Petroleum Reservoir Engineering Practice

Petroleum Reservoir Engineering Practice

Nnaemeka Ezekwe

PRENTICE HALL

Upper Saddle River, NJ • Boston • Indianapolis • San Francisco
New York • Toronto • Montreal • London • Munich • Paris • Madrid
Capetown • Sydney • Tokyo • Singapore • Mexico City

The publisher offers excellent discounts on this book when ordered in quantity for bulk purchases or special sales, which may include electronic versions and/or custom covers and content particular to your business, training goals, marketing focus, and branding interests. For more information, please contact:

U.S. Corporate and Government Sales
(800) 382-3419
corpsales@pearsontechgroup.com

For sales outside the United States please contact:

International Sales
international@pearson.com

Visit us on the Web: informit.com/ph

Library of Congress Cataloging-in-Publication Data

Ezekwe, Nnaemeka.
Petroleum reservoir engineering practice / Nnaemeka Ezekwe.
p. cm.
Includes bibliographical references and index.
ISBN 0-13-715283-3 (hardcover : alk. paper) 1. Oil reservoir engineering. 2. Petroleum—Geology. I. Title.
TN870.57.E94 2011
622'.3382—dc22

2010024160

ISBN-13: 978-0-13-715283-4
ISBN-10: 0-13-715283-3

Text printed in the United States on recycled paper at Courier in Westford, Massachusetts.
First printing, September 2010

To the memory of my parents, Vincent Nweke and Rosaline Oriaku, for their sacrifices, commitment, courage, and unwavering support towards my education and those of my brother and sisters. This book is the fruit of the seed they planted and nurtured.

Contents

Preface

In writing any book, there is always a conflict in deciding the topics to include and the ones to leave out. Of course there is the desire to present as many topics as possible, but that is not practical due to limitations on the physical size of the book. The scope and organization of this book were conceived to cover petroleum reservoir engineering topics which will provide strong fundamentals to an engineer-in-training in a classroom setting, and at the same time be useful as a handbook to a practicing engineer.

Chapters 1 to 5 are devoted to discussing the sources and applications of basic rock and fluid properties data which are the bedrock for all petroleum reservoir engineering calculations. In Chapter 1, the porosity of reservoir rocks are presented. Chapter 2 discusses rock permeability and relative permeability, and Chapter 3 discusses determination of fluid saturations and classification of formation intervals of reservoir rocks. These chapters treat these topics at the introductory to intermediate levels. They are designed to emphasize the importance of these types of data as basic input for reservoir engineering calculations. Correlations and methods for calculation of PVT data for reservoir fluids are presented in Chapter 4. Chapter 5 discusses reservoir fluid sampling methods and laboratory measurements of PVT data on reservoir fluid samples. Chapter 6 presents the prediction of PVT properties with equations of state. This topic is presented at the intermediate-to-advance level because many engineers conduct compositional reservoir simulation work which requires equations of state in most cases.

Basic reservoir engineering fundamentals are covered in Chapters 7 to 9. In Chapter 7, the general material balance equation is presented as a fundamental tool which could be used for basic reservoir analysis. Chapter 8 discusses different types of gas reservoirs and calculation of gas-in-place volumes with volumetric and graphical methods. Similar treatment of oil reservoirs are presented in Chapter 9. Case histories of gas and oil reservoirs are presented in Chapters 8 and 9 for discussion and review purposes. I consider the use of case histories as more realistic examples and effective tools for teaching of reservoir engineering fundamentals and practices than sanitized problems included in many traditional petroleum engineering textbooks. Most real-life reservoir engineering problems are usually not encountered in these neatly encapsulated problem formats. For these reasons, I have not presented end-of-chapter problems in this book. Instead, teachers are encouraged to use case histories as effective vehicles for teaching fundamental reservoir engineering principles to students.

An introduction to fluid flow in petroleum reservoirs is presented in Chapter 10 with the derivation of the continuity equation and the radial diffusivity equation. This chapter lays the foundation for the fundamental equations used in the development of well test analysis by straight-line methods presented in Chapter 11. In Chapter 12, the application of type curves, especially Gringarten and Bourdet type curves, are presented with emphasis on procedures for type-curve matching. The importance of hydraulically fractured wells and naturally fractured reservoirs in the production of many reservoirs around the world is recognized by the presentation of well test analysis methods for these well and reservoir types in Chapter 13. Chapter 14 presents deconvolution in a rather rudimentary form to acquaint the reader with this method for well test analysis.

The book introduces basic concepts in immiscible fluid displacement in Chapter 15, including the derivation of the fractional flow equation, the Buckley-Leverett equation, and the Welge method. This lays the foundation for the introduction of secondary recovery methods in Chapter 16. In Chapter 17, enhanced oil recovery methods are presented. Special emphasis is placed on screening criteria and field implementation of enhanced oil recovery processes. The main purpose of Chapters 16 and 17 is to introduce the engineer to the fundamentals of secondary and enhanced oil recovery processes, and also provide practical procedures for field implementation of these processes.

The availability of high speed computers has placed very powerful engineering tools in the hands of modern petroleum engineers. These tools are readily available in applications of geologic modeling, reservoir characterization, and reservoir simulation. Every practicing engineer has been exposed to these tools in the form of commercial software readily available in the petroleum industry. In this book, I present introductory fundamentals on geologic modeling, reservoir characterization, and reservoir simulation in Chapters 18 and 19. On these topics, considerable emphasis is placed on procedures for using these tools rather than an in-depth presentation of the theoretical basis of the methods.

Finally in Chapter 20, I present fundamental principles of petroleum reservoir management based on my experience. These principles are simple, practical, and can be applied in the management of reservoirs around the world. I encourage readers of this book to adopt these principles to review current reservoir management strategies and in the implementation of new strategies in the management of old and new reservoirs.

This book came into form with encouragement from Mr. Stuart Filler, who read the early drafts of many chapters and gave me many useful suggestions. I thank him for his advice and encouragement that compelled me to continue work on the book. I also give thanks to my publisher, Mr. Bernard Goodwin, for his unwavering support that made this book possible. Finally, I express special thanks to my wife, Anulika, and my children (Nkem, Emeka, Chioma, Ifeoma, Obinna, and Eze) for their patience, comfort, and the emotional support they provided to me during the long hours and many years it took to write this book. This book would not have been completed without their love and support.

Acknowledgments

The author gratefully thanks the Society of Petroleum Engineers for permissions to reprint the following tables and figures in this text:

Chapter 1

Figure 1.1

Al-Ruwaili, S. A., and Al-Waheed, H. H.: "Improved Petrophysical Methods and Techniques for Shaly Sands Evaluation," paper SPE 89735 presented at the 2004 SPE International Petroleum Conference in Puebla, Mexico, November 8–9, 2004.

Chapter 3

Figure 3.4

Worthington, P.F., and Cosentino, L.: "The Role of Cutoffs in Integrated Reservoir Studies," *SPEREE* (August 2005) 276–290.

Chapter 4

Figures 4.9, 4.10

Sutton, R.P.: "Compressibility Factors for High-Molecular-Weight Reservoir Gases," paper SPE 14265 presented at the 60th Annual Technical Conference and Exhibition, Las Vegas, NV (Sept. 22–25, 1985).

Figures 4.11, 4.12

Mattar, L., Brar, G.S., and Aziz, K.: "Compressibility of Natural Gases," *J. Cdn. Pet. Tech.* (Oct-Dec. 1975) 77–80.

Chapter 5

Figure 5.1

Moffatt, B.J. and Williams, J.M.: "Identifying and Meeting the Key Needs for Reservoir Fluid Properties—A Multi-Disciplinary Approach," paper SPE 49067 presented at the 1998 SPE Annual Technical Conference and Exhibition, New Orleans, Louisiana, Sept. 27–30, 1998.

Figures 5.4, 5.5, 5.6, 5.7, 5.9

Elshahawi, H., Samir, M., and Fathy, K.: "Correcting for Wettability and Capillary Pressure Effects on Formation Tester Measurements," paper SPE 63075 presented at the 2000 SPE Annual Technical Conference and Exhibition, Dallas, Texas, Oct. 1–4, 2000.

Chapter 6

Tables 6.1, 6.2

Jhaveri, B.S., and Youngren, G.K.: "Three-Parameter Modification of the Peng-Robinson Equation of State to Improve Volumetric Predictions," *SPERE* (August 1988) 1033–1040.

Figures 6.1, 6.2

Wang, P., and Pope, G.A.: "Proper Use of Equations of State for Compositional Reservoir Simulation," SPE 69071, Distinguished Author Series, (July 2001) 74–81.

Chapter 8

Figure 8.4

Havlena, D., and Odeh, A.S.: "The Material Balance as an Equation of a Straight Line," *JPT* (August 1963) 896–900.

Figure 8.11

Hammerlindl, D.J.: "Predicting Gas Reserves in Abnormally Pressured Reservoirs," paper SPE 3479 presented at the 1971 SPE Fall Meeting, New Orleans, October 3–5, 1971.

Chapter 9

Figures 9.1, 9.4, 9.5

Havlena, D., and Odeh, A.S.: "The Material Balance as an Equation of a Straight Line," *JPT* (August 1963) 896–900.

Chapter 10

Appendices 10B, 10C, 10D, 10E, 10F

van Everdingen, A.F., and Hurst, W.: "The Application of the Laplace Transformation to Flow Problems in Reservoirs," *Trans.*, AIME (1949) 186, 305–324.

Chapter 11

Table 11.2

Tiab, D., Ispas, I.N., Mongi, A., and Berkat, A.: "Interpretation of Multirate Tests by the Pressure Derivative. 1. Oil Reservoirs," Paper SPE 53935 presented at the 1999 SPE Latin American and Caribbean Petroleum Engineering Conference, Caracas, Venezuela, April 21–23, 1999.

Chapter 12

Figure 12.1

Agarwal, R.G., Al-Hussainy, R., and Ramey, H.J., Jr.: "An Investigation of Wellbore Storage and Skin Effect in Unsteady Liquid Flow: I. Analytical Treatment," *SPEJ* (September 1970) 279–290.

Figure 12.2

Gringarten, A.C., Bourdet, D.P., Landel, P.A., and Kniazeff, V.J.: "A Comparison Between Different Skin and Wellbore Storage Type-Curves for Early-Time Transient Analysis," Paper SPE 8205 presented at the SPE Annual Technical Conference and Exhibition, Las Vegas, Nevada, September 23–26, 1979.

Chapter 13

Figures 13.1, 13.3, 13.4, 13.5, 13.6, 13.7

Cinco-Ley, H., and Samaniego-V, F.: "Transient Pressure Analysis for Fractured Wells," *JPT,* (September 1981) 1749–1766.

Figures 13.8, 13.9, 13.11

Cinco, H.: "Evaluation of Hydraulic Fracturing By Transient Pressure Analysis Methods," paper SPE 10043 presented at the International Petroleum Exhibition and Technical Symposium, Beijing, China, March 18–26, 1982.

Figure 13.10

Cinco-L, H., Samaniego-V, F., and Dominguez-A, N.: "Transient Pressure Behavior for a Well With a Finite-Conductivity Vertical Fracture," *SPEJ* (August 1978) 253–264.

Figures 13.16, 13.17a, 13.17b, 13.18, 13.19a, 13.19b, 13.20, 13.21, 13.22, 13.23, 13.24, 13.25

Cinco-Ley, H.: Well-Test Analysis for Naturally Fractured Reservoirs," *JPT* (January 1996) 51–54.

Figure 13.26

Warren, J.E., and Root, P.J.: "The Behavior of Naturally Fractured Reservoirs," *SPEJ* (September 1963) 245–255.

Figure 13.27

Serra, K., Reynolds, A.C., and Raghavan, R.: "New Pressure Transient Analysis Methods for Naturally Fractured Reservoirs," *JPT* (December 1983) 2271–2283.

Figure 13.32, 13.33, 13.34

Gringarten, A.C.: "Interpretation of Tests in Fissured and Multilayered Reservoirs with Double-Porosity Behavior: Theory and Practice," *JPT* (April 1984) 549–564.

Chapter 14

Figures 14.1, 14.2, 14.3, 14.4, 14.5, 14.6, 14.7, 14.8

Levitan, M.M., Crawford, G.E., and Hardwick, A.: "Practical Considerations for Pressure-Rate Deconvolution of Well-Test Data," *SPEJ* (March 2006) 35–46.

Chapter 19

Figures 19.2, 19.7, 19.9, 19.10

Aziz, K.: "Reservoir Simulation Grids: Opportunities and Problems," *JPT* (July 1993) 658–663.

Chapter 20

Figures 20.16, 20.17, 20.18, 20.21, 20.22

Salamy, S.P., Al-Mubarak, H.K., Ghamdi, M.S., and Hembling, D.: "Maximum-Reservoir-Contact-Wells Performance Update: Shaybah Field, Saudi Arabia," *SPE Production & Operation* (November 2008) 439–443.

Figures 20.19, 20.20

Saleri, N.G., Salamy, S.P., Mubarak, H.K., Sadler, R.K., Dossary, A.S., and Muraikhi, A.J.: "Shaybah-220: A Maximum-Reservoir-Contact (MRC) Well and its Implications for Developing Tight-Facies Reservoirs," *SPEREE* (August 2004) 316–321.

About the Author

Nnaemeka Ezekwe holds B.S., M.S., and Ph.D. degrees in chemical and petroleum engineering, and an MBA, all from the University of Kansas. For many years, he worked in several supervisory roles including manager of reservoir evaluation and development for Bechtel Petroleum Operations. As a senior petroleum engineer advisor for Pennzoil and later Devon Energy, he performed reservoir engineering analyses on many domestic and worldwide projects. Nnaemeka was an SPE Distinguished Lecturer in 2004–2005, during which he spoke on reservoir management strategies and practices to audiences in 33 countries in Africa, Asia, Europe, Middle East, and North and South America. He has published numerous technical papers on chemical and petroleum engineering topics. Nnaemeka is a registered professional engineer in California and Texas.

CHAPTER 1

Porosity of Reservoir Rocks

1.1 Introduction

Porosity is defined as a measure of the capacity of reservoir rocks to contain or store fluids. The fluids stored in the pore spaces within the reservoir rocks could be gas, oil, and water. High porosity values indicate high capacities of the reservoir rocks to contain these fluids, while low porosity values indicate the opposite. Consequently, porosity data are routinely used qualitatively and quantitatively to assess and estimate the potential volume of hydrocarbons contained in a reservoir. For instance, in a discovery well that shows the presence of hydrocarbons in the reservoir rocks, the set of data that is reviewed at least qualitatively to evaluate reservoir potential is porosity data acquired with either logging-while-drilling (LWD) tools or by running wireline tools. Porosity data are obtained from direct measurements on core samples and/or indirectly from well logs. In most cases, porosity data from core samples are used to validate or calibrate porosity data from well logs. Porosity data are also used in reservoir characterization for the classification of lithological facies, and the assignment of permeabilities using porosity-permeability transforms. Since porosity data are very important in many reservoir engineering calculations, this book begins by reviewing basic concepts in the determination of rock porosities. This review is concise and serves to refresh the reader with the many sources of porosity data that exist through applications of different formation evaluation tools.

1.2 Total Porosity and Effective Porosity

The porosity of a rock is a measure of its capacity to contain or store fluids. Porosity is calculated as the pore volume of the rock divided by its bulk volume.

$$\text{Porosity} = \frac{\text{Pore Volume}}{\text{Bulk Volume}} \tag{1.1}$$

Expressed in terms of symbols, Eq. (1.1) is represented as:

$$\phi = \frac{V_P}{V_B} \tag{1.2}$$

In Eq. (1.2), ϕ = porosity; V_p = pore volume; and V_B = bulk volume. Pore volume is the total volume of pore spaces in the rock, and bulk volume is physical volume of the rock, which includes the pore spaces and matrix materials (sand and shale, etc.) that compose the rock.

Two types of porosities can exist in a rock. These are termed primary porosity and secondary porosity. **Primary porosity** is described as the porosity of the rock that formed at the time of its deposition. **Secondary porosity** develops after deposition of the rock. Secondary porosity includes vugular spaces in carbonate rocks created by the chemical process of leaching, or fracture spaces formed in fractured reservoirs. Porosity is further classified as total porosity and effective porosity. **Total porosity** is defined as the ratio of the entire pore space in a rock to its bulk volume. **Effective porosity** is the total porosity less the fraction of the pore space occupied by shale or clay. In very clean sands, total porosity is equal to effective porosity. As shown in Figure 1.1, effective porosity represents pore space that contains hydrocarbon and non-clay water.[1] Free formation water that is neither bound to clay nor to shale is called non-clay water. An accurate definition of effective porosity is total porosity minus volume of clay-bound water (Figure 1.1). The relationship between total porosity and effective porosity can be represented for a shaly sand model as:

$$\phi_t = \phi_e + V_{sh} \times \phi_{sh} \tag{1.3}$$

In Eq. (1.3), ϕ_t = total porosity, fraction; ϕ_e = effective porosity, fraction; V_{sh} = volume of shale, fraction; and ϕ_{sh} = shale porosity, fraction. The determination of shale porosity from

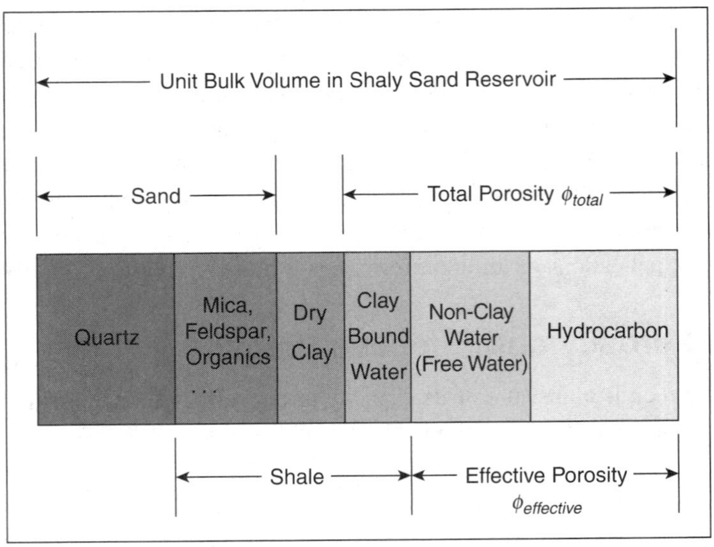

Figure 1.1 Porosity model for a shaly sand reservoir (from Al-Ruwaili et al.[1] © 2004 SPE, Reproduced with permission).

well logs can be difficult and erroneous because the selection of the 100% shale section can be wrong and subjective.[1] For this reason, an approximate form of Eq. (1.3) is obtained by replacing shale porosity ϕ_{sh} with total porosity ϕ_t to get:

$$\phi_t = \phi_e + V_{sh} \times \phi_t \tag{1.4}$$

For a clay model, effective porosity is represented as:

$$\phi_t = \phi_e + V_{cbw} \tag{1.5}$$

In Eq. (1.5), V_{cbw} = volume of clay-bound water, fraction. The application of Eq. (1.5) for calculation of accurate effective porosity depends on accurate quantification of the volume of clay-bound water. This can be determined from an elemental capture spectroscopy (ECS) well logs.[1]

1.3 Sources of Porosity Data

Rock porosity data are obtained by direct or indirect measurements. Laboratory measurements of porosity data on core samples are examples of direct methods. Determinations of porosity data from well log data are considered indirect methods.

1.3.1 Direct Methods for Measurement of Porosity

Direct measurements of porosity data on core samples in a laboratory typically require measurements of bulk and pore volumes of the core samples. For irregular-shaped core samples, the bulk volume is determined by gravimetric or volumetric methods. In gravimetric methods, the apparent loss in weight of the sample when immersed completely in a liquid of known density is measured. Volumetric methods measure the volume of liquid displaced by the rock sample when completely immersed in the liquid. These methods use specially designed equipment so that the liquid is not absorbed by the rock sample. For regular-shaped samples, the bulk volume is calculated from physical measurement of the dimensions of the core sample. For instance, if the core plug is cylindrical in shape, the bulk volume is calculated as:

$$V_B = \pi r^2 l \tag{1.6}$$

In Eq. (1.6), V_B = bulk volume; r = radius of the core plug; and l = length of the core plug.

Other direct methods for measuring the porosity of a rock sample include use of mercury porosimeter or gas expansion porosimeter. The use of mercury porosimeter or gas expansion porosimeter for measurement of porosity is not presented in this book because they are described in many introductory textbooks[2] on petroleum reservoir engineering.

Most laboratory routines based on direct methods measure total porosity. It is important to remember to distinguish between total porosity data obtained from core samples and porosity data derived from well logs, which may include effective porosities. Porosity data obtained from core samples using direct methods are generally considered to be accurate and reliable. They are used to calibrate and validate log-derived porosity data which are based on indirect methods.

Example 1.1 Calculation of Porosity from Gravimetric Data

Problem

The dimensions of a cylindrical core sample are 10.16 cm long and 3.81 cm in diameter after it was thoroughly cleaned and dried. The dried core sample weighed 365.0 g. The core sample was then completely (100%) saturated with brine that has specific gravity of 1.04. The weight of the saturated core sample is 390.0 g. Calculate the porosity of the core sample.

Solution

Using Eq. (1.6), the bulk volume of the core sample is:

$$V_B = \pi r^2 l$$

$$= \pi\left(\frac{3.81}{2}\right)^2 \times 10.16 = 115.8333 \text{ cm}^3$$

The pore volume of the core sample is given by:

$$V_p = \frac{\text{wt. of saturated core} - \text{wt. of dried core}}{\text{specific gravity of brine}}$$

$$= \frac{390.0 - 365.0}{1.0400} = 24.0385 \text{ cm}^3$$

Using Eq. (1.2), porosity of the core sample is:

$$\phi = \frac{V_P}{V_B}$$

$$= \frac{24.0385}{115.8333}$$

$$= 0.2075 \text{ or } 20.75\%$$

1.3.2 Indirect Methods for Derivation of Porosity

Indirect methods for derivation of porosity data are based on well log data. The well logs generally used for this purpose are density, sonic, neutron, and nuclear magnetic response (NMR) logs. In most formation evaluation programs, density, sonic, and neutron logs are routinely acquired. The NMR log is frequently run in many wells because of its capability of providing other data for formation evaluation, in addition to porosity data. In most deepwater wells, it is common practice to run NMR logs, in addition to density, sonic, and neutron logs. It is important to note that density, sonic, and neutron logs are lithology-dependent, while the NMR logs are lithology-independent for derivation of porosity.[3] NMR data are very sensitive to environmental conditions. It is recommended that NMR tools should be run together with conventional logs, such as density logs or neutron logs for quality control and validation of the NMR data. A summary of the basic principles, data requirements, advantages, and disadvantages of all the porosity tools is provided in Table 1.1.

Table 1.1 Summary of Principles, Advantages, and Disadvantages of Porosity Tools

	Type of Porosity Log			
Attributes	**Density**	**Neutron**	**Sonic**	**NMR**
Basic principle	Gamma ray attenuation	"Slowed" neutrons or Gamma ray capture	Transit times	Excitation of hydrogen in pore spaces
Required data	Matrix and fluid densities	Calibration	Matrix and fluid transit times	Hydrogen index
Advantages	Little effect of presence of gas in formation	Ability to detect presence of gas in formation; can be used in cased hole	Good compensation for environmental effects; combinable with induction logs	Lithology independent
Disadvantages	Shallow depth of investigation; affected by wellbore washouts	Sensitive to irregular borehole; requires calibration	Depth of investigation dependent on type of formation	Environmental corrections; tool run speed affects results

1.3.2.1 Density Logs

Density logs are based on the attenuation of gamma rays in the formation.[3] Density logging tools measure the attenuation of gamma rays produced by a gamma source of known strength. The attenuation caused by the interaction between gamma ray photons and electrons on the outer shell of electrons (called Compton scattering) is directly proportional to the bulk density (ρ_b) of the formation. The formation bulk density is related to formation matrix density (ρ_{ma}) and formation fluid density (ρ_f) as:

$$\rho_b = (1 - \phi)\rho_{ma} + \phi\rho_f \tag{1.7}$$

Re-arranging Eq. (1.7), density-derived porosity is given by:

$$\phi_d = \frac{\rho_{ma} - \rho_b}{\rho_{ma} - \rho_f} \tag{1.8}$$

In Eq. (1.8), ϕ_d = density-derived porosity; ρ_{ma} = matrix density; ρ_b = bulk density; and ρ_f = fluid density. The porosity data obtained from density logs are considered to be total porosity. This relationship can be represented as:

$$\phi_d = \phi_t \tag{1.9}$$

For density logs, effective porosity is derived from Eq. (1.3) as:

$$\phi_e = \phi_d - V_{sh} \times \phi_{dsh} \tag{1.10}$$

In Eq. (1.10), ϕ_{dsh} is the shale porosity derived from the density logs. The depth of investigation of density logging tools is shallow and typically within the zone invaded by mud filtrate. For this reason, it is sometimes appropriate to assume that the density of formation fluid is equal to the density of the mud filtrate. However, this assumption may cause errors in the density-derived porosity data, if virgin formation fluid remains within the depth of investigation of the density tool.[4] The matrix density can be determined from elemental capture spectroscopy (ECS) log, if available.

Example 1.2 Calculation of Porosity from Density Logs

Problem

The bulk density of a clean, sandy interval saturated with water was measured by the density logging tool to be 2.4 g/cm^3. Assuming that the density of the formation water is 1.04 g/cm^3 and the density of the matrix is 2.67 g/cm^3, calculate the density porosity of this interval.

Solution

Using Eq. (1.8), density porosity is calculated to be:

$$\begin{aligned}\phi_d &= \frac{\rho_{ma} - \rho_b}{\rho_{ma} - \rho_f}\\ &= \frac{2.67 - 2.4}{2.67 - 1.04}\\ &= 0.166 \text{ or } 16.6\%\end{aligned}$$

1.3.2.2 Sonic (acoustic) Logs

In sonic (acoustic) logging, the formation is probed with sound waves. The time it takes the sound waves to travel a given distance is measured. This interval transit time depends on the elastic properties of the rock matrix, the properties of the fluid in the rock, and the porosity of the rock. Wyllie et al.[5] proposed that the interval transit time (Δt) can be represented as the sum of the transit time in the matrix fraction (Δt_{ma}) and the transit time in the liquid fraction (Δt_f) thus:

$$\Delta t = (1 - \phi)\Delta t_{ma} + \phi \Delta t_f \tag{1.11}$$

Re arranging Eq. (1.11), sonic-derived porosity is given by:

$$\phi_s = \frac{\Delta t - \Delta t_{ma}}{\Delta t_f - \Delta t_{ma}} \tag{1.12}$$

In Eq. (1.12), ϕ_s = sonic-derived porosity; Δt = transit time; Δt_f = fluid transit time; and Δt_{ma} = transit time for the rock matrix. Total porosity is related to porosity derived from sonic logs as:

$$\phi_s = \phi_t + V_{clay} \times \phi_{scl} \tag{1.13}$$

In Eq. (1.13), V_{clay} = the volume of clay; and ϕ_{scl} = sonic porosity derived in the clay. Effective porosity as calculated from sonic logs as:

$$\phi_e = \phi_s - V_{sh} \times \phi_{ssh} \tag{1.14}$$

In Eq. (1.14), V_{sh} = volume of shale; and ϕ_{ssh} = sonic porosity derived for shale. Analysis of sonic logs based on Eq. (1.12) gives reliable porosity data only for consolidated formations. For unconsolidated sandstones and carbonates, Eq. (1.12) gives porosity values that are too high. Other equations similar to Eq. (1.12) have been proposed for calculation of porosity for unconsolidated formations and carbonates by Raymer et al.[6] These equations should be used for calculations of sonic porosities on unconsolidated formations and carbonates. Note that sonic logs are well-compensated for environmental effects such as mud velocity, borehole diameter, etc. and that its depth of investigation is dependent on the compactness of the formation.

Example 1.3 Calculation of Porosity from Sonic Logs

Problem

The transit time for a well-consolidated sandstone interval saturated with brine was measured to be 82×10^{-6} sec/ft. The matrix transit time is 55.5×10^{-6} sec/ft and the brine transit time is 189×10^{-6} sec/ft. Calculate the sonic porosity for the interval.

Solution

Applying Eq. (1.12), sonic porosity is calculated to be:

$$\begin{aligned}\phi_s &= \frac{\Delta t - \Delta t_{ma}}{\Delta t_f - \Delta t_{ma}} \\ &= \frac{82 \times 10^{-6} - 55.5 \times 10^{-6}}{189 \times 10^{-6} - 55 \times 10^{-6}} \\ &= 0.199 \text{ or } 19.9\%\end{aligned}$$

1.3.2.3 Neutron Porosity Logs

The first logging tool that was used for the estimation of formation porosity is the neutron logging tool, which was introduced around 1940. The neutron porosity logging tool consists of either a chemical source or an electrical source of fast neutrons, and detectors located some distance from the source. The fast neutrons from the neutron source are slowed down by successive collisions with individual nuclei in the rock, thereby losing most of their energy. The detectors in the neutron tool record either the "slowed" down neutrons directly or capture gamma radiation generated when the neutrons are captured by nuclei. The neutron porosity log is sensitive to the amount of hydrogen in the formation because the neutrons interact most effectively with hydrogen due to the closeness of their masses. Neutron logs estimate the amount of hydrogen in the rock, and relate it to the amount of fluid in the formation. From the amount of fluid in the formation, the porosity of the rock is estimated after calibration for different lithologies (sandstone, dolomite, and limestone). Neutron porosity tools are sensitive to borehole conditions, especially variations in the size of the borehole. In combination with density porosity logs, neutron porosity logs can be used to detect the presence of gas in some formations. This known crossover of density porosity log and neutron porosity log in gas-filled formation intervals results from the apparent increase of density-derived porosities and apparent decrease of neutron-derived porosities in gas-filled formation intervals (Figure 1.2).

For neutron porosity logs, correction for total porosity is applied as:

$$\phi_n = \phi_t + V_{clay} \times \phi_{ncl} \quad \textbf{(1.15)}$$

In Eq. (1.15), ϕ_n = porosity from neutron logs; and ϕ_{ncl} = neutron porosity for clay. Effective porosity is defined as:

$$\phi_e = \phi_n - V_{sh} \times \phi_{nsh} \quad \textbf{(1.16)}$$

In Eq. (1.16), ϕ_{nsh} is the neutron porosity for shale.

1.3.2.4 Nuclear Magnetic Resonance (NMR) Porosity Logs

Nuclear magnetic resonance (NMR) porosity tools have a clear advantage over other porosity tools (density, sonic, and neutron) because their determination of porosity is independent of lithology of the rock. Porosities calculated from density, sonic, and neutron logs depend on "knowing" or estimating the properties of the rock matrix. NMR porosities are calculated from the number of hydrogen atoms in the fluids (hydrocarbon and water) within a specific measurement volume of the tool, and are independent of the lithology of the rock formation.[7] For reservoirs with highly heterogeneous rocks consisting of mixed or unknown lithology, porosity data derived from NMR logs are more consistent and reliable than porosity data from the other porosity tools.[8] NMR logs report porosities in terms of total porosity, bound-fluid porosity, and free-fluid porosity (Figure 1.1). **Free-fluid porosity** (also termed free-fluid index) is a qualitative measure of effective porosity and is linked to the hydrocarbon storage potential of the formation. A comparison of porosities measured with NMR in the laboratory to porosities measured by direct methods on core samples from a reservoir is shown in Figure 1.3. The porosity data plotted in Figure 1.3 show close agreements between NMR porosities and core porosities measured on core samples. It demonstrates a method for calibrating NMR porosities with core porosities, which can then be used to calculate porosities from NMR data in other wells. In addition to measurement of porosity, NMR tools are used for determination of pore size distributions, measurements of permeability (Chapter 2), and fluid saturations (Chapter 3). NMR tools have become standard in most wireline logging operations because they can quickly provide qualitative data on formation porosity, permeability, pore size distributions, and fluid saturations.[9,10] These data are very valuable and useful. They are frequently used to make decisions on selection of fluid sampling points and formation intervals to be tested in discovery, appraisal, and development wells.

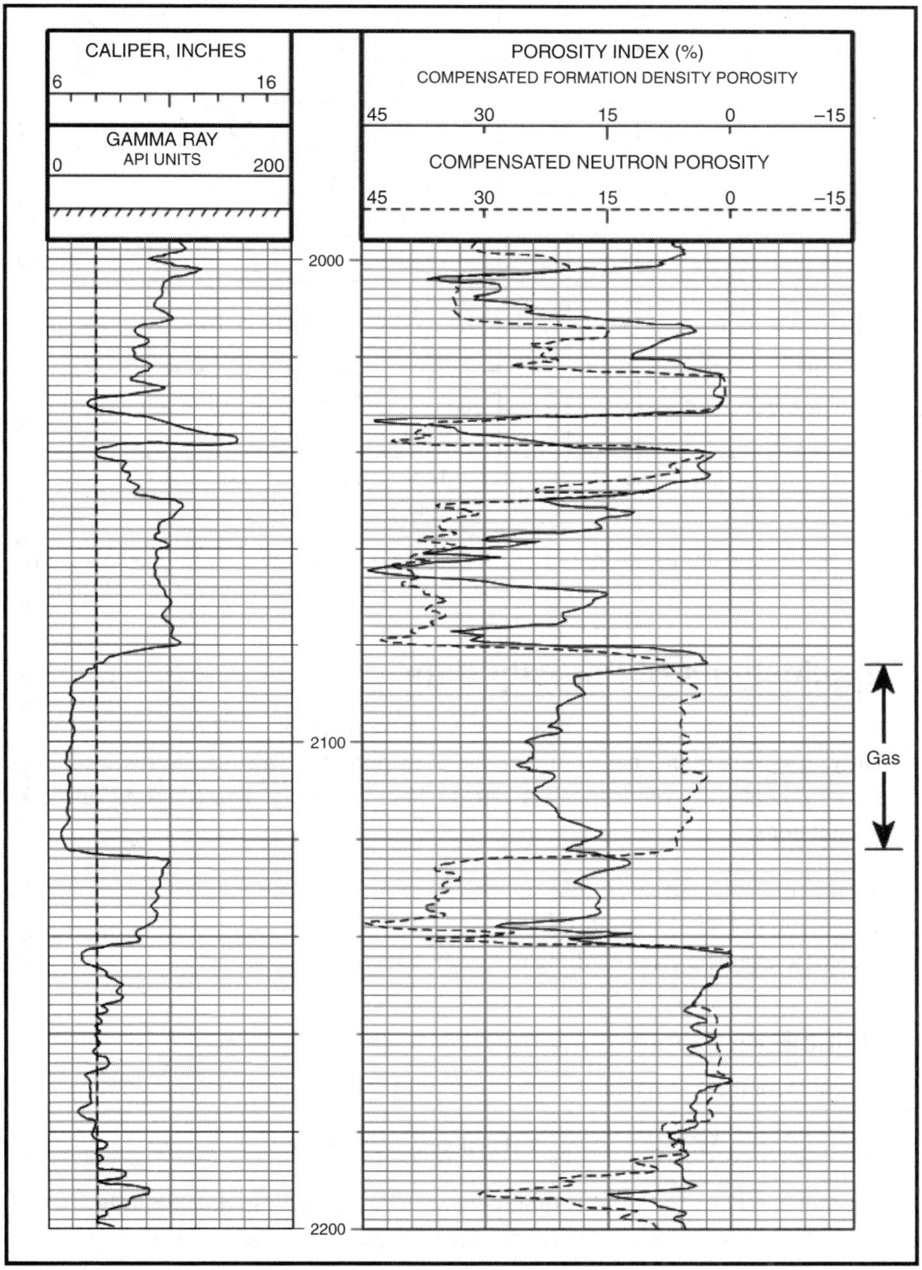

Figure 1.2 Density and neutron well logs showing crossover in a gas interval.

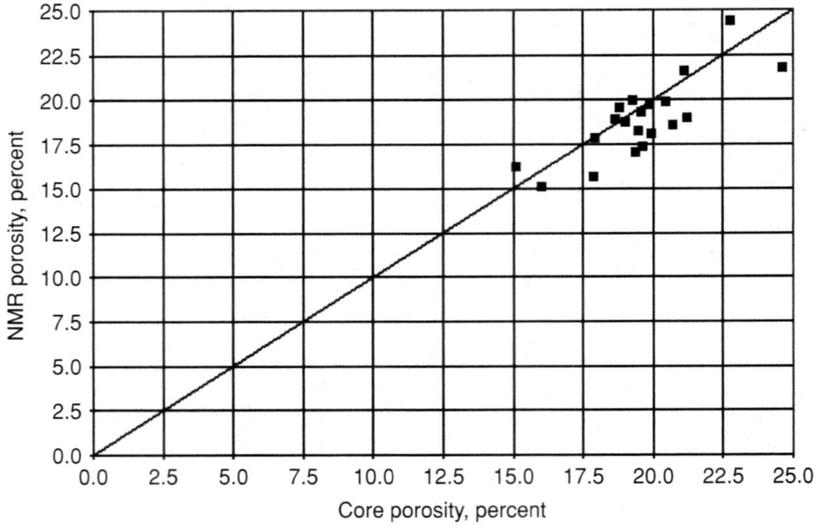

Figure 1.3 NMR porosity versus core porosity based on core samples from a reservoir.

1.4 Applications of Porosity Data

One of the primary rock property data used in most reservoir evaluation is porosity data. Consequently, it is important that accurate values of porosity data for the reservoir rocks are measured and validated by other independent methods. Porosity data are used in these basic reservoir evaluations:

1. Volumetric calculation of fluids in the reservoir
2. Calculation of fluid saturations
3. Geologic characterization of the reservoir

1.4.1 Volumetric Calculation

A general formula for the calculation of the volume of hydrocarbons in a reservoir is represented as:

$$\text{HCPV} = \text{Area} \times \text{Thickness} \times \phi \times (1 - S_w) \quad \textbf{(1.17)}$$

In Eq. (1.17), HCPV = hydrocarbon pore volume; Area = hydrocarbon-bearing area of the reservoir; Thickness = net productive thickness or pay of the reservoir; ϕ = porosity, fraction; and S_w = water saturation, fraction. The hydrocarbon volumes of specific types of fluids (oil and/or gas) in the reservoir can be calculated with minor modifications of Eq. (1.17) as demonstrated in Chapter 8 for Gas Reservoirs, and Chapter 9 for Oil Reservoirs. Note the prominence of porosity in Eq. (1.17) for the calculation of volumes of hydrocarbons present in a reservoir. It is evident from Eq. (1.17) that inaccurate porosity data can directly cause underestimation or overestimation of the hydrocarbon volumes in the reservoir. For marginal reservoirs, underestimation of in-place hydrocarbon volumes may contribute to a decision not to pursue development of the reservoir. Overestimation of in-place hydrocarbon volumes may lead to economic losses, if projected reserves estimated prior to development are far below actual reservoir performance. Note that there are other geologic and reservoir factors (such as permeability barriers, faults, compartments, recovery mechanisms) which can also cause reservoir performance to be below projected levels. The impact of these other factors on reservoir performance are presented and discussed in more details in several chapters in this book.

1.4.2 Calculation of Fluid Saturations

For clean, non-shaly rocks, water saturations can be calculated from the Archie equation[1] as:

$$C_t = \phi_t^m \times S_w^n \times C_w \tag{1.18}$$

In Eq. (1.18), C_t = formation conductivity; ϕ_t = total porosity; S_w = water saturation; C_w = formation water conductivity; m = cementation factor; and n = saturation exponent. The parameters m and n are also called the electrical properties of the rock.

For shaly sands, water saturations can be calculated from modified forms[1] of the Archie equation, which are shown as:

$$C_t = \phi_t^{m^v} \times S_{wt}^{n^v} \times C_{we} \tag{1.19}$$

$$C_t = \phi_t^{m^v} \times S_{wt}^{n^v} \times (C_w + X) \tag{1.20}$$

In Eqs. (1.19) and (1.20), C_{we} = effective conductivity; and m^v, n^v are general forms of the electrical properties. In Eq. (1.19), C_{we} is expressed in terms of C_w and a function of shale (in the shale model) or a function of clay (in the clay model). In Eq. (1.20), X is a function that accounts for the conductivity caused by shale or clay that occur in shaly sands. Note that in Eq. (1.20) as X approaches zero, Eq. (1.20) becomes equivalent to Eq. (1.18).

The main point to note from Eqs. (1.18), (1.19), and (1.20) is that total porosity is an important data input for calculation of water saturation with water saturation models. If errors exist in the calculations of total porosity, these errors will be transferred to the calculation of water saturations. This could ultimately lead to errors in the estimation of reservoir in place hydrocarbon volumes as shown in Eq. (1.17). The calculation of water (fluid) saturation is presented in more detail in Chapter 3.

1.4.3 Reservoir Characterization

Porosities can be measured directly from cores or indirectly determined from well logs as discussed previously in this chapter. On the one hand, rock permeability can be measured most reliably from cores or in aggregate sense from well tests. Indirect methods for acquiring permeability data are discussed in Chapter 2. There are usually more porosity data than permeability data available on a reservoir. A cross-plot of permeability versus porosity data (Figure 1.4) to create a porosity-permeability transform is sometimes used to assign permeability values to areas of the

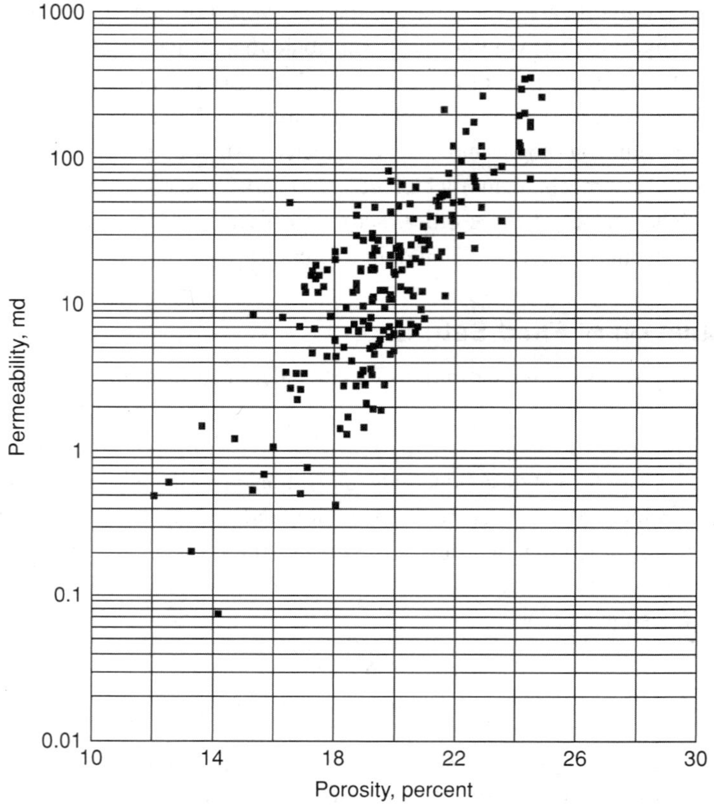

Figure 1.4 Porosity-permeability cross-plot based on core samples from a reservoir.

reservoir where permeability data do not exist. The practice of using porosity-permeability transforms in reservoir characterization is presented in Chapter 18.

Facies or rock types can be defined or assigned to parts of a reservoir by using porosity values as part of a system of criteria for rock classification. This process of classifying reservoir rock in terms of facies or rock types is useful in the process of reservoir characterization. This is also presented in Chapter 18.

Nomenclature

C_t	formation conductivity
C_{we}	effective conductivity
C_w	formation water conductivity
l	length of core plug
m	cementation factor

n	saturation exponent
r	radius of core plug
S_w	water saturation, fraction
V_{cbw}	volume of clay-bound water
V_{clay}	volume of clay
V_B	bulk volume
V_p	pore volume
V_{sh}	shale volume
X	function in Eq. 1.20
ϕ	porosity
ϕ_d	density-derived porosity
ϕ_{dsh}	density-derived shale porosity
ϕ_e	effective porosity
ϕ_n	neutron-derived porosity
ϕ_{ncl}	neutron-derived porosity in clay
ϕ_{nsh}	neutron-derived porosity in shale
ϕ_s	sonic-derived porosity
ϕ_{sh}	shale porosity
ϕ_{scl}	sonic-derived porosity in clay
ϕ_{ssh}	sonic-derived porosity in shale
ϕ_t	total porosity
ρ_b	bulk density
ρ_f	fluid density
ρ_{ma}	rock matrix density
Δt	formation interval transit time
Δt_f	fluid transit time
Δt_{ma}	rock matrix transit time

Abbreviations

LWD	Logging-While-Drilling
NMR	Nuclear Magnetic Resonance
ECS	Elemental Capture Spectroscopy
HCPV	Hydrocarbon Pore Volume

References

1. Al-Ruwaili, S.A., and Al-Waheed, H.H.: "Improved Petrophysical Methods and Techniques for Shaly Sands Evaluation," paper SPE 89735 presented at the 2004 SPE International Petroleum Conference in Puebla, Mexico, November 8–9, 2004.

2. Amyx, J.W., Bass, D.M., Jr., and Whiting, R.L.: *Petroleum Reservoir Engineering, Physical Properties,* McGraw-Hill, New York, 1960.

3. Coates, G.R., Menger, S., Prammer, M., and Miller, D.: "Applying NMR Total and Effective Porosity to Formation Evaluation," paper SPE 38736 presented at the 1997 SPE Annual Technical Conference and Exhibition, San Antonio, Texas, October 5–8, 1997.

4. Ellis, D.: "Formation Porosity Estimation from Density Logs," *Petrophysics,* (September–October, 2003) 306–316.

5. Wyllie, M.R.J., Gregory, A.R., and Gardner, G.H.F.: "Elastic Wave Velocity in Heterogeneous and Porous Media," *Geophysics,* (1956) 41–70.

6. Raymer, L.L., Hunt, E.R., and Gardner, J.S.: "An Improved Sonic Transit Time-To-Porosity Transform," SPWLA Twenty-First Annual Logging Symposium, July 8–11, 1980.

7. Freedman, R.: "Advances in NMR Logging," SPE 89177, Distinguished Author Series, *JPT,* (January 2006) 60–66.

8. Bachman, H.N., Crary, S., Heidler, R., LaVigne, J., and Akkurt, R.: "Porosity Determination from NMR Log Data: The Effects of Acquisition Parameters," paper SPE 110803 presented at the 2007 SPE Annual Technical Conference and Exhibition, Anaheim, California, November 11–14, 2007.

9. Akkurt, R., Kersey, D.G., and Zainalabedin, K.: "Challenges for Everyday NMR: An Operator's Perspective," paper SPE 102247 presented at the 2006 SPE Annual Technical Conference and Exhibition, San Antonio, Texas, September 24–27, 2006.

10. Seifert, D.J., Akkurt, R., Al-Dossary, S., Shokeir, R., and Ersoz, H.: "Nuclear Magnetic Resonance Logging: While Drilling, Wireline, and Fluid Sampling," paper SPE 105605 presented at the 15th SPE Middle East Oil & Gas Show and Conference, Bahrain, Kingdom of Bahrain, March 11–14, 2007.

General Reading

- Towler, B.F.: *Fundamental Principles of Reservoir Engineering,* SPE Textbook Series Vol. 8 (2002).
- Mezzatesta, A., Méndez, F., and Rodriguez, E.: "Effective and Total Porosities: Their Reconciliation in Carbonate and Shaly-Sand Systems," paper SPE 102811 presented at the 2006 SPE Annual Technical Conference and Exhibition, San Antonio, Texas, September 24–27, 2006.
- Wiltgen, N.: "Shale Volumes Have Large Uncertainties in the Lower Tertiary Deepwater," paper SPE 114775 presented at the 2008 SPE Annual Technical Conference and Exhibition, Denver, Colorado, September 21–24, 2008.
- Wiltgen, N.: "The Contradiction in the Lower Tertiary Deepwater GoM," paper SPE 114758 presented at the 2008 SPE Annual Technical Conference and Exhibition, Denver, Colorado, September 21–24, 2008.

CHAPTER 2

Permeability and Relative Permeability

2.1 Introduction

Permeability is a measure of the ability of a porous medium, such as reservoir rock, to transmit fluids through its system of interconnected pore spaces. If the porous medium is completely saturated (100% saturated) with a single fluid, the permeability measured is the absolute permeability. **Absolute permeability** is an intrinsic property of the porous medium, and the magnitude of absolute permeability is independent of the type of fluid in the pore spaces. When the pore spaces in the porous medium are occupied by more than one fluid, the permeability measured is the effective permeability of the porous medium to that particular fluid. For instance, the effective permeability of a porous medium to oil is the permeability to oil when other fluids, including oil, occupy the pore spaces. **Relative permeability** is defined as the ratio of effective permeability to absolute permeability of a porous medium. The relationship for relative permeability is represented as:

$$k_{ri} = \frac{k_i}{k_a} \tag{2.1}$$

In Eq. (2.1), k_{ri} = relative permeability of the porous medium to fluid i; k_i = effective permeability of the porous medium for fluid i; and k_a = absolute permeability of the porous medium. For instance, the relative permeability of a porous medium to oil is expressed in a form similar to Eq. (2.1) as:

$$k_{ro} = \frac{k_o}{k_a} \tag{2.2}$$

In Eq. (2.2), k_{ro} = relative permeability of the porous medium to oil; k_o = effective permeability of the porous medium to oil; and k_a = absolute permeability of the porous medium. Similarly, the relative permeability of the porous medium to water and gas are expressed in Eqs. (2.3) and (2.4), respectively, as:

$$k_{rw} = \frac{k_w}{k_a} \tag{2.3}$$

$$k_{rg} = \frac{k_g}{k_a} \tag{2.4}$$

In Eq. (2.3), k_{rw} = relative permeability of the porous medium to water; and k_w = effective permeability of the porous medium to water. In Eq. (2.4), k_{rg} = relative permeability of the porous medium to gas; and k_g = effective permeability of the porous medium to gas. Absolute permeability is measured in a laboratory by flowing a fluid of known viscosity through a core sample from a porous medium while measuring the flow rates and pressure differences across the core sample. The core sample must be totally saturated (100%) with the fluid. By definition, any fluid can be used to measure absolute permeability. In practice, absolute permeability is measured by flowing air through a core sample that has been completely dried. Core permeability measured with air or any other gas must be reviewed and corrected for Klinkenberg[1] effect, if necessary. At low pressures, gas permeability can be higher than absolute permeability due to Klinkenberg effect, as shown in Eq. (2.5):

$$k_{air} = k_a\left(1 + \frac{b}{\overline{P}}\right) \tag{2.5}$$

In Eq. (2.5), k_{air} = air (gas) permeability of core sample as measured; k_a = absolute permeability of the core sample; b = Klinkenberg constant for a given gas in a given core sample; and $\overline{P}$ = mean flowing pressure of the gas at which the gas permeability was measured. Klinkenberg[1] found that at low pressures gas permeability could be higher than absolute permeability of the porous medium due to interactions between gas molecules and the walls of the pore spaces. This interaction is referred to as "slippage," and is related to the molecular size and kinetic energy of the gas. The apparent increase in core air permeabilities at low measurement pressures due to this slippage is called the Klinkenberg effect. The standard unit of measure for permeability in the petroleum industry is the darcy (d). The darcy is the permeability of a porous medium with cross-sectional area of 1 cm^2 through which a fluid with a viscosity of 1 cp is flowing at the volumetric rate of 1 cc/sec under a pressure gradient of 1 atm per 1 cm of length in the direction of flow (see Eq. (2.6)). The millidarcy (md) is 1/1000th of a darcy.

2.2 Sources of Permeability Data

The main sources of permeability data are:

1. Core samples
2. Pressure transient tests
3. Well logs

These sources of permeability data sample volumes of the reservoir at different scales. Permeability data measured on core samples are at a different scale from permeability data calculated from pressure transient tests or well logs. A typical core plug is about 2 inches in diameter by 2 inches long, and is taken from a full-diameter core sample at every half-foot to 1 foot frequency, depending on the sampling program. This represents a sample volume of 0.0036 ft^3 for each core plug. In contrast, a typical well log, such as an NMR (Nuclear Magnetic Resonance) log, has a vertical resolution of 2 to 5 ft, depending on the configuration of the tool. Consequently, the volume of reservoir rock sampled by an NMR log is several orders of magnitude larger than the volume of reservoir rock sampled by core plugs. Pressure transient tests are capable of achieving radii of investigation that extend several hundred feet away from the wellbore, depending on rock and fluid properties of the reservoir and duration of the test. Clearly, the volume of reservoir rock sampled by pressure transient tests is much larger than the volume of reservoir rock sampled by NMR logs. The differences in scale among the various sources of permeability data should be considered before the data are integrated and used for reservoir analysis. Many techniques have been reported in the literature for integrating permeability data with different scales. A method based on calculation of running averages was reported by Worthington.[2] This method was applied to Saudi Arabian reservoirs by Al-Ali and Worthington.[3] Shafer and Ezekwe[4] demonstrated that core and well log permeability data on different scales could be integrated by applying regression analysis based on the Coates[5] equation.

2.2.1 Permeability from Core Samples

The primary source of reliable permeability data is core samples. Ample permeability data are measured on core plugs taken from whole core samples. In the absence of whole cores, permeability data can be obtained from sidewall core samples. Generally, permeability data from sidewall cores are not as reliable as those measured on core plugs taken from whole core samples and should be used with caution. Core plugs are usually taken parallel to the plane of deposition of the formation. Consequently, the permeability data measured on core plugs are horizontal permeability data. To measure vertical permeability, the core plugs should be taken perpendicular to the plane of deposition. In practice, vertical permeability data are seldom measured. However, if the dominant oil displacement mechanism (such as gravity drainage discussed in Chapter 9) is dependent on vertical permeability, extra effort should be made to obtain vertical permeability data from laboratory core measurements. The typical permeability data measured under a routine core analysis (RCAL) program is air permeability. Air permeability is calculated using the Darcy equation Eq. (2.6) from data obtained by placing the core sample in a chamber and measuring the pressure differential across it at stabilized air flow rates under steady-state conditions. Air permeability should be measured at the net confining stress of the reservoir. **Net confining stress** is the pressure difference between the overburden pressure and the pore pressure at the depth of the reservoir. Air permeability data should be corrected for the Klinkenberg effect using Eq. (2.5). As part of a quality control program, the core samples used in core permeability measurements should be examined for the presence of induced fractures, fissures, or any deformations which could influence the magnitude of the permeability data.

2.2.2 Permeability from Pressure Transient Tests

Pressure transient tests are very important sources of permeability data for reservoirs. There are various forms of pressure transient tests that are used to calculate permeability data. These tests can be categorized in two groups, namely: wireline or tubing-conveyed openhole tests and traditional wellhead or down-hole well tests. The wireline or tubing-conveyed openhole tests include wireline formation testers (WFT) (see Chapter 5) and drillstem tests. Traditional wellhead or down-hole well tests comprise of drawdown or buildup tests conducted on active wells, as discussed in Chapter 11. Generally, permeability data from WFT or traditional well tests are at different scales from permeability data measured from core samples or obtained from well logs. The volume of reservoir rock sampled by pressure transient tests depends on the extent of the interval tested, rock and fluid properties of the interval, and duration of the test. Pressure transient tests measure effective permeability, and should be used to condition permeability data from core samples and well logs as necessary to achieve observed well and reservoir performance.

2.2.3 Permeability from Well Logs Based on Empirical Correlations

The concept of permeability as a proportionality constant that governs the rate of fluid flow is demonstrated by Darcy's equation for single phase, linear, horizontal flow in a porous medium as:

$$k = \frac{q \times \mu \times L}{A \times \Delta P} \tag{2.6}$$

In Eq. (2.6), k = permeability, darcy; q = flow rate, cc/sec; μ = viscosity, cp; L = length, cm; A = cross-sectional area, cm^2; and ΔP = pressure difference, atm. If a dimensionless analysis is performed on Eq. (2.6) by substituting for the dimensions of permeability, the result is:

$$\begin{aligned} k &= \frac{rate \times viscosity \times length}{area \times pressure} \\ &= \frac{\left(\frac{L^3}{time}\right)\left(\frac{force \times time}{L^2}\right) \times L}{L^2 \times \left(\frac{force}{L^2}\right)} \\ &= L^2 \end{aligned} \tag{2.7}$$

Eq. (2.7) shows that permeability has dimensions of area (L^2) which are physically related to the surface area of grains exposed to flow. The size and distribution of rock grains in a porous medium are related to surface area, which affects the permeability of the porous medium. This physical aspect of permeability has been used to create empirical equations for prediction of permeabiliy.[6]

An equation that predicts permeability from the porosity and surface area of grains exposed to fluid flow was proposed by Kozeny[7] and Carman[8] as:

$$k = \frac{\phi^3}{F_s\tau^2 A_g^2(1 - \phi)^2} \tag{2.8}$$

Eq. (2.8) is called the Kozeny-Carman equation. In Eq. (2.8), k = permeability, μm^2; ϕ = porosity, fraction; F_s = shape factor; τ = tortuosity; and A_g = surface area per unit grain volume, μm^{-1}. The term, $F_s\tau^2$, is called the Kozeny constant. The Kozeny constant is variable and has been shown by Rose and Bruce[9] to vary from 5 to 100 for reservoir rocks. A modification of the Kozeny-Carman equation was proposed by Wyllie and Rose[10] by substituting irreducible water saturation for specific surface area to get:

$$k^{1/2} = \frac{C_1\phi}{S_{wir} - C_2} \tag{2.9}$$

In Eq. (2.9), C_1 and C_2 are constants that vary depending on the rock and fluid properties of the formation; and S_{wir} = irreducible water saturation. A more generalized form of the Wyllie-Rose correlation was proposed by Timur[11] as:

$$k = \frac{P\phi^Q}{S_{wir}^R} \tag{2.10}$$

In Eq. (2.10), P, Q, and R are regression parameters to be determined by fitting the equation to core data. A form of Eq. (2.10) was developed for sandstones by Timur[11] as:

$$k = \frac{0.136\phi^{4.4}}{S_{wir}^2} \tag{2.11}$$

In Eq. (2.11), k = permeability, md; ϕ = porosity, percent; and S_{wir} = irreducible water saturation, percent. The Timur equation is based on 155 sandstone core samples from three different oil fields in North America. A modified form[12] of the Timur equation which was in general use is represented as:

$$k = \left[\frac{100\phi^{2.25}}{S_{wir}}\right]^2 \tag{2.12}$$

In Eq. (2.12), k = permeability, md; ϕ = porosity, fraction; and S_{wir} = irreducible water saturation, fraction. Eq. (2.12) was used to correlate actual data by adjusting the constant, 100, and the exponent, 2.25, as necessary. Coates and Denoo[13] developed an equation that ensured that permeability is zero at zero effective porosity and 100% irreducible water saturation. The Coates-Denoo[13] equation is represented as:

$$k = \left[\frac{100\phi_e^2(1 - S_{wir})}{S_{wir}}\right]^2 \tag{2.13}$$

In Eq. (2.13), k = permeability, md; ϕ_e = effective porosity, fraction; and S_{wir} = irreducible water saturation, fraction. Bulk volume irreducible water can be represented as:

$$V_{bwi} = \phi_t S_{wir} \tag{2.14}$$

In Eq. (2.14), V_{bwi} = bulk volume irreducible water, fraction; and ϕ_t = total porosity, fraction. Eq. (2.13) can be represented in terms of bulk-volume irreducible water by multiplying the numerator and denominator of the equation with total porosity to get:

$$k = \left[\frac{100\phi_e^2(\phi_t - V_{bwi})}{V_{bwi}}\right]^2 \tag{2.15}$$

The improved reliability of nuclear magnetic resonance (NMR) logs led to the reformulation of Eq. (2.15) by Coates et al.[5] in terms of free-fluid index (*FFI*) and bulk volume irreducible (*BVI*) saturations. *FFI* is the fractional part of the pore volume occupied by fluids that are free to flow and *BVI* is the fractional part of the pore volume occupied by immobile bound water. The Coates equation for calculation of permeability from NMR logs is expressed as:

$$k = \left(\frac{\phi}{C}\right)^m \left(\frac{FFI}{BVI}\right)^n \tag{2.16}$$

In Eq. (2.16), k = permeability, md; ϕ = porosity, percent; C = 10; m = 4; and n = 2. The values given for C, m, and n parameters are default values. Values of these parameters are dependent on the core samples and can be regressed from actual data. *FFI* is obtained by summing the T_2 distribution over T_2 values greater than the T_2 cutoff. The transverse relaxation time, T_2, is the characteristic time constant that represents the envelope of the spin-echo signal decay in a NMR logging sequence.[14] The main difficulty of using NMR logs to determine *FFI* is calibration of T_2 cutoff. One of the procedures of calibrating T_2 cutoff is to match irreducible water saturations (S_{wir}) obtained from an independent method, such as the Dean-Starks method or Centrifugal method, with irreducible water saturations derived from NMR data.[15] The irreducible water saturations of core samples shown in Figure 2.1 were measured by the centrifugal method and matched against irreducible water saturations from NMR core laboratory data based on T_2 cutoffs. The T_2 cutoffs were then used to determine *FFI* and *BVI* for the core samples. NMR permeability for each core sample was then calculated from these data using Eq. (2.16). The comparison of calculated NMR permeability versus core permeability measured by routine core analysis method is shown in Figure 2.2. The approach discussed here is a method regularly used to calibrate NMR calculated permeability with core permeability obtained through routine

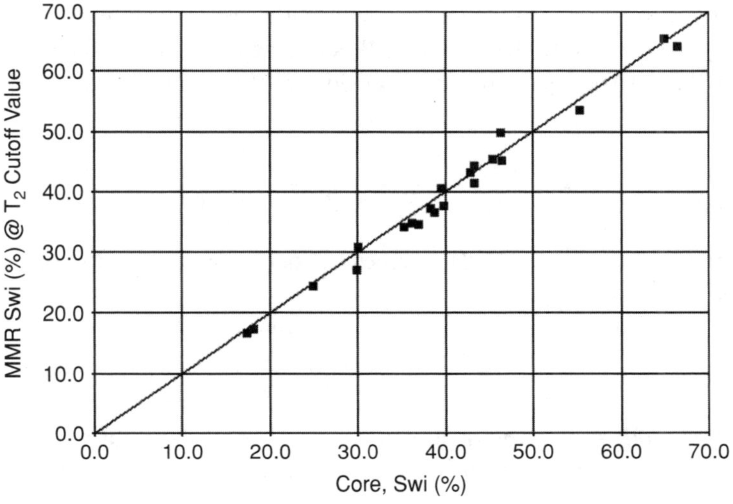

Figure 2.1 Determination of T_2 cutoff from irreducible water saturation comparison.

core analysis methods. The parameters for the Coates equation obtained from regression of the calibrated permeability data are then used to calculate permeability data from NMR data measured in other wells. The Coates equation is widely used for the calculation of permeability data as discussed, if NMR logs are available. This is the method recommended for the calculation of permeability data from NMR and porosity logs.

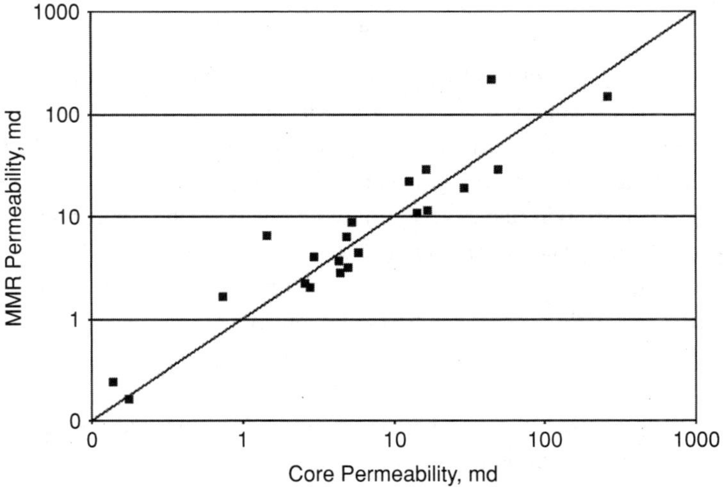

Figure 2.2 NMR permeability calculated from Coates equation versus core permeability.

An important correlation for predicting permeability was developed by Amaefule et al.[16] from the Kozeny-Carman equation (Eq. (2.8)). By dividing both sides of Eq. (2.8) with effective porosity, ϕ_e, and taking the square root of both sides of the equation, the resulting equation is expressed as:

$$\sqrt{\frac{k}{\phi_e}} = \left[\frac{\phi_e}{1 - \phi_e}\right]\left[\frac{1}{\sqrt{F_s \tau A_g}}\right] \tag{2.17}$$

The reservoir quality index (RQI) parameter is defined in terms of permeability in millidarcies (md) as:

$$RQI = 0.0314\sqrt{\frac{k}{\phi_e}} \tag{2.18}$$

In Eq. (2.18), RQI = reservoir quality index, μm; k = permeability, md; and ϕ_e = effective porosity, fraction. The pore volume-to-grain volume ratio, ϕ_z, is defined as:

$$\phi_z = \frac{\phi_e}{1 - \phi_e} \tag{2.19}$$

The flow zone indicator (FZI) parameter is designated by the following expression as:

$$\begin{aligned} FZI &= \frac{1}{\sqrt{F_s \tau A_g}} \\ &= \frac{RQI}{\phi_z} \end{aligned} \tag{2.20}$$

In Eq. (2.20), FZI = flow zone indicator, μm. Taking the logarithm of both sides of Eq. (2.20) gives:

$$\log RQI = \log FZI + \log \phi_z \tag{2.21}$$

As demonstrated by Amaefule et al.,[16] a log-log plot of RQI versus ϕ_z will give a straight line with unit slope for all samples with similar FZI values. Data from core samples with different FZI values will lie on separate, parallel lines. The value of the constant for FZI is determined from the intercept of the unit slope straight line at $\phi_z = 1$. As stated by Amaefule et al.,[16] core data that lie on the same straight line have similar pore throat attributes, and thereby can be categorized as a hydraulic unit. The classification of reservoir rock as hydraulic units is useful as a means of characterizing the reservoir as discussed in Chapter 18. Numerous reservoir studies[17–19] have been conducted by application of the hydraulic unit concepts. However, note that the application of this method for prediction of permeability data requires the availability of significant quantities of core permeability data.

A common empirical relationship that is widely used to correlate permeability and porosity data is a crossplot of the data on a semi-log scale as represented by the equation:

$$\log k = a\phi + b \tag{2.22}$$

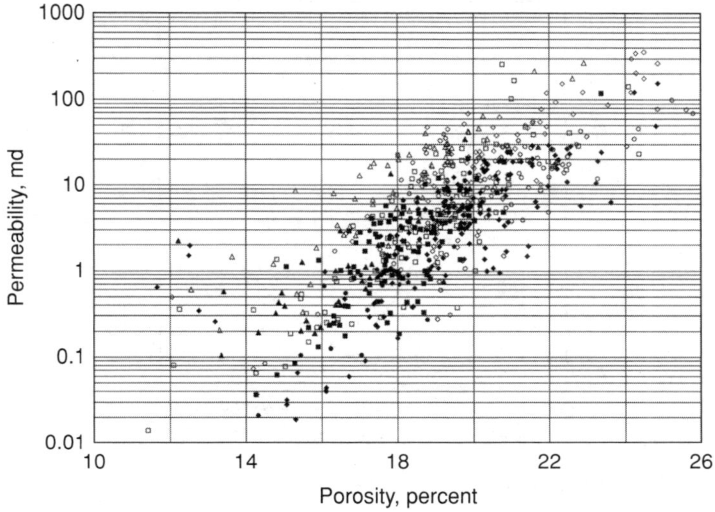

Figure 2.3 Porosity-permeability crossplot based on core samples from a reservoir.

In Eq. (2.22), a and b are regression constants that depend on the data set. Eq. (2.22) has no theoretical basis but has been adopted because the distribution of permeability data appears to be log-normal. An example of a permeability-porosity ($\phi - k$) crossplot for core samples from a reservoir is shown in Figure 2.3. The $\phi - k$ crossplot is a convenient method for displaying core or log data but is generally misused as a correlation for predicting permeability data from porosity data. Worthington[2] and Delfiner[20] reported that indiscriminate use of the $\phi - k$ crossplot for predicting permeability data could cause the underestimation of permeability data by at least a factor of 2. The $\phi - k$ crossplot is not recommended as a method for predicting permeability data in the form of a correlation. The crossplot can be used qualitatively to review permeability data predicted using Eq. (2.16) or Eq. (2.21). The application of the $\phi - k$ crossplots in the form of cloud transforms could be used for reservoir characterization, as described in Chapter 18.

2.3 Relative Permeability

The relative permeability of a fluid flowing in a porous medium is the ratio of its effective permeability to the absolute permeability of the porous medium as represented in Eq. (2.1). Relative permeability data can be presented graphically in plots called relative permeability curves. A typical relative permeability curve for an oil-water system is shown in Figure 2.4. In Figure 2.4, the range of water saturation is from the initial water saturation, S_{wi} to water saturation at residual oil saturation, $1 - S_{orw}$. Oil relative permeability, k_{ro}, is highest at S_{wi} and declines to zero at $1 - S_{orw}$. Water relative permeability, k_{rw}, increases from zero at S_{wi} to its highest value at $1 - S_{orw}$. Figure 2.4 shows the location of critical water saturation, S_{wc}. Critical water saturation is the level of water saturation at which water starts to flow in the reservoir. The relative

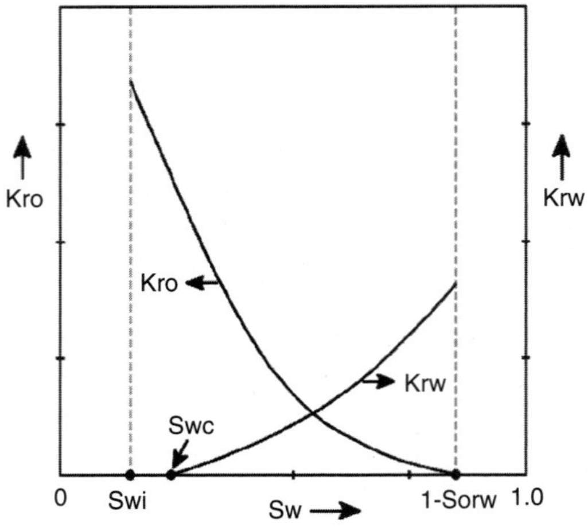

Figure 2.4 Oil-water relative permeability curves.

permeability curve for a gas-oil system is shown in Figure 2.5. The range of gas saturation is from initial gas saturation, S_{gi}, to gas saturation at residual oil saturation, $1 - S_{org}$. Oil relative permeability, k_{ro}, is highest at S_{gi} and declines to zero at $1 - S_{org}$. Gas relative permeability, k_{rg}, increases from zero at S_{gi} to its highest value at $1 - S_{org}$. The critical gas saturation, S_{gc}, is the level of gas saturation at which gas starts to flow in the reservoir.

The relative permeability of a fluid in a multi-fluid system is a function of the saturation level of the fluid as shown in Figures 2.4 and 2.5. However, many distributions of the fluid are possible at the same saturation level, depending on the wettability and the direction of saturation changes (or saturation history) of the porous media. For this reason, permeability data measured

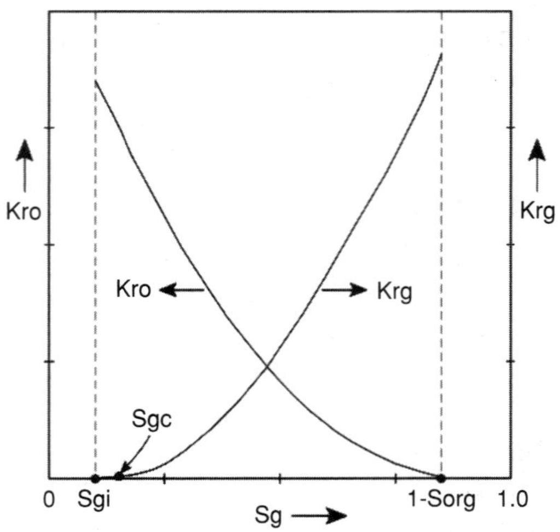

Figure 2.5 Gas-oil relative permeability curves.

for drainage displacements (decrease in saturation of the wetting phase) are usually different from permeability data measured for imbibition displacement (increase in saturation of the wetting phase). This difference is described as the **hysteresis** of the permeability curves. It is similar to the hysteresis of the capillary pressure curves discussed in Chapter 15, Section 15.2.2. Basic concepts on rock wettability are presented in Chapter 15, Section 15.2.1.

2.4 Sources of Relative Permeability Data

Relative permeability data are obtained from many sources for hydrocarbon reservoirs. These sources include laboratory measurements on core samples, calculations from field production data, and empirical correlations. The most reliable source of relative permeability data are laboratory measurements on core samples.[21] Some success has been reported on estimating relative permeability data from field production data.[22] In the absence of laboratory data, relative permeability data can be calculated from empirical correlations.

2.4.1 Laboratory Measurements of Relative Permeability Data

Laboratory methods used in measuring relative permeability data can be classified under two groups, namely: steady-state methods and unsteady-state methods. Steady-state methods are more widely used and are considered more reliable than unsteady-state methods.[21] However, steady-state methods require more experimental time and more elaborate equipment than unsteady-state methods. Some steady-state measurements may require days of experimental time in comparison to hours for unsteady-state methods. Both methods are affected by capillary end effects. **Capillary end effect** is a phenomenon that causes the saturation of the wetting phase to be high close to the inlet and outlet ends of the core sample. The higher saturations at the ends of the cores are caused by the tendency of the wetting phase to remain in the core capillary spaces rather than exit into a noncapillary space. Several measures are employed in relative permeability measurements to reduce or eliminate capillary end effects in core samples. One of the methods used to eliminate capillary end effects is the Hassler's technique that involves the placement of porous plates in contact with both ends of the core.[21] Both steady-state and unsteady-state measurements should be conducted at reservoir conditions.

2.4.1.1 Steady-State Methods

Steady-state methods yield the most reliable relative permeability data because capillary equilibrium is achieved in the method, fluid saturations can be measured directly, and the calculation procedure is based on the Darcy equation (Eq. (2.6)). In the steady-state method, two or three fluids are injected simultaneously at constant rates or pressures to achieve equilibrium inside the porous medium.[21] After equilibrium is achieved, the saturations, flow rates, and pressure gradients of each phase are measured. Using the Darcy equation, the effective permeability of each phase is then calculated. The entire saturation range can be measured in a stepwise fashion by changing the ratio of injection rates and repeating the procedure.[23] Steady-state measurement requires many hours or even days to complete because equilibrium conditions must be attained at each saturation level.

2.4.1.2 Unsteady-State Methods

Unlike steady-state methods, relative permeability data can be obtained with unsteady-state methods very quickly. Because unsteady-state methods do not require the attainment of equilibrium in the displacement process, many relative permeability data can be acquired in a few hours. In a typical unsteady-state method, the in-situ fluids are displaced by injection of the displacing fluid at constant rate or pressure, while measuring the produced fluids continuously. The relative permeability data are calculated from the production data on the basis of the Buckley-Leverett equation.[24,25] A widely used unsteady-state method is the use of centrifuge for measurement of relative permeability data. In this method, saturated core samples are rotated at high angular speed, creating a known centrifugal force. The rate of fluid production is measured as a function of time. Relative permeability data are then calculated from the measurements.[26] The centrifuge method is extremely fast and is not affected by the viscous fingering problems associated with other unsteady-state methods. Permeability data from unsteady-state methods are affected by displacement problems, such as viscous fingering and channeling, including instability of the displacement process.[27] The method is recommended to be used to measure relative permeability at the saturation endpoints, while steady-state methods are used to infill the data between the endpoints.[28]

2.4.2 Estimations from Field Data

Relative permeability data can be estimated through history-matching of field production data using reservoir simulators. The generation of relative permeability data from historical production data have been reported by many authors, including Kulkarni and Datta-Gupta,[22] Watson et al.,[29] Yang and Watson,[30] Reynolds et al.,[31] and Eydinov et al.[32] The agreement between permeability data measured on core samples and those estimated from production data are generally not good. This should be expected since cores are samples of very small volumes of the reservoir (microscopic scale), while production data are based on much larger volumes of the reservoir (macroscopic scale). Relative permeability data calculated from field data should be used to adjust laboratory-derived relative permeability data for reservoir simulation applications.

2.4.3 Empirical Correlations

Several correlations are available for estimation of relative permeability data. These correlations should be used in situations where measured relative permeability data are not available. Reasonable caution should be exercised in applying these correlations because they may not be representative of the rock systems that exist in the reservoir. The most widely used correlations for the estimation of relative permeability data are modifications of the Corey equations.[33]

The basic Corey equations are represented as:

$$k_{rwp} = \left(S_{wp}^*\right)^4 \tag{2.23}$$

$$k_{rnwp} = \left(1 - S_{wp}^*\right)^2 \left[1 - \left(S_{wp}^*\right)^2\right] \tag{2.24}$$

In Eqs. (2.23) and (2.24), k_{rwp} = relative permeability of the wetting phase; k_{rnwp} = relative permeability of the non-wetting phase; and S_{wp}^* = normalized saturation of the wetting phase. For a drainage process, the normalized saturation of the wetting phase is expressed as:

$$S_{wp}^* = \frac{S_{wp} - S_{wpr}}{1 - S_{wpr}} \tag{2.25}$$

For an imbibition process, the normalized saturation of the wetting phase is expressed as:

$$S_{wp}^* = \frac{S_{wp} - S_{wpr}}{1 - S_{wpr} - S_{nwpr}} \tag{2.26}$$

In Eqs. (2.25) and (2.26), S_{wp} = saturation of the wetting phase; S_{wpr} = residual saturation of the wetting phase; and S_{nwpr} = residual saturation of the non-wetting phase.

Applying the Corey equations to a water-wet, two-phase oil-water system, and assuming oil displacement process is by imbibition, substituting Eq. (2.26) gives:

$$S_w^* = \frac{S_w - S_{wi}}{1 - S_{wi} - S_{orw}} \tag{2.27}$$

From Eq. (2.27),

$$1 - S_w^* = \frac{1 - S_w - S_{orw}}{1 - S_{wi} - S_{orw}} \tag{2.28}$$

In Eqs. (2.27) and (2.28), S_w = water saturation, fraction; S_{orw} = residual oil saturation to water, fraction; and S_{wi} = initial water saturation, fraction. Substituting Eq. (2.27) into Eq. (2.23) gives:

$$k_{rw} = \left(\frac{S_w - S_{wi}}{1 - S_{wi} - S_{orw}}\right)^4 \tag{2.29}$$

In Eq. (2.29), k_{rw} = relative permeability of water. A generalized form of the equation can be written by expressing the Corey exponent as N and including the endpoint of relative permeability of water at residual oil saturation, $k_{rw@S_{orw}}$, such that Eq. (2.29) becomes:

$$k_{rw} = k_{rw@S_{orw}}\left(\frac{S_w - S_{wi}}{1 - S_{wi} - S_{orw}}\right)^N \tag{2.30}$$

Eq. (2.24) can be simplified by dropping the second term in the equation to get:

$$k_{rnwp} = \left(1 - S_{wp}^*\right)^2 \tag{2.31}$$

Substituting Eq. (2.28) into Eq. (2.31) yields:

$$k_{ro} = \left(\frac{1 - S_w - S_{orw}}{1 - S_{wi} - S_{orw}}\right)^2 \tag{2.32}$$

In Eq. (2.32), k_{ro} = relative permeability of oil. A generalized form of the equation can be written by expressing the Corey exponent as N and including the endpoint of relative permeability of oil at initial water saturation, $k_{ro@S_{wi}}$, such that Eq. (2.32) becomes:

$$k_{ro} = k_{ro@S_{wi}}\left(\frac{1 - S_w - S_{orw}}{1 - S_{wi} - S_{orw}}\right)^N \tag{2.33}$$

Corey equations can also be derived for a water-wet, gas-oil system as follows. From Eq. (2.27), it can be deduced that:

$$S_w^* = \frac{1 - S_g - S_{wi} - S_{org}}{1 - S_{wi} - S_{org}} \tag{2.34}$$

Substituting Eq. (2.34) into Eq. (2.23) gives:

$$k_{ro} = \left(\frac{1 - S_g - S_{wi} - S_{org}}{1 - S_{wi} - S_{org}}\right)^4 \tag{2.35}$$

A generalized form of the equation can be written by expressing the Corey exponent as N and including the endpoint of relative permeability of oil at initial gas saturation, $k_{ro@S_{gi}}$, such that Eq. (2.35) becomes:

$$k_{ro} = k_{ro@S_{gi}}\left(\frac{1 - S_g - S_{wi} - S_{org}}{1 - S_{wi} - S_{org}}\right)^N \tag{2.36}$$

The gas relative permeability, k_{rg}, can be derived by substituting Eq. (2.28) into Eq. (2.31) to get:

$$\begin{aligned} k_{rg} &= \left(\frac{1 - S_w - S_{org} - S_{gc}}{1 - S_{wi} - S_{org} - S_{gc}}\right)^2 \\ &= \left(\frac{S_g - S_{gc}}{1 - S_{wi} - S_{org} - S_{gc}}\right)^2 \end{aligned} \tag{2.37}$$

A generalized form of the equation can be written by expressing the Corey exponent as N and including the endpoint of relative permeability of gas at residual oil saturation, $k_{rg@S_{org}}$, such that Eq. (2.37) becomes:

$$k_{rg} = k_{rg@S_{org}}\left(\frac{S_g - S_{gc}}{1 - S_{wi} - S_{org} - S_{gc}}\right)^N \tag{2.38}$$

Note that similar modified Corey equations can be derived for an oil-wet system assuming drainage displacement process from Eqs. (2.23), (2.24), and (2.25).

In the absence of data, Corey exponent can be estimated for different rock formations as follows: well consolidated sandstones, $N = 3.0$; poorly consolidated sandstones, $N = 3.5$; cemented sandstone and limestones, $N = 4.0$; and fractured formations, $N = 1.0$.

Example 2.1 Calculation of Oil-Water and Gas-Oil Relative Permeability Curves Using the Modified Corey Equations

Problem

Calculate the oil-water and gas-oil relative permeability data for a water-wet, poorly consolidated rock with the following rock and fluid properties:

Initial water saturation, S_{wi}	0.2
Residual oil saturation to water, S_{orw}	0.2
Residual oil saturation to gas, S_{org}	0.2
Critical gas saturation, S_{gc}	0.05
Oil curve endpoint, $k_{ro@S_{wi}}$	0.9
Oil curve endpoint, $k_{ro@S_{gi}}$	0.9
Water curve endpoint, $k_{rw@S_{orw}}$	0.15
Gas curve endpoint, $k_{rg@S_{org}}$	1.0
Corey exponent	3.5

Solution

Using Eq. (2.30), the relative permeability of water at water saturation, $S_w = 0.3$ is calculated as:

$$\begin{aligned} k_{rw} &= k_{rw@S_{orw}}\left(\frac{S_w - S_{wi}}{1 - S_{wi} - S_{orw}}\right)^N \\ &= 0.15\left(\frac{0.3 - 0.2}{1 - 0.2 - 0.2}\right)^{3.5} \\ &= 0.000284 \end{aligned}$$

Similarly, the remaining water relative permeabilities at other water saturations were calculated as shown in Table 2.1.

Using Eq. (2.33), the relative permeability of oil at water saturation, $S_w = 0.3$ is calculated as:

$$\begin{aligned} k_{ro} &= k_{ro@S_{wi}}\left(\frac{1 - S_w - S_{orw}}{1 - S_{wi} - S_{orw}}\right)^N \\ &= 0.9\left(\frac{1 - 0.3 - 0.2}{1 - 0.2 - 0.2}\right)^{3.5} \\ &= 0.4755 \end{aligned}$$

Similarly, the remaining oil relative permeabilities at other water saturations were calculated as shown in Table 2.1. The calculated relative permeability data for oil and water are shown in Figure 2.6.

Table 2.1 Oil-water relative permeabilities calculated for Example 2.1

S_w	k_{rw}	k_{ro}
0.20	0.0000	0.9000
0.25	0.0000	0.6637
0.30	0.0003	0.4755
0.35	0.0012	0.3288
0.40	0.0032	0.2177
0.45	0.0070	0.1364
0.50	0.0133	0.0795
0.55	0.0227	0.0420
0.60	0.0363	0.0192
0.65	0.0548	0.0070
0.70	0.0792	0.0017
0.75	0.1106	0.0002
0.80	0.1500	0.0000

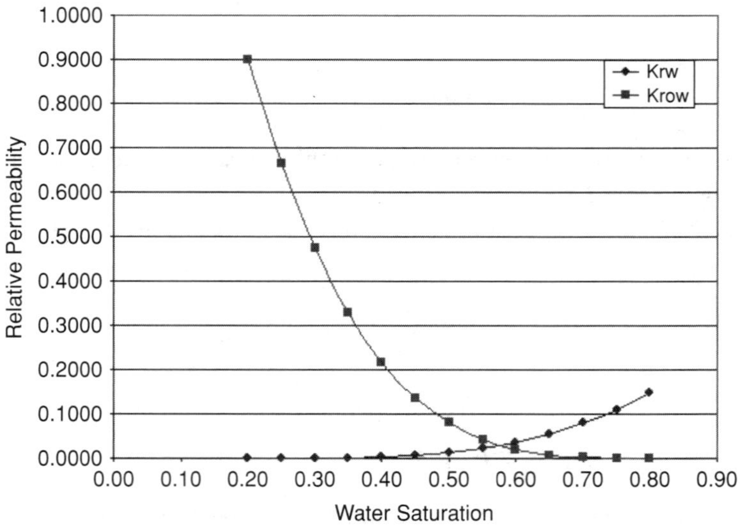

Figure 2.6 Oil-water relative permeability plot for Example 2.1.

Using Eq. (2.36), the relative permeability of oil at gas saturation, $S_g = 0.4$ is calculated as:

$$k_{ro} = k_{ro@S_{gi}}\left(\frac{1 - S_g - S_{wi} - S_{org}}{1 - S_{wi} - S_{org}}\right)^N$$

$$= 0.9\left(\frac{1 - 0.4 - 0.2 - 0.2}{1 - 0.2 - 0.2}\right)^{3.5}$$

$$= 0.01925$$

Similarly, the remaining oil relative permeabilities at other gas saturations were calculated as shown in Table 2.2.

Using Eq. (2.38), the relative permeability of gas at gas saturation, $S_g = 0.4$ is calculated as:

$$k_{rg} = k_{rg@S_{org}}\left(\frac{S_g - S_{gc}}{1 - S_{wi} - S_{org} - S_{gc}}\right)^N$$

$$= 1.0 \times \left(\frac{0.4 - 0.05}{1 - 0.2 - 0.2 - 0.05}\right)^{3.5}$$

$$= 0.2056$$

Table 2.2 Gas-oil relative permeabilities calculated for Example 2.1

S_g	k_{ro}	k_{rg}
0.05	0.6637	0.0000
0.10	0.4755	0.0002
0.15	0.3288	0.0026
0.20	0.2177	0.0106
0.25	0.1364	0.0290
0.30	0.0795	0.0633
0.35	0.0420	0.1199
0.40	0.0192	0.2056
0.45	0.0070	0.3281
0.50	0.0017	0.4954
0.55	0.0002	0.7164
0.60	0.0000	1.0000

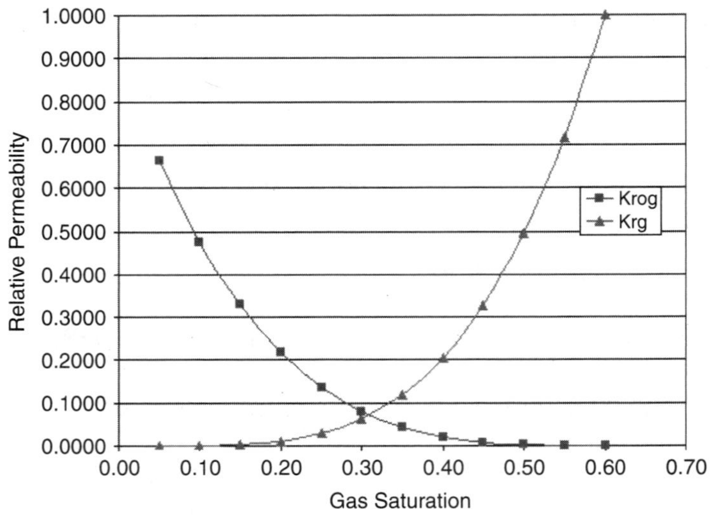

Figure 2.7 Gas-oil relative permeability plot for Example 2.1.

The remaining gas relative permeabilities at other gas saturations were calculated as shown in Table 2.2. The calculated relative permeability data for gas and oil are shown in Figure 2.7.

2.5 Three-Phase Relative Permeability

The relative permeability curves shown in Figures 2.4 and 2.5 are based on two-phase systems. However, in many reservoirs, the flow of three phases (oil, water, and gas) is present during primary, secondary, and enhanced oil recovery processes. The simultaneous flow of three phases requires the use of three-phase relative permeability data to represent fluid flow in the reservoir. Three-phase relative permeability data are usually not measured in the laboratory because there are expensive, tedious, and in some cases impractical. The current accepted engineering practice for prediction of three-phase relative permeability is based on empirical models. Most of the empirical models use data from two-phase relative permeabilities, capillary pressure, and saturation history to predict three-phase relative permeability data. Stone's empirical models (Stone I[34] and Stone II[35]) are the two empirical models widely used in the petroleum industry for the prediction of three-phase relative permeability data. Numerous other empirical models that appear to be superior to Stone's models have been published in the literature.[36–38] However, reviews of these models available in the literature indicate that the empirical models should be tested for accuracy before using them for a specific application.[36–38]

2.6 Applications of Permeability and Relative Permeability Data

Permeability and relative permeability data are two of the most important data that determine the flow of fluids in the reservoir. Consequently, permeability and relative permeability data are very important input in flow calculations, including reservoir simulation. As discussed in Chapter 19, it is common practice in reservoir simulation to adjust permeability and relative permeability data to match performance history, achieve production targets and projected reservoir performance. This practice should be conducted with careful consideration of the reasonableness and soundness of the data generated by this process. Permeability and relative permeability data should not be adjusted to the extent that the resulting data violate known physical properties of the rocks and fluids that are present in the reservoir.[39,40]

Nomenclature

A	Area, cm^2
A_g	surface area
b	Klinkenberg constant
F_s	shape factor
k	permeability
k_a	absolute permeability
k_{air}	permeability measured with air
k_g	effective permeability for gas
k_i	effective permeability for fluid, i
k_o	effective permeability for oil
k_w	effective permeability for water
k_{rg}	relative permeability for gas
k_{ri}	relative permeability for fluid, i
k_{ro}	relative permeability for oil
k_{rw}	relative permeability for water
k_{rwp}	relative permeability of the wetting phase
k_{rnwp}	relative permeability of the non-wetting phase
L	length, cm
N	Corey's exponent
$\overline{p}$	mean flowing pressure
Δp	pressure difference, atm
q	flow rate, cc/sec
S_g	gas saturation, fraction
S_{gc}	critical gas saturation, fraction
S_w	water saturation, fraction
S_w^*	normalized water saturation
S_{wc}	critical water saturation, fraction
S_{wi}	initial water saturation, fraction
S_{wir}	irreducible water saturation, fraction
S_{wp}	saturation of the wetting phase
S_{wpr}	residual saturation of the wetting phase
S_{nwpr}	residual saturation of the non-wetting phase

S_{wp}^*	normalized saturation of the wetting phase
S_{org}	residual oil saturation to gas
S_{orw}	residual oil saturation to water
T_2	NMR characteristic time constant, milliseconds
V_{bwi}	bulk volume irreducible water
ϕ	porosity
ϕ_e	effective porosity
ϕ_t	total porosity
ϕ_z	pore volume-to-grain volume ratio
μ	viscosity, cp
τ	tortuosity

Abbreviations

BVI	Bulk Volume Irreducible
FFI	Free Fluid Index
FZI	Flow Zone Indicator
NMR	Nuclear Magnetic Resonance
RCAL	Routine Core Analysis
RQI	Reservoir Quality Index
WFT	Wireline Formation Tester

References

1. Klinkenberg, L.J.: "The Permeability of Porous Media to Liquids and Gases," *Drilling and Production Practices*, American Petroleum Institute, Washington, DC, 1941, 200–213.
2. Worthington, P.F.: "The Effect of Scale on the Petrophysical Estimation of Intergranular Permeability," *Petrophysics* (January–February 2004) 45, 1, 59–72.
3. Al-Ali, H.A., and Worthington, P.F.: "Applications of Petrophysical Scale Reconciliation to Saudi Arabian Reservoirs," paper IPTC 10726 presented at the International Petroleum Technology Conference, Doha, Qatar, November 21–23, 2005.
4. Shafer, J.L., and Ezekwe, N.J.: "Methods for Upscaling Diverse Rock Permeability Data for Reservoir Characterization and Modeling," paper SPE 109128 presented at the Annual Technical Conference and Exhibition, Anaheim, California, November 11–14, 2007.
5. Coates, G.R., Miller, M., and Henderson, G.: "The MRIL in Conoco 33—1: An Investigation of a New Magnetic Resonance Imaging Log," paper DD presented at the 32nd SPWLA Annual Logging Symposium, Midland, Texas, June 16–19, 1991.
6. Shang, B.Z., Hamman, J.G., Chen, H., and Caldwell, D.H.: "A Model to Correlate Permeability with Efficient Porosity and Irreducible Water Saturation," paper SPE 84303 presented at SPE Annual Technical Conference and Exhibition, Denver, Colorado, October 5–8, 2003.

7. Kozeny, J., 1927, Uber Kapillare Leitung des Wassers im Boden: Sitzungsber. Akad. Wiss. Wien, 136, 271–306.

8. Carman, P. C.: "Fluid Flow through a Granular Bed," *Trans. Inst. Chem. Eng. London,* 1937, 15, 150–156.

9. Rose, W., and Bruce, W.A.: "Evaluation of Capillary Character in Petroleum Reservoir Rock," *Trans., AIME* (1949) 127–142.

10. Wyllie, M.R.J., and Rose, W.D.: "Some Theoretical Considerations Related to the Quantitative Evaluation of the Physical Characteristics of Reservoir Rock from Electrical Log Data," *Trans., AIME* (1950) 105–118.

11. Timur, A: "An Investigation of Permeability, Porosity, and Residual Water Saturation Relationship for Sandstone Reservoirs," *The Log Analyst* (July–August), 1968.

12. Coates, G.R., and Dumanoir, J.L.: "A New Approach to Improved Log-Derived Permeability," *Proceedings,* SPWLA 14th Annual Logging Symposium, Lafayette, Louisiana (May 6–9, 1973).

13. Coates, G., and Denoo, S.: "The Producibility Answer Product," *The Technical Review,* Schlumberger, Houston, Texas (June 1981) 29, 2, 55–63.

14. Freedman, R.: "Advances in NMR Logging," Distinguished Author Series, *JPT* (January 2006) 60–66.

15. Shafer, J., Chen, S., Georgi, D., and Chen, J.: "Methods and Procedures for Calibrating NMR Log Derived Permeabilities," The 11th Formation Evaluation Symposium of Japan, Chiba, Japan (October 5–6, 2005) 1–15.

16. Amaefule, J.O., Altunbay, M., Tiab, D., Kersey, D.G., and Keelan, D.K.: "Enhanced Reservoir Description: Using Core and Log Data to Identify Hydraulic (Flow) Units and Predict Permeability in Uncored Intervals/Wells," paper SPE 26436 presented at the SPE 68th Annual Technical Conference and Exhibition, Houston, Texas, October 3–6, 1993.

17. Abbaszadeh, M., Fujii, H., and Fujimoto, F.: "Permeability Prediction by Hydraulic Flow Units—Theory and Applications," *SPEFE* (December 1996) 263–271.

18. Soto B.R., Garcia, J.C., Torres, F., and Perez, G.S.: "Permeability Prediction Using Hydraulic Flow Units and Hybrid Soft Computing Systems," paper SPE 71455 presented at the 2001 SPE Annual Technical Conference and Exhibition, New Orleans, Louisiana, September 30–October 3, 2001.

19. Shenawi, S.H., White, J.P., Elrafie, E.A., and Kilany, A.: "Permeability and Water Saturation Distribution by Lithologic Facies and Hydraulic Units: A Reservoir Simulation Case Study," paper SPE 105273 presented at the 15th SPE Middle East Oil & Gas Show and Conference, Bahrain, March 11–14, 2007.

20. Delfiner, P.: "Three Statistical Pitfalls of Phi-K Transforms," *SPERE&E* (December 2007) 609–617.

21. Honapour, M., and Mahmood, S.M.: "Relative Permeability Measurements: An Overview," *JPT* (August 1988) 963–966.

22. Kulkarni, K.N., and Datta-Gupta, A.: "Estimating Relative Permeability From Production Data: A Streamline Approach," *SPEJ* (December 2000) 402–411.

23. Braun, E.M., and Blackwell, R.J.: "A Steady-State Technique for Measuring Oil-Water Relative Permeability Curves at Reservoir Conditions," paper SPE 10155 presented at the 56th Annual Technical Conference and Exhibitions, San Antonio, Texas, October 5–7, 1981.

24. Johnson, E.F., Bossler, D.P., and Naumann, V.O.: "Calculation of Relative Permeability from Displacement Experiments," SPE 1023G, *Trans. AIME* (1959) 370–372.

25. Batycky, J.P., McCaffery, F.G., Hodgins, P.K., and Fisher, D.B.: "Interpreting Relative Permeability and Wettability from Unsteady-State Displacement Measurements," *SPEJ* (June 1981) 296–308.

26. Spronsen, E.V.: "Three-Phase Relative Permeability Measurements Using Centrifuge Method," paper SPE/DOE 10688 presented at the SPE/DOE Third Joint Symposium on Enhanced Oil Recovery, Tulsa, Oklahoma, April 4–7, 1982.

27. Peters, E.J., and Khataniar, S.: "The Effect of Instability on Relative Permeability Curves Obtained by the Dynamic-Displacement Method," *SPEFE* (December 1987) 469–474.

28. Shafer, J.L., Braun, E.M., Wood, A.C., and Wooten, J.M.: "Obtaining Relative Permeability Data Using a Combination of Steady-State and Unsteady-State Corefloods," paper 9009 presented at the 1990 SCA Conference, Houston, Texas, August 14–16, 1990.

29. Watson, A.T, Seinfeld, J.H., Gavalas, G.R., and Woo, P.T.: "History Matching in Two-Phase Petroleum Reservoirs," *SPEJ* (December 1980) 521–532.

30. Yang, P-H., and Watson, A.T.: "A Bayesian Methodology for Estimating Relative Permeability Curves," *SPERE* (May 1991) 259–265.

31. Reynolds, A.C., Li, R., and Oliver, D.S.: "Simultaneous Estimation of Absolute and Relative Permeability by Automatic History Matching of Three-Phase Flow Production Data," *JCPT* (2004) 43, 3, 37–46.

32. Eydinov, D., Gao, G., Li, G., and Reynolds, A.C.: "Simultaneous Estimation of Relative Permeability and Porosity/Permeability Fields by History Matching Production Data," *JCPT* (December 2009) 48, 12, 13–25.

33. Corey, A.T.: "The Interrelation between Gas and Oil Relative Permeabilities," *Production Monthly* (1954) 19, 1, 38–41.

34. Stone, H.L.: "Probability Model for Estimating Three-Phase Relative Permeability," *JPT* (February 1970) 214–218.

35. Stone, H. L.: "Estimation of Three-Phase Relative Permeability and Residual Oil Data," *JCPT* (1973) 12, 53–61.

36. Manjnath, A., and Honarpour, M.M.: "An Investigation of Three-Phase Relative Permeability," paper SPE 12915 presented at the 1984 Rocky Mountain Regional Meeting, Casper, Wyoming, May 21–23, 1984.

37. Blunt, M.J.: "An Empirical Model for Three-Phase Relative Permeability," *SPEJ* (December 2000) 435–445.

38. Ahmadloo, F., Asghari, K., and Jamaloei, B.Y.: "Experimental and Theoretical Studies of Three-Phase Relative Permeability," paper SPE 124538 presented at the 2009 SPE Annual Technical Conference and Exhibition, New Orleans, Louisiana, October 4–7, 2009.

39. Coats, K.H.: "Use and Misuse of Reservoir Simulation Models," *JPT* (1969) 183–190.

40. Wang, K., Killough, J.E., and Sepehrnoori, K.: "A New Upscaling Method of Relative Permeability Curves for Reservoir Simulation," paper SPE 124819 presented at the 2009 SPE Annual Technical Conference and Exhibition, New Orleans, Louisiana, October 4–7, 2009.

General Reading

- Babadagli, T., and Al-Salmi, S.: "A Review of Permeability-Prediction Methods for Carbonate Reservoirs Using Well-Log Data," *SPEREE* (April 2004) 75–88.
- Nelson, P.H.: "Permeability-Porosity Relationships in Sedimentary Rocks," *The Log Analyst* (May–June 1994) 38–62.
- Ahmed, U., Crary, S.F., and Coates, G.R.: "Permeability Estimation: The Various Sources and Their Interrelationships," *JPT* (May 1991) 578–587.
- Haddad, S., Cribbs, M., Sagar, R., Tang, Y., Viro, E., and Castelijins, K.: "Integrating Permeabilities from NMR, Formation Tester, Well Test and Core Data," paper SPE 71722 presented at the 2001 SPE Annual Technical Conference and Exhibition, New Orleans, Louisiana, September 30–October 3, 2001.
- Haddad, S., Cribbs, M., Sagar, R., Viro, E., Castelijins, K., and Tang, Y.: "So What is the Reservoir Permeability?" paper SPE 63138 presented at the 2000 SPE Annual Technical Conference and Exhibition, Dallas, Texas, October 1–4, 2000.
- Heaviside, J., Black, C.J.J., and Berry, J.F.: "Fundamentals of Relative Permeability: Experimental and Theoretical Considerations," paper SPE 12173 presented at the 58th Annual Technical Conference and Exhibition, San Francisco, California, October 5–8, 1983.
- Jones, S.C., and Roszelte, W.O.: "Graphical Techniques for Determining Relative Permeability From Displacement Experiments," *JPT* (May 1978) 807–817.
- Honarpour, M., Koederitz, L.F., and Harvey, A.H.: "Empirical Equations for Estimating Two-Phase Relative Permeability in Consolidated Rock," *JPT* (December 1982) 2905–2908.

- Johnson, C.E., Jr.: "Graphical Determination of the Constants in the Corey Equation for Gas-Oil Relative Permeability Ratio," SPE 2346, *JPT Forum* (October 1968) 1111–1113.
- Grattoni, C.A., and Bidner, M.S.: "History Matching of Unsteady-State Corefloods for Determining Capillary Pressure and Relative Permeabilities," paper SPE 21135 presented at the SPE Latin American Petroleum Engineering Conference, Rio de Janeiro, Brazil, October 14–19, 1990.
- Amabeoku, M.O., Kersey, D.G., BinNasser, R.H., and Belowi, A.R.: "Relative Permeability Coupled Saturation-Height Models Based on Hydraulic (Flow) Units in a Gas Field," *SPEREE* (December 2008) 1013–1028.
- App, J.F., and Mohanty, K.K.: "The Benefit of Local Saturation Measurements in Relative Permeability Estimation from Centrifuge Experiments," *SPEJ* (September 2002) 288–298.
- Labban, Z.S., and Horne, R.N.: "Estimating Relative Permeabilities by Automated History Matching Using Linear Modeling Techniques," paper SPE 21206 presented at the 11th SPE Symposium on Reservoir Simulation, Anaheim, California, February 17–20, 1991.
- Blunt, M.J.: "Effects of Heterogeneity and Wetting on Relative Permeability Using Pore Level Modeling," *SPEJ* (March 1997) 70–87.
- Kjosavik, A., Ringen, J.K., and Skjaeveland, S.M.: "Relative Permeability Correlation for Mixed-Wet Reservoirs," *SPEJ* (March 2002) 49–58.
- Corey, A.T., Rathjens, C.H., Henderson, J.H., and Wyllie, M.R.J.: "Three-Phase Relative Permeability," *JPT* (November 1956) 63–65.

CHAPTER 3

Reservoir Fluid Saturations

3.1 Introduction

The pore spaces in reservoir rocks are occupied by fluids. In petroleum reservoirs, the fluids are usually water and hydrocarbons. The relative volumes of water and hydrocarbons in the pore volume of the reservoir rock are designated as saturations. **Water saturation** in the reservoir rock is the fraction of the pore volume occupied by water. By the same definition, **hydrocarbon saturation** in the reservoir rock is the fraction of the pore volume occupied by hydrocarbons. The sum of the water and hydrocarbon saturations in the reservoir rock is equal to unity. This relationship can be expressed simply as:

$$S_h + S_w = 1 \tag{3.1}$$

In Eq. (3.1), S_h = hydrocarbon saturation, fraction; and S_w = water saturation, fraction. If the hydrocarbon in the reservoir exists in oil and gas phases, Eq. (3.1) can be written as:

$$S_o + S_g + S_w = 1 \tag{3.2}$$

In Eq. (3.2), S_o = oil saturation, fraction; and S_g = gas saturation, fraction. The importance of determining accurate data for water saturations in the reservoir, especially at discovery, is evident from either Eq. (3.1) or Eq. (3.2). The presence of low water saturations in the reservoir indicates the presence of high hydrocarbon saturations. Conversely, high water saturations are interpreted as representing the presence of low hydrocarbon saturations. This rule-of-thumb is routinely applied qualitatively to assess the potential hydrocarbon contents of a reservoir after it has been penetrated with a well, especially at discovery. For calculation of the volume of hydrocarbons in a reservoir, a general equation that applies is expressed as:

$$HCPV = Area \times Thickness \times \phi \times (1 - S_w) \tag{3.3}$$

In Eq. (3.3), $HCPV$ = hydrocarbon pore volume; $Area$ = hydrocarbon-bearing area of the reservoir; $Thickness$ = net productive thickness or pay of the reservoir; ϕ = porosity, fraction; and S_w = water saturation, fraction. Water saturation is one of the key data required in Eq. (3.3). If the water saturation data are incorrect, it could result in the over- or under-estimation of the volume of hydrocarbon present in the reservoir. The economic impact of erroneous calculation of in-place-hydrocarbon volume can not be overstated. It could lead to execution of uneconomic projects with erroneously high estimated in-place-hydrocarbon volumes or lead to the abandonment of projects with erroneously low estimates of in-place-hydrocarbon volumes.

In this chapter, various models and equations used for the calculation of water saturations in clean sands, shaly sands, and carbonate reservoirs are presented. The widely used Archie equation for clean sands is presented. This is followed with the presentation of models for shaly sands, such as the Waxman-Smits model, Simandoux equation, Poupon-Leveaux equation, and the Dual-Water model. The application of the Archie equation to carbonate reservoirs is discussed. This section ends with a presentation on the use of nuclear magnetic resonance (NMR) logs for the determination of in-situ fluid saturations for most types of reservoirs.

Also in this chapter, the use of cutoffs in the determination of net pay thickness is discussed. As Eq. (3.3) shows, net pay thickness is a very important data required in the calculation of in-place-hydrocarbon volumes. The impact of net pay thickness data on the calculation of hydrocarbon volumes is as important as those of porosity and water saturation. The criteria generally used in the application of cutoffs are based on porosity, permeability, and water saturation limiting values. There are no accepted industry-wide limits on the criteria used for the determination of net pay based on these petrophysical properties. This is because the application of cutoff criteria is by necessity reservoir-specific, since it is rare to encounter two reservoirs that have identical petrophysical properties. Even in the same reservoir owned jointly by different companies, it is not uncommon to find that the cutoff criteria applied by each company are different. This situation has resulted in partners on the same reservoir using different values of net pay in calculating the in-place-hydrocarbon volumes. In this chapter, common practices and definitions used in the application of cutoff criteria to determine net pay are presented.

3.2 Determination of Water Saturations

Many models and equations have been developed over many years for the determination of water saturations in hydrocarbon-bearing formations. The models and equations vary in terms of complexity from the Archie equation developed for clean sands to more elaborate models developed for shaly sands. In this section, the Archie equations for clean sands are presented followed by a selection of models developed for shaly sands. The application of the Archie equations on carbonate reservoirs is discussed. Finally, the utility of nuclear magnetic resonance (NMR) logs in the evaluation of water saturation for most types of reservoirs is discussed.

3.2.1 Clean Sands

Clean sands are classified as sands that satisfy the assumptions used in the development of the Archie equations.[1] These are sands that do not contain clays or clay minerals. In such sands, conduction of electricity occurs only through free ions within the formation water. There is an absence of the "shale effect." Archie equations are assumed to apply at these "perfect" rock conditions. In practice, Archie equations are generally applied under rock conditions that do not meet these ideal conditions.

3.2.1.1 Archie Equations

The derivation of Archie equations[1] follows from the definition of **resistivity index** (I_R) as the ratio of the resistivity of a reservoir rock partially saturated with water (R_t) to the resistivity of the rock if fully saturated with water (R_o). Thus, the resistivity index is defined as:

$$I_R = \frac{R_t}{R_o} \tag{3.4}$$

In Eq. (3.4), I_R = resistivity index; R_t = formation resistivity or resistivity of partially water-saturated rock, $\Omega \cdot \text{m}$; and R_o = resistivity of fully (100%) water-saturated rock, $\Omega \cdot \text{m}$. Resistivity index can also be related to water saturation as:

$$I_R = S_w^{-n} \tag{3.5}$$

In Eq. (3.5), S_w = water saturation, fraction; and n = saturation exponent. Combining Eqs. (3.4) and (3.5) gives:

$$S_w^{-n} = \frac{R_t}{R_o} \tag{3.6}$$

For a clean rock that is fully saturated with water, **formation resistivity factor** (sometimes called **formation factor**), F, is defined as:

$$F = \frac{R_o}{R_w} \tag{3.7}$$

In Eq. (3.7), F = formation factor; and R_w = resistivity of formation water, $\Omega \cdot \text{m}$. Substituting Eq. (3.7) into Eq. (3.6) gives:

$$S_w^{-n} = \frac{R_t}{FR_w} \tag{3.8}$$

Archie[1] expressed the formation factor, F, as a function of total porosity, ϕ_t. thus:

$$F = \phi_t^{-m} \tag{3.9}$$

In Eq. (3.9), ϕ_t = total porosity, fraction; and m = cementation exponent. Winsauer et al.[2] modified Eq. (3.9) later by inserting the tortuosity factor, a, to get:

$$F = a\phi_t^{-m} \tag{3.10}$$

Substituting Eq. (3.10) into Eq. (3.8) gives:

$$S_w^{-n} = \frac{R_t}{a\phi_t^{-m} R_w} \tag{3.11}$$

Eq. (3.11) can be rearranged to give:

$$S_w = \left(\frac{a}{\phi_t^m} \frac{R_w}{R_t}\right)^{1/n} \tag{3.12}$$

Eq. (3.12) is the Archie equation for the calculation of water saturation in reservoir rocks. Electrical conductivity of a material is the reciprocal of its resistivity. For example, the conductivity of a reservoir rock is the reciprocal of its resistivity. This can be expressed as:

$$C_t = \frac{1}{R_t} \tag{3.13}$$

In Eq. (3.13), C_t = conductivity of the reservoir rock, Sm^{-1}. Expressing Archie equation (Eq. (3.12)) in terms of conductivities yields:

$$S_w = \left(\frac{a}{\phi_t^m} \frac{C_t}{C_w}\right)^{1/n} \tag{3.14}$$

In Eq. (3.14), C_w = conductivity of the formation water, Sm^{-1}. If it is assumed that the parameter $a = 1$, then Eq. (3.14) can be rearranged as:

$$C_t = \phi_t^m \times S_w^n \times C_w \tag{3.15}$$

3.2.1.2 Determination of Archie Parameters *n, m,* and *a*

The Archie parameters n, m, and a can be determined by applying the expression derived by rearranging Eq. (3.12) as:

$$\log R_t = -n \log S_w + \log\left(\frac{aR_w}{\phi_t^m}\right) \tag{3.16}$$

A least-squares straight line fitted through a log-log plot of R_t versus S_w laboratory data from core samples has a negative slope equal to the parameter, n. An alternative method that is equivalent to Eq. (3.16) can be derived from Eq. (3.5) as:

$$\log I_R = -n \log S_w \tag{3.17}$$

A least-squares straight line fitted through a log-log plot of I_R versus S_w laboratory data from core samples yields the same value of parameter, n.

The parameters m and a can be determined by application of Eq. (3.10) in the form:

$$\log F = -m \log \phi_t + \log a \tag{3.18}$$

A least-squares straight line drawn through a log-log plot of F versus ϕ_t data based on core samples has a negative slope equal to the parameter m, and an intercept equal to the parameter a. Note that the entire core data used in this plot must be obtained at $S_w = 1$ as required by the definition of formation factor in Eq. (3.7).

A mathematically rigorous method for determining Archie parameters was proposed by Maute et al.[3] This method determines Archie parameters n, m, and a by minimizing the mean-square water saturation error, ε, represented as:

$$\varepsilon = \sum_j \sum_i \left[S_{wij} - \left(\frac{a}{\phi_{tj}^m} \frac{R_{wj}}{R_{tij}} \right)^{1/n} \right]^2 \quad (3.19)$$

In Eq. (3.19), ε = water saturation error, fraction; j = core sample index; i = index for each core sample, j, data; S_{wij} = ith laboratory-measured water saturation for core sample, j, fraction; R_{tij} = ith laboratory-measured resistivity for core sample, j, $\Omega \cdot$m; R_{wj} = resistivity of water used in core sample, j, $\Omega \cdot$m; and ϕ_{tj} = total porosity of core sample, j, fraction. Maute et al.[3] method uses *all* available core sample data in the regression of the Archie parameters n, m, and a. Detailed procedure for the implementation of this method was provided by Maute et al.[4]

Example 3.1 Calculation of the Water Saturation Using the Archie Equation

Problem

Calculate the water saturation of clean consolidated sandstone given the following rock properties and Archie parameters:

Total porosity, ϕ_t	0.26
Resistivity of sandstone, R_t	22.5 $\Omega \cdot$m
Resistivity of formation water, R_w	0.265 $\Omega \cdot$m
Archie parameter, m	1.953
Archie parameter, n	1.861
Archie parameter, a	0.838

Solution

Substituting the given data and Archie parameters into Eq. (3.12), water saturation for the clean sandstone is calculated as:

$$\begin{aligned} S_w &= \left(\frac{a}{\phi_t^m} \frac{R_w}{R_t} \right)^{1/n} \\ &= \left[\left(\frac{0.838}{(0.26)^{1.953}} \right) \left(\frac{0.265}{22.5} \right) \right]^{1/1.861} \\ &= 0.344 \text{ or } 34.4\% \end{aligned}$$

3.2.2 Shaly Sands

Shaly sands can be described as reservoir rocks that contain shales. These sands are sometimes described as "non-Archie" rocks, and exhibit the effects of the presence of shales on the electrical conductivity of rocks. The presence of shales causes reservoir rocks to become conductive, which adds to the conductivity of the formation water. Archie equation assumes that the formation water is the only conductive phase in the formation and the reservoir rock is non-conductive. To account for the effects of shales on the extra conductivity of reservoir rocks, many shaly sand models have been proposed in the literature.[5] Most of the shaly sand saturation models have the following forms:[5,6]

$$C_t = \phi_t^{m^v} \times S_w^{n^v} \times C_{we} \tag{3.20}$$

$$C_t = \phi_t^{m^v} \times S_w^{n^v} \times (C_w + X) \tag{3.21}$$

The forms of Eqs. (3.20) and (3.21) are similar to the Archie equation as shown in Eq. (3.15). In Eq. (3.20), C_{we} = effective conductivity that accounts for the extra conductivity caused by the presence of shales in the rocks. In Eq. (3.21), X = a function that accounts for the extra conductivity caused by the presence of shales in the rocks. In Eqs. (3.20) and (3.21), m^v and n^v are general forms of the electrical properties of the rock that correspond to the Archie parameters of m and n. They reduce to Archie parameters as the shale function, $X \rightarrow 0$.

3.2.2.1 Waxman-Smits Model

The Waxman-Smits[7,8] model is based on the results of an extensive experimental study on the effects of shales on the conductivity of shaly sands. The model takes the form of Eq. (3.21) and is represented as:

$$C_t = \phi_t^{m^*} \times S_w^{n^*} \times \left(C_w + \frac{BQ_v}{S_w} \right) \tag{3.22}$$

In Eq. (3.22), B = specific conductivity of exchangeable cations, mho/m per meq/cc. B is also called counterion equivalent conductance; Q_v = clay cation-exchange-capacity (CEC) in milliequivalents per unit pore volume, meq/cc; The parameters m^* and n^* are the electrical parameters for the Waxman-Smits equation. The experimental procedure for determining the parameters m^* and n^* is very complicated.[6] An approximate approach is to use Archie parameters m and n to replace Waxman-Smits parameters m^* and n^* in Eq. (3.22).[6]

3.2.2.2 Simandoux Equation

The Simandoux[9] equation takes the form of Eq. (3.20), and is represented as:

$$C_t = C_w \phi_t^m S_w^n + \varepsilon V_{sh} C_{sh} \tag{3.23}$$

In Eq. (3.23), V_{sh} = wetted shale volume of rock, fraction; C_{sh} = conductivity of wetted shale, Sm^{-1}. The shale-term coefficient, ε, has a range of $0 \leq \varepsilon \leq 1$. As $S_w \rightarrow$ high, $\varepsilon \rightarrow 1$. The parameters m and n are the same as Archie parameters.

3.2.2.3 Poupon-Leveaux Equation

The Poupon-Leveaux[10] equation is also of the form of Eq. (3.20). It is expressed as:

$$C_t = C_w \phi_t^m S_w^n + C_{sh} S_w^n V_{sh}^{2-V_{sh}} \tag{3.24}$$

This basic equation has been modified by adding additional term in the development of the Indonesian saturation equation represented as:[6]

$$C_t = C_w \phi_t^m S_w^n + C_{sh} S_w^n V_{sh}^{2-V_{sh}} + S_w^n \sqrt{C_w \phi_t^m V_{sh}^{2-V_{sh}} C_{sh}} \tag{3.25}$$

It is also the precursor of the Nigerian saturation equation expressed as:[6]

$$C_t = \left[\sqrt{C_w \phi_t^m} + \sqrt{C_{sh} V_{sh}^{2.8}}\right]^2 S_w^n \tag{3.26}$$

The parameters *m* and *n* are the same as Archie parameters.

3.2.2.4 The Dual-Water Model

The Dual-Water model proposed by Clavier et al.[11] assumes that the clay-bound water and the free non-clay water act as two parallel conductive layers that contribute to the total conductivity, C_t, measured in the formation. The Dual-Water model is expressed as:

$$C_t = \phi_t^m S_w^n \left[C_w + \frac{v_Q Q_v (C_{cbw} - C_w)}{S_w} \right] \tag{3.27}$$

In Eq. (3.27), C_{cbw} = conductivity of clay-bound water, Sm^{-1}; v_Q = amount of clay water associated with milliequivalents of clay counterions, $meq^{-1}cm^3$; and Q_v = cation-exchange-capacity per unit pore volume, meq cm^{-3}. The parameters *m* and *n* can be approximated with the Archie parameters.

3.2.3 Carbonate Rocks

Carbonate rocks, unlike sandstones, have complex pore systems. These pore systems may have bi- or trimodal pore size distributions. Pore sizes may range from less than an inch to feet. The pore geometry of carbonate rocks is very heterogeneous and variable. The texture and structure of carbonate rocks are further rendered more complex by the diagenesis caused by chemical dissolution, precipitation, dolomitization, leaching, and fracturing. Due to these reasons, petrophysical models comparable in terms of simplicity to the Archie equation have not been developed for carbonate rocks. In some petrophysical analyses, the Archie equation is used to calculate water saturation in carbonate rocks. This approach could lead to significant errors. An alternative approach is to base the calculation of water saturation of carbonate rocks on data from NMR logs.

3.2.4 Water Saturations from Nuclear Magnetic Resonance Logs

The use of nuclear magnetic resonance (NMR) logging tools for determination of porosity and permeability was described in Chapters 1 and 2, respectively. Data from NMR logging tools are also used to determine water saturations. Water saturations in reservoir rocks are determined on the basis of T_2 distribution measured with NMR logging tools.[12] The transverse relaxation time, which is named T_2, is the characteristic time constant that represents the envelope of the spin-echo signal decay in a NMR response sequence.[12] T_2 distributions computed from NMR echo data are used to determine total porosity, effective porosity, permeability, pore size distribution, and water saturation. An example of a T_2 distribution curve for typical fluid-saturated shaly sandstone is shown in Figure 3.1. The T_2 distribution curve can be divided into two regions representing bound water and free fluids by application of a T_2 cutoff (see Figure 3.1). For shaly sandstones, bound water includes clay- and capillary-bound water, and free fluids include non-clay water (or free water) and hydrocarbons as shown in Figure 3.2. A default value of 33 milliseconds is sometimes applied as the T_2 cutoff for sandstones in the absence of laboratory data. For carbonates, the default T_2 cutoff is 100 milliseconds or higher, depending on the heterogeneity of the rock. More accurate T_2 cutoffs for specific rock samples can be measured in the laboratory with NMR tools after the samples are centrifuged. Another method for determining T_2 cutoffs is by matching irreducible water saturation obtained from an independent source (such as using the Dean-Stark method on core samples) with the irreducible water saturation derived from NMR data.[13] The irreducible water saturations of core samples shown in Figure 3.3 were measured by the centrifugal method and matched against irreducible water saturations from NMR core laboratory data for determination of T_2 cutoffs. In many formation evaluation processes, water saturations calculated from Archie-type equations are compared with water saturations derived from NMR data. In highly heterogeneous formations, such as carbonates, NMR-derived water saturations are considered more reliable in comparison with saturations calculated with Archie-type equations.

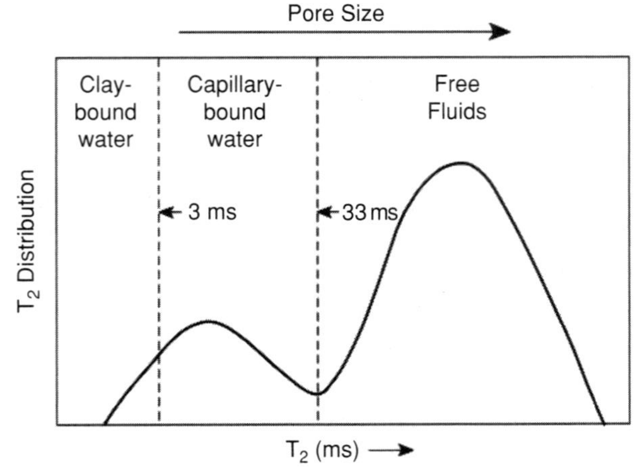

Figure 3.1 T_2 distribution curve for typical fluid-saturated shaly sandstone.

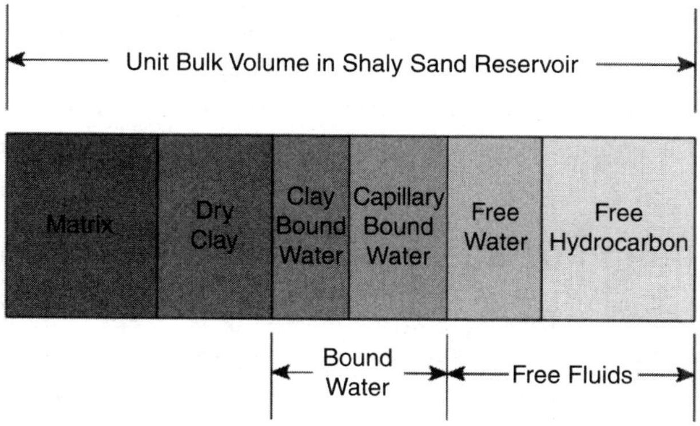

Figure 3.2 Fluid saturation distribution in a shaly sand reservoir.

3.2.5 Uncertainties in Estimation of Water Saturation

The estimation of water saturation for Archie and non-Archie rocks require input data that have considerable uncertainties associated with them. For instance, calculation of water saturations for Archie rocks (clean sands) using the Archie equation require the following input data: porosity, ϕ; cementation exponent, m; saturation exponent, n; the tortuosity factor, a; formation conductivity, C_t; and formation water conductivity, C_w. For the non-Archie rocks (shaly sands), additional input data required include shale conductivity, C_{sh}; and volume of shale, V_{sh}. In similar manner, water saturations calculated from NMR data are affected by uncertainties related to the computation of the T_2 distribution and accurate values for T_2 cutoffs. The uncertainties associated with

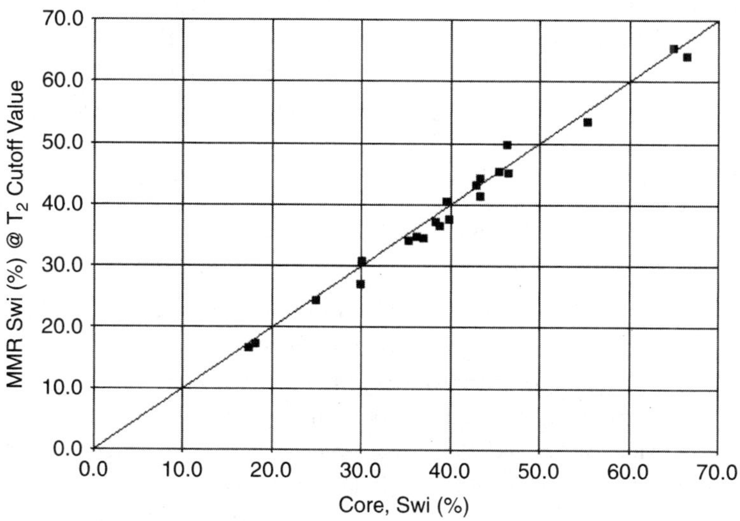

Figure 3.3 Determination of T_2 cutoff from irreducible water saturation comparison.

these input data have direct impact on the calculated water saturations.[14,15] The degree of uncertainty in calculated water saturations can be minimized by using regression analysis on core data to generate variable parameters (*m*, *n*, and *a*) for Archie or Archie-type equations. The core database can also be classified in terms of facies to improve the correlation of the parameters to rock type.[16]

3.3 Determination of Reservoir Productive Intervals

In Chapter 1, the importance of porosity data as a means of defining the storage capacity of reservoir rocks was discussed. In Chapter 2, permeability was presented as a measure of the ability of reservoir rocks to transmit fluids. Earlier in this chapter, various equations for the calculation of water saturation in different types of formations were presented. As shown in Eq. (3.1), the hydrocarbon saturation is the formation pore volume not occupied by water. As a result, the level of water saturation in a formation is a direct measure of the level of hydrocarbon saturation present in the formation. Porosity, permeability, and water saturation are the three petrophysical properties that constitute the criteria for the application of cutoffs in the determination of productive intervals in a reservoir. Generally, formation cutoffs are applied as limiting values that are used to eliminate formation intervals that appear not to have the potential to contribute significantly to the storage and productive capacity of the reservoir.[17] With this description of cutoffs, it is evident that there are no generally accepted criteria for the application of cutoffs in the petroleum industry.[18] Even on the same reservoirs, petroleum companies have used different cutoff criteria to obtain widely different estimation of productive intervals. And in many organizations, subsurface specialists (engineers, geologists, petrophysicists, etc.) frequently disagree on the cutoffs to be used in the determination of productive intervals. The lack of general criteria for the application of cutoffs can be accommodated by recognition of the fact that reservoirs are rarely identical in terms of their petrophysical properties. Consequently, it is impractical to expect that general cutoff criteria or guidelines can be adequate or applicable to all reservoirs. Due to their arbitrary nature, cutoff criteria should be devised for each reservoir to accommodate the objectives of the reservoir evaluation process in terms of in-place-hydrocarbon volumes, displacement mechanisms, and projected hydrocarbon recoveries.[19]

3.3.1 Net Sands, Net Reservoir, and Net Pay

Net sands, net reservoir, and net pay are terms commonly used to represent different categories of the formation intervals. These terms are defined in this book in accordance with the definitions proposed by Worthington.[17,20] The definitions of gross rock, net sand, net reservoir, and net pay are illustrated in Figure 3.4. **Gross rock thickness** represents the entire thickness of the formation interval. **Net sand** is the summation of the intervals whose sand content is greater than or equal to a limiting value set by the shale cutoff. Shale cutoff represents the limiting fractional volume of the formation that is considered to be shale and non-productive. In some reservoirs, it is preferable to use net sands in the characterization of reservoir models because it includes all the potential sources of drive energy that are present in the reservoir. **Net reservoir** comprises of

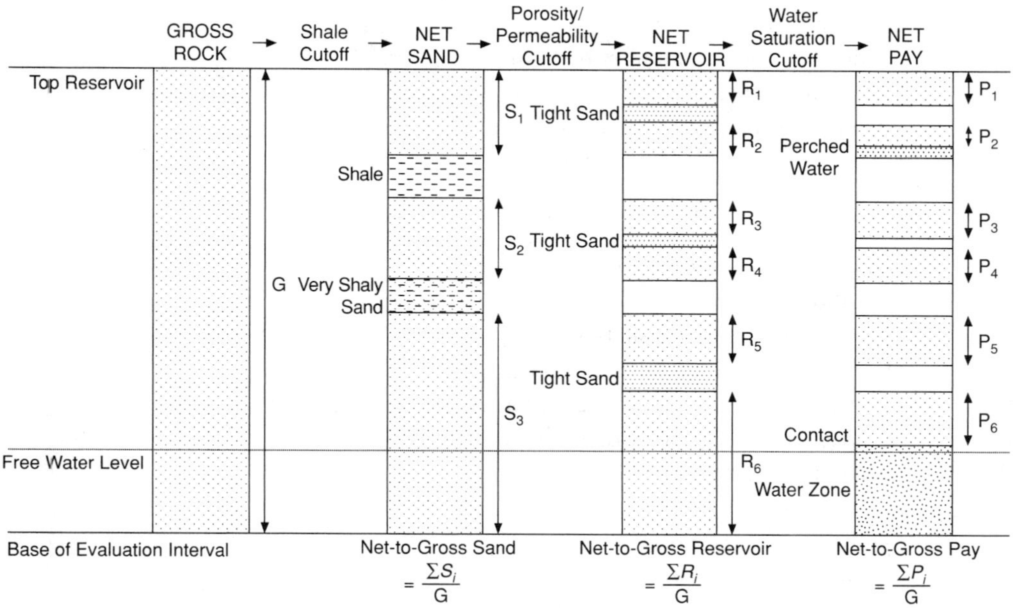

Figure 3.4 Definitions of net sand, net reservoir, and net pay with cutoff criteria based on petrophysical properties (from Worthington and Cosentino[17] © 2005 SPE, Reproduced with permission).

intervals of net sand that have the capacity to both store and transmit reservoir fluids. This is generally determined by the application of porosity and permeability cutoffs. **Net pay** is the summation of the net reservoir intervals that contain hydrocarbons above the limit set by the water saturation cutoff. Note the definitions of net-to-gross sand, net-to-gross reservoir, and net-to-gross pay ratios as shown in Figure 3.4. Net pay thickness is typically used as shown in Eq. (3.3) for the calculation of in-place-hydrocarbon volumes. But observe that since net pay is derived from the application of arbitrarily set cutoff criteria, it may lead to under- or over-estimation of in-place-hydrocarbon volumes. By extension, this could cause under- or over estimation of reserves. It is recommended that net pay thicknesses or net-to-gross pay ratios should be subjected to uncertainty analysis, as applied to porosity and water saturation data, in the evaluation of ranges of in-place-hydrocarbon volumes and potential recoveries.[21]

Nomenclature

a	tortuosity factor
B	specific conductivity of exchangeable cations, mho/m per meq/cc
C_{cbw}	conductivity of clay-bound water, Sm^{-1}
C_o	conductivity of fully water-saturated formation or rock, Sm^{-1}

C_{sh}	conductivity of shale, Sm^{-1}
C_t	conductivity of formation or rock, Sm^{-1}
C_w	conductivity of formation water, Sm^{-1}
C_{we}	effective conductivity of shaly sands, Sm^{-1}
F	formation factor
I_R	resistivity index, ratio
m	cementation factor
m^v	cementation factor for shaly sands
n	saturation exponent
n^v	saturation exponent for shaly sands
Q_v	clay cation-exchange-capacity, meq/cc
R_o	resistivity of fully water-saturated formation or rock, $\Omega \cdot m$
R_t	resistivity of formation or rock, $\Omega \cdot m$
R_w	resistivity of formation water, $\Omega \cdot m$
S_g	gas saturation, fraction
S_h	hydrocarbon saturation, fraction
S_o	oil saturation, fraction
S_w	water saturation, fraction
T_2	NMR characteristic time constant, milliseconds
V_{sh}	rock shale volume, fraction
v_Q	clay water associated with clay counterions, $meq^{-1}cm^3$
X	function to account for extra conductivity of shale
ε	mean-square in water saturation regression
ϕ	porosity
ϕ_e	effective porosity
ϕ_t	total porosity

Abbreviations

CEC	Cation-Exchange-Capacity
HCPV	Hydrocarbon Pore Volume
NMR	Nuclear Magnetic Resonance

References

1. Archie, G.E.: "The Electrical Resistivity Log as an Aid in Determining Some Reservoir Characteristics," *Trans., AIME* (1942) 146, 54–62.
2. Winsauer, W.O., Shearin, Jr., H.M., Masson, P.H., and Williams, M.: "Resistivity of Brine-Saturated Sands in Relation to Pore Geometry," *AAPG Bulletin* (1952) 36, 2, 253–277.

3. Maute, R.E., Lyle, W.D., and Sprunt, E.S.: "Improved Data-Analysis Method Determines Archie Parameters from Core Data," *JPT* (January 1992) 103–107.

4. Maute, R.E., Lyle, W.D., and Sprunt, E.S.: "Supplement to SPE 19399, Improved Data-Analysis Method Archie Parameters from Core Data," paper SPE 24223.

5. Worthington, P.F.: "The Evolution of Shaly-Sand Concepts in Reservoir Evaluation," *The Log Analyst* (January–February, 1985) 26, 1, 23–40.

6. Al-Ruwali, S.B., and Al-Waheed, H.H.: "Improved Petrophysical Methods and Techniques for Shaly Sands Evaluation," paper SPE 89735 presented at the 2004 SPE International Petroleum Conference, Puebla, Mexico, November 8–9, 2004.

7. Waxman, M.H., and Smits, L.J.M.: "Electrical Conductivities in Oil-Bearing Shaly Sands," *SPEJ* (June 1968) 107–122.

8. Waxman, M.H., and Thomas, E.C.: "Electrical Conductivities in Shaly Sands—1. The Relation Between Hydrocarbon Saturation and Resistivity Index; II. The Temperature Coefficient of Electrical Conductivity," *JPT* (February 1974) 213–223.

9. Simandoux, P.: "Dielectric Measurements on Porous Media: Application to the Measurement of Water Saturations: Study of the Behaviour of Argillaceous Formations," *Revue de l'Institu Francais du Petrole* 18, supplementary issue (1963) 193–215 (Translated text in Shaly Sand Reprint Volume, SPWLA, Houston, Texas, pp. IV 97–124).

10. Poupon, A., and Leveaux, J.: "Evaluation of Water Saturation in Shaly Formations," *The Log Analyst* (July–August, 1971) 3–8.

11. Clavier, C., Coates, G., and Dumanoir, J.: "Theoretical and Experimental Bases for the Dual-Water Model for Interpretation of Shaly Sands," *SPEJ* (April 1984) 153–168.

12. Freedman, R.: "Advances in NMR Logging," SPE 89177, Distinguished Author Series, *JPT* (January 2006) 60–66.

13. Shafer, J., Chen, S., Georgi, D., and Chen, J.: "Methods and Procedures for Calibrating NMR Log Derived Permeabilities," The 11th Formation Evaluation Symposium of Japan, Chiba, Japan (October 5–6, 2005) 1–15.

14. Worthington, P.F., and Pallatt, N.: "Effect of Variable Saturation Exponent on the Evaluation of Hydrocarbon Saturation," *SPEFE* (December 1992) 331–336.

15. Zeybek, A.D. Onur, M., Tureyen, O.I., Ma, S.M., Al-Shahri, A.M., and Kuchuk, F.J.: "Assessment of Uncertainty in Saturation Estimated From Archie's Equation," paper SPE 120517 presented at the 2009 SPE Middle East Oil & Gas Show and Conference, Bahrain, Kingdom of Bahrain, March 15–18, 2009.

16. Worthington, P.F.: "Improved Quantification of Fit-for-Purpose Saturation Exponents," *SPEREE* (August 2004) 270–284.

17. Worthington, P.F., and Cosentino, L.: "The Role of Cutoffs in Integrated Reservoir Studies," *SPEREE* (August 2005) 276–290.

18. Caldwell, R.H., and Heather, D.I.: "Characterizing Uncertainty in Oil and Gas Evaluations," paper SPE 68592 presented at the SPE Hydrocarbon Economics and Evaluation Symposium, April 2–3, 2001, Dallas, Texas.

19. Worthington, P.F.: "The Application of Cutoffs in Integrated Reservoir Studies," *SPEREE* (December 2008) 968–975.

20. Worthington, P.F.: "Net Pay: What is it? What does it do? How do we quantify it? How do we use it?" paper SPE 123561 presented at the 2009 SPE Asia Pacific Oil and Gas Conference and Exhibition, August 4–6, 2009, Jakarta, Indonesia.

21. Egbele, E., Ezuka, I., and Onyekonwu, M.: "Net-To-Gross Ratios: Implications in Integrated Reservoir Management Studies," paper SPE 98808 presented at the Nigeria Annual International Conference and Exhibition, August 1–3, 2005, Abuja, Nigeria.

General Reading

- Poupon, A., Clavier, C., Dumanoir, J., Gaymard, R., and Misk, A.: "Log Analysis of Sand-Shale Sequences—A Systematic Approach," *JPT* (July 1970) 867–881.
- Juhász, I.: "The Central Role of Q_v and Formation-Water Salinity in the Evaluation of Shaly Formations," *The Log Analyst* (July–August, 1979) 3–13.
- Worthington, P.F.: "Petrophysical Type Curves for Identifying the Electrical Character of Petroleum Reservoirs," *SPEREE* (December 2007) 711–729.
- Worthington, P.F.: "Identifying the Potential Upside of Hydrocarbon Saturation from Electric Logs," *SPEREE* (February 2009) 53–67.

CHAPTER 4

Pressure-Volume-Temperature (PVT) Properties of Reservoir Fluids

4.1 Introduction

Fluids exist in reservoirs as mixtures of gas, oil, and water. Some reservoirs may contain only gas and water, only oil and water, or mixtures of gas, oil, and water. Irrespective of the proportions of these fluids present in a reservoir, obtaining fluid samples and studying their phase behavior in a laboratory are necessary for establishing reservoir type, devising strategies for reservoir management, and estimating expected hydrocarbon recovery. The importance of collecting representative reservoir fluid samples (preferably early in the life of the reservoir) and having the samples analyzed in a reputable laboratory can not be over emphasized. It is one of the essential functions of the engineers working on new or existing reservoirs as an integral part of a comprehensive data collection program. Sampling methods for reservoir fluids and types of laboratory measurements are presented in Chapter 5. This chapter begins by discussing basic concepts of phase diagrams as tools for understanding the phase behavior of reservoir fluids. The chapter provides many correlations for calculation of properties of reservoir fluids especially in circumstances where laboratory data measured on reservoir fluid samples may not be available.

4.2 Phase Diagrams

The PVT properties of reservoir fluids are introduced by reviewing the basic concepts of phase diagrams. **Phase diagrams** are graphical representations that relate the properties of a fluid system. The properties used in the representation could be intensive or extensive properties. **Intensive** properties (pressure, temperature, density) are independent of the extent of the system.

Extensive properties (mass, volume, heat) depend on the extent of the system. Common forms of phase diagrams show pressure-temperature (P-T), pressure-volume (P-V), and temperature-volume (T-V) relationships. The basic concepts of phase behavior and phase diagrams are discussed in this book by the use of P-T diagrams.

4.2.1 Single Component Systems

The P-T diagram of a single component system such as pure water is shown in Figure 4.1. In this diagram, the pure substance can exist as three phases—solid, liquid, and vapor depending on the temperature and pressure of the system. A **phase** is a continuous, homogeneous portion of the system separated by a physical interface if another phase is present.

The three curves in Figure 4.1 represent states of phase equilibrium. The line ST is the sublimation curve. Along the **sublimation curve,** solid and vapor phases are in equilibrium. The line MT is the **melting curve,** along which solid and liquid phases are in equilibrium. Line TC is the **vapor pressure curve,** which represents the locus of equilibrium between liquid and vapor phases. The point T, at which the three curves meet, is the triple point. At the **triple point,** the three phases are in equilibrium. Along the vapor pressure curve, the terminal point C is called the **critical point.** Above the critical point, the liquid and vapor phases are indistinguishable. The physical interface between them disappears. The concept of the critical point and its location on the phase diagram are used later in this chapter to classify the phase behavior of petroleum systems.

The phases within a system composed of one or more components are considered to be in **thermodynamic equilibrium** when the chemical potentials of each component in the phases are equal. Equilibrated phases within a system have the same temperatures and pressures, and there

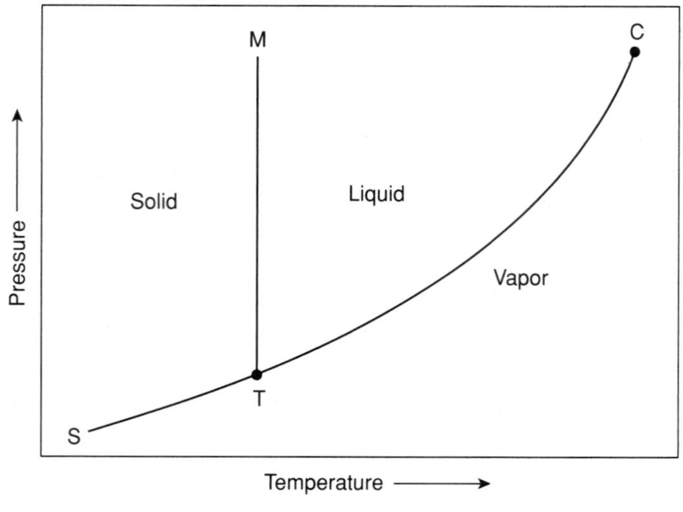

Figure 4.1 P-T phase diagram of a single-component system.

are no net mass transfers between the phases. If a system has two phases denoted as i and j, the condition of equilibrium is denoted as:

$$\begin{aligned} \psi_i &= \psi_j \\ T_i &= T_j \text{ and} \\ p_i &= p_j \end{aligned} \tag{4.1}$$

In Eq.(4.1), ψ is chemical potential, T is temperature, and p is pressure in each phase.

The number of intensive properties required to define a system in equilibrium can be determined by using Gibbs' phase rule.[1] If C is the number of components in a system and P is the number of phases, then the number of intensive properties, F (or degrees of freedom), that must be specified to define its equilibrated state is:

$$F = C - P + 2 \tag{4.2}$$

Applying the Gibbs' phase rule to Figure 4.1, it can be deduced that for a single component system with two phases in coexistence (such as a liquid phase and a vapor phase), if the temperature is known, then the pressure is fixed by the vapor pressure curve because the number of degrees of freedom, F, is equal to 1.

Another form of a phase diagram representing a single component system is shown in Figure 4.2. This is the pressure-volume (P-V) phase diagram of a single-component system. The solid curve ACB in Figure 4.2 is called the **phase envelope.** Within the phase envelope, vapor and liquid phases coexist in equilibrium. Outside the phase envelope (left of segment AC in Figure 4.2) is only the liquid phase, and to the right of segment BC is the vapor phase. Line AC

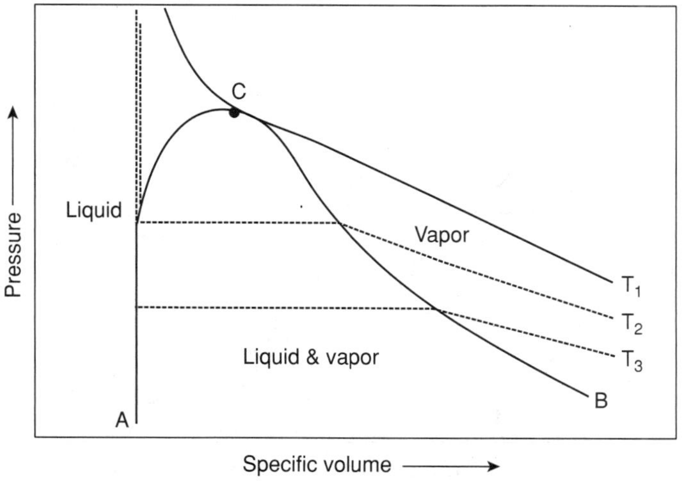

Figure 4.2 P-V phase diagram of a single-component system.

is the bubble point line, and line BC is the dew point line. The **bubble point** for any system is the condition of temperature and pressure at which the liquid phase is in equilibrium with an infinitesimal amount of the vapor phase. Similarly, the **dew point** is the temperature and pressure at which the vapor phase is in equilibrium with an infinitesimal amount of the liquid phase. In Figure 4.2, the critical point is represented as C. As was the case in Figure 4.1, beyond the critical point, liquid and vapor phases overlap and cannot be distinguished from each other.

An important feature of P-V phase diagrams is the use of **constant temperature lines,** or **isotherms.** These are shown as T_1, T_2, and T_3 in Figure 4.2. Along isotherms T_2 and T_3, the specific volume decreases with an increase in pressure within the vapor region. Within the two-phase region, the specific volume decreases at constant pressure. Across the bubble point line, the pressure increases rapidly with a relatively small reduction in the specific volume, because the liquid phase is significantly less compressible than the vapor phase. This concept is used in the measurement of bubble point pressures of most fluid systems. Note that the phase behavior along the T_1 isotherm is significantly different from the phase behavior along the isotherm T_2 or T_3 because it does not traverse the two-phase region within the phase envelope.

A temperature-volume (T-V) phase diagram of a single-component system is shown in Figure 4.3. This diagram is very similar to the P-V diagram shown in Figure 4.2. The main difference is that instead of isotherms shown in Figure 4.2, **constant pressure lines, or isobars,** are shown in Figure 4.3.

4.2.2 Binary Systems

Binary systems are mixtures of two components. Figure 4.4 is the P-T phase diagram of a binary mixture of components A and B. The vapor pressure curve of pure component A is shown terminating at the critical point C_A. Similarly, the vapor pressure curve of pure component B

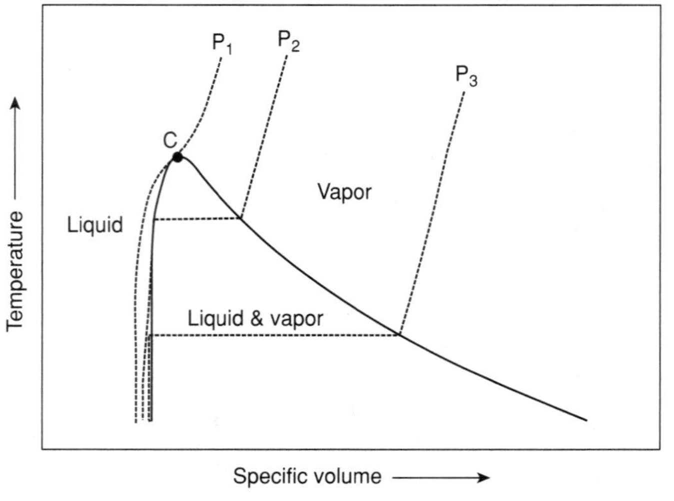

Figure 4.3 T-V phase diagram of a single-component system.

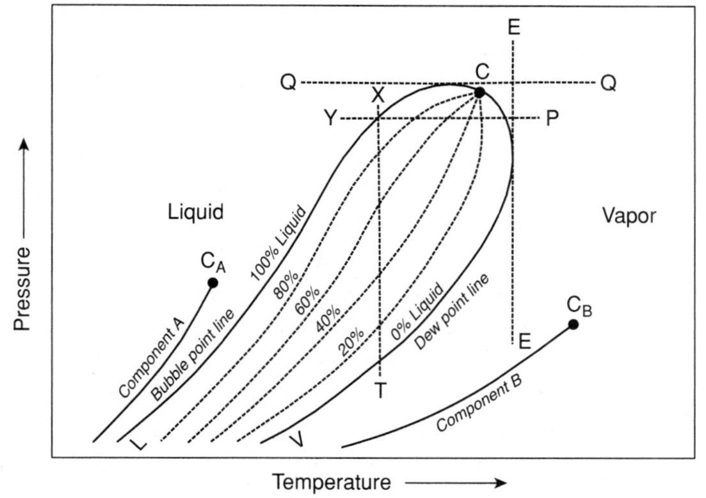

Figure 4.4 P-T phase diagram of a binary system.

terminates at the critical point C_B. The phase diagrams of hypothetical mixtures of components A and B will lie between the vapor pressure curves of the pure components, as shown in Figure 4.4 for the 50-50 mole percent mixture. The area in Figure 4.4 bounded by the solid curve LCV is the phase envelope, which encloses conditions of temperatures and pressures where liquid and vapor phases can coexist in equilibrium. The curved line LC defines the bubble point line, and curved line VC is the dew point line. The bubble point line LC and the dew point line VC meet at point C, which is defined as the critical point. Unlike the critical point that terminates the vapor pressure curve of a pure single component system (Figure 4.1), the critical point of a binary mixture does not define the highest temperature and pressure at which two phases can coexist. The highest temperature in Figure 4.4 at which two phases can coexist is shown by line EE. This is called the **cricondentherm.** Similarly, the highest pressure at which two phases can coexist is shown by line QQ. This is called the **cricondenbar.** Note that the cricondentherm and the cricondenbar are higher than the critical temperature and pressure of the mixture, respectively. The cricondentherm lies between the critical temperatures of components A and B. The cricondenbar is higher than the critical pressure of either component A or component B.

The dashed lines within the phase envelope in Figure 4.4 denote lines of constant volume percentage liquid drawn from the critical point. At the bubble point line, the mixture is 100% liquid. At the dew point line, the mixture is 100% vapor. Consider a process in which the mixture is taken at constant temperature along the line XT. At point X, the mixture is an undersaturated liquid. As the pressure is reduced, the liquid expands until it reaches the bubble point line, where an infinitesimal amount of bubble will form. This is the bubble point pressure of the mixture at the isotherm XT. As the pressure is reduced along XT, the volume of liquid phase in the mixture will decrease until the dew point line is reached. At the dew point, an infinitesimal amount of liquid will remain. This is the dew point pressure of the mixture at the isotherm XT. Further pressure reduction to point T will create a system composed entirely of the vapor phase.

A process similar to the isothermal expansion of the mixture along XT can be repeated by expanding the mixture at constant pressure (isobaric) along YP. At point Y, the mixture is undersaturated liquid, just as at point X. At the bubble point line, an infinitesimal amount of bubble will form. Further expansion along YP will result in the reduction of the volume of liquid present in the mixture. At the dew point line, a very small amount of liquid will remain. Further temperature increase to point P will create a vapor system.

The P-T phase diagram of ethane/n-heptane mixtures of varying composition of ethane is shown in Figure 4.5. This figure illustrates that mixtures of the same components having varying compositions will have varying phase envelopes. For instance, an ethane/n-heptane mixture with 90.22% ethane has a phase envelope shown as C_1. The shape of this phase envelope is different from the phase envelope of the C_2 mixture (50.25% ethane) or that of the C_3 mixture (9.78% ethane). In Figure 4.5 the critical points of the three mixtures are connected with a dashed line, which is the locus of the critical points of all potential mixtures of ethane and n-heptane at temperatures and pressures between the critical points of pure ethane and pure n-heptane.

4.2.3 Multicomponent Systems

The phase diagram of a binary system (Figure 4.4) introduces several useful concepts, such as the phase envelope, critical point, cricondentherm, and cricondenbar. Figure 4.5 (phase diagrams

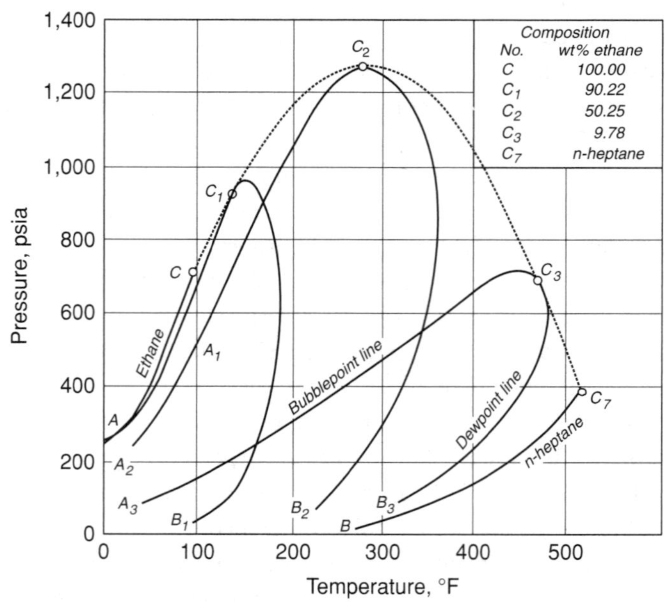

Figure 4.5 P-T phase diagram for ethane and *n*-heptane system at various compositions of ethane (from Kay[3]).

of ethane/*n*-heptane at various compositions) also illustrates the effect of the composition of the mixture on the shape of the phase envelope. The shape and nature of phase diagrams for multicomponent systems are similar to those of binary systems. However, it is much more difficult to map out the phase envelope of a multicomponent system. These concepts are used in this section to discuss the phase envelopes of various fluid types that will be encountered in reservoirs.

The phase envelopes of common reservoir fluid types are shown in Figure 4.6. The reservoir fluid types shown are for a gas, a gas condensate, a volatile oil, and a black oil reservoir. In Figure 4.6, the letter C denotes the location of the critical point on the phase envelopes. Consider a reservoir whose initial pressure and temperature are denoted as p_i,T_i, as shown in Figure 4.6. Note the path of the solid line starting at p_i,T_i, which represents isothermal depletion by reduction in reservoir pressure in relation to the phase envelopes that represent a gas, a gas condensate, a volatile oil, and a black oil reservoir. The phase envelope (A) for the gas reservoir is clearly to the left of the solid line. The critical temperature (point C) and cricondentherm of the phase envelope are lower than reservoir temperature, T_i. As the reservoir pressure is reduced at constant reservoir temperature, there is no chance that the solid line will cross any part of the phase envelope that defines the gas reservoir. Consequently, no liquid will form inside the gas reservoir as its pressure is reduced by production.

Gas reservoirs are further classified as dry or wet. This classification is based on the pressure and temperature of the facilities used in handling and processing the produced fluids. If the pressure and temperature of the flow and processing facilities fall within the two-phase region defined by the phase envelope, then liquids could form within the wellbore, flowlines, and separators. This type of gas is called a **wet gas** (Figure 4.7). Describing the gas as wet does not mean that some of the gas exists as liquid within the reservoir. If the temperature and pressure of the

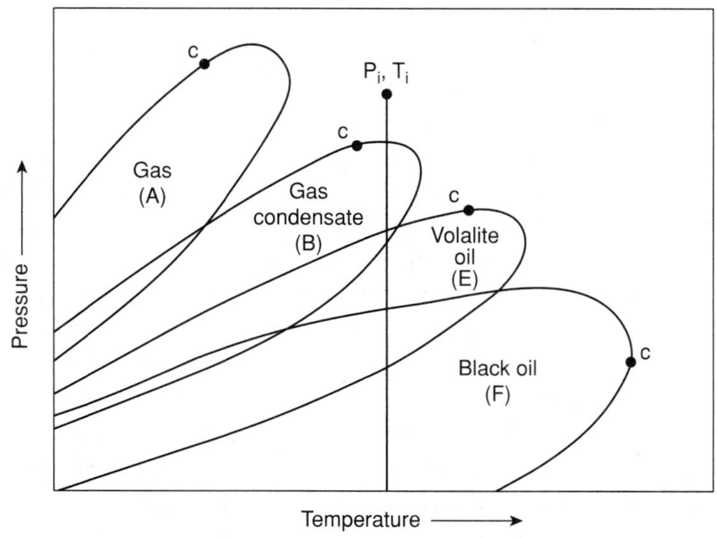

Figure 4.6 Phase diagram showing various reservoir fluid types.

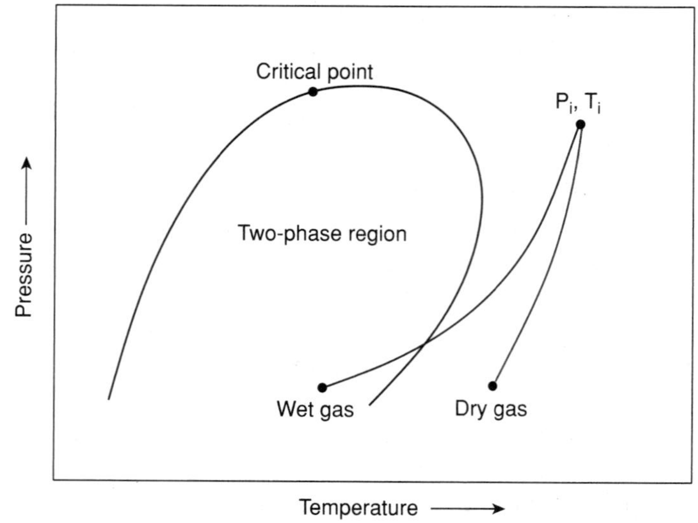

Figure 4.7 Phase diagram showing dry gas and wet gas behavior based on the phase envelope.

facilities fall outside the two-phase region of the phase envelope, no liquid will condense in the wellbore, flowlines, and separators. This type of gas is described as a **dry gas.**

The phase envelope (B) for a gas-condensate reservoir intersects the solid line from p_i, T_i. For the gas-condensate phase envelope, the reservoir temperature, T_i, is higher than the critical temperature (point C) but lower than the cricondentherm. Note that the solid line crosses the phase envelope on the side of the dew point line. As reservoir pressure is reduced along the solid line, a liquid phase will condense inside the reservoir when the dew point line is reached. The nature and behavior of the condensation process are presented in more detail in Section 4.2.4 for retrograde gas-condensate reservoirs.

Volatile oil reservoirs are represented in Figure 4.6 by the phase envelope, E. The critical temperature (point C) and the cricondentherm of the phase envelope are higher than the reservoir temperature, T_i. At initial reservoir pressure and temperature, p_i, T_i, the fluid in the reservoir is a compressed and undersaturated liquid. When reservoir pressure is reduced at constant temperature along the solid line, the reservoir fluid will expand as a single phase until it reaches the bubble point line of the phase envelope. Further reduction of reservoir pressure along the solid line will cause a gas phase to form in the reservoir. Initially, the gas phase exists as isolated bubbles within the liquid phase. As reservoir pressure continues to decline, more gas will evolve until the scattered bubbles will coalesce to form a continuous phase. Generally, the stock-tank oil gravities are higher than 35° API, for volatile oils, resulting in gas-oil ratios that range between 1000 and 3000 scf/STB, and oil formation volume factors that are higher than 1.5 RB/STB. Black oil reservoirs are represented by phase envelope, F, in Figure 4.6. Note the location of the critical point on the phase envelope for black oils compared with its location on the phase envelope for volatile oils. The reservoir temperature, T_i, is much farther from the critical temperature for black oils in comparison with volatile oils. As is the case for volatile oils, a gas phase will form in black oil reservoirs when the reservoir pressure reaches the bubble point pressure on the bubble point line along the solid line of isothermal depletion (p_i,T_i). Generally, the stock-tank oil gravities of black oil are less than API, producing gas-oil ratios are less than 2000 scf/STB, and oil formation volume factors are less than 2 RB/STB.

It is apparent from the stated rules of thumb for volatile and black oils that there is significant overlap in properties between the two types of oils. Actual designation of a particular oil type should be determined from careful PVT laboratory measurements.

4.2.4 Retrograde Behavior of Gas-Condensate Systems

The phase envelope of a gas-condensate reservoir is represented in Figure 4.6 by phase envelope B. A similar phase envelope has been expanded in Figure 4.8 to illustrate the phenomenon of retrograde condensation. The dashed lines within the phase envelope are lines of constant liquid volume varying from 100% liquid volume at the bubble point line to 0% liquid volume at the dew point line. The reservoir temperature, T_i, is higher than the critical temperature, C, but lower than the cricondentherm of the phase envelope. This is a necessary condition for the occurrence of retrograde behavior.

Consider an experiment in which reservoir pressure is reduced at constant reservoir temperature, T_i, along the path defined by the solid line ABDE from initial conditions at p_i,T_i. At point

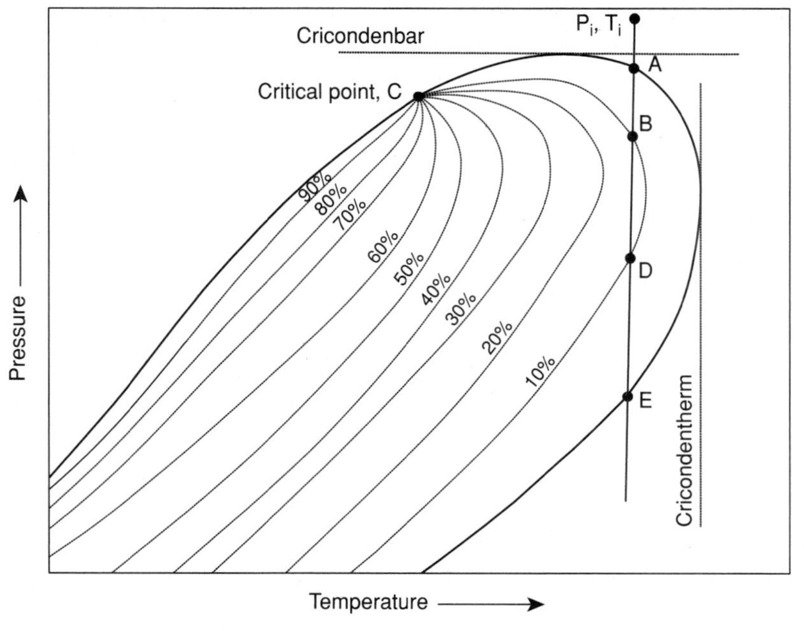

Figure 4.8 Phase diagram showing retrograde behavior.

A on the dew point line, a liquid phase will begin to form in the reservoir. As reservoir pressure is reduced, more liquid will condense in the reservoir such that at point B, 10% of the reservoir volume is occupied by the liquid phase. Further reduction in reservoir pressure will increase the percentage of liquid volume to a maximum of about 12%. Beyond this maximum, as pressure is reduced, the volume of liquid phase in the reservoir will begin to shrink until it reaches 10% again at point D. From point D to point E, liquid volume will continue to shrink and will disappear completely at the lower dew point at point E. This condensation behavior, which begins with increasing liquid volume followed by a reversal of liquid volume with pressure reduction at constant temperature in the reservoir, is described as **retrograde condensation.**

Generally, the liquid phase that forms in the reservoir rarely reaches the critical liquid saturation necessary to become mobile. The condensed liquid may appear as isolated "islands" of the liquid phase scattered throughout the reservoir. In a strict sense, the process described along the isotherm ABDE never occurs in an actual reservoir. As condensation occurs in a reservoir, the liquid phase will contain the heavier components, and the vapor phase will contain fewer of the heavier components. Because the liquid phase is not mobile, none of these heavier components will be produced. Consequently, the composition of the fluid system remaining in the reservoir is not constant but changes with production. As the composition changes, so does the phase envelope, resulting in retrograde behavior that is different from the process described here, which assumes constant composition of the fluid system in the reservoir.

Proper management of retrograde gas-condensate reservoirs is very important if the goal is to maximize hydrocarbon recovery. The key steps are identifying the gas-condensate system as exhibiting retrograde behavior and quantifying the extent of retrograde condensation that could occur either in the reservoir or within the production facilities. This can be done by taking several representative samples of reservoir fluid as close to initial reservoir conditions as possible. The samples should be analyzed by competent PVT laboratories, especially covering the range of operating conditions expected during depletion. The PVT data should be modeled with an equation of state. Further reservoir work should involve performing compositional simulation by using a geological model of the reservoir. Geologic modeling and reservoir simulation are presented in Chapters 18 and 19, respectively.

Management of gas-condensate reservoirs may require pressure maintenance, usually by gas injection, either to maintain reservoir pressure or to evaporate some of the condensed liquid. Ideally, it will be more beneficial in terms of recovery to maintain reservoir pressure higher than the dew point so that no condensation occurs within the reservoir. This will require initiating the gas injection project as early as possible in the recovery process. The decision to undertake a pressure maintenance or liquid recovery project depends on a number of factors, such as the location of the reservoir, the availability of gas for injection, the cost of injected gas, the availability of facilities for recovery of liquids, and expected recovery from the project. Each of these factors should be evaluated and combined in a comprehensive economic analysis of the project as discussed in Chapter 20 on Petroleum Reservoir Management Strategies. More details in designing gas cycling for gas-condensate reservoirs are presented in Chapter 16, Section 16.3 under Gasflooding.

4.3 Gas and Gas-Condensate Properties

This section presents equations and correlations that can be used to calculate thermodynamic properties of dry, wet, or condensate gases. Recall that a dry gas is one in which no liquid forms either in the reservoir, in flowlines, or in the separators during the entire depletion process (Figure 4.7). A wet gas is one in which a liquid phase forms in the flowlines or separators, but no liquid phase forms in the reservoir. For condensate gas, a liquid phase can precipitate in the reservoir, the flowlines, and the separators (Figure 4.8).

4.3.1 Ideal Gas Equation

The ideal gas equation is derived from the combination of Boyle's law and Charles' law for ideal gases. Boyle's law states that the pressure of a given mass of gas is inversely proportional to its volume if its temperature is kept constant:

$$pV = k, \text{ if } T \text{ is constant} \tag{4.3}$$

In Eq. (4.3), p is pressure, V is volume, T is temperature, and k is constant. Another form of Eq. (4.3) that is sometimes useful is:

$$p_1V_1 = p_2V_2, \quad \text{at constant temperature, } T \tag{4.4}$$

In Eq. (4.4), p_1V_1 and p_2V_2 are pressures and volumes at conditions 1 and 2, respectively.

Charles' law states that the pressure of a given mass of gas is directly proportional to its temperature if its volume is kept constant:

$$\frac{p}{T} = k, \quad \text{if } V \text{ is constant} \tag{4.5}$$

Another way of expressing Charles' law is:

$$\frac{p_1}{T_1} = \frac{p_2}{T_2} \text{ at constant volume, } V \tag{4.6}$$

Combining Boyle's and Charles' laws gives the ideal gas equation:

$$pV = nRT \tag{4.7}$$

In Eq. (4.7), n is the number of moles, and R is the gas constant. If p is in psia, V is in ft^3, n is in lbmol, and $T$is in °R, then the gas constant $R = 10.732\ (\text{psia} \cdot \text{ft}^3)/(\text{lbmol} \cdot °\text{R})$. The ideal gas equation can be used to represent real gas behavior at low temperatures and pressures. The ideal gas equation can be expressed in another useful form by combining Eqs. (4.4) and (4.6) or derived from Eq. (4.7) as:

$$\frac{p_1V_1}{T_1} = \frac{p_2V_2}{T_2} \tag{4.8}$$

In Eq. (4.8), pressures, volumes, and temperatures are denoted at two thermodynamic states, 1 and 2, respectively.

4.3.2 Real Gas Equation

Gases, especially those found at reservoir conditions, do not exhibit ideal behavior. In fact, it can be said that no gases exhibit true ideal behavior, primarily because of the conditions of the kinetic theory (conditions that define ideal gas behavior). At low temperatures and pressures, low molecular weight gases approach ideal behavior. The thermodynamic behavior of real gases can be represented with the real gas equation as:

$$pV = znRT \tag{4.9}$$

In Eq. (4.9), z is sometimes called the **gas factor, gas deviation factor,** or **gas compressibility factor.** The gas compressibility factor, z, is a modifying factor introduced to account for the non-ideal behavior of real gases. When the compressibility factor, z, is equal to unity, the gas behaves as an ideal gas. Eqs. (4.7) and (4.9) become identical. Values of the gas compressibility factor vary over a wide range, depending on the pressure, temperature, and composition of the gas. The gas compressibility factor is one of the key properties necessary to represent the thermodynamic behavior of a gas. In this chapter, methods of calculating the gas compressibility factor using correlations are presented. Calculation of gas compressibility factors using equations of state is treated in Chapter 6.

4.3.3 Gas Gravity

The specific gravity of gas, known simply as **gas gravity,** is defined with reference to air. It is defined as the ratio of the density of the gas to the density of air at the same temperature and pressure:

$$\gamma_g = \frac{\rho_g}{\rho_{air}} \tag{4.10}$$

In Eq. (4.10), γ_g is gas gravity, ρ_g is gas density, and ρ_{air} is density of air.

Most of the petroleum industry has adopted the temperature of 60°F and the pressure of 14.65 psia as standard conditions. Unless otherwise stated, these standard conditions are adopted in this book.

Density is defined as mass per unit volume. Hence, gas density is defined as:

$$\rho_g = \frac{m_g}{V_g} \tag{4.11}$$

In Eq. (4.11), m_g is the mass of the gas, and V_g is the volume of the gas. $m_g = nM_g$, where n is moles of gas and M_g is its molecular weight. If we assume ideal gas behavior at standard conditions ($T_{sc} = 60°F$ and $p_{sc} = 14.65$ psia), then we can apply Eq. (4.7) such that Eq. (4.11) becomes:

$$\rho_g = \frac{pM_g}{RT} \tag{4.12}$$

Similarly, the density of air is given by:

$$\rho_{air} = \frac{pM_{air}}{RT} \tag{4.13}$$

In Eq. (4.13), M_{air} is the molecular weight of air, which is equal to 28.9586 (see Table 4.1). Combining Eqs. (4.12) and (4.13) gives:

$$\begin{aligned} \gamma_g &= \frac{M_g}{M_{air}} \\ &= \frac{M_g}{28.9586} \end{aligned} \tag{4.14}$$

4.3.4 Reduced Temperature and Pressure

The **principle of corresponding states** provides a method for correlating the properties of pure substances and mixtures. It states that two substances of similar chemical nature at the same conditions will exhibit similar physical properties when referenced to the same properties, such as critical temperatures and pressures.[2] The conditions of temperature and pressure referenced to critical temperature and critical pressure are called **reduced temperature** (T_r) and **reduced pressure** (p_r).

$$T_r = \frac{T}{T_c} \tag{4.15}$$

$$p_r = \frac{p}{p_c} \tag{4.16}$$

In Eqs. (4.15) and (4.16), T_c is critical temperature and p_c is critical pressure.

For gas mixtures, pseudo-reduced temperature (T_{pr}) and pseudo-reduced pressure (p_{pr}) are represented as:

$$T_{pr} = \frac{T}{T_{pc}} \tag{4.17}$$

$$p_{pr} = \frac{p}{p_{pc}} \tag{4.18}$$

In Eqs. (4.17) and (4.18), T_{pc} is the pseudo-critical temperature and p_{pc} is the pseudo-critical pressure for gas mixtures. Pseudo-critical properties for mixtures do not have real physical meaning as critical properties for pure substances. The main use of pseudo-critical properties for mixtures is as a means of correlating the properties of the mixtures. Section 4.4 presents various methods for estimating pseudo-critical properties for gas mixtures.

Table 4.1 Physical Properties of Compounds

Compound	Formula	Molecular Weight	Critical Pressure (psia)	Critical Temperature (°F)	Critical Volume (ft³/lb)	Liquid Specific Gravity (Water = 1)	Gas Specific (Air = 1)	Acentric Factor
Methane	CH_4	16.042	667.0	–116.66	0.0985	(0.3)*	0.55400	0.0115
Ethane	C_2H_6	30.069	706.6	89.92	0.0775	0.35643	1.03830	0.0994
Propane	C_3H_8	44.096	615.5	205.92	0.0728	0.50738	1.52270	0.1529
Isobutane	C_4H_{10}	58.122	527.9	274.41	0.0715	0.56295	2.00710	0.1865
n-Butane	C_4H_{10}	58.122	550.9	305.55	0.0703	0.58408	2.00710	0.2003
Isopentane	C_5H_{12}	72.149	490.4	369	0.0685	0.62460	2.49140	0.2284
n-Pentane	C_5H_{12}	72.149	488.8	385.8	0.0676	0.63113	2.49140	0.2515
n-Hexane	C_6H_{14}	86.175	436.9	453.8	0.0688	0.66404	2.97580	0.2993
n-Heptane	C_7H_{16}	100.202	396.8	512.9	0.0682	0.68819	3.46020	0.3483
n-Octane	C_8H_{18}	114.229	360.7	564.2	0.0673	0.70698	3.94450	0.3977
n-Nonane	C_9H_{20}	128.255	330.7	610.8	0.0693	0.72186	4.42890	0.4421
n-Decane	$C_{10}H_{22}$	142.282	304.6	652.2	0.0703	0.73406	4.91330	0.4875
Carbon Monoxide	CO	28.01	506.7	–220.63	0.0527	0.79265	0.96720	0.0510
Carbon Dioxide	CO_2	44.01	1070.0	87.76	0.0343	0.82203	1.51970	0.2239
Hydrogen Sulfide	H_2S	34.082	1306.5	212.81	0.0462	0.80269	1.17690	0.1010
Air	—	28.9586	551.9	–220.97	0.0458	0.87603	1.00000	———
Hydrogen	H_2	2.0159	190.7	–399.9	0.5319	0.07087	0.06961	–0.2140
Oxygen	O_2	31.9988	731.4	–181.43	0.0367	1.14230	1.10500	0.0222
Nitrogen	N_2	28.0135	492.5	–232.53	0.0511	0.80687	0.96740	0.0372
Water	H_2O	18.0153	3200.1	705.1	0.04975	1.00000	0.62210	0.3443

4.4 Pseudo-critical Properties of Gas Mixtures

The methods for calculating the pseudo-critical properties of gas mixtures are presented under two groups. These groups are:

1. Composition of gas mixtures known
2. Composition of gas mixtures unknown

4.4.1 Composition of Gas Mixtures Known

If the composition of the mixture is known, the pseudo-critical properties of gas mixtures can be calculated by using various mixing rules. The simplest mixing rule was proposed by Kay.[3] Kay's mixing rule is the molar weighted average of the critical properties of the pure components of the gas mixture. It is written as:

$$p_{pc} = \sum y_i p_{ci} \tag{4.19}$$

$$T_{pc} = \sum y_i T_{ci} \tag{4.20}$$

In Eqs. (4.19) and (4.20), y_i is the mole fraction of component i; p_{ci} and T_{ci} are the critical pressure and critical temperature, respectively, of component i. Eqs. (4.19) and (4.20) are suitable for low-molecular-weight, homologous gas mixtures with specific gravity less than 0.75. These equations can be used for quick estimation of pseudo-critical properties and ultimately for calculation of gas compressibility (z) factors. However, it should be noted that this method under estimates gas compressibility factors, with errors as high as 15%.[2]

The mixing rules proposed by Stewart, Burkhart, and Voo (termed SBV)[4] and modified by Sutton (termed SSBV)[2] improved the accuracy of calculating the pseudo-critical properties of high molecular weight gas mixtures. The mixing rules proposed by Stewart et al.[4] for high molecular weight gas mixtures are:

$$J = \frac{1}{3}\sum_{i=1}^{n_c}\left(\frac{yT_c}{p_c}\right)_i + \frac{2}{3}\left[\sum_{i=1}^{n_c}\left(y\sqrt{\frac{T_c}{p_c}}\right)_i\right]^2 \tag{4.21}$$

$$K = \sum_{i=1}^{n_c}\left(\frac{yT_c}{\sqrt{p_c}}\right)_i \tag{4.22}$$

$$T_{pc} = \frac{K^2}{J} \tag{4.23}$$

$$p_{pc} = \frac{T_{pc}}{J} \tag{4.24}$$

In Eqs. (4.21) and (4.22), y = mole fraction of component, i, in mixture; T_c = critical temperature of component, i, in °R; and p_c = critical pressure of component, i, in psia.

Sutton[2] modified the SBV equations (Eqs. (4.21) to (4.24)) for gas mixtures that contain high concentrations of heptane plus fraction (up to 14.27 mole %). The SSBV modifications are:

$$F_j = \frac{1}{3}\left(\frac{yT_c}{p_c}\right)_{C7+} + \frac{2}{3}\left(\frac{y^2T_c}{p_c}\right)_{C7+} \tag{4.25}$$

$$\varepsilon_j = 0.6081F_j + 1.1325F_j^2 - 14.004F_jy_{C7+} + 64.434F_jy_{C7+}^2 \tag{4.26}$$

$$\varepsilon_k = \left(\frac{T_c}{\sqrt{p_c}}\right)_{C7+} \times \left(0.3129y_{C7+} - 4.8156y_{C7+}^2 + 27.3751y_{C7+}^3\right) \tag{4.27}$$

$$J' = J - \varepsilon_j \tag{4.28}$$

$$K' = K - \varepsilon_k \tag{4.29}$$

$$T_{pc} = \frac{K'^2}{J'} \tag{4.30}$$

$$p_{pc} = \frac{T_{pc}}{J'} \tag{4.31}$$

The critical properties of pure components in a mixture are provided in Table 4.1. The critical properties of the heptanes-plus fraction must be determined from correlations. Sutton[2] reported that the Kessler-Lee[5] equations provide the lowest error in comparison with other methods. The Kessler-Lee[5] equations are:

$$p_{pc,C7+} = \exp\left[\begin{matrix} 8.3634 - \dfrac{0.0566}{\gamma_{C7+}} - \left(0.24244 + \dfrac{2.2898}{\gamma_{C7+}} + \dfrac{0.11857}{\gamma_{C7+}^2}\right)\dfrac{T_{b,C7+}}{10^3} + \\ \left(1.4685 + \dfrac{3.648}{\gamma_{C7+}} + \dfrac{0.47227}{\gamma_{C7+}^2}\right)\dfrac{T_{b,C7+}^2}{10^7} - \left(0.42019 + \dfrac{1.6977}{\gamma_{C7+}^2}\right)\dfrac{T_{b,C7+}^3}{10^{10}} \end{matrix}\right] \tag{4.32}$$

$$\begin{aligned} T_{pc,C7+} &= 341.7 + 811\gamma_{C7+} + \left(0.4244 + 0.1174\gamma_{C7+}\right)T_{b,C7+} \\ &\quad + \left(0.4669 - 3.2623\gamma_{C7+}\right)\frac{10^5}{T_{b,C7+}} \end{aligned} \tag{4.33}$$

The boiling point, $T_{b,C7+}$, is estimated from the Whitson[6] equation:

$$T_{b,C7+} = \left(4.5579 \times M_{C7+}^{0.15178} \times \gamma_{C7+}^{0.15427}\right)^3 \tag{4.34}$$

The application of the SSBV equations for calculation of pseudo-critical properties of a mixture is shown in Example 4.3. The units of the terms in all the equations are as shown in the example.

4.4.2 Correction for Non-Hydrocarbon Gas Impurities

Gas mixtures may contain varying amounts of hydrogen sulfide, carbon dioxide, nitrogen, and water vapor as impurities. The critical properties of hydrocarbon mixtures should be corrected for the presence of these substances. Sutton[7] reported that the correlations proposed by Wichert and Aziz[8] for correcting for the presence of hydrogen sulfide and carbon dioxide are superior to other published methods. The Wichert and Aziz correlations are:

$$\xi = 120\left[(y_{CO_2} + y_{H_2S})^{0.9} - (y_{CO_2} + y_{H_2S})^{1.6}\right] + 15\left(y_{H_2S}^{0.5} - y_{H_2S}^{4}\right) \tag{4.35}$$

$$T_{pc}^{*} = T_{pc} - \xi \tag{4.36}$$

$$p_{pc}^{*} = \frac{p_{pc}(T_{pc} - \xi)}{T_{pc} + y_{H_2S}(1 - y_{H_2S})\xi} \tag{4.37}$$

At present, there are no satisfactory published correlations for correcting for the presence of nitrogen and water vapor. In Section 4.4.3, a method proposed by Standing[9] that could be used to correct for the presence of water and nitrogen in hydrocarbon gas mixtures is presented. The units of the terms in all the equations are as shown in Example 4.3.

4.4.3 Composition of Gas Mixture Unknown

The critical properties of gas mixtures can be calculated from Sutton[7] correlations when the composition of the mixture is not known. Sutton[7] based the correlation on hydrocarbon gas gravity, which is calculated using the method proposed by Standing:[9]

$$y_{HC} = 1 - y_{H_2S} - y_{CO_2} - y_{N_2} \tag{4.38}$$

The original Standing equation (Eq. (4.38)) can be modified slightly to account for the presence of water vapor as with the other impurities:

$$y_{HC} = 1 - y_{H_2S} - y_{CO_2} - y_{N_2} - y_{H_2O} \tag{4.39}$$

The hydrocarbon gas gravity, $\gamma_{g,HC}$, is calculated as:

$$\gamma_{g,HC} = \frac{\gamma_g - (y_{H_2S}M_{H_2S} + y_{CO_2}M_{CO_2} + y_{N_2}M_{N_2} + y_{H_2O}M_{H_2O})/M_{AIR}}{y_{HC}} \tag{4.40}$$

According to Standing,[9] the mixture's pseudo-critical properties are calculated as follows:

$$p_{pc} = y_{HC}p_{pc,HC} + y_{H_2S}p_{c,H_2S} + y_{CO_2}p_{c,CO_2} + y_{N_2}p_{c,N_2} + y_{H_2O}p_{c,H_2O} \tag{4.41}$$

$$T_{pc} = y_{HC}T_{pc,HC} + y_{H_2S}T_{c,H_2S} + y_{CO_2}T_{c,CO_2} + y_{N_2}T_{c,N_2} + y_{H_2O}T_{c,H_2O} \tag{4.42}$$

Note that if hydrogen sulfide and carbon dioxide are present in the gas mixture, then p_{pc} and T_{pc} calculated from Eqs. (4.41) and (4.42) should be corrected by using the Wichert and Aziz correlations (Eqs. (4.35), (4.36), and (4.37)).

Sutton[7] proposed two sets of correlations for calculating hydrocarbon pseudo-critical properties based on hydrocarbon gas gravity. The first set of correlations is for associated gas from oil reservoirs. The second set of correlations is to be used for gas condensates.

The correlations to be used for associated hydrocarbon gas gravity as proposed by Sutton are:

$$p_{pc,HC} = 671.1 + 14\gamma_{g,HC} - 34.3\gamma_{g,HC}^2 \quad (4.43)$$

$$T_{pc,HC} = 120.1 + 429\gamma_{g,HC} - 62.9\gamma_{g,HC}^2 \quad (4.44)$$

Equations (4.43) and (4.44) are applicable over a hydrocarbon gas gravity range of $0.554 \leq \gamma_{g,HC} \leq 1.862$. The equations apply if the gas mixture contains less than 10 mol % hydrogen sulfide, less than 55.8 mol % carbon dioxide, and less than 21.7 mol % nitrogen. The correlations were derived from a database that contained 3256 compositions with 4817 gas compressibility factor measurements. The average absolute error reported by Sutton[7] for compressibility factors calculated with the correlations is 0.80%.

The correlations for condensate hydrocarbon gas gravity proposed by Sutton[7] are:

$$p_{pc,HC} = 744 - 125.4\gamma_{g,HC} + 5.9\gamma_{g,HC}^2 \quad (4.45)$$

$$T_{pc,HC} = 164.3 + 357.7\gamma_{g,HC} - 67.7\gamma_{g,HC}^2 \quad (4.46)$$

Equations (4.45) and (4.46) are applicable over a hydrocarbon gas gravity range of $0.554 \leq \gamma_{g,HC} \leq 2.819$. The composition of the gas mixture should contain less than 90 mole % hydrogen sulfide, less than 89.9 mole % carbon dioxide and less than 33.3 mole % nitrogen. The database for these correlations contained 2264 compositions with 10177 compressibility factor measurements. The average absolute error reported by Sutton[7] for compressibility factors calculated with the correlations is 1.11%.

4.5 Wet Gas and Gas Condensate

In this section, methods for calculating the specific gravities of wet and condensate gases are presented. As defined previously in Section 4.3, a wet gas forms no liquids inside the reservoir but condenses liquids in the separators and flowlines. A condensate gas will condense liquids inside the reservoir if reservoir pressure falls below the dew point. The methods presented here are not applicable to a condensate reservoir if reservoir pressure falls below the dew point. Before these methods are applied, it is important to ensure that excessive liquid condensation has not occurred in the flowlines before the sampling point at the separators. The first method is a recombination method that is used when the compositions of the gas and liquid phases sampled at the separators are determined from laboratory measurements. The second method is based on correlations when the compositions of the produced phases are not known.

4.5.1 Recombination Method

The recombination method is illustrated with a two-stage separation system consisting of a separator and a stock tank. The compositions of the separator gas, stock-tank gas, and stock-tank oil are known.

Example 4.1 Calculation of Gravity of a Wet Gas by the Recombination Method

Problem

Suppose a gas reservoir produces wet gas through a two-stage separation system consisting of a separator and a stock tank. The gas-oil ratio (GOR) of the separator is given as 62,000 scf/STB. The stock-tank GOR is 350 scf/STB. Stock-tank oil gravity is 50.6°API. Calculate the composition of the reservoir gas given the compositions of the separator and stock-tank fluids as shown in Table 4.2.

Solution

The specific gravity of oil is defined as the density of oil relative to the density of water at the same temperature and pressure:

$$\gamma_o = \frac{\rho_o}{\rho_w} \tag{4.47}$$

In Eq. (4.47), ρ_o and ρ_w are densities of oil and water, respectively, in pounds per cubic foot. The specific gravity of oil can be related to API gravity (°API) as:

$$\gamma_o = \frac{141.5}{131.5 + {}^\circ API} \tag{4.48}$$

From Eq. (4.48), stock-tank oil specific gravity,

$$\gamma_o = \frac{141.5}{131.5 + 50.6} = 0.7770.$$

Molecular weight of stock-tank oil from Table 4.2 is $\sum x_i M_i = 105.7146$ lbm/lb-mole.

Step 1: Convert separator GOR to lb-mole/STB.

Assuming standard conditions of $T_{sc} = 60$ °F and $p_{sc} = 14.65$ psia:

Molar volume, $V_M = RT_{sc}/p_{sc} = 10.732(60 + 459.67)/14.65 = 380.69$ scf/lb-mole

$$\begin{aligned}\text{Pound moles of separator gas per STB} &= (62000\ \text{scf/STB}) \times (\text{lb-mole}/380.69\ \text{scf}) \\ &= 162.8622\ \text{lb-mole/STB}\end{aligned}$$

Step 2: Convert stock-tank GOR to lb-mole/STB.

$$\begin{aligned}\text{Pound moles of stock tank gas per STB} &= (350\ \text{scf/STB}) \times (\text{lb-mole}/380.69\ \text{scf}) \\ &= 0.9194\ \text{lb-mole/STB}\end{aligned}$$

Step 3: Convert stock-tank liquid to lb-mole/STB.

$$\text{Pound moles of stock tank liquid per STB} = (0.777 \times 62.4 lbm/ft^3) \times (5.615 ft^3/STB) \times (lb\text{-}mole/105.7146\ lbm) = 2.5753\ \text{lb-mole/STB}$$

The composition of the reservoir gas is computed as shown in Table 4.3 using the values calculated in steps 1–3. Because the reservoir gas composition has been determined, calculation

Table 4.2 Compositions of Separator and Stock-Tank Fluids for Example 4.1

Component	Vapor Composition Separator, Mole Fraction, y_i	Vapor Composition Stock-Tank, Mole Fraction, y_i	Liquid Composition Stock-Tank, Mole Fraction, X_i	Molecular Weight of Components, M_i	X_iM_i
N_2	0.0030	0.0010	0.0000	28.0134	0.0000
CO_2	0.0042	0.0014	0.0000	44.0100	0.0000
CH_4	0.7905	0.2700	0.0015	16.0430	0.0241
C_2H_6	0.1230	0.2221	0.0072	30.0700	0.2165
C_3H_8	0.0563	0.2812	0.0328	44.0970	1.4464
i-C_4H_{10}	0.0052	0.0426	0.0126	58.1230	0.7323
n-C_4H_{10}	0.0118	0.1107	0.0138	58.1230	0.8021
i-C_5H_{12}	0.0020	0.0232	0.0409	72.1500	2.9509
n-C_5H_{12}	0.0023	0.0285	0.0267	72.1500	1.9264
C_6H_{14}	0.0014	0.0171	0.0894	86.1770	7.7042
C_{7+}	0.0003	0.0022	0.7751	116.0000	89.9116
Total	1.0000	1.0000	1.0000	—	105.7146
Properties of the heptanes plus stock tank liquid:					
Specific gravity		0.802			
Molecular weight		116 lb/lb-mole			

Table 4.3 Calculation of Gravity of a Wet Gas by the Recombination Method with Two-Stage Separation

Component	Vapor Composition Separator, Mole Fraction, y_{SP}	Vapor Composition Stock Tank, Mole Fraction, y_{ST}	Liquid Composition Stock Tank, Mole Fraction, x_{ST}	Component Moles Separator Gas y_{SP} × 162.8622	Component Moles Stock Tank Gas y_{ST} × 0.9194	Component Moles Stock Tank Liquid x_{ST} × 2.5753	Total	Composition of Reservoir Gas
N_2	0.0030	0.0010	0.0000	0.4886	0.0009	0.0000	0.4895	0.0029
CO_2	0.0042	0.0014	0.0000	0.6840	0.0013	0.0000	0.6853	0.0041
CH_4	0.7905	0.2700	0.0015	128.7426	0.2482	0.0039	128.9947	0.7754
C_2H_6	0.1230	0.2221	0.0072	20.0321	0.2042	0.0185	20.2548	0.1218
C_3H_8	0.0563	0.2812	0.0328	9.1691	0.2585	0.0845	9.5121	0.0572
i-C_4H_{10}	0.0052	0.0426	0.0126	0.8469	0.0392	0.0324	0.9185	0.0055
n-C_4H_{10}	0.0118	0.1107	0.0138	1.9218	0.1018	0.0355	2.0591	0.0124
i-C_5H_{12}	0.0020	0.0232	0.0409	0.3257	0.0213	0.1053	0.4524	0.0027
n-C_5H_{12}	0.0023	0.0285	0.0267	0.3746	0.0262	0.0688	0.4695	0.0028
C_6H_{14}	0.0014	0.0171	0.0894	0.2280	0.0157	0.2302	0.4740	0.0028
C_{7+}	0.0003	0.0022	0.7751	0.0489	0.0020	1.9961	2.0470	0.0123
Total	1.0000	1.0000	1.0000	—	—	—	166.3569	1.0000

Basis: 1 bbl of stock-tank liquid

of pseudo-critical properties can proceed as illustrated in Example 4.3. Note that a similar procedure can be used to calculate the specific gravity of the reservoir gas if fluid sampling was conducted only at the primary separator.

4.5.2 Correlation Method

The gas specific gravity of wet or condensate reservoirs can be calculated from production data if the compositions of the fluid phases at the separators are not known. The production data required for the calculation are the specific gravities of the fluids plus gas and liquid production data at the separators. For instance, for a three-stage separation system, the gas gravities and gas-liquid ratios from the primary separator, secondary separator, and stock tank, as well as the stock-tank liquid gravity, are required. Thus, the expression for the gas specific gravity for a three-stage separation system can be derived as reported by Gold et al.:[10]

$$\gamma_w = \frac{R_1\gamma_1 + 4602\gamma_o + R_2\gamma_2 + R_3\gamma_3}{R_1 + (133{,}316\gamma_o/M_o) + R_2 + R_3} \tag{4.49}$$

In Eq. (4.49), γ_w is wellstream gas gravity ($air = 1$); R_1, R_2, and R_3 are the gas-liquid ratios in scf/STB for the primary separator, secondary separator, and stock tank, respectively; γ_1, γ_2, and γ_3 are gas gravities for the primary separator, secondary separator, and stock tank, respectively; M_o and γ_o are the molecular weight and specific gravity, respectively, of the stock-tank liquid.

For a two-stage separation system, the expression for the specific gravity is derived as:

$$\gamma_w = \frac{R_1\gamma_1 + 4602\gamma_o + R_3\gamma_3}{R_1 + (133{,}316\gamma_o/M_o) + R_3} \tag{4.50}$$

Note the absence of the terms R_2 and γ_2 to represent the secondary separator in Eq. (4.50) because the primary separator and stock tank constitute a two-stage separation system. Eqs. (4.49) and (4.50) can be used to calculate gas gravities accurately if all the terms in the equations are available from production data. In most cases, only the primary separator and stock-tank production data are measured. From the production data, the primary separator gas-liquid ratio, R_1, the primary separator gas gravity, γ_1, and the gravity of the stock-tank liquid, γ_o can be obtained. All the other terms in Eqs. (4.49) and (4.50) are unknown. The gas produced at the secondary separator and stock tank (for a three-stage separation system) or the stock tank (for a two-stage separation system) is accounted for as additional gas produced, G_{pa}. Thus, for a three-stage separation system:[10]

$$G_{pa} = R_2\gamma_2 + R_3\gamma_3 \tag{4.51}$$

For a two-stage separation system:

$$G_{pa} = R_3\gamma_3 \tag{4.52}$$

Similarly, the vapor equivalent of the primary separator liquid, V_{eq}, accounts for the stock-tank liquid production and gas production beyond the primary separator. This is represented for a three-stage separation system as:

$$V_{eq} = (133{,}316\gamma_o/M_o) + R_2 + R_3 \tag{4.53}$$

For a two-stage separation system:

$$V_{eq} = (133{,}316\gamma_o/M_o) + R_3 \tag{4.54}$$

Using Eqs. (4.51) and (4.53), Eq. (4.49) can be rewritten for a three-stage separation system as:

$$\gamma_w = \frac{R_1\gamma_1 + 4602\gamma_o + G_{pa}}{R_1 + V_{eq}} \tag{4.55}$$

The expression for a two-stage separation system is identical to Eq. (4.55) except that G_{pa} refers to Eq. (4.52), and V_{eq} refers to Eq. (4.54).

Gold et al.[10] proposed a correlation for estimating molecular weight, M_o, of the stock-tank liquid as:

$$M_o = \frac{5954}{°API - 8.811} = \frac{42.43\gamma_o}{1.008 - \gamma_o} \tag{4.56}$$

The additional gas produced, G_{pa}, and vapor equivalent, V_{eq}, for a three-stage or a two-stage separation system can be estimated with correlations proposed by Gold et al.[10] The correlation for additional gas produced in a three-stage separation system is:

$$G_{pa} = \beta_1(p_{s1} - 14.65)^{\beta_2}\gamma_1^{\beta_3}\gamma_{API}^{\beta_4}T_{s1}^{\beta_5}T_{s2}^{\beta_6} \tag{4.57}$$

In Eq. (4.57), $\beta_1 = 2.9922$, $\beta_2 = 0.97050$, $\beta_3 = 6.8049$, $\beta_4 = 1.0792$, $\beta_5 = -1.1960$, and $\beta_6 = 0.55367$.

The correlation for the vapor equivalent in a three-stage separation system is:

$$V_{eq} = \beta_0 + \beta_1 p_{s1}^{\beta_2}\gamma_1^{\beta_3}\gamma_{API}^{\beta_4}T_{s1}^{\beta_5}T_{s2}^{\beta_6} \tag{4.58}$$

For Eq. (4.58), $\beta_0 = 535.92$, $\beta_1 = 2.6231$, $\beta_2 = 0.79318$, $\beta_3 = 4.6612$, $\beta_4 = 1.2094$, $\beta_5 = -0.84911$, and$\beta_6 = 0.26987$. In Eqs. (4.57) and (4.58), p_{s1} is the primary separator pressure in psia, γ_1 is the primary separator gas gravity, T_{s1} is the primary separator temperature in °F, γ_{API} is the API gravity of the stock-tank liquid, and T_{s2} is the secondary separator temperature in °F.

Similar equations for a two-stage separation system were also correlated by Gold et al.[10] The correlation for additional gas produced in a two-stage separation system is:

$$G_{pa} = \beta_1(p_{s1} - 14.65)^{\beta_2}\gamma_1^{\beta_3}\gamma_{API}^{\beta_4}T_{s1}^{\beta_5} \tag{4.59}$$

In Eq. (4.59), $\beta_1 = 1.4599$, $\beta_2 = 1.3394$, $\beta_3 = 7.0943$, $\beta_4 = 1.1436$, and $\beta_5 = -0.93446$.

The correlation for vapor equivalent in a two-stage separation system is:

$$V_{eq} = \beta_0 + \beta_1 p_{s1}^{\beta_2} \gamma_1^{\beta_3} \gamma_{API}^{\beta_4} T_{s1}^{\beta_5} \tag{4.60}$$

In Eq. (4.60), $\beta_0 = 635.53$, $\beta_1 = 0.36182$, $\beta_2 = 1.0544$, $\beta_3 = 5.0831$, $\beta_4 = 1.5812$, and $\beta_5 = -0.79130$. Other terms in Eqs. (4.59) and (4.60) are as defined previously in Eq. (4.58).

According to Gold et al.,[10] the average absolute error calculated for G_{pa} estimated with Eq. (4.57) is 10.5%, and with Eq. (4.59) is 14.9%. The average absolute error calculated for V_{eq} estimated with Eq. (4.58) is 5.8%, and with Eq. (4.60) is 6.3%. Eqs. (4.57) to (4.60) can be used for reservoir gas with total non-hydrocarbon content up to 20 mole %. For reservoir gas with non-hydrocarbon content between 5 and 20 mole %, the average absolute error of gas gravity calculated with the correlations is 0.95%, and the maximum error is 8.1%. The presence of non-hydrocarbon content up to 20 mole % increased the average absolute error of the V_{eq} correlation by 15 percentage points from 6% to 21%.

The total reservoir fluid production rate for a wet gas or condensate reservoir can be estimated if we have the total gas production rate (separator plus stock-tank gas), stock-tank oil production rate, and properties of the stock-tank oil with Eq. (4.61).

$$q = q_g + 133.316(\gamma_o q_o / M_o) \tag{4.61}$$

In Eq. (4.61), q = total reservoir fluid rate, Mscf/D; q_g = gas production rate (separator plus stock-tank gas), Mscf/D; γ_o = stock-tank oil gravity; q_o = stock-tank oil rate, STB/D; M_o = molecular weight of stock-tank oil, lbm/lb-mole. Alternatively, if the vapor equivalent of the primary separator liquid, V_{eq}, is calculated from prior equations or correlations, then the total reservoir fluid production rate can be estimated by using Eq. (4.62).

$$q = q_{s1}(1 + V_{eq}/R_1) \tag{4.62}$$

In Eq. (4.62), q = total reservoir fluid rate, Mscf/D; q_{s1} = gas rate of the primary separator, Mscf/D; and R_1 = gas-liquid ratio of the primary separator, scf/STB.

Example 4.2 Calculation of Gas Gravity and Total Reservoir Fluid Production Rate

Problem

The separator and stock tank conditions and production data of a condensate gas reservoir in the Gulf of Mexico, USA, are as follows:

Primary separator pressure	1155 psia
Primary separator temperature	92°F
Primary separator gas rate	49,545 Mcf/D
Primary separator GOR	28,721 scf/STB
Primary separator gas gravity	0.6346 (air = 1)
Stock-tank gas rate	778 Mcf/D

Stock-tank GOR	451 scf/STB
Stock-tank gas gravity	0.9394 (air = 1)
Stock-tank liquid gravity	44.8°API
Stock-tank pressure	15.025 psia
Stock-tank temperature	60°F
Condensate production rate	1725 STB/D

Solution

The surface rates and fluid properties of the primary separator and stock-tank are known. The gas gravity can be calculated by using the recombination equation for two-stage separation shown previously as Eq. (4.50):

$$\gamma_w = \frac{R_1\gamma_1 + 4602\gamma_o + R_3\gamma_3}{R_1 + (133{,}316\gamma_o/M_o) + R_3}$$

$$\text{From Eq. (4.48), } \gamma_o = \frac{141.5}{131.5 + °API} = \frac{141.5}{131.5 + 44.8} = 0.8026$$

$$\text{From Eq. (4.56), } M_o = \frac{42.43\gamma_o}{1.008 - \gamma_o} = \frac{42.43(0.8026)}{1.008 - 0.8026} = 165.7951$$

Substituting,

$$\gamma_w = \frac{28721(0.6346) + 4602(0.8026) + 451(0.9394)}{28721 + (133316) \times 0.8026/165.7951 + 451}$$
$$= 0.7493$$

Actual reported reservoir fluid gas gravity is 0.7437. The calculated error is 0.75%.

Assuming that surface rates and fluid properties are not available, gas gravity can still be estimated from the operating parameters of the primary separator by using the correlations from Gold et al.[10]

From Eq. (4.59),

$$G_{pa} = \beta_1(p_{s1} - 14.65)^{\beta_2}\gamma_1^{\beta_3}\gamma_{API}^{\beta_4}T_{s1}^{\beta_5}$$
$$= 1.4599(1155 - 14.65)^{1.3394}(0.6346)^{7.0943}(44.8)^{1.1436}(92)^{-0.93446}$$
$$= 814.9231 \text{ scf/STB}$$

From Eq. (4.60),

$$V_{eq} = \beta_0 + \beta_1 p_{s1}^{\beta_2}\gamma_1^{\beta_3}\gamma_{API}^{\beta_4}T_{s1}^{\beta_5}$$
$$= 635.53 + 0.36182(1155)^{1.0544}(0.6346)^{5.0831}(44.8)^{1.5812}(92)^{-0.79130}$$
$$= 1328.673 \text{ scf/STB}$$

Substituting Eq. (4.55):

$$\gamma_w = \frac{R_1\gamma_1 + 4{,}602\gamma_o + G_{pa}}{R_1 + V_{eq}} = \frac{28721(0.6346) + 4601(0.8026) + 814.9231}{28721 + 1328.673}$$
$$= 0.7566$$

Actual reported reservoir fluid gas gravity is 0.7437. The calculated error is 1.7%.

The total reservoir fluid production rate is estimated by using Eq. (4.62):

$$\begin{aligned} q &= q_{s1}(1 + V_{eq}/R_1) \\ &= 49545(1 + 1328.673/28721) \\ &= 51{,}837 \text{ Mscf/D} \end{aligned}$$

4.6 Correlations for Gas Compressibility Factor

Standing and Katz (SK)[11] developed a generalized chart for predicting the compressibility factors of gases using parameters of reduced pressure, p_r, and reduced temperature, T_r, based on the principle of corresponding states. The SK chart with extended ranges has been reproduced numerically by many authors for computer applications. Takacs[12] reviewed the accuracies of these correlations for numerical representation of the SK chart. On the basis of the errors reported by Takacs, the correlations of Dranchuk and Abou-Kassem (DAK)[13] and Hall and Yarborough (HY)[14] give the most accurate representation of the SK chart over a wide range of temperatures and pressures. Sutton[7] also compared several methods developed by various authors for numerical representation of the SK chart. According to Sutton,[7] the Dranchuk et al. (DPR) [15] method shows the lowest average error but higher standard deviation, whereas the DAK method has comparably low average error and low standard deviation. On the basis of these studies reported by Takacs[12] and Sutton,[7] the DAK method was selected for the calculations of gas compressibility factors (z factor).

The DAK correlation is based on the 11-parameter Starling equation of state. Using nonlinear regression methods, Dranchuk and Abou-Kassem[13] fitted the Starling equation of state to 1500 data points from the SK chart, with an average absolute error of 0.486%. The DAK correlation for calculating z factors is represented as follows:

$$z = 1 + c_1(T_{pr})\rho_r + c_2(T_{pr})\rho_r^2 - c_3(T_{pr})\rho_r^5 + c_4(\rho_r,T_{pr}) \tag{4.63}$$

In Eq. (4.63):

$$\rho_r = 0.27p_{pr}/(z \times T_{pr}) \tag{4.64}$$

$$c_1(T_{pr}) = A_1 + A_2/T_{pr} + A_3/T_{pr}^3 + A_4/T_{pr}^4 + A_5/T_{pr}^5 \tag{4.65}$$

$$c_2(T_{pr}) = A_6 + A_7/T_{pr} + A_8/T_{pr}^2 \tag{4.66}$$

$$c_3(T_{pr}) = A_9(A_7/T_{pr} + A_8/T_{pr}^2) \tag{4.67}$$

$$c_4(\rho_r,T_{pr}) = A_{10}(1 + A_{11}\rho_r^2)(\rho_r^2/T_{pr}^3)\exp(-A_{11}\rho_r^2) \tag{4.68}$$

The constants A_1 through A_{11} are as follows:

$A_1 = 0.3265$ $A_2 = -1.0700$ $A_3 = -0.5339$ $A_4 = 0.01569$
$A_5 = -0.05165$ $A_6 = 0.5475$
$A_7 = -0.7361$ $A_8 = 0.1844$ $A_9 = 0.1056$ $A_{10} = 0.6134$ $A_{11} = 0.7210$

The DAK correlation is applicable over the following ranges of reduced pressure and temperature: $0.2 \leq p_{pr} < 30$, $1.0 < T_{pr} \leq 3.0$ and $p_{pr} < 1.0$; $0.7 < T_{pr} \leq 1.0$. The correlation gives poor results for $T_{pr} = 1.0$ and $p_{pr} > 1.0$.

Eq. (4.63) can be solved for z by using iterative techniques such as the Newton-Raphson method. The form of the Newton-Raphson method is:

$$x_{n+1} = x_n - \frac{f_n}{f'_n} \tag{4.69}$$

By rearranging Eq. (4.63), f_n is:

$$F(z) = z - (1 + c_1(T_{pr})\rho_r + c_2(T_{pr})\rho_r^2 - c_3(T_{pr})\rho_r^5 + c_4(\rho_r, T_{pr})) = 0 \tag{4.70}$$

The derivative, f'_n, is the derivative of Eq. (4.70) with respect to z at constant reduced temperature:

$$\begin{aligned}(\partial F(z)/\partial z)_{T_{pr}} &= 1 + c_1(T_{pr})\rho_r/z + 2c_2(T_{pr})\rho_r^2/z - 5c_3(T_{pr})\rho_r^5/z \\ &+ \frac{2A_{10}\rho_r^2}{T_{pr}^3 z}\left(1 + A_{11}\rho_r^2 - \left(A_{11}\rho_r^{2}\right)^2\right)\exp\left(-A_{11}\rho_r^2\right)\end{aligned} \tag{4.71}$$

The DAK correlations are also provided in Figures 4.9 and 4.10, which express z factors as a function of reduced temperature and pressure. These figures should be used to obtain z factors in manual calculations as shown in Example 4.4.

4.7 Gas Formation Volume Factor (FVF)

The z factor calculated from Eq. (4.63) numerically or by using the graphs in Figures 4.9 and 4.10 can be used to calculate the gas **formation volume factor** (FVF). Gas FVF is defined as its volume at reservoir conditions of temperature and pressure relative to its volume at standard temperature and pressure. Gas FVF, represented as B_g, is thus:

$$B_g = \frac{V_R}{V_{sc}} \tag{4.72}$$

In Eq. (4.72), V_R is the gas volume at reservoir temperature and pressure, and V_{sc} is the gas volume at standard temperature (60°F) and standard pressure (14.65 psia).

Applying Eq. (4.9) for real gases,

$$V_R = \frac{z_R n R T_R}{p_R} \tag{4.73}$$

$$V_{sc} = \frac{z_{sc} n R T_{sc}}{p_{sc}} \tag{4.74}$$

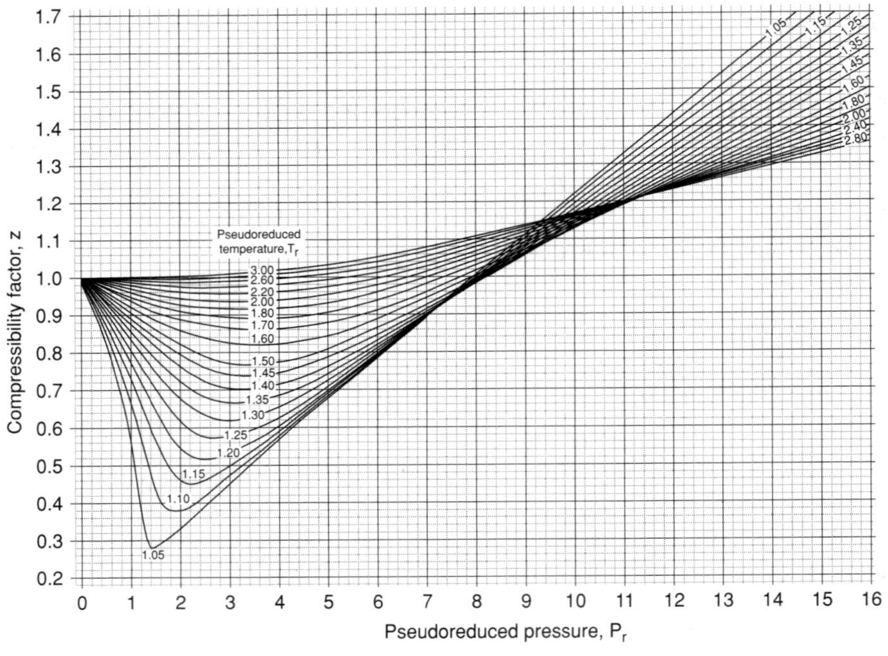

Figure 4.9 Gas compressibility factor chart for $0 \leq p_r \leq 16$ (from Sutton[7] © 1985 SPE, Reproduced with permission).

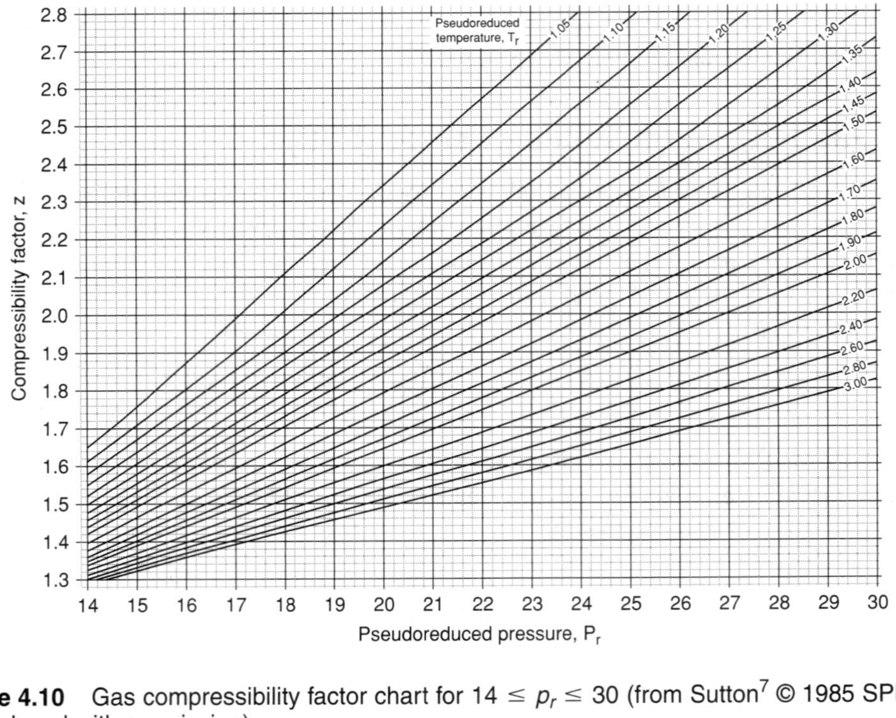

Figure 4.10 Gas compressibility factor chart for $14 \leq p_r \leq 30$ (from Sutton[7] © 1985 SPE, Reproduced with permission).

At standard conditions, ideal gas behavior is assumed for most gases, and hence $z_{sc} = 1.0$. Substituting Eqs. (4.73) and (4.74) into Eq. (4.72) gives:

$$\begin{aligned} B_g &= \frac{z_R T_R (14.65)}{p_R (519.67)} \\ &= \frac{0.02819 z_R T_R}{p_R}, \ \frac{\text{ft}^3}{\text{scf}} \\ &= \frac{0.005021 z_R T_R}{p_R}, \ \frac{\text{RB}}{\text{scf}} \end{aligned} \tag{4.75}$$

Note that in Eq. (4.75), T_{sc} = absolute temperature in °R. This is calculated as: $T_{sc} = 60°\text{F} + T_{abs} = 60 + 459.67 = 519.67°\text{R}$. In this book, absolute zero temperature, $T_{abs} = 459.67°\text{R}$ unless otherwise stated.

4.8 Gas Density

Density is mass per unit volume. Gas density is:

$$\rho_g = \frac{m}{V} \tag{4.76}$$

Assuming real gas behavior, gas volume, V, is given by Eq. (4.9) as:

$$V = \frac{znRT}{p} \tag{4.77}$$

Substituting Eq. (4.77) into Eq. (4.76) gives:

$$\rho_g = \frac{mp}{znRT} \tag{4.78}$$

Because the molecular weight of gas, $M_g = m/n$, where m is the mass of gas, and n is the number of moles of gas, Eq. (4.78) can be written as:

$$\rho_g = \frac{M_g p}{zRT} \tag{4.79}$$

From Eq. (4.14), gas gravity, $\gamma_g = M_g/28.9586$. Expressing Eq. (4.79) in terms of gas gravity:

$$\rho_g = \frac{28.9586 \gamma_g p}{zRT} \tag{4.80}$$

4.9 Gas Viscosity

Many correlations are available in the literature for estimation of gas viscosity. These include correlations reported by Carr et al.,[16] Jossi et al.,[17] Dean and Stiel,[18] Lohrenz et al.,[19] Lee et al.,[20] Lucas,[21] Londono et al.,[22] and Sutton.[7] The correlations reported by Lee et al[20] and Sutton[7] for calculations of gas viscosities are recommended.

The expression for the Lee et al.[20] correlation is:

$$\mu_g = 1 \times 10^{-4} K \exp\left(X \rho_g^Y\right) \tag{4.81}$$

where
$$K = \frac{(9.379 + 0.01607 M_g) T^{1.5}}{209.2 + 19.26 M_g + T} \tag{4.82}$$

and
$$X = 3.448 + \left(\frac{986.4}{T}\right) + 0.01009 M_g \tag{4.83}$$

and
$$Y = 2.447 - 0.2224 X \tag{4.84}$$

Note that in Eqs. (4.81) to (4.84), ρ_g is gas density in g/cm^3; M_g is gas molecular weight in lbm/lb-mole; T is temperature in °R, and μ_g is gas viscosity in centipoise (cp).

The Lee et al.[20] correlation should be used for gases having a specific gravity less than 0.77. The ranges of applicability are as follows: $100 \leq p\,(\text{psia}) \leq 8000$; $200 \leq T(°\text{F}) \leq 340$; $0.55 \leq N_2(\text{mol}\%) \leq 4.8$; and $0.90 \leq CO_2(\text{mol}\%) \leq 3.20$. The reported accuracy of the Lee et al[20] correlation was a standard deviation of $\pm 2.69\%$ and a maximum deviation of 8.99%.

The Sutton[7] correlation should be used for gases having gravity as high as 1.861. It is a hybrid correlation that uses the Lucas[21] method for low-pressure viscosity, and the Lee et al.[20] method for the viscosity ratio. The Sutton[7] correlation is expressed as follows:

$$\mu_g = \mu_{gsc} \exp\left[X \rho_g^Y\right] \tag{4.85}$$

In Eq. (4.85),

$$X = 3.47 + \frac{1588}{T} + 0.0009 M_g \tag{4.86}$$

$$Y = 1.66378 - 0.04679 X \tag{4.87}$$

$$\mu_{gsc}\xi = 10^{-4}\begin{bmatrix} 0.807 T_{pr}^{0.618} - 0.357 \exp(-0.449 T_{pr}) \\ + 0.340 \exp(-4.058 T_{pr}) + 0.018 \end{bmatrix} \tag{4.88}$$

$$\xi = 0.9490 \left(\frac{T_{pc}}{M_g^3 P_{pc}^4}\right)^{1/6} \tag{4.89}$$

In Eqs. (4.85) to (4.89), ρ_g is gas density in g/cm^3; M_g is gas molecular weight in lbm/lb-mole; T is temperature in °R; ξ is viscosity normalizing parameter; μ_{gsc} is low-pressure gas viscosity in centipoise (cp); and μ_g is gas viscosity in centipoise (cp). The Sutton[7] correlation applies under the following conditions: $0 \leq H_2S(\text{mol }\%) \leq 1.7$; $0 \leq CO_2(\text{mol }\%) \leq 8.9$; $0 \leq N_2(\text{mol }\%) \leq 5.2$;

$0 \leq$ Heptanes-plus (mol%) ≤ 24.3; $0.554 \leq \gamma_g \leq 1.861$; $14.7 \leq p$(psia) ≤ 20305; $-45.7 \leq T°R \leq 1112$; and $0.008 \leq \mu_g$ (cp) ≤ 0.435. The average error of the correlation is minus 0.5%, with a standard deviation of 4.2%.

4.10 Gas Coefficient of Isothermal Compressibility

The **coefficient of isothermal compressibility** of gas (or simply **gas compressibility**) is defined as the change in gas volume per change in pressure at constant temperature. Gas compressibility, c_g, is expressed as:

$$c_g = -\frac{1}{V}\left(\frac{\partial V}{\partial p}\right)_T \tag{4.90}$$

For an ideal gas, $V = nRT/p$, and hence:

$$\left(\frac{\partial V}{\partial p}\right)_T = -\frac{nRT}{p^2} \tag{4.91}$$

Substituting Eq. (4.91) into Eq. (4.90) gives:

$$c_g = \frac{1}{p} \tag{4.92}$$

Thus, the units for c_g are psi^{-1}. Eq. (4.92) is useful for an estimate of the value of gas compressibility. But remember that to arrive at this equation, ideal gas behavior was assumed. The true value of gas compressibility will differ substantially depending on the conditions of pressure and temperature of the gas system being evaluated.

For a real gas, $V = \frac{znRT}{p}$, and hence:

$$\begin{aligned}\left(\frac{\partial V}{\partial p}\right)_T &= \frac{nRT}{p}\left(\frac{\partial z}{\partial p}\right)_T - \frac{znRT}{p^2}\\ &= \left(\frac{znRT}{p}\right)\frac{1}{z}\left(\frac{\partial z}{\partial p}\right)_T - \left(\frac{znRT}{p}\right)\left(\frac{1}{p}\right)\end{aligned} \tag{4.93}$$

Dividing both sides of Eq. (4.93) by $1/V$, we get:

$$\frac{1}{V}\left(\frac{\partial V}{\partial p}\right)_T = \frac{1}{z}\left(\frac{\partial z}{\partial p}\right)_T - \frac{1}{p} \tag{4.94}$$

From Eqs. (4.90) and (4.94), then,

$$c_g = \frac{1}{p} - \frac{1}{z}\left(\frac{\partial z}{\partial p}\right)_T \tag{4.95}$$

The similarity between Eq. (4.92) for an ideal gas and Eq. (4.95) for a real gas is clearly evident. By definition for an ideal gas, $z = 1$, Eq. (4.95) reduces to Eq. (4.92).

The principle of corresponding states can be used to transform Eq. (4.95) into a reduced form. From Eq. (4.18), $p = p_{pr} \times p_{pc}$ and $dp = p_{pc}dp_{pr}$. Substituting in Eq. (4.95) gives:

$$c_g = \frac{1}{p_{pr}p_{pc}} - \frac{1}{zp_{pc}}\left(\frac{\partial z}{\partial p_{pr}}\right)_{T_{pr}} \tag{4.96}$$

Trube[23] introduced the term **pseudo-reduced compressibility,** c_r, which is defined as:

$$c_r = c_g \times p_{pc} \tag{4.97}$$

Multiplying Eq. (4.96) by p_{pc} gives:

$$c_r = \frac{1}{p_{pr}} - \frac{1}{z}\left(\frac{\partial z}{\partial p_{pr}}\right)_{T_{pr}} \tag{4.98}$$

Mattar et al.[24] applied Eq. (4.63) to obtain Figures 4.11 and 4.12, in which the product, $c_r T_{pr}$, is plotted as a function of reduced pseudo-temperature, T_{pr}, and reduced pseudo-pressure, p_{pr}. The use of Figures 4.11 and 4.12 to estimate gas compressibility is demonstrated in Example 4.4.

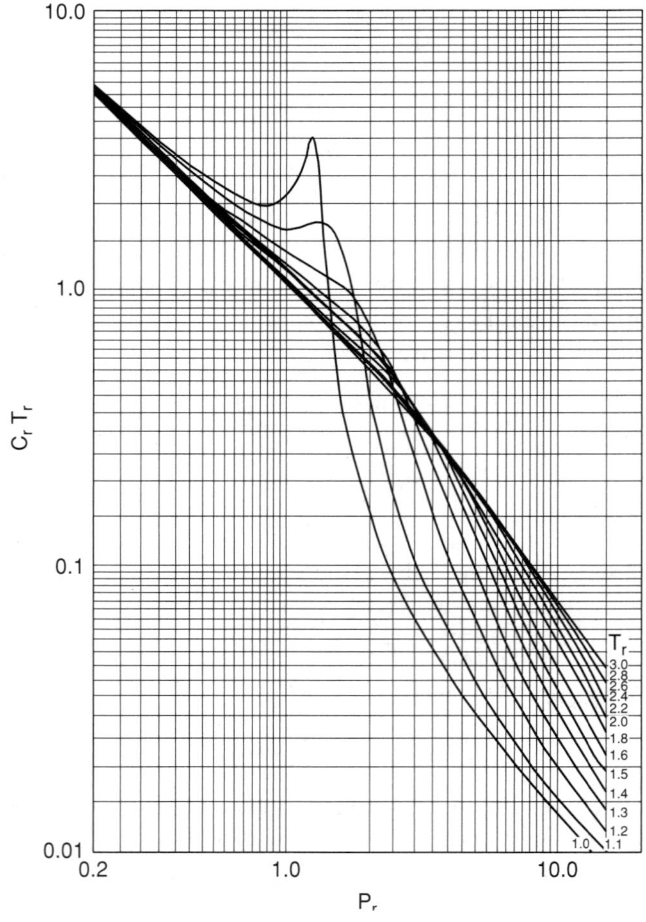

Figure 4.11 Pseudo-reduced compressibility chart for $1.05 \leq T_r \leq 3.0$ and $0.2 \leq p_r \leq 15.0$ (from Mattar et al.[24] © 1975 SPE, Reproduced with permission).

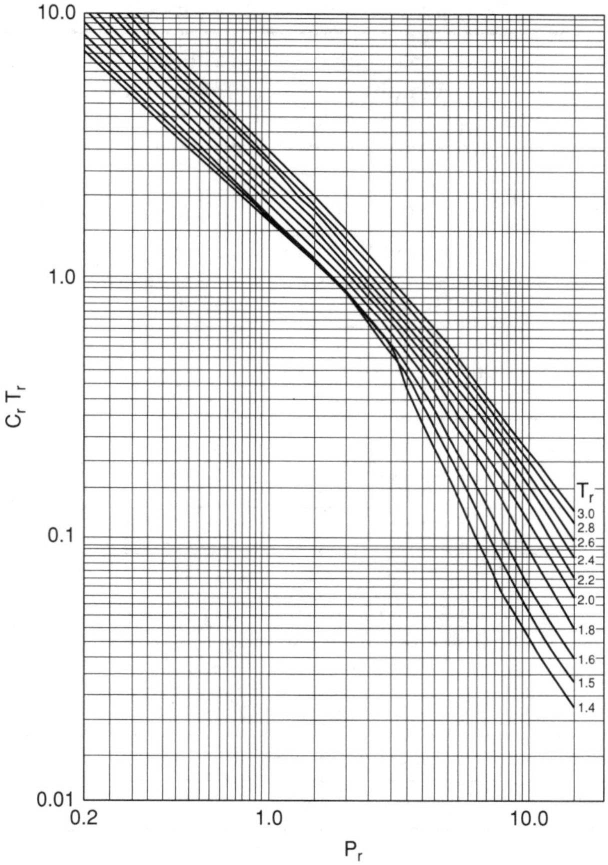

Figure 4.12 Pseudo-reduced compressibility chart for $1.4 \leq T_r \leq 3.0$ and $0.2 \leq p_r \leq 15.0$ (from Mattar et al.[24] © 1975 SPE, Reproduced with permission).

Example 4.3 Calculation of Pseudo-critical Properties of a Gas Mixture Using SSBV Mixing Rules

Problem

Calculate the pseudo-critical properties of a gas mixture with composition shown in Table 4.4 using the SSBV mixing rules

Solution

Step 1: Calculate the boiling point of C_{7+} using Eq. (4.34) from Whitson:[6]

$$\begin{aligned} T_{b,C7+} &= \left(4.5579 \times M_{C7+}^{0.15178} \times \gamma_{C7+}^{0.15427}\right)^3 \\ &= (4.5579 \times 193.72^{0.15178} \times 0.816^{0.15427})^3 \\ &= 948.13^\circ\text{R} \end{aligned}$$

Table 4.4 Properties and Compositions of Gas Mixture for Example 4.3

Component	Mole Fraction	Molecular Weight lbm/lb-mole	Critical Temperature °R	Critical Pressure psia
N_2	0.0030	28.013	227.16	493.1
CO_2	0.1016	44.010	547.58	1071.0
H_2S	0.0504	34.080	672.12	1300.0
CH_4	0.6632	16.043	343.00	666.4
C_2H_6	0.0387	30.070	549.49	706.5
C_3H_8	0.0283	44.097	665.73	616.0
i-C_4H_{10}	0.0067	58.123	734.13	527.9
n-C_4H_{10}	0.0130	58.123	765.29	550.6
i-C_5H_{12}	0.0060	72.150	828.77	490.4
n-C_5H_{12}	0.0066	72.150	845.47	488.6
C_6H_{14}	0.0096	86.177	913.27	436.9
C_{7+}	0.0729	193.72	—	—
Total	1.0000	—	—	—
Molecular weight, C7+ =		193.72		
Specific gravity, C7+ =		0.816		

Step 2: Calculate the critical pressure and temperature of C_{7+} using the Kessler-Lee[5] equations (Eqs. (4.32) and (4.33)):

$$p_{pc,C7+} = \exp\left[\begin{array}{l} 8.3634 - \dfrac{0.0566}{\gamma_{C7+}} - \left(0.24244 + \dfrac{2.2898}{\gamma_{C7+}} + \dfrac{0.11857}{\gamma_{C7+}^2}\right)\dfrac{T_{b,C7+}}{10^3} \\ + \left(1.4685 + \dfrac{3.648}{\gamma_{C7+}} + \dfrac{0.47227}{\gamma_{C7+}^2}\right)\dfrac{T_{b,C7+}^2}{10^7} - \left(0.42019 + \dfrac{1.6977}{\gamma_{C7+}^2}\right)\dfrac{T_{b,C7+}^3}{10^{10}} \end{array}\right]$$

$$= \exp\left[\begin{array}{l} 8.3634 - \dfrac{0.0566}{0.816} - \left(0.24244 + \dfrac{2.2898}{0.816} + \dfrac{0.11857}{0.816^2}\right)\dfrac{948.13}{10^3} \\ + \left(1.4685 + \dfrac{3.648}{0.816} + \dfrac{0.47227}{0.816^2}\right)\dfrac{948.13^2}{10^7} - \left(0.42019 + \dfrac{1.6977}{0.816^2}\right)\dfrac{948.13^3}{10^{10}} \end{array}\right]$$

$$= 264.88 \text{ psia}$$

$$
\begin{aligned}
T_{pc,C7+} &= 341.7 + 811\gamma_{C7+} + (0.4244 + 0.1174\gamma_{C7+})T_{b,C7+} \\
&\quad + (0.4669 - 3.2623\gamma_{C7+})\frac{10^5}{T_{b,C7+}} \\
&= 341.7 + 811 \times 0.816 + (0.4244 + 0.1174 \times 0.816)948.13 \\
&\quad + (0.4669 - 3.2623 \times 0.816)\frac{10^5}{948.13} \\
&= 1265.17°\text{R}
\end{aligned}
$$

Step 3: Calculate the terms in SBV correlations (Eqs. (4.21) to (4.24)) as shown in Table 4.5:

$$
\begin{aligned}
J &= \frac{1}{3}\sum_{i=1}^{n_c}\left(\frac{yT_c}{p_c}\right)_i + \frac{2}{3}\left[\sum_{i=1}^{n_c}\left(y\sqrt{\frac{T_c}{p_c}}\right)_i\right]^2 \\
&= \frac{1}{3}[0.8986] + \frac{2}{3}[0.8632]^2 = 0.7963
\end{aligned}
$$

$$
K = \sum_{i=1}^{n_c}\left(\frac{yT_c}{\sqrt{p_c}}\right)_i = 20.2428
$$

$$
T_{pc} = \frac{K^2}{J} = \frac{(20.2428)^2}{0.7963} = 514.59°R
$$

$$
p_{pc} = \frac{T_{pc}}{J} = \frac{514.59}{0.7963} = 646.23 \text{ psia}
$$

Step 4: Apply Sutton's corrections for high concentrations of heptane plus fraction (Eqs. (4.25) to (4.31)):

$$
F_j = \frac{1}{3}\left(\frac{yT_c}{p_c}\right)_{C7+} + \frac{2}{3}\left(\frac{y^2T_c}{p_c}\right)_{C7+} = \frac{1}{3}(0.3482) + \frac{2}{3}(0.0254) = 0.1330
$$

$$
\begin{aligned}
\varepsilon_j &= 0.6081F_j + 1.1325F_j^2 - 14.004F_jy_{C7+} + 64.434F_jy_{C7+}^2 \\
&= 0.6081(0.1330) + 1.1325(0.1330)^2 - 14.004(0.1330)(0.0729) + 64.434(0.1330)(0.0729)^2 \\
&= 0.0107
\end{aligned}
$$

$$
\begin{aligned}
\varepsilon_k &= \left(\frac{T_c}{\sqrt{p_c}}\right)_{C7+} \times (0.3129y_{C7+} - 4.8156y_{C7+}^2 + 27.3751y_{C7+}^3) \\
&= \left(\frac{1265.17}{\sqrt{264.88}}\right) \times [0.3129(0.0729) - 4.8156(0.0729)^2 + 27.3751(0.0729)^3] = 0.6082
\end{aligned}
$$

$$
J' = J - \varepsilon_j = 0.7963 - 0.0107 = 0.7856
$$

$$
K' = K - \varepsilon_k = 20.2428 - 0.6082 = 19.6346
$$

Table 4.5 Calculation of Pseudo-critical Properties of Gas Mixture Using SSBV Mixing Rules in Example 4.3

Component	Mole Fraction y_i	Molecular Weight (lbm/lb-mole) M_i	Critical Tempe-rature(°R) T_{ci}	Critical Pressure (psia) p_{ci}	y_iM_i	y_iT_{ci}	$(T_{ci}/p_{ci})^{0.5}$	$(p_{ci})^{0.5}$	$(y_iT_{ci})/p_{ci}$	$y_i(T_{ci}/p_{ci})^{0.5}$	$y_iT_{ci}/\sqrt{p_{ci}}$
N_2	0.0030	28.013	227.16	493.1	0.0840	0.6815	0.6787	22.2059	0.0014	0.0020	0.0307
CO_2	0.1016	44.010	547.58	1071.0	4.4714	55.6341	0.7150	32.7261	0.0519	0.0726	1.7000
H_2S	0.0504	34.080	672.12	1300.0	1.7176	33.8748	0.7190	36.0555	0.0261	0.0362	0.9395
CH_4	0.6632	16.043	343.00	666.4	10.6397	227.4776	0.7174	25.8147	0.3414	0.4758	8.8119
C_2H_6	0.0387	30.070	549.49	706.5	1.1637	21.2653	0.8819	26.5801	0.0301	0.0341	0.8000
C_2H_6	0.0283	44.097	665.73	616.0	1.2479	18.8402	1.0396	24.8193	0.0306	0.0294	0.7591
i-C_4H_{10}	0.0067	58.123	734.13	527.9	0.3894	4.9187	1.1793	22.9761	0.0093	0.0079	0.2141
n-C_4H_{10}	0.0130	58.123	765.29	550.6	0.7556	9.9488	1.1789	23.4649	0.0181	0.0153	0.4240
i-C_5H_{12}	0.0060	72.150	828.77	490.4	0.4329	4.9726	1.3000	22.1450	0.0101	0.0078	0.2245
n-C_5H_{12}	0.0066	72.150	845.47	488.6	0.4762	5.5801	1.3154	22.1043	0.0114	0.0087	0.2524
C_6H_{14}	0.0096	86.177	913.27	436.9	0.8273	8.7674	1.4458	20.9022	0.0201	0.0139	0.4194
C_{7+}	0.0729	193.72	1265.17	264.88	14.1222	92.2309	2.1855	16.2751	0.3482	0.1593	5.6670
Total	1.0000	—	—	—	36.3281	—	—	—	0.8986	0.8632	20.2428

$$T_{pc} = \frac{K'^2}{J'} = \frac{(19.6346)^2}{0.7856} = 490.73°R$$

$$p_{pc} = \frac{T_{pc}}{J'} = \frac{490.73}{0.7856} = 624.66 \text{ psia}$$

Step 5: Adjust pseudo-critical properties from step 4 for the presence of hydrogen sulfide and carbon dioxide in the mixture by using the Wichert and Aziz[8] correlation (Eqs. (4.35) to (4.37)):

$$\xi = 120\left[\left(y_{CO_2} + y_{H_2S}\right)^{0.9} - \left(y_{CO_2} + y_{H_2S}\right)^{1.6}\right] + 15\left(y_{H_2S}^{0.5} - y_{H_2S}^{4}\right)$$

$$= 120\left[(0.1520)^{0.9} - (0.1520)^{1.6}\right] + 15\left[(0.0504)^{0.5} - (0.0504)^{4}\right] = 19.50$$

$$T^*_{pc} = T_{pc} - \xi = 490.73 - 19.50 = 471.23 \text{ °}R$$

$$p^*_{pc} = \frac{p_{pc}(T_{pc} - \xi)}{T_{pc} + y_{H_2S}(1 - y_{H_2S})\xi} = \frac{624.66(471.23)}{490.73 + 0.0504(1 - 0.0504)(19.50)}$$

$$= 598.70 \text{ psia}$$

Example 4.4 Calculation of Gas Compressibility Factor, Gas Formation Volume Factor, Gas Gravity, Gas Density, Gas Viscosity, and Gas Compressibility for the Gas Mixture in Example 4.3

Given:
Reservoir pressure, $p = 9620$ psia
Reservoir temperature, $T = 146°F$

Solution

From Example 4.3, pseudo-critical temperature, $T_{pc} = 471.23$ °R and pseudo-critical pressure, $p_{pc} = 598.70$ psia.

From Eq. (4.17), $T_{pr} = \frac{T}{T_{pc}} = \frac{459.67 + 146}{471.23} = 1.28$

From Eq. (4.18), $p_{pr} = \frac{p}{p_{pc}} = \frac{9620}{598.70} = 16.07$

Using Figure 4.9, gas compressibility factor, $z = 1.67$

Gas formation volume factor, B_g, is expressed in Eq. (4.75) as:

$$B_g = \frac{0.005021 z_R T_R}{p_R}$$

$$= \frac{0.005021(1.67)(605.67)}{9620}$$

$$= 0.00053 \text{ RB/scf}$$

From Table 4.5, the molecular weight of the gas mixture, M_g = 36.328 lbm/lb-mole. Gas gravity, γ_g, is shown in Eq. (4.14) as

$$\begin{aligned}\gamma_g &= M_g/28.9586 \\ &= 36.328/28.9586 \\ &= 1.254\end{aligned}$$

Gas density, ρ_g, is shown in Eq. (4.80) as:

$$\begin{aligned}\rho_g &= 28.9586\gamma_g p/zRT \\ &= (28.9586 \times 1.254 \times 9620)/(1.67 \times 10.732 \times 605.67) \\ &= 32.182 \text{ lbm/ft}^3 \\ &= 0.516 \text{ gm/cm}^3\end{aligned}$$

Gas viscosity is calculated using the Sutton[7] correlation because gas gravity is 1.254, which is greater than the limit of 0.77 for the Lee et al.[20] correlation.

From Eqs. (4.85) to (4.89):

$$\xi = 0.9490\left(\frac{T_{pc}}{M_g^3 p_{pc}^4}\right)^{1/6} = 0.9490\left[471.44/(36.328)^3(598.74)^4\right]^{1/6} = 0.006183$$

$$\begin{aligned}\mu_{gsc}\xi &= 10^{-4}\begin{bmatrix}0.807T_{pr}^{0.618} - 0.357\exp(-0.449T_{pr}) \\ + 0.340\exp(-4.058T_{pr}) + 0.018\end{bmatrix} \\ &= 10^{-4}[0.807(1.28)^{0.618} - 0.357\exp(-0.449 \times 1.28) + 0.340\exp(-4.058 \times 1.28) + 0.018] \\ &= 0.000076\end{aligned}$$

$$\mu_{gsc} = 0.000076/0.006183 = 0.01229 \text{ cp}$$

$$X = 3.47 + 1588/T + 0.0009M_g = 3.47 + 1588/605.67 + 0.0009 \times 36.328 = 6.1246$$

$$Y = 1.66378 - 0.04679X = 1.66378 - 0.04679 \times 6.1246 = 1.3772$$

$$\begin{aligned}\mu_g &= \mu_{gsc}\exp\left[X\rho_g^Y\right] = 0.01229\exp\left[6.1246(0.5161)^{1.3772}\right] \\ &= 0.1443 \text{ cp}\end{aligned}$$

The gas compressibility is calculated with the aid of Figure 4.11:

For p_{pr} = 16.07 and T_{pr} =1.28, $c_r T_{pr}$ = 0.02 (Fig. 4.11), c_r = 0.02/1.28 = 0.01563

$$\begin{aligned}c_r &= c_g \times p_{pc} \text{ or } c_g = c_r/p_{pc} = 0.01563/598.74 \\ &= 26 \times 10^{-6} \text{ psi}^{-1} \text{ or 26 microsips}\end{aligned}$$

Example 4.5 Calculation of Pseudo-critical Properties Using Kay's Mixing Rules

Problem

Calculate the pseudo-critical properties and gas compressibility factor of the gas mixture in Table 4.4 using Kay's mixing rules

Solution

Step 1: Calculate the boiling point of C_{7+} using Eq. (4.34) from Whitson:[6]
From Example 4.3, $T_{b,C7+} = 948.13°R$.

Step 2: Calculate the critical pressure and temperature of C_{7+} using the Kessler-Lee[5] equations (Eqs. (4.32) and (4.33)):
From Example 4.3, $P_{pc,C7+} = 264.88$ psia and $T_{pc,C7+} = 1265.17°R$.

Step 3: Calculate pseudo-critical temperature and pressure for mixture.

Using Kay's mixing rules (Eqs. (4.19) and (4.20)):

$$p_{pc} = \sum y_i p_{ci} = 702.9 \text{ psia and } T_{pc} = \sum y_i T_{ci} = 484.34°R \text{ as shown in Table 4.6.}$$

Step 4: Apply corrections for hydrogen sulfide and carbon dioxide.
From Example 4.3, $\xi = 19.50°R$. $T^*_{pc} = T_{pc} - \xi = 484.34 - 19.50 = 464.84°R$.

$$p^*_{pc} = \frac{p_{pc}(T_{pc} - \xi)}{T_{pc} + y_{H_2S}(1 - y_{H_2S})\xi} = \frac{702.9(464.69)}{484.34 + 0.0504(1 - 0.0504)(19.50)} = 673.3 \text{ psia}$$

Table 4.6 Calculation of Pseudo-critical Properties of Gas Mixture in Table 4.4 Using Kay's Mixing Rules in Example 4.5

Component	Mole Fraction y_i	Molecular Weight (lbm/lb-mole) M_i	Critical Temperature (°R) T_{ci}	Critical Pressure (psia) p_{ci}	y_iM_i	y_iT_{ci}	(y_iT_{ci})
N_2	0.0030	28.013	227.16	493.1	0.084	0.68	1.5
CO_2	0.1016	44.010	547.58	1071.0	4.471	55.63	108.8
H_2S	0.0504	34.080	672.12	1300.0	1.718	33.87	65.5
CH_4	0.6632	16.043	343.00	666.4	10.640	227.48	442.0
C_2H_6	0.0387	30.070	549.49	706.5	1.164	21.27	27.3
C_3H_8	0.0283	44.097	665.73	616.0	1.248	18.84	17.4
i-C_4H_{10}	0.0067	58.123	734.13	527.9	0.389	4.92	3.5
n-C_4H_{10}	0.0130	58.123	765.29	550.6	0.756	9.95	7.2
i-C_5H_{12}	0.0060	72.150	828.77	490.4	0.433	4.97	2.9
n-C_5H_{12}	0.0066	72.150	845.47	488.6	0.476	5.58	3.2
C_6H_{14}	0.0096	86.177	913.27	436.9	0.827	8.77	4.2
C_{7+}	0.0729	193.72	1267.17	264.88	14.122	92.38	19.3
Total	1.0000	—	—	—	36.328	484.34	702.9

Step 5: Calculate the gas compressibility factor.
Reservoir pressure = 9620 psia Reservoir temperature = 146°F

$$T_{pr} = \frac{T}{T_{pc}} = \frac{459.67 + 146}{464.84} = 1.30 \text{ and } p_{pr} = \frac{p}{p_{pc}} = \frac{9620}{673.3} = 14.29$$

Using Fig. 4.9, gas compressibility factor, $z = 1.50$.

The actual measured z factor for this mixture is 1.67. This was the value calculated using SSBV mixing rules in Example 4.4. The error introduced by using Kay's mixing rules is 10%. This example demonstrates that the SSBV method is more accurate than Kay's method. However, Kay's method is relatively simple and can be used for quick estimation.

Example 4.6 Calculation of Pseudo-critical Properties and Gas Compressibility Factor of a Gas Mixture if the Composition is Unknown

Problem

The gas gravity of the gas mixture in Example 4.3 was calculated to be 1.254. Assuming that the gas composition is not known, calculate the pseudo-critical properties of the mixture using Sutton's correlation and the gas compressibility factor.

Solution

Step 1: Calculate the hydrocarbon gas composition.
From Eq. (4.39):

$$y_{HC} = 1 - y_{H_2S} - y_{CO_2} - y_{N_2} - y_{H_2O} = 1 - 0.0504 - 0.1016 - 0.003 = 0.8450$$

Step 2: Calculate the hydrocarbon gas gravity.
From Eq. (4.40):

$$\gamma_{g,HC} = \frac{\gamma_g - (y_{H_2S}M_{H_2S} + y_{CO_2}M_{CO_2} + y_{N_2}M_{N_2} + y_{H_2O}M_{H_2O})/M_{AIR}}{y_{HC}}$$

$$= \frac{1.254 - (0.0504 \times 34.08 + 0.1016 \times 44.01 + 0.003 \times 28.013)/28.963}{0.845}$$

$$= 1.228$$

Step 3: Calculate the hydrocarbon pseudo-critical properties.
Using Eqs. (4.45) and (4.46):

$$p_{pc,HC} = 744 - 125.4\gamma_{g,HC} + 5.9\gamma_{g,HC}^2 = 744 - 125.4 \times 1.228 + 5.9(1.228)^2 = 598.91 \text{ psia}$$

$$T_{pc,HC} = 164.3 + 357.7\gamma_{g,HC} - 67.7\gamma_{g,HC}^2 = 164.3 + 357.7 \times 1.228 - 67.7(1.228)^2$$

$$= 501.47\ °R$$

Step 4: Calculate the pseudo-critical properties for the gas mixture.
Using Eqs. (4.41) and (4.42):

$$\begin{aligned} p_{pc} &= y_{HC}p_{pc,HC} + y_{H_2S}p_{c,H_2S} + y_{CO_2}p_{c,CO_2} + y_{N_2}p_{c,N_2} + y_{H_2O}p_{c,H_20} \\ &= 0.845 \times 598.91 + 0.0504 \times 1300 + 0.1016 \times 1071 + 0.003 \times 493.1 = 681.89 \text{ psia} \end{aligned}$$

$$\begin{aligned} T_{pc} &= y_{HC}T_{pc,HC} + y_{H_2S}T_{c,H_2S} + y_{CO_2}T_{c,CO_2} + y_{N_2}T_{c,N_2} + y_{H_2O}T_{c,H_2O} \\ &= 0.845 \times 501.47 + 0.0504 \times 672.12 + 0.1016 \times 547.58 + 0.003 \times 227.16 = 513.93°\text{R} \end{aligned}$$

Step 5: Apply corrections for hydrogen sulfide and carbon dioxide.
From Example 4.3, $\xi = 19.50°\text{R}$. $T^*_{pc} = T_{pc} - \xi = 513.93 - 19.50 = 494.43°\text{R}$.

$$p^*_{pc} = \frac{p_{pc}(T_{pc} - \xi)}{T_{pc} + y_{H_2S}(1 - y_{H_2S})\xi} = \frac{681.89(494.43)}{513.93 + 0.0504(1 - 0.0504)(19.50)} = 654.83 \text{ psia}$$

Step 6: Calculate the z factor.
Reservoir pressure = 9620 psia Reservoir temperature = 146°F

$$T_{pr} = \frac{T}{T_{pc}} = \frac{459.67 + 146}{494.43} = 1.22 \text{ and } p_{pr} = \frac{p}{p_{pc}} = \frac{9620}{654.83} = 14.69$$

Using Figure 4.10, the gas compressibility factor is calculated to be, $z = 1.56$. The actual measured z factor for this mixture is 1.67. The error from using Sutton's correlation for gases with unknown composition is calculated to be 6.6% for this example.

4.11 Correlations for Calculation of Oil PVT Properties

The most reliable source of PVT data for oil mixtures is to conduct laboratory measurements on fluid samples taken from the reservoir. Chapter 5 describes the methods used to obtain reservoir fluid samples and the types of laboratory measurements that are conducted. However, in many circumstances, such as new reservoir discoveries, laboratory PVT data may not be available. To calculate early estimates of oil-in-place volumes, oil reserves, oil production rates, recovery factors, etc, the engineer needs empirical correlations to estimate the PVT properties of the reservoir oil. In this section, these empirical correlations are presented. Many empirical correlations for calculation of oil properties exist in the literature.[25,26] Some of the correlations are based on oil samples from specific geographical regions: Africa,[27,28] Gulf of Mexico,[29,30] Middle East,[31,32] North Sea,[33] and South America.[34] According to the analyses conducted by Al-Shammasi[25] and supported by the work of Valko and McCain,[35] correlations based on specific geographical regions are not better than correlations based on a global data set. Consequently, the correlations recommended in this section may have been derived from either a geographical or a global data set but have been deemed most accurate based on published results. However, the user is encouraged to search for other correlations if not satisfied with the estimates calculated from the recommended correlations.

4.11.1 Bubble Point Pressure

Valko and McCain[35] developed correlations for oil bubble point pressures at reservoir conditions. The correlating parameters are solution gas-oil ratio at bubble point, R_{sb}; oil API gravity, γ_{API}; separator gas gravity, γ_{sp}; and reservoir temperature, T_F, in °F.

The Valko and McCain[35] correlations for bubble point pressure are as follows:

$$\ln p_b = 7.475 + 0.713Z + 0.0075Z^2 \tag{4.99}$$

$$Z = \sum_{n=1}^{4} Z_n \tag{4.100}$$

$$Z_1 = -5.48 - 0.0378\ln R_{sb} + 0.281(\ln R_{sb})^2 - 0.0206(\ln R_{sb})^3 \tag{4.101}$$

$$Z_2 = 1.27 - 0.0449\gamma_{API} + 4.36 \times 10^{-4}(\gamma_{API})^2 - 4.76 \times 10^{-6}(\gamma_{API})^3 \tag{4.102}$$

$$Z_3 = 4.51 - 10.84\gamma_{gsp} + 8.39(\gamma_{gsp})^2 - 2.34(\gamma_{gsp})^3 \tag{4.103}$$

$$Z_4 = -0.7835 + 6.23 \times 10^{-3}T_F - 1.22 \times 10^{-5}(T_F)^2 + 1.03 \times 10^{-8}(T_F)^3 \tag{4.104}$$

In Eqs. (4.99) to (4.104), p_b = bubble point pressure, psia; R_{sb} = solution gas-oil ratio at bubble point, scf/STB; γ_{API} = stock-tank oil gravity, °API; γ_{gsp} = separator gas gravity (air = 1); and T_F = reservoir temperature, °F.

The average relative error and the absolute relative error reported for the Valko and McCain[35] correlations are 0.0% and 10.9%, respectively. The global data set (1745 records) used to obtain these correlations has the following range:

$$10 \leq R_{sb} \leq 2216 \text{ scf/STB}$$
$$6.0 \leq \gamma_{API} \leq 63.7\ °API$$
$$0.555 \leq \gamma_{gsp} \leq 1.685 \text{ (air = 1)}$$
$$60 \leq T_F \leq 342\ °\text{F}$$
$$82 \leq p_b \leq 6700 \text{ psia}$$

A simpler but less accurate correlation developed by Standing[9] can be used to calculate oil bubble point pressures at reservoir conditions. The Standing correlation has wide acceptance in the petroleum industry. It was developed from data based on samples from California reservoirs. The Standing correlation (valid to 325 °F) is as follows:

$$p_b = 18.2(C_{pb} - 1.4) \tag{4.105}$$

$$\text{and } C_{pb} = (R_{sb}/\overline{\gamma}_g)^{0.83} \times 10^{(0.00091T_F - 0.0125\gamma_{API})} \tag{4.106}$$

In Eqs. (4.105) and (4.106), $\overline{\gamma}_g$ = average gas gravity (separator plus stock-tank vent gas). Average gas gravity can be calculated as shown in Eq. (4.110) or estimated as shown in Eq. (4.111). All other terms have the same units as in the Valko and McCain[35] correlations. The average relative error and the average absolute relative error of the Standing[9] correlation are −2.1% and 12.7%, respectively, when applied to the same global data set used for the Valko and McCain[35] correlations.

The solution gas-oil ratio at bubble point, R_{sb}, in Eqs. (4.101) and (4.106) is the sum of the producing gas-oil ratio from the separators and the stock-tank for a three or a two-stage separation system. Thus, for a two-stage separation system:

$$R_{sb} = R_{sp} + R_{st} \tag{4.107}$$

In most field operations, the separator gas and stock-tank oil production rates are measured. The stock-tank vent gas is rarely measured. As a result, the total solution gas-oil ratio as represented in Eq. (4.107) is underreported. If the stock-tank vent gas is not included in the reported total solution gas-oil ratio, then a simple correlation reported by Valko and McCain[35] can be used to account for the deficiency. The correlation is:

$$R_{sb} = 1.1618 R_{sp} \tag{4.108}$$

Equation (4.108) was based on a data set having 881 data points, with an average absolute relative error of 9.9%.

4.11.2 Solution Gas-Oil Ratio (GOR)

At reservoir pressures equal to or less than bubble point pressures, solution gas-oil ratios can be estimated by using Eqs. (4.105) and (4.106). The errors associated with using these equations are the same as reported for bubble point pressure estimation using the same equations.

4.11.3 Oil Formation Volume Factor (FVF)

Oil formation volume factors (FVF) at or less than bubble point pressures can be estimated by using the correlation obtained by Petrosky and Farshad[29] from regression of data:

$$B_o = 1.0113 + 7.2046 \times 10^{-5}\left[R_s^{0.3738}\left(\frac{(\overline{\gamma}_g)^{0.2914}}{\gamma_o^{0.6265}}\right) + 0.24626 T_F^{0.5371}\right]^{3.0936} \tag{4.109}$$

In Eq. (4.109), B_o = FVF, RB/STB; R_s = total solution gas-oil ratio (separator plus stock-tank gas), scf/STB; $\overline{\gamma}_g$ = average gas gravity (separator plus stock tank), air = 1; γ_o = oil specific gravity (water = 1); and T_F = temperature, °F.

The Petrosky and Farshad[29] correlation was regressed from data taken from Gulf of Mexico oil samples. However, the correlation was applied by Al-Shammasi[25] to a global data set and was found to have the lowest average absolute relative error and standard deviation when compared with all other available correlations. Al-Shammasi[25] reported average absolute relative error of 1.728% and standard deviation of 1.92% for the Petrosky and Farshad correlation. The global data set used by Al-Shammasi[25] has 1345 points and the following ranges:

$$6.0 \leq R_s \leq 3298.6 \text{ scf/STB}$$

$$0.51 \leq \gamma_g \leq 3.44 \text{ (air} = 1)$$

$$6.0 \leq \gamma_{API} \leq 63.7°\text{API}$$

$$74 \leq T_F \leq 341.6°\text{F}$$

$$1.02 \leq B_o \leq 2.916 \text{ RB/STB}$$

$$31.7 \leq p_b \leq 7127.0 \text{ psi}$$

The average gas gravity in Eq. (4.109) is calculated as a weighted average as follows:

$$\overline{\gamma}_g = \frac{\gamma_{gsp}R_{sp} + \gamma_{gst}R_{st}}{R_{sp} + R_{st}} \tag{4.110}$$

In Eq. (4.110), $\overline{\gamma}_g$ = average gas gravity; γ_{gsp} = separator gas gravity; γ_{gst} = stock-tank gas gravity; R_{sp} = separator GOR, scf/STB; and R_{st} = stock-tank GOR, scf/STB. In most field operations, stock-tank gas gravity and stock-tank GOR are not measured. In such circumstances, average gas gravity can be estimated by using a correlation suggested by Valko and McCain:[35]

$$\overline{\gamma}_g = 1.066\gamma_{gsp} \tag{4.111}$$

Eq. (4.111) was based on 626 data points, with an average absolute error of 3.8%. Note that Eq. (4.110) or (4.111) is used to estimate average gas gravity where required in this book.

For pressures greater than bubble point pressure, oil FVF is calculated by using:

$$B_o = B_{ob} \exp\left[c_o(p_b - p)\right] \tag{4.112}$$

Oil FVF at bubble point pressure, B_{ob}, is calculated from Eq. (4.109). The coefficient of oil isothermal compressibility, c_o, is calculated by using the correlation shown in Eq. (4.113).

4.11.4 Coefficient of Isothermal Compressibility of Oil

The coefficient of isothermal compressibility of oil (or simply oil compressibility) at bubble point pressure or greater than bubble point pressure can be calculated by using correlations developed by Spivey et al.:[36]

$$\ln c_{ob} = 2.434 + 0.475Z + 0.048Z^2 \tag{4.113}$$

$$Z = \sum_{n=1}^{6} Z_n \tag{4.114}$$

$$Z_1 = 3.011 - 2.6254 \ln \gamma_{API} + 0.497(\ln \gamma_{API})^2 \tag{4.115}$$

$$Z_2 = -0.0835 - 0.259 \ln \gamma_{gsp} + 0.382(\ln \gamma_{gsp})^2 \tag{4.116}$$

$$Z_3 = 3.51 - 0.0289 \ln p_b - 0.0584(\ln p_b)^2 \tag{4.117}$$

$$Z_4 = 0.327 - 0.608 \ln (p/p_b) + 0.0911[\ln (p/p_b)]^2 \tag{4.118}$$

$$Z_5 = -1.918 - 0.642 \ln R_{sb} + 0.154(\ln R_{sb})^2 \tag{4.119}$$

$$Z_6 = 2.52 - 2.73 \ln T_F + 0.429(\ln T_F)^2 \tag{4.120}$$

In Eqs. (4.113) to (4.120), c_{ob} = oil compressibility in microsips; γ_{API} = stock-tank oil gravity, °API; γ_{gsp} = separator gas gravity (air = 1); p_b = bubble point pressure, psia; p = reservoir pressure, psia; R_{sb} = solution gas-oil ratio at bubble point, scf/STB; and T_F = reservoir temperature, °F.

The average relative error and average absolute relative error of the Spivey et al.[36] correlations are 0.1% and 7.1%, respectively, for a global data set having 2433 data points from 369 reservoir samples. The applicable ranges of the data set are as follows:

$$11.6 \leq \gamma_{API} \leq 57.7\ °\text{API}$$
$$0.561 \leq \gamma_{gsp} \leq 1.798\ (\text{air} = 1)$$
$$120.7 \leq p_b \leq 6658.7\ \text{psia}$$
$$414.7 \leq p \leq 8014.7\ \text{psia}$$
$$12 \leq R_{sb} \leq 1808\ \text{scf/STB}$$
$$70.7 \leq T_F \leq 320°\text{F}$$
$$3.6 \leq c_o \leq 50.3\ \text{microsips}$$

Note that the units of oil compressibility calculated by using Eq. (4.113) are in microsips, or 10^{-6} psi^{-1}.

Al-Marhoun[37] developed a simpler but less accurate correlation for oil compressibility at pressures greater than the bubble point pressure. The Al-Marhoun[37] correlation, which is based on oil samples from the Middle East, is:

$$\ln c_o = -14.1042 + \frac{2.7314}{\gamma_{ob}} - 56.0605 \times 10^{-6} \frac{(p - p_b)}{\gamma_{ob}^3} - \frac{580.8778}{(T_F + 460)} \tag{4.121}$$

$$\gamma_{ob} = \frac{(\gamma_o + 2.18 \times 10^{-4} R_{sb} \overline{\gamma}_g)}{B_{ob}} \tag{4.122}$$

In Eqs. (4.121) and (4.122), c_o = oil compressibility, psi^{-1}; and $\overline{\gamma}_g$ = average gas gravity, air = 1. All other terms are as previously defined for Eq. (4.113).

Spivey et al.[36] reported average relative error and average absolute error of −3.3% and 12.2%, respectively, when the Al-Marhoun[37] correlation was applied to the same global data set used in their work.

For pressures less than bubble point pressure, McCain et al.[38] developed several correlations for oil compressibility depending on available data. If bubble point pressure and solution gas-oil ratio are available, then oil compressibility can be calculated by using:

$$\ln c_o = -7.573 - 1.450 \ln p - 0.383 \ln p_b + 1.402 \ln T + 0.256 \ln \gamma_{API} + 0.449 \ln R_{sb} \tag{4.123}$$

If the bubble point pressure is not available, oil compressibility can be calculated by using:

$$\ln c_o = -7.633 - 1.497 \ln p + 1.115 \ln T + 0.533 \ln \gamma_{API} + 0.184 \ln R_{sb} \tag{4.124}$$

If bubble point pressure and solution GOR are not available, oil compressibility can be calculated by using:

$$\ln c_o = -7.114 - 1.394 \ln p + 0.981 \ln T + 0.770 \ln \gamma_{API} + 0.446 \ln \overline{\gamma}_g \tag{4.125}$$

In Eqs. (4.123) to (4.125), c_o = oil compressibility, psi^{-1}; p = reservoir pressure, psia; p_b = bubble point pressure, psia; T = reservoir temperature, °R; γ_{API} = stock-tank oil gravity, °API; R_{sb} = solution gas-oil ratio at bubble point, scf/STB; and $\overline{\gamma}_g$ = weighted average of separator and stock-tank gas gravity.

The reported maximum absolute deviation for all the equations is 10.2% for pressures greater than 500 psia. The ranges of data (2500 data points) used to generate these correlations are:

$$500 \leq p \leq 5300 \text{ psia}$$

$$763 \leq p_b \leq 5300 \text{ psia}$$

$$78 \leq T_F \leq 330°\text{F}$$

$$0.58 \leq \overline{\gamma}_g \leq 1.2 \text{ (air = 1)}$$

$$18 \leq \gamma_{API} \leq 52°\text{API}$$

$$15 \leq R_{sb} \leq 1947 \text{ scf/STB}$$

$$31 \leq c_o \leq 6600 \text{ microsips}$$

4.11.5 Oil Viscosity

The calculation of the viscosity of live oil (gas-saturated oil) at bubble point pressure or at pressures less than bubble point pressure is a two-step process. In the first step, the viscosity of the dead oil (no dissolved gas) is calculated by using the correlation by Ng and Egbogah.[39] In the second step, the viscosity of the live oil is calculated by using the correlation by Beggs and Robinson.[40] The Ng-Egbogah[39] dead oil viscosity correlation is:

$$\log_{10}[\log_{10}(\mu_{od} + 1)] = 1.8653 - 0.025086\gamma_{API} - 0.5644 \log_{10}(T_F) \quad \textbf{(4.126)}$$

For Eq. (4.126), μ_{od} = dead oil viscosity, cp; γ_{API} = stock-tank oil gravity, °API; and T_F = temperature, °F.

The average absolute relative error reported by Ng and Egbogah[39] for the correlation is 6.6% for 394 samples with the following property ranges:

$$59 \leq T_R \leq 176°\text{F}$$

$$-58 \leq T_{pour} \leq 59°\text{F}$$

$$5 \leq \gamma_{API} \leq 58°\text{API}$$

The Beggs and Robinson[40] correlation for gas-saturated oil at or less than bubble point pressures is:

$$\begin{aligned} \mu_{ob} &= a(\mu_{od})^b \\ \text{where } a &= 10.715(R_s + 100)^{-0.515} \\ \text{and } b &= 5.44(R_s + 150)^{-0.338} \end{aligned} \quad \textbf{(4.127)}$$

In Eq. (4.127), μ_{ob} = viscosity of gas-saturated oil, cp; and R_s = solution gas-oil ratio, scf/STB.

The average error reported by Beggs and Robinson[40] for the correlation is −1.83%, with standard deviation of 27.25%. The correlation was developed from a database of 2073 points from 600 samples having the following ranges:

$$20 \le R_s \le 2070 \text{ scf/STB}$$
$$16 \le \gamma_{API} \le 58 \text{ °API}$$
$$70 \le T_F \le 295 \text{ °F}$$
$$0 \le p \le 5250 \text{ psig}$$

For pressures greater than the bubble point, Bergman and Sutton[41] developed the following correlation:

$$\begin{aligned} \mu_o &= \mu_{ob} e^{\alpha(p-p_b)^\beta} \\ \text{where } \alpha &= 6.5698 \times 10^{-7} \ln(\mu_{ob})^2 - 1.48211 \times 10^{-5} \ln(\mu_{ob}) + 2.27877 \times 10^{-4} \\ \text{and } \beta &= 2.24623 \times 10^{-2} \ln(\mu_{ob}) + 0.873204 \end{aligned} \tag{4.128}$$

In Eq. (4.128), μ_o = undersaturated oil viscosity, cp; p = reservoir pressure, psia; and p_b = bubble point pressure, psia.

The Bergman and Sutton[41] correlation is remarkable because of its applicability to severe conditions of high pressures and high temperatures typically encountered in deepwater reservoirs. The database consisted of 1399 oil samples and 10,248 data points having the following ranges:

$$115 \le p \le 25{,}015 \text{ psia}$$
$$14.5 \le p_b \le 11{,}195 \text{ psia}$$
$$32 \le T_F \le 425 \text{ °F}$$
$$6 \le \gamma_{API} \le 61.2 \text{ °API}$$
$$0 \le R_s \le 4630.6 \text{ scf/STB}$$
$$0.204 \le \mu_{od} \le 14{,}200 \text{ cp}$$
$$0.063 \le \mu_{ob} \le 14{,}200 \text{ cp}$$
$$0.067 \le \mu_o \le 24{,}180 \text{ cp}$$

The average absolute relative error reported by Berman and Sutton[41] for the correlation is 3.88%, with standard deviation of 4.97%.

Example 4.7 Calculation of Bubble Point Pressure (p_b), Oil FVF at Bubble Point (B_{ob}), Oil Compressibility at Bubble Point (c_{ob}), and Oil Viscosity at Bubble Point (μ_{ob})

Problem

Consider an oil sample from a reservoir that has the following properties:

Solution gas-oil ratio at bubble point, R_{sb}	805 scf/STB
Average gas gravity, $\overline{\gamma}_g$	0.929 (air = 1)
Stock-tank oil gravity, γ_{API}	39.1°API
Reservoir temperature, T_R	200°F

Calculate bubble point pressure, oil formation volume factor at bubble point, oil compressibility at bubble point, and oil viscosity at bubble point.

Solution

To calculate bubble point pressure for the oil sample, use the Valko and McCain correlation given in Eqs. (4.99) to (4.104):

$$\begin{aligned} Z_1 &= -5.48 - 0.0378 \ln R_{sb} + 0.281(\ln R_{sb})^2 - 0.0206(\ln R_{sb})^3 \\ &= -5.48 - 0.0378 \ln 805 + 0.281(\ln 805)^2 - 0.0206(\ln 805)^3 = 0.6764 \end{aligned}$$

$$\begin{aligned} Z_2 &= 1.27 - 0.0449\gamma_{API} + 4.36 \times 10^{-4}(\gamma_{API})^2 - 4.76 \times 10^{-6}(\gamma_{API})^3 \\ &= 1.27 - 0.0449(39.1) + 4.36 \times 10^{-4}(39.1)^2 - 4.76 \times 10^{-6}(39.1)^3 = -0.1036 \end{aligned}$$

$$\begin{aligned} Z_3 &= 4.51 - 10.84\gamma_{gsp} + 8.39(\gamma_{gsp})^2 - 2.34(\gamma_{gsp})^3 \\ &= 4.51 - 10.84(0.929) + 8.39(0.929)^2 - 2.34(0.929)^3 = -0.1956 \end{aligned}$$

$$\begin{aligned} Z_4 &= -0.7835 + 6.23 \times 10^{-3}T_F - 1.22 \times 10^{-5}(T_F)^2 + 1.03 \times 10^{-8}(T_F)^3 \\ &= -0.7835 + 6.23 \times 10^{-3}(200) - 1.22 \times 10^{-5}(200)^2 + 1.03 \times 10^{-8}(200)^3 = 0.0569 \end{aligned}$$

$$Z = \sum_{n=1}^{n} Z_n = Z_1 + Z_2 + Z_3 + Z_4 = 0.6764 - 0.1036 - 0.1956 + 0.0569 = 0.4341$$

$$\begin{aligned} \ln p_b &= 7.475 + 0.7132Z + 0.0075Z^2 \\ &= 7.475 + 0.7132(0.4341) + 0.0075(0.4341)^2 = 7.7860 \end{aligned}$$

$$\begin{aligned} p_b &= \exp(7.7860) \\ &= 2406.7 \text{ psia} \end{aligned}$$

For comparison, the bubble point pressure calculations were repeated using the Standing correlation given in Eqs. (4.105) to (4.106):

$$\begin{aligned} C_{pb} &= (R_{sb}/\overline{\gamma}_g)^{0.83} \times 10^{(0.00091T_F - 0.0125\gamma_{API})} \\ &= (805/0.929)^{0.83} \times 10^{(0.00091(200) - 0.0125(39.1))} = 135.3965 \end{aligned}$$

$$\begin{aligned} P_b &= 18.2(C_{pb} - 1.4) = 18.2(135.3965 - 1.4) \\ &= 2438.7 \text{ psia} \end{aligned}$$

Actual measured bubble point pressure for the oil sample is 2392 psia. The estimation error for the Valko-McCain[35] correlation for this example is 0.61%, and the error for the Standing[9] correlation is 1.95%.

To calculate formation volume factor at bubble point, the correlation by Petrosky and Farshad[29] given in Eq. (4.109) is used:

$$B_o = 1.0113 + 7.2046 \times 10^{-5}\left[R_s^{0.3738}\left(\frac{(\overline{\gamma}_g)^{0.2914}}{\gamma_o^{0.6265}}\right) + 0.24626T_F^{0.5371}\right]^{3.0936}$$

$$= 1.0113 + 7.2046 \times 10^{-5}\left[(805)^{0.3738}\left(\frac{(0.929)^{0.2914}}{(0.8294)^{0.6265}}\right) + 0.24626(200)^{0.5371}\right]^{3.0936}$$

$$= 1.5304 \text{ RB/STB}$$

Note that oil specific gravity, $\gamma_o = 141.5/(131.5 + \gamma_{API}) = 141.5/(131.5 + 39.1) = 0.8294$. The actual measured formation volume factor for the oil sample is 1.4790 rb/STB. The error calculated for the Petrosky-Farshad[29] correlation for this example is 3.5%.

The coefficient of isothermal compressibility of oil is calculated by using the Spivey et al.[36] correlations shown in Eqs. (4.113) to (4.120):

$$Z_1 = 3.011 - 2.6254 \ln \gamma_{API} + 0.497(\ln \gamma_{API})^2$$
$$= 3.011 - 2.6254 \ln 39.1 + 0.497(\ln 39.1)^2 = 0.0659$$

$$Z_2 = -0.0835 - 0.259 \ln \gamma_{gsp} + 0.382(\ln \gamma_{gsp})^2$$
$$= -0.0835 - 0.259 \ln 0.929 + 0.382(\ln 0.929)^2 = -0.0624$$

$$Z_3 = 3.51 - 0.0289 \ln p_b - 0.0584(\ln p_b)^2$$
$$= 3.51 - 0.0289 \ln 2392 - 0.0584(\ln 2392)^2 = -0.2496$$

$$Z_4 = 0.327 - 0.608 \ln (p/p_b) + 0.0911[\ln (p/p_b)]^2$$
$$= 0.327 - 0.608 \ln (2392/2392) + 0.0911[\ln (2392/2392)]^2 = 0.3270$$

$$Z_5 = -1.918 - 0.642 \ln R_{sb} + 0.154(\ln R_{sb})^2$$
$$= -1.918 - 0.642 \ln 805 + 0.154(\ln 805)^2 = 0.6807$$

$$Z_6 = 2.52 - 2.73 \ln T_F + 0.429(\ln T_F)^2$$
$$= 2.52 - 2.73 \ln 200 + 0.429(\ln 200)^2 = 0.0986$$

$$Z = \sum_{n=1}^{6} Z_n = Z_1 + Z_2 + Z_3 + Z_4 + Z_5 + Z_6 = 0.0659 - 0.0624 - 0.2496 + 0.3270$$
$$+ 0.6807 + 0.0986 = 0.8602$$

$$\ln c_{ob} = 2.434 + 0.475Z + 0.048Z^2$$
$$= 2.434 + 0.475(0.8602) + 0.048(0.8602)^2 = 2.8781$$

$$c_{ob} = \exp(2.8781) = 17.78 \times 10^{-6} \text{ psi}^{-1}$$
$$= 17.78 \text{ microsips}$$

Alternatively, oil compressibility can be calculated by using the Al-Marhoun[37] correlation shown in Eqs. (4.121) and (4.122):

$$\gamma_{ob} = \frac{(\gamma_o + 2.18 \times 10^{-4} R_{sb}\overline{\gamma}_g)}{B_{ob}} = \frac{(0.8294 + 2.18 \times 10^{-4}(805 \times 0.929))}{1.479} = 0.6710$$

$$\ln c_o = -14.1042 + \frac{2.7314}{\gamma_{ob}} - 56.0605 \times 10^{-6} \frac{(p - p_b)}{\gamma_{ob}^3} - \frac{580.8778}{(T_F + 460)}$$
$$= -14.1042 + \frac{2.7314}{0.6710} - 56.0605 \times 10^{-6} \frac{(2392 - 2392)}{(0.6710)^3} - \frac{580.8778}{(200 + 460)} = -10.9137$$

$$c_{ob} = \exp(-10.9137) = 18.21 \times 10^{-6} \text{ psi}^{-1}$$
$$= 18.21 \text{ microsips}$$

For this example, there is no significant difference between the oil compressibility at bubble point calculated with the two correlations. But note the relative ease of using the Al-Marhoun correlation.

Calculation of oil viscosity for the sample at bubble point pressure is a two-step process. In step 1, dead oil viscosity is calculated by using the Ng-Egbogah[39] correlation. In step 2, oil viscosity at bubble point is calculated by using the Beggs-Robinson[40] correlation.

Step 1: Calculate dead oil viscosity using the Ng-Egbogah correlation.
From Eq. (4.126):

$$\log_{10}\left[\log_{10}(\mu_{od} + 1)\right] = 1.8653 - 0.025086\gamma_{API} - 0.5644 \log_{10}(T_F)$$
$$= 1.8653 - 0.025086 \times 39.1 - 0.5644 \log_{10}(200) = -0.4143$$

$$\mu_{od} = 1.4278 \text{ cp}$$

Step 2: Calculate oil viscosity at bubble point by using the Beggs-Robinson correlation.
From Eq. (4.127):

$$a = 10.715(R_s + 100)^{-0.515} = 10.715(805 + 100)^{-0.515} = 0.3216$$
$$b = 5.44(R_s + 150)^{-0.338} = 5.44(805 + 150)^{-0.338} = 0.5350$$
$$\mu_{ob} = 0.3216(1.4278)^{0.5350}$$
$$= 0.3891 \text{ cp}$$

An additional exercise is to calculate oil viscosity for the sample at 4000 psia reservoir pressure using the Bergman-Sutton[41] correlation in Eq. (4.128).

$$\alpha = 6.5698 \times 10^{-7} \ln(\mu_{ob})^2 - 1.48211 \times 10^{-5} \ln(\mu_{ob}) + 2.27877 \times 10^{-4}$$
$$= 6.5698 \times 10^{-7} \ln(0.3891)^2 - 1.48211 \times 10^{-5} \ln(0.3891) + 2.27877 \times 10^{-4}$$
$$= 0.00024245$$

$$\beta = 2.24623 \times 10^{-2} \ln(\mu_{ob}) + 0.873204$$
$$= 2.24623 \times 10^{-2} \ln(0.3891) + 0.873204 = 0.8520$$

Note that $\mu_{ob} = 0.3891$ cp from the Beggs-Robinson correlation, as shown earlier.

$$\mu_o = \mu_{ob} e^{\alpha(p-p_b)^\beta} = 0.3891 \exp^{0.00024245(4000-2392)^{0.8520}}$$
$$= 0.4434 \text{ cp}$$

4.12 Correlations for Calculation of Water PVT Properties

4.12.1 Water Formation Volume Factor (FVF)

The formation volume factor of water, B_w, can be calculated by using a correlation reported by McCain:[26]

$$B_w = (1 + \Delta V_{wp})(1 + \Delta V_{wT}) \quad \textbf{(4.129)}$$

where $\Delta V_{wT} = -1.0001 \times 10^{-2} + 1.33391 \times 10^{-4} T + 5.50654 \times 10^{-7} T^2$ **(4.130)**

and $\Delta V_{wp} = -1.95301 \times 10^{-9} pT - 1.72834 \times 10^{-13} p^2 T - 3.58922 \times 10^{-7} p - 2.25341 \times 10^{-10} p^2$ **(4.131)**

For Eqs. (4.129) to (4.131), p = pressure in psia, and T = temperature in °F. The correlation is valid for temperatures up to 260°F and pressures up to 5000 psia. The reported error for the correlation is within 2%.

4.12.2 Density of Formation Water

The density of formation water at standard conditions is calculated by using the McCain correlation:[26]

$$\rho_w = 62.368 + 0.438603S + 1.60074 \times 10^{-3} S^2 \quad \textbf{(4.132)}$$

In Eq. (4.132), ρ_w = density in lbm/ft^3, and S = salinity in weight percent. Water density at reservoir temperature and pressure is calculated by dividing the density at standard conditions by the formation volume factor at the same reservoir conditions.

4.12.3 Coefficient of Isothermal Compressibility of Formation Water

At pressures greater than bubble point, water compressibility can be calculated by using a correlation from Osif:[42]

$$\frac{1}{c_w} = 7.033p + 0.5415C - 537T + 403{,}300 \quad \textbf{(4.133)}$$

In Eq. (4.133), c_w = water compressibility, psi^{-1}; p = pressure, psi; C = salinity, mg/L; and T = temperature, °F. Equation (4.133) is valid for temperatures between 200 °F and 270°F, pressures of 1000 to 20,000 psi, and salinities up to 200,000 mg/L. The accuracy of this correlation is not reported. Osif[42] reported that water compressibility was independent of dissolved gas.

4.12.4 Viscosity of Formation Water

Water viscosity at reservoir temperature and 1 atm can be calculated by using a correlation from McCain:[26]

$$\mu_{w1} = AT^{-B} \tag{4.134}$$

where

$$A = 109.574 - 8.40564S + 0.313314S^2 + 8.72213 \times 10^{-3}S^3 \tag{4.135}$$

and

$$B = 1.12166 - 2.63951 \times 10^{-2}S + 6.79461 \times 10^{-4}S^2 + 5.47119 \times 10^{-5}S^3 - 1.55586 \times 10^{-6}S^4 \tag{4.136}$$

In Eqs. (4.134) to (4.136), μ_{w1} = water viscosity at reservoir temperature and 1 atm, cp; T = temperature, °F; and S = salinity in weight percent. McCain[26] reported that the correlation has an error of 5% when compared with graphical correlations shown by Matthews and Russell[43] for temperatures between 100°F and 200°F and salinities up to 26 weight percent.

Water viscosity at reservoir pressure is calculated by adjusting water viscosity at 1 atm using a correlation from Mathews and Russell:[43]

$$\frac{\mu_w}{\mu_{w1}} = 0.9994 + 4.0295 \times 10^{-5}p + 3.1062 \times 10^{-9}p^2 \tag{4.137}$$

McCain[26] reported that Eq. (4.137) had an error of 4% for data in the temperature range of 86°F to 167°F and pressures less than 10,000 psia. The reported error is within 7% for pressures between 10,000 and 15,000 psia.

Nomenclature

B_g	gas formation volume factor, RB/STB
B_o	oil formation volume factor, RB/STB
B_{ob}	oil formation volume factor at bubble point pressure, RB/STB
B_w	water formation volume factor, RB/STB
c_o	isothermal oil compressibility, psi^{-1}
c_{ob}	isothermal oil compressibility at bubble point, psi^{-1}
c_g	isothermal gas compressibility, psi^{-1}
c_r	pseudo-reduced isothermal compressibility
c_w	isothermal water compressibility, psi^{-1}

C	salinity, mg/L
m	mass, lbm or gm
M	molecular weight, lbm/lb-mole
M_i	molecular weight of component i, lbm/lb-mole
n	moles of component
p	pressure
p_b	bubble point pressure, psia
p_c	critical pressure
p_i	initial reservoir pressure
p_{pc}	pseudo-critical pressure
p_{pr}	reduced pseudo-pressure, dimensionless
p_r	reduced pressure, dimensionless
P	phases
R	gas constant
R_s	solution gas-oil ratio, scf/STB
R_{sb}	solution gas-oil ratio at bubble point, scf/STB
S	salinity, wt % solids
T	temperature
T_c	critical temperature
T_i	initial reservoir temperature
T_F	reservoir temperature, °F
T_{pc}	pseudo-critical temperature
T_{pr}	reduced pseudo-temperature, dimensionless
T_r	reduced temperature, dimensionless
x_i	mole fraction of component i in the liquid phase
y_i	mole fraction of component i in the vapor phase
z	gas compressibility factor
Z	correlation parameter in Eq. (4.99) or (4.113)
γ_{API}	stock-tank oil gravity, °API
γ_g	gas gravity, air = 1
$\overline{\gamma}_g$	average gas gravity, air = 1
γ_o	stock-tank oil specific gravity, water = 1
γ_w	wellstream gas gravity, air = 1
μ_g	gas viscosity, cp
μ_o	oil viscosity, cp
μ_{od}	dead-oil viscosity, cp
μ_w	water viscosity, cp
μ_{w1}	water viscosity at 1 atm and reservoir temperature, cp
ρ	density, mass per unit volume
ρ_r	reduced density
ψ	chemical potential

Subscripts

b	bubble point
g	gas
HC	hydrocarbon
o	oil
sc	standard conditions
sp	separator
st	stock tank
R	reservoir
w	water

References

1. Gibbs, J.W.: "The Collected Works of J. Willard Gibbs," Yale U. Press, New Haven, CT (1948).
2. Sutton, R.P.: "Compressibility Factors for High-Molecular-Weight Reservoir Gases," paper SPE 14265 presented at the 60th Annual Technical Conference and Exhibition, Las Vegas, NV (Sept. 22–25, 1985).
3. Kay, W.B.: "Density of Hydrocarbon Gases and Vapors at High Temperature and Pressure," *Ind. Eng. Chem.* (Sept. 1936) 1014–1019.
4. Stewart, W.F., Burkhart, S.F., and Voo, D.: "Prediction of Pseudocritical Parameters for Mixtures," paper presented at the AICHE Meeting, Kansas City, MO (May 18, 1959).
5. Kessler, M.G., and Lee, B.I.: "Improve Prediction of Enthalpy of Fractions," *Hyd. Proc.* (March 1976) 153–158.
6. Whitson, C.H.: "Effect of C_{7+} Properties on Equation-of-State Predictions," *SPEJ* (Dec. 1987) 685–696.
7. Sutton, R.P.: "Fundamental PVT Calculations for Associated and Gas/Condensate Natural Gas Systems," paper SPE 97099 presented at the 2005 SPE Annual Technical Conference and Exhibition, Dallas, TX (Oct. 9–12, 2005).
8. Wichert, E., and Aziz, K.: "Calculate Z's for Sour Gases," *Hyd. Proc.* (May 1972) 119–122.
9. Standing, M.B.: *Volumetric and Phase Behavior of Oil Hydrocarbon Systems*, 9th Printing, Society of Petroleum Engineers of AIME, Dallas, TX (1981).
10. Gold, D.K., McCain, W.D., and Jennings, J.W.: "An Improved Method for the Determination of the Reservoir-Gas Specific Gravity for Retrograde Gases," *JPT* (July 1989) 747–752.
11. Standing, M.B., and Katz, D.L.: "Density of Natural Gases," *Trans. AIME* (1942) 146, 140.
12. Takacs, G.: "Comparisons Made for Computer Z-Factor Calculations," *Oil and Gas J.* (Dec. 20, 1976) 64–66.
13. Dranchuk, P.M., and Abou-Kassem, J.H.: "Calculation of Z Factors for Natural Gases Using Equations of State," *J. Cdn. Pet. Tech.* (July-Sept. 1975) 34–36.

14. Hall, K.R., and Yarborough, L.: "A New Equation of State for Z-Factor Calculations," *Oil and Gas J.* (June 18, 1973) 82–92.

15. Dranchuk, P.M., Purvis, R.A., and Robinson, D.B.: "Computer Calculation of Natural Gas Compressibility Factors Using the Standing and Katz Correlations," Institute of Petroleum Technical Series, No. IP74-008 (1974) 1–13.

16. Carr, N.L., Kobayashi, R., and Burrows, D.B.: "Viscosity of Hydrocarbon Gases Under Pressure," *Trans. AIME* (1954) 201, 264–272.

17. Jossi, J.A., Stiel, L.I., and Thodos, G.: "The Viscosity of Pure Substances in the Dense Gaseous and Liquid Phases," *AICHE Journal* (Mar. 1962) Vol. 8, No. 1, 59–62.

18. Dean, D.E., and Stiel, L.I.: "The Viscosity of Nonpolar Gas Mixtures at Moderate and High Pressures," *AICHE Journal* (May 1965) 526–532.

19. Lohrenz, J., Bray, B.G., and Clark, C.R.: "Calculating Viscosities of Reservoir Fluids from Their Compositions," *JPT* (Oct. 1964) 1171–1176.

20. Lee, A.L, Gonzalez, M.H., and Eakin, B.E.: "The Viscosity of Natural Gas," *Trans. AIME* (1966) Vol. 237, 997–1000.

21. Poling, B.E., Prausnitz, J.M., and O'Connell, J.P.: *The Properties of Gases and Liquids*, Fifth Edition, McGraw-Hill (2001) Chapters 5 and 9.

22. Londono, F.E., Archer, R.A., and Blasingame, T.A.: "Correlations for Hydrocarbon-Gas Viscosity and Gas Density-Validation and Correlation of Behavior Using a Large-Scale Database," *SPEREE* (Dec. 2005) 561–572.

23. Trube, A.S. "Compressibility of Natural Gases," *Trans. AIME* (1957) 210, 61.

24. Mattar, L., Brar, G.S., and Aziz, K.: "Compressibility of Natural Gases," *J. Cdn. Pet. Tech.* (Oct-Dec. 1975) 77–80.

25. Al-Shammasi, A.A.: "A Review of Bubblepoint Pressure and Oil Formation Volume Factor Correlations," *SPEREE* (April 2001) 146–160.

26. McCain, W.D. Jr.: "Reservoir-Fluid Property Correlations: State of the Art," *SPERE* (May 1991) 266.

27. Labedi, R.: "Use of Production Data to Estimate Volume Factor Density and Compressibility of Reservoir Fluids," *J. Pet. Sci. Eng.* (1990) 4, 357.

28. Obomanu, D.A., and Okpobori, G.A.: "Correlating the PVT Properties of Nigerian Crudes," *Trans. ASME* (1987) 109, 214.

29. Petrosky, G.E. Jr., and Farshad, F.: "Pressure-Volume-Temperature Correlations for Gulf of Mexico Crude Oils," *SPEREE* (Oct. 1998) 416–420.

30. Sutton, R.P., and Farshad, F.: "Evaluation of Empirically Derived PVT Properties for Gulf of Mexico Crude Oils," *SPERE* (Feb. 1990) 79–86.

31. Al-Marhoun, M.A.: "PVT Correlations for Middle East Crude Oils," *JPT* (May 1988) 650–666.

32. Dokla, M., and Osman, M.: "Correlation of PVT Properties for UAE Crudes," *SPEFE* (March 1992) 41.

33. Glaso, O.: "Generalized Pressure-Volume-Temperature Correlations," *JPT* (May 1980) 785.

34. Farshad, F., LeBlanc, J.L., Garber, J.D., and Osorio, J.G.: "Empirical PVT Correlations for Colombian Crude Oils," paper SPE 36105 presented at the 1996 SPE Latin American and Caribbean Petroleum Engineering Conference, Port of Spain, Trinidad, and Tobago, April 23–26, 1996.

35. Valko, P.P., and McCain, W.D. Jr.: "Reservoir oil bubblepoint pressures revisited; solution gas-oil ratios and surface gas specific gravities," *J. Pet. Sci. Eng*. 37 (2003) 153–169.

36. Spivey, J.P., Valko, P.P., and McCain, W.D. Jr.: "Applications of the Coefficient of Isothermal Compressibility to Various Reservoir Situations with New Correlations for Each Situation," paper SPE 96415 presented at the 2005 SPE Annual Technical Conference and Exhibition, Dallas, TX (Oct. 9–12, 2005).

37. Al-Marhoun, M.A.: "The Coefficient of Isothermal Compressibility of Black Oils," paper SPE 81432 presented at the SPE 13th Middle East Oil Show & Conference, Bahrain, June 9–12, 2003.

38. McCain, W.D. Jr., Rollins, J.B., and Villena Lanzi, A.J.: "The Coefficient of Isothermal Compressibility of Black Oils at Pressures Below the Bubblepoint," *SPEFE* (Sept. 1988) 659–662.

39. Ng, J.T.H., and Egbogah, E.O.: "An Improved Temperature-Viscosity Correlation for Crude Oil Systems," paper CIM 83-34-32 presented at the 1983 Annual Technical Meeting of the Petroleum Soc. Of CIM, Banff, Alta. (May 10–13, 1983).

40. Beggs, H.D., and Robinson, J.R.: "Estimating the Viscosity of Crude Oil Systems," *JPT* (Sept. 1975) 1140–1141.

41. Bergman, D.F., and Sutton, R.P.: "Undersaturated Oil Viscosity Correlation for Adverse Conditions," paper SPE 103144 presented at the 2006 SPE Annual Technical Conference and Exhibition, San Antonio, TX (Sept. 24–27, 2006).

42. Osif, T.L.: "The Effects of Salt, Gas, Temperature, and Pressure on the Compressibility of Water," *SPERE* (Feb. 1988) 175–181.

43. Matthews, C.S., and Russell, D.G.: *Pressure Buildup and Flow Tests in Wells*, Monograph Series, SPE, Richardson, TX (1967) 1.

General Reading

- Borges, P.R.: "Correction Improves Z-Factor Values for High Gas Density," *Oil and Gas J.* (Mar. 4, 1991) 55.
- Eilerts, C.K.: *Phase Relations of Gas-Condensate Fluids, Vol. II,* Monograph 10, Bureau of Mines, American Gas Association (1959) 764–770.

- Elsharkawy, A.M., Hashem, Y.Kh., and Alikhan, A.A.: "Compressibility Factor for Gas Condensates," paper SPE 59702 presented at the SPE 2000 Permian Basin Oil and Gas Conference, Midland, TX (Mar. 21–23, 2000).
- Elsharkawy, A.M., and Elkamel, A.: "Compressibility Factor for Sour Gas Reservoirs," paper SPE 64284 presented at the 2000 SPE Asia Pacific Oil & Gas Conference, Brisbane, Australia (Oct. 16–18, 2000).
- Elsharkawy, A.M.: "Efficient Methods for Calculations of Compressibility, Density, and Viscosity of Natural Gases," *Fluid Phase Equilibria* 218 (2004) 1–13.
- Kay, W.B.: "The Ethane-Heptane System," *Ind. & Eng. Chem.* (1938) 30, 459.
- Lee, J., and Wattenbarger, R.A.: *Gas Reservoir Engineering*, Soc. of Pet. Engrs., Richardson, TX (1996).
- McCain, W.D. Jr.: *The Properties of Petroleum Fluids*, 2nd Ed., PennWell Publishing Co., Tulsa, OK (1990).
- Piper, L.D., McCain, W.D., and Corredor, J.H.: "Compressibility Factors for Naturally Occurring Petroleum Gases," paper SPE 26668 presented at the 68th Annual Technical Conference and Exhibition, Houston, TX (Oct. 3–6, 1993).
- Whitson, C., and Brule, M.: *Phase Behavior, Monograph Vol. 20*, Soc. of Pet. Engrs., Richardson, TX (2000).
- Yarborough, L., and Hall, K.R.: "How to Solve Equation of State for Z-Factors," *Oil and Gas J.* (Feb. 18, 1974) 86–88.
- Bergman, D.F., and Sutton, R.P.: "An Update to Viscosity Correlations for Gas-Saturated Crude Oils," paper SPE 110195 presented at the SPE Annual Technical Conference and Exhibition, Anaheim, CA (Nov. 11–14, 2007).
- Bergman, D.F., and Sutton, R.P.: "A Consistent and Accurate Dead-Oil-Viscosity Method," *SPEREE* (December 2009) 815–840.
- *Engineering Data Book,* 12th Edition, Gas Processors Suppliers Assn., Tulsa (2004).

Reservoir Fluid Sampling and PVT Laboratory Measurements

5.1 Overview of Reservoir Fluid Sampling

One of the important activities in reservoir management is to ensure that representative reservoir fluid samples are obtained by sampling and proper laboratory measurements are conducted on the samples. It is also equally important that reservoir fluid samples should be stored in containers and maintained under conditions that retain the composition of the original sample over time. Laboratory measurements on reservoir fluid samples that are not representative of original or current state of reservoir fluids at the time of sampling could yield erroneous data which could adversely impact all engineering calculations on the reservoir. Considerable care and attention should be spent in planning the sampling program, quality checking the samples at the site and laboratory, and monitoring the actual laboratory process of obtaining data from the samples. This chapter is devoted to the entire process of fluid sampling and PVT data measurements on fluid samples to emphasize their importance in the overall process of reservoir management. A summary of guidelines on fluid sampling methods and PVT laboratory measurements for specific reservoir condition is provided in Table 5.1.

The primary objective of sampling fluids from a reservoir is to obtain a sample that is representative of fluids existing in the reservoir at the time the samples were taken. Reservoir fluid samples are used in one or combinations of the following important functions:

1. PVT studies under reservoir and surface operating conditions.
2. Reservoir fluid characterization or modeling.
3. Economic evaluations (studies of in-place volumes, reserves, etc.).

Table 5.1 Guidelines on Fluid Sampling Methods and PVT Laboratory Measurements

Reservoir			Sampling Method				PVT LAB	
			Surface	Subsurface				
Fluid Type	Fluid State	Life Cycle	Separator	Bottomhole	WFT	QA/QC Issues	Test Types	QA/QC Issues
Dry/Wet Gas	**Single-Phase**	Expl/App			X	Ensure monophasic samples	Composition, CCE, Viscosity, Separator tests	Verify opening pressures Check lab procedures/protocol Check data for consistency
		Prod	X			Ensure monophasic samples Avoid condensation in tubing, flowlines etc.		
Gas Condensate	**Undersaturated**	Expl/App	X	X	X	Ensure monophasic samples Minimize well drawdown Check liquids drop-out in tubing, flowlines. & separators Check separator condition, calibration, & efficiency	Composition, Sat Press., CCE, CVD, Separator tests	Verify opening pressures Check lab procedures/protocol Check data for consistency
		Prod	X	X				
	Saturated	Expl/App	X		X	Minimize well drawdown Check liquids drop-out in tubing, flowlines, & separators Check separator condition, calibration & efficiency Well conditioning at stable rates crucial		
		Prod	X					

Volatile Oil	Undersaturated	Expl/App	X	X	X	Minimize well drawdown Check separator condition, calibration & efficiency Well conditioning at stable low rates crucial	Composition, Sat. Press., CCE, DL, Viscosity, Separator tests	Verify opening pressures Check lab procedures/protocol Check data for consistency
		Prod	X	X				
	Saturated	Expl/App	X		X			
		Prod	X					
Black Oil	Undersaturated	Expl/App	X	X	X	Minimize well drawdown Check separator condition, calibration & efficiency Well conditioning at stable rates crucial	Composition, Sat. Press., CCE, DL, Viscosity, Separator tests	Verify opening pressures Check lab procedures/protocol Check data for consistency
		Prod	X	X				
	Saturated	Expl/App	X					
		Prod	X		X			

Expl/App = Exploration/Appraisal

Prod = Production

WFT = Wireline Formation Tester

4. Geochemical analyses to determine fluid sources.
5. Areal and vertical stratification or compartmentalization issues (or reservoir continuity).
6. Evaluation of reservoir performance or management strategies.
7. Special core analyses or core flooding.
8. Design of gathering and separation facilities.
9. Flow assurance studies (flow potential, precipitation of wax, asphaltenes, hydrates etc.).
10. Processing plant design.
11. Early identification of corrosive and dangerous fluid components such as hydrogen sulfide, and/or carbon dioxide. These will also influence the choice of materials used for well tubulars (casing, tubing, etc.), surface equipment (pipelines, separators, tanks, etc.), and processing plants.
12. Crude assays for refining plants.

Reservoir fluid sampling methods are generally divided into two categories: subsurface sampling and surface sampling. Subsurface sampling (sometimes called bottom-hole sampling) describes the collection of fluid samples downhole from cased or open-hole wellbores. In surface sampling, fluid samples are collected at various surface locations such as wellhead, pipelines, separators, stock-tank, etc. The method selected for collection of reservoir fluid samples is determined mainly by the type of reservoir. In Chapter 4, five reservoir types were identified based on their initial conditions of temperature and pressure relative to their phase diagrams. The five reservoir types were identified as: dry gas, wet gas, gas-condensate, volatile oil, and black oil reservoirs. In general, dry or wet gas reservoirs can be sampled with subsurface or surface methods. Gas-condensate and volatile oil reservoirs are typically sampled with surface methods. Black oil reservoirs are sampled with subsurface and surface methods depending on the state of the reservoir. Note that the sampling methods prescribed at this point for these reservoir types are based on gross generalizations. There are numerous other factors that influence the selection of sampling method for any reservoir, such as potential uses of the samples, the production characteristics of the wells, the type of mechanical equipment in the wellbore, the cost of preparing the well for sampling, mechanical condition of surface separators, availability of sampling equipment, and safety issues. All these factors, in addition to the state of the reservoir, influence the selection of the sampling method for any reservoir.

The entire process of collecting reservoir fluid samples either by surface or subsurface methods is beset with opportunities for errors at every stage. The sources of these potential errors which can affect the quality of the samples are illustrated in Figure 5.1. Common errors associated with subsurface sampling are:

1. Contamination from drilling fluids, mud filtrates, and completion fluids.
2. Near wellbore phase separation due to excessive drawdown.
3. Commingled reservoir fluids.
4. Two-phase flow in the wellbore.
5. Intermittent flow or "heading" at low flow rates.

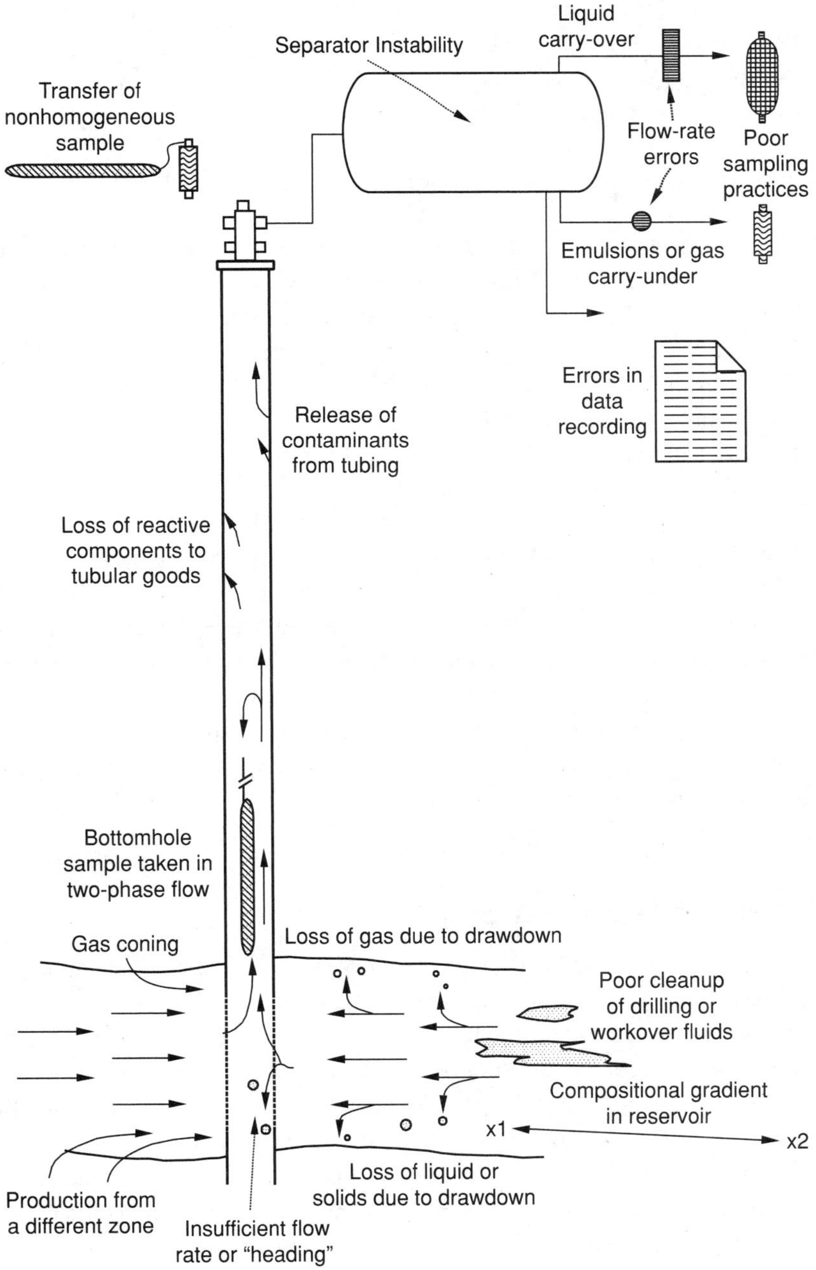

Figure 5.1 A schematic view of wellsite sampling and measurement errors (from Moffatt et al.[1] © 1998 SPE, Reproduced with permission).

6. Loss of reactive components, such as sulfur compounds, to downhole tubular equipment.
7. Separation of phases in static well fluid columns.
8. Transfer and transportation of samples.
9. Errors in recording of subsurface conditions.

Common errors that are encountered in surface sampling are:

1. Fluid stream not equilibrated in the separator due to improper separator size, high throughput rates, insufficient residence time, improper operation, or poor mechanical condition.
2. Entrainment of liquid in the separator exit gas stream.
3. Gas "carry-under" in the separator exit liquid stream.
4. Emulsions in the separator exit liquid stream.
5. Poor sampling practices at the separator.
6. Transfer and transportation of samples.
7. Recording of data at the separator.

The most effective means of eliminating these common errors present in subsurface and surface sampling methods is to devise a comprehensive and detailed sampling program with specific procedures at each major step, identification of well-trained personnel responsible for monitoring activities at the major steps, and verifiable means of quality checking data and samples collected as prescribed in the sampling program.

The probability of capturing a representative reservoir fluid sample is largely influenced by the type and state of the candidate reservoir, the procedures used in well conditioning, sampling methods, and type of sampling tools used. The roles played by these factors are discussed in the following sections.

5.2 Reservoir Type and State

The five reservoir types identified in Chapter 4 are dry gas, wet gas, gas-condensate, volatile oil, and black oil reservoirs. The states of gas-condensate, volatile, and black oil reservoirs are further classified as undersaturated or saturated reservoirs on the basis of initial or current reservoir pressure and temperature. For undersaturated reservoirs, the initial or current reservoir pressure and temperature are above the saturation conditions. In saturated reservoirs, initial or current reservoir pressure and temperature are equal to or less than saturation conditions. It is very important to emphasize that, irrespective of the type and state of a reservoir, the best opportunity to collect the most representative reservoir fluid sample exists early in the life of the reservoir before substantial production has occurred.

5.2.1 Undersaturated Oil Reservoirs

In undersaturated oil reservoirs, initial or current pressures are higher than bubble point pressures of the fluids at reservoir temperature. If the oil reservoir is highly undersaturated, then reservoir pressures are considerably higher than bubble point pressures. For fluids to flow toward a well in a reservoir, a pressure difference must be maintained between the wellbore and the drainage boundary of the well. This pressure difference is called the **drawdown**. High flow rates can be achieved at high drawdowns. Conversely, low flow rates can be maintained at low drawdowns. This concept is utilized in the process of conditioning the well for sampling (termed "well conditioning") and is generally applicable to all reservoir types.

High oil flow rates at high drawdown may cause flowing pressures within the drainage area of the well to fall below bubble point pressure. If this happens, gas bubbles will form within the oil phase in the drainage area. The formation of this gas phase within the drainage area of the well could distort the true gas-oil ratio of the reservoir fluid either by promoting the flow of gas or limiting the flow of oil. In either case, the two-phase system created within the drainage area of the well must be removed before sampling can take place. This is done through well conditioning by systematically reducing the flow rate of the well while monitoring the producing gas-oil ratio. The well is ready for sampling when the producing gas-oil ratio is stable (not changing) after an extended period of flow at approximately constant rate. If there are no other adverse factors such as mechanical condition of the wellbore, prohibitive sampling costs, safety issues, etc., the most representative fluid samples of this type of reservoirs are captured with subsurface samplers.

5.2.2 Undersaturated Gas Condensate Reservoirs

In undersaturated gas condensate reservoirs, initial or current pressures are higher than dew point pressures of the fluids at reservoir temperature. As illustrated in Figure 4.8, a liquid phase will form in the reservoir when reservoir pressure falls below the dew point pressure. This generally will occur if the drawdown applied to the well causes the flowing bottomhole pressure to fall below the dew point pressure. This liquid phase in the near wellbore region must be removed or stabilized by well conditioning prior to sampling. To obtain a sample from an undersaturated gas condensate reservoir that is representative of the original reservoir fluid, it is important that the following conditions are observed:

1. Reservoir pressure within the drainage area of the well must be above dew point.
2. Samples should be collected as early as possible before substantial production has occurred.
3. Well conditioning should be conducted as long as possible until stable rates (gas and condensate) are achieved at the separator.

The following procedure for sampling gas-condensate reservoirs proposed by McCain and Alexander[2] is recommended:

1. Produce at lowest initial rates to remove liquids from the production string.
2. Maintain that rate reasonably constant until well cleans up.
3. Stabilize separator gas and condensate rates.
4. Then sample.

McCain and Alexander[2] recommended that condensate reservoirs with high permeability which had been produced at high rates should be stabilized over a period of days before sampling to ensure that the spike in heptanes-plus fraction has subsided.

Sampling of gas-condensate reservoirs is typically conducted on the surface at the primary separator. Gas and liquid samples obtained from the separator are recombined at the producing gas-liquid ratio to obtain the composition of a representative reservoir fluid sample as shown in Example 4.1.

5.2.3 Saturated Oil Reservoirs

Reservoir pressures in saturated oil reservoirs are below saturation or bubble point pressures at reservoir temperature. In saturated oil reservoirs, a gas cap exists in equilibrium with an oil zone. The pressure at the gas-oil contact is the saturation pressure of the system. It is questionable whether a true representative sample of the original reservoir fluid could be collected from saturated oil reservoirs because of the relative mobilities of gas and oil, the potential effects of gas cusping in the reservoir, and additional evolution of solution gas due to drawdown. Many procedures,[3–6] which are heavily dependent on well conditioning, have been devised for sampling saturated oil reservoirs. These procedures, when properly administered, eliminate gas cusping, replace altered near-wellbore fluids with virgin reservoir fluids, and ensure that oil and gas are flowing at rates that correspond to the gas-oil ratio of the virgin reservoir fluid. The main features of most well conditioning procedures for saturated oil reservoirs are:

1. Reduce well flow rate in a step-wise fashion.
2. Stabilize well at each flow rate and measure producing gas-oil ratio until relatively constant.
3. Continue rate reduction until gas-oil ratio no longer changes with rate reduction.
4. Then sample.

It is considered best practice to sample saturated oil reservoirs at the primary separator. If subsurface sampling is to be conducted, the properly conditioned well should be produced at very low rates during the sampling process.

5.2.4 Saturated Gas Condensate Reservoirs

In saturated gas condensate reservoirs, reservoir pressures are or have declined below dew point pressures to form a mobile or immobile liquid phase. As explained in Chapter 4, retrograde behavior may occur as reservoir pressures continue to decline. The formation of a liquid phase in a gas condensate reservoir continuously alters the composition of fluids remaining in the reservoir because the condensed liquid is composed mostly of heavier hydrocarbons. Since the liquid phase is usually less mobile than the gas phase, the fluids remaining in the reservoir will be richer with heavier hydrocarbons. Consequently, it is doubtful that well conditioning will ever produce a sample that is representative of the original reservoir fluid. Sampling of saturated gas condensate reservoirs is not recommended if the objective of the sampling program is to capture a sample of the original reservoir fluid. However, a sampling program may be devised for other reservoir management purposes, such as monitoring the progress of reservoir depletion.

5.3 Well Conditioning

Well conditioning is the process of preparing a well for sampling by removing near-wellbore fluids which have been altered from virgin reservoir fluids by excessive drawdowns, or by invasion from drilling and remedial fluids. Well conditioning procedures usually (not always) involve step-wise reduction of well flow rates accompanied with measurements of pressures, temperatures, and production rates after stabilization at each step. The well is considered to be properly conditioned when producing gas-oil ratios as measured at appropriate surface locations do not change appreciably with rate change. Generalized well conditioning procedures have been recommended earlier for undersaturated and saturated reservoirs. These generalized procedures must be adapted and possibly revised depending on the specific conditions of the reservoir and the well to be sampled. Detailed well conditioning procedures are also available in *API Recommended Practice 44: Sampling Petroleum Reservoir Fluids.*[3]

Getting a well to be properly conditioned takes time and patience. Conditioning wells in low permeability reservoirs may take several weeks. In some cases, a combination of factors such as well location, lost revenues, and rig cost may prevent adequate conditioning to be achieved on a well. These adverse conditions should be taken into consideration in the preparation of the sampling program with adjustments made to obtain the best possible samples given the prevailing conditions.

The most important aspect of well conditioning is collection of data on the state of the reservoir. The data to be collected should include bottomhole pressures and temperatures, wellhead pressures and temperatures, separator pressures and temperatures, separator gas and liquid flow rates, and stock tank oil, gas and water production rates at each step change. These data are necessary to assess the quality of the conditioning process and hence the quality of the samples that were collected. It is recommended that the data collected during well conditioning should be reviewed thoroughly to ensure that stabilized rates were achieved and actual sampling was initiated at an appropriate point in the well conditioning process. Specially prepared tables for data collection are normally provided by the service company. Other formats of tables that could be used for data collection during well conditioning and sampling are provided in *API Recommended Practice 44: Sampling Petroleum Reservoir Fluids.*[3]

5.4 Subsurface Sampling Methods and Tools

The subsurface sampling tools that are discussed in this section are:

1. Conventional bottomhole samplers
2. Pistonned bottomhole samplers
3. Single-phase samplers
4. Exothermic samplers

5.4.1 Conventional Bottomhole Samplers

Conventional bottomhole samplers are chambers with valves that are used to capture a sample of fluids, typically in cased holes at bottomhole conditions. They are usually run on a wireline. These samplers use either the flow-through or evacuated chamber technique to capture samples. For the flow-through technique, the tool is lowered into the well with the valves open. As the tool descends downhole, the sample chamber is thoroughly flushed with wellbore fluids. At the selected sampling depth, the valves are triggered to close by timing, or mechanical or electrical mechanisms, depending on the type of tool. For the evacuated chamber technique, a vacuumed chamber is positioned on a wireline at the desired sampling depth. The valves are opened and the chamber is filled with fluids present in the wellbore at that depth. In either technique, the sampler filled with fluids is retrieved to the surface and the fluid sample transferred to another vessel for transportation to the laboratory. In most samplers, the displacing fluid for the transfer has been mercury, water, and glycol. The use of mercury has declined considerably due to safety and environmental concerns. An alternative practice is to transport the sampler directly to the laboratory where better equipment exists for reconstituting the sample before transferring it to another container. This option is suitable for gas condensate samples or oil samples that contain asphaltenes or waxes which can precipitate due to changes in pressures and temperatures.

5.4.2 Pistonned Bottomhole Samplers

Pistonned bottomhole samplers are a newer generation of conventional bottomhole samplers. They are equipped with a displaceable piston in the sample chamber that separates the reservoir fluid samples from a hydraulic fluid on the other side of the piston.[7] This eliminates the use of mercury or any other fluid for transfer of the sample into cylinders for transportation. Piston-type samplers also allow the rate of fluid sampling to be controlled by applying a back-pressure during the sampling process. This enhances the chances of collecting the sample in a single-phase. However, it is important to point out that this tool can leak hydraulic fluid past the piston and contaminate the fluid sample. This is a potential risk of contamination posed by piston-type bottomhole samplers.

5.4.3 Single-Phase Samplers

Single-phase samplers are used to capture reservoir fluids which may contain asphaltenes in solution.[5] Asphaltenes may precipitate out of solution if the sampling conditions (temperature and pressure) are reduced. Single-phase samplers are pressure-compensated by applying a back-pressure with nitrogen against a piston in the sample chamber. By maintaining the back pressure much higher than reservoir pressure, it is expected that monophasic conditions will be retained in the sample chamber. However, since the sampler will undergo temperature reduction as it is retrieved from the well, it is likely that some precipitation of asphaltenes will occur, depending on the concentration of asphaltenes in the sample. The standard practice in most laboratories is to restore the sample by reheating it and maintaining it at reservoir temperature and pressure over several days with agitation to achieve thorough mixing of its contents. The process of asphaltene precipitation may not be completely reversible even after extended period of reheating and mixing. If asphaltenes precipitation could become critical for process design, it might be preferable to collect the sample with exothermic samplers which are temperature compensated.

5.4.4 Exothermic Samplers

Exothermic samplers are similar to single-phase samplers except that they are also designed to maintain the temperature of the sample. They are used mainly for samples that may contain asphaltenes. The goal is to prevent asphaltenes from precipitating by maintaining temperatures and pressures as close as possible to sampling conditions. Exothermic samplers are kept hot with battery-operated heating jackets.

5.5 Wireline Formation Testers

Wireline Formation Tester (WFT) is the generic name used to describe wireline tools run usually in an open-hole to measure formation pressures and temperatures, and collect fluid samples. WFT in their current configurations are very technologically advanced tools. They have become very useful tools for reservoir characterization. In their various industry configurations, the tool consists of a probe and seal assembly that can be extended against the wellbore to create a flow path between the formation and the tool which is isolated from drilling or completion fluids in the wellbore. Fluids can then flow from the formation into several chambers within the tool that can be selectively opened and closed by remote control from the surface. Most modern WFT are equipped with samplers that can apply back pressure on the samples to maintain them in a single-phase if desired.

5.5.1 Oil-Based Mud Contamination of WFT Samples

Oil-based mud (OBM) systems are increasingly being used to drill wells because they yield better wellbore quality, increased penetration rate, and reduced drilling time. Drilling of wells with OBM typically causes the invasion of the near wellbore region by OBM filtrates. The depth and extent of the zone invaded by OBM filtrates depend on a number of factors, such as type of rock, formation permeability, buildup of mudcake, drilling operations, etc. Within the invaded zone, formation fluids are contaminated with OBM filtrate. When WFT are used in wells drilled with OBM, contamination of samples by OBM filtrates should be expected. This is the main risk of running WFT in wellbores drilled with OBM.

Several techniques are employed to minimize the level of contamination in fluid samples collected from wellbores drilled with OBM. One of these techniques is the use of pump-out modules in WFT. Pump-out modules provide the option of dumping as much fluids as possible from the invaded zone into the wellbore before samples are collected. The objective of the pump-out process is to replace the contaminated fluid in the invaded zone as much as possible with virgin fluid from the uninvaded parts of the formation. The degree to which the clean-up process is achieved depends on the extent of the invaded zone, the duration of the pump-out period, and mobilities of the fluids in the formation. In many cases, the pump-out module is used in conjunction with an Optical Fluid Analyzer (OFA) to monitor the quality of the fluid before sampling. The OFA monitors fluid flow in the flowline with two sensor systems: an optical spectrometer and an optical gas detector.[8–10] The combined application of pump-out and OFA modules on a sampling operation can be used to reduce contamination in fluid samples to very low levels. Considerable skill is required to decipher the output on OFA time log. The output on OFA time log is subject to a wide range of interpretations. The interpretation of OFA time log is best done by a specialist. Recent advances in OFA technology are leading to in-situ determination of hydrocarbon properties during wireline fluid sampling.[11] This technology will be useful for the in-situ measurement of the properties of fluids in a formation.

Fluid samples collected in wells drilled with OBM should be expected to contain some level of contamination with OBM filtrate. OBM filtrate is miscible with reservoir fluids in most cases. Hence, it is not possible to decontaminate the samples with physical methods. Contamination of fluid samples by OBM filtrates significantly changes the composition and phase behavior of the original reservoir fluids. Even at low levels of contamination, PVT properties of the reservoir fluids such as saturation pressure, formation volume factor, gas-liquid ratio, viscosity, density, and molecular weight can be affected. Because accurate values of these properties are needed in assessment of reserves, reservoir development, and process design, it is extremely important that reliable PVT data of reservoir fluids should be calculated from decontaminated samples.

The two simplest methods for determining the original composition of the reservoir fluid from contaminated samples are called the **Subtraction method** and the **Skimming method.** Both methods are based on the observation that there is an exponential relationship between concentrations of components in the C_{n+} of real reservoir fluids and the corresponding carbon number or molecular weight.[12,13] A plot of the composition of the C_{n+} fraction of the contaminated sample against carbon number or molecular weight on a semi-logarithmic scale will not be linear. The departure or "bump" from the straight line is due to contribution from the OBM filtrate. For the Skimming method, the composition of uncontaminated reservoir fluid is determined from a straight line drawn through the plot. This method does not require data on the composition of the OBM or OBM filtrate. An example showing the application of the Skimming method to an OBM contaminated sample from a deepwater reservoir in the Gulf of Mexico is shown in Figure 5.2. The Subtraction method requires knowledge of the composition of the OBM or OBM filtrate. The contamination level of the sample is determined on-site or at the laboratory by gas chromatography. The composition of the reservoir fluid is determined by subtracting the mass of OBM or OBM filtrate from the composition of the contaminated fluid sample.

There are other decontamination methods, such as the Experimental method, Scaling method, and Statistical method. These methods are more complicated for routine applications but could yield more accurate results. These methods are discussed in detail by Hy-Billiot et al.[13]

5.5.2 Formation Pressures from WFT

One of the primary uses of WFT is to measure formation pressures. Formation pressures can be used in a pressure-depth plot to calculate the density of the continuous fluid phase in the formation since:

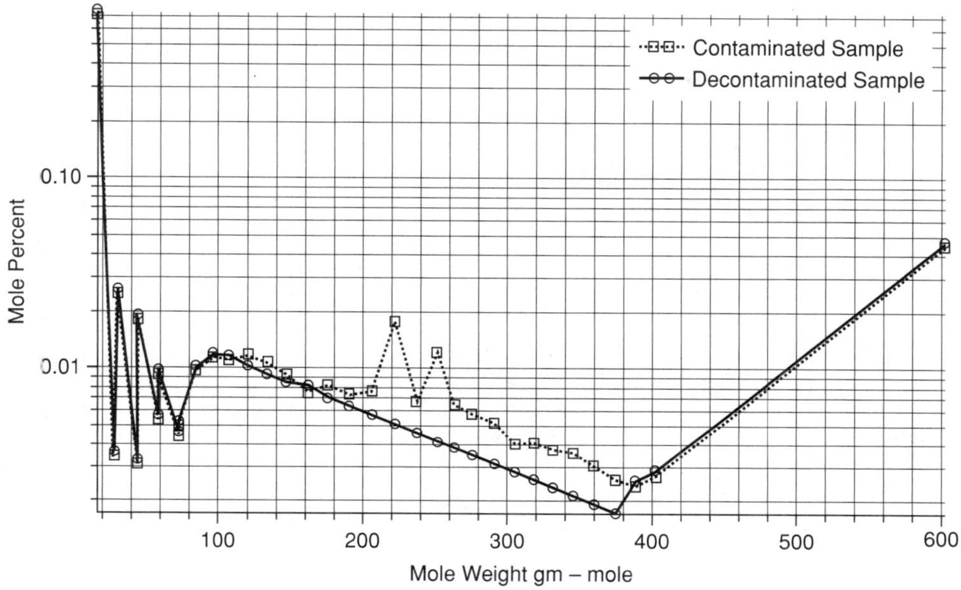

Figure 5.2 Determination of the composition of the uncontaminated reservoir fluid by the Skimming method.

$$\frac{dp}{dz} = \frac{\rho_f g \cos\theta}{144} \tag{5.1}$$

In Eq. (5.1), p = pressure, psia; z = depth, ft; ρ_f = fluid density, lbm/ft^3; g = acceleration due to gravity, ft/sec^2; and θ = well bore deviation, degrees. An example of pressure-depth plot is shown in Figure 5.3 for a deepwater Gulf of Mexico reservoir. In addition to formation fluid densities, pressure-depth plots can be used to determine fluid contacts, identify existence of reservoir heterogeneities or permeability barriers, and presence of reservoir compartments, and/or multiple reservoirs. However, before formation pressures from WFT are used for these determinations, it is important to recognize that WFT formation pressures are affected by capillary pressure changes and the phenomenon of supercharging within the invaded zone close to the wellbore.

5.5.3 Capillary Effects on WFT Formation Pressures

Capillary pressure is defined as the pressure difference between the non-wetting phase and the wetting phase:

$$p_c = p_{nw} - p_w \tag{5.2}$$

In Eq. (5.2), p_c is capillary pressure; p_{nw} is pressure in the non-wetting phase; and p_w is pressure in the wetting phase. The magnitude of capillary pressure depends on the saturation of each

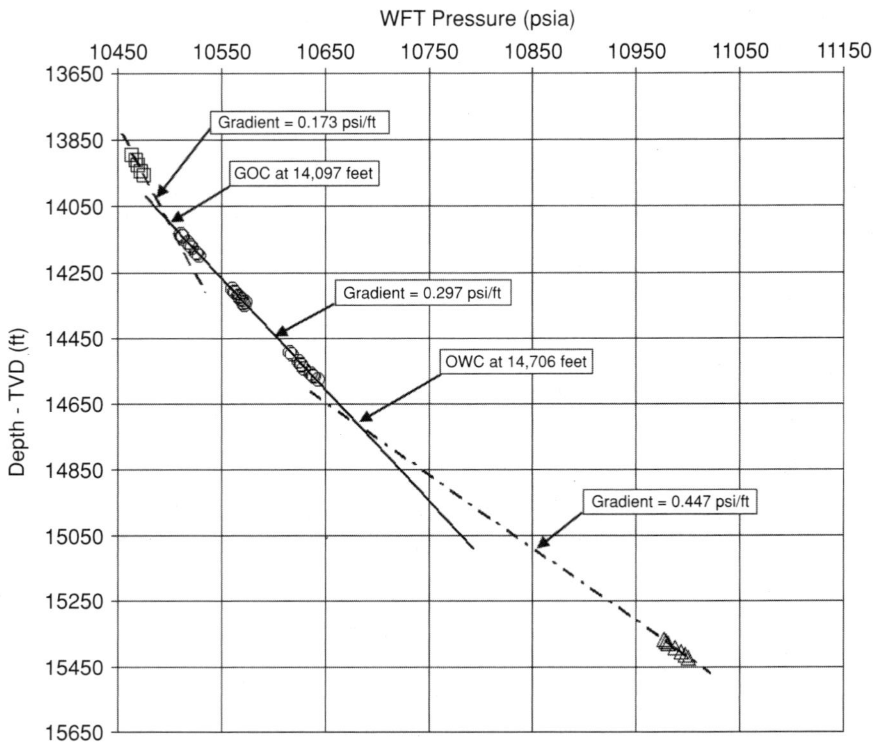

Figure 5.3 Pressure-depth plot for a deepwater Gulf of Mexico reservoir.

phase, on the nature of the continuous phase, and on the distribution, shape, and size of the pores and pore throats.[14] For a capillary tube, capillary pressure is represented as:

$$p_c = \frac{2\sigma \cos \theta}{r} \tag{5.3}$$

In Eq. (5.3), σ is the interfacial tension between the two fluids; θ is the angle of contact which denotes the wettability of the capillary tube; and r is the radius of the capillary tube. **Rock wettability** is described as the preference of either the water phase or the oil phase to adhere to the surface of the rock. The wettability of a rock is denoted by a measure of its contact angle. Values of contact angles less than 90 degrees indicate a water-wet system, while contact angles greater than 90 degrees indicate an oil-wet system. The capillary pressure profile of a water-wet system is different from that of an oil-wet system as defined by Eq. (5.3). Additional discussion of rock wettability and capillary pressure concepts are presented in Chapter 15, Sections 15.2.1 and 15.2.2, respectively.

The distribution of fluid saturation in a reservoir is governed by capillary pressure, which is dependent on depth. The **Free Water Level** (FWL) in a reservoir is the depth at which oil-water capillary pressure is non-existent, and **Oil-Water Contact** (OWC) is defined as the depth at which oil saturation starts to increase from some level of minimum saturation.[15] The FWL and OWC are not generally at the same depth but are separated by a distance related to the capillary displacement pressure. Capillary displacement is the threshold or entry capillary pressure needed for the non-wetting phase to displace the wetting phase from the largest pores.[14] Capillary displacement pressure is calculated as:

$$p_d = p_{nw} - p_w = \frac{2\sigma \cos\theta}{r_L} \tag{5.4}$$

In Eq. (5.4), p_d is capillary displacement pressure; and r_L is the radius of the largest pore throat. In a water-wet reservoir, the FWL exists at a depth, d_{owc}, below the OWC given by:[15]

$$d_{owc} = \frac{144 \times p_d}{g(\rho_{nw} - \rho_w)} \tag{5.5}$$

In Eq. (5.5), ρ_{nw} and ρ_w = densities of the non-wetting and wetting phases in lbm/ft^3, respectively; p_d = displacement pressure in psi; and d_{owc} is in feet. The location of the OWC and FWL for a water-wet reservoir as illustrated by Elshahawi et al.[15] is shown in Figure 5.4. Note that in Figure 5.4, the FWL is at some distance, d_{owc}, below the OWC. The transition zone as shown in Figure 5.4 is the region above the OWC where water saturation will decrease from a maximum value at the OWC to an irreducible value at some distance above the OWC. The height of the transition zone depends on reservoir wettability, pore size distribution, fluid density difference, and fluid interfacial tension.

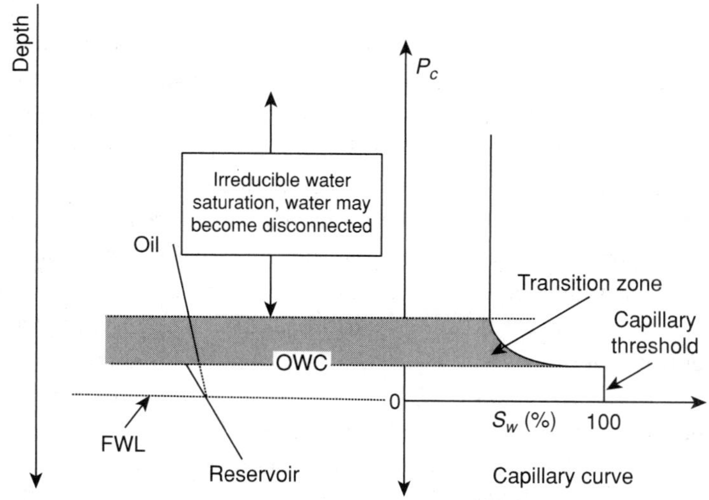

Figure 5.4 Location of OWC and FWL in a water-wet reservoir (from Elshahawi et al.[15] © 2000 SPE, Reproduced with permission).

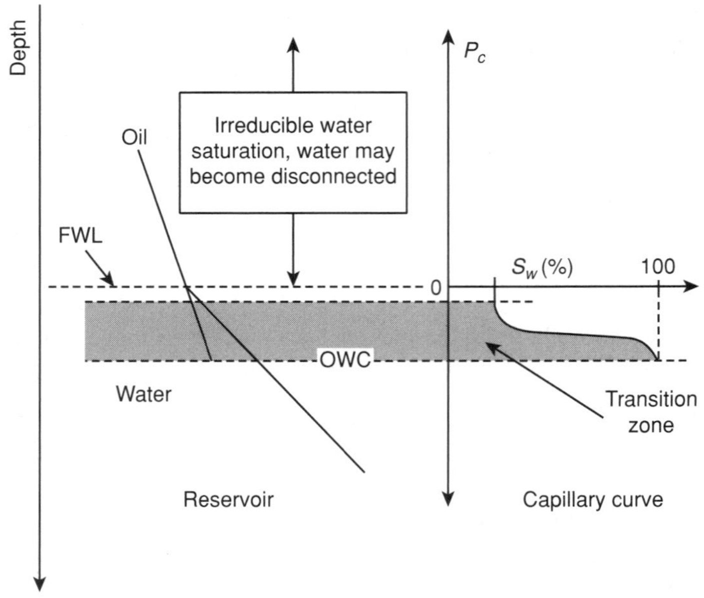

Figure 5.5 Location of OWC and FWL in an oil-wet reservoir (from Elshahawi et al.[15] © 2000 SPE, Reproduced with permission).

The location of the OWC and FWL for an oil-wet reservoir as illustrated by Elshahawi et al.[15] is reproduced in Figure 5.5. Note that for an oil-wet reservoir, the FWL is above the OWC, in contrast to a water-wet reservoir. This is due to the profile of capillary pressure for an oil-wet sand, which is generally negative. The locations of OWC and FWL in Figure 5.5 become evident when the concepts of these fluid levels as defined earlier is applied for an oil-wet reservoir.

The intersection of the pressure gradient lines drawn through hydrocarbon-bearing, and water-filled (aquifer) sections of a reservoir as measured by WFT indicates an apparent location of the FWL. In general, the apparent FWL location will differ from the true FWL in a direction that reflects the wettability of the rock and by an amount that is dependent on the degree of wettability, magnitude of capillary pressure, and type of mud used in drilling the well.[15] The differences between the apparent FWL from WFT measurements and the true FWL is illustrated in Figures 5.6 and 5.7 for water-wet and oil-wet reservoirs, respectively, drilled with water-based mud (WBM) or oil-based mud (OBM). In Figure 5.6, the pressure gradient plot on the left represents water-wet reservoir drilled with WBM. As the plot shows, formation pressures measured by WFT in the oil zone are less (to the left) than true formation pressures because capillary pressure effects in the invaded zone have been eliminated by WBM filtrate. In the water zone, capillary pressure is zero, so WFT measures true formation pressures. Also in Figure 5.6, the pressure gradient plot on the right denotes a water-wet reservoir drilled with OBM. In this case, WFT measures true formation pressures in the oil zone. In the water zone, WFT pressures are higher than the true formation pressures because OBM filtrates will introduce capillary effects.

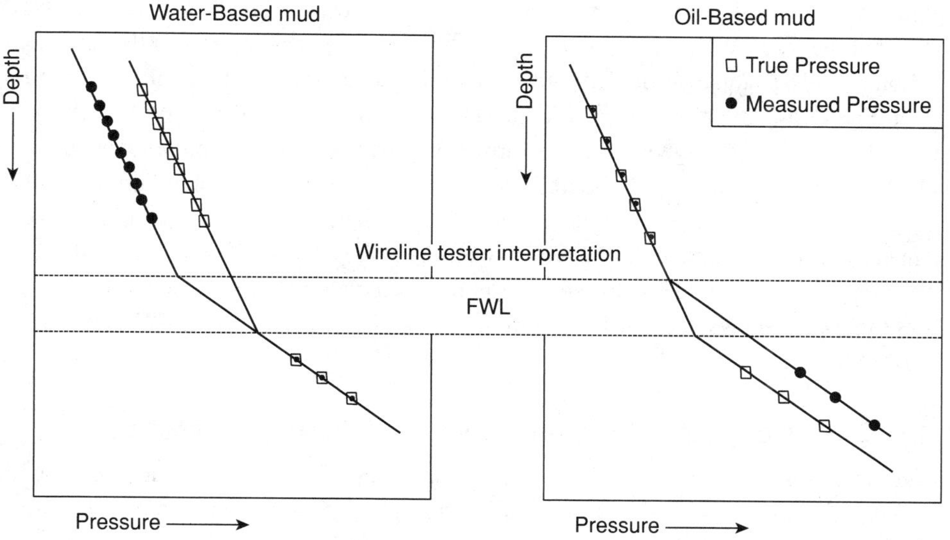

Figure 5.6 Pressure versus depth plot in a water-wet reservoir drilled with a WBM (left) and with OBM (right) (from Elshahawi et al.[15] © 2000 SPE, Reproduced with permission).

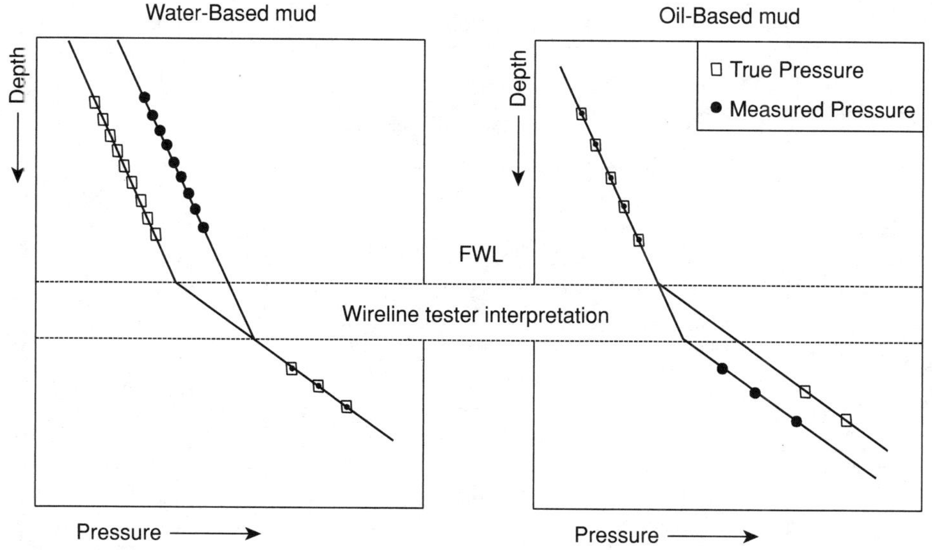

Figure 5.7 Pressure versus depth plot in an oil-wet reservoir drilled with WBM (left) and with OBM (right) (from Elshahawi et al.[15] © 2000 SPE, Reproduced with permission).

As Figure 5.6 shows, the summary effect for a water-wet reservoir drilled with either WBM or OBM is that the true FWL is at some level lower than that determined from WFT gradients.

Comparable results can be deduced for an oil-wet reservoir drilled with either WBM or OBM. In Figure 5.7, the plot to the left is for an oil-wet reservoir drilled with WBM. In the oil zone, WFT pressures are higher than true formation pressures due to capillary pressure created by WBM filtrate. In the water zone, there is no change in capillarity. Again in Figure 5.7, the plot to the right is for an oil-wet reservoir drilled with OBM. There is no difference in formation pressures measured in the oil zone. In the water zone, WFT pressures will be less than the true formation pressures because OBM filtrate will eliminate capillary effects. Generally for oil-wet sands as Figure 5.7 shows, the true FWL will be higher than the WFT based FWL level for either WBM or OBM.

5.5.4 Effects of Supercharging on WFT Formation Pressures

Mud filtrate invasion in the near-wellbore region may create formation pressures that are higher than the true formation pressures. This is called **supercharging**, which is more evident in low permeability reservoirs. High formation pressures due to supercharging may dissipate over time due to deposition of mud cake in the wellbore, which will reduce or eliminate further invasion of mud filtrate. Even if a mud cake is established in the wellbore, higher formation pressures may still exist when WFT is run. Data points affected by supercharging in either the hydrocarbon or water zones will generally be higher and appear to the right side of the true formation pressure line on a pressure-depth plot. This is shown in Figure 5.8 for a deepwater reservoir in the Gulf

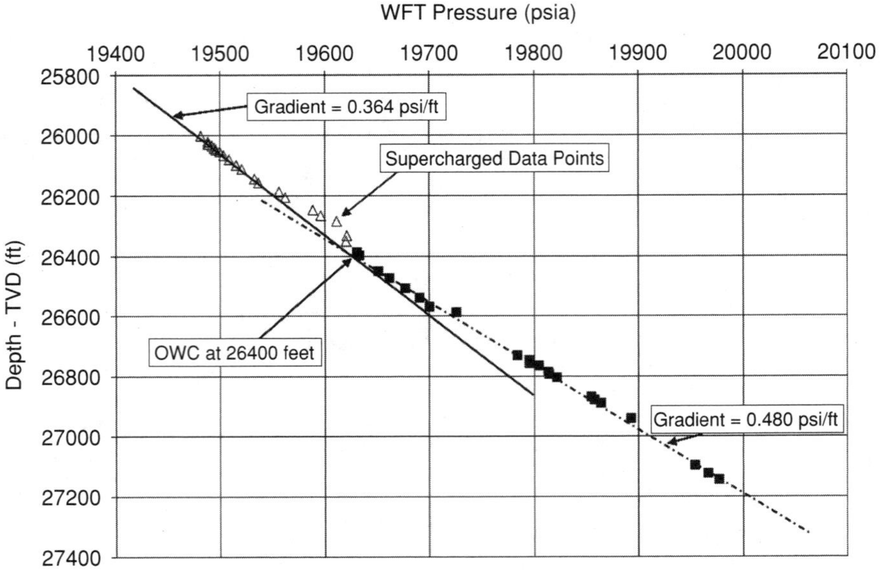

Figure 5.8 Pressure-depth plot for a deepwater Gulf of Mexico reservoir with supercharged data points.

of Mexico. The primary factors controlling supercharging are pressure differential across the mud cake, properties of the mud cake, and total mobility of the formation.[8] The effects of supercharging can be reduced by delaying running WFT as much as possible, especially after a wiper trip, and by using WFT with pump-out modules to drain the mud filtrate in the invaded zone. A graphical method has been developed that can be used to correct formation pressures affected by supercharging. This is shown in Figure 5.9. This method requires measuring borehole pressures (mud pressures) and formation pressures several times at approximately the same depth. By plotting the formation pressures against borehole pressures, as shown in Figure 5.9, the true formation pressure can be determined at the intersection of a line drawn through the data points and the 45 degree line.

5.5.5 Comments on Applications of WFT Pressure Data

WFT pressure data are routinely used in pressure-depth plots to determine the following:

1. Calculation of fluid densities.
2. Estimation of reservoir fluid contact levels.
3. Identification of the existence of reservoir heterogeneities or permeability barriers.
4. Identification of the presence of reservoir compartments and/or multiple reservoirs.

Formation pressures measured with WFT can be affected by capillary pressures, and supercharging as discussed earlier, especially for low permeability reservoirs. Capillary effects

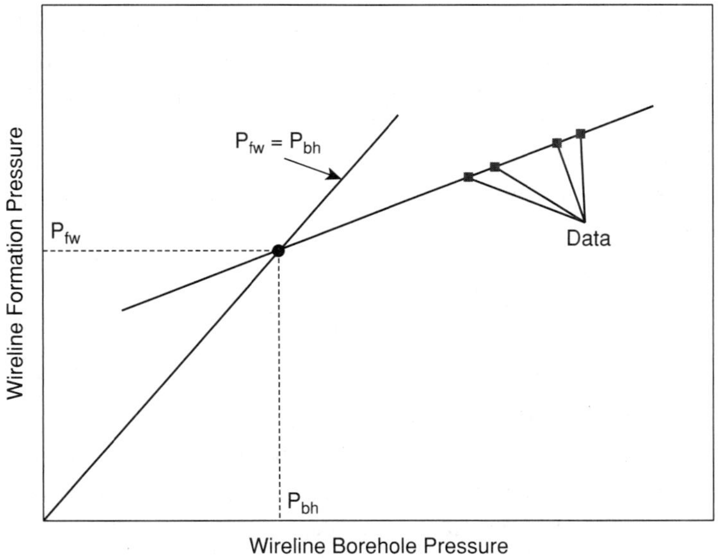

Figure 5.9 Correction of supercharged formation pressures (from Elshahawi et al.[15] © 2000 SPE, Reproduced with permission).

can be corrected as suggested by Elshahawi et al.[15] by using core capillary pressure data or capillary pressure data determined from nuclear magnetic resonance log (NMR). These methods may not be available or at best may be difficult to apply. Supercharging effects can be corrected as shown in Figure 5.9 if required data are available.

Capillary effects and supercharging are factors that should be considered when using WFT data for interpretative purposes. Potential errors in actual measurements (from leaking seals, different hole conditions, accuracy of the pressure gauges, time allowed for pressure stabilization, depth corrections, etc.) must also be taken into consideration. The engineer should avoid the temptation to interpret every deviation from the apparent fluid gradient line as evidence for multiple fluid contacts, reservoir heterogeneities, reservoir compartments or multiple reservoirs, etc. Interpretation of these geologic features from WFT data must be preceded by elimination of potential effects of capillary pressure, supercharging, measurement errors, depth corrections, etc. It is recommended practice to use independent geological interpretation as supporting evidence and guide in the interpretation of WFT data especially with respect to the presence of compartments, and permeability barriers in the reservoir, or the existence of multiple reservoirs.

5.6 PVT Laboratory Measurements

After a representative sample of the reservoir fluid has been collected, the next step is measurement of the fluid properties in series of Pressure-Volume-Temperature (PVT) experiments. This is a very important operation because the engineer should ascertain that the PVT laboratory chosen for the measurements has the necessary experimental equipment and proper procedures for the prescribed reservoir fluid studies. For measurement of some critical data, it might be advisable to engage two separate PVT laboratories to conduct fluid properties studies on the same samples for verification and quality checking of data reported on the samples.

Reservoir fluid studies typically involve combinations of the following six procedures:

1. Fluid composition.
2. Constant Composition Expansion (CCE).
3. Differential Liberation (DL).
4. Constant Volume Depletion (CVD).
5. Separator tests.
6. Viscosity measurements.

5.6.1 Fluid Composition

The composition of the reservoir fluid sample is usually measured to various levels of detail, depending on the laboratory or as requested. Typically, the mole percent of each component in the C_1 to C_9 range is measured. The average molecular weight and specific gravity of the decane-plus (C_{10}^+) fraction are reported. In some cases, the mole percent of the components in the C_1 to C_6 hydrocarbon range are reported with the molecular weight and specific gravity of the heptane-plus (C_7^+) fraction. The composition of the fluid sample can be used for phase calculations with an equation of state, if needed. The composition of a black oil sample from a typical reservoir fluid study is shown in Appendix 5A. Note that for this study, component composition was determined for hydrocarbons in the C_1 to C_{29} range.

5.6.2 Constant Composition Expansion (CCE)

This laboratory procedure is called various names, such as constant composition expansion, constant mass expansion, flash expansion, flash vaporization or pressure-volume relations, etc. Irrespective of what it is called, the actual laboratory procedure is the same. The reservoir fluid sample is placed in a windowed PVT cell maintained at reservoir temperature and pressure. The pressure of the PVT cell is reduced to a predetermined level by increasing the volume of the PVT cell. The contents of the PVT cell are equilibrated at this pressure and volume by agitation. After equilibration, the cell volume is increased until the next predetermined pressure level is reached. This is then followed with equilibration at this pressure level. This procedure is followed until all the pressure stages have been tested. At each equilibrated stage, the pressure and volume of the PVT cell are measured and recorded. Note that no gas or liquid is removed from the PVT cell at any stage. This process is illustrated in Figure 5.10. A typical plot of pressure versus volume that results from this procedure is shown in Figure 5.11. In this case, the bubble point is at the intersection of the compressed liquid line and the two-phase line. Constant composition expansion experiments are conducted for gas condensate and oil samples. Typical reports for CCE experiments are shown in Appendix 5A for an oil sample, and Appendix 5B for a gas condensate sample.

5.6.3 Differential Liberation (DL)

Differential liberation is also called differential vaporization or differential expansion. In this experiment procedure, the reservoir fluid sample is equilibrated in a windowed PVT cell at its bubble point pressure and reservoir temperature. The pressure inside the cell is reduced by increasing its volume. Since the cell pressure is now less than bubble point pressure, a gas phase will form. The gas is equilibrated with the liquid in the cell by agitation. Once equilibrium has been established

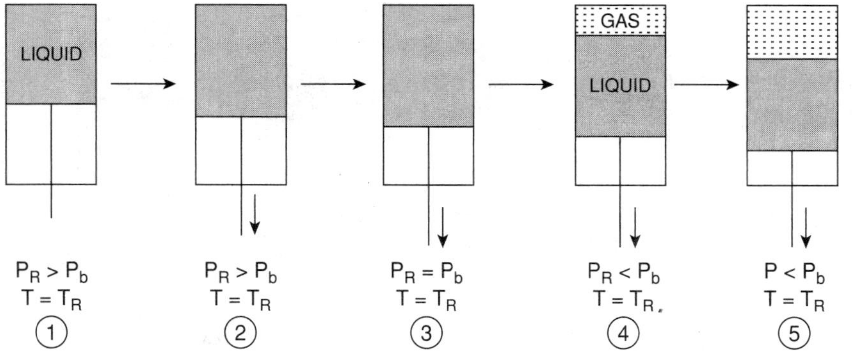

Figure 5.10 Constant composition expansion (CCE) experimental diagram.

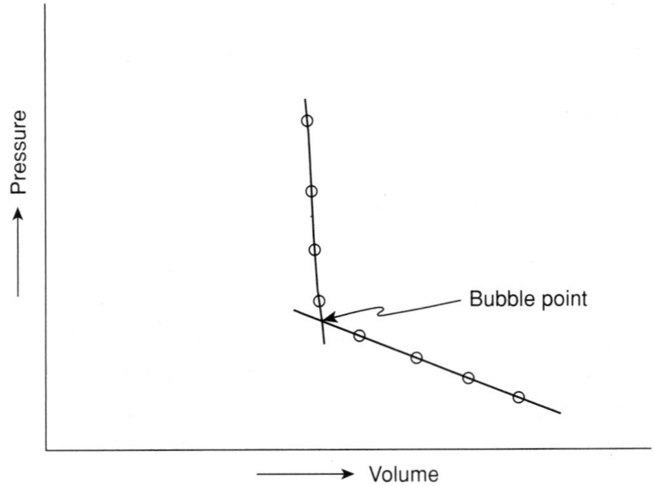

Figure 5.11 Pressure versus volume plot for determination of bubble point pressure from CCE data.

between the fluid phases, the gas is completely displaced from the cell at constant pressure by slowly reducing the volume of the cell. The volume and specific gravity of the gas expelled from the cell is measured. The volume of liquid left in the cell is also measured. This represents Stage 1, as illustrated in Figure 5.12. The procedure used in Stage 1 is repeated for the next pressure level at Stage 2, as illustrated in Figure 5.12. These stages are repeated until atmospheric pressure is reached as the final stage with the cell still at reservoir temperature. The temperature of the cell is further reduced to 60°F to determine residual oil volume from differential liberation. This is the volume of liquid remaining in the cell at 60°F. A typical report of data from differential liberation is shown in Appendix 5A. From differential liberation data, fluid properties such as oil formation volume factor, gas compressibility factor, gas formation volume factor, solution gas-oil

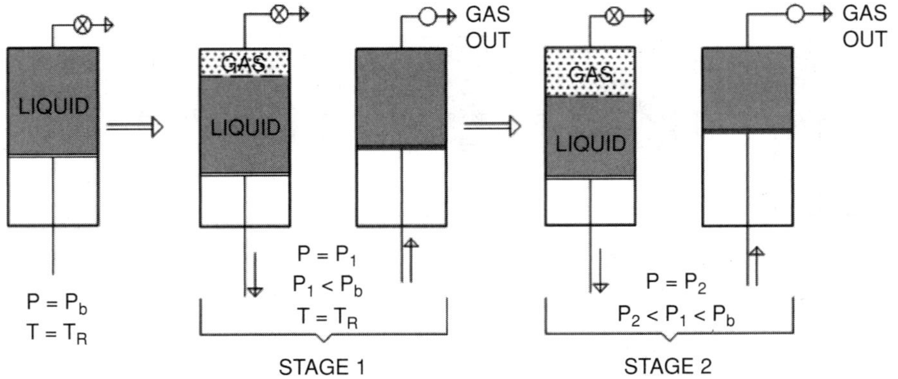

Figure 5.12 Differential liberation experimental diagram.

ratio, and oil compressibility are calculated. These data are further adjusted for reservoir engineering applications, as shown in Section 5.7.

5.6.4 Constant Volume Depletion (CVD)

Constant volume depletion, generally called depletion study, is an experimental procedure used to study gas condensate samples. The fluid sample is equilibrated in a PVT cell at reservoir temperature and pressure. The cell volume is increased, thereby reducing its pressure to some predetermined level. The fluid sample is thoroughly mixed at this lower pressure by agitation. Then a portion of the fluid in the cell is withdrawn slowly while maintaining its pressure constant at the current level until the original cell volume is reached. The composition and specific gravity of the fluid removed from the cell is measured. This procedure is represented as Stage 1 in Figure 5.13. Subsequent stages at lower pressure levels are repeated as described for Stage 1. A typical report from a depletion study is shown in Appendix 5B. The data from a depletion study can be used to fit parameters of any preferred equation of state. The equation of state can then be used in a reservoir simulator to model depletion of a condensate reservoir.

5.6.5 Separator Tests

A separator test is a flash vaporization process. A fluid sample at reservoir temperature and bubble point pressure in a PVT cell is displaced at bubble point pressure through two or more stages of separation. The pressure in the PVT cell is held constant by slowly reducing its volume. The pressures and temperatures of the separators in the laboratory are selected to approximate as closely as possible the expected separator conditions in the field. The stock tank (final stage) is at atmospheric pressure. A three-stage separator test representing this procedure is shown in Figure 5.14. An alternative procedure for separator tests which is used in some laboratories is very similar to the differential liberation experiment. In this procedure, the temperature of fluid sample initially at bubble point pressure and reservoir temperature is reduced to the temperature

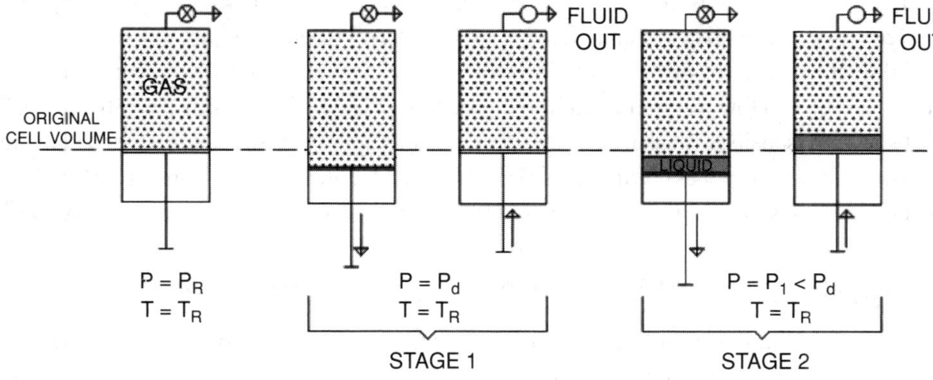

Figure 5.13 Constant volume depletion experimental diagram.

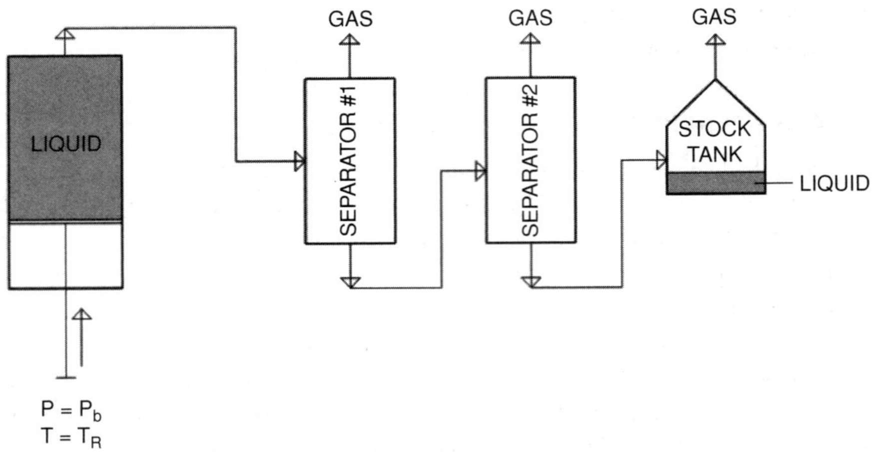

Figure 5.14 A three-stage separator test experimental diagram.

of the first stage separator. After equilibration, the pressure of the sample is reduced to the pressure set for the first stage separator. The liberated gas and the remaining liquid phase are allowed to equilibrate. Then the liberated gas is displaced at constant pressure from the PVT cell into an evacuated chamber for density measurements and compositional analysis. This process is repeated for other separator stages down to stock tank conditions. An example of gas-oil ratio (R_{sSb}) and formation oil volume factor (B_{oSb}) measured for a four-stage separation is shown in Appendix 5A. These data are used to adjust fluid properties measured in CCE and DL tests, as demonstrated in Section 5.7.

5.6.6 Viscosity Measurements

Oil viscosities are measured at the same pressure levels as the differential liberation tests by obtaining liquid samples devoid of gas at each pressure level in a separate test from the DL test. The equipment used to measure oil viscosities is either a capillary viscometer or electromagnetic (EM) viscometer. Viscosities are calculated from capillary viscometers directly from measured data based on the principles of fluid flow in a tube. Viscosities from the EM viscometer are based on indirect methods. However, either equipment requires viscosity standards for calibration. The reliability of oil viscosities measured with either EM or capillary viscometer depends considerably on the calibration process. Oil viscosities measured for the example fluid study are shown in Appendix 5A. Gas viscosities are calculated from correlations such as Lee et al.[16] correlation.

5.7 Applications of Laboratory PVT Measurements

In laboratory PVT studies of volatile and black oil samples, flash vaporization (CCE) data, differential liberation (DL) data, and separator test data are reported. An example of a reservoir fluid study report is provided in Appendix 5A for a black oil sample. Differential liberation procedures in the laboratory are designed to represent the processes that occur in the reservoir, in which gas released from oil is presumed to migrate away from the oil because of higher mobility of the gas phase. In the reservoir, some of the liberated gas remains in contact with the oil. In the laboratory differential liberation procedure, all the liberated gas is removed from the oil. The actual process that occurs in the reservoir can be reproduced in the laboratory with a procedure called "composite liberation."[17] This procedure is superior to the differential liberation procedure, but it requires a large amount of fluid sample and takes a long time to complete. McCain[18] showed that the differences between oil properties calculated with composite liberation data and differential liberation data are within experimental accuracy. In most applications, differential liberation data in conjunction with separator test data are used for calculation of oil PVT data.

5.7.1 Calculation of Oil FVF and Solution GOR

Oil formation volume factor (FVF) that can be used for material balance and reserves calculations is calculated from flash vaporization, differential liberation, and separator test data. At pressures above bubble point,

$$B_o = \left(\frac{V}{V_{\text{sat}}}\right)_{\text{CCE}} \times B_{oSb} \quad \text{for } p > p_b \tag{5.6}$$

In Eq. (5.6), B_o = oil FVF; $(V/V_{SAT})_{CCE}$ = relative volume from constant composition expansion; and B_{oSb} = oil FVF at bubble point from separator test.

For pressures below bubble point,

$$B_o = B_{oD}\left(\frac{B_{oSb}}{B_{oDb}}\right) \quad \text{for } p < p_b \tag{5.7}$$

In Eq. (5.7), B_{oD} = oil FVF from differential liberation; and B_{oDb} is FVF at bubble point from differential liberation.

For pressures below bubble point,

$$R_s = R_{sD}\left(\frac{R_{sSb}}{R_{sDb}}\right) \quad \text{for } p < p_b \tag{5.8}$$

In Eq. (5.8), R_s = oil solution GOR; R_{sD} = oil solution GOR by differential liberation; R_{sSb} = oil solution GOR at bubble point from separator test; and R_{sDb} = oil solution GOR at bubble point from differential liberation.

5.7.2 Calculation of Gas Compressibility Factor, Gas FVF, and Total FVF

The volume of gas evolved at each stage of the differential liberation process is measured at the temperature and pressure of the PVT cell (reservoir conditions). The volume of the gas is also measured at standard conditions (atmospheric pressure and 60°F). Gas compressibility factor is calculated from these volumes as:

$$z_R = \frac{p_R V_R T_{sc}}{p_{sc} V_{sc} T_R} \tag{5.9}$$

In Eq. (5.9), p_R = pressure of the cell; V_R = volume of the gas at cell conditions; T_R = temperature of the cell; p_{sc} = pressure at standard conditions; V_{sc} = volume of gas at standard conditions; and T_{sc} = temperature at standard conditions (or 60°F).

Gas formation volume factor is calculated from Eq. 4.75 as:

$$\begin{aligned} B_g &= \frac{z_R T_R (14.65)}{p_R (519.67)} \\ &= \frac{0.02819 z_R T_R}{p_R}, \ \frac{\text{ft}^3}{\text{scf}} \\ &= \frac{0.005021 z_R T_R}{p_R}, \ \frac{\text{RB}}{\text{scf}} \end{aligned}$$

Note that in Eq. 4.75, pressure at standard conditions is set at 14.65 psia.

Total formation volume factor for differential liberation data is calculated as:

$$B_t = B_{oD} + R_l B_g \tag{5.10}$$

In Eq. (5.10), B_t = total formation volume factor, RB/STB; B_{oD} = oil FVF from differential liberation, RB/STB; R_l = total liberated gas, scf/STB; and B_g = gas FVF, RB/scf.

The calculation of gas properties from differential liberation data is shown in Appendix 5A.

5.7.3 Calculation of Oil Compressibility Factor

If oil compressibility factor is not reported with differential liberation data in Appendix 5A, it can be calculated from constant composition data for pressures above bubble point and differential liberation data for pressures below bubble point.

For pressures above bubble point, oil compressibility is represented as:

$$c_o = -\frac{1}{V}\left(\frac{\partial V}{\partial p}\right)_T \tag{5.11}$$

By integrating Eq. (5.11), assuming c_o is constant as pressure changes, it can be shown that:

$$c_o(p_2 - p_1) = -\ln\left(\frac{V_2}{V_1}\right) \tag{5.12}$$

Substituting relative volume (V/V_{SAT}) from CCE for volume in Eq. (5.12), we obtain:

$$c_o = \frac{\ln \dfrac{(V/V_{sat})_2}{(V/V_{sat})_1}}{p_1 - p_2} \quad \textbf{(5.13)}$$

Applying Eq. (5.13) to CCE data in Appendix 5A, $(V/V_{sat})_{8000} = 0.9899$ and $(V/V_{sat})_{9000} = 0.9797$. Hence,

$$c_o = \frac{\ln\left(\dfrac{0.9899}{0.9797}\right)}{9000 - 8000} = 10.4 \times 10^{-6}\text{psi}^{-1} \text{ for 9000–8000 psia range}$$

For pressures below bubble point, it was shown by McCain[19] that oil compressibility can be calculated as:

$$c_o = -\frac{1}{B_o}\left[\left(\frac{\partial B_o}{\partial p}\right)_T - B_g\left(\frac{\partial R_s}{\partial p}\right)_T\right] \quad \text{for } p < p_b \quad \textbf{(5.14)}$$

Eq. (5.14) can be rearranged as:

$$c_o = \frac{1}{B_o}\left(\frac{\partial R_s}{\partial p}\right)_T\left[B_g - \left(\frac{\partial B_o}{\partial R_s}\right)_T\right] \quad \textbf{(5.15)}$$

Eq. (5.15) can be rewritten in terms of differential liberation data as:

$$c_o = \frac{1}{B_{oD}}\left(\frac{\partial R_{sD}}{\partial p}\right)_T\left[B_g - \left(\frac{\partial B_{oD}}{\partial R_{sD}}\right)_T\right] \quad \textbf{(5.16)}$$

As an example, Eq. (5.16) can be used to estimate oil compressibility at 5600 psia from differential data in Appendix 5A. From Appendix 5A, $B_{oD} = 1.4232$ RB/bbl and $B_g = 649 \times 10^{-6}$ RB/scf at 5600 psia. From a plot of B_o versus R_s, $\dfrac{\partial B_o}{\partial R_s} = 0.000405$ RB/scf. From a plot of R_{sD} versus P, $\dfrac{\partial R_{\text{sD}}}{\partial p} = 0.1000$ scf/(bbl = psi). Note that the plots are not shown but can be easily generated from the data in Appendix 5A. Substituting in Eq. (5.16),

$$\begin{aligned} c_o &= \frac{1}{1.4232}(0.1000)[649 \times 10^{-6} - 0.000405] \\ &= 17.1 \times 10^{-6}\ \text{psi}^{-1} \end{aligned}$$

Note that the calculated oil compressibility is higher than the value reported at the same pressure in Appendix 5A. The difference can be traced to the errors associated with calculating the slope of the tangent line at 5600 psia for the $\dfrac{\partial R_{sD}}{\partial P}$ term. But this method could be used to estimate oil compressibility below bubble point pressures from differential liberation data if actual data are not reported.

Nomenclature

B_g gas formation volume factor, RB/STB
B_o oil formation volume factor, RB/STB
B_{oDb} oil formation volume factor at P_b for DL, RB/residual bbl
B_{oD} oil formation volume factor at $P<P_b$ for DL, RB/residual bbl
B_{oSb} oil formation volume factor at P_b in separator test, RB/STB
B_w water formation volume factor, RB/STB
c_o isothermal oil compressibility, psi^{-1}
c_{ob} isothermal oil compressibility at bubble point, psi^{-1}
d_{owc} depth below OWC
p pressure
p_b bubble point pressure, psia
p_c capillary pressure
p_d capillary displacement pressure or dew point pressure
R gas constant, (psia $\times$ ft^3)/(lb-mol $\times$ °R)
R_l total liberated gas in DL, scf/residual bbl
R_s solution gas-oil ratio, scf/STB
R_{sSb} solution gas-oil ratio at P_b in separator test, scf/STB
R_{sD} solution gas-oil ratio at $P<P_b$ for DL, scf/residual bbl
R_{sDb} solution gas-oil ratio at P_b for DL, scf/residual bbl
V volume
V_{sat} volume at saturated pressure
r radius of capillary tube
r_L largest pore throat radius
T temperature
z gas compressibility factor
ρ density, mass per unit volumel
σ interfacial tension
θ angle of contact

Subscripts

b bubble point
f fluid
n wetting phase
nw non-wetting phase
o oil
sc standard conditions
sp separator

st stock-tank
R reservoir

Abbreviations

App appraisal (Table 5.1)
CCE constant composition expansion
CVD constant volume depletion
DL differential liberation
EM electromagnetic viscometer
Expl exploration (Table 5.1)
FVF formation volume factor
FWL free water level
OBM oil-based mud
OFA optical fluid analyzer
OWC oil-water contact
PVT pressure-volume-temperature
Prod production (Table 5.1)
WBM water-based mud
WFT wireline formation tester

References

1. Moffatt, B.J., and Williams, J.M.: "Identifying and Meeting the Key Needs for Reservoir Fluid Properties—A Multi-Disciplinary Approach," paper SPE 49067 presented at the 1998 SPE Annual Technical Conference and Exhibition, New Orleans, Louisiana, Sept. 27–30, 1998.
2. McCain Jr., W.D., and Alexander, R.A.: "Sampling Gas-Condensate Wells," *SPERE* (August 1992) 358–362.
3. *RP 44, Recommended Practice for Sampling Petroleum Reservoir Fluids*, Second Edition, API, Washington, DC (April 2003).
4. Williams, J.M., and Kokal, S.L.: *Fluid Sampling,* Petroleum Engineering Handbook, General Engineering, Vol. 1, Chapter 4, SPE (2006).
5. Bon, J., Sarma, H.K., Rodrigues, J.T., and Bon, J.G.: "Reservoir Fluid Sampling Revisted—A Practical Perspective," paper SPE 101037 presented at the 2006 SPE Asia Pacific Oil & Gas Conference and Exhibition, Adelaide, Australia, Sept. 11–13, 2006.
6. Towler, B.F.: "Reservoir Engineering Aspects of Bottomhole Sampling of Saturated Oils for PVT Analysis," paper SPE 19438, 1990.
7. Proett, M.A., Gilbert, G.N., Chin, W.C., and Monroe, M.L.: "New Wireline Formation Testing Tool with Advanced Sampling Technology," paper SPE 56711 presented at the 1999 SPE Annual Technical Conference and Exhibition, Houston, Texas, Oct. 3–6, 1999.

8. *Fundamentals of Formation Testing*, Schlumberger, Sugar Land, Texas (2006).

9. Morris, C.W., Felling, M.M., Butsch, R.J., and Sass, W.: "Using Optical Fluid Analysis to Evaluate Downhole Fluid Sample Contamination," paper SPE 50603 presented at the 1998 SPE European Petroleum Conference, The Hague, The Netherlands, Oct. 20–22, 1998.

10. Smits, A.R., Fincher, D.V., Nishida, K., Mullins, O.C., Schroeder, R.J., and Yamate, T.: "In-situ Optical Fluid Analysis as an Aid to Wireline Formation Sampling," *SPEFE* (June 1995) 91–98.

11. Van Dusen, A., Williams, S., Fadnes, F.H., and Irvine-Fortescue, J.: "Determination of Hydrocarbon Properties by Optical Analysis During Wireline Fluid Sampling," *SPEREE* (August 2003) 286–292.

12. Gozalpour, F., Danesh, A., Tehrani, D.-H., Todd, A.C., and Tohidi, B.: "Predicting Reservoir Fluid Phase and Volumetric Behavior from Samples Contaminated with Oil-Based Mud," *SPEREE* (June 2002) 197–205.

13. Hy-Billiot, J., Bickert, J., Montel, F., and Segalini, G.: "Getting the Best From Formation Tester Sampling," paper SPE 77771 presented at the 2002 SPE Annual Technical Conference and Exhibition, San Antonio, Texas, Sept. 29–Oct. 2, 2002.

14. Elshahawi, H., Fathy, K., and Hiekel, S.: "Capillary Pressure and Rock Wettability Effects on Wireline Formation Tester," paper SPE 56712 presented at the 1999 SPE Annual Technical Conference and Exhibition, Houston, Texas, Oct. 3–6, 1999.

15. Elshahawi, H., Samir, M., and Fathy, K.: "Correcting for Wettability and Capillary Pressure Effects on Formation Tester Measurements," paper SPE 63075 presented at the 2000 SPE Annual Technical Conference and Exhibition, Dallas, Texas, Oct. 1–4, 2000.

16. Lee, A.L, Gonzalez, M.H., and Eakin, B.E.: "The Viscosity of Natural Gas," *Trans. AIME* (1966) Vol. 237, 997–1000.

17. Moses, P.L.: "Engineering Applications of Phase Behavior of Crude Oil and Condensate Systems," *JPT* (July 1986) 38, 715–723.

18. McCain, Jr., W.D.: "Analysis of Black Oil PVT Reports Revisited," paper SPE 77386 presented at the 2002 SPE Annual Technical Conference and Exhibition, San Antonio, Texas, Sept. 29–Oct. 2, 2002.

19. McCain, Jr., W.D.: *The Properties of Petroleum Fluids*, Second Edition, Pennwell Books, Tulsa, OK, 1990.

General Reading

- Pham, T.R., Al-Afaleg, N.I., Kelder, O., Al-Otaibi, U.F., and Zeybek, M.: "Field Example of Capillary Pressure Effects on Wireline Formation Tester Measurements and OWC Estimation in a Mixed-Wettability Oil Reservoir," paper SPE 93262 presented at the SPE Middle East Oil & Gas Show and Conference, Bahrain, March 12–15, 2005.

- Nagarajan, N.R, Honarpour, M.M., and Sampath, K.: "Reservoir Fluid Sampling and Characterization—Key to Efficient Reservoir Management," paper SPE 101517 presented at the 2006 Abu Dhabi International Petroleum Exhibition and Conference, Abu Dhabi, U.A.E. November 5–8, 2006.
- Michaels, J., Moody, M., and Shwe, T.: "Wireline Fluid Sampling," paper SPE 30610 presented at the SPE Annual Technical Conference & Exhibition, Dallas, Texas, Oct. 22–25, 1995.
- Williams, J.M.: "Getting the Best Out of Fluid Samples," *JPT* (September 1994) 752.
- El-Banbi, A., and McCain, Jr., W.D.: "Sampling Volatile Oil Wells," paper SPE 67232 presented at the SPE Production and Operations Symposium, Oklahoma City, Oklahoma, March 24–27, 2001.
- Felling, M.M. and Morris, C.W.: "Characterization of In-Situ Fluid Responses by Use of Optical Fluid Analysis," *SPEREE* (August 1998) 297–302.

APPENDIX 5A

Typical Reservoir Fluid Study for a Black Oil Sample

5A.1 Reservoir Fluid Summary

Reservoir temperature, °F	212
Saturation pressure at 212°F, psia	7153
Compressibility of reservoir oil at 212°F, psi^{-1}:	
At static bottomhole pressure of 10,640 psia	8.25E-06
At bubble point pressure of 7153 psia	11.96E-06
Saturated oil at 7153 psia, 212°F:	
Density, gm. per cc.	0.6690
Density, lb. per bbl.	234.5
Specific volume, ft^3 per lb.	0.02394
Viscosity, cp.	0.467
Formation Volume Factor (FVF):	
*FVF at p_b, rb/bbl. residual oil at 60°F	1.5572
**FVF at p_b, rb/stock tank oil at 60°F	1.5186
Solution gas-oil ratio:	
*Standard cu. ft. per bbl. residual oil at 60°F	1153
**Standard cu. ft. per bbl. stock tank oil at 60°F	1049
Reservoir oil at 10460 psia, 212°F:	
Density, gm. per cc.	0.6931
Density, lb. per bbl.	242.9

Specific volume, cu. ft. per lb.	0.02311
Viscosity, cp.	0.676
Formation volume factor:	
* Bbl. per bbl. residual oil at 60°F	1.5031
**Bbl. per bbl. stock tank oil at 60°F	1.4658

* Differential Liberation.

** Flash Separation at field separator conditions.

5A.2 Calculated Analysis of Reservoir Fluid

Component	Mole %
Hydrogen Sulfide	0.00
Carbon Dioxide	0.23
Nitrogen	0.30
Methane	67.51
Ethane	2.25
Propane	1.87
i-Butane	0.61
n-Butane	1.00
i-Pentane	0.50
n-Pentane	0.55
Hexanes	1.10
Heptanes	1.28
Octanes	1.26
Nonanes	1.38
Decanes	1.28
Undecanes	1.05
Dodecanes	0.99
Tridecanes	0.99
Tetradecanes	0.89
Pentadecanes	0.93
Hexadecanes	1.19
Heptadecanes	1.05
Octadecanes	0.62
Nonadecanes	1.21
Eicosanes	0.69
Heneicosanes	0.58
Docosanes	0.53
Tricosanes	0.49
Tetracosanes	0.44
Pentacosanes	0.43

Hexacosanes	0.32
Heptacosanes	0.35
Octacosanes	0.32
Nonacosanes	0.30
Triacontanes Plus	5.51
Total	100.00

Basis of Recombination = 1167 scf/bbl
Total molecular weight = 88.37 lb/lb-mole
Tricontanes Plus molecular weight = 610.6
Tricontanes Plus specific gravity = 0.9847

5A.3 Pressure-Volume Properties at 212°F (Constant Composition Expansion)

Pressure (psia)	Relative Volume (V/V_{sat})	Density (gm/cc)	Compressibility ($\Delta V/V/\Delta psi \times 10^6$)	"Y" Function $(p_{sat} - p)/p(B_t - 1)$
10640*	0.9652	0.6931	8.25	
10000	0.9706	0.6893	8.61	
9000	0.9797	0.6829	9.27	
8000	0.9899	0.6759	10.29	
7153**	1.0000	0.6690	11.96	
TWO-PHASE REGION DATA BELOW				
7000	1.0035	0.6667		6.203
6300	1.0230	0.6539		5.876
5600	1.0500	0.6372		5.548
4900	1.0881	0.6148		5.216
4200	1.1437	0.5850		4.893
3500	1.2287	0.5445		4.563
2800	1.3671	0.4894		4.235
2100	1.6155	0.4141		3.909
1400	2.1474	0.3115		3.581
700	3.8323	0.1746		3.255

* Static Bottomhole Pressure
** Bubble point, p_b

5A.4 Differential Liberation at 212°F

Pressure (psia)	Density (gm/cc)	FVF B_{oD} RB/bbl. residual oil at 60°F	FVF B_o† (RB/STB)	Compressibility $\frac{\Delta V}{V\Delta p} \times 10^{-6}$ (psi^{-1})	Liberated Gas R_l (scf/bbl)	Solution GOR R_{sD} (scf/bbl)	Solution GOR R_s† (scf/bbl)
10640*	0.6931	1.5031	1.4658	8.25	0	1153	1049
10000	0.6893	1.5114	1.4739	8.61	0	1153	1049
9000	0.6829	1.5256	1.4877	9.27	0	1153	1049
8000	0.6759	1.5414	1.5032	10.29	0	1153	1049
7153**	0.6690	1.5572	1.5186	11.96	0	1153	1049
TWO-PHASE REGION DATA BELOW							
7000	0.6728	1.5436	1.5053	10.64	24	1129	1027
6300	0.6881	1.4803	1.4435	9.49	176	977	889
5600	0.7033	1.4232	1.3879	8.90	310	843	767
4900	0.7175	1.3725	1.3385	8.53	434	719	654
4200	0.7312	1.3263	1.2934	8.20	549	604	550
3500	0.7448	1.2836	1.2517	7.90	656	497	452
2800	0.7584	1.2434	1.2126	7.61	755	398	362
2100	0.7718	1.2056	1.1757	7.35	852	301	274
1400	0.7856	1.1684	1.1394	7.09	947	206	187
700	0.7993	1.1313	—	6.86	1041	112	—
15	0.8215	1.0728	—	—	1153	0	—
15@ 60°F	0.8813	1.0000	—	—	1153	0	—

* Static Bottomhole Pressure

** Bubble point Pressure, p_b

† Adjusted for field separator conditions per bbl. of stock tank oil at 60°F.

5A.5 Gas Differentially Liberated at 212°F

Pressure (psia)	Specific Gravity (Air = 1.00)		Compressibility Factor (z)	Gas FVF, B_g (RB/MMscf)	Total FVF B_t (RB/STB)	Gas Viscosity (Calculated)* (cp)
	Incremental	Accumulative				
7153	(Saturation Pressure at 212 °F)					
7000	0.6030	0.6030	1.133	560	1.5571	0.0309
6300	0.5950	0.5961	1.091	599	1.5853	0.0285
5600	0.5900	0.5935	1.051	649	1.6242	0.0262
4900	0.5861	0.5914	1.013	715	1.6824	0.0241
4200	0.5840	0.5898	0.980	807	1.7688	0.0221
3500	0.5811	0.5884	0.952	941	1.8999	0.0202
2800	0.5820	0.5875	0.935	1154	2.1151	0.0183
2100	0.5832	0.5871	0.928	1528	2.5064	0.0167
1400	0.5920	0.5876	0.936	2313	3.3580	0.0153
700	0.6540	0.5935	0.962	4753	6.0777	0.0140
15	0.9210	0.6254	1.000	230,176	266.4026	0.0122

$B_t = B_o + [R_l B_g] \times 10^{-6}$

* Based on Lee et al.[16]

5A.6 Viscosity Data at 212°F

Pressure (psia)	Oil Viscosity (cp)	Gas Viscosity (Calculated)† (cp)	Oil/Gas Viscosity Ratio
10640*	0.676		
10000	0.637		
9000	0.579		
8000	0.517		
7153**	0.467		
TWO-PHASE REGION DATA BELOW			
7000	0.471	0.0309	15.23
6300	0.507	0.0285	17.81
5600	0.548	0.0262	20.89
4900	0.598	0.0241	24.80
4200	0.671	0.0221	30.36
3500	0.772	0.0202	38.31
2800	0.944	0.0183	51.47
2100	1.286	0.0167	77.05
1400	1.774	0.0153	116.18
700	2.435	0.0140	174.43
15	3.436	0.0122	282.80

* Static Bottomhole Pressure

** Bubble point, p_b

† Based on Lee et al.[16]

5A.7 Comparison of Reservoir Oil Flash Liberation Tests

Liberation Type At 212°F	Gas-Oil Ratio*	Gas Gravity	Stock Tank Oil Gravity (°API @ 60°F)	FVF (V_{sat}/V_{st})**
	Total Solution Gas			
Flash Separation	1049	0.6079	30.2	1.5186
Differential	1153	0.6254	28.9	1.5572

* Gas-Oil Ratio is cubic feet of gas at 15.025 psia and 60°F per bbl of stock tank oil at 60°F.

** V_{sat}/V_{st} is the volume of reservoir oil at the saturation pressure and reservoir temperature relative to stock tank oil volume at 60°F.

APPENDIX 5B

Typical Reservoir Fluid Study for a Gas Condensate Sample

5B.1 Summary of Reservoir Data and Surface Sampling Conditions

Reservoir Data	
Formation	Unknown
Perforated interval	15,390–15,516 feet
Flowing bottomhole pressure	Not available
Static bottomhole pressure	12,688 psia
Reservoir temperature	267°F

Surface Sampling Summary	
Well status	On Production
Choke size	42/64 inch
Date and time sampled	March 2, 1997; 1200 hours
Flowing tubing pressure	9020 psig
Primary separator temperature	92°F
Meter run temperature	92°F
Primary separator pressure	1140 psig
Primary separator gas rate (field)	48,453 Mcf/D
Primary separator gas rate (lab)	49,544.8 Mcf/D
Oil rate measured at primary separator	2120 B/D
Water rate measured at primary separator	40 B/D
Gas-oil ratio	23,370 scf 1st stage gas per bbl. separator liquid
Gas-oil ratio	28,721 scf 1st stage gas per bbl. stock tank liquid
Standard pressure and temperature	15.025 psia, 60°F

5B.2 Chromatograph Analysis of Separator Gas at 1140 psig and 92°F

Component	Mole (%)	Weight (%)	GPM* @15.025 psia
Nitrogen	0.20	0.30	0.000
Hydrogen Sulfide	0.00	0.00	0.000
Carbon Dioxide	0.94	2.25	0.000
Methane	90.44	79.13	0.000
Ethane	5.28	8.65	1.440
Propane	1.70	4.09	0.478
i-Butane	0.36	1.15	0.122
n-Butane	0.41	1.30	0.132
i-Pentane	0.16	0.65	0.062
n-Pentane	0.12	0.45	0.043
Hexanes	0.16	0.72	0.063
Heptanes Plus	0.23	1.31	0.103
Totals	100.00	100.00	2.443

* GPM is Gallons per Mcf

Calculated Properties of Gas:

Gas specific gravity (Air = 1.00) = 0.6346

Gas Heat of Combustion (Btu/cu. Ft. @ 15.025 psia & 60°F) Dry = 1133.7

Gas Heat of Combustion (Btu/cu. Ft. @ 15.025 psia & 60°F) Wet = 1113.9 water saturated

Gas compressibility factor, z (at 1 atmosphere, 60°F) = 0.9973

5B.3 Chromatograph Analysis of Separator Liquid at 1140 psig and 92°F

Component	Mole (%)	Liquid Volume (%)	Weight (%)	Mole Weight	Specific Gravity
Hydrogen Sulfide	0.00	0.00	0.00	34.076	0.8014
Carbon Dioxide	0.59	0.23	0.26	44.010	0.8180
Nitrogen	0.04	0.01	0.01	28.013	0.8094
Methane	28.16	9.94	4.49	16.043	0.3324
Ethane	5.89	2.63	1.76	30.070	0.4935
Propane	4.74	3.01	2.08	44.097	0.5070
i-Butane	1.76	1.33	1.02	58.123	0.5629
n-Butane	2.75	2.00	1.59	58.123	0.5840
i-Pentane	2.04	1.72	1.46	72.150	0.6247
n-Pentane	1.84	1.54	1.32	72.150	0.6311
Hexanes	2.08	1.98	1.78	86.177	0.6644
Heptanes Plus	50.11	75.61	84.23	169.100	0.8197
Totals	100.00	100.00	100.00	—	—

Properties	Measured for Heptanes Plus	Calculated for Liquid
Specific Gravity	0.8197	0.7358 @ 60/60°F
Molecular Weight	169.10	100.60
Cu. Ft/Gal.	15.00	22.63 @ 15.025 psia & 60°F

5B.4 Composition of Reservoir Fluid (Calculated)

Component	Mole (%)
Hydrogen Sulfide	0.00
Carbon Dioxide	0.93
Nitrogen	0.19
Methane	88.01
Ethane	5.30
Propane	1.82
i-Butane	0.42
n-Butane	0.50
i-Pentane	0.23
n-Pentane	0.19
Hexanes	0.24
Heptanes Plus	2.17
Totals	100.00

Basis of recombination calculation (1)	= 23,370 scf separator gas per bbl. separator liquid
Basis of recombination calculation (2)	= 42.79 bbl. separator liquid per MMscf separator gas
Total molecular weight	= 21.5 lb/lb-mole
Total specific gravity (Air = 1.0)	= 0.7437
Heptanes Plus molecular weight	= 162.1 lb/lb-mole
Heptanes Plus specific gravity (water = 1.0)	= 0.8125

5B.5 Measured Saturation Pressures from Stepwise Recombinations at 267°F

Gas-Liquid Ratio (scf 1st stage gas) / (bbl. 1st stage liquid)	Saturation Pressure (Dew Point) (psia)
40,000	9304
30,000	9960
23,000	9912

5B.6 Pressure-Volume Properties of Reservoir Fluid at 267°F (or CCE)

Pressure (psia)	Relative Volume (V/V_{Rcs})	Specific Volume (cu. ft./lb)	Reservoir bbl/MMcf	Reservoir Liquid Vol. (%)†	Deviation Factor (z)	Gas Viscosity (cp)	Percent Wellstream Produced††
12668*	1.0000	0.04698			1.6441	0.0503	0.00
12000	1.0208	0.04796			1.5898	0.486	2.04
11500	1.0392	0.04883			1.5511	0.0471	3.78
11000	1.0547	0.04955			1.5058	0.0460	5.19
10750	1.0643	0.05000			1.4849	0.0454	6.04
10500	1.0745	0.05048			1.4643	0.0447	6.93
10000	1.0946	0.05143			1.4207	0.0434	8.65
9912**	1.0990	0.05164	0.00	0.00	1.4138	0.0432	9.01
9500	1.1187	0.05256	0.16	0.03	1.3793		10.61
9000	1.1450	0.05379	0.21	0.04	1.3374		12.66
8500	1.1743	0.05517	0.42	0.09	1.2955		14.84
8000	1.2085	0.05678	0.77	0.16	1.2547		17.25
7500	1.2471	0.05859	1.34	0.28	1.2139		19.81
7000	1.2921	0.06071	1.86	0.38	1.1739		22.61
6000	1.4078	0.06614	3.05	0.63	1.0963		28.97
5000	1.5838	0.07441	4.30	0.89	1.0278		36.86
4000	1.8703	0.08787	5.86	1.21	0.9710		46.53
3000	2.3994	0.11273	7.16	1.48	0.9342		58.32
2000	3.5510	0.16684	7.33	1.51	0.9217		71.84
1000	7.2273	0.33957	6.92	1.43	0.9380		86.16

* Static bottomhole pressure

** Saturation pressure (dew point)

† Liquid volume is calculated as percent of hydrocarbon pore space at reservoir volume occupied by liquid.

†† Wellstream produced is calculated as volume percent of original reservoir fluid.

5B.7 Depletion Study at 267°F: Hydrocarbon Analyses of Produced Wellstream (Mole %)

Component	Reservoir Pressure (psia)						
	Reservoir Fluid						Abandonment Liquid
	12,668	9912	8000	6000	4000	2000	2000
Hydrogen Sulfide	0.00	0.00	0.00	0.00	0.00	0.00	0.00
Nitrogen	0.19	0.19	0.19	0.20	0.19	0.20	0.04
Carbon Dioxide	0.93	0.93	0.94	0.94	0.95	0.93	0.41
Methane	88.01	88.01	88.78	89.29	89.62	88.81	32.10
Ethane	5.30	5.30	5.27	5.25	5.27	5.34	4.41
Propane	1.82	1.82	1.81	1.81	1.81	1.81	2.68
i-Butane	0.42	0.42	0.41	0.40	0.40	0.40	0.88
n-Butane	0.50	0.50	0.49	0.49	.049	0.48	1.26
i-Pentane	0.23	0.23	0.22	0.21	0.21	0.23	0.88
n-Pentane	0.19	0.19	0.18	0.18	0.18	0.17	0.74
Hexanes	0.24	0.24	0.23	0.22	0.21	0.23	1.64
Heptanes Plus	2.17	2.17	1.48	1.01	0.67	1.40	54.96
Totals	100.00	100.00	100.00	100.00	100.00	100.00	100.00
Specific Gravity	0.7437	0.7437	0.6976	0.6729	0.6585	0.6942	—
Properties of Heptanes Plus							
Specific Gravity	0.8125	0.8125	0.7890	0.7745	0.7699	0.7908	0.8300
Molecular Weight	162.1	162.1	142.4	132.0	128.9	142.5	181.1
Liquid Content of Well Stream (Gallons/Mscf)	3.914	3.914	3.349	3.026	2.833	3.314	—
Reservoir Fluid Deviation Factor, *z*							
Two-Phase Fluid	—	1.414	1.256	1.101	0.974	0.929	—
Gas Phase Produced	1.644	1.414	1.267	1.110	0.993	0.940	—
Gas FVF, B_g (Res. cu. ft./Mcf)	2.726	2.996	3.326	3.886	5.214	9.871	—
Gas Viscosity (cp)	0.0503	0.0432	0.0348	0.0285	0.0225	0.0173	—
Initial Fluid Produced Cumulative (%)	0.00	9.02	16.53	27.56	45.69	72.05	—

5B.8 Retrograde Condensation During Gas Depletion at 267°F

Pressure (psia)	Reservoir Liquid (Volume %)*
9912	0.00
8000	0.15
6000	0.56
4000	0.82
2000	0.89

* Percent of reservoir hydrocarbon pore space at reservoir pressure.

PVT Properties Predictions from Equations of State

6.1 Historical Introduction to Equations of State (EOS)

The most common and familiar equation of state (EOS) is the ideal gas EOS. This was derived in Chapter 4 from the combination of Boyle's and Charles' Laws on gases. The ideal gas EOS was represented in Eq. (4.7) as $pV = nRT$. The EOS for real gases was represented in Eq. (4.9) as $pV = znRT$. As discussed in Chapter 4, the gas compressibility factor, z, is used to account for the non-ideality of real gases. The gas compressibility factor is not constant but varies with the temperature and pressure of the system. This imposes mathematical limitations on the direct application of real gas EOS for representation of thermodynamic state of gases, since the gas compressibility factor must be evaluated at each thermodynamic condition.

In 1873, van der Waals[1] proposed an EOS that represented non-ideal behavior by accounting for finite volumes occupied by molecules of substances and the repulsive and attractive forces between these molecules. The van der Waals EOS is shown as:

$$\left(p + \frac{a}{V^2}\right)(V - b) = RT \tag{6.1}$$

In Eq. (6.1), p is pressure; V is volume; T is temperature; a is the attraction parameter to correct pressures for attraction between molecules; b is the effective molecular volume to correct for volume occupied by the molecules; and R is the universal gas constant.

Many EOS of varying degrees of complexity have been proposed since the emergence of the van der Waals EOS. A summary of early equations of state as documented by Walas[2] shows the rich history of advancement in the development of equations of state.

The presentation of equations of state is limited to van der Waals–type EOS in this book. These are generally called two-parameter equations of state because they have the form of the original van der Waals EOS. In addition to the van der Waals (vdW) EOS, two other equations of state presented in this book are the Soave-Redlich-Kwong (SRK) EOS and the Peng-Robinson (PR) EOS. The reasons for the presentation of only these two other equations of state are that most publications, research work, and software in the petroleum industry are based on the SRK and PR equations of state or variants of these equations of state.

6.2 van der Waals (vdW) EOS

As shown in Eq. (6.1), the van der Waals (vdW) EOS is represented as:

$$\left(p + \frac{a}{V^2}\right)(V - b) = RT$$

In cubic forms, it can be represented as:

$$V^3 - \left(b + \frac{RT}{p}\right)V^2 + \frac{a}{p}V - \frac{ab}{p} = 0 \tag{6.2}$$

$$z^3 - z^2\left(\frac{bp}{RT} + 1\right) + z\frac{ap}{(RT)^2} - \frac{abp^2}{(RT)^3} = 0 \tag{6.3}$$

The above representations of vdW EOS in cubic forms are the bases that these types of equations of state are often referred to as cubic equations of state.

To derive formulas for the constants a and b in vdW EOS (Eq. (6.1)), it is useful to observe that at the critical temperature, the first and second derivative of pressure with respect to volume is zero for a pure substance. This means that for a pure substance, the critical temperature isotherm on a pressure-volume (P-V) diagram has a zero gradient as it passes through the critical point. Thus,

$$\left(\frac{\partial p}{\partial V}\right)_{T_c} = -\frac{RT_c}{(V_c - b)^2} + \frac{2a}{V_c^3} = 0 \tag{6.4}$$

$$\left(\frac{\partial^2 p}{\partial V^2}\right)_{T_c} = \frac{2RT_c}{(V_c - b)^3} - \frac{6a}{V_c^4} = 0 \tag{6.5}$$

Also, vdW EOS can be expressed in terms of critical properties as:

$$\left(p_c + \frac{a}{V_c^2}\right)(V_c - b) = RT_c \tag{6.6}$$

Eqs. (6.4), (6.5), and (6.6) can be solved simultaneously for a and b as:

$$a = 3p_c V_c^2 = \frac{27R^2T_c^2}{64p_c} \tag{6.7}$$

$$b = \frac{V_c}{3} = \frac{RT_c}{8p_c} \tag{6.8}$$

$$z_c = 0.375 \tag{6.9}$$

For mixtures, the following mixing rules based on the pure component parameters may be applied for vdW EOS:

$$a_m = \sum_i^{N_C} \sum_j^{N_C} y_i y_j \sqrt{a_i a_j} \tag{6.10}$$

$$b_m = \sum_i^{N_C} y_i b_i \tag{6.11}$$

In Eqs. (6.10) and (6.11), a_m and b_m are mixture parameters; y_i and y_j are compositions for mixture components, i and j; a_i, a_j, and b_i are parameters for the pure components, i and j; and N_C is the total number of components in the mixture.

6.3 Soave-Redlich-Kwong (SRK) EOS

The Soave-Redlich-Kwong (SRK) EOS is a modification of the Redlich-Kwong (RK) EOS. The equations representing RK EOS are presented first followed by the modifications introduced in the SRK EOS.

The RK EOS was proposed in 1949.[3] It is a modification of the vdW EOS with the introduction of a temperature correction for the attraction parameter, a, in the original vdW EOS. The equations representing the RK EOS are as follows:

$$p + \frac{a}{\sqrt{T}V(V+b)} = \frac{RT}{V-b} \tag{6.12}$$

$$a = \Omega_a \frac{R^2T_c^{2.5}}{p_c} = 0.42748\frac{R^2T_c^{2.5}}{p_c} \tag{6.13}$$

$$b = \Omega_b \frac{RT_c}{p_c} = 0.08664\frac{RT_c}{p_c} \tag{6.14}$$

$$A = \frac{ap}{R^2T^{2.5}} = 0.42748\frac{p_r}{T_r^{2.5}} \tag{6.15}$$

$$B = \frac{bp}{RT} = 0.08664\frac{p_r}{T_r} \tag{6.16}$$

In cubic forms, the RK EOS is shown as:

$$V^3 - V^2\frac{RT}{p} + \frac{V}{p}\left(\frac{a}{\sqrt{T}} - bRT - pb^2\right) - \frac{ab}{p\sqrt{T}} = 0 \tag{6.17}$$

$$z^3 - z^2 + z(A - B - B^2) - AB = 0 \tag{6.18}$$

For mixtures, the following mixing rules for the parameters are applicable:

$$a_m = \sum_i^{N_C}\sum_j^{N_C} y_i y_j a_{ij} \text{ where } a_{ij} = \sqrt{a_i a_j} \tag{6.19}$$

$$b_m = \sum_i^{N_C} y_i b_i \tag{6.20}$$

$$A_m = \sum_i^{N_C}\sum_j^{N_C} y_i y_j A_{ij} \text{ where } A_{ij} = \sqrt{A_i A_j} \tag{6.21}$$

$$B_m = \sum_i^{N} y_i B_i \tag{6.22}$$

In Eqs. (6.19) to (6.22), a_m, b_m, A_m, and B_m are parameters for the mixture; y_i and y_j are compositions of mixture components; a_i, b_i, A_i, and B_i are parameters for the pure components; a_{ij} and A_{ij} are the cross parameter terms between components in the mixture.

In 1972, Soave[4] modified the RK EOS by replacing the temperature-dependent term $a/\sqrt{T}$ in the RK EOS with another term $a(T,\omega)$ which defined the attraction term, a, as a function of temperature, T, and acentric factor, ω. The SRK EOS is:

$$p + \frac{a(T,\omega)}{V(V + b)} = \frac{RT}{V - b} \tag{6.23}$$

Using the coefficients published by Graboski and Daubert,[5,6] the parameters in the SRK EOS are defined as:

$$a(T,\omega) = a\alpha = 0.42747\frac{R^2T_c^2}{p_c}\alpha \tag{6.24}$$

$$\alpha = \left[1.0 + m\left(1.0 - T_r^{0.5}\right)\right]^2 \tag{6.25}$$

$$m = 0.48508 + 1.55171\omega - 0.15613\omega^2 \tag{6.26}$$

$$b = 0.08664\frac{RT_c}{p_c} \tag{6.27}$$

$$A = \frac{\alpha a p}{R^2 T^2} = \frac{0.42747 \alpha p_r}{T_r^2} \tag{6.28}$$

$$B = \frac{bp}{RT} = \frac{0.08664 p_r}{T_r} \tag{6.29}$$

In Eq. (6.26), ω is the Pitzer acentric factor. Pitzer et al.[7] defined acentric factor as:

$$\omega = -\left(\log\left(\frac{p_v^*}{p_c}\right) + 1\right) \text{ at } T_r = \frac{T}{T_c} = 0.7 \tag{6.30}$$

In Eq. (6.30), p_v^* =vapor pressure at $T = 0.7\,T_c$; $T_c =$ critical temperature, and $p_c =$ critical pressure. Values of acentric factors for pure substances are given in Table 4.1.

In cubic forms, SRK EOS is represented as:

$$V^3 - V^2\frac{RT}{p} + \frac{V}{p}(a\alpha - bRT - pb^2) - \frac{a\alpha b}{p} = 0 \tag{6.31}$$

$$z^3 - z^2 + (A - B - B^2)z - AB = 0 \tag{6.32}$$

For mixtures, as was the case for RK EOS, these mixing rules are applicable:

$$(a\alpha)_m = \sum_i^{N_C}\sum_j^{N_C} y_i y_j (a\alpha)_{ij} \tag{6.33}$$

$$b_m = \sum y_i b_i \tag{6.34}$$

$$A_m = \sum_i^{N_C}\sum_j^{N_C} y_i y_j A_{ij} \tag{6.35}$$

$$B_m = \sum_i^{N_C} y_i B_i \tag{6.36}$$

The cross parameter terms are expressed as:

$$(a\alpha)_{ij} = (1 - k_{ij})\sqrt{(a\alpha)_i (a\alpha)_j} \tag{6.37}$$

$$A_{ij} = (1 - k_{ij})\sqrt{A_i A_j} \tag{6.38}$$

The term k_{ij} is called the binary interaction parameter (BIP). By definition, $k_{ii} = 0$ and $k_{ij} = k_{ji}$. Binary interaction parameters are sometimes adjusted during "tuning" of cubic equations of state to match experimental phase equilibria data.

6.4 Peng-Robinson (PR) EOS

In 1976, Peng and Robinson[8] introduced a new two-parameter equation of state developed primarily to improve calculations in the vicinity of the critical region, and also improve calculation of liquid densities. The PR EOS is similar in form to the familiar RK and SRK equations of state. The PR EOS is expressed as:

$$p + \frac{a(T, \omega)}{V(V + b) + b(V - b)} = \frac{RT}{V - b} \tag{6.39}$$

The parameters for the PR EOS are as follows:

$$a(T, \omega) = a\alpha = 0.45724\frac{R^2T_c^2}{p_c}\alpha \tag{6.40}$$

$$\alpha = \left[1.0 + m\left(1.0 - T_r^{0.5}\right)\right]^2 \tag{6.41}$$

$$m = 0.37464 + 1.54226\omega - 0.26992\omega^2 \tag{6.42}$$

$$b = 0.07780\frac{RT_c}{p_c} \tag{6.43}$$

$$A = \frac{\alpha a p}{R^2T^2} = \frac{0.4572\alpha p_r}{T_r^2} \tag{6.44}$$

$$B = \frac{bp}{RT} = \frac{0.07780p_r}{T_r} \tag{6.45}$$

For heavier components, $\omega > 0.49$:[9,10]

$$m = 0.3796 + 1.485\omega - 0.1644\omega^2 + 0.01667\omega^3 \tag{6.46}$$

In cubic form, PR EOS is represented as:

$$z^3 - (1 - B)z^2 + (A - 3B^2 - 2B)z - (AB - B^2 - B^3) = 0 \tag{6.47}$$

For mixtures, the same mixing rules presented in Eqs. (6.33) to (6.38) for SRK EOS are applicable to the PR EOS.

6.5 Phase Equilibrium of Mixtures

It was stated in Chapter 4 that the phases in a system composed of one or more components are considered to be in equilibrium if the chemical potential of each component in all the phases are equal at the same temperature and pressure. The chemical potential, μ, of a pure substance is related to temperature and pressure by the expression:

$$d\mu = -sdT + vdp \tag{6.48}$$

In Eq. (6.48), s is the molar entropy, and v is the molar volume. Chemical potential is sometimes called Gibbs energy. For an ideal gas at constant temperature, Eq. (6.48) reduces to:

$$d\mu = vdp = RTd\ln p \tag{6.49}$$

Eq. (6.49) can be applied to a real fluid by replacing the pressure term, p, with a new term, f, called fugacity such that:

$$d\mu = RTd\ln f = vdp \tag{6.50}$$

For an ideal gas, fugacity is equal to pressure and has pressure units. For a real fluid, fugacity approaches pressure at low pressure as defined by the limit:

$$\lim_{p\to 0} f = p \tag{6.51}$$

The ratio of fugacity to pressure for a pure substance is called fugacity coefficient:

$$\phi = \frac{f}{p} \tag{6.52}$$

The fugacity coefficient of a pure substance is expressed by the general thermodynamic relationship:

$$\ln\phi = \ln\frac{f}{p} = \frac{1}{RT}\int_0^p \left(V - \frac{RT}{p}\right)dp \tag{6.53}$$

By substitution of the SRK EOS (Eq. (6.23)) into Eq. (6.53), the fugacity coefficient of a pure substance is expressed as:

$$\ln\phi = z - 1 - \ln(z - B) - \frac{A}{B}\ln\left(\frac{z + B}{z}\right) \tag{6.54}$$

Similarly for the PR EOS, by substituting Eq. (6.39) into Eq. (6.53), the fugacity coefficient for a pure substance is expressed as:

$$\ln\phi = z - 1 - \ln(z - B) - \frac{A}{2\sqrt{2}B}\ln\left(\frac{z + 2.414B}{z - 0.414B}\right) \tag{6.55}$$

For a component in a mixture, the rigorous thermodynamic relationship for the fugacity coefficient is:

$$RT\ln\phi_i = \int_V^\infty \left[\frac{\partial P}{\partial n_i} - \frac{RT}{V}\right]dV - RT\ln z \tag{6.56}$$

In Eq. (6.56), ϕ_i is the partial fugacity coefficient of component i in the mixture where:

$$\phi_i = \frac{f_i}{x_i p} \tag{6.57}$$

In Eq. (6.57), f_i is the partial fugacity of the component in the mixture; x_i is the mole fraction of the component in the mixture; and p is the total pressure of the mixture.

For the SRK EOS, the thermodynamic expression for partial fugacity coefficient of a component in a mixture as derived from Eq. (6.56) is:

$$\ln \phi_i = \frac{B_i}{B}(z - 1) - \ln (z - B) + \frac{A}{B}\left[\frac{B_i}{B} - \frac{2}{a\alpha}\sum_j x_j(a\alpha)_{ij}\right]\ln\left(1 + \frac{B}{z}\right) \tag{6.58}$$

A similar expression for the PR EOS is:

$$\ln \phi_i = \frac{B_i}{B}(z - 1) - \ln (z - B) + \frac{A}{2.828B}\left[\frac{B_i}{B} - \frac{2}{a\alpha}\sum_j x_j(a\alpha)_{ij}\right]\ln\left[\frac{z + 2.414B}{z - 0.414B}\right] \tag{6.59}$$

It was stated earlier that the chemical potentials of all components in all coexisting phases at equilibrium are equal. From Eq. (6.50), chemical potential can be related directly to fugacity. It thus follows that the fugacity of a component in a mixture is equal in all phases of the mixture coexisting at equilibrium. For instance, if the mixture is composed only of a vapor phase and a liquid phase, it means that the fugacity of a component in the vapor phase is equal to its fugacity in the liquid phase. This statement can be expressed simply as:

$$f_i^V = f_i^L \tag{6.60}$$

In Eq. (6.60), f_i is the partial fugacity of component i; V and L denote the vapor and liquid phases, respectively. Eq. (6.60) is the fundamental equation that represents the condition of equilibrium in a mixture. If the mixture consisted of N_C components with N phases in equilibrium, Eq. (6.60) can be extended to illustrate that the partial fugacity of a component in all the N phases are equal:

$$f_i^k = f_i^j, i = 1, \ldots, N_C; k, j = 1, \ldots, N \tag{6.61}$$

Applying Eq. (6.57), Eq. (6.60) can be written as:

$$y_i p \phi_i^V = x_i p \phi_i^L \tag{6.62}$$

In Eq. (6.62), p is the pressure of the mixture; y_i and x_i are vapor and liquid phase compositions of component i in equilibrium; and ϕ_i is the corresponding partial fugacity coefficient of the component in each phase. From Eq. (6.62), it is deduced that:

$$\frac{y_i}{x_i} = \frac{\phi_i^L}{\phi_i^V} = K_i \tag{6.63}$$

K_i is the equilibrium constant of component i.

6.6 Roots from Cubic EOS

The roots of the cubic forms of vdW, RK, SRK, and PR equations of state can be determined by analytical or iterative methods. The solution may yield one or three real roots. When three real roots are obtained, the smallest root is selected for the liquid phase and the largest root is selected for the vapor phase. The middle root is discarded because it has no physical significance. Michelsen[11] devised a method for choosing the correct root for mixtures based on the phase with the lowest normalized Gibbs energy. For the vapor phase, normalized Gibbs energy, g_y^* is defined as:

$$g_y^* = \sum_{i=1}^{N} y_i \ln f_i(y) \tag{6.64}$$

For the liquid phase, normalized Gibbs energy, g_x^* is defined as:

$$g_x^* = \sum_{i=1}^{N} x_i \ln f_i(x) \tag{6.65}$$

In Eqs. (6.64) and (6.65), y_i and x_i are mole fractions of vapor and liquid, respectively, and f_i is the fugacity of the component in the mixture.

6.7 Volume Translation

In a very exhaustive effort to improve volumetric predictions with generalized cubic equations of state, Martin[12] introduced the concept of volume translation. He proved that no other two-parameter cubic equation of state can be developed to be superior to a two-parameter cubic equation with volume translation. Working independently, Peneloux et al.[13] applied volume translation to improve volumetric predictions with SRK EOS with an approach that was directly applicable to phase equilibria and volumetric calculations in the petroleum industry. In 1988, Jhaveri and Youngren[14] extended the work of Peneloux et al.[13] to the Peng-Robinson EOS. The concepts of volume translation are presented as described by Peneloux et al.[13] and extended by Jhaveri and Youngren.[14]

Peneloux et al.[13] introduced a third term, c, into the two-parameter cubic equations of state to correct the molar volume calculated with the cubic equations such that:

$$\widetilde{v} = v - c \tag{6.66}$$

In Eq. (6.66), $\widetilde{v}$ = corrected (translated) molar volume; v = molar volume calculated with EOS; and c = third (correction) component dependent parameter. For mixtures, the parameter, c, is related to the third parameter, c_i, for each component through a linear mixing rule such that:

$$c = \sum_{i} c_i z_i \tag{6.67}$$

In Eq. (6.67), z_i is the mole fraction of the component in the mixture. For multicomponent systems, Peneloux et al.[13] showed that volume correction can be applied to the liquid and vapor phases (assuming a biphasic system) as:

$$\widetilde{v}_L = v_L - \sum_{i}^{N_C} c_i x_i \tag{6.68}$$

$$\widetilde{v}_V = v_V - \sum_{i}^{N_C} c_i y_i \tag{6.69}$$

In Eqs. (6.68) and (6.69), $\widetilde{v}_L$ and $\widetilde{v}_V$ are corrected molar volume for the liquid and vapor phases, respectively; v_L and v_V are EOS calculated molar volumes for the same phases; c_i is the third parameter for component i; and x_i and y_i are mole fractions of the component in the liquid and vapor phases, respectively.

The important contribution made by Peneloux et al.[13] is that they showed that vapor-liquid equilibrium conditions determined with the unmodified SRK EOS are not altered by the introduction of volume translation parameters. This is seen readily by relating the fugacity of a component of a mixture in the modified EOS to the original EOS in the vapor and liquid phases yielding:

$$\widetilde{f}_i^V = f_i^V \exp\left(-\frac{c_i p}{RT}\right) \tag{6.70}$$

$$\widetilde{f}_i^L = f_i^L \exp\left(-\frac{c_i p}{RT}\right) \tag{6.71}$$

In Eqs. (6.70) and (6.71), $\widetilde{f}_i^V$ and $\widetilde{f}_i^L$ are the modified fugacities for the vapor and liquid phase, respectively, and f_i^V and f_i^L are the original fugacities for the vapor and liquid phase. From Eqs. (6.70) and (6.71), the fugacity ratio of the modified EOS is equal to the fugacity ratio of the original EOS not modified with volume shift:

$$\frac{\widetilde{f}_i^V}{\widetilde{f}_i^L} = \frac{f_i^V}{f_i^L} \tag{6.72}$$

The equivalence of fugacity ratios in Eq. (6.72) demonstrate that equilibrium conditions determined with the original EOS are not affected by volume translation.

For pure components, Peneloux et al.[13] determined the volume shift parameter, c, by calculating the saturated liquid molar volume, v, at the reduced temperature of 0.7 with the unmodified two-parameter EOS and matching it against the experimental volume, v_E, so that:

$$c = v - v_E \tag{6.73}$$

In Eq. (6.73), $c =$ volume shift parameter; $v =$ saturated liquid molar volume calculated with the unmodified two-parameter EOS; $v_E =$ experimentally measured saturated liquid molar volume.

For correlational purposes, Jhaveri and Youngren[14] introduced a dimensionless shift parameter, s_i, for any component defined as:

$$s_i = \frac{c_i}{b_i} \tag{6.74}$$

In Eq. (6.74), $s_i =$ dimensionless shift parameter for component i; $c_i =$ volume shift parameter for component i; and $b_i =$ second parameter for component i in the unmodified PR EOS. Values of s_i for light hydrocarbons generated by Jhaveri and Youngren[14] are shown in Table 6.1 for PR EOS.

Table 6.1 Volume Shift Parameter for Hydrocarbons (from Jhaveri and Youngren[14] © 1988 SPE, Reproduced with permission)

Component	s_i
Methane	–0.15400
Ethane	–0.10020
Propane	–0.08501
Iso-Butane	–0.07935
n-Butane	–0.06413
Iso-Pentane	–0.04350
n-Pentane	–0.04183
n-Hexane	–0.01478

For heavier hydrocarbons (heptanes$^+$), Jhaveri and Youngren[14] developed a correlation that is dependent on molecular weight. The correlation is:

$$s_i = 1 - \frac{d}{M_i^e} \tag{6.75}$$

In Eq. (6.75), $M_i =$ molecular weight of component i; and d and e are positive correlation coefficients. Values of d and e for paraffins, naphthenes, and aromatics are shown in Table 6.2 for the PR EOS.

In effect, application of volume translation converts two-parameter cubic equations of state to essentially three-parameter equations. This has been shown by Peneloux et al.[13] and Jhaveri and Youngren[14] to improve volumetric calculations with SRK and PR EOS, respectively.

Table 6.2 Volume Shift Parameter Correlation Coefficients for Hydrocarbons Heavier Than Hexane (from Jhaveri and Youngren[14] © 1988 SPE, Reproduced with permission)

	Correlation Coefficient		
Component Type	***d***	***e***	**Average Error (%)**
Paraffins	2.258	0.1823	0.19
Napthenes	3.004	0.2324	0.28
Aromatics	2.516	0.2008	0.24

6.8 Two-Phase Flash Calculation

Consider a biphasic mixture consisting of vapor and liquid phases at equilibrium. Suppose n_t is the total moles in the mixture, with n_V moles in the vapor phase and n_L moles in the liquid phase. If $\mathbb{Z}_i$ is the mole fraction of component i in the mixture, y_i is its mole fraction in the vapor phase, and x_i is its mole fraction in the liquid phase, the following equations represent the material balance of the component in the mixture:

$$n_t = n_V + n_L \quad \textbf{(6.76)}$$

$$n_t\mathbb{Z}_i = n_V y_i + n_L x_i \quad \textbf{(6.77)}$$

For a system in equilibrium, the equilibrium constant, K_i, of a component i is the ratio of the mole fraction of the component in the vapor phase to its mole fraction in the liquid phase, as designated in Eq. (6.77). Thus as previously shown in Eq. (6.63),

$$K_i = \frac{y_i}{x_i} \quad \textbf{(6.78)}$$

Substituting for y_i in Eq. (6.77) with Eq. (6.78) and re arranging, gives:

$$x_i = \frac{n_t\mathbb{Z}_i}{n_V K_i + n_L} \quad \textbf{(6.79)}$$

Substituting for n_L in Eq. (6.79) with Eq. (6.76), we have:

$$x_i = \frac{n_t\mathbb{Z}_i}{n_V(K_i - 1) + n_t} \quad \textbf{(6.80)}$$

Dividing Eq. (6.80) by n_t yields:

$$x_i = \frac{\mathbb{Z}_i}{F_v(K_i - 1) + 1} \quad \textbf{(6.81)}$$

In Eq. (6.81), F_V is the mole fraction of vapor in the mixture where $F_V = n_V/n_t$. Similarly, it can be shown that the mole fraction, y_i, of the same component in the vapor phase is represented by:

$$y_i = \frac{K_i\mathbb{Z}_i}{F_V(K_i - 1) + 1} \quad \textbf{(6.82)}$$

By definition, the sum of the mole fractions of the equilibrated phases and the entire mixture must be equal to one. Hence for N_C components in the mixture:

$$\sum_{i=1}^{N_C}\mathbb{Z}_i = \sum_{i=1}^{N_C} y_i = \sum_{i=1}^{N_C} x_i = 1.0 \quad \textbf{(6.83)}$$

From Eq. (6.83), a constraining equation can be obtained such that:

$$\sum_{i=1}^{N_C} y_i - \sum_{i=1}^{N_C} x_i = 0 \tag{6.84}$$

Substituting Eqs. (6.81) and (6.82) into Eq. (6.84) yields the Rachford-Rice[15] equation:

$$f(F_v) = \sum_{i=1}^{N_C} \frac{z_i(K_i - 1)}{1 + F_v(K_i - 1)} = 0. \tag{6.85}$$

Eq. (6.85) is a monotonically decreasing convergence function that can be solved by root solving techniques, such as successive substitution or the Newton-Raphson method.

6.8.1 Generalized Procedure for Two-Phase Flash Calculations

Step 1: Conduct phase stability check on the mixture.

Phase stability of the mixture is based primarily on the minimization of Gibbs free energy as suggested by Michelsen.[11] A procedure for phase stability check based on the Michelsen's method was presented by Whitson and Brule.[16] If a phase is determined not to be stable, then a flash calculation is initiated. Initial estimates of K-values can be obtained from this step. This step may not be necessary if temperature and pressure conditions of the mixture suggest the existence of a biphasic mixture.

Step 2: Estimate initial K-values.

Initial K-values can be obtained from the Wilson[17] equation:

$$K_i = \frac{\exp\left[5.37\left(1 + \omega_i(1 - T_{ri}^{-1})\right)\right]}{p_{ri}} \tag{6.86}$$

This step can be skipped if K-values were obtained from Step 1.

Step 3: Calculate F_v from Eq. (6.85).

Solving Eq. (6.85) involves iteration, either by successive substitution or Newton-Raphson method, until convergence is reached at a set tolerance. A faster method for performing flash calculations has been proposed by Li and Johns.[18]

Step 4: Compute phase compositions from Eqs. (6.81) and (6.82).

Step 5: Calculate phase molar volumes from Eqs. (6.32) or (6.47).

Step 6: Compute partial component fugacities from Eqs. (6.58) or (6.59).

Step 7: Check for equivalence of partial component fugacities.

In this step, the equality of partial component fugacity in each phase is checked to a set tolerance level.

Step 8: Convergence check.

If convergence is reached in Step 7, stop. If not, update K-values using Eq. (6.63) for each component. Repeat Steps 3 to 7 until convergence is achieved.

6.9 Bubble Point and Dew Point Pressure Calculations

The bubble point pressure or the dew point pressure of a mixture is defined as the pressure at constant temperature at which the mixture is in equilibrium with an infinitesimal amount of the second phase, assuming a biphasic mixture. At bubble point, the mixture consists entirely of liquid except for an infinitesimal amount of gas (vapor). Conversely, at the dew point, the mixture is mostly gas (vapor) with an infinitesimal amount of liquid. At equilibrium, the sum of the mole fractions of the infinitesimal phase is equal to one. Thus, at bubble point:

$$\sum_{i=1}^{N_C} y_i = 1.0 \tag{6.87}$$

And at dew point:

$$\sum_{i=1}^{N_C} x_i = 1.0 \tag{6.88}$$

For Eqs. (6.87) and (6.88), y_i and x_i are mole fractions of component i in the vapor and liquid phases, respectively, and N_C is the total number of components in the mixture.

Substituting Eq. (6.78) into Eqs. (6.87) and (6.88), gives:
For bubble point:

$$\sum_{i=1}^{N_C} y_i = \sum_{i=1}^{N_C} K_i x_i = \sum_{i=1}^{N_C} K_i \mathbb{Z}_i = 1.0 \tag{6.89}$$

In Eq. (6.89), $\mathbb{Z}_i$ is the mole fraction of component i in the mixture, and $x_i \approx \mathbb{Z}_i$ at the bubble point.
Similarly at dew point:

$$\sum_{i=1}^{N_C} x_i = \sum_{i=1}^{N_C} \frac{y_i}{K_i} = \sum_{i=1}^{N_C} \frac{\mathbb{Z}_i}{K_i} = 1.0 \tag{6.90}$$

In Eq. (6.90), $y_i \approx \mathbb{Z}_i$ at the dew point.
Eqs. (6.89) and (6.90) can be converted to a form similar to Eq. (6.85) for convergence calculations. Thus, Eqs. (6.89) and (6.90) become, respectively:

$$\sum_{i=1}^{N_C} K_i \mathbb{Z}_i - 1.0 = 0 \tag{6.91}$$

$$\sum_{i=1}^{N_C} \frac{\mathbb{Z}_i}{K_i} - 1.0 = 0. \tag{6.92}$$

6.10 Characterization of Hydrocarbon Plus Fractions

In many laboratory analyses of reservoir fluid samples, the heavy hydrocarbons are typically lumped together and classified as C_n^+ fraction. In some PVT laboratory reports, the C_n^+ fraction is represented as C_7^+ fraction, while higher carbon number such as C_{10}^+ may be chosen to represent the hydrocarbon plus fractions in other reports. Typically, in most of these PVT reports, only the molecular weight and specific gravity of the C_n^+ fraction are reported. Ideally, a complete True-Boiling-Point (TBP) analyses by distillation should be performed on fluid samples to yield boiling points, specific gravities, and molecular weights for Single-Carbon-Number (SCN) groups, which could be used to characterize the fluid samples. TBP data are rarely available in most PVT reports because of the prohibitive expense and time required to measure them.

Although the mole fraction of the hydrocarbon plus fraction in a fluid sample is relatively small, it has a huge impact on the phase behavior of the fluid. In particular, the mole fractions and properties assigned to pseudo-components that result from splitting the plus fraction can significantly influence the phase behavior predicted for the fluid sample. Numerous methods have been used to split the hydrocarbon plus fraction into pseudo-components. In general, these can be grouped into two methods:

Method 1: The hydrocarbon plus fraction is split into many SCN groups using a probability distribution function or some other criteria. The SCN groups are assigned properties from correlations and regrouped into a few Multiple-Carbon-Number groups (MCN) with various averaging techniques. This method was proposed and applied by Whitson.[19,20]

Method 2: The hydrocarbon plus fraction is split into pseudo-components using semi-continuous thermodynamics as proposed by Behrens and Sandler[21] and adapted by Whitson and Brule.[16] The pseudo-components are assigned properties from correlations.

The splitting of the hydrocarbon plus fraction with semi-continuous thermodynamics has wide appeal in the petroleum industry and is available in most commercial PVT software. Since most engineers will most likely encounter this technique in commercial PVT software, the theoretical basis of the method is presented in this book. Semi-continuous thermodynamics is based on representing a fluid mixture with identifiable, discrete components, and using a continuous distribution function to represent the remaining components in the mixture. Behrens and Sandler[21] used a truncated exponential function as the continuous distribution function in their work. The three-parameter gamma function that was first proposed by Whitson[19] is the probability density function used for the description of the method. Note that any other continuous distribution function could be used in place of the three-parameter gamma function.

The three-parameter gamma function is represented as:

$$h(M) = \frac{(M - \eta)^{(\alpha-1)} \exp\left[\dfrac{M - \eta}{\beta}\right]}{\beta^{\alpha}\Gamma(\alpha)} \tag{6.93}$$

For Eq. (6.93), the distribution is defined by the three parameters, α,β, and η. The parameter, α, represents the shape of the distribution. If $\alpha = 1$, Eq. (6.93) reduces to an exponential distribution. The gamma function is represented as Γ. The parameter, β, is the normalized condition, which is given by:

$$\alpha\beta = M_{C^+_n} - \eta \tag{6.94}$$

In Eq. (6.94), $M_{C_n^+}$ is the average molecular weight of the plus fraction; and η is the lowest molecular weight in the plus fraction. Generally, η is approximated from the empirical relation:

$$\eta = 14n - 6 \tag{6.95}$$

In Eq. (6.95), n is the first SCN group in the plus fraction.
The cumulative distribution function $H(x)$ is the integral of $h(x)$ from η to x:

$$H(x) = \int_{\eta}^{x} h(x)\, dx` \tag{6.96}$$

The numerical form of Eq. (6.96) is given by:

$$H(x) = e^{-y}\sum_{j=0}^{\infty}\left[\frac{y^{\alpha+j}}{\Gamma(\alpha + j + 1)}\right], \quad \text{where } y = (x - \eta)/\beta \tag{6.97}$$

The frequency, ξ_i, of a component having molecular weight between boundaries M_{i-1} and M_i is given by the integral:

$$\xi_i = \int_{M_{i-1}}^{M_i} h(M)dM = H(M_i) - H(M_{i-1}) \tag{6.98}$$

The mole fraction, z_i, is calculated by multiplying the mole fraction of the plus fraction, z_n^+, by the respective frequency from Eq. (6.98):

$$z_i = z_n^+ \times \xi_i \tag{6.99}$$

The average molecular weight in the same interval is given by:

$$\overline{M_i} = \eta + \alpha\beta\frac{H(M_i,\alpha + 1) - H(M_{i-1},\alpha + 1)}{H(M_i,\alpha) - H(M_{i-1},\alpha)} \tag{6.100}$$

The above equations represent the basic application of the three-parameter gamma function for splitting of plus fraction into many SCN groups.

Whitson et al.[22] adapted the gamma distribution model with Gaussian quadrature using the principles of semi-continuous thermodynamics to characterize multiple fluid samples from the same reservoir simultaneously. Each fluid sample can have its own molecular weight and specific gravity for the C_n^+ fraction before the split. After the split, each split fraction (pseudo-component) will have the same molecular weight and specific gravity while the mole fraction distribution will be different for each fluid sample. The procedure for applying Gaussian quadrature to the gamma distribution was modified from the procedure proposed by Whitson and Brule.[16]

1. Determine the number of C_n^+ split fractions, N. Obtain the quadrature values X_i and W_i from Table 6.3.
2. Specify η and α. Assume $\eta = 92$ and $\alpha = 1$.
3. Specify the heaviest molecular weight of fraction N as M_N. A good estimate is $M_N = 2.5M_{C_n^+}$.
4. Calculate the parameter δ, where $\delta = \exp\left(\dfrac{\alpha\beta^*}{M_{C_n^+} - \eta} - 1\right)$ and $\beta^* = (M_N - \eta)/X_N$
5. Calculate the, C_n^+, mole fraction, z_i, and molecular weight, M_i, for each fraction

$$z_i = z_{C_n^+} \times \left[W_i f(X_i)\right] \tag{6.101}$$

$$M_i = \eta + \beta^* X_i \tag{6.102}$$

$$f(X) = \frac{(X)^{\alpha-1}(1 + \ln \delta)^{\alpha}}{\Gamma(\alpha) \qquad \delta^X} \tag{6.103}$$

Table 6.3 Gaussian Quadrature Variables and Weight Factors

Integration Points (Pseudo-components)	Function Variables X_i	Weight Factors W_i
	Two Quadrature Points	
1	0.5858	0.8536
2	3.4142	0.1464
	Three Quadrature Points	
1	0.4158	0.7111
2	2.2943	0.2785
3	6.2899	0.0104
	Five Quadrature Points	
1	0.2636	0.5218
2	1.4134	0.3987
3	3.5964	0.07594
4	7.0858	0.003612
5	12.6408	0.00002337

6. Check whether the condition $\sum_{i=1}^{N} z_i M_i = M_{C_n^+}$ is satisfied. Modify the value of δ and repeat Steps 5 and 6 until a reasonable match is achieved.

7. For each split fraction, calculate specific gravity assuming an average Watson Characterization Factor, $\overline{K}_W$. From Whitson:[20]

$$\gamma_i = 6.0108 M_i^{0.17947} K_W^{-1.18241} \tag{6.104}$$

In Eq. (6.104), K_W is obtained from an approximation by Whitson:[19]

$$K_W = 4.5579 M_{C_n^+}^{0.15178} \gamma_{C_n^+}^{-0.84573} \tag{6.105}$$

8. Estimate boiling points for each split fraction from the equation:

$$T_{bi} = (K_w \gamma_i)^3 \tag{6.106}$$

9. With known boiling points, specific gravities, and molecular weights for each split fraction (pseudo-component), the characterization of the hydrocarbon plus fractions is completed by calculating critical properties for each pseudo-component using the Kessler-Lee correlations (Eqs. 4.32 and 4.33). The Riazi-Daubert[23] correlations can also be used. From Riazi-Daudert correlations, the correlation for estimating critical pressure is:

$$p_{ci} = (3.12281 \times 10^9) \times T_{bi}^{-2.3125} \times \gamma_i^{2.3201} \tag{6.107}$$

For critical temperature, the correlation is:

$$T_{ci} = 24.27871 \times T_{bi}^{0.58848} \times \gamma_i^{0.3596} \tag{6.108}$$

In Eqs. (6.107) and (6.108), T_{bi} is the boiling point of the pseudo-component in degrees Rankine; and γ_i is the specific gravity of the pseudo-component.

The above procedure is superior to other methods of splitting plus hydrocarbon fractions because multiple fluid samples from the same reservoir could be split together to produce a common set of pseudo-components with the same molecular weights and critical properties but different mole fractions representing each fluid sample. This is essential for the generation of a single "tuned" equation of state model for compositional simulation.

6.11 Phase Equilibrium Predictions with Equations of State

This section discusses the use of equations of state to predict phase equilibrium of reservoir fluids for compositional simulation. The PR EOS and SRK EOS were presented earlier in this chapter because of their simplicity, robustness, and popularity in the petroleum industry, especially in applications that require compositional simulations of petroleum reservoirs. Many commercial software programs that are available for analyses of fluid samples and generation of PVT data input for simulators usually have options for selection of either equation of state. Fundamental topics that laid the foundation for the use of PR and SRK equations of state in phase equilibrium calculations were presented in this chapter to improve understanding of commercial PVT software programs. The key topics that were reviewed are as follows:

1. The concept of fugacity and equivalence of partial fugacity of all components in all phases of a mixture in equilibrium.
2. The significance of calculation of roots of cubic equations of state.
3. Improvement of volumetric predictions with volume shifts from cubic equations of state.
4. The basic equations for flash calculations in compositional simulations.
5. Characterization of the hydrocarbon plus fractions to improve phase equilibrium predictions with equations of state.

Most equations of state used in compositional reservoir simulation are cubic in form. These types of equations of state have been shown to adequately represent the phase behavior of reservoir fluids over *normal* ranges of temperatures and pressures encountered in reservoir processes. But before equations of state can be used in compositional reservoir simulations, they must be adjusted to represent the reservoir fluids under varying processes as temperatures and pressures change. The adjustments, sometimes called "tuning," involve changing the physical properties and other empirical parameters of the components of the fluid system that represent the reservoir fluid mixture. But before getting to "tuning" of equations of state, there are basic steps necessary in preparing an equation of state for use in analyses of PVT data. These steps are summarized as follows:

1. Assemble the PVT data to be used in the analyses.
   a. Evaluate the quality of the PVT data for consistency, errors, and poor laboratory practice.
   b. Check compositions of fluid samples for contamination with mud filtrate, especially if the sample was collected from a wellbore drilled with oil-based mud. Decontaminate the fluid samples if necessary using the techniques described in Chapter 5.
   c. Tabulate the type of PVT data measured for each fluid sample, such as constant composition expansion (CCE) data, constant volume depletion (CVD) data, differential liberation (DL) data, separator tests data, etc.
   d. Decide the sets of data to be used in "tuning" the EOS, and the sets of data to be used in validating the tuned EOS.
2. Estimate the total number of components to be used to model the reservoir fluid. An initial estimate of twelve components is recommended. Note that the number of components could be increased or reduced during the process of matching experimental PVT data by adjusting the properties of the components as represented in the equation of state fluid model. The effects of number of components on phase equilibria predictions are illustrated in Figures 6.1 and 6.2.

a. Select the number of pure hydrocarbons to represent the lighter hydrocarbons such as methane, ethane, propane, butane, pentane, and hexane. If non-hydrocarbon gases such as carbon dioxide, nitrogen, and hydrogen sulfide are present, these must be represented as components depending on their concentrations. The intermediate components (C_2 to C_6) can be grouped into two pseudo-components if predictions of liquid production from these groups of hydrocarbons by surface separation are not required.

b. Select the number of pseudo-components to represent the hydrocarbon plus fraction. This depends on the extent of the compositional analyses of the fluid samples. For instance, if the composition, molecular weight, and specific gravity of C_{10+} were reported, the splitting of the hydrocarbon plus fraction into pseudo-components could be started from this point. Generally, the hydrocarbon plus fraction is split into three or more pseudo-components. For this example with C_{10+} fraction, the hydrocarbons in the C_7 to C_9 range could be grouped into a pseudo-component. It is important to note that selection and grouping of components depend on the surface and reservoir processes that are modeled with compositional simulation. The number of components selected from the non-hydrocarbon group, the pure hydrocarbon group, and the pseudo-components group should be designed to represent the processes that will be modeled.

c. The technique of semi-continuous thermodynamics with Gaussian quadrature is recommended for splitting the hydrocarbon plus fraction into pseudo-components.

3. Compute the properties (T_c, P_c, ω) for the components.

a. For the pure hydrocarbons, the properties are taken from published data such as Table 4.1.

b. For the pseudo-components obtained from splitting of the hydrocarbon-plus fractions, the properties are calculated from correlations as discussed in Section 6.10.

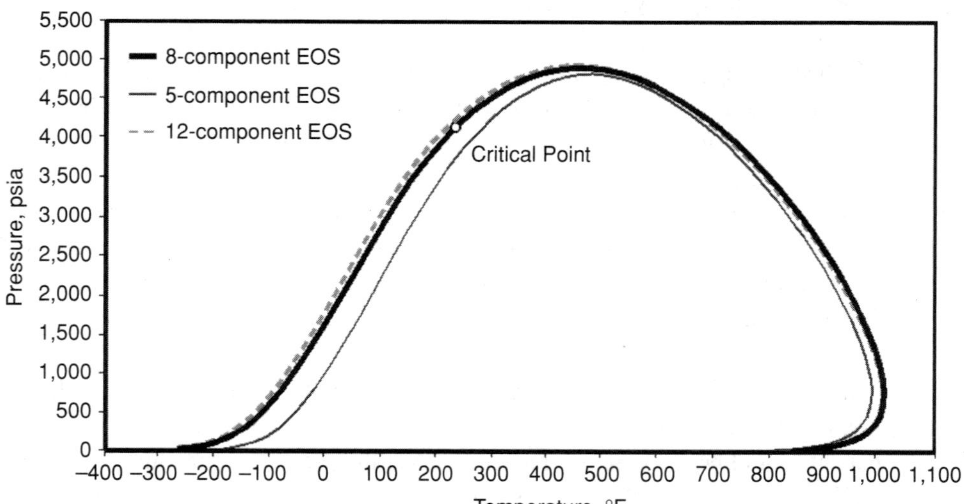

Figure 6.1 Effects of number of components on phase envelope of an actual gas condensate (from Wang and Pope[28] © 2001 SPE, Reproduced with permission).

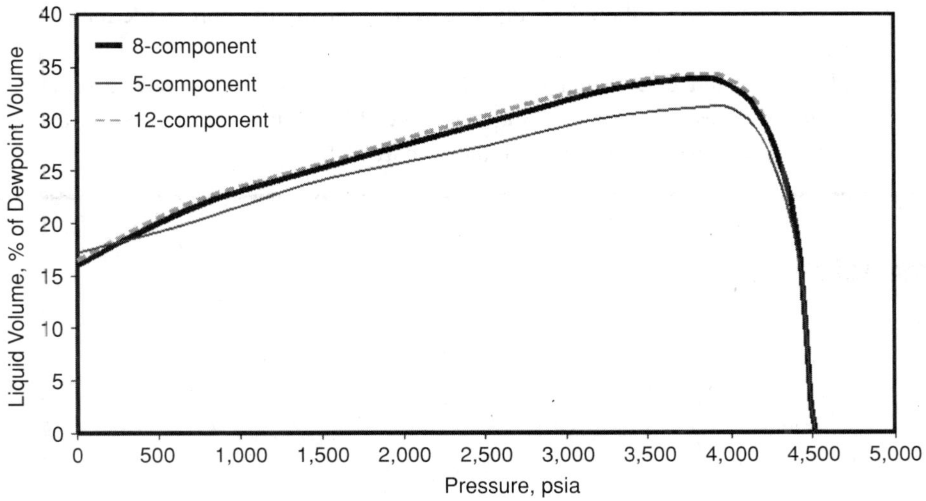

Figure 6.2 Effects of number of components on liquid dropout (from Wang and Pope[28] © 2001 SPE, Reproduced with permission).

c. For the grouped components, the properties are calculated from application of one of three averaging methods as presented by Joergensen and Stenby:[24]
- Weight averaging
- Molar averaging
- Mixing rule

d. Note that the averaging step is embedded in all commercial software and performed almost seamlessly. However, it may be necessary to select the correlations you prefer for calculation of the properties in Step 3b and the type of averaging method to apply in Step 3c.

4. Use the EOS model in current form to compute experimental PVT data and compare results. Determine if tuning EOS parameters is necessary.

5. Tune EOS parameters. This process is generally based on experience since there are many parameters that can be adjusted to tune an EOS.[25,26] The two main approaches are:

a. Adjustment of critical properties of the pseudo-components.[27,28]

b. Adjustment of the binary interaction parameters between the light hydrocarbons and the pseudo-components. Also, the binary interaction parameters between the non-hydrocarbons and the pseudo-components may be adjusted.[29,30]

The process of tuning an EOS should proceed in a systematic and organized manner to avoid generating an EOS model that has limited predictive capacity. Most commercial software programs provide computational facilities for performing non-linear regression on the adjusted parameters to minimize the differences between measured and predicted PVT data. The output from the regression analyses should be examined to determine the group of parameters that are most sensitive to adjustments. The most sensitive parameters should be adjusted systematically to improve the match between measured and predicted PVT data.

6. Validate the tuned EOS with other PVT data not used in the tuning process. It is important to test the tuned EOS against other PVT data such as swelling test, multi-contact, and slim-tube data if such data are available.[31]

Nomenclature

c	volume shift parameter
f	fugacity
$\tilde{f}$	modified fugacity
F_v	mole fraction of vapor in mixture
g^*	normalized Gibbs energy
K	equilibrium constant
K_W	Watson characterization factor
k_{ij}	binary interaction parameter
M	molecular weight
N	number of phases
N_C	number of components
n	moles of substance
p	pressure
p_c	critical pressure
p_r	reduced pressure
p_v	vapor pressure
R	universal gas constant
s	molar entropy or dimensionless shift parameter in volume shift
T	temperature
T_b	boiling point temperature
T_c	critical temperature
T_r	reduced temperature
V	volume
V_c	critical volume
v	molar volume
$\overline{v}$	corrected molar volume
x	liquid phase mole fraction
y	vapor phase mole fraction
z	gas compressibility factor
$\mathbb{Z}$	total mole fraction of component in mixture
z_c	critical compressibility factor

α	shape parameter in gamma function
β	nomalization parameter in gamma function
β^*	modified nomalization parameter in gamma function
η	parameter in gamma function
Γ	gamma function
ϕ	fugacity coefficient
μ	chemical potential
ω	acentric factor
γ	specific gravity
ξ	frequency of a component having molecular weight between certain boundaries

Subscripts

E	experimental data
i,j	components in mixture
C	component
c	critical property
V	vapor phase
L	liquid phase

Superscripts

V	vapor phase
L	liquid phase

Abbreviations

BIP	Binary Interaction Parameter
EOS	Equation of State
MCN	Multiple Carbon Number
PR	Peng-Robinson
PVT	Pressure-Volume-Temperature
RK	Redlich-Kwong
SCN	Single Carbon Number
SRK	Soave-Redlich-Kwong
TBP	True-Boiling-Point
vdW	van der Waals

References

1. van der Waals, J.D.: *On the Continuity of the Gas and Liquid State*. Dissertation, Leiden, The Netherlands (1873).
2. Walas, S.M.: *Phase Equilibria in Chemical Engineering,* Butterworth Publishers, Stoneham, MA (1985).
3. Redlich, O., and Kwong, J.N.S.: "On the Thermodynamics of Solutions: V. An Equation of State. Fugacities of Gaseous Solutions," *Chem. Review* (1949) 44, 233–244.
4. Soave, G.: "Equilibrium Constants from a modified Redlich-Kwong Equation of State," *Chem. Eng. Sci.* (1972) 27, 6, 1197–1203.
5. Graboski, M.S., and Daubert, T.E.: "A Modified Soave Equation of State for Phase Equilibrium Calculations. 1. Hydrocarbon Systems," *Ind. Eng. Chem. Process Des. Dev.* (1978) 17, 4, 443–448.
6. Graboski, M.S., and Daubert, T.E.: "A Modified Soave Equation of State for Phase Equilibrium Calculations. 2. Systems Containing CO_2, H_2S, N_2 and CO," *Ind. Eng. Chem. Process Des. Dev.* (1978) 17, 4, 448–454.
7. Pitzer, K.S., Lippmann, D.Z., Curl, R.F., Jr., Huggins, C.M., and Peterson, D.E.: "The Volumetric and Thermodynamic Properties of Fluids. II. Compressibility Factor, Vapor Pressure and Entropy of Vaporization," *J. Am. Chem. Soc.* (1955) 77, 13, 3433–3440.
8. Peng, D.-Y., and Robinson, D.B.: "A New Two-Constant Equation of State," *Ind. Eng. Chem. Fundamentals* (1976) 15, 1, 59–64.
9. Robinson, D.B., Peng, D.-Y., and Ng, H.-Y: "Capabilities of the Peng-Robinson Programs, Part 2: Three-Phase and Hydrate Calculations," *Hydrocarbon Proc.* (1979) 58, 269.
10. Robinson, D.B., and Peng, D.-Y.: "The Characterizations of the Heptanes and Heavier Fractions," Research Report 28, Gas Producers Assn., Tulsa, Oklahoma (1978).
11. Michelsen, M.L.: "The Isothermal Flash Problem. Part II. Phase-Split Calculation," *Fluid Phase Equilibria* (1982) 9, 21.
12. Martin, J.J.: "Cubic Equations of State-Which?" *Ind. & Eng. Chem. Fund.* (1979) 18, 2, 81–97.
13. Peneloux, A., Rauzy, E., and Freze, R: "A Consistent Correction for Redlich-Kwong-Soave Volumes," *Fluid Phase Equilibria* (1982) 8, 7–23.
14. Jhaveri, B.S., and Youngren, G.K.: "Three-Parameter Modification of the Peng-Robinson Equation of State to Improve Volumetric Predictions," *SPERE* (August 1988) 1033–1040.
15. Rachford, H.H., and Rice, J.D.: "Procedure for Use of Electrical Digital Computers in Calculating Flash Vaporization Hydrocarbon Equilibrium," *JPT* (October 1952) 19; *Trans.*, AIME, 195.
16. Whitson, C.H., and Brule, M.R.: *Phase Behavior,* SPE Monograph Volume 20, Society of Petroleum Engineers, Richardson, Texas, 2000.

17. Wilson, G.M.: "A Modified Redlich-Kwong Equation of State, Application to General Physical Data Calculations," paper 15c presented at the 1969 AICHE Natl. Meeting, Cleveland, Ohio.

18. Li, Y., and Johns, R.T.: "Rapid Flash Calculations for Compositional Simulation," *SPEREE* (October 2006) 521–529.

19. Whitson, C.H.: "Characterizing Hydrocarbon Plus Fractions," *SPEJ.* (August 1983) 683–694.

20. Whitson, C.H.: "Effect of C_{7+} Properties on Equation-of-State Predictions," *SPEJ* (December 1984) 685–696.

21. Behrens, R.A., and Sandler, S.I.: "The Use of Semicontinuous Description to Model the C_{7+} Fraction in Equation of State Calculations," *SPERE* (August 1988) 1041–1047.

22. Whitson, C.H., Andersen, T.F., and Soreide, I.: "C_{7+} Characterization of Related Equilibrium Fluids Using the Gamma Distribution," *C_{7+} Fraction Characterization*, L.G. Chorn, and G.A. Mansoori (eds.), Advances in Thermodynamics, Taylor and Francis, New York (1989) 1, 35–56.

23. Riazi, M.R., and Daubert, T.E.: "Simplify Property Predictions," *Hydro. Proc.* (March 1980) 115.

24. Joergensen, M., and Stenby, E.H.: "Optimization of Pseudo-component Selection for Compositional Studies of Reservoir Fluids," paper SPE 30789 presented at the 1995 SPE Annual Technical Conference and Exhibition, Dallas, Texas, October 22–25.

25. Agarwal, R.K, Li, Y.-K., and Nghiem, L.: "A Regression Technique with Dynamic Parameter Selection for Phase-Behavior Matching," *SPERE*, (February 1990) 115–120.

26. Merrill, R.C., Hartman, K.J., and Creek, J.L.: "A comparison of Equation of State Tuning Methods," paper SPE 28589 presented at the 1994 SPE Annual Technical Meeting and Exhibition, New Orleans, September 26–28.

27. Coats, K.H., and Smart, G.T.: "Application of a Regression-Based EOS PVT Program to Laboratory Data," *SPERE* (May 1986) 277–299.

28. Wang, P., and Pope, G.A.: "Proper Use of Equations of State for Compositional Reservoir Simulation," SPE 69071, Distinguished Author Series, (July 2001) 74–81.

29. Slot-Petersen, C: "A Systematic and Consistent Approach to Determine Binary Interaction Coefficients for the Peng-Robinson Equation of State," *SPERE* (November 1989) 488–494.

30. Ezekwe, J.N.: "Effect of Paraffinic, Naphthenic and Aromatic Distribution in the Hydrocarbon Mixture and Water on the Phase Equilibria of Carbon Dioxide-Hydrocarbon Systems Over the Temperature Range from 333 K TO 366 K," Ph.D. Dissertation, University of Kansas, 1982.

31. Stalkup, F., and Yuan, H.: "Effect of EOS Characterization on Predicted Miscibility Pressure," paper SPE 95332 presented at the 2005 SPE Annual Technical Conference and Exhibition, Dallas, October 9–12.

CHAPTER 7

The General Material Balance Equation

7.1 Introduction

The General Material Balance Equation (GMBE) as developed in this book is based on an oil reservoir with a primary gas cap at initial conditions and reservoir pressure designated as p_i. At a later time, t, reservoir pressure is assumed to have been reduced from p_i to p from production of oil, water, and gas. During the production period, it is assumed that there was water influx into the reservoir from an aquifer. It is also assumed that water and/or gas was injected into the reservoir. These conditions are illustrated in Figure 7.1.

The description of the equation that is developed as a material balance equation is in a strict sense not accurate. Actually, it is a volumetric balance based on reservoir barrels instead of a material balance based on mass. But in this book, it is called the GMBE to conform to the generally accepted terminology used in the petroleum industry. The GMBE is derived from the basis that the reservoir pore volume as existing at its initial conditions is constant. Thus, the sum of the volume changes of gas, oil, water, and the formation rock caused by fluid production is equal to zero.

7.2 Derivation of the General Material Balance Equation (GMBE)

The GMBE can be represented volumetrically in reservoir barrels as:

$$\begin{aligned}\text{Net change in oil and water production volume} = {} & \text{Gas cap volume change} + \text{Released solution gas} \\ & + \text{Oil volume change} + \text{Connate water expansion} \\ & + \text{Rock expansion}\end{aligned} \tag{7.1}$$

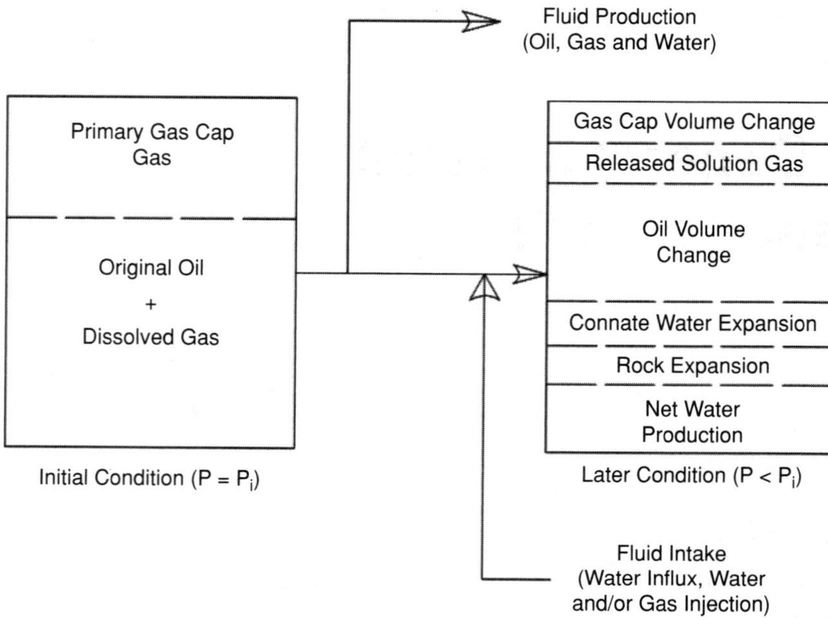

Figure 7.1 Material balance model for an oil reservoir.

Before developing the terms of the expression for the GMBE in Eq. (7.1), it is necessary to derive other expressions that apply to oil reservoirs with primary gas caps. The ratio of original reservoir gas cap volume and the original reservoir oil zone volume is defined as:

$$m = \frac{\text{Original reservoir gas cap volume}}{\text{Original reservoir oil zone volume}} = \frac{GB_{gi}}{NB_{oi}} \tag{7.2}$$

In Eq. (7.2), G = original gas in place (OGIP) in the gas cap, scf; N = original oil in place (OOIP) in the oil zone, STB; B_{gi} = initial gas formation volume factor (FVF), RB/scf; B_{oi} = initial oil FVF, RB/STB. From Eq. (7.2), $NmB_{oi} = GB_{gi}$.

7.2.1 Development of Terms in the Expression of Equation (7.1)

1. Net change in oil production volume:

$$\text{Net change in oil production volume (RB)} = N_pB_o \tag{7.3}$$

In Eq. (7.3), N_p = cumulative oil production, STB; and B_o = oil FVF, RB/STB, at current reservoir pressure, p.

2. Net change in water production volume:

$$\text{Net change in water production volume (RB)} = W_p B_w - W_e - W_{inj} B_w \tag{7.4}$$

In Eq. (7.4), W_p = cumulative water production, STB; W_e = cumulative water influx, RB; W_{inj} = cumulative water injection, STB; and B_w = water FVF, RB/STB, at current reservoir pressure, p.

3. Gas cap volume change;

$$\text{Gas cap volume change (RB)} = (G - G_p + G_{inj})B_g - GB_{gi} \tag{7.5}$$

In Eq. (7.5), G_p = cumulative gas production, scf; G_{inj} = cumulative gas injection, scf; and B_g = gas FVF, RB/scf. Expressing Eq. (7.5) in terms of N by using Eq. (7.2) and the relationship, $G_p = N_p R_p$:

$$\text{Gas cap volume change (RB)} = \left(\frac{NmB_{oi}}{B_{gi}} - N_p R_p + G_{inj}\right)B_g - NmB_{oi} \tag{7.6}$$

In Eq. (7.6), R_p = cumulative production gas-oil ratio, scf/STB.

4. Solution gas released into the gas cap:

$$\begin{aligned}\text{Released solution gas(RB)} &= [NR_{si} - (N - N_p)R_s]B_g \\ &= [N(R_{si} - R_s) + N_p R_s]B_g\end{aligned} \tag{7.7}$$

In Eq. (7.7), R_{si} = initial solution gas-oil ratio (GOR), scf/STB; and R_s = solution GOR, scf/STB, at current reservoir pressure, p.

5. Oil volume change:

$$\text{Oil volume change } (RB) = NB_o - NB_{oi} \tag{7.8}$$

6. Connate water and rock expansion:

$$\text{Reduction in Hydrocarbon Pore Volume (HCPV)} = \text{Connate water and rock expansion} \tag{7.9}$$

$$-d(\text{HCPV}) = (c_w V_w + c_f V_p)(p_i - p) \tag{7.10}$$

$$V_w = V_p S_{wi} \tag{7.11}$$

In Eqs. (7.10) and (7.11), c_w = water compressibility, psi^{-1}; c_f = rock compressibility, psi^{-1}; V_w = connate water volume, RB; V_p = total pore volume, RB; p_i = initial reservoir pressure, psi; p = reservoir pressure in psi at time, t; and S_{wi} = initial connate water saturation, fraction.

$$V_p = \frac{\text{HCPV}}{1 - S_{wi}} \tag{7.12}$$

$$\begin{aligned}\text{HCPV} &= \text{initial oil reservoir volume} + \text{initial reservoir gas volume} \\ &= NB_{oi} + NmB_{oi} = (1 + m)NB_{oi}\end{aligned} \tag{7.13}$$

From Eqs. (7.12) and (7.13),

$$V_p = \frac{(1 + m)NB_{oi}}{1 - S_{wi}} \tag{7.14}$$

$$V_w = \frac{(1 + m)NB_{oi}S_{wi}}{1 - S_{wi}} \tag{7.15}$$

Substituting Eqs. (7.14) and (7.15) into Eq. (7.10) give;

$$\begin{aligned} -d(\text{HCPV}) &= (1 + m)NB_{oi}\left[\frac{c_w S_{wi} + c_f}{1 - S_{wi}}\right](p_i - p) \\ &= (1 + m)NB_{oi}\left[\frac{c_w S_{wi} + c_f}{1 - S_{wi}}\right]\Delta p \end{aligned} \tag{7.16}$$

Substituting Eqs. (7.3), (7.4), (7.6), (7.7), (7.8) and (7.16) into Eq. (7.1) gives:

$$\begin{aligned} N_pB_o + W_pB_w - W_e - W_{inj}B_w = {} & \left(\frac{NmB_{oi}}{B_{gi}} - N_pR_p + G_{inj}\right)B_g - NmB_{oi} \\ & + [N(R_{si} - R_s) + N_pR_s]B_g + NB_o - NB_{oi} \\ & + (1 + m)NB_{oi}\left[\frac{c_w S_{wi} + c_f}{1 - S_{wi}}\right]\Delta p \end{aligned} \tag{7.17}$$

By expanding and arranging like terms, Eq. (7.17) becomes:

$$\begin{aligned} & NB_{oi} - NB_o + N_pB_o + NmB_{oi} - \frac{NmB_{oi}B_g}{B_{gi}} + N_pR_pB_g - NR_{si}B_g + NR_sB_g - N_pR_sB_g \\ & = G_{inj}B_g + W_{inj}B_w + W_e - W_pB_w + (1 + m)NB_{oi}\left[\frac{c_w S_{wi} + c_f}{1 - S_{wi}}\right]\Delta p \end{aligned} \tag{7.18}$$

Adding the expression $N_pB_gR_{si}$ to both sides of Eq. (7.18) gives:

$$\begin{aligned} & NB_{oi} - NB_o + N_pB_o + NmB_{oi} - \frac{NmB_{oi}B_g}{B_{gi}} + N_pR_pB_g - NR_{si}B_g + NR_sB_g - N_pR_sB_g + N_pB_gR_{si} \\ & = G_{inj}B_g + W_{inj}B_w + W_e - W_pB_w + (1 + m)NB_{oi}\left[\frac{c_w S_{wi} + c_f}{1 - S_{wi}}\right]\Delta p + N_pB_gR_{si} \end{aligned} \tag{7.19}$$

Grouping terms in Eq. (7.19) yields:

$$\begin{aligned} & NB_{oi} + NmB_{oi} - N[B_o + (R_{si} - R_s)B_g] + N_p[B_o + (R_{si} - R_s)B_g] + N_p(R_p - R_{si})B_g \\ & - \frac{NmB_{oi}B_g}{B_{gi}} = G_{inj}B_g + W_{inj}B_w + W_e - W_pB_w + (1 + m)NB_{oi}\left[\frac{c_w S_{wi} + c_f}{1 - S_{wi}}\right]\Delta p \end{aligned} \tag{7.20}$$

The two-phase FVF, B_t, is defined as:

$$B_t = B_o + (R_{si} - R_s)B_g$$
$$\text{and } B_{ti} = B_{oi} \text{ at initial conditions} \tag{7.21}$$

Substituting Eq. (7.21) into Eq. (7.20) gives:

$$N(B_{ti} - B_t) + N_p[B_t + (R_p - R_{si})B_g] + NmB_{ti}\left[1 - \frac{B_g}{B_{gi}}\right]$$
$$= G_{inj}B_g + W_{inj}B_w + W_e - W_pB_w + (1 + m)NB_{oi}\left[\frac{c_wS_{wi} + c_f}{1 - S_{wi}}\right]\Delta p \tag{7.22}$$

Eq. (7.22) is the General Material Balance Equation (GMBE). Further re-arrangement of Eq. (7.22) gives a useful form of the GMBE:

$$N(B_t - B_{ti}) + \frac{NmB_{ti}}{B_{gi}}[B_g - B_{gi}] + (1 + m)NB_{ti}\left[\frac{c_wS_{wi} + c_f}{1 - S_{wi}}\right]\Delta p + G_{inj}B_g + W_{inj}B_w + W_e$$
$$= N_p[B_t + (R_p - R_{si})B_g] + W_pB_w \tag{7.23}$$

Another useful form of Eq. (7.22) is re-arranged as:

$$N_p[B_t + (R_p - R_{si})B_g] + W_pB_w - G_{inj}B_g - W_{inj}B_w$$
$$= N\left[(B_t - B_{ti}) + (1 + m)B_{ti}\left(\frac{c_wS_{wi} + c_f}{1 - S_{wi}}\right)\Delta p + \frac{mB_{ti}}{B_{gi}}(B_g - B_{gi})\right] + W_e \tag{7.24}$$

7.3 The GMBE for Gas Reservoirs

The GMBE was developed for a saturated oil reservoir with a primary gas cap. However, it can be modified to apply to gas reservoirs (dry gas, wet gas, and gas condensate reservoirs). For gas condensate reservoirs, there is the requirement that the gas initially exists as a single phase (that means no condensation has occurred in the reservoir) before the GMBE can be applied. Starting from Eq. (7.24) and expanding all terms, the result is:

$$N_pB_t + N_pR_pB_g - N_pR_{si}B_g + W_pB_w - G_{inj}B_g - W_{inj}B_w$$
$$= NB_t - NB_{ti} + (NB_{ti} + NmB_{ti}) \times \left[\frac{c_wS_{wi} + c_f}{1 - S_{wi}}\right]\Delta p + \frac{NmB_{ti}}{B_{gi}}(B_g - B_{gi}) + W_e \tag{7.25}$$

From Eq. (7.2), $NmB_{ti} = GB_{gi}$. Also, $G_p = N_pR_p$. Substituting in Eq. (7.25) gives:

$$N_pB_t + G_pB_g - N_pR_{si}B_g + W_pB_w - G_{inj}B_g - W_{inj}B_w$$
$$= NB_t - NB_{ti} + (NB_{ti} + GB_{gi}) \times \left[\frac{c_wS_{wi} + c_f}{1 - S_{wi}}\right]\Delta p + G(B_g - B_{gi}) + W_e \tag{7.26}$$

Since at initial conditions, it is assumed there is no oil in the gas reservoir, $N = 0$ and $N_p = 0$. Also, it is assumed that there is no water or gas injection. Thus, Eq. (7.26) reduces to:

$$G_pB_g + W_pB_w = GB_{gi}\left(\frac{c_wS_{wi} + c_f}{1 - S_{wi}}\right)\Delta p + G(B_g - B_{gi}) + W_e \quad \textbf{(7.27)}$$

Eq. (7.27) is the general material balance equation for gas reservoirs. It is applied in Chapter 8 to derive equations for volumetric and geopressured gas reservoirs.

7.4 Discussion on the Application of the GMBE

Schilthius[1] published the first GMBE in 1936. In 1953, van Everdingen et al.[2] applied a linear form of the GMBE to a partial water-drive reservoir. The full potential of the GMBE as a straight line equation was developed by Havlena and Odeh[3,4] in classical papers published in 1963 and 1964. Since then, there has been a large body of work on the applications of the GMBE on many reservoir engineering problems published in the literature.

The GMBE is a valuable, analytical tool for evaluation of reservoir drive mechanisms. Even though it is zero dimensional, it offers tools for gaining diagnostic insights into the processes occurring in the reservoir. Some of these diagnostic applications of the GMBE are presented in Chapter 9 using the straight line method of Havlena and Odeh.[3,4] Computerized applications of the Havlena and Odeh[3] method have been published in the literature by Wang and Teasdale[5] for gas reservoirs, and Wang et al.[6] for oil reservoirs. Computer software for GMBE applications are also available commercially. The applications provide the reservoir engineer with powerful tools to gain in-depth knowledge of the key processes influencing reservoir performance before embarking on more rigorous analyses.

The most rigorous method for reservoir evaluation and analyses is reservoir modeling (also literally termed reservoir simulation). The application of reservoir modeling is widespread in the petroleum industry. In contrast to GMBE, which is zero dimensional, reservoir modeling is multi-dimensional. The application of reservoir modeling to petroleum reservoirs is presented in Chapters 18 and 19.

With the availability of fast, powerful digital computers with large storage capacity for reservoir modeling, questions have been raised on the utility of material balance methods for reservoir analyses. The debate must be engaged bearing in mind that material balance methods were mostly developed before the advent of modern computers. Furthermore, all the analyses that can be performed with material balance methods can be replicated in many cases more rigorously with reservoir modeling.

However, material balance methods are still very useful and are incredibly simple tools for gaining an understanding of the reservoir processes before undertaking the more elaborate and time-consuming task of reservoir modeling. In practical terms, material balance methods and reservoir modeling should be viewed as complementary tools in the toolkit of the engineer for reservoir analyses. If the reservoir engineer is unsure of the reservoir drive mechanisms, it is advisable to start the process of analyzing the performance of the reservoir with material balance methods, and later advance to reservoir modeling as more knowledge of the reservoir drive mechanisms is gained. In practice, material balance methods can be of assistance in determining the type of reservoir model to construct, and assessing the quality of reservoir performance data to use in conducting reservoir modeling.

Nomenclature

B_g	gas formation volume factor, RB/scf
B_o	oil formation volume factor, RB/STB
B_t	two-phase formation volume factor, RB/STB
B_w	water formation volume factor, RB/STB
c_f	formation compressibility factor, psi^{-1}
c_w	water compressibility factor, psi^{-1}
G	original gas in place, scf
G_{inj}	cumulative gas injection, scf
G_p	cumulative gas produced, scf
m	ratio original reservoir gas volume to original reservoir oil volume
N	original oil in place, STB
N_p	cumulative oil production, STB
p	pressure, psi
R_p	cumulative production gas-oil ratio, scf/STB
R_s	solution gas-oil ratio, scf/STB
S_w	water saturation, fraction
V_p	total pore volume, RB
V_w	connate water volume, RB
W_e	cumulative water influx, RB
W_{inj}	cumulative water injection, STB
W_p	cumulative water production, STB

Subscripts

i	initial
o	oil or condensate
p	produced

Abbreviations

GMBE	General Material Balance Equation
HCPV	Hydrocarbon pore volume
OGIP	Original gas in place

References

1. Schilthius, R.J.: "Active Oil and Reservoir Energy," *Trans.* AIME (1936) 148, 33–51.
2. van Everdingen, A.F, Timmerman, E.H., and McMahon, J.J: "Application of the Material Balance Equation to a Partial Water-Drive Reservoir," *Trans.* AIME (1953) 198, 51–60.
3. Havlena, D., and Odeh, A.S.: "The Material Balance as an Equation of a Straight Line," *JPT* (August 1963) 896–900.
4. Havlena, D., and Odeh, A.S.: "The Material Balance as an Equation of a Straight Line—Part II, Field Cases," *JPT* (July 1964) 815–822.
5. Wang, B., and Teasdale, T.S.: "GASWAT-PC: A Microcomputer Program for Gas Material Balance With Water Influx," paper SPE 16484 presented at the 1987 SPE Petroleum Industry Applications of Microcomputers, Lake Conroe, Montgomery, Texas, June 23–26, 1987.
6. Wang, B., Litvak, B.L., and Bowman, G.W.: "OILWAT: Microcomputer Program for Oil Material Balance With Gas Cap and Water Influx," paper SPE 24437 presented at the 1992 SPE Petroleum Computer Conference, Houston, Texas, July 19–22, 1992.

CHAPTER 8

Gas Reservoirs

8.1 Introduction

This chapter presents methods for calculations of in-place volumes and reserves for gas reservoirs. Gas reservoirs are classified as dry, wet or condensate depending on path of depletion (in terms of pressure and temperature) relative to the phase envelope of the reservoir fluids. In Chapter 4, the characteristics of dry, wet, and retrograde condensate gas reservoirs were presented. For dry gas reservoirs, no liquids are formed either in the reservoir or production facilities because the path of depletion is completely outside the phase envelope (Figure 4.7). For wet gas reservoirs, no liquids are condensed within the reservoir, but liquids may condense in the production facilities because temperature and pressure conditions in the production facilities may fall within the two-phase region (Figure 4.7). Retrograde gas condensate reservoirs may initially exist as single-phase fluid systems but condense liquid phases in the reservoir as pressure is reduced by production (Figure 4.8). In this chapter, gas reservoirs are further classified as volumetric or non-volumetric reservoirs. Volumetric gas reservoirs are defined as completely isolated, closed systems with approximately constant hydrocarbon pore volumes. Volumetric gas reservoirs are presumed not to gain significant pressure support or fluid influx from outside sources, such as water influx from aquifers or neighboring shale (non-reservoir) layers. On the other hand, non-volumetric gas reservoirs exhibit evidence of pressure support or influx of fluids (mostly water) from outside sources, such as aquifers or neighboring shale intervals. A special group of gas reservoirs are classified as overpressured gas reservoirs. These reservoirs are sometimes called geopressured or abnormally pressured gas reservoirs. The term geopressured gas reservoir is used to represent such reservoirs in this book. Geopressured gas reservoirs have pressure gradients that are sometimes approximately twice as high as those of normally pressured gas reservoirs.

8.2 Volumetric Gas Reservoirs

Volumetric gas reservoirs are considered as totally isolated, closed systems with approximately constant hydrocarbon pore volume. For volumetric gas reservoirs, it is assumed that the reservoir does not receive significant pressure support or fluid from outside sources, such as water influx from aquifers or neighboring shale (non-reservoir) layers. In practice, there are relatively few gas reservoirs that are truly volumetric as defined. The classification of a gas reservoir as volumetric allows the application of simplifying assumptions for the assessment of its in-place-hydrocarbon volumes and prediction of reservoir performance. In this section, volumetric calculations for dry gas, wet gas, and retrograde gas condensate reservoirs are presented. This is then followed by material balance on volumetric gas reservoirs.

8.2.1 Volumetric Calculations for Dry Gas Reservoirs

The original gas in place (OGIP) denoted as G in a dry gas reservoir can be calculated in terms of standard cubic feet (scf) with a simple volumetric equation:

$$G = \frac{7758 \times A \times h \times \phi \times (1 - S_{wi})}{B_{gi}} \tag{8.1}$$

In Eq. (8.1), A = reservoir area in acres; h = net sand thickness (or net pay), feet; ϕ = porosity, fraction; S_{wi} = initial water saturation, fraction (and $S_{gi} = 1 - S_{wi}$, where S_{gi} is initial gas saturation, fraction) ; and B_{gi} = initial gas formation volume factor (FVF), RB/scf. In the units of RB/scf, initial gas FVF is shown in Eq. (4.75) as:

$$B_{gi} = \frac{0.005021 z_i T_R}{p_i} \tag{8.2}$$

Methods for calculating B_{gi} for a gas mixture from its composition or gas gravity are shown in example calculations in Chapter 4.

At any intermediate pressure, the cumulative volume of gas produced from the reservoir is the difference between the OGIP and the volume of gas remaining in the reservoir. This is represented as:

$$G_p = G - G_r \tag{8.3}$$

In Eq. (8.3), G_p is the cumulative volume of produced gas, and G_r is the volume of gas remaining in the reservoir at the intermediate pressure. From Eq. (8.1),

$$G_r = \frac{7758 \times A \times h \times \phi \times (1 - S_{wi})}{B_g} \tag{8.4}$$

Note that in Eq. (8.4), B_g is the gas FVF at intermediate pressure, p. At an abandonment pressure, p_a, Eq. (8.3) is represented as:

$$G_p = G - G_a \tag{8.5}$$

In Eq. (8.5), G_a is the gas remaining in the reservoir at abandonment pressure, p_a. Substituting Eqs. (8.1) and (8.4) into Eq. (8.5) gives:

$$G_p = 7758 \times A \times h \times \phi \times (1 - S_{wi})\left[\frac{1}{B_{gi}} - \frac{1}{B_{ga}}\right] \quad (8.6)$$

The ultimate recovery factor, R, for a volumetric gas reservoir is defined as:

$$R = \frac{G - G_a}{G} = \frac{G_P}{G} \quad (8.7)$$

Substituting Eqs. (8.1) and (8.6) into Eq. (8.7), the ultimate gas recovery factor, R, is:

$$\begin{aligned} R &= 1 - \frac{B_{gi}}{B_{ga}} \\ &= 1 - \frac{p_a z_i}{p_i z_a} \end{aligned} \quad (8.8)$$

Note that the basic assumption in developing the equations for a volumetric dry gas reservoir is that initial water saturation does not change over the productive life of the reservoir. Gas recovery factors for volumetric dry gas reservoirs are usually high, approaching 90% in many cases. Even at low reservoir pressures, recovery from volumetric gas reservoirs can be improved by utilizing compressors to reduce well head pressures. However, it is important to be aware that there are few true volumetric gas reservoirs. As reservoir pressures are reduced through production, limited migration of fluids, especially water from neighboring shale layers or aquifers, can occur. These migrations may increase water saturations, and pressures in gas reservoirs. But if fluid influx is insignificant relative to the size of the reservoir, the assumption of volumetric behavior is considered valid and applicable even in such cases.

8.2.2 Volumetric Calculations for Wet Gas and Retrograde Gas Condensate Reservoirs

The original gas in place for a wet or a retrograde gas condensate reservoir can be calculated with the same equations as for a dry gas reservoir, but an important difference should be noted. The equations that follow will apply to a retrograde condensate reservoir if reservoir pressure is **above** its dew point pressure and **no condensation** has occurred in the reservoir. If the pressure of a retrograde gas condensate reservoir is below its dew point pressure, volumetric calculation of in-place volumes and reserves will be in error since condensation must have taken place in the reservoir. Liquid condensation changes the volume and composition of gas remaining in the reservoir. This affects the uniformity of the composition of gas in the reservoir, which is the basic assumption in volumetric calculations. The best approach for estimating reserves for condensate reservoirs in this condition is to conduct thorough constant volume depletion (CVD) experiments (described in Chapter 5) with fluid samples obtained from the reservoir, and by performing compositional simulation of the depletion process with the CVD data.

The original wet gas or retrograde condensate gas in place, G_c, can be calculated volumetrically as:

$$G_c = \frac{7758 \times A \times h \times \phi \times (1 - S_{wi})}{B_{gi}} \quad \textbf{(8.9)}$$

All the terms in Eq. (8.9) are as previously defined for Eq. (8.1). The main difference between the application of the volumetric equation to a dry gas reservoir versus a wet gas or a retrograde gas condensate reservoir is calculation of the initial gas FVF, B_{gi}. For wet or retrograde condensate gases, calculation of wellstream compositions should include all the liquids condensed in the production facilities. The best approach is to sample the gas and liquid streams at the separators and stock tank. The total composition of the reservoir gas (wet or condensate) is then determined by the recombination method. This method was demonstrated in Example 4.1 in Chapter 4. If the compositions of the streams at the separators and stock tank are not available, the correlation method proposed by Gold et al.[1] should be applied. This method was used in Example 4.2 in Chapter 4 to calculate the gas gravity of a retrograde gas condensate reservoir. With the gas gravity calculated from either example, the gas compressibility factor can be calculated as shown in Example 4.6 in Chapter 4. The cumulative wet or retrograde gas condensate produced, G_{cp}, is given by:

$$G_{cp} = G_p + \frac{133{,}316(\gamma_o N_p)}{M_o} \quad \textbf{(8.10)}$$

In Eq. (8.10), G_p is cumulative dry gas produced in scf; γ_o is the specific gravity of stock tank oil (condensate); N_p is cumulative stock tank oil (condensate) produced in STB; and M_o is the molecular weight of stock tank oil (condensate), lbm/lb-mole. Equations for calculating specific gravity of the oil and molecular weight of the oil are presented in Chapter 4 as Eqs. (4.48) and (4.56), respectively.

Example 8.1 Calculation of Gas in Place for the Condensate Reservoir in Example 4.2

Problem

Calculate the gas in place for the condensate reservoir in Example 4.2. Additional data for the condensate reservoir in Example 4.2 are as follows:

Initial reservoir pressure, p_i	12,688 psia
Reservoir temperature, T_R	267°F
Area, A	297 acres
Net thickness, h	79 feet
Average porosity, ϕ	21%
Initial connate water saturation, S_{wi}	0.301
Well stream gas gravity, γ_g	0.7437

Solution

Step 1: Calculate pseudo-critical pressure and temperature of the gas.
From Eqs. 4.45 and 4.46 in Chapter 4:

$$\begin{aligned} p_{pc,HC} &= 744 - 125.4\gamma_{g,HC} + 5.9\gamma_{g,HC}^2 \\ &= 744 - 125.4 \times 0.7437 + 5.9 \times (0.7437)^2 \\ &= 654.0 \text{ psia} \end{aligned}$$

$$\begin{aligned} T_{pc,HC} &= 164.3 + 357.7\gamma_{g,HC} - 67.7\gamma_{g,HC}^2 \\ &= 164.3 + 357.7 \times 0.7437 - 67.7 \times (0.7437)^2 \\ &= 392.9°R \end{aligned}$$

Compositional analysis of the gas shows negligible amounts of nitrogen and carbon dioxide. Consequently, no corrections are applied to the calculated critical properties due to these non-hydrocarbons.

Step 2: Calculate pseudo-reduced pressure and temperature.
The pseudo-reduced pressure and temperature are calculated as:

$$p_{pr} = \frac{p}{p_{pc}} = \frac{12{,}688}{654} = 19.4$$

$$T_{pr} = \frac{T}{T_{pc}} = \frac{459.67 + 267}{392.9} = 1.85$$

Step 3: Calculate gas compressibility factor.
From Figure 4.9, gas compressibility factor, $z = 1.66$. The initial gas formation volume factor is calculated from Eq. (8.2):

$$\begin{aligned} B_{gi} &= \frac{0.005021 z_i T_R}{p_i} \\ &= \frac{0.005021 \times 1.66 \times (459.67 + 267)}{12{,}688} \\ &= 4.77 \times 10^{-4} \text{ RB/scf} \end{aligned}$$

Step 4: Calculate condensate gas in place.
The condensate gas in place is calculated from Eq. (8.9) as:

$$\begin{aligned} G_c &= \frac{7758 \times A \times h \times \phi \times (1 - S_{wi})}{B_{gi}} \\ &= \frac{7758 \times 297 \times 79 \times 0.21 \times (1 - .301)}{4.77 \times 10^{-4}} \\ &= 56.0 \times 10^9 \text{ scf or 56 Bcf} \end{aligned}$$

8.2.3 Material Balance for Volumetric Dry Gas, Wet Gas, and Retrograde Gas Condensate Reservoirs

The material balance equations developed in this section will apply to volumetric dry gas, wet gas, and retrograde gas condensate reservoirs. It is important to note that these equations are not applicable to a retrograde gas condensate reservoir if reservoir pressure is below dew point pressure and condensation has occurred in the reservoir. In addition, for wet and retrograde gas condensate reservoirs, extreme care must be taken to ensure that the total gas produced from the reservoir includes all the liquids condensed and produced from the production facilities and converted to its vapor equivalent volume. The equations for converting produced liquid volumes to their vapor equivalent volumes were presented as Eqs. (4.53) and (4.54) in Chapter 4.

For any volumetric gas reservoir, a simple material balance equation for the reservoir can be written as:

$$\text{Initial Gas Moles} = \text{Gas Moles Remaining} + \text{Gas Moles Produced} \tag{8.11}$$

Equation (8.11) can be written as:

$$n_i = n_r + n_p \tag{8.12}$$

In Eq. (8.12), n = moles of gas; and the subscripts i = initial, r = remaining, and p = produced. Re-arranging Eq. (8.12), gives:

$$n_p = n_i - n_r \tag{8.13}$$

In terms of surface gas units (scf), Eq. (8.13) becomes:

$$G_p = G - G\left(\frac{B_{gi}}{B_g}\right) \tag{8.14}$$

Alternatively, Eq. (8.14) can be derived from the General Material Balance Equation (GMBE) for gas reservoirs developed in Chapter 7. The GMBE for gas reservoirs is shown in Eq. (7.27) as:

$$G_pB_g + W_pB_w = GB_{gi}\left(\frac{c_wS_{wi} + c_f}{1 - S_{wi}}\right)\Delta p + G(B_g - B_{gi}) + W_e \tag{8.15}$$

Since for volumetric reservoirs, water influx and water production are assumed to be zero, and water and formation compressibility are considered negligible, Eq. (8.15) can be reduced to Eq. (8.14). From Eq. (8.2):

$$\frac{B_{gi}}{B_g} = \frac{z_ip}{zp_i} \tag{8.16}$$

Substituting Eq. (8.16) into Eq. (8.14), yields:

$$G_p = G - G\left(\frac{z_ip}{zp_i}\right) \tag{8.17}$$

From Eq. (8.17), it can be derived easily that:

$$\frac{p}{z} = \frac{p_i}{z_i} - \frac{p_i}{z_i}\left(\frac{G_p}{G}\right) \tag{8.18}$$

Equation (8.18) is the basic equation derived from application of material balance to volumetric gas reservoirs. A plot of p/z vs. G_p yields a straight line. When $G_p = 0$, the intercept on the y-axis is equal to p_i/z_i from which initial pressure can be estimated if not available. When $p/z = 0$, the intercept on the x-axis is equal to the original gas in place, G. The p/z vs. G_p plot is widely misapplied to all types of gas reservoirs. It is important to remember that it applies in a strict sense only to volumetric gas reservoirs. Volumetric reservoirs by definition have no fluid influxes either from aquifers or any other external sources. However, the p/z vs. G_p plot can be used in a diagnostic sense to qualitatively demonstrate the existence of potential energy support for the reservoir from external sources. Some of the characteristic shapes of the plot for different reservoir drive mechanisms are shown in Figure 8.1 for volumetric, geopressured, weak, and strong water-drive gas reservoirs. It is sometimes permissible to apply the p/z vs. G_p plot to a retrograde gas condensate reservoir at pressures below the dew point if the amount of liquid condensation in the reservoir is presumed to be less than 10 percent of the hydrocarbon pore volume.

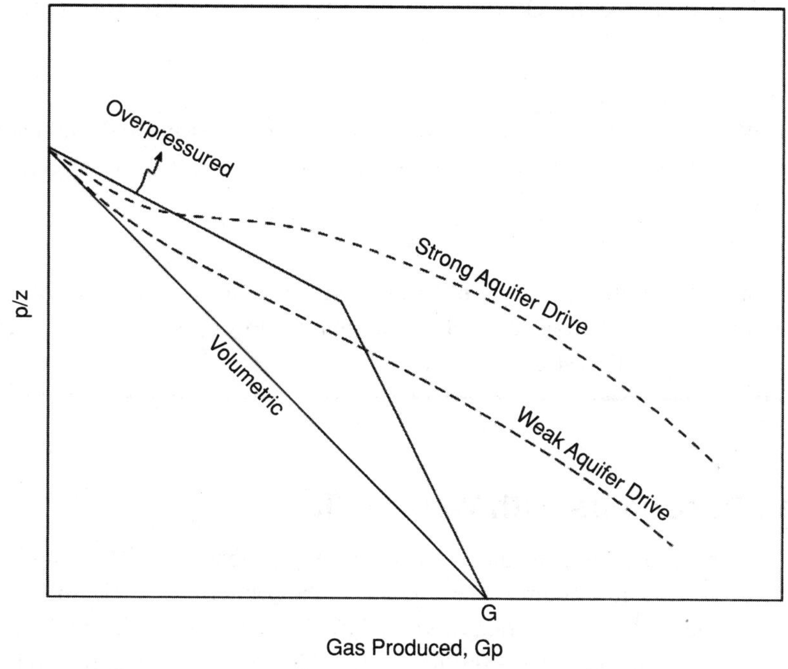

Figure 8.1 Sketch of p/z plots for various gas reservoir drive mechanisms.

Table 8.1 Data for p/z vs. G_p Plot for a Volumetric, Dry Gas Reservoir

Pressure (psia)	Compressibility Factor, z	Cum. Gas Production G_p (MMscf)	p/z
3600	0.85	0.0	4235.3
3450	0.83	47.8	4156.6
3300	0.82	126.5	4024.4
3150	0.81	204.8	3888.9
2850	0.79	382.5	3607.6
2685	0.77	440.0	3487.0

However, limited valid results can be derived from the application of this plotting technique on a retrograde gas condensate reservoir below dew point pressures since in a practical sense, it is difficult to determine with sufficient degree of accuracy, the actual amount of condensation that may have occurred in the reservoir at these pressures.

Example 8.2 Calculation of the OGIP for a Volumetric, Dry Gas Reservoir Using the p/z vs. G_p Plot

Problem

Calculate the OGIP for a volumetric, dry gas reservoir from the pressure and cumulative gas production data in Table 8.1. The gas compressibility factors at each reservoir pressure are also shown in Table 8.1.

Solution

From the data in Table 8.1, the p/z vs. G_p plot can be readily performed on a spreadsheet and plotted as shown in Figure 8.2. Extrapolation of the straight line to $p/z = 0$ gives the original gas in place (OGIP) as: $G = 2400$ MMscf or 2.4 Bcf.

8.3 Gas Reservoirs with Water Influx

Gas reservoirs with water influx are non-volumetric reservoirs that have water invading the reservoir mainly from an adjoining aquifer. The strength of the aquifer can be classified qualitatively as weak or strong depending on the rate of water influx into the reservoir. Weak aquifer influx indicates that relatively the rate of water influx is low, while strong aquifer influx indicates a high rate of water influx. Two methods are presented in this section for assessing the performance of

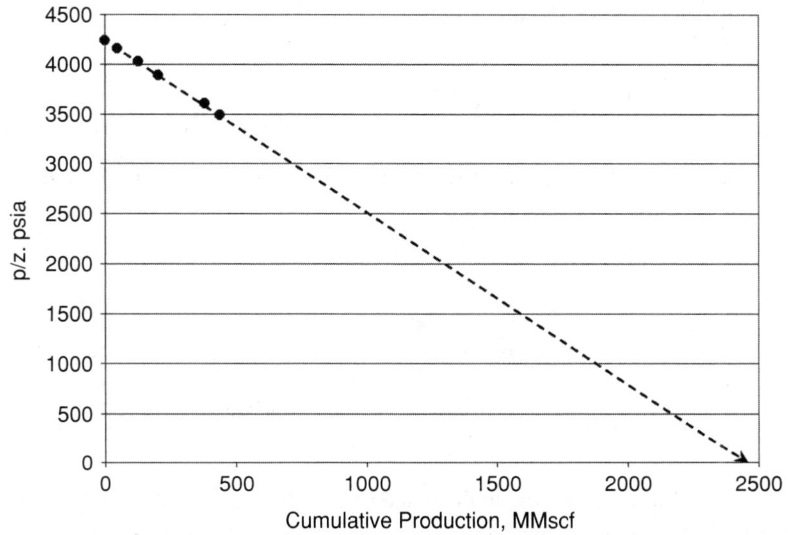

Figure 8.2 p/z vs. G_p plot for Example 8.2.

gas reservoirs with water influx. These methods are the volumetric approach and the material balance approach. The volumetric method is presented first followed by the material balance method.

8.3.1 Volumetric Approach

Suppose water had invaded a gas reservoir from an adjoining aquifer. Let us assume that the hydrocarbon pore volume of the gas reservoir had been occupied by the volume of water that invaded the reservoir and was not produced. On the basis of these assumptions, volumetric methods can be used to estimate the total volume of gas produced from the reservoir and its recovery factor. From Eq. (8.1), the original volume of gas in the reservoir is:

$$G = \frac{7758 \times A \times h \times \phi \times (1 - S_{wi})}{B_{gi}} \tag{8.19}$$

At abandonment, the volume of gas remaining in the reservoir, G_a is given as:

$$G_a = \frac{7758 \times A \times h \times \phi \times S_{gr}}{B_{ga}} \tag{8.20}$$

In Eq. (8.20), S_{gr} = residual gas saturation at abandonment; and B_{ga} = gas FVF at abandonment pressure. The cumulative gas produced, G_p, is equal to:

$$G_p = G - G_a \tag{8.21}$$

Substituting with Eqs. (8.19) and (8.20):

$$G_p = 7758 \times A \times h \times \phi \times \left[\frac{S_{gi}}{B_{gi}} - \frac{S_{gr}}{B_{ga}}\right] \tag{8.22}$$

In Eq. (8.22), $S_{gi} = 1 - S_{wi}$ from Eq. (8.19). Gas recovery factor, R, is given by:

$$R = \frac{G_p}{G} = \frac{\dfrac{S_{gi}}{B_{gi}} - \dfrac{S_{gr}}{B_{ga}}}{\dfrac{S_{gi}}{B_{gi}}} = \left[1 - \frac{S_{gr}B_{gi}}{S_{gi}B_{ga}}\right] \tag{8.23}$$

Residual gas saturations are difficult to determine either from laboratory measurements or correlations. At best, it can be estimated from laboratory core flood experiments. In absence of any laboratory data, it can be estimated from Agarwal[2] correlations provided in Appendix 8A. These correlations should be used with extreme caution since they may not be representative of the particular reservoir under analysis. Recoveries from gas reservoirs with water influx are generally less than recoveries from volumetric gas reservoirs. The recovery factors for gas reservoirs with water influx range from 50 to 70 percent, while volumetric gas reservoirs could have recovery factors as high as 90 percent. The lower recoveries from gas reservoirs with water influx are due to trapped or bypassed gas as water invades the reservoir. The general strategy for managing a gas reservoir with strong water drive is to produce the reservoir at high gas production rates in an attempt to "out run" the water. In some cases, this strategy may require producing gas wells at high gas-water ratios until the wells can no longer flow on their own energy. Even under such circumstances, de-watering practices could be used to prolong the productive life of gas wells.

8.3.2 Material Balance Approach

The material balance equation for a gas reservoir with water influx can be derived readily from the GMBE shown as Eq. (8.15). Assuming water and formation compressibilities are negligible, the material balance equation for a gas reservoir with water influx can be re-arranged from Eq. (8.15) as:

$$GB_{gi} - (G - G_p)B_g = W_e - W_pB_w \tag{8.24}$$

8.3.3 The Cole Plot

Equation (8.24) can be re-arranged into the following form:

$$\frac{G_pB_g}{B_g - B_{gi}} = G + \frac{W_e - W_pB_w}{B_g - B_{gi}} \tag{8.25}$$

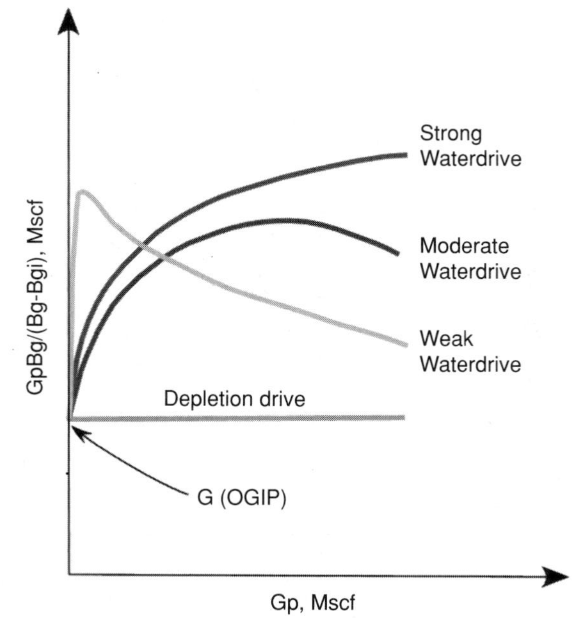

Figure 8.3 Sketch of the Cole Plot for gas reservoirs.

Cole[3] plotted the expression $G_pB_g/(B_g - B_{gi})$ on the y-axis, and cumulative gas production, G_p, to develop a graphic representation called the Cole plot. A sketch of the Cole plot for various reservoir drive mechanisms is shown in Figure 8.3. As shown in Figure 8.3, the Cole plot for a volumetric depletion drive gas reservoir is a horizontal line. The characteristic shapes of the Cole plot for weak, moderate, and strong water-drive reservoirs are also shown in Figure 8.3. The Cole plot is most useful as a diagnostic plot that will indicate the presence of water influx much earlier in the production life of the reservoir than the p/z plot.[4]

8.3.4 The Havlena-Odeh Straight Line Method

Havlena and Odeh[5,6] re-arranged Eq. (8.24) in the form of a straight line equation as follows:

$$\frac{G_pB_g - W_pB_w}{B_g - B_{gi}} = G + \frac{W_e}{B_g - B_{gi}} \tag{8.26}$$

Equation (8.26) can be written in terms of aquifer function such that:

$$\frac{G_pB_g - W_pB_w}{B_g - B_{gi}} = G + C_{aq}\frac{\Phi(p,t)}{B_g - B_{gi}} \tag{8.27}$$

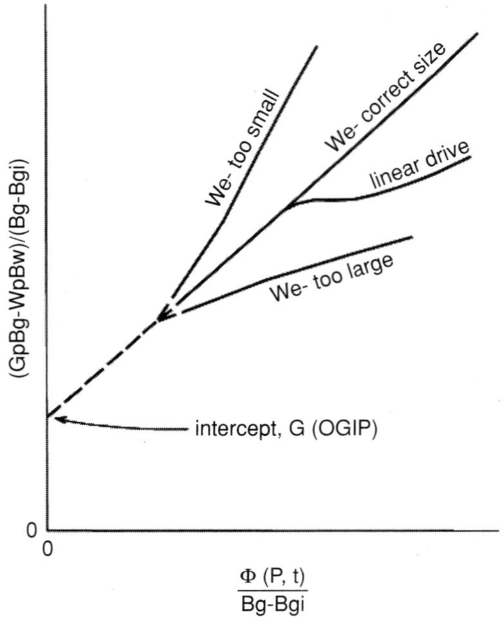

Figure 8.4 Sketch of the Havlena-Odeh[5] plot for a gas reservoir (©1963 SPE, Reproduced with permission).

In Eq. (8.27), $W_e = C_{aq}\Phi(p,t)$ where C_{aq} = aquifer constant; and $\Phi(p,t)$ = aquifer function. The definition of the aquifer constant and aquifer function depends on the type of aquifer. A plot of $(G_pB_g - W_pB_w)/(B_g - B_{gi})$ on the y-axis and $\Phi(p,t)/(B_g - B_{gi})$ on the x-axis yields a straight line with the intercept on the y-axis equal to G and the slope equal to the aquifer constant C_{aq}. Figure 8.4 is a representation of potential shapes of curves that may result from the Havlena and Odeh (HO) plot. As shown in Figure 8.4, if the size of aquifer is correct, the resulting plot will be a straight line. If the aquifer size is too small, the plot will curve upward. If the size of the aquifer is too large, the plot will curve downward. An S-shaped curve indicates that a linear aquifer might be more appropriate. The Havlena-Odeh Straight Line method is an important tool that can be used to assess the type and strength of the aquifer, and gain knowledge on the performance of the reservoir before using more sophisticated techniques, such as reservoir modeling.

8.4 Water Influx Models

Water influx into reservoirs (oil and gas) can be estimated with steady-state, unsteady-state, pseudosteady-state, and approximation models. The Schilithuis[7] aquifer model is an example of a steady-state model. Steady-state models assume that the rate of water influx is directly proportional to the pressure drop between the original oil-water contact and the external aquifer boundary. For steady-state models, the pressure at the external aquifer boundary is assumed to be constant. Steady-state models are not presented in this book because they have limited applications for calculating water influx, especially for reservoirs with short production history. The van Everdingen and Hurst[8] aquifer model is an unsteady-state model, which is based on the radial diffusivity equation. (The derivation and applications of the radial diffusivity equation are presented in Chapter 10.) The van Everdingen-Hurst aquifer model uses superposition to calculate cumulative water influx. The process of calculating water influx using the van Everdingen-Hurst method is also presented in Chapter 10. The Fetkovich[9] aquifer model is a pseudosteady-state model. The Fetkovich aquifer model was developed from a combination of an inflow equation and a material balance model based on the aquifer. The Carter-Tracy[10] aquifer model is an approximate model which is based on the van Everdingen-Hurst unsteady-state model. The Fetkovich and Carter-Tracy aquifer models do not require superposition to calculate cumulative water influx. Both methods provide simple and more direct methods for calculating cumulative water influx with accuracy that is comparable to the more tedious van Everdingen-Hurst method. Furthermore, most of the data on the aquifer (such as porosity, net thickness, permeability, compressibility, pressure, etc.) required for calculation of water influx are usually not measured or known with any reasonable level of accuracy. This is mainly due to common industry practice of **not** drilling wells intentionally into the aquifer to acquire these data. The rock properties of the aquifer are often based on the rock properties of the reservoir. Consequently, calculation procedures such as Fetkovich and Carter-Tracy methods are adequate for predicting cumulative water influx for most engineering applications.

The Fetkovich model applies to finite-acting aquifers, while the Carter-Tracy aquifer models can be applied to both finite and infinite-acting aquifers. Both models are applicable to both radial and linear aquifers. The geometrical representations of radial and linear aquifer geometries are shown in Figures 8.5 and 8.6, respectively. The Fetkovich aquifer model applies to

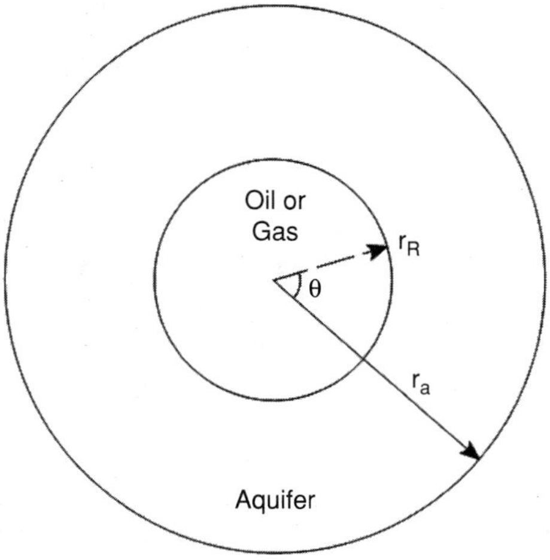

Figure 8.5 Sketch of a reservoir with radial aquifer.

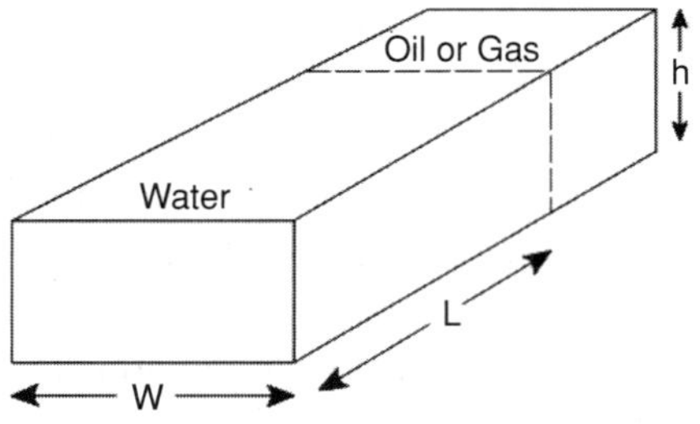

Figure 8.6 Sketch of a reservoir with linear aquifer.

edge-water and bottom-water drive reservoirs, while the Carter-Tracy aquifer model applies to edge-water drive reservoirs. In edge-water drive, water influx occurs around the flanks of the reservoir. In bottom-water drive, the reservoir is underlain by the aquifer which influxes vertically into the reservoir. Figures 8.7 and 8.8 are sketches of edge-water drive and bottom-water drive reservoirs, respectively.

8.4.1 Fetkovich Aquifer Model

The Fetkovich[9] aquifer model for calculating water influx is based on equating an inflow equation for the aquifer to a material balance model on the aquifer. Because pseudosteady state is assumed in developing this model, early transient effects on cumulative water influx are ignored. Note that for this reason, cumulative water influx calculated with the Fetkovich method is usually less than cumulative water influx calculated with either the van Everdingen-Hurst or Carter-Tracy methods. However, Fetkovich[9] demonstrated in his paper that this "simplified approach is accurate enough for engineering purposes."

The generalized inflow equation for an aquifer can be written as:

$$q_w = J(\overline{p}_a - p_R)^n = \frac{dW_e}{dt} \tag{8.28}$$

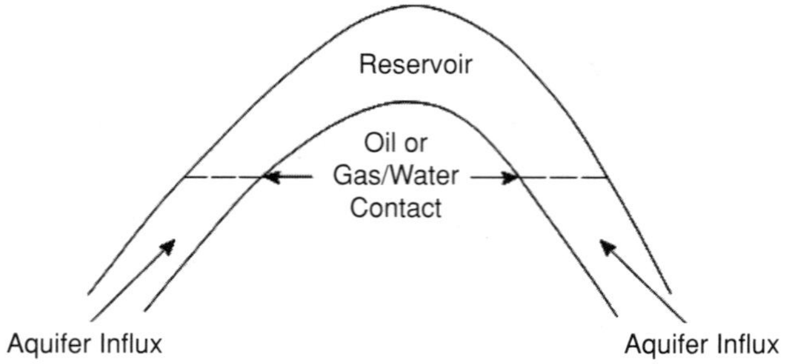

Figure 8.7 Sketch of edge-water drive reservoir.

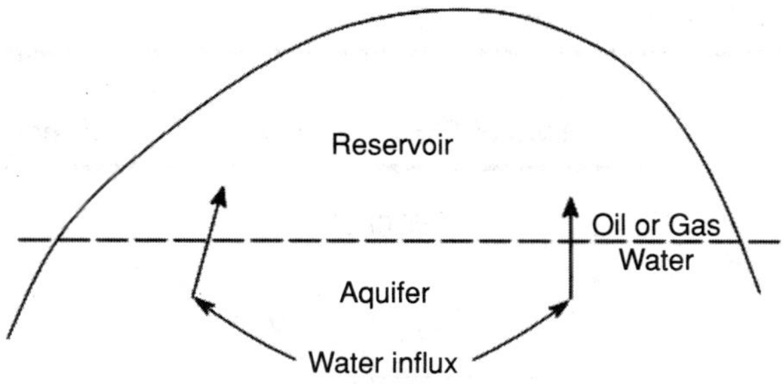

Figure 8.8 Sketch of bottom-water drive reservoir.

In Eq. (8.28), q_w = water influx rate; J = aquifer productivity index; $\overline{p}_a$ = average aquifer pressure; p_R = pressure at the reservoir/aquifer boundary (or original hydrocarbon/aquifer contact); n = inflow equation exponent. For Darcy flow at pseudosteady-state or steady-state conditions, $n = 1$, and for turbulent flow conditions, $n = 0.5$; and W_e = cumulative water influx.

The cumulative water influx can be represented in a simple material balance equation based on the expansion of the aquifer as:

$$W_e = c_t W_i (p_{a,i} - \overline{p}_a) \tag{8.29}$$

In Eq. (8.29), c_t = total aquifer compressibility; W_i = initial aquifer water volume; $p_{a,i}$ = initial aquifer pressure; and $\overline{p}_a$ = average aquifer pressure. Initial aquifer pressure is assumed equal to initial reservoir pressure in most cases. Eq. (8.29) can be expressed in an equivalent form as:

$$\overline{p}_a = p_{a,i}\left(1 - \frac{W_e}{c_t W_i p_{a,i}}\right) = p_{a,i}\left(1 - \frac{W_e}{W_{ei}}\right) \tag{8.30}$$

In Eq. (8.30), $W_{ei} = c_t W_i p_{a,i}$. W_{ei} is the maximum amount of influx water at initial conditions. Differentiating Eq. (8.30) with respect to time, t, and re-arranging gives:

$$\frac{dW_e}{dt} = -\frac{W_{ei}}{p_{a,i}}\frac{d\overline{p}_a}{dt} \tag{8.31}$$

Equating Eqs. (8.28) and (8.31) followed by integrating between initial conditions and time, t, gives:

$$\int_{p_{a,i}}^{p_a} \frac{d\overline{p}_a}{\overline{p}_a - p_R} = -\int_0^t \frac{J p_{a,i}}{W_{ei}} dt \tag{8.32}$$

After integration of Eq. (8.32), the result is:

$$\overline{p}_a - p_R = (p_{a,i} - p_R)\exp\left(\frac{-J p_{a,i} t}{W_{ei}}\right) \tag{8.33}$$

Substituting Eq. (8.28) into Eq. (8.33) and rearranging results in:

$$\frac{dW_e}{dt} = J(p_{a,i} - p_R)\exp\left(\frac{-J p_{a,i} t}{W_{ei}}\right) \tag{8.34}$$

Table 8.2 Aquifer Productivity Indices

Type of Outer Aquifer Boundary	Radial Flow* (STB/D-psi)	Linear Flow* (STB/D-psi)
Finite- no flow	$J = \dfrac{0.00708kh\left(\dfrac{\theta}{360}\right)}{\mu\left[\ln\left(\dfrac{r_a}{r_R}\right) - 0.75\right]}$	$J = \dfrac{0.003381kwh}{\mu L}$
Finite- constant pressure	$J = \dfrac{0.00708kh\left(\dfrac{\theta}{360}\right)}{\mu\left[\ln\left(\dfrac{r_a}{r_R}\right)\right]}$	$J = \dfrac{0.001127kwh}{\mu L}$

* All variables are in field units.

Integrating Eq. (8.34) for cumulative water influx yields:

$$W_e = \frac{W_{ei}}{p_{a,i}}(p_{a,i} - p_R)\left[1 - \exp\left(\frac{-Jp_{a,i}t}{W_{ei}}\right)\right] \tag{8.35}$$

Fetkovich[9] approximated constant pressure at the reservoir/aquifer boundary by assuming that the pressure history at the boundary can be divided into small finite number of time intervals such that the incremental water influx at the nth interval can be represented thus:

$$\Delta W_{en} = \frac{W_{ei}}{p_{a,i}}(\overline{p}_{a,n-1} - \overline{p}_{R,n})\left[1 - \exp\left(\frac{-Jp_{a,i}\Delta t_n}{W_{ei}}\right)\right] \tag{8.36}$$

For Eq. (8.36),

$$\overline{p}_{a,n-1} = p_{a,i}\left(1 - \frac{W_{e,n-1}}{W_{ei}}\right) \tag{8.37}$$

$$\overline{p}_{R,n} = \frac{p_{R,n-1} + p_{R,n}}{2} \tag{8.38}$$

The Fetkovich method was developed for finite aquifer systems. The productivity indices for various aquifer flow geometries and boundary conditions are provided in Table 8.2.

Example 8.3 Calculation of Water Influx Using the Fetkovich Method

Problem

The properties of a radial aquifer are shown in Table 8.3. The pressure history at the reservoir-aquifer boundary is also provided in Table 8.3. Calculate the water influx using the Fetkovich method assuming a finite aquifer with no flow at the outer aquifer boundary.

Table 8.3 Aquifer/Reservoir Data and Pressure History for Example 8.3

Initial reservoir pressure	3665 psia
Porosity, ϕ	0.225
Permeability, k	669 md
Aquifer thickness, h	45 feet
Water viscosity, μ	0.35 cp
Total aquifer compressibility, c_t	6×10^{-6} psia^{-1}
Reservoir radius, r_R	8210 feet
Aquifer radius, r_a	123,150 feet
Encroachment angle, θ	360°

Pressure History at Reservoir/Aquifer Boundary

Time, *t* (days)	**Pressure, *p* (psia)**
0	3665
365	3586
730	3543
1095	3520
1460	3491

Solution

Step 1: Calculate the maximum water influx from the aquifer, W_{ei}.
Initial volume of water in the aquifer, W_i, is calculated as:

$$W_i = \frac{\pi\left(r_a^2 - r_R^2\right)h\phi\left(\frac{\theta}{360}\right)}{5.615}$$

$$= \frac{\pi[(123150)^2 - (8210)^2] \times 45 \times 0.225 \times 1}{5.615}$$

$$= 85.53 \times 10^9 \text{ RB}$$

$$W_{ei} = c_t W_i p_{a,i} = 6 \times 10^{-6} \times 85.53 \times 10^9 \times 3665$$

$$= 1.88 \times 10^9 \text{ RB}$$

Step 2: Calculate aquifer productivity index, J.
From Table 8.2,

$$J = \frac{0.00708kh\dfrac{\theta}{360}}{\mu\left[\ln\left(\dfrac{r_a}{r_R}\right) - 0.75\right]} = \frac{0.00708 \times 669 \times 45}{0.35\left[\ln\left(\dfrac{123150}{8210}\right) - 0.75\right]}$$
$$= 311.0141 \text{ STB/D-psi}$$

Step 3: Calculate incremental water influx for each time period, n.
From Eq. (8.36):

$$\Delta W_{en} = \frac{W_{ei}}{p_{a,i}}\left(\overline{p}_{a,n-1} - \overline{p}_{R,n}\right)\left[1 - \exp\left(\frac{-Jp_{a,i}\Delta t_n}{W_{ei}}\right)\right]$$

From Eq. (8.37):

$$\overline{p}_{a,n-1} = p_{a,i}\left(1 - \frac{W_{e,n-1}}{W_{ei}}\right)$$

And from Eq. (8.38):

$$\overline{p}_{R,n} = \frac{p_{R,n-1} + p_{R,n}}{2}$$

Substituting above equations at $n = 1$ gives:

$$\overline{p}_{a,0} = 3665\left(1 - \frac{0}{1.88 \times 10^9}\right) = 3665 \text{ psia}$$
$$\overline{p}_{R,1} = \frac{3665 + 3586}{2} = 3625.5 \text{ psia}$$
$$\overline{p}_{a,0} - \overline{p}_{R,1} = 3665 - 3625.5 = 39.5 \text{ psia}$$
$$\Delta W_{e1} = \frac{1.88 \times 10^9}{3665} \times 39.5 \times \left[1 - \exp\left(\frac{-311.0141 \times 3665 \times 365}{1.88 \times 10^9}\right)\right]$$
$$= 4.023 \times 10^6 \text{ RB}$$

At $n = 2$:

$$\overline{p}_{a,1} = 3665\left(1 - \frac{4.023 \times 10^6}{1.88 \times 10^9}\right) = 3657.16 \text{ psia}$$
$$\overline{p}_{R,2} = \frac{3586 + 3543}{2} = 3564.5 \text{ psia}$$
$$\overline{p}_{a,1} - \overline{p}_{R,2} = 3657.16 - 3564.5 = 92.66 \text{ psia}$$
$$\Delta W_{e2} = \frac{1.88 \times 10^9}{3665} \times 92.66 \times \left[1 - \exp\left(\frac{-311.0141 \times 3665 \times 365}{1.88 \times 10^9}\right)\right]$$
$$= 9.436 \times 10^6 \text{ RB}$$

Table 8.4 Water Influx Calculations Using the Fetkovich Method for Example 8.3

n	t (days)	$p_{R,n}$ (psia)	$\bar{p}_{R,n}$ (psia)	$\bar{p}_{a,n}$ (psia)	$\bar{p}_{a,n-1} - \bar{p}_{R,n}$ (psia)	J (STB/D-psi)	ΔW_e (RB)	W_e (RB)
0	0	3665	3665.0	3665.0	0.0	311.0141	0	0
1	365	3586	3625.5	3657.2	39.5	311.0141	4.02E+6	4.02E+6
2	730	3543	3564.5	3638.8	92.7	311.0141	9.44E+6	1.35E+7
3	1095	3520	3531.5	3617.5	107.3	311.0141	1.09E+7	2.44E+7
4	1460	3491	3505.5	3595.2	112.0	311.0141	1.14E+7	3.58E+7

Cumulative water influx at $n = 2$ is:

$$W_{e2} = W_{e1} + \Delta W_{e2} = 4.023 \times 10^6 + 9.436 \times 10^6$$
$$= 1.346 \times 10^7 \text{ RB}$$

The calculations for the remaining time periods are summarized in Table 8.4.

8.4.2 Carter-Tracy Aquifer Model

The Carter-Tracy[10] method is an approximate method that calculates aquifer influx with results that are comparable to the exact solutions of the van Everdingen-Hurst method. The Carter-Tracy method is recommended in preference to the van Everdingen-Hurst method because it does not use superposition which makes the calculations less tedious especially if performed manually on a calculator.

The equations for the Carter-Tracy[10] method were developed by expressing cumulative water influx as a function of varying pressure with the aid of a convolution integral:

$$W_e(t_{Dj}) = B \int_0^{t_{Dj}} \Delta p(\lambda) Q'_D(t_{Dj} - \lambda) d\lambda \qquad (8.39)$$

In Eq. (8.39), W_e = cumulative water influx; B = constant; p = pressure; Q_D = dimensionless cumulative water influx; and t_D = dimensionless time. In terms of field units,

$$B = 1.119 \phi c_t h r_R^2 \left(\frac{\theta}{360} \right) \quad \text{for radial flow} \qquad (8.40)$$

$$B = 0.178 \phi c_t h L \quad \text{for linear flow} \qquad (8.41)$$

Also, in terms of field units, dimensionless time, t_D, is defined as:

$$t_D = \frac{0.00633kt}{\phi\mu c_t r_R^2} \quad \text{for radial flow} \tag{8.42}$$

$$t_D = \frac{0.00633kt}{\phi\mu c_t L^2} \quad \text{for linear flow} \tag{8.43}$$

In Eqs. (8.40) to (8.43), ϕ = porosity, fraction; c_t = total aquifer compressibility, psi^{-1}; h = net aquifer thickness, feet; r_R = reservoir radius, feet; L = length of reservoir, feet; θ = angle of encroachment, degrees; k = permeability, md; and t = time, days.

By approximating water influx by a series of constant rate intervals, Eq. (8.39) can be represented with the following alternate forms:

$$W_e(t_{Dj}) = \sum_{n=0}^{j-1} q_{Dn}(t_{Dn+1} - t_{Dn}) \tag{8.44}$$

$$W_e(t_{Dj}) = W_e(t_{Di}) + \sum_{n=i}^{j-1} q_{Dn}(t_{Dn+1} - t_{Dn}) \tag{8.45}$$

Combining Eqs. (8.39) and (8.45), Laplace transform techniques can be used to solve for cumulative water influx in terms of pressure drop, Δp_n:

$$W_e(t_{Dn}) = W_e(t_{Dn-1}) + \left[\frac{B\Delta p(t_{Dn}) - W_e(t_{Dn-1})p'_D(t_{Dn})}{p_D(t_{Dn}) - t_{Dn-1}p'_D(t_{Dn})}\right] \times (t_{Dn} - t_{Dn-1}) \tag{8.46}$$

The intermediate equations before arriving at Eq. (8.46) are shown in the paper by Carter and Tracy.[10] In Eq. (8.46),

$$\Delta p(t_{Dn}) = p_0 - p(t_{Dn}) \tag{8.47}$$

In Eq. (8.46), n and $n - 1$ represent current and previous time intervals, respectively. Also $p(t_{Dn})$ represents pressure at dimensionless time interval, t_{Dn}, and $W_e(t_{Dn-1})$ represents cumulative water influx at dimensionless time interval, t_{Dn-1}. In Eq. (8.47), p_0 is initial aquifer pressure at $t = 0$. In Eq. (8.46), $p_D(t_{Dn})$ represents dimensionless pressure as a function of dimensionless time, t_{Dn}, and p'_D represents the derivative of p_D. p_D and p'_D for finite and infinite acting aquifers can be obtained from polynomials developed by Fanchi.[11] The polynomials are provided in Appendix 8B. Similar polynomials for infinite acting aquifers only were published by Edwardson et al.[12] These polynomials are also presented in Appendix 8C. Klins et al.[13] published sets of polynomials for finite and infinite acting reservoirs. These were not provided in this book. For infinite acting aquifers, the polynomials reported by Edwardson et al.[12] are recommended because the reported average error of 0.2% is less than the average error of 1.5% reported by Fanchi.[11] However, for consistency and digital applications, the polynomials presented by Fanchi[11] may be preferable since the polynomials are also applicable to finite aquifers.

Example 8.4 Calculation of Water Influx Using the Carter-Tracy Method

Problem

Using the same data presented in Table 8.3 for the example on application of the Fetkovich method, calculate the water influx using the Carter-Tracy method.

Solution

Step 1: Calculate the parameter B for radial flow.
From Eq. (8.40):

$$B = 1.119\phi c_t h r_R^2\left(\frac{\theta}{360}\right) = 1.119 \times 0.225 \times 6 \times 10^{-6} \times 45 \times (8210)^2 \times \left(\frac{360}{360}\right)$$
$$= 4582.08 \text{ RB/psi}$$

Step 2: Calculate dimensionless time, t_D.
From Eq. (8.42):

$$t_D = \frac{0.00633kt}{\phi\mu c_t r_R^2} = \frac{0.00633 \times 669 \times t}{0.225 \times 0.35 \times 6 \times 10^{-6} \times (8210)^2}$$
$$= 0.1323t$$

Step 3: Calculate pressure change, $\Delta p(t_{Dn})$.
From Eq. (8.47):

$$\Delta p(t_{Dn}) = p_0 - p(t_{Dn})$$

At $n = 1$, $t_{D1} = 48.30$. $\Delta p(t_{D1}) = 3665 - 3586 = 79$ psia.
At $n = 2$, $t_{D2} = 96.62$. $\Delta p(t_{D2}) = 3665 - 3543 = 122$ psia.

Step 4: Calculate dimensionless pressure for infinite acting aquifers at $n = 1$.
From Appendix 8C using the correlations by Edwardson et al.[12] at $t_{D1} = 48.3$;

$$p_D = \frac{370.529\sqrt{t_D} + 137.582t_D + 5.69549t_D\sqrt{t_D}}{328.834 + 265.488\sqrt{t_D} + 45.2157t_D + t_D\sqrt{t_D}}$$
$$= \frac{370.529\sqrt{48.3} + 137.582 \times 48.3 + 5.69549 \times 48.3\sqrt{48.3}}{328.834 + 265.488\sqrt{48.3} + 45.2157 \times 48.3 + 48.3\sqrt{48.3}}$$
$$= 2.3718$$

Step 5: Calculate dimensionless pressure derivative for infinite acting aquifers at $n = 1$
From Appendix 8C using the correlations by Edwardson et al.[12] at $t_{D1} = 48.3$;

$$p'_D = \frac{716.441 + 46.7984\sqrt{t_D} + 270.038t_D + 71.0098t_D\sqrt{t_D}}{1269.86\sqrt{t_D} + 1204.73t_D + 618.618t_D\sqrt{t_D} + 538.072t_D^2 + 142.410t_D^2\sqrt{t_D}}$$
$$= \frac{716.441 + 46.7984\sqrt{48.3} + 270.038 \times 48.3 + 71.0098 \times 48.3\sqrt{48.3}}{1269.86\sqrt{48.3} + 1204.73 \times 48.3 + 618.618 \times 48.3\sqrt{48.3} + 538.072(48.3)^2 + 142.410(48.3)^2\sqrt{48.3}}$$
$$= 0.009878$$

Step 6: Calculate water influx at $n = 1$.
From Eq. (8.46):

$$W_e(t_{Dn}) = W_e(t_{Dn-1}) + \left[\frac{B\Delta p(t_{Dn}) - W_e(t_{Dn-1})p'_D(t_{Dn})}{p_D(t_{Dn}) - t_{Dn-1}p'_D(t_{Dn})}\right] \times (t_{Dn} - t_{Dn-1})$$

At $t_{D1} = 48.3$;

$$W_e(t_{D1}) = 0 + \left[\frac{4582.08 \times 79 - 0}{2.3718 - 0}\right] \times (48.3 - 0)$$
$$= 7{,}371{,}829 \text{ RB}$$

Step 7: Repeat Steps 4, 5, 6 at $n = 2$ and $t_{D2} = 96.61$.

$$p_D = \frac{370.529\sqrt{96.61} + 137.582 \times 96.61 + 5.69549 \times 96.61\sqrt{96.61}}{328.834 + 265.488\sqrt{96.61} + 45.2157 \times 96.61 + 48.3\sqrt{96.61}}$$
$$= 2.7061$$

$$p'_D = \frac{716.441 + 46.7984\sqrt{96.61} + 270.038 \times 96.61 + 71.0098 \times 96.61\sqrt{96.61}}{1269.86\sqrt{96.61} + 1204.73 \times 96.61 + 618.618 \times 96.61\sqrt{96.61} + 538.072(96.61)^2 + 142.410(96.61)^2\sqrt{96.61}}$$
$$= 0.005036$$

$$W_e(t_{D2}) = 7{,}371{,}829 + \left[\frac{4582.08 \times 122 - 7{,}371{,}829 \times 0.005036}{2.7061 - 48.3 \times 0.005036}\right] \times (96.61 - 48.3)$$
$$= 17{,}607{,}516 \text{ RB}$$

The calculations for the remaining time periods are summarized in Table 8.5. Note that cumulative water influx predicted with the Fetkovich method is less than the influx calculated with the Carter-Tracy method by comparing Tables 8.4 and 8.5. This is because the Fetkovich method does not include transient effects for unsteady-state conditions in early time periods, and is limited in its application to infinite-acting aquifers.

Table 8.5 Water Influx Calculations Using the Carter-Tracy Method for Example 8.4

n	t (days)	$p_{R,n}$ (psia)	t_D	$\Delta p(t_D)$	B (RB/psi)	p_D	P'_D	W_e (RB)
0	0	3665	0.00	0	4582.08	0.0000	0.0000	0
1	365	3586	48.30	79	4582.08	2.3718	0.0099	7.37E+6
2	730	3543	96.61	122	4582.08	2.7061	0.0050	1.76E+7
3	1095	3520	144.91	145	4582.08	2.9043	0.0034	2.89E+7
4	1460	3491	193.21	174	4582.08	3.0457	0.0025	4.20E+7

Generally, it is recommended practice that a water-drive gas reservoir should be modeled with a numerical aquifer in a reservoir simulator. In constructing the reservoir model, the aquifer size should reflect the assumed reservoir-aquifer ratio depending on whether the aquifer is considered to be finite- or infinite-acting. The Havlena-Odeh plot can assist in the decision pertaining to the size and strength of the aquifer. If sufficient production history is available, a reasonable aquifer size will give an acceptable history match. This aquifer size should be reviewed to ensure that it is reasonable and fits with the known geology of the area. Construction of gas reservoir models with numerical aquifers reduces the difficulties associated with determining the rate of water influx using analytical models such as Fetkovich, and Carter Tracy models. In addition, it provides a versatile tool that can be used to investigate many possible water drive mechanisms that may be occurring in the reservoir. The applications of reservoir simulation in evaluation of reservoir performance are discussed in Chapter 19.

8.5 Geopressured Gas Reservoirs

Geopressured gas reservoirs are abnormally pressured reservoirs which have been encountered all over the world, including the Gulf Coast region of the United States. The initial pressure gradients in normally pressured reservoirs range between 0.43 psi/ft and 0.5 psi/ft. In geopressured reservoirs, initial pressure gradients are between 0.6 psi/ft and 1.0 psi/ft. According to Poston and Berg,[14] the higher pressures in geopressured reservoirs are probably caused by inability of excess fluids to leak-off after major tectonic events of compressional folding or rapid deposition of thick sediments in young sedimentary sections. Figure 8.9 shows the occurrences of geopressured zones along the Gulf Coast of the United States of America.

In normally pressured gas reservoirs, gas compressibility is dominant in comparison to water and formation (rock) compressibility as the source of energy for gas production. For this reason, water and formation compressibility were considered negligible in the development of the material balance equation (Eq. (8.14)) for volumetric, normally pressured gas reservoirs. In geopressured reservoirs, water and formation compressibility can be almost as high as gas compressibility at initial pressures. Consequently, it is necessary to include water and formation compressibility in developing material balance equations for geopressured reservoirs. If the conventional p/z versus G_p plot developed for normally pressured gas reservoirs (Eq. (8.18) is applied to geopressured gas reservoirs, the original gas in place will be overestimated at early stages of reservoir depletion. The typical shape of p/z vs. G_p plot for a geopressured gas reservoir is shown in Figure 8.10. Note the typical dual slope of the plot, representing the flatter slope during the early production history when water and rock compressibility effects are significant and the steeper slope at later stages of depletion when gas compressibility is dominant. Extrapolation of the early straight line will give an apparent gas in place (AGIP), which is significantly higher than the actual original gas in place (OGIP). This could cause significant errors in the estimation of OGIP and reserves for geopressured reservoirs. To avoid this error, two methods are presented for calculation of OGIP for geopressured gas reservoirs. The methods account for the effects of water and formation compressibility in the development of material balance equations for geopressured reservoirs.

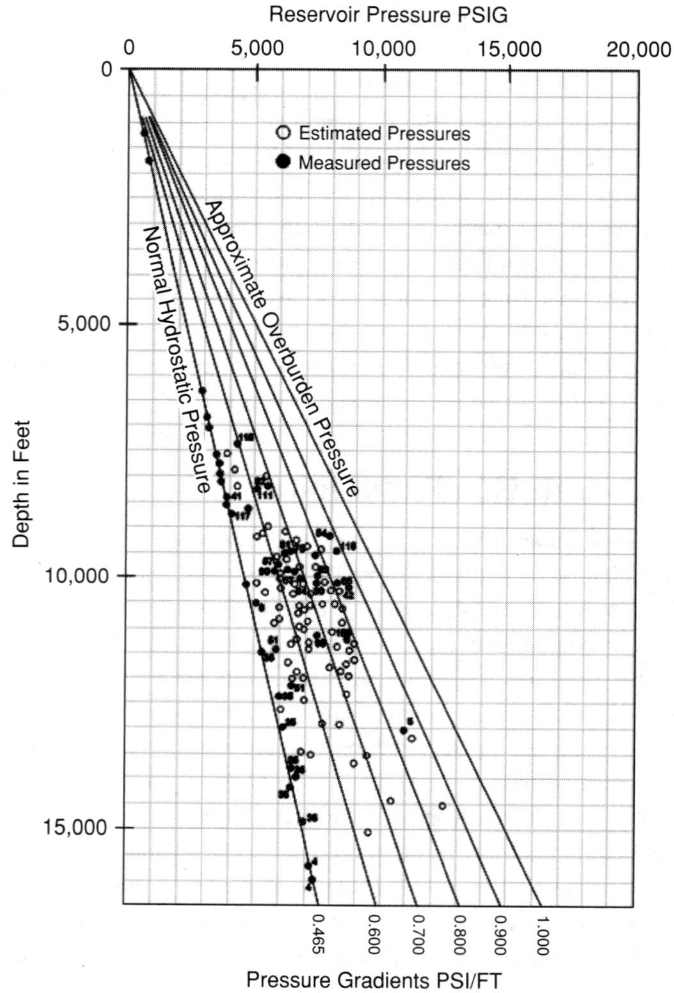

Figure 8.9 Geopressured zones along Gulf Coast, USA. (from Dickinson[15]).

8.5.1 The Ramagost and Farshad Method

The GMBE for a geopressured gas reservoir, assuming no water influx and negligible water production, can be derived from Eq. (8.15) as:

$$G_pB_g = GB_{gi}\left(\frac{c_wS_{wi} + c_f}{1 - S_{wi}}\right)\Delta p + G(B_g - B_{gi}) \tag{8.48}$$

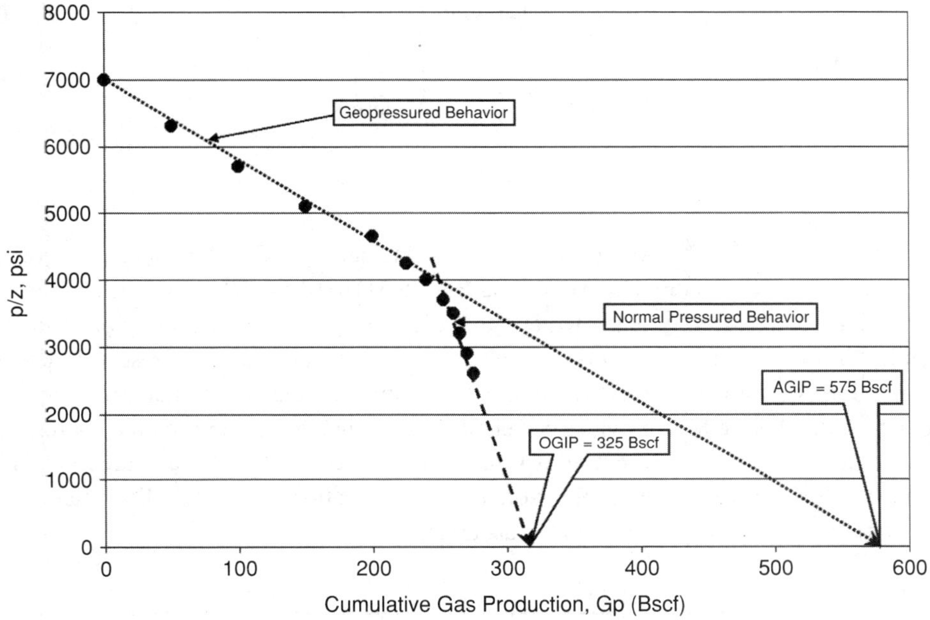

Figure 8.10 Typical p/z vs. G_p plot for a geopressured gas reservoir.

Equation (8.48) can be rearranged to get:

$$G = \frac{G_p B_g}{B_g - B_{gi} + \left[\frac{B_{gi}(p_i - p) \times (c_f + c_w S_{wi})}{1 - S_{wi}}\right]} \tag{8.49}$$

Equation (8.49) was further rearranged by Ramagost and Farshad[16] to develop a useful form of the material balance equation for geopressured reservoirs which is similar in form to Eq. (8.18) for normally pressured gas reservoirs:

$$\frac{p}{z}\left[1 - \left(\frac{c_f + c_w S_{wi}}{1 - S_{wi}}\right)(p_i - p)\right] = \frac{p_i}{z_i} - \left(\frac{p_i}{z_i G}\right)G_p \tag{8.50}$$

Equation (8.50) is different from Eq. (8.18) only by the expression: $[1 - \{(c_f + c_w S_{wi})/(1 - S_{wi})\}(p_i - p)]$. This expression accounts for the effects of water and rock compressibility on the p/z ratio of geopressured reservoirs. Eq. (8.50) is in the form of the equation for a straight line where:

$$Y \text{ variable} = \frac{p}{z}\left[1 - \left(\frac{c_f + c_w S_{wi}}{1 - S_{wi}}\right)(p_i - p)\right] \tag{8.51}$$

$$X \text{ variable} = G_p \quad (8.52)$$

$$\text{slope, } m = -\left(\frac{p_i}{z_i G}\right) \quad (8.53)$$

$$Y \text{ intercept} = \frac{p_i}{z_i} \quad (8.54)$$

The Ramagost and Farshad method can be used to calculate OGIP by making a plot of the production data as shown in Example 8.5 using Eqs. (8.51) to (8.54). This approach assumes that the formation compressibility is known and that it is a single constant value. Formation compressibility is not constant and may be variable, especially at early stages of reservoir depletion. The variability of formation compressibility has been reported by Poston and Chen[17] and Fetkovich et al.[18] Figure 8.11 provides correlations for estimation of formation compressibility. These correlations should only be used if formation compressibility data measured in a laboratory for the reservoir are not available. Note that the formation compressibility obtained from these correlations is most likely not highly accurate.

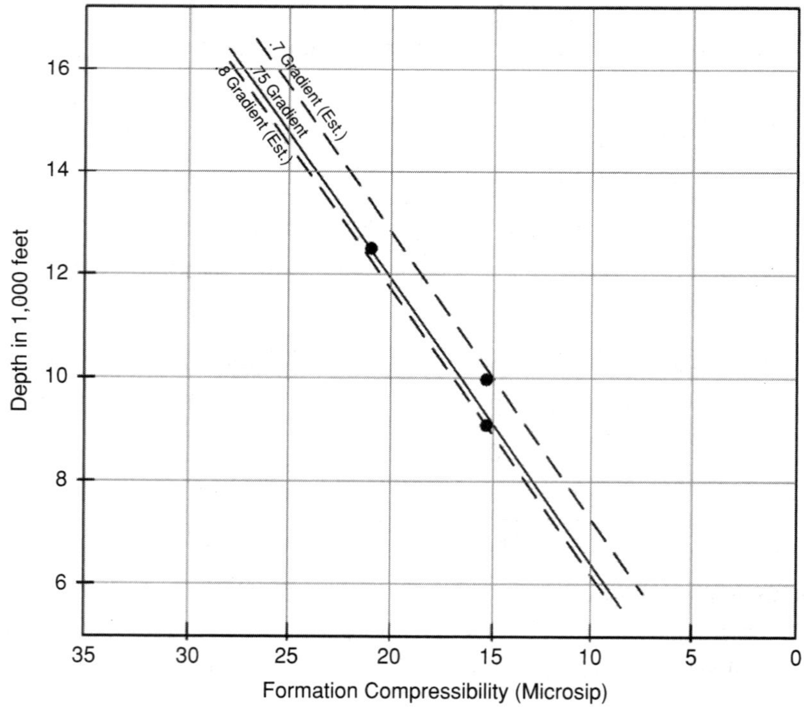

Figure 8.11 Correlation for estimation of formation compressibility (from Hammerlindl[19] © 1971 SPE, Reproduced with permission).

Example 8.5 Calculation of Original Gas in Place Using the Ramagost-Farshad Method

Problem

Calculate the OGIP with the Ramagost-Farshad method for a geopressured reservoir using the reservoir data and production history of the Anderson "L" reservoir as reported by Duggan.[20] The data summary for the Anderson "L" reservoir is as follows:

Depth	11,167 ft. (subsea)
Porosity, ϕ	24%
Initial Water Saturation, S_{wi}	35%
Permeability, k	0.7 to 791 md
Estimated Dew Point Pressure	6118 psia
Initial Bottomhole Pressure	9507 psia
Initial Pressure Gradient	0.843 psi/ft.
Bottom-hole Temperature	266°F
Volumetric OGIP	2240 Mcf/ac-ft. or 69.552 Bcf

Solution

The formation compressibility was estimated from Figure 8.11 as 19.3×10^{-6} psi^{-1} and the compressibility of water was estimated as 3.2×10^{-6} psi^{-1}. The calculations were performed on a spreadsheet as shown in Table 8.6 using Eqs. (8.51) to (8.54). The data in Table 8.6 were used to plot Figure 8.12. Note in Figure 8.12, the OGIP was shown to be 68 Bcf. This is comparable to volumetrically calculated OGIP of 69.6 Bcf reported by Duggan.[20] The AGIP was shown to be 107 Bcf based on the p/z plot. This represents an error of about 54% if the p/z method was used early in this reservoir to estimate OGIP.

8.5.2 The Roach Method

Roach[21] re-arranged the familiar material balance equation for geopressured reservoir shown in Eq. (8.49) in an alternative form that treats the formation compressibility and original gas in place as the unknown variables. The Roach[21] form of the material balance equation is as follows:

$$\frac{1}{(p_i - p)}\left(\frac{p_i z}{p z_i} - 1\right) = \frac{1}{G}\left[\frac{G_p}{(p_i - p)}\left(\frac{p_i z}{p z_i}\right)\right] - \left[\frac{c_f + c_w S_{wi}}{1 - S_{wi}}\right] \tag{8.55}$$

Again, Eq. (8.55) is in the form of equation for a straight line where:

$$Y \text{ variable} = \frac{1}{(p_i - p)}\left(\frac{p_i z}{p z_i} - 1\right) \tag{8.56}$$

Table 8.6 Data and Calculations for Example 8.5 Using the Ramagost-Farshad method

DATE	Bottomhole Pressure (psia)	Compressi-bility Factor, *z*	Cumulative Gas Production (MMscf)	*p/z*	Y Variable
12/22/65	9507	1.440	0.0	6602.1	6602.1
03/01/66	9292	1.418	392.5	6552.9	6508.6
06/22/66	8970	1.387	1642.2	6467.2	6358.1
09/29/66	8595	1.344	3225.8	6395.1	6211.9
11/17/66	8332	1.316	4260.3	6331.3	6097.6
12/30/66	8009	1.282	5503.5	6247.3	5953.3
03/23/67	7603	1.239	7538.1	6136.4	5769.4
05/15/67	7406	1.218	8749.2	6080.5	5679.1
07/31/67	7002	1.176	10509.3	5954.1	5485.5
09/14/67	6721	1.147	11758.9	5859.6	5346.8
10/19/67	6535	1.127	12789.2	5798.6	5257.2
03/05/68	5764	1.048	17262.5	5500.0	4853.3
09/04/68	4766	0.977	22890.8	4878.2	4151.6
03/19/69	4295	0.928	28144.6	4628.2	3870.4
09/29/69	3750	0.891	32566.7	4208.8	3447.6
03/31/70	3247	0.854	36819.9	3802.1	3054.4

$$X \text{ variable} = \left[\frac{G_p}{(p_i - p)}\left(\frac{p_i z}{p z_i}\right)\right] \tag{8.57}$$

$$\text{slope, } m = \frac{1}{G} \tag{8.58}$$

$$Y \text{ intercept} = -\left[\frac{c_f + c_w s_{wi}}{1 - s_{wi}}\right] \tag{8.59}$$

Example 8.6 is the application of the Roach method to the same data used in Example 8.5. Note that this method assumes that formation compressibility is constant although unknown. The strength of the Roach method is that it can be applied without prior knowledge of the data for formation compressibility.

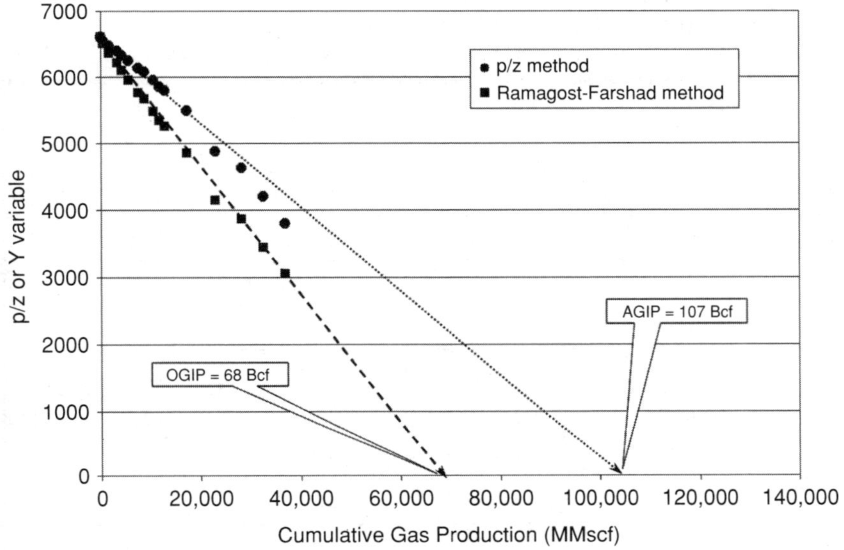

Figure 8.12 Comparison of p/z method and Ramagost-Farshad method for the Anderson "L" Reservoir in Example 8.5.

Example 8.6 Calculation of the Original Gas in Place Using the Roach Method

Problem

Calculate the OGIP with the Roach method using the Anderson "L" reservoir data in Example 8.5.

Solution

The terms for the x-axis and the y-axis were calculated in a spreadsheet as shown in Table 8.7. Figure 8.13 is the plot of the Roach method based on the data in Table 8.7. The slope of the straight line in Figure 8.13 is $m = 1.4082 \times 10^{-5}$ MMscf^{-1}. From Eq. (8.58):

$$\text{slope, } m = \frac{1}{G}$$

$$\text{or } G = \frac{1}{m} = \frac{1}{1.4082 \times 10^{-5}}$$

$$= 71 \text{ Bcf}$$

The OGIP calculated with the Roach method compares favorably to the volumetric OGIP (69 Bcf) reported by Duggan[20] and the OGIP of 68 Bcf calculated in Example 8.5 using the Ramagost-Farshad method.

Table 8.7 Data and Calculations for Example 8.6 Using the Roach Method

DATE	Bottomhole Pressure (psia)	Compressibility Factor, *z*	Cumulative Gas Production (MMscf)	X Variable	Y Variable
12/22/65	9507	1.440	0.0	0.00	0.00E+00
03/01/66	9292	1.418	392.5	1.84	3.49E–05
06/22/66	8970	1.387	1642.2	3.12	3.88E–05
09/29/66	8595	1.344	3225.8	3.65	3.55E–05
11/17/66	8332	1.316	4260.3	3.78	3.64E–05
12/30/66	8009	1.282	5503.5	3.88	3.79E–05
03/23/67	7603	1.239	7538.1	4.26	3.99E–05
05/15/67	7406	1.218	8749.2	4.52	4.08E–05
07/31/67	7002	1.176	10509.3	4.65	4.34E–05
09/14/67	6721	1.147	11758.9	4.76	4.55E–05
10/19/67	6535	1.127	12789.2	4.90	4.66E–05
03/05/68	5764	1.048	17262.5	5.54	5.35E–05
09/04/68	4766	0.977	22890.8	6.53	7.45E–05
03/19/69	4295	0.928	28144.6	7.70	8.18E–05
09/29/69	3750	0.891	32566.7	8.87	9.88E–05
03/31/70	3247	0.854	36819.9	10.21	1.18E–04

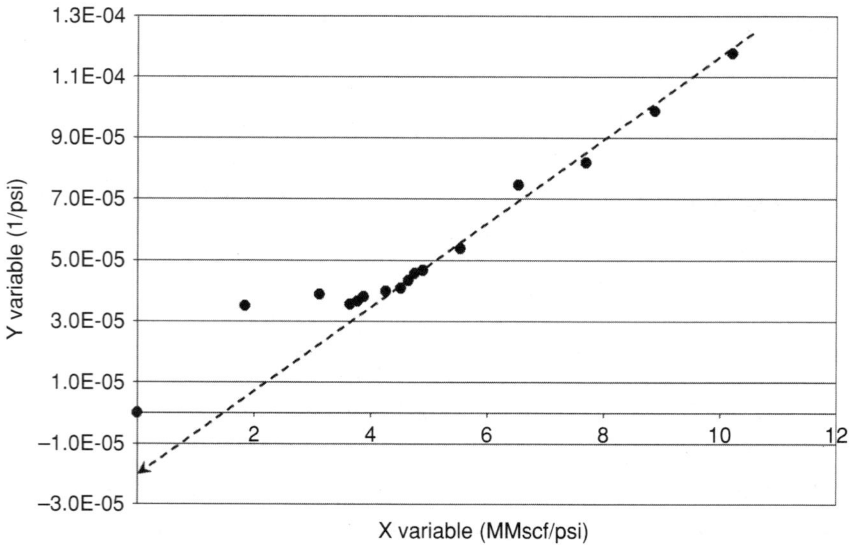

Figure 8.13 Plot of the Roach method for the Anderson "L" reservoir in Example 8.6.

The y-intercept in Figure 8.13 is -2.2×10^{-5} psi^{-1}. From the y-intercept, the formation compressibility factor can be estimated from Eq. (8.59):

$$
\begin{aligned}
Y_{\text{intercept}} &= -\left[\frac{c_f + c_w s_{wi}}{1 - s_{wi}}\right] \\
-2.2 \times 10^{-5} &= -\left[\frac{c_f + 3.2 \times 10^{-6} \times 0.35}{1 - 0.35}\right] \\
c_f &= (1 - 0.35) \times 2.2 \times 10^{-5} - 3.2 \times 10^{-6} \times 0.35 \\
&= 13.18 \times 10^{-6}\ \text{psi}^{-1}
\end{aligned}
$$

Note that the formation compressibility estimated from Figure 8.11 in Example 8.5 is 19.3 (10^{-6} psi^{-1}). The difference illustrates the fact that formation compressibility data exhibit a wide range of values and should be used with caution.

8.6 Case Histories of Two Gas Reservoirs

The case histories of two gas reservoirs are presented to serve as examples for discussions on performance of gas reservoirs. These reservoirs were selected because they have been economically depleted and no additional production is expected. One of the reservoirs is classified as a dry gas reservoir, and the other is classified as a retrograde gas condensate reservoir. The case histories have been condensed from all the activities that occurred in these reservoirs and only main events that could have affected reservoir performance are presented. The data presented are considered sufficient for routine reservoir analyses and should provide some insights on the performance of the reservoirs. The case histories are presented under the following categories: reservoir geology, reservoir rock and fluid properties, and reservoir pressure and production data. For each case history, some review questions are suggested to assist in the discussion and analysis of reservoir performance.

8.6.1 The Case History of Red Hawk Reservoir

The Red Hawk gas reservoir is located in Garden Banks (GB) Block 877 in the Gulf of Mexico (GOM), USA as shown in Figure 8.14. The water depth at GB Block 877 is about 5300 ft, which is classified as deep water in the GOM. The Red Hawk gas reservoir was discovered in August 2001. Gas production began in July 2004 through the first cell spar facility built in the world. The two main productive intervals in the Red Hawk reservoir were depleted in April 2008.

8.6.1.1 Reservoir Geology

The Red Hawk reservoir is composed of sands trapped on the west flank of a large salt dome. The Upper Pliocene sands can be described as slope fans with imbedded channel/levee complex. The three productive intervals present in the reservoir are named the Brouweri, the Surculus, and the Tamalis, after the nanofossils found in each interval. The type log for the Red Hawk reservoir

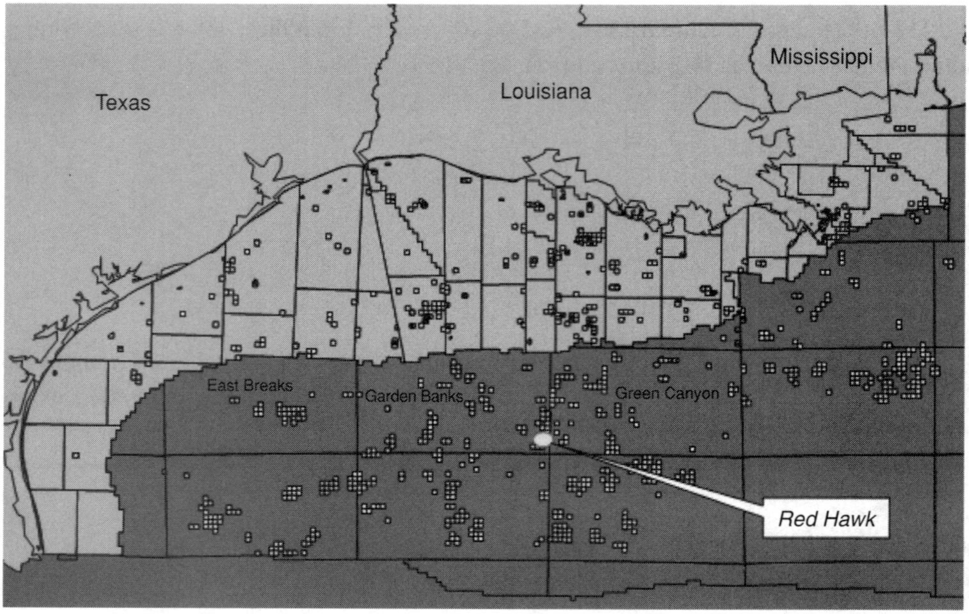

Figure 8.14 Location of Red Hawk reservoir in Gulf of Mexico, USA.

is shown in Figure 8.15. The Brouweri sands appear to contain limited reserves and were not part of the initial development. Structure maps on the Surculus and the Tamalis intervals are shown as Figure 8.16 and Figure 18.17, respectively. As shown in the structure maps, the two productive intervals are rotated at steep angles near the salt dome. The structure maps show the presence of faults which appear to isolate portions of the reservoir. The Red Hawk reservoirs were penetrated with seven boreholes during discovery, appraisal, and development phases. The locations of the penetrations are shown in Figure 18.18 with the observations at the locations summarized in Table 8.8. The gas-water contact was located at 18,250 feet TVD subsea for the Surculus and Tamalis intervals on the basis on wireline formation tester data.

8.6.1.2 Reservoir Rock and Fluid Properties

The average net pay thickness determined for the Surculus and Tamalis intervals are 45 feet and 99 feet, respectively. The average rock properties for each interval are summarized in Table 8.9. The composition of the reservoir fluid from a typical bottomhole sample obtained from laboratory analysis is shown in Table 8.10. The gas-liquid ratio for the sample is 234,000 scf/STB. This low liquid yield essentially led to the classification of Red Hawk as a dry gas reservoir. Also, note the high composition of methane in the sample.

8.6.1.3 Reservoir Pressure and Production Data

The initial reservoir pressures of the Surculus and Tamalis intervals are 8923 psia and 9000 psia, respectively. The gas reserves in the Surculus and Tamalis intervals were recovered through two wells.

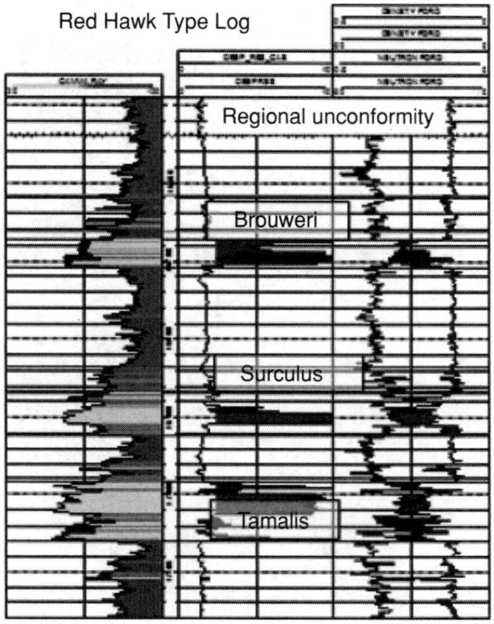

Figure 8.15 Type log for Red Hawk reservoir.

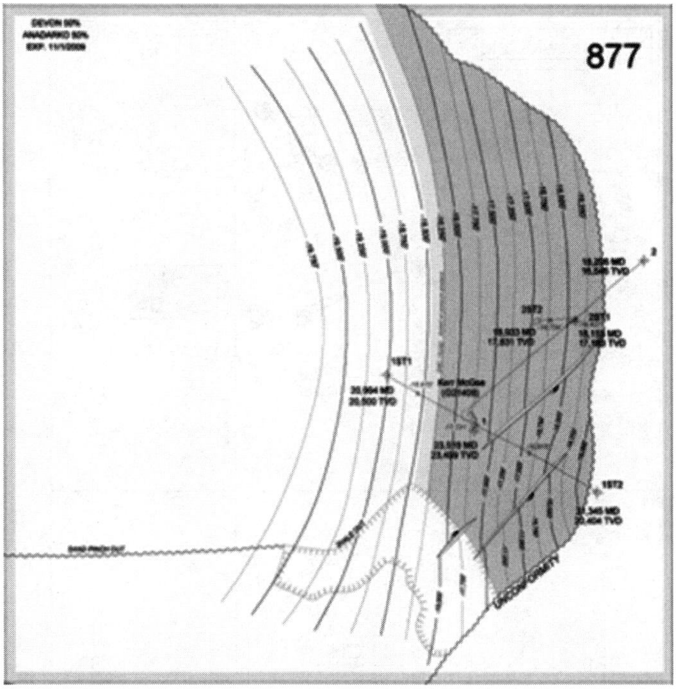

Figure 8.16 Structure map for Surculus interval.

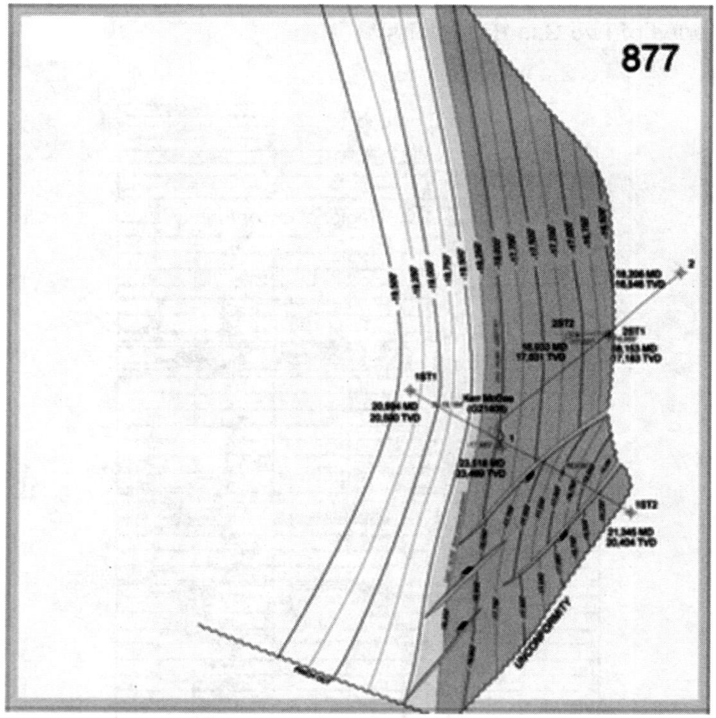

Figure 8.17 Structure map for Tamalis interval.

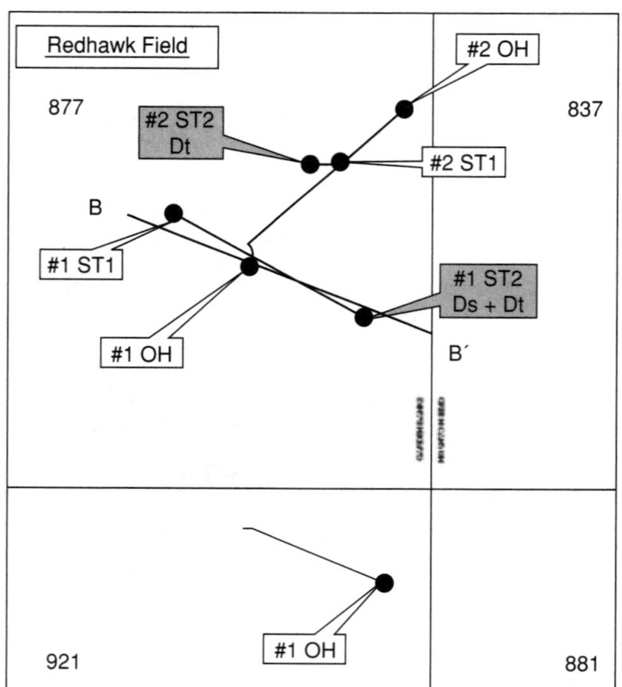

Figure 8.18 Location of wellbores in Red Hawk reservoir.

Table 8.8 Summary of Red Hawk Well Drilling History

Block Number	Borehole #	Date Drilled	Measured Depth (ft)	True Vertical Depth (ft)	Intervals Penetrated	Classification
877	1 OH	August 2001	23,518	23,500	D. Brouweri	Wet
					D. Surculus	Pay
					D. Tamalis	Pay
877	1 ST1	Sept. 2001	20,994	20,500	All 3 Sands	Wet
877	1 ST2	Oct. 2001	21,345	20,419	All 3 Sands	Pay
921	1	Dec. 2001	17,814	16,600	D. Brouweri	Pay
					D. Surculus	Wet
					D. Tamalis	Wet
877	2 OH	Sept. 2002	18,206	16,551	D. Brouweri	Wet
					D. Surculus	Absent
					D. Tamalis	Absent
877	2 ST1	Jan. 2003	18,153	17,186	D. Brouweri	Absent
					D. Surculus	Pay
					D. Tamalis	Faulted out
877	2 ST2	Jan. 2003	18,933	17,832	D. Brouweri	Absent
					D. Surculus	Pay
					D. Tamalis	Pay

Table 8.9 Average Rock Properties of Intervals in Red Hawk Reservoir

Rock Property	Intervals	
	Surculus	Tamalis
Porosity (%)	24	23
Permeability (md)	165	190
Water Saturation (%)	33	26
Total Compressibility (1/psi)	2.35E–04	2.35E–04

Table 8.10 Composition of Red Hawk Reservoir Fluid

Sample Depth 17,340 Feet Reservoir Fluid Flashed from 11,000 psia and 137°F to 15.025 psia and 60°F		
Sampling Conditions:	11,000 psia at 137°F	
Gas-Liquid Ratio	234,016	ft^3 of Stock Tank Vapors/bbl Stock Tank Liquid
API Gravity of Liquid	43.7	@60°F
Color of Stock Tank Liquid:	Clear	
Gas-liquid ratio is cubic feet of gas at 15.025 psia and 60°F per barrel of stock tank liquid at 60°F.		

Chromatograph Analysis of Flash Gas

Component	Mole %	GPM @ 15.025 psia	wt%	Mole Weight
Nitrogen	0.249	0.000	0.428	28.013
Carbon Dioxide	0.026	0.000	0.071	44.010
Hydrogen Sulfide	0.000	0.000	0.000	34.076
Methane	99.258	0.000	97.648	16.043
Ethane	0.150	0.041	0.277	30.070
Propane	0.067	0.019	0.182	44.097
Iso-Butane	0.026	0.009	0.094	58.123
N-Butane	0.032	0.010	0.116	58.123
Iso-Pentane	0.022	0.008	0.098	72.150
N-Pentane	0.021	0.008	0.092	72.150
Hexanes	0.034	0.014	0.179	85.817
Heptanes	0.036	0.015	0.208	96.798
Octanes	0.038	0.019	0.258	111.708
Nonanes	0.021	0.011	0.160	120.668
Decane Plus	0.020	0.014	0.189	142.285
Totals	100.000	0.168	100.000	

Calculated Properties of Gas

Gas Specific Gravity	(Air = 1.00)	= 0.5641
Net Heat of Combustion	(Btu/Cu.Ft. @15.025 Psia @ 60°F)	Dry = 940.8 Real
Gross Heat of Combustion	(Btu/Cu.Ft. @15.025 Psia @ 60°F)	Dry = 1044.4 Real
Gross Heat of Combustion—Sat.	(Btu/Cu.Ft. @ 15.025 Psia @ 60°F)	Wet = 1026.1 Water Sat.
Gas Compressibility	(@ 1 Atm. @ 60°F)	Z = 0.9978

Well 1ST2 was completed in the Surculus and Tamalis intervals, and Well 2ST2 was completed only in the Tamalis intervals (Figures 8.16, 8.17, and 8.18). A permanent downhole pressure gauge was installed in each well. These gauges were used to measure flowing bottomhole pressures during production, and static bottomhole pressures during periodic shut-ins. From these sources, historical static bottomhole pressures (SBHP) were estimated. The historical SBHP and monthly gas production rates are presented in Table 8.11 and Figure 8.19 for Well 1ST2, and Table 8.12 and Figure 8.20 for Well 2ST2. Gas production was initiated in the two wells in mid-July 2004 and climbed quickly to average above 56 MMCFPD in one month (Figures 8.19 and 8.20). These high gas production rates were maintained in the both wells until mid-September 2005, when both wells were shut-in due to damage to production facilities caused by the hurricane named Rita. The wells were returned to production in mid-May 2006 after eight months of shut-in. There were signs of aquifer influx into the two producing intervals during the shut-in period. These signs were reflected mostly in higher flowing bottomhole pressures measured after the wells were returned to production. However, no sharp increases in water production rates were observed in the two wells (Figures 8.21 and 8.22) after the shut-in period. Well 1ST2 showed steeper decline in reservoir pressures and productivity following the shut-in period due to Rita. This was attributed to the possibility that the aquifer moved updip and isolated Well 1ST2 within a compartment created by faults (Figure 8.16). In effect, the updip movement of the aquifer potentially reduced the effective drainage area of the well. There was no evidence of communication between Well 1ST2 and 2ST2 based on interference tests. A compressor was installed in October 2007 to assist production from both wells. Gas production from Well 1ST2 in the Surculus and Tamalis intervals was terminated in mid-February 2008. Cumulative gas production from Well 1ST2 is 49.84 Bcf. At this time also, Well 2ST2 was showing increased water production from the Tamalis interval. The processing facilities were limited to handling no more than 250 BWPD. In March 2008, the Surculus interval was tested in Well 2ST2. The tests showed that the Surculus interval in this well was completely swept by aquifer influx because it produced only water. Attempts to close the valve across the Surculus interval and re-open the valve across the Tamalis interval were not successful. Gas production from Well 2ST2 in the Tamalis interval was terminated in mid-April 2008. Cumulative gas production from Well 2ST2 is 73.02 Bcf.

8.6.1.4 Review Questions

The data presented in the case history of the Red Hawk reservoir could be used to assess the performance of the reservoir. Some review questions that could be addressed in the assessment are as follows:

1. Is the p/z method applicable to this reservoir?
2. How can the OGIP be assessed using the Havlena-Odeh method?
3. Can a simple tank simulation model be used to evaluate the performance of the reservoir, as discussed in Chapter 19?
4. What are the potential effects of shutting-in a gas reservoir under an active water drive?

Table 8.11 Historical Performance Data for Well 1ST2

DATE	SBHP (PSIA)	GAS RATE (MCFPD)	LIQUIDS RATE (BOPD)	WATER RATE (BWPD)
Jun-04	8923	0	0	0
Jul-04	8903	15,843	30	0
Aug-04	8640	55,955	34	0
Sep-04	8354	52,822	63	0
Oct-04	8100	56,730	68	0
Nov-04	7829	59,449	78	0
Dec-04	7585	60,677	80	0
Jan-05	7403	60,111	71	0
Feb-05	7265	62,611	71	0
Mar-05	6994	64,462	80	0
Apr-05	6805	62,332	102	5
May-05	6693	62,059	94	23
Jun-05	6515	59,756	126	29
Jul-05	6411	63,145	141	21
Aug-05	6221	53,395	181	27
Sep-05	6105	37,224	134	20
Oct-05	6105	0	0	0
Nov-05	6105	0	0	0
Dec-05	6105	0	0	0
Jan-06	6105	0	0	0
Feb-06	6105	0	0	0
Mar-06	6105	0	0	0
Apr-06	6105	0	0	0
May-06	6090	34,297	157	12
Jun-06	6075	63,666	176	31
Jul-06	6050	65,891	175	18
Aug-06	5778	64,536	161	16
Sep-06	5553	57,591	148	15
Oct-06	5350	55,544	136	14
Nov-06	5083	52,000	150	15

DATE	SBHP (PSIA)	GAS RATE (MCFPD)	LIQUIDS RATE (BOPD)	WATER RATE (BWPD)
Dec-06	4903	50,711	163	16
Jan-07	4563	49,775	220	20
Feb-07	4339	53,606	235	18
Mar-07	4090	52,675	217	17
Apr-07	3767	45,654	176	13
May-07	3347	38,639	120	9
Jun-07	2969	31,274	79	7
Jul-07	2642	25,619	50	5
Aug-07	2396	20,082	34	4
Sep-07	2171	15,767	29	3
Oct-07	2019	11,031	23	3
Nov-07	1915	7614	21	2
Dec-07	1815	6597	18	2
Jan-08	1730	6921	9	1
Feb-08	1654	2414	8	3
Mar-08	1685	0	0	0

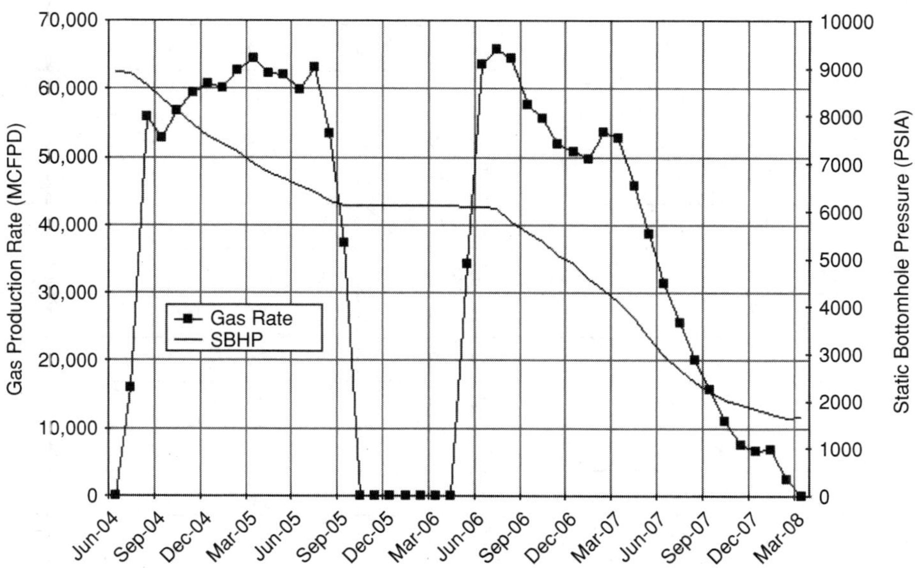

Figure 8.19 Historical pressure and gas production for Well 1ST2.

Table 8.12 Historical Performance Data for Well 2ST2

DATE	SBHP (PSIA)	GAS RATE (MCFPD)	LIQUIDS RATE (BOPD)	WATER RATE (BWPD)
Jun-04	9000	0	0	0
Jul-04	8950	14,816	45	0
Aug-04	8900	56,565	19	0
Sep-04	8608	53,749	37	0
Oct-04	8461	60,109	49	0
Nov-04	8238	66,777	58	0
Dec-04	8096	66,458	64	0
Jan-05	7966	63,298	77	0
Feb-05	7872	66,243	82	0
Mar-05	7754	64,965	87	0
Apr-05	7643	64,211	60	3
May-05	7541	64,386	65	10
Jun-05	7449	61,170	28	15
Jul-05	7396	59,994	11	6
Aug-05	7302	49,311	19	15
Sep-05	7250	39,395	2	1
Oct-05	7249	0	0	0
Nov-05	7249	0	0	0
Dec-05	7249	0	0	0
Jan-06	7249	0	0	0
Feb-06	7249	0	0	0
Mar-06	7249	0	0	0
Apr-06	7249	660	0	0
May-06	7200	41,773	29	22
Jun-06	7150	63,575	13	27
Jul-06	7076	64,454	8	3
Aug-06	6928	62,573	9	5
Sep-06	6866	58,353	16	8
Oct-06	6777	60,198	18	9
Nov-06	6665	64,363	12	6
Dec-06	6520	65,995	14	7
Jan-07	6378	62,319	19	10

DATE	SBHP (PSIA)	GAS RATE (MCFPD)	LIQUIDS RATE (BOPD)	WATER RATE (BWPD)
Feb-07	6249	73,354	18	9
Mar-07	6138	76,566	21	11
Apr-07	6056	75,150	19	10
May-07	5950	75,687	22	12
Jun-07	5859	73,029	18	10
Jul-07	5764	72,957	21	11
Aug-07	5670	73,819	21	12
Sep-07	5589	74,352	20	11
Oct-07	5557	70,249	18	10
Nov-07	5487	69,730	18	10
Dec-07	5415	74,225	17	9
Jan-08	5331	71,106	20	22
Feb-08	5345	62,926	30	155
Mar-08	5334	46,391	31	224
Apr-08	5317	12,992	24	271
May-08	5317	0	0	0

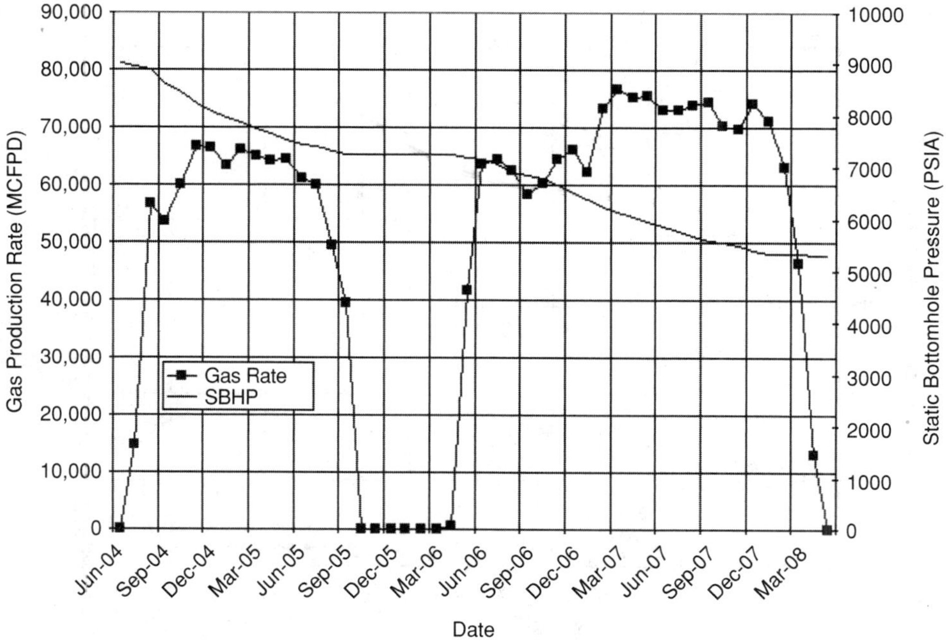

Figure 8.20 Historical pressure and gas production for Well 2ST2.

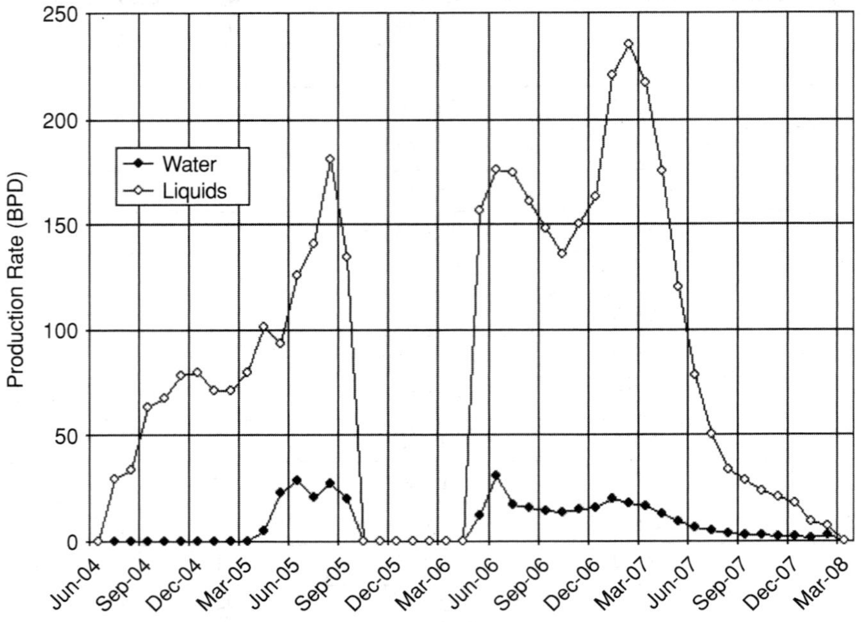

Figure 8.21 Historical liquids and water production for Well 1ST2.

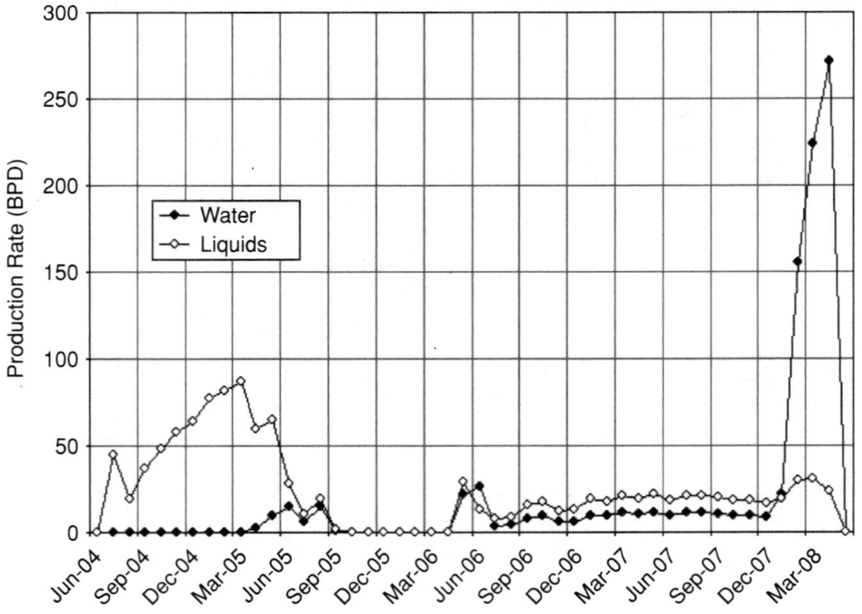

Figure 8.22 Historical liquids and water production for Well 2ST2.

8.6.2 The Case History of West Cameron 580 Reservoir

The West Cameron Block 580 (WC 580) field consists of mid-sized, over-pressured, and moderately rich retrograde gas condensate reservoirs located in the Gulf of Mexico at a water depth of 240 feet. The reservoirs are located mostly within WC 580 but extend partly into East Cameron Block 334 (EC 334). These blocks are located about 140 miles south of Lake Charles in Louisiana, U.S.A. The WC 580 reservoir consists of the M-4/M-4A and the M-6 reservoirs that extend over WC 580 and EC 334 blocks. These reservoirs were produced through separate wells.

8.6.2.1 Reservoir Geology

The sequences of sands that constitute the M-4/M-4A and the M-6 reservoirs are illustrated in a cross section drawn in Figure 8.23. The M-4/M-4A reservoir is structurally higher than the M-6 reservoir. The structure map of the M-4/M-4A reservoir is shown as Figure 8.24. The structure map of the M-6 reservoir is shown as Figure 8.25. Both reservoirs were trapped against a fault. A summary of the structural data for the M-4/M-4A and M-6 reservoirs is provided in Table 8.13.

8.6.2.2 Reservoir Rock and Fluid Properties

The average rock properties for the M-4/M-4A and M-6 reservoirs are presented in Table 8.14. The fluid properties data for a recombined sample taken at the separator from the M-4/M-4A

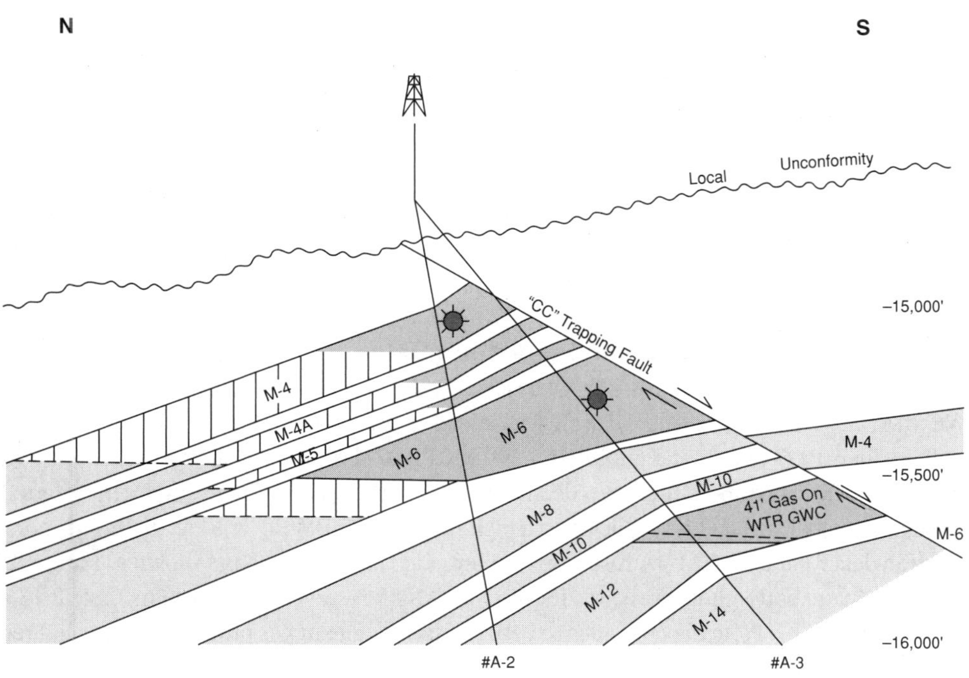

Figure 8.23 Cross-section of WC 580 reservoirs.

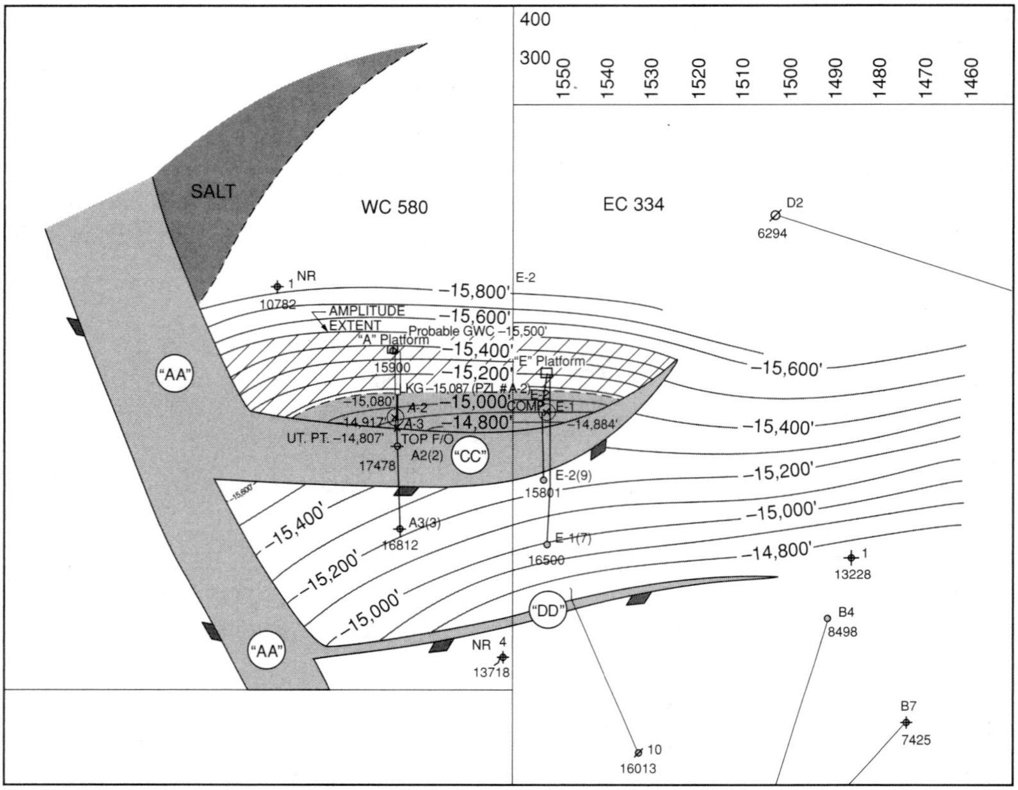

Figure 8.24 Structure map of the M-4/M-4A reservoirs.

reservoir is given in Table 8.15. Similar data for a recombined sample obtained at the separator for the M-6 reservoir are shown in Table 8.16. Note the high yield of condensate liquids obtained from both samples. The two reservoirs are classified as retrograde gas condensate reservoirs.

8.6.2.3 Reservoir Pressure and Production Data

Initial production started in the M-4/M-4A reservoirs in February 1997 through one well located in WC 580. In April 1997, a second well was added in EC 334. Peak production was achieved quickly in June 1997 at the total rate of 166,299 MCFPD. Gas production declined rapidly from the peak rate with corresponding steep decline in reservoir pressure. Water production started to increase in January 1998 and rose quickly to a high rate of 6883 BWPD in May 1999. The monthly production data for the M-4/M-4A reservoirs are provided in Table 8.17 and shown as Figure 8.26. The average static bottomhole pressures for the M-4/M-4A reservoirs are given in Table 8.18 and shown in Figure 8.27. Note the coincidence between the decline in gas production rates and reservoir pressures, and the rise in water production. Cumulative gas, liquids, and water production from the M-4/M-4A reservoirs are 56.8 Bcf, 1.9 MMBO, and 3.7 MMBW, respectively.

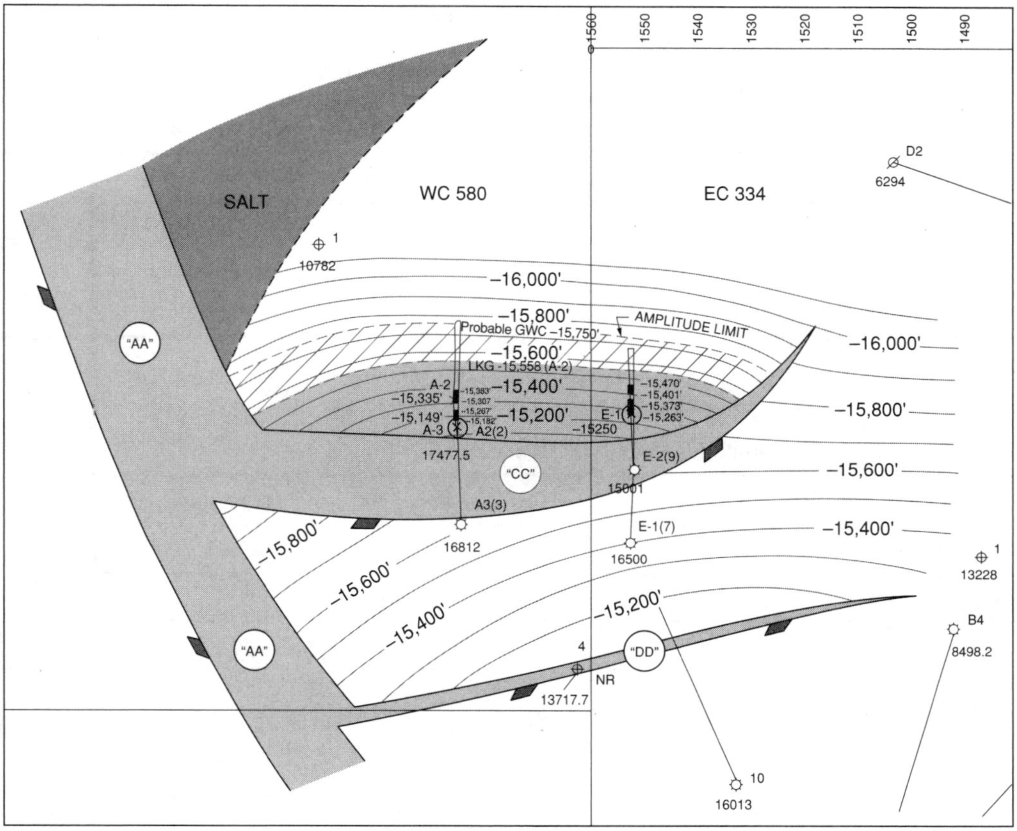

Figure 8.25 Structure map of the M-6 reservoir.

Table 8.13 Summary of Reservoir Structure Data for WC 580 Reservoirs

	Reservoir Name	
Structure Parameter	**M-4/M-4A**	**M-6**
Reservoir average TVD (ft, ss)	15,283	15,525
Lowest known gas (ft, ss)	15,087	15,558
Gas-water contact (ft, ss)	15,450	15,590
Total productive volume (ac.-ft)	31,956	35,886

Table 8.14 Average Rock Properties for WC 580 Reservoirs

Rock Property	Reservoir	
	M-4/M-4A	M-6
Porosity (%)	21.3	23.0
Permeability (md)	58.6	117.5
Initial water saturation (%)	28.6	20.0
Rock compressibility (1/psi)	7E-06	7E-06

Table 8.15 Fluid Properties for M-4/M-4A Reservoirs

Sample Type:	Recombined fluid sample at primary separator
Separator Temperature:	92°F
Separator Pressure:	1155 psia
GOR at separator:	23,370 scf/bbl separator liquid
Stock tank pressure:	15.025 psia
Stock tank temperature:	60°F
GOR at stock tank:	451 scf/STB
Condensate yield:	36.43 STB/MMscf
Condensate gravity:	44.8° API
Initial reservoir pressure:	12,721 psia
Reservoir temperature:	267°F
Dewpoint pressure at 267°F:	9912 psia
Composition of Reservoir Fluid	
Component	**Mole (%)**
Hydrogen Sulfide	0.00
Carbon Dioxide	0.93
Nitrogen	0.19
Methane	88.01
Ethane	5.30

Component	Mole (%)
Propane	1.82
i-Butane	0.42
n-Butane	0.50
i-Pentane	0.23
n-Pentane	0.19
Hexanes	0.24
Heptanes Plus	2.17
Total	**100.00**
Specific gravity of reservoir fluid: 0.7437 (Air = 1.0)	
Heptanes Plus molecular weight: 162.1 lb/lbmol	
Heptanes Plus specific gravity: 0.8125 (Water = 1.0)	

Table 8.16 Fluid Properties for M-6 Reservoir

Sample Type:	Recombined fluid sample at primary separator
Separator Temperature:	70°F
Separator Pressure:	1075 psia
GOR at separator:	17,899 scf/bbl separator liquid
Stock tank pressure:	15.025 psia
Stock tank temperature:	60°F
GOR at stock tank:	417 scf/STB
Condensate yield:	3.48 STB/MMscf
Condensate gravity:	45.2° API
Initial reservoir pressure:	12,825 psia
Reservoir temperature:	272°F
Dewpoint pressure at 267°F:	9850 psia

(*continued*)

Table 8.16 Fluid Properties for M-6 Reservoirs (continued)

Composition of Reservoir Fluid	
Component	**Mole (%)**
Hydrogen Sulfide	0.00
Carbon Dioxide	0.74
Nitrogen	0.17
Methane	87.85
Ethane	4.84
Propane	1.93
i-Butane	0.47
n-Butane	0.57
i-Pentane	0.25
n-Pentane	0.23
Hexanes	0.38
Heptanes Plus	2.57
Total	**100.00**
Specific gravity of reservoir fluid:	0.7714 (Air = 1.0)
Heptanes Plus molecular weight:	166.9 lb/lb-mol
Heptanes Plus specific gravity:	0.8140 (Water = 1.0)

Initial production in the M-6 reservoir started in October 1996 through a single well located in WC 580. A second well was put in production in EC 334 in April 1997. Peak production at the rate of 165,663 MCFPD was achieved in September 1997. The performance of the M-6 reservoir is very similar to the performance of the M-4/M-4A reservoirs. The M-6 reservoir also shows rapid decline in gas production which is accompanied with rapid rise in water production. The monthly production data for the M-6 reservoir are provided in Table 8.19 and shown as Figure 8.28. The average static bottomhole pressures for the M-6 reservoir are given in Table 8.20 and shown in Figure 8.29. Cumulative gas, liquids, and water production from the M-6 reservoirs are 74.2 Bcf, 3.3 MMBO, and 13.0 MMBW, respectively.

Table 8.17 Historical Production Data for M-4/M-4A Reservoirs

DATE	Gas Rate (MCFPD)	Liquids Rate (BOPD)	Water Rate (BWPD)
Feb-97	12,381	577	1
Mar-97	42,633	1,784	1
Apr-97	137,111	5,054	9
May-97	163,106	6,409	4
Jun-97	166,299	5,790	20
Jul-97	163,112	5,586	27
Aug-97	148,001	4,829	24
Sep-97	128,068	4,216	17
Oct-97	81,525	2,859	87
Nov-97	73,940	2,573	1
Dec-97	64,649	2,229	48
Jan-98	76,924	2,284	363
Feb-98	45,577	1,307	352
Mar-98	75,475	2,348	885
Apr-98	71,929	1,962	1,018
May-98	73,953	1,896	1,115
Jun-98	57,066	1,460	1,419
Jul-98	27,560	856	1,691
Aug-98	26,414	810	1,908
Sep-98	22,468	651	2,063
Oct-98	23,469	683	2,535
Nov-98	22,937	726	3,026
Dec-98	21,803	691	3,437
Jan-99	19,751	649	3,797
Feb-99	18,721	636	4,345
Mar-99	19,374	612	5,402
Apr-99	18,666	599	6,415
May-99	14,438	584	6,883
Jun-99	6,493	231	3,758
Jul-99	3,144	140	3,848

(continued)

Table 8.17 Historical Production Data for M-4/M-4A Reservoirs (continued)

DATE	Gas Rate (MCFPD)	Liquids Rate (BOPD)	Water Rate (BWPD)
Sep-99	2,845	80	3,751
Sep-99	3,149	117	3,781
Nov-99	3,994	62	4,629
Nov-99	2,480	65	4,489
Jan-00	1,970	68	5,012
Feb-00	2,136	30	5,626
Feb-00	2,076	42	5,607
Mar-00	1,952	42	4,423
Apr-00	1,803	33	3,922
May-00	2,475	43	4,259
Jun-00	2,444	37	4,105
Jul-00	1,348	13	2,835
Aug-00	2,131	13	3,442
Sep-00	2,146	11	3,570
Oct-00	1,792	10	3,933
Nov-00	0	12	4,418

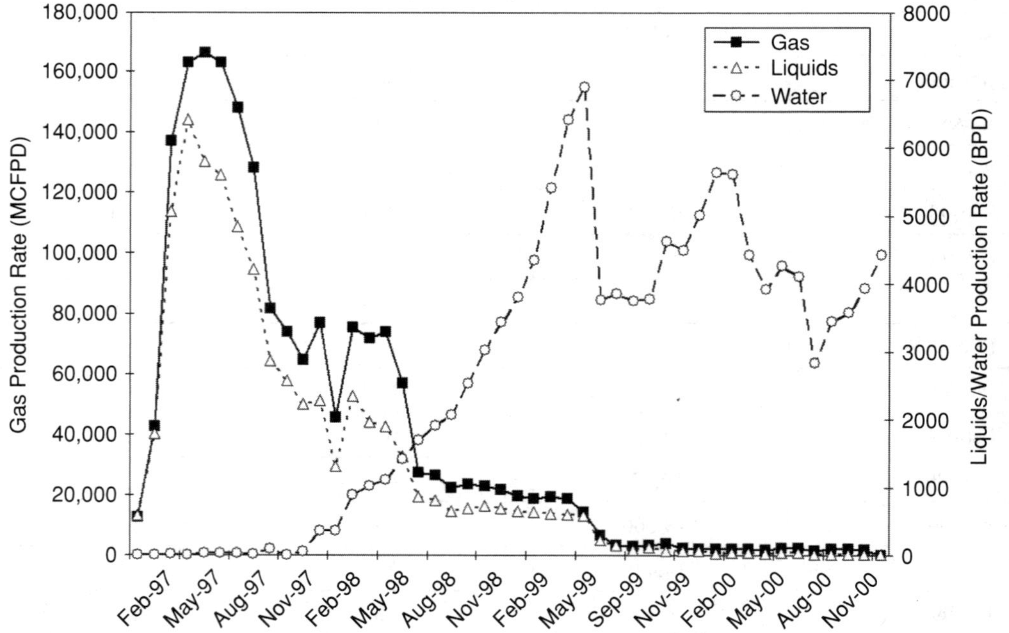

Figure 8.26 Historical production data for M-4/M-4A reservoirs.

Table 8.18 Average Reservoir Pressures for M-4/M-4A Reservoirs

DATE	Pressure (PSIA)	DATE	Pressure (PSIA)
Oct-96	12,721	Dec-98	4,032
Nov-96	12,721	Jan-99	3,925
Dec-96	12,721	Feb-99	3,823
Jan-97	12,721	Mar-99	3,736
Feb-97	12,555	Apr-99	3,634
Mar-97	12,327	May-99	3,532
Apr-97	11,710	Jun-99	3,456
May-97	10,396	Jul-99	3,452
Jun-97	9,366	Aug-99	3,459
Jul-97	8,492	Sep-99	3,467
Aug-97	7,654	Oct-99	3,476
Sep-97	6,819	Nov-99	3,486
Oct-97	6,455	Dec-99	3,496
Nov-97	6,211	Jan-00	3,506
Dec-97	5,866	Feb-00	3,516
Jan-98	5,751	Mar-00	3,526
Feb-98	5,916	Apr-00	3,538
Mar-98	5,620	May-00	3,550
Apr-98	5,259	Jun-00	3,563
May-98	4,867	Jul-00	3,576
Jun-98	4,593	Aug-00	3,590
Jul-98	4,583	Sep-00	3,603
Aug-98	4,467	Oct-00	3,615
Sep-98	4,345	Nov-00	3,627
Oct-98	4,251	Dec-00	3,640
Nov-98	4,140		

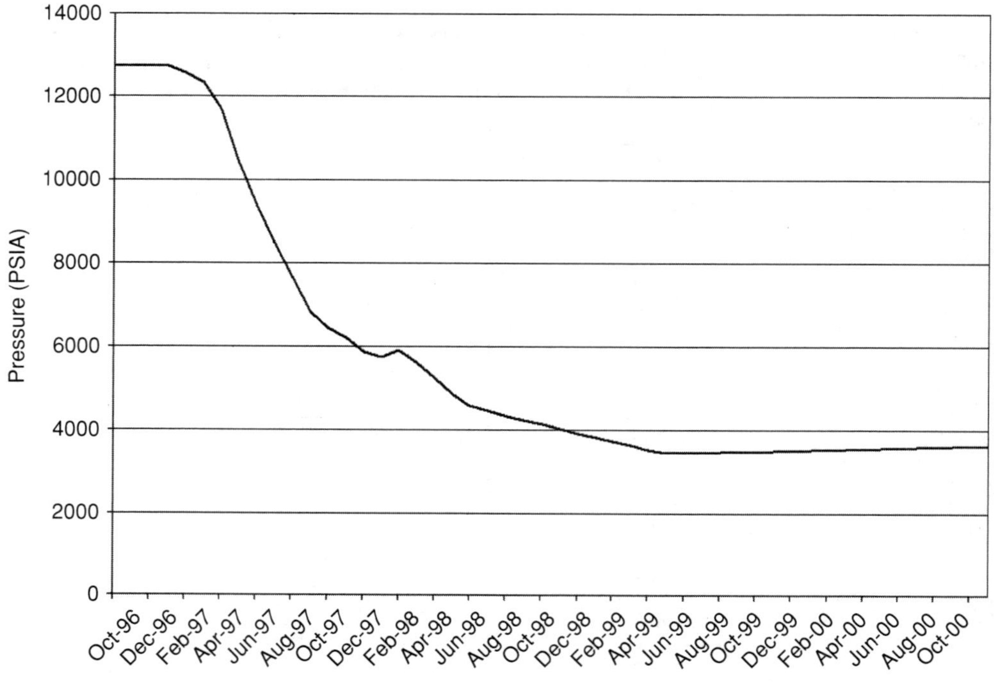

Figure 8.27 Average reservoir pressures for M-4/M-4A reservoirs.

8.6.2.4 Review Questions

The data presented in the case history of the WC 580 reservoirs could be used to assess the performance of the reservoirs. Some review questions that could be addressed in the assessment are as follows:

1. Is the p/z method applicable to these reservoirs?
2. Compare the OGIP calculated volumetrically to those determined by other methods such as the Ramagost and Farshad and Roach methods. Are these methods applicable to these reservoirs?
3. Can a simple tank simulation model be used to evaluate the performance of the reservoir as discussed in Chapter 19?
4. What are the potential effects of shutting-in a gas reservoir under an active water drive?

Table 8.19 Historical Production Data for M-6 Reservoir

DATE	Gas Rate (MCFPD)	Liquids Rate (BOPD)	Water Rate (BWPD)
Oct-96	44,452	3,035	1
Nov-96	60,025	3,561	1
Dec-96	0	0	1
Jan-97	0	0	1
Feb-97	0	0	1
Mar-97	28,766	1,584	1
Apr-97	131,272	6,503	3
May-97	149,188	7,101	11
Jun-97	155,106	7,129	36
Jul-97	154,536	7,292	35
Aug-97	154,004	7,056	33
Sep-97	165,663	7,585	32
Oct-97	161,093	7,353	176
Nov-97	157,661	6,998	746
Dec-97	132,358	5,947	1,669
Jan-98	116,736	5,231	4,607
Feb-98	55,396	2,180	6,583
Mar-98	111,881	4,841	7,570
Apr-98	103,241	4,440	7,432
May-98	71,407	3,044	7,996
Jun-98	37,024	1,362	9,612
Jul-98	32,449	1,191	10,735
Aug-98	31,924	1,157	12,116
Sep-98	12,540	482	7,702
Oct-98	21,303	1,015	11,901
Nov-98	30,851	1,118	14,402
Dec-98	19,748	792	13,352
Jan-99	19,449	878	13,417
Feb-99	19,993	812	13,087

(continued)

Table 8.19 Historical Production Data for M-6 Reservoir (continued)

DATE	Gas Rate (MCFPD)	Liquids Rate (BOPD)	Water Rate (BWPD)
Mar-99	19,843	758	12,955
Apr-99	17,947	571	11,458
May-99	9,385	282	7,330
Jun-99	8,290	283	10,078
Jul-99	14,590	467	12,401
Aug-99	15,939	628	13,335
Sep-99	6,290	212	10,315
Oct-99	13,161	393	12,253
Nov-99	13,263	395	12,590
Jan-00	12,864	413	13,099
Feb-00	12,606	396	13,203
Mar-00	7,462	272	7,307
Apr-00	6,537	245	7,246
May-00	6,461	235	7,622
Jun-00	6,349	232	7,359
Jul-00	7,431	266	8,336
Aug-00	7,025	246	8,222
Sep-00	6,470	224	8,300
Oct-00	6,745	225	8,266
Nov-00	6,240	207	7,770
Dec-00	2,765	93	4,400
Jan-01	4,144	143	8,868
Feb-01	6,047	228	9,023
Mar-01	5,889	218	9,108
Apr-01	5,402	199	8,701
May-01	5,345	204	9,728
Jun-01	4,748	159	9,057
Jul-01	4,144	125	7,101

Table 8.20 Average Reservoir Pressures for M-6 Reservoir

DATE	Pressure (PSIA)	DATE	Pressure (PSIA)
Sep-96	12,825	Mar-99	5,793
Oct-96	12,584	Apr-99	5,722
Nov-96	12,533	May-99	5,665
Dec-96	12,482	Jun-99	5,669
Jan-97	12,482	Jul-99	5,668
Feb-97	12,482	Aug-99	5,598
Mar-97	12,417	Sep-99	5,545
Apr-97	12,099	Oct-99	5,501
May-97	11,582	Nov-99	5,463
Jun-97	10,622	Dec-99	5,435
Jul-97	10,188	Jan-00	5,412
Aug-97	9,734	Feb-00	5,390
Sep-97	9,183	Mar-00	5,404
Oct-97	8,614	Apr-00	5,416
Nov-97	8,227	May-00	5,425
Dec-97	7,903	Jun-00	5,433
Jan-98	7,636	Jul-00	5,440
Feb-98	7,528	Aug-00	5,446
Mar-98	7,053	Sep-00	5,451
Apr-98	6,719	Oct-00	5,455
May-98	6,499	Nov-00	5,459
Jun-98	6,404	Dec-00	5,463
Jul-98	6,296	Jan-01	5,466
Aug-98	6,188	Feb-01	5,469
Sep-98	6,185	Mar-01	5,471
Nov-98	6,119	Apr-01	5,474
Dec-98	5,997	May-01	5,476
Jan-99	5,927	Jun-01	5,478
Feb-99	5,858	Jul-01	5,480

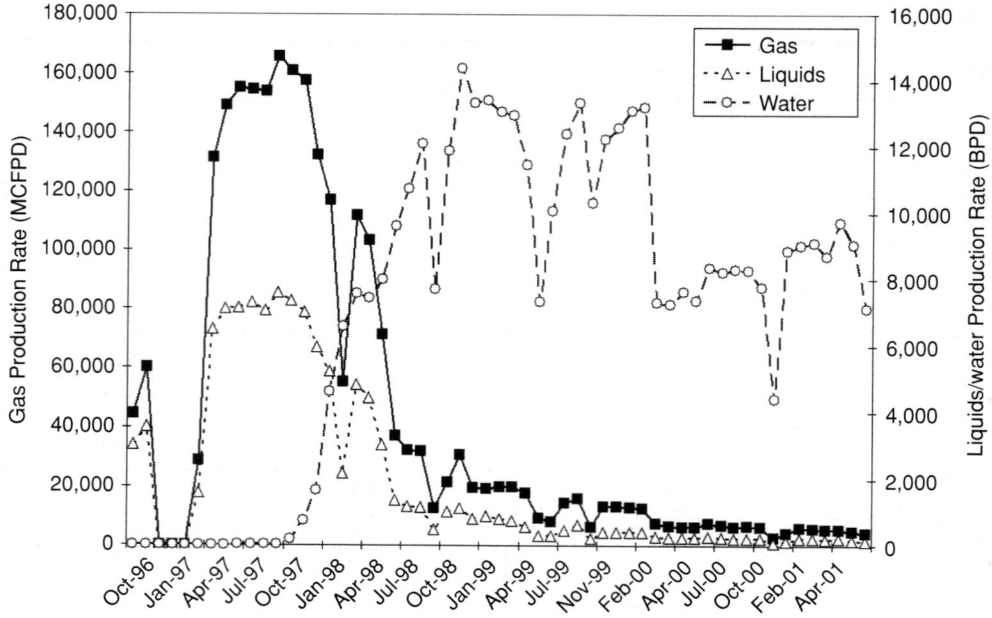

Figure 8.28 Historical production data for M-6 reservoir.

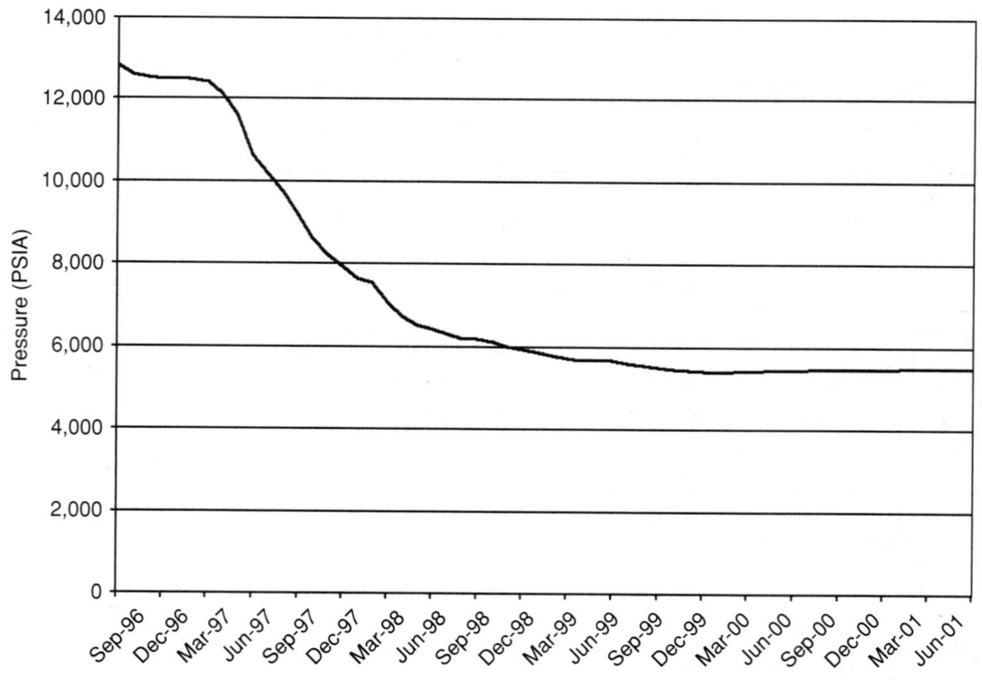

Figure 8.29 Average reservoir pressures for M-6 reservoir.

Nomenclature

A	reservoir area, acres
B	constant in Eq. 8.39, RB/psi
B_g	gas formation volume factor, RB/scf
B_w	water formation volume factor, RB/STB
c_f	formation compressibility, psi^{-1}
c_w	water compressibility, psi^{-1}
c_t	total compressibility, psi^{-1}
G	original gas in place, scf
G_{cp}	cumulative wet or condensate gas produced
G_p	cumulative gas produced, scf
G_r	gas remaining in reservoir, scf
h	net sand thickness, feet
k	permeability, md
J	aquifer productivity index, STB/day-psi
L	length of linear reservoir or aquifer, feet
M_o	molecular weight of stock tank oil, lbm/lb-mole
N_p	cumulative condensate produced, STB
n	moles of gas
p	pressure, psia
$p_{a,i}$	initial aquifer pressure, psia
$\overline{p}_a$	average aquifer pressure, psia
p_D	dimensionless pressure
p'_D	derivative of p_D
p_0	initial aquifer pressure at $t = 0$, psia
p_R	pressure at reservoir/aquifer boundary, psia
q_w	water influx rate, STB/day
Q_D	dimensionless cumulative water influx
q_D	dimensionless aquifer rate
R	recovery factor
r	radius, feet
S_g	gas saturation, fraction
S_{gi}	initial gas saturation, fraction
S_{gr}	residual gas saturation, fraction
S_w	initial water saturation, fraction
t	time, days
T_R	reservoir temperature, $^\circ R$
t_D	dimensionless time
W_e	cumulative water influx, RB

W_{ei}	maximum volume of encroachable water at initial conditions, RB
W_i	initial aquifer water volume, RB
W_p	cumulative water produced, STB
w	width of linear reservoir or aquifer, feet
z	gas compressibility factor
θ	aquifer encroachment angle
ϕ	porosity, fraction
μ	viscosity, cp
γ_o	specific gravity of stock tank oil

Subscripts

a	abandonment or aquifer
g	gas
i	initial
c	condensate
o	oil or condensate
p	produced
r	remaining or residual
R	reservoir
w	water

Abbreviations

AGIP	Apparent Gas In Place
GOR	Gas-Oil Ratio
GMBE	General Material Balance Equation
OGIP	Original Gas In Place
HO	Havlena-Odeh
TVD	True Vertical Depth
SS	Sub-sea

References

1. Gold, D.K., McCain, W.D., and Jennings, J.W.: "An Improved Method for the Determination of the Reservoir-Gas Specific Gravity for Retrograde Gases," *JPT* (July 1989) 747–752.
2. Agarwal, R.G.: "Unsteady-State Performance of Water-Drive Gas Reservoir," PhD dissertation, Texas A&M University, College Station, Texas (1967).
3. Cole, F.W.: *Reservoir Engineering Manual,* Gulf Publishing Co., Houston, Texas (1969) 285.
4. Pletcher, J.L.: "Improvements to Reservoir Material-Balance Methods," *SPEREE* (February 2002) 49–59.

5. Havlena, D., and Odeh, A.S.: "The Material Balance as an Equation of a Straight Line," *JPT* (August 1963) 896–900.

6. Havlena, D., and Odeh, A.S.: "The Material Balance as an Equation of a Straight Line- Part II, Field Cases," *JPT* (July 1964) 815–822.

7. Schilthuis, R.J.: "Active Oil and Reservoir Energy," *Trans.* AIME (1936) 148, 33–51.

8. van Everdingen, A.F., and Hurst, W.: "The Application of the Laplace Transformation to Flow Problems in Reservoirs," *Trans.* AIME (1949) 186, 305–324.

9. Fetkovich, M.J.: "A Simplified Approach to Water Influx Calculations-Finite Aquifer Systems," *JPT* (July 1971) 814–828.

10. Carter, R.D., and Tracy, G.W.: "An Improved Method for Calculating Water Influx," *Trans.*, AIME (1960), 219, 415–417.

11. Fanchi, J.R.: "Analytical Representation of the van Everdingen-Hurst Aquifer Influence Functions for Reservoir Simulation," *SPEJ* (June 1985) 405–406.

12. Edwardson, M.J., Girner, H.M., Parkison, H.R., Williams, C.D., and Matthews, C.S.: "Calculation of Formation Temperature Disturbances Caused by Mud Circulation," *JPT* (April 1962) 416–426.

13. Klins, M.A., Bouchard, A.J., and Cable, C.L.: "A Polynomial Approach to the van Everdingen-Hurst Dimensionless Variables for Water Encroachment," *SPERE* (February 1988) 320–326.

14. Poston, S.W., and Berg, R.R.: *Overpressured Gas Reservoirs*, SPE, Richardson, Texas (1997).

15. Dickinson, G.: "Geological Aspects of Abnormal Reservoir Pressures in Gulf Coast Louisiana," *AAPG Bull.* (February 1953), 37, 2, 410.

16. Ramagost, B.P., and Farshad, F.F.: "P/Z Abnormally Pressured Gas Reservoirs," paper SPE 10125 presented at the SPE 56th Annual Fall Technical Conference and Exhibition, San Antonio, Texas, October 5–7, 1981.

17. Poston, S.W., and Chen, H.Y.: "Case History Studies: Abnormal Pressured Gas Reservoirs," paper SPE 18857 presented at the SPE Production Operations Symposium, Oklahoma City, Oklahoma, March 13–14, 1989.

18. Fetkovich, M.J., Reese, D.E., and Whitson, C.H.: "Application of a General Material Balance for High-Pressure Gas Reservoirs," *SPEJ* (March 1998) 3–13.

19. Hammerlindl, D.J.: "Predicting Gas Reserves in Abnormally Pressured Reservoirs," paper SPE 3479 presented at the 1971 SPE Fall Meeting, New Orleans, Louisiana, October 3–5, 1971.

20. Duggan, J.O.: "The Anderson "L"—An Abnormally Pressured Gas Reservoir in South Texas," *JPT* (February, 1972) 132–138.

21. Roach, R.H.: "Analyzing Geopressured Reservoirs—A Material Balance Technique," paper SPE 9968 available from SPE, Richardson, Texas, August 1981.

General Reading

- Ambastha, A.K.: "Evaluation of Material Balance Analysis Methods for Volumetric, Abnormally-pressured Gas Reservoirs," *JCPT* (October 1993) 32, 8, 19–24.
- Bruns, J.R., Fetkovich, M.J., and Meitzen, V.C.: "The Effect of Water Influx on p/z-Cumulative Gas Production Curves," *JPT* (March 1965) 287–291.
- Bass, D.M.: "Analysis of Abnormally Pressured Gas Reservoirs with Partial Water Influx," paper SPE 3850 presented at the Abnormal Subsurface Pressure Symposium, Baton Rouge, Louisiana, May 15–16, 1972.
- Bernard, W.J.: "Gulf Coast Geopressured Gas Reservoirs: Drive Mechanism and Performance Prediction," paper SPE 14362 presented at the 1985 Annual Technical Conference, Las Vegas, Nevada, September 22–25, 1985.
- Begland, T.F., and Whitehead, W.R.: "Depletion Performance of Volumetric High-Pressured Gas Reservoirs," *SPERE* (August 1989) 279–282.
- Lee, J., and Wattenbarger, R.A.: *Gas Reservoir Engineering*, SPE Textbook Series, Vol. 5, SPE, Richardson, Texas (1996).
- Wang, B., and Teasdale, T.S.: "GASWAT-PC: A Microcomputer Program for Gas Material Balance with Water Influx," paper SPE 16484 presented at the 1987 SPE Petroleum Industry Applications of Microcomputers, Lake Conroe, Montgomery, Texas, June 23–26, 1987.

APPENDIX 8A

Correlations for Estimating Residual Gas Saturations for Gas Reservoirs under Water Influx*

Agarwal[2] developed correlations for the estimation of residual gas saturations in gas reservoirs under water influx. Note that these correlations are to be used if no laboratory data are available for the reservoir.

For unconsolidated sandstones:

$$S_{gr} = \frac{a_1\left(10^2 S_{gi}\right) + a_2\left(10^4 \phi S_{gi}\right) + a_3(10^2\phi) + a_4}{100} \tag{8A.1}$$

In Eq. (8A.1),

$a_1 = -0.51255987$, $a_2 = 0.026097212$, $a_3 = -0.26769575$, $a_4 = 14.796539$.

For consolidated sandstones:

$$S_{gr} = \frac{a_1\left(10^2 S_{gi}\right) + a_2\left(10^2 S_{gi}\right)^2}{100} \tag{8A.2}$$

In Eq. (8A.2), $a_1 = 0.80841168$, $a_2 = -0.0063869116$

For limestones:

$$S_{gr} = \frac{a_1(10^2\phi) + a_2\log(k) + a_3\left(10^2 S_{gi}\right) + a_4}{100} \tag{8A.3}$$

In Eq. (8A.3),

$a_1 = -0.53482234$, $a_2 = 3.3555165$, $a_3 = 0.15458573$, $a_4 = 14.403977$.

*From Agarwal

APPENDIX 8B

Dimensionless Pressure for Finite and Infinite Aquifers*

The regression equation is given as:

$$p_D = a_0 + a_1 t_D + a_2 \ln t_D + a_3(\ln t_D)^2 \quad \textbf{(8B.1)}$$

The derivative p'_D can be obtained from differentiation of Eq. (8B.1).

Case	Regression Coefficients				Correlation Range of	Standard Error of Estimate	Average Deviation From Actual
r_a/r_R	a_0	a_1	a_2	a_3	t_D	σ	(%)
1.5	0.10371	1.66657	−0.04579	−0.01023	0.06 to 0.6	0.00032	0.06
2.0	0.30210	0.68178	−0.01599	−0.01356	0.22 to 5.0	0.00140	0.16
3.0	0.51243	0.29317	0.01534	−0.06732	0.52 to 5.0	0.00091	0.09
4.0	0.63656	0.16101	0.15812	−0.09104	1.5 to 10.0	0.00058	0.03
5.0	0.65106	0.10414	0.30953	−0.11258	3.0 to 15.0	0.00041	0.02
6.0	0.63367	0.06940	0.41750	−0.11137	4.0 to 30.0	0.00081	0.04
8.0	0.40132	0.04104	0.69592	−0.14350	8.0 to 45.0	0.00047	0.02
10.0	0.14386	0.02649	0.89646	−0.15502	12.0 to 70.0	0.00043	0.01
∞	0.82092	−3.68E-4	0.28908	0.02882	0.01 to 1000.0	0.01710	1.50

*From Fanchi

Appendix 8C

Dimensionless Pressure for Infinite Aquifers*

For $0.01 < t_D < 500$;

$$p_D = \frac{370.529\sqrt{t_D} + 137.582t_D + 5.69549t_D\sqrt{t_D}}{328.834 + 265.488\sqrt{t_D} + 45.2157t_D + t_D\sqrt{t_D}} \tag{8C.1}$$

$$p'_D = \frac{716.441 + 46.7984\sqrt{t_D} + 270.038t_D + 71.0098t_D\sqrt{t_D}}{1269.86\sqrt{t_D} + 1204.73t_D + 618.618t_D\sqrt{t_D} + 538.072t_D^2 + 142.410t_D^2\sqrt{t_D}} \tag{8C.2}$$

*From Edwardson et al

CHAPTER 9

Oil Reservoirs

9.1 Introduction

One of the important functions of a reservoir engineer is the estimation of oil-in-place (OIP) volumes, and the calculation of oil reserves thereafter. In this chapter, methods for estimating OIP volumes are presented for volumetric reservoirs, reservoirs with primary gas caps, and reservoirs with water influx. As was the case for gas reservoirs (Chapter 8), volumetric oil reservoirs are assumed to have constant hydrocarbon pore volume. Oil reservoirs are discovered at undersaturated or saturated conditions. As discussed in Chapter 4, an undersaturated oil reservoir has no initial gas cap at discovery, and reservoir pressure is at or above bubble point pressure. A saturated oil reservoir has a primary gas cap at discovery, and reservoir pressure is below bubble point pressure. The behavior of undersaturated reservoirs is remarkably different from those of saturated reservoirs in terms of performance and oil recovery. For these reasons, oil reservoirs are classified as undersaturated or saturated for purposes of the calculation methods presented in the chapter. However, it is worthy to note that most undersaturated reservoirs will ultimately become saturated reservoirs possibly with secondary gas caps, if reservoir pressure is allowed to fall below bubble point pressure as a result of production of reservoir fluids. Reservoir fluids are produced either through the release of energy stored in the fluids and the formation or supply of external energy from influx of fluids, represented as water influx from aquifers or injection of fluids like water and/or gas injection. The relative importance of these sources of drive energy varies over the life of the reservoir. The different drive mechanisms that exist in an oil reservoir are presented in the chapter.

9.2 Oil Reservoir Drive Mechanisms

Reservoir drive mechanism is a term that is used to describe the predominant force or forces that push the production of reservoir fluids. For volumetric, undersaturated reservoirs, the predominant forces that drive production are the expansion of the fluids and the formation rocks. For volumetric, saturated reservoirs, the predominant forces are the expansion of the gas caps (primary or secondary), and the expansion of the gas in solution in the oil (or solution gas). For non-volumetric reservoirs, the strength of the aquifer determines relative contributions from expansion of the solution gas, the gas cap, and water influx. These drive mechanisms have been reduced to indices as a means of quantifying their contributions to the production of fluids from the reservoir. Using the scheme devised by Pirson,[1] and the General Material Balance Equation (GMBE), four drive indices are defined for an oil reservoir. The GMBE as derived in Chapter 7 for an oil reservoir (Eq. 7.23) is:

$$N(B_t - B_{ti}) + \frac{NmB_{ti}}{B_{gi}}[B_g - B_{gi}] + (1 + m)NB_{ti}\left[\frac{c_w S_{wi} + c_f}{1 - S_{wi}}\right] \times \Delta p + G_{inj}B_g + W_{inj}B_w + W_e = N_p[B_t + (R_p - R_{si})B_g] + W_pB_w \tag{9.1}$$

Assuming no gas or water injection and rearranging, Eq. (9.1) becomes:

$$N(B_t - B_{ti}) + \frac{NmB_{ti}}{B_{gi}}[B_g - B_{gi}] + (1 + m)NB_{ti}\left[\frac{c_w S_{wi} + c_f}{1 - S_{wi}}\right]\Delta p + W_e - W_pB_w = N_p[B_t + (R_p - R_{si})B_g] \tag{9.2}$$

Dividing Eq. (9.2) by the term on the right-side of the equation gives:

$$\frac{N(B_t - B_{ti})}{N_p[B_t + (R_p - R_{si})B_g]} + \frac{\frac{NmB_{ti}}{B_{gi}}[B_g - B_{gi}]}{N_p[B_t + (R_p - R_{si})B_g]} + \frac{(1 + m)NB_{ti}\left[\frac{c_w S_{wi} + c_f}{1 - S_{wi}}\right]\Delta p}{N_p[B_t + (R_p - R_{si})B_g]} + \frac{(W_e - W_pB_w)}{N_p[B_t + (R_p - R_{si})B_g]} = 1 \tag{9.3}$$

The first term in Eq. (9.3), which represents drive by expansion of oil, is the oil drive index, I_{OD}:

$$I_{OD} = \frac{N(B_t - B_{ti})}{N_p[B_t + (R_p - R_{si})B_g]} \tag{9.4}$$

The second term in Eq. (9.3), which represents expansion of the gas cap, is the gas cap drive index, I_{GCD}:

$$I_{GCD} = \frac{\frac{NmB_{ti}}{B_{gi}}[B_g - B_{gi}]}{N_p[B_t + (R_p - R_{si})B_g]} \tag{9.5}$$

The third term in Eq. (9.3) is the formation and connate water compressibility drive index, I_{FWCD}:

$$I_{FWCD} = \frac{(1 + m)NB_{ti}\left[\frac{c_w S_{wi} + c_f}{1 - S_{wi}}\right]\Delta p}{N_p[B_t + (R_p - R_{si})B_g]} \tag{9.6}$$

The fourth term in Eq. (9.3) is the water drive index, I_{WD}:

$$I_{WD} = \frac{(W_e - W_p B_w)}{N_p[B_t + (R_p - R_{si})B_g]} \tag{9.7}$$

As shown in Eq.(9.3), the sum of the indices is equal to one. Hence,

$$I_{OD} + I_{GCD} + I_{FWCD} + I_{WD} = 1 \tag{9.8}$$

For oil reservoirs with gas cap and water influx, the formation and water compressibility drive index is negligible. For such reservoirs, Eq. (9.8) reduces to:

$$I_{OD} + I_{GCD} + I_{WD} = 1 \tag{9.9}$$

Note that it is usually necessary to include formation and water compressibility drive index in calculations involving undersaturated reservoirs, especially high pressure, undersaturated oil reservoirs.

9.3 Gravity Drainage Mechanism

The gravity drainage mechanism is an effective and efficient means of oil production in many reservoirs with steep dip angles of the formation. In this process, oil drains downdip from the gas cap and gas migrate updip from the oil column in a system of counter-flow assuming the reservoir has sufficient vertical permeability for vertical flow. Gravity drainage mechanism is an effective means of oil production if most of the following conditions exist in the reservoir:

1. High formation dip angle
2. High effective oil permeability
3. Low oil viscosity
4. Low residual oil saturation
5. Large gas cap

Reservoirs under gravity drainage mechanism can achieve very high oil recoveries. Dykstra[2] reported an oil recovery of 64% for the Lakeview Pool in Midway Sunset Field after 40 years of production. Oil production from the Lakeview Pool was by free-fall gravity drainage because the gravity head was the only remaining source of energy after extensive blow-out of the discovery well. The key reservoir and fluid data for the Lakeview Pool are as follows: average formation dip angle = 24 degrees; effective oil permeability = 1100 md; average oil viscosity = 19 cp; and estimated residual oil saturation = 0.10. The gas cap was considered to be large. Dykstra[2] also reported oil recovery of 58.5% for the Mile Six Pool. Reservoir pressure was maintained at about 800 psi during the reported 25-year history of the reservoir. The key reservoir and fluid data for the Mile Six Pool are as follows: formation dip angle = 17.5 degrees; oil permeability = 300 md; oil viscosity = 1.05 cp; residual oil saturation = 0.20; and a large gas cap. Wei et al.[3] estimated ultimate oil recovery for the 26R Reservoir in Kern County,

California, at 50%. The 26R reservoir is a gravity drainage reservoir with pressure maintenance by crestal gas injection. At the top of the 26R Reservoir, the average formation dip is 23 degrees and increases to 62 degrees at the oil-water contact. Other key reservoir and fluid properties for the 26R Reservoir are as follows: average formation permeability = 88 md; oil viscosity = 0.42 cp; and residual oil saturation = 0.49. The 26R reservoir had a very large gas cap. The 26R Reservoir is an example of forced or assisted gravity drainage because of gas injection at the crest of the structure for pressure maintenance.

The rate of gravity drainage is affected by reservoir pressure since fluid properties depend on pressure. Essley et al.[4] suggest that the benefits of gravity drainage can be improved by maintaining reservoir pressure through gas injection in the crestal areas of the reservoir. This mode of operation can be described as gravity drainage with complete pressure maintenance. Gravity drainage reservoirs can also be operated with declining gas-cap pressure either due to no gas injection or insufficient gas injection into the gas-cap to maintain pressure (Hall[5]). Analyses of the performance of gravity drainage reservoirs should consider the influence of pressure maintenance by gas injection on oil recoveries.

Many analytical models have been published on gravity drainage mechanism (Li and Horne[6]). Most of these analytical models are complicated and do not adequately represent oil production by gravity drainage. For these reasons, none of these analytical models is recommended or presented in this book. Rather, it is considered best practice to analyze the performance of reservoirs whose recovery mechanism is dominated by gravity drainage with reservoir models on numerical simulators. Simulation of reservoirs is presented in Chapter 19.

9.4 Volumetric Undersaturated Oil Reservoirs

Volumetric undersaturated oil reservoirs are defined as reservoirs that are completely isolated with constant hydrocarbon pore volume and whose pressures are above bubble point pressures. Volumetric reservoirs may be considered closed systems that do not receive significant pressure support or fluid influx from outside sources, such as water influx from aquifers. In this section, the discussions and equations that follow apply only to undersaturated, volumetric reservoirs.

9.4.1 Volume Calculations Above Bubble Point Pressure

The original oil in place (OOIP), represented as N in an undersaturated oil reservoir, can be calculated in terms of stock tank barrels (STB) with the equation:

$$N = \frac{7758 \times A \times h \times \phi \times (1 - S_{wi})}{B_{oi}} \tag{9.10}$$

In Eq. (9.10), A = area in acres; h = average net thickness in feet; ϕ = porosity, fraction; S_{wi} = water saturation, fraction; and B_{oi} = initial formation volume factor (FVF) in RB/STB. Suppose the reservoir is depleted to a lower pressure which is still above bubble point pressure. The cumulative oil production, N_p, is the difference between the OOIP and the oil remaining in the reservoir, N_r:

$$N_p = N - N_r \tag{9.11}$$

The oil remaining in the reservoir at this lower pressure is given by:

$$N_r = \frac{7758 \times A \times h \times \phi \times (1 - S_{wi})}{B_o} \tag{9.12}$$

In Eq. (9.12), B_o is the FVF at the lower reservoir pressure. Note that the lower reservoir pressure is still above bubble point pressure for this derivation. Substituting Eqs. (9.10) and (9.12) into Eq. (9.11) gives:

$$N_p = 7558 \times A \times h \times \phi \times (1 - S_{wi})\left[\frac{1}{B_{oi}} - \frac{1}{B_o}\right] \tag{9.13}$$

The oil recovery factor, R, at pressures above bubble point is given by:

$$\begin{aligned} R &= \frac{N_p}{N} \\ &= \left[1 - \frac{B_{oi}}{B_o}\right] \end{aligned} \tag{9.14}$$

Equation (9.14) can also be derived starting from the GMBE shown as Eq. (9.2). Since this is a volumetric reservoir, there is no water influx ($W_e = 0$), and water production is negligible ($W_p = 0$). Also assume that water and formation compressibilities are negligible ($c_w = 0$, and $c_f = 0$). (It is demonstrated later in this chapter that this assumption is not recommended, especially for reservoirs with high formation compressibility.) With these assumptions, the GMBE reduces to:

$$N(B_t - B_{ti}) = N_p[B_t + (R_p - R_{si})B_g] \tag{9.15}$$

Above bubble point pressure, $R_p = R_{si}$ since all produced gas comes from the gas in solution. Also, above bubble point pressure, $B_{ti} = B_{oi}$ and $B_t = B_o$ since the amount of solution gas is constant. Thus, Eq. (9.15) becomes:

$$\begin{aligned} N(B_t - B_{ti}) &= N_p B_t \text{ and transforms to:} \\ N(B_o - B_{oi}) &= N_p B_o \end{aligned} \tag{9.16}$$

From Eq. (9.16), oil recovery factor, R, above bubble point pressure is:

$$R = \left[1 - \frac{B_{oi}}{B_o}\right] \text{ same as Eq. (9.14)}$$

To estimate oil recovery above bubble point pressure based on oil expansion alone, contributions to oil recovery caused by formation and water expansion were ignored. This could lead to large errors in the estimation of oil recovery at pressures above bubble point. Inclusion of formation and water compressibilities is necessary in the calculation of oil recovery above bubble point pressure, especially for reservoirs with high formation compressibility. This type of reservoir is encountered frequently in high pressure, deepwater environments. The expression for oil recovery above bubble point pressure that includes formation and water compressibilities can be derived from the GMBE. Again since there is no water influx (volumetric reservoir), and water production is negligible, Eq. (9.2) becomes:

$$N(B_t - B_{ti}) + NB_{ti}\left[\frac{c_w S_{wi} + c_f}{1 - S_{wi}}\right]\Delta p = N_p[B_t + (R_p - R_{si})B_g] \tag{9.17}$$

Since the reservoir is above bubble point, Eq. (9.17) can be written in equivalent form as:

$$N(B_o - B_{oi}) + NB_{oi}\left[\frac{c_w S_{wi} + c_f}{1 - S_{wi}}\right]\Delta p = N_p B_o \tag{9.18}$$

Oil compressibility, c_o, can be calculated from the expression:

$$c_o = \frac{B_o - B_{oi}}{B_{oi}\Delta p} \tag{9.19}$$

Substituting Eq. (9.19) into Eq. (9.18) gives:

$$Nc_o B_{oi}\Delta p + NB_{oi}\left[\frac{c_w S_{wi} + c_f}{1 - S_{wi}}\right]\Delta p = N_p B_o \tag{9.20}$$

Since for volumetric, undersaturated reservoirs $S_o = 1 - S_{wi}$, Eq. (9.20) can be written as:

$$NB_{oi}\left[\frac{c_o S_o + c_w S_{wi} + c_f}{1 - S_{wi}}\right]\Delta p = N_p B_o \tag{9.21}$$

The reservoir effective compressibility, c_e, is defined as:

$$c_e = \left[\frac{c_o S_o + c_w S_{wi} + c_f}{1 - S_{wi}}\right] \tag{9.22}$$

Substituting Eq. (9.22) into Eq. (9.21) gives:

$$NB_{oi}c_e\Delta p = N_p B_o \tag{9.23}$$

By including formation and water compressibilities, oil recovery factor is expressed as:

$$\begin{aligned} R &= \frac{N_p}{N} \\ &= c_e\Delta p\left[\frac{B_{oi}}{B_o}\right] \end{aligned} \tag{9.24}$$

Example 9.1 Calculation of Oil Recovery Factor for an Undersaturated Reservoir

Problem

Estimate oil recovery factors for a super-undersaturated reservoir in ultra-deepwater Gulf of Mexico, USA by neglecting formation and water compressibility, and including formation and water compressibility. The volumetric reservoir was produced from initial pressure and abandoned at average reservoir pressure of 12000 psia. The average reservoir properties are as follows:

Initial reservoir pressure, p	19600 psia
Temperature, T	252 °F
Porosity, ϕ	0.22
Initial water saturation, S_{wi}	0.38
Permeability, k	42 md
Water compressibility, c_w	4×10^{-6} psi^{-1}
Formation compressibility, c_f	15×10^{-6} psi^{-1}
Bubble point pressure, p_b	1200 psia

The oil PVT properties are presented in Table 9.1.

Solution

Using Eq. (9.14), which neglects formation and water compressibility, and PVT data in Table 9.1:

$$R = \left[1 - \frac{B_{oi}}{B_o}\right] = 1 - \frac{1.123}{1.159}$$
$$= 0.031 \text{ or } 3.1\%$$

Using Eq. (9.24), which includes formation and water compressibility:

$$R = c_e \Delta p \left[\frac{B_{oi}}{B_o}\right]$$

From Eq. (9.22), the reservoir effective compressibility is given by:

$$c_e = \left[\frac{c_o S_o + c_w S_{wi} + c_f}{1 - S_{wi}}\right]$$

From Eq. (9.19), oil compressibility is estimated as:

$$c_o = \frac{B_o - B_{oi}}{B_{oi}\Delta p} = \frac{1.159 - 1.123}{1.123 \times (19600 - 12000)}$$
$$= 4.218 \times 10^{-6} \text{ psi}^{-1}$$

Table 9.1 PVT Data of the Reservoir in Example 9.1

Pressure (psia)	FVF (RB/STB)	Rs (scf/STB)
19564	1.123	254
15000	1.143	254
14000	1.148	254
13000	1.153	254
12000	1.159	254
11000	1.165	254
10000	1.172	254
9000	1.178	254
8000	1.186	254
7000	1.193	254
6000	1.201	254
5000	1.210	254
4000	1.219	254
3000	1.229	254
2000	1.240	254
1240	1.248	254

By substituting with reservoir data, the reservoir effective compressibility is:

$$c_e = \left[\frac{c_o S_o + c_w S_{wi} + c_f}{1 - S_{wi}}\right] = \frac{4.218 \times 10^{-6} \times 0.62 + 4 \times 10^{-6} \times 0.38 + 15 \times 10^{-6}}{(1 - 0.38)}$$
$$= 3.086 \times 10^{-5} \text{ psi}^{-1}$$

The recovery factor is:

$$R = c_e \Delta p \left[\frac{B_{oi}}{B_o}\right] = 3.086 \times 10^{-5} \times (19600 - 12000) \times \left[\frac{1.123}{1.159}\right]$$
$$= 0.23 \text{ or } 23\%$$

This rather simple example illustrates the impact of formation and fluid compressibilities on oil recovery, especially for reservoirs with high formation compressibility. These types of reservoirs are encountered in high pressure conditions, which are prevalent in deepwater environments. The effect of formation compressibility on oil recovery is comparable to the effect it has on the performance of geopressured gas reservoirs. This was discussed in Chapter 8.

The intent of this example is to emphasize the importance of including formation and fluid compressibilities in the estimation of oil recovery for **all** undersaturated reservoirs.

9.4.2 Volume Calculations Below Bubble Point Pressure

Suppose the pressure of an oil reservoir at initial conditions was above bubble point pressure. The OOIP can be calculated with Eq. (9.10). If the reservoir was produced until the pressure falls below bubble point, gas will evolve and form a gas phase dispersed within the oil phase. As reservoir pressure continues to fall due to production, gas saturation will continue to increase. When the critical gas saturation is reached, gas will begin to flow either toward the producing wells or migrate toward the top of the structure to form a secondary gas cap. Assuming the reservoir is volumetric (no water influx), the oil remaining in the reservoir at any pressure below the bubble point can be estimated as:

$$N_r = \frac{7758 \times A \times h \times \phi \times (1 - S_{wi} - S_g)}{B_o} \tag{9.25}$$

In Eq. (9.25), S_g is the gas saturation at any pressure below the bubble point. From Eqs. (9.10), (9.11), and (9.25), the oil recovery factor for the reservoir below bubble point is:

$$\begin{aligned} R &= \frac{N_p}{N} \\ &= \frac{\left[\left(\frac{1 - S_{wi}}{B_{oi}}\right) - \left(\frac{1 - S_{wi} - S_g}{B_o}\right)\right]}{\left[\frac{1 - S_{wi}}{B_{oi}}\right]} \\ &= \left[1 - \left(\frac{1 - S_{wi} - S_g}{1 - S_{wi}}\right)\left(\frac{B_{oi}}{B_o}\right)\right] \end{aligned} \tag{9.26}$$

Equation (9.26) is not in a very useful form because of the difficulty of determining gas saturation in the reservoir at any pressure below the bubble point. In some cases, gas saturations can be estimated from core studies or material balance calculations. Note, however, that these sources for gas saturation are unreliable and may lead to large errors in the application of Eq. (9.26).

The oil recovery factor for a reservoir below bubble point pressure can be determined from the GMBE. Again, assuming the reservoir is volumetric (no water influx), and water production is negligible, Eq. (9.2) reduces to Eq. (9.17):

$$N(B_t - B_{ti}) + NB_{ti}\left[\frac{c_w S_{wi} + c_f}{1 - S_{wi}}\right]\Delta p = N_p[B_t + (R_p - R_{si})B_g]$$

Since gas is present in the reservoir, the compressibility of gas is much higher than the combined compressibility of the formation and water. Hence, formation and water compressibility are considered negligible and Eq. (9.17) becomes:

$$N(B_t - B_{ti}) = N_p[B_t + (R_p - R_{si})B_g] \tag{9.27}$$

The oil recovery factor, R, can be derived from Eq. (9.27) as:

$$R = \frac{(B_t - B_{ti})}{[B_t + (R_p - R_{si})B_g]} \tag{9.28}$$

In Eq. (9.28), R_p is the cumulative produced gas-oil ratio. R_p is calculated as:

$$R_p = \frac{G_p}{N_p} \tag{9.29}$$

Also, in Eq. (9.29), G_p = cumulative gas production, scf; and N_p = cumulative oil production, STB. Note that all the other terms in Eq. (9.28) except R_p are based on the properties of the reservoir fluid and are dependent on the pressure of the reservoir. In contrast to Eq. (9.26), it is much easier to calculate the recovery factor for a reservoir below its bubble point by using Eq. (9.28). The cumulative production gas-oil ratio can be calculated from production data, although serious errors might occur if accurate production data are not maintained. The other terms in Eq. can be determined from PVT data on the reservoir fluids. Again, if representative samples were not taken from the reservoir and reliable PVT data were not measured, application of Eq. (9.28) may lead to considerable errors.

By examining Eq. (9.28), it can be readily deduced that reducing the cumulative produced gas-oil ratio, R_p, can lead to higher recovery factors. The cumulative produced gas-oil ratio can be reduced in practice by isolating high gas producing intervals with packers, partially shutting-off high gas producing intervals with cement, or simply shutting in high gas-oil ratio wells. If the facilities are available, some or all of the produced gas can be returned to the reservoir through the process of gas re-injection. The practice of reservoir management by gas re-injection for pressure maintenance is discussed in Chapter 20.

9.5 Undersaturated Oil Reservoirs with Water Influx

Undersaturated oil reservoirs with water influx are treated differently from volumetric, undersaturated oil reservoirs. Unlike volumetric reservoirs, pressures in reservoirs with water influx could be supported completely or partially, depending on the strength of the aquifer. Fluid saturations will also change within areas of the reservoir that had been invaded by water influx from the aquifer.

9.5.1 Volume Method

At initial conditions, the OOIP in undersaturated reservoirs with water influx is determined with Eq. (9.10). Suppose the aquifer influx is strong, and reservoir pressure is maintained at initial levels, the oil remaining in the reservoir after a period of production is given by:

$$N_r = \frac{7758 \times A \times h \times \phi \times (1 - S_{wi} - S_{or})}{B_{oi}} \tag{9.30}$$

Combining Eqs. (9.10) and (9.30), oil recovery factor for this special case of no pressure decline in the undersaturated reservoir is given by:

$$R = \frac{N_p}{N} = \frac{N - N_r}{N} = 1 - \frac{N_r}{N} \\ = 1 - \left[\frac{1 - S_{wi} - S_{or}}{1 - S_{wi}}\right] \tag{9.31}$$

In Eqs. (9.30) and (9.31), S_{or} is the residual oil saturation in the areas of the reservoir invaded by water. Residual oil saturation can be determined from special core analyses. If it is assumed that reservoir pressure declined from initial levels but still was maintained above the bubble point, the oil remaining in the reservoir after a period of production will be:

$$N_r = \frac{7758 \times A \times h \times \phi \times (1 - S_{wi} - S_{or})}{B_o} \tag{9.32}$$

For this case, the oil recovery factor is derived as:

$$R = 1 - \frac{N_r}{N} = 1 - \left[\frac{1 - S_{wi} - S_{or}}{1 - S_{wi}}\right]\left[\frac{B_{oi}}{B_o}\right] \tag{9.33}$$

Equations (9.31) and (9.33) are to be used only for estimation of oil recovery factors in undersaturated reservoirs with water influx. The reasons are due to difficulties of obtaining reliable values for residual oil saturations, and the likelihood that some oil bearing intervals may have been bypassed due to poor sweep efficiency. Consequently, these equations may yield high estimates of oil recovery factors.

9.5.2 Material Balance Method

The GMBE can be applied to undersaturated oil reservoirs with water influx. Starting with Eq. (9.2), the GMBE for an undersaturated reservoir with water influx is:

$$N(B_t - B_{ti}) + NB_{ti}\left[\frac{c_w S_{wi} + c_f}{1 - S_{wi}}\right]\Delta p + W_e - W_p B_w = N_p[B_t + (R_p - R_{si})B_g] \tag{9.34}$$

Formation and water compressibility are considered negligible for undersaturated reservoirs with strong water influx. Hence, Eq. (9.34) becomes:

$$N(B_t - B_{ti}) + W_e - W_p B_w = N_p[B_t + (R_p - R_{si})B_g] \tag{9.35}$$

Re-arranging Eq. (9.35) gives:

$$N_p[B_t + (R_p - R_{si})B_g] + W_p B_w = N(B_t - B_{ti}) + W_e \tag{9.36}$$

The left side of Eq. (9.36) represents cumulative voidage from the reservoir, and the right side represents energy from oil expansion and water influx. If cumulative reservoir voidage is denoted as F, then:

$$F = N_p[B_t + (R_p - R_{si})B_g] + W_pB_w \tag{9.37}$$

Also, if oil expansion is represented as E_o, then:

$$E_o = B_t - B_{ti} \tag{9.38}$$

Substituting Eqs. (9.37) and (9.38) into Eq. (9.36) yields:

$$F = NE_o + W_e \tag{9.39}$$

Dividing both sides of Eq. (9.39) by E_o gives:

$$\frac{F}{E_o} = N + \frac{W_e}{E_o} \tag{9.40}$$

Equation (9.40) is the Havlena-Odeh[7] straight line form of the material balance equation for an undersaturated oil reservoir with water influx. A plot of F/E_o vs. W_e/E_o will give a straight line with the intercept on the y-axis equal to N (the OOIP), and the slope equal to unity. Note that for the special case in which reservoir pressure is still above bubble point, $F = N_pB_t + W_pB_w$ in Eq. (9.37). The shape of F/E_o vs. W_e/E_o plot depends on the values of W_e. If correct values of W_e are used, the plot will be a straight line. If values of W_e are too small, the plot will curve upward. If the values are too large, the plot will curve downward. A linear aquifer will yield an S-shaped curve. These possible shapes of the F/E_o vs. W_e/E_o plot are shown in Figure 9.1. Note that Figure 9.1 is similar to Figure 8.4 for a gas reservoir with water influx.

It is important to note that Eq. (9.40) was derived by assuming that formation and water compressibility are negligible. This will apply to relatively low pressure reservoirs with low formation and water compressibility. However, this assumption will not apply to high pressure reservoirs with high formation and water compressibility, such as reservoirs encountered in a deepwater environment. For reservoirs with high formation and water compressibility, Eq. (9.40) will not be applicable. To derive an equation similar in form to Eq. (9.40), the GMBE for such reservoirs is the same as Eq. (9.34):

$$N(B_t - B_{ti}) + NB_{ti}\left[\frac{c_wS_{wi} + c_f}{1 - S_{wi}}\right]\Delta p + W_e - W_pB_w = N_p[B_t + (R_p - R_{si})B_g]$$

As shown previously in Eqs. (9.37) and (9.38):

$$F = N_p[B_t + (R_p - R_{si})B_g] + W_pB_w$$

$$E_o = B_t - B_{ti}$$

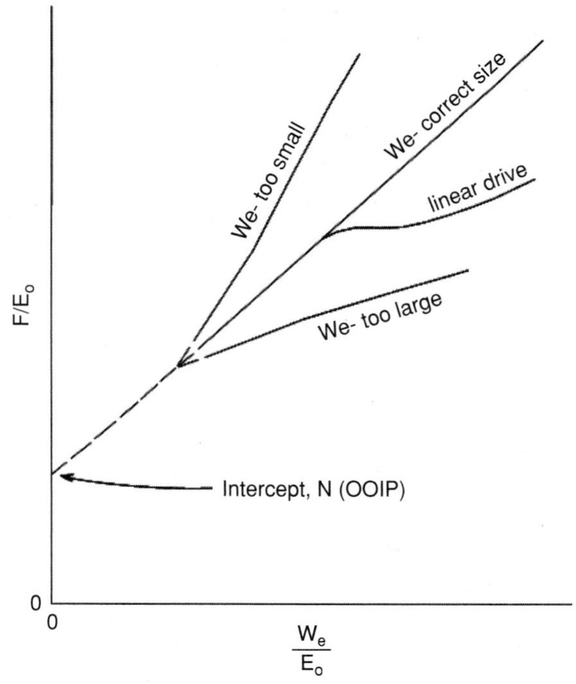

Figure 9.1 Oil reservoir with water influx (from Havlena and Odeh[7] © 1963 SPE, Reproduced with permission).

A new term is defined to represent formation and water expansion as:

$$E_{fw} = B_{ti}\left[\frac{c_w S_w + c_f}{1 - S_{wi}}\right]\Delta p \tag{9.41}$$

Substituting these terms into Eq. (9.34) and re-arranging gives:

$$\frac{F}{E_o + E_{fw}} = N + \frac{W_e}{E_o + E_{fw}} \tag{9.42}$$

Note that the only difference between Eqs. (9.40) and (9.42) is the addition of the term, E_{fw}, for formation and water expansion. A plot of $F/(E_o + E_{fw})$ vs. $W_e/(E_o + E_{fw})$ will give a straight line with the intercept on the y-axis equal to N (the OOIP), and the slope equal to unity. The characteristics and shapes of this line will be the same as described previously for the plot of F/E_o vs. W_e/E_o in Figure 9.1.

Example 9.2 Calculation of OOIP for an Undersaturated Reservoir with Water Influx

Problem

Calculate the OOIP and water influx volumes for an undersaturated reservoir with water influx. Calculate water influx volumes with the Carter-Tracy method assuming an infinite-acting aquifer. The properties of the aquifer and reservoir are as follows:

Initial pressure, p_i	10,535 psia
Current reservoir pressure, p	9806 psia
Temperature, T	212°F
Porosity, ϕ	0.275
Permeability, k	36.5 md
Net thickness, h	12.5 ft
Water compressibility, c_w	3×10^{-6} psi^{-1}
Water viscosity, μ_w	0.588 cp
Water formation volume factor, B_w	1.02 RB/STB
Formation compressibility, c_f	20×10^{-6} psi^{-1}
Total compressibility, c_t	23×10^{-6} psi^{-1}
Aquifer encroachment angle, θ	180°
Reservoir Properties:	
Area of reservoir, A	1022 acres
Radius of reservoir, r_R	3764.39 ft
Initial water saturation, S_{wi}	0.30
Bubble point pressure, p_b	8104 psia

The PVT properties of the oil above bubble point pressure are presented in Table 9.2 and Figure 9.2. The production and pressure history of the reservoir are presented in Table 9.3.

Table 9.2 PVT Data of the Reservoir in Example 9.2

Pressure (psia)	FVF (RB/STB)	Rs (scf/STB)
10535	1.552	1417
10000	1.561	1417
9500	1.568	1417
9000	1.576	1417
8500	1.584	1417
8104	1.592	1417

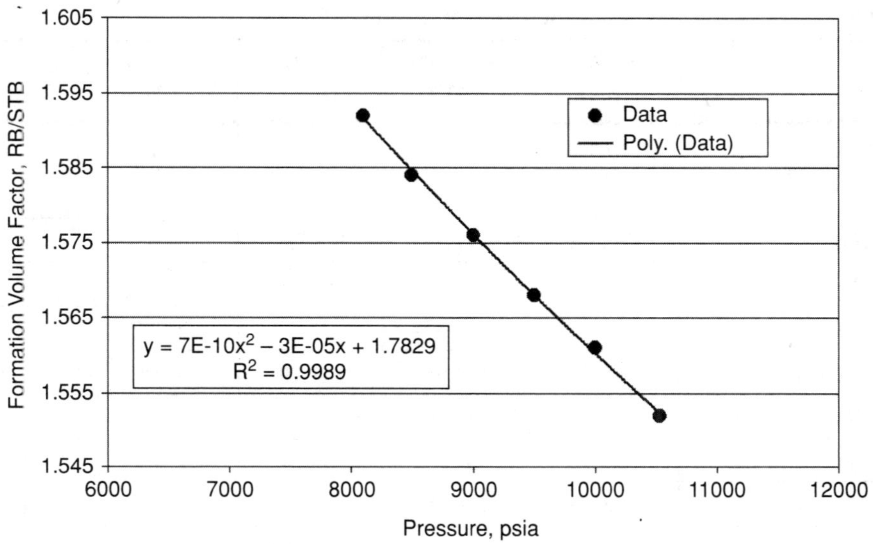

Figure 9.2 PVT data for the reservoir in Example 9.2.

Solution

Step 1: Calculate the parameter B for the Carter-Tracy method (see Example 8.4 in Chapter 8).

$$B = 1.119\phi c_t h r_R^2\left(\frac{\theta}{360}\right) = 1.119 \times 0.275 \times 23 \times 10^{-6} \times 12.5 \times (3764.39)^2 \times \left(\frac{180}{360}\right)$$
$$= 626.8446 \text{ RB/psi}$$

Step 2: Calculate the dimensionless time, t_D:

$$t_D = \frac{0.00633kt}{\phi\mu c_t r_R^2} = \frac{0.0063 \times 36.5 \times t}{0.275 \times 0.588 \times 23 \times 10^{-6} \times (3764.39)^2}$$
$$= 0.00436t$$

Step 3: Calculate water influx volumes using the Carter-Tracy method as shown in Example 8.4 (Chapter 8):

The remaining procedure for calculating the water influx volumes are shown in Example 8.4 in Chapter 8. The water influx volumes are calculated as shown in Table 9.4. The results in Table 9.4 were generated in a spreadsheet and are similar to Table 8.5 in Chapter 8.

Step 4: Calculation of the terms in Eq. (9.42):

For the production data of November 1, 2003 as example:

From Eq. (9.38):

$$E_o = B_t - B_{ti}$$
$$= 1.5560 - 1.5445 = 0.0115 \text{ RB/STB}$$

Table 9.3 Production Data for the Reservoir in Example 9.2

Date	Cum (days)	Res. Press (psia)	Cum Oil (STB)	Cum Gas (Mcf)	Cum Water (STB)	Cum GOR (scf/STB)
1-May-2003	0	10,535	0	0	0	0
1-Jun-2003	30	10,527	75,926	46,419	1,907	611
1-Jul-2003	61	10,519	151,993	123,750	1,913	814
1-Aug-2003	92	10,502	211,109	188,814	1,922	894
1-Sep-2003	122	10,293	363,314	360,419	1,927	992
1-Oct-2003	153	10,076	544,988	580,480	1,939	1,065
1-Nov-2003	183	9,806	719,095	802,639	1,956	1,116
1-Dec-2003	214	9,784	902,185	1,044,303	2,018	1,158
1-Jan-2004	245	9,637	1,084,186	1,290,928	2,078	1,191
1-Feb-2004	274	9,497	1,264,710	1,539,785	2,124	1,218
1-Mar-2004	305	9,356	1,453,826	1,803,340	2,137	1,240
1-Apr-2004	335	9,142	1,638,530	2,092,009	2,167	1,277
1-May-2004	366	9,079	1,803,146	2,353,667	2,195	1,305
1-Jun-2004	396	9,035	1,951,444	2,589,641	2,206	1,327
1-Jul-2004	427	9,003	2,097,122	2,822,767	2,228	1,346
1-Aug-2004	458	9,013	2,226,579	3,024,247	2,241	1,358
1-Sep-2004	488	9,023	2,306,688	3,152,648	2,241	1,367
1-Oct-2004	519	8,912	2,458,169	3,387,957	2,248	1,378
1-Nov-2004	549	8,800	2,611,623	3,640,493	2,248	1,394
1-Dec-2004	580	8,689	2,774,991	3,908,663	2,248	1,409
1-Jan-2005	611	8,697	2,940,599	4,174,634	2,248	1,420
1-Feb-2005	639	8,704	3,089,323	4,417,571	2,270	1,430
1-Mar-2005	670	8,712	3,215,581	4,613,268	2,435	1,435
1-Apr-2005	700	8,709	3,369,394	4,860,906	2,588	1,443
1-May-2005	731	8,705	3,471,539	5,103,481	5,254	1,470
1-Jun-2005	761	8,597	3,581,969	5,275,471	11,974	1,473
1-Jul-2005	792	8,488	3,660,120	5,394,232	23,754	1,474
1-Aug-2005	823	8,380	3,731,048	5,498,702	36,340	1,474

Table 9.4 Calculation of Water Influx Volumes for the Reservoir in Example 9.2 Using the Carter-Tracy Method

n	t (days)	$p_{R,n}$ (psia)	t_D	$\Delta p\ (t_D)$ (psia)	$\sqrt{t_D}$	p_D	p'_D	W_e (RB)
0	0	10535	0.00	0	0.0000	0.0000	0.0000	0
1	30	10527	0.13	8	0.3617	0.3535	1.1766	1,856
2	61	10519	0.27	16	0.5157	0.4780	0.7451	4,927
3	92	10502	0.40	33	0.6333	0.5652	0.5644	10,756
4	122	10293	0.53	242	0.7293	0.6320	0.4635	53,778
5	153	10076	0.67	459	0.8167	0.6897	0.3943	128,834
6	183	9806	0.80	729	0.8932	0.7379	0.3461	235,206
7	214	9784	0.93	751	0.9659	0.7820	0.3082	335,610
8	245	9637	1.07	898	1.0335	0.8216	0.2784	448,551
9	274	9497	1.19	1038	1.0930	0.8553	0.2558	564,958
10	305	9356	1.33	1179	1.1532	0.8885	0.2356	699,870
11	335	9142	1.46	1393	1.2086	0.9182	0.2191	850,070
12	366	9079	1.60	1456	1.2632	0.9468	0.2044	1,004,142
13	396	9035	1.73	1500	1.3140	0.9727	0.1921	1,150,891
14	427	9003	1.86	1532	1.3644	0.9979	0.1810	1,299,186
15	458	9013	2.00	1522	1.4131	1.0217	0.1712	1,439,851
16	488	9023	2.13	1512	1.4587	1.0435	0.1627	1,569,719
17	519	8912	2.26	1623	1.5043	1.0650	0.1548	1,712,000
18	549	8800	2.39	1735	1.5471	1.0847	0.1479	1,857,497

(continued)

Table 9.4 Calculation of Water Influx Volumes for the Reservoir in Example 9.2 Using the Carter-Tracy Method (continued)

n	t (days)	$p_{R,n}$ (psia)	t_D	$\Delta p\ (t_D)$ (psia)	$\sqrt{t_D}$	p_D	p'_D	W_e (RB)
19	580	8689	2.53	1846	1.5902	1.1043	0.1414	2,015,367
20	611	8697	2.66	1838	1.6322	1.1230	0.1355	2,167,624
21	639	8704	2.79	1831	1.6691	1.1392	0.1306	2,301,018
22	670	8712	2.92	1823	1.7092	1.1565	0.1256	2,444,072
23	700	8709	3.05	1826	1.7470	1.1727	0.1211	2,579,621
24	731	8705	3.19	1830	1.7853	1.1887	0.1168	2,716,984
25	761	8597	3.32	1938	1.8215	1.2038	0.1130	2,857,732
26	792	8488	3.45	2047	1.8583	1.2188	0.1092	3,010,987
27	823	8380	3.59	2155	1.8943	1.2333	0.1058	3,171,731

The values for B_t and B_{ti} were calculated from a polynomial fit of the data shown in Table 9.2 and Figure 9.2. The equation for the polynomial is: $FVF = 7E - 10p^2 - 3E - 05p + 1.7829$.
From Eq. (9.41):

$$E_{fw} = B_{ti}\left[\frac{c_w S_w + c_f}{1 - S_{wi}}\right]\Delta p = 1.5445\left[\frac{3 \times 10^{-6} \times 0.3 + 20 \times 10^{-6}}{1 - 0.3}\right] \times (10535 - 9806) = 0.0336 \text{ RB/STB}$$

$W_e = 235{,}206$ RB, as shown in Table 9.4.

From Eq. (9.37):

$$F = N_p B_t + W_p B_w = 719{,}095 \times 1.5560 + 1956 \times 1.02 = 1{,}120{,}928 \text{ RB}$$

Values for the $y -$ axis and the $x -$ axis were calculated as follows:

$$y_{axis} = \frac{F}{(E_o + E_{fw})} = \frac{1{,}120{,}928}{(0.0115 + 0.0336)} = 24{,}850{,}381 \text{ STB}$$

$$x_{axis} = \frac{W_e}{(E_o + E_{fw})} = \frac{235{,}206}{(0.0115 + 0.0336)} = 5{,}214{,}383 \text{ STB}$$

A plot of y_{axis} vs. x_{axis} for the results in Table 9.5 is shown in Figure 9.3. The intercept on the y_{axis} shows the OOIP is estimated to be 21.5 MMBO. The volumetrically calculated OOIP for this example is 21 MMBO. Thus, the OOIP calculated from material balance, and volumetrically from geologic maps are remarkably close for this example. It is important to note that the material balance calculations required many iterations on the estimated cumulative volume of water influx from the aquifer. The cumulative water influx volume was the key variable that was systematically changed iteratively before achieving the solution presented for this example. These iterations were not shown in Table 9.4 and Table 9.5 due to limitations of space. Note that the early data points (June to October 2003) were not included in Figure 9.3 because of the large variability of the data. These might have been caused by instability of production data, especially during the early stages of initiating production from the reservoir. Note also that toward the end of the production history (May to August 2005), the plotted data points appear to have decreased. This may have been caused by using constant formation compressibility for the entire calculations. Formation compressibility is not typically constant, but varies with depletion of the reservoir. The variability of formation compressibility was reported by Poston and Chen[8] and Fetkovich et al.[9] An important requirement for using the material balance method for estimating OOIP and W_e is that the plotted points must fall on the unit slope line as evident in Eq. (9.42). This condition is a necessary check at each iteration step.

Table 9.5 Tabulation of the Results from Eq. (9.42) Used in Plotting Figure 9.3 for Example 9.2

Date	Cum Oil (STB)	Cum Water (STB)	FVF (RB/STB)	Δp (psia)	E_o (RB/STB)	E_{fw} (RB/STB)	W_e (RB)	F (RB)	$\frac{W_e}{(E_o - E_{fw})}$ (STB)	$\frac{F}{(E_o + E_{fw})}$ (STB)
01-May-2003	0	0	1.5445	0	0.0000	0.0000	0	0	0	0
01-Jun-2003	75,926	1,907	1.5447	8	0.0001	0,0004	1,856	119,225	3,779,867	242,837,944
01-Jul-2003	151,993	1,913	1.5448	16	0.0002	0.0007	4,927	236,747	5,016,846	241,082,255
01-Aug-2003	211,109	1,922	1.5450	33	0.0005	0,0015	10,756	328,133	5,309,683	161,976,375
01-Sep-2003	363,314	1,927	1.5483	242	0.0037	0,0112	53,778	564,474	3,611,365	37,906,243
01-Oct-2003	544,988	1,939	1.5517	459	0.0071	0,0212	128,834	847,629	4,550,169	29,936,727
01-Nov-2003	719,095	1,956	1.5560	729	0.0115	0.0336	235,206	1,120,928	5,214,383	24,850,381
01-Dec-2003	902,185	2,018	1.5564	751	0.0118	0.0346	335,610	1,406,209	7,220,536	30,254,112
01-Jan-2004	1,084,186	2,078	1.5588	898	0.0143	0.0414	448,551	1,692,149	8,057,286	30,395,922
01-Feb-2004	1,264,710	2,124	1.5611	1,038	0.0166	0,0479	564,958	1,976,537	8,765,694	30,667,239
01-Mar-2004	1,453,826	2,137	1.5635	1,179	0.0190	0.0544	699,870	2,275,228	9,545,107	31,030,480
01-Apr-2004	1,638,530	2,167	1.5671	1,393	0.0226	0.0642	850,070	2,570,022	9,788,953	29,594,997
01-May-2004	1,803,146	2,195	1.5682	1,456	0.0237	0,0671	1,004,142	2,829,987	11,055,009	31,156,483
01-Jun-2004	1,951,444	2,206	1.5690	1,500	0.0245	0.0692	1,150,891	3,064,051	12,292,894	32,727,716
01-Jul-2004	2,097,122	2,228	1.5695	1,532	0.0250	0.0706	1,299,186	3,293,807	13,582,123	34,434,558
01-Aug-2004	2,226,579	2,241	1.5694	1,522	0.0248	0.0702	1,439,851	3,496,622	15,153,289	36,799,156

01-Sep-2004	2,306,688	2,241	1.5692	1,512	0.0247	0.0697	1,569,719	3,621,942	16,631,162	38,374,457
01-Oct-2004	2,458,169	2,248	1.5711	1,623	0.0266	0.0748	1,712,000	3,864,413	16,877,095	38,095,822
01-Nov-2004	2,611,623	2,248	1.5731	1,735	0.0286	0.0800	1,857,497	4,110,659	17,107,893	37,859,934
01-Dec-2004	2,774,991	2,248	1.5751	1,846	0.0305	0.0851	2,015,367	4,373,124	17,424,141	37,808,471
01-Jan-2005	2,940,599	2,248	1.5749	1,838	0.0304	0.0848	2,167,624	4,633,550	18,823,757	40,237,983
01-Feb-2005	3,089,323	2,270	1.5748	1,831	0.0303	0.0844	2,301,018	4,867,418	20,060,119	42,433,819
01-Mar-2005	3,215,581	2,435	1.5747	1,823	0.0301	0.0841	2,444,072	5,065,961	21,402,668	44,362,478
01-Apr-2005	3,369,394	2,588	1.5747	1,826	0.0302	0.0842	2,579,621	5,308,500	22,551,799	46,408,458
01-May-2005	3,471,539	5,254	1.5748	1,830	0.0303	0.0844	2,716,984	5,472,317	23,699,689	47,733,888
01-Jun -2005	3,581,969	11,974	1.5767	1,938	0.0322	0.0894	2,857,732	5,659,996	23,509,887	46,563,459
01-Jul-2005	3,660,120	23,754	1.5787	2,047	0.0342	0.0944	3,010,987	5,802,432	23,423,186	45,138,495
01-Aug-2005	3,731,048	36,340	1.5807	2,155	0.0361	0.0994	3,171,731	5,934,574	23,408,921	43,800,051

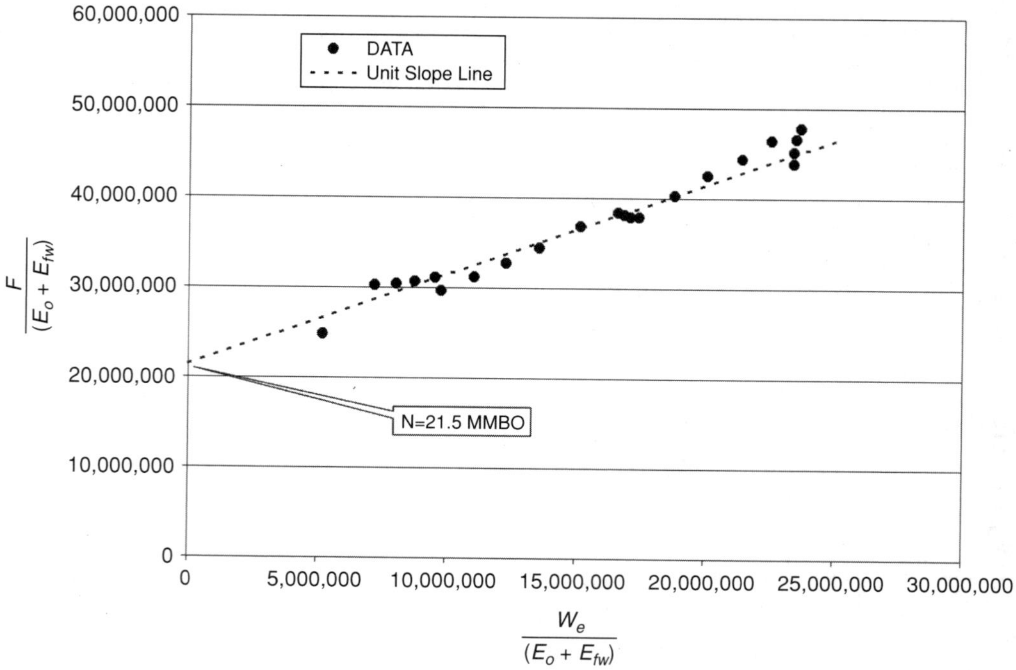

Figure 9.3 Plot of material balance calculations from Table 9.5 for Example 9.2.

9.6 Volumetric Saturated Oil Reservoirs

A volumetric saturated oil reservoir is described as a reservoir with a primary gas cap in contact with the oil zone at discovery and negligible influence from external energy sources, such as an aquifer. At initial conditions, the oil zone is assumed to be in thermodynamic equilibrium with the gas cap so that gas saturation within the oil zone is essentially zero. Note that the concept of volumetric reservoirs described previously as isolated, closed systems for other types of reservoirs is applicable to saturated oil reservoirs.

9.6.1 Volume Method

The OGIP in the primary gas cap of a saturated oil reservoir can be calculated as:

$$G = \frac{7758 \times A_{gc} \times h_{gc} \times \phi_{gc} \times (1 - S_{wigc})}{B_{gi}} \tag{9.43}$$

For Eq. (9.43), A_{gc} = area of gas cap, acres; h_{gc} = average net thickness of gas cap, feet; ϕ_{gc} = average porosity of gas cap, fraction; and S_{wigc} = initial average connate water saturation in the gas cap, fraction. Similarly, the OOIP in the oil zone of a saturated reservoir can be calculated as:

$$N = \frac{7758 \times A_{oz} \times h_{oz} \times \phi_{oz} \times (1 - S_{wioz})}{B_{oi}} \tag{9.44}$$

For Eq. (9.44), A_{oz} = area of the oil zone, acres; h_{oz} = average net thickness of the oil zone, feet; ϕ_{oz} = average porosity in oil zone, fraction; and S_{wioz} = average connate water saturation in the oil zone. For simplicity, it can be assumed that the initial connate water saturations in the gas cap and oil zone are equal. Also, it can be assumed that the porosity of the formation is the same in the gas cap and in the oil zone. With both assumptions, the ratio of the initial gas cap volume to the initial oil volume can be derived from Eqs. (9.43) and (9.44) as:

$$\begin{aligned} m &= \frac{A_{gc}h_{gc}}{A_{oz}h_{oz}} \\ &= \frac{GB_{gi}}{NB_{oi}} \end{aligned} \tag{9.45}$$

9.6.2 Material Balance Method

The GMBE for a volumetric saturated oil reservoir is obtained from Eq. (9.2) as:

$$\begin{aligned} &N(B_t - B_{ti}) + \frac{NmB_{ti}}{B_{gi}}[B_g - B_{gi}] + (1 + m)NB_{ti}\left[\frac{c_wS_{wi} + c_f}{1 - S_{wi}}\right]\Delta p_e - W_pB_w \\ &= N_p[B_t + (R_p - R_{si})B_g] \end{aligned} \tag{9.46}$$

For reservoirs with gas caps, formation and water compressibility are generally negligible because of the higher compressibility of the gas in the gas cap. By re-arrangement, Eq. (9.46) becomes:

$$N_p[B_t + (R_p - R_{si})B_g] + W_pB_w = N(B_t - B_{ti}) + \frac{NmB_{ti}}{B_{gi}}[B_g - B_{gi}] \tag{9.47}$$

As in Eq. (9.37), the cumulative voidage $F = N_p[B_t + (R_p - R_{si})B_g] + W_pB_w$. The oil expansion is represented in Eq. (9.38) as $E_o = B_t - B_{ti}$. The gas expansion term is expressed as:

$$E_g = B_g - B_{gi} \tag{9.48}$$

Substituting Eq. (9.47) becomes:

$$\begin{aligned} F &= NE_o + Nm\left(\frac{B_{ti}}{B_{gi}}\right)E_g \\ &= N\left[E_o + m\left(\frac{B_{ti}}{B_{gi}}\right)E_g\right] \end{aligned} \tag{9.49}$$

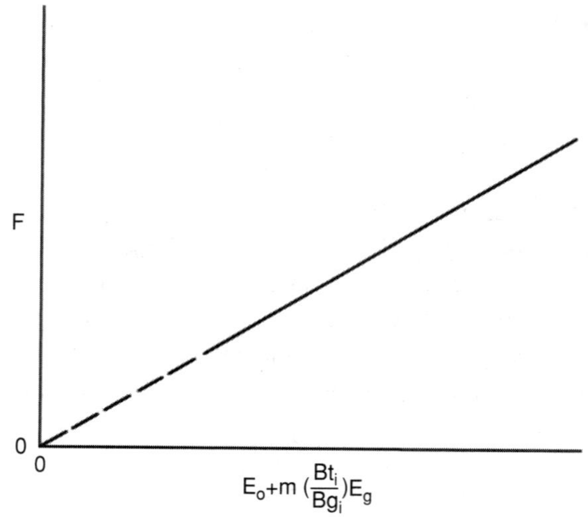

Figure 9.4 Volumetric saturated oil reservoir with *m* known and *N* unknown (from Havlena and Odeh[7] © 1963 SPE, Reproduced with permission).

As demonstrated by Havlena and Odeh,[7] if *m* is known, a plot of *F* vs. $[E_o + m(B_{ti}/B_{gi})E_g]$ will give a straight line passing through the origin with the slope equal to *N*. This is shown as Figure 9.4. If *N* and *m* are unknown, a plot of *F* vs. $[E_o + m(B_{ti}/B_{gi})E_g]$ for various values of *m* will produce a plot similar to Figure 9.5. If *m* is too small, the line will curve upward. If *m* is too large, the line will curve downward. If the correct value of *m* is assumed, a straight line that passes through the origin will be obtained, as shown in Figure 9.5.

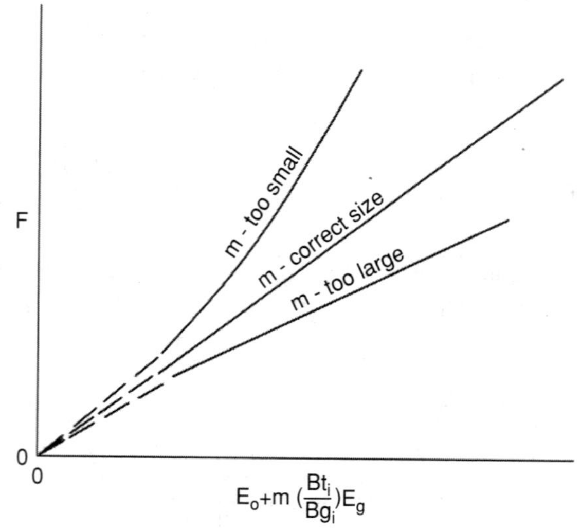

Figure 9.5 Volumetric saturated oil reservoir with *m* and *N* as unknowns (from Havlena and Odeh[7] © 1963 SPE, Reproduced with permission).

9.7 Material Balance Approach for Saturated Oil Reservoirs with Water Influx

The GMBE for a saturated oil reservoir with water influx is as shown in Eq. (9.2):

$$N(B_t - B_{ti}) + \frac{NmB_{ti}}{B_{gi}}[B_g - B_{gi}] + (1 + m)NB_{ti}\left[\frac{c_w S_{wi} + c_f}{1 - S_{wi}}\right]\Delta p + W_e - W_p B_w$$
$$= N_p[B_t + (R_p - R_{si})B_g]$$

Since the reservoir has a gas cap and water influx, the effects of formation and water compressibility are negligible. Thus, Eq. (9.2) can be reduced and re-arranged into the following form:

$$N_p[B_t + (R_p - R_{si})B_g] + W_p B_w = N(B_t - B_{ti}) + \frac{NmB_{ti}}{B_{gi}}[B_g - B_{gi}] + W_e \quad \textbf{(9.50)}$$

Using the symbols F, E_o, E_g as defined in Eqs. (9.37), (9.38) and (9.48), in Eq. (9.50) gives:

$$F = N\left[E_o + m\left(\frac{B_{ti}}{B_{gi}}\right)E_g\right] + W_e \quad \textbf{(9.51)}$$

Eq. (9.51) can be re-arranged in the form of a straight line equation for application of the Havlena-Odeh[7] method:

$$\frac{F}{\left[E_o + m\left(\frac{B_{ti}}{B_{gi}}\right)E_g\right]} = N + \frac{W_e}{\left[E_o + m\left(\frac{B_{ti}}{B_{gi}}\right)E_g\right]} \quad \textbf{(9.52)}$$

A plot of $F/[E_o + m(B_{ti}/B_{gi})E_g]$ on the y-axis against $W_e/[E_o + m(B_{ti}/B_{gi})E_g]$ on the x-axis will give a straight line with the intercept of the y-axis equal to N, and slope equal to unity if the correct values of W_e are used for known gas cap size, m. If W_e is too small, the line will curve upward. If W_e is too large, the line will curve downward. A linear aquifer will yield an S-shaped curve. The shapes of the curves that could result from this plot are as shown in Figure 9.1. Note the similarity of the plots resulting from applying the Havlena-Odeh[7] method to an undersaturated reservoir with water influx, and a saturated reservoir with water influx. The main difference between Eqs. (9.40) and (9.52) is the inclusion of the gas expansion term in Eq. (9.52) for the gas cap in saturated reservoirs.

9.8 Case History of Manatee Reservoirs

The Manatee reservoirs are located in the Gulf of Mexico, USA about 120 miles southwest of New Orleans, Louisiana, in Green Canyon Block 155 (GC 155) and in the western part of the same basin that contain the Angus and Troika reservoirs (Figure 9.6). The water depth at this location is 1900 feet. There are three reservoirs that constitute the Manatee reservoirs. These are the S10A1, S10A2, and S10B reservoirs. The S10A1 reservoir is relatively small in terms of

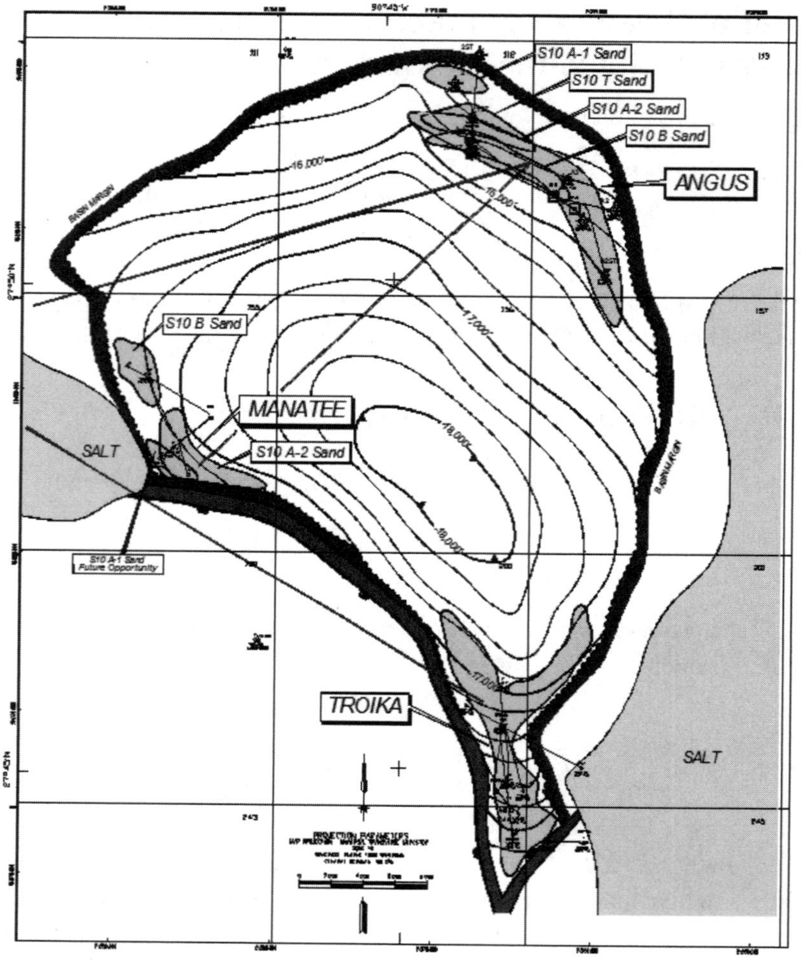

Figure 9.6 Basin map showing Manatee, Angus, and Troika reservoirs.

oil-in-place volumes and was not part of initial development. The type log of the sands in the Manatee reservoirs is shown as Figure 9.7. This case history is based on the S10A2 and S10B reservoirs.

9.8.1 Reservoir Geology

The S10A2 and S10B reservoirs consist of turbidite sands trapped against marl and salt in a mini-basin setting (Figure 9.6). The age of the pay intervals is lower Pliocene. The structure and net pay isopach maps of the S10A2 reservoir are shown as Figures 9.8 and 9.9, respectively. The planimetered reservoir volume of the S10A2 reservoir is 12,140 acre-ft. The structure and net pay isopach maps of the S10B reservoir are shown as Figures 9.10 and 9.11, respectively.

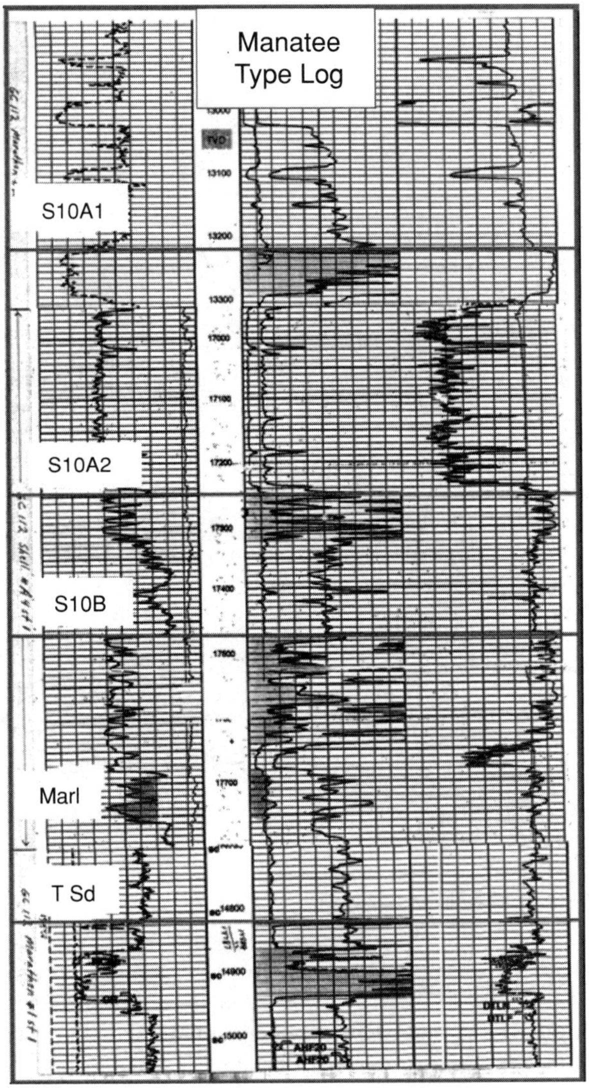

Figure 9.7 Type log of Manatee reservoirs.

The planimetered reservoir volume of the S10B reservoir is 13,707 acre-ft. The Manatee sands have five well penetrations (Wells #1, #1ST1, #1ST2, #1ST3, and #2) as shown in the maps. The oil-water contacts in the S10A2 and S10B reservoirs were determined to be at 17,000 feet and 17,100 feet, respectively. In June 1998, original pressures measured in the S10A2 and S10B reservoirs at the oil-water contact were 11,225 psia and 11,237 psia, respectively.

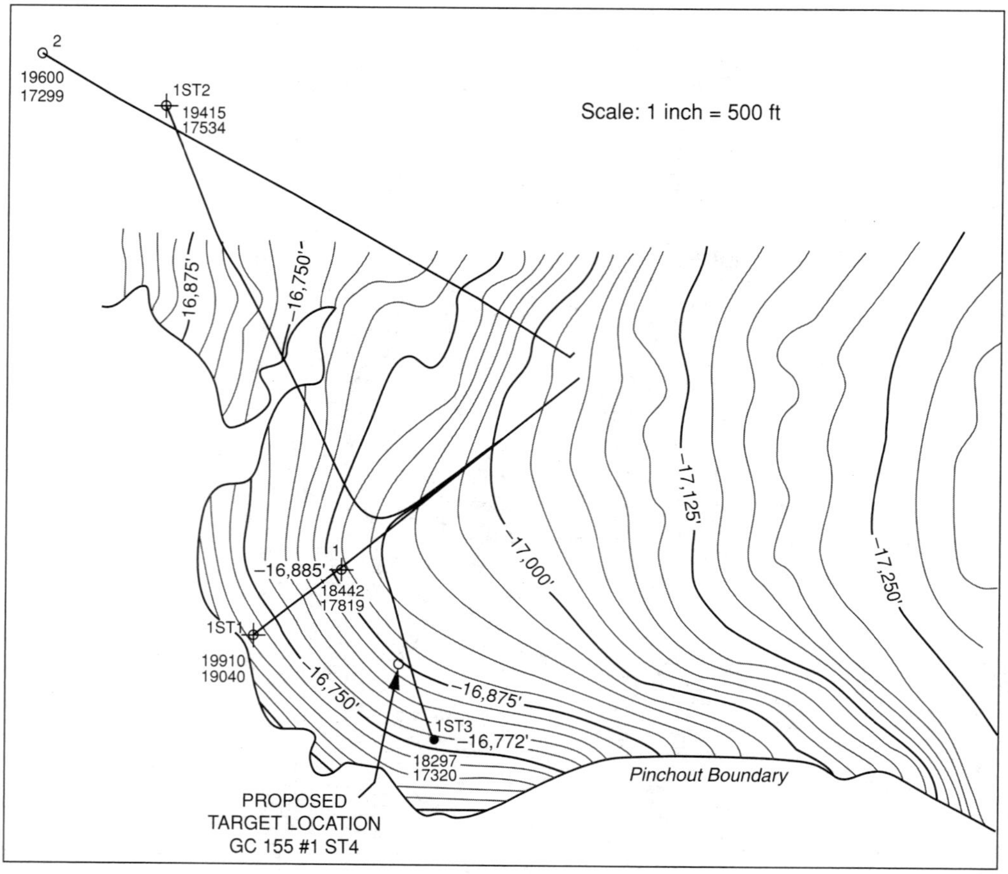

Figure 9.8 Structure map of S10A2 reservoir.

9.8.2 Rock and Fluid Properties

The rock properties of the S10A2 and S10B sands are summarized in Table 9.6. These averages are based on three well penetrations of the S10A2 sands, and two well penetrations of the S10B sands. The PVT properties of reservoir fluids for the S10A2 and S10B reservoirs are given in Tables 9.7 for oil and Table 9.8 for associated gas.

9.8.3 Reservoir Pressure and Production Data

The Manatee Field was discovered in June 1998 and oil production started in July 2002. The S10A2 reservoir was produced from a single well designated as Well #1ST3 (Figure 9.8). The S10B reservoir was also produced from a single well shown as Well #2 in Figure 9.10. The production history of the S10A2 reservoir is shown in Table 9.9 and Figure 9.12. The production history of the S10B reservoir is shown in Table 9.10 and Figure 9.13. The historical average reservoir pressures are also provided in Table 9.9 and Table 9.10 for S10A2 and S10B reservoirs, respectively.

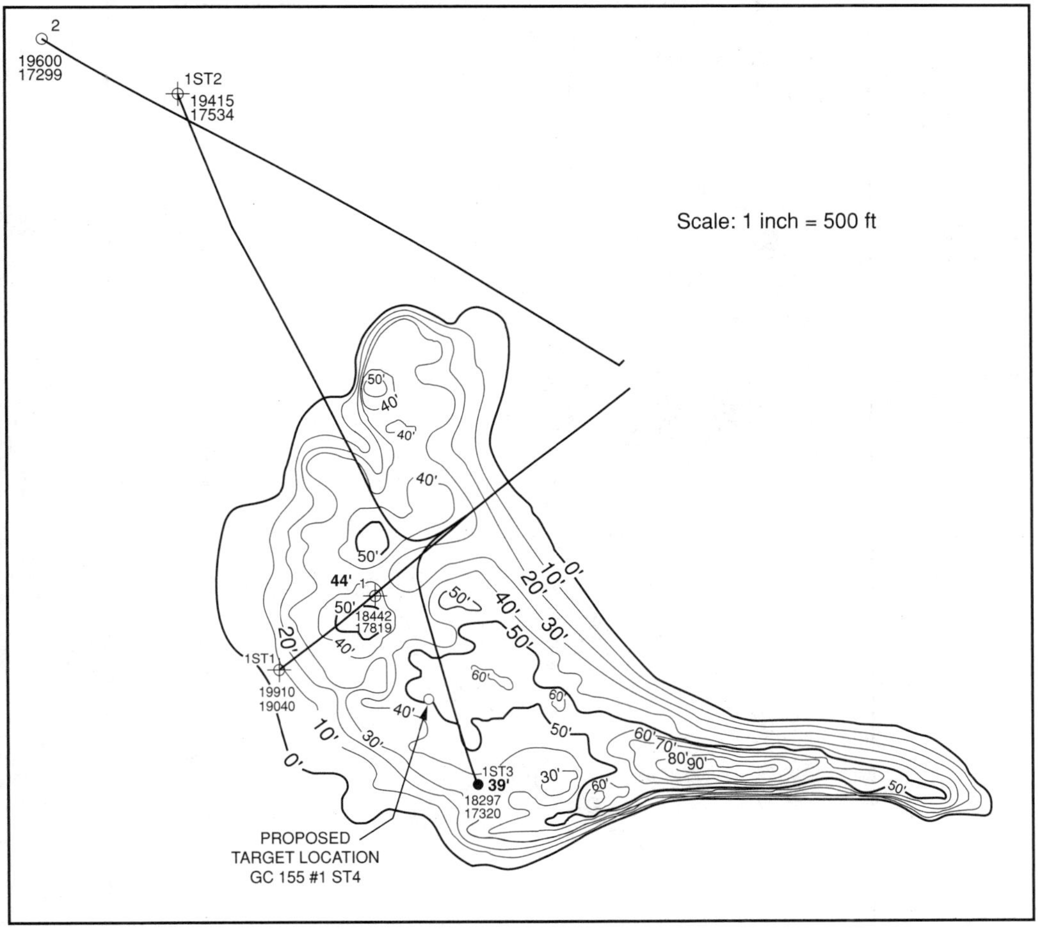

Figure 9.9 Net pay isopach map for S10A2 reservoir.

9.8.4 Review Questions

The case histories of the two reservoirs in the Manatee Field could be used as primers for discussions on the performance of oil reservoirs. The following review questions are suggested to aid in the discussions:

1. The cumulative production from the S10A2 reservoir appears to be substantially less than potential resources present in the reservoir. What are the potential reservoir factors that may have limited the productivity of the S10A2 reservoir?
2. Is the S10A2 reservoir a good candidate for a sidetrack well?
3. Is the S10B reservoir depleted? Is there potential for additional recovery?
4. What is the potential impact of the location of the Manatee reservoirs in the same basin as the Angus and Troika reservoirs? Could the three reservoirs be in hydraulic communication?

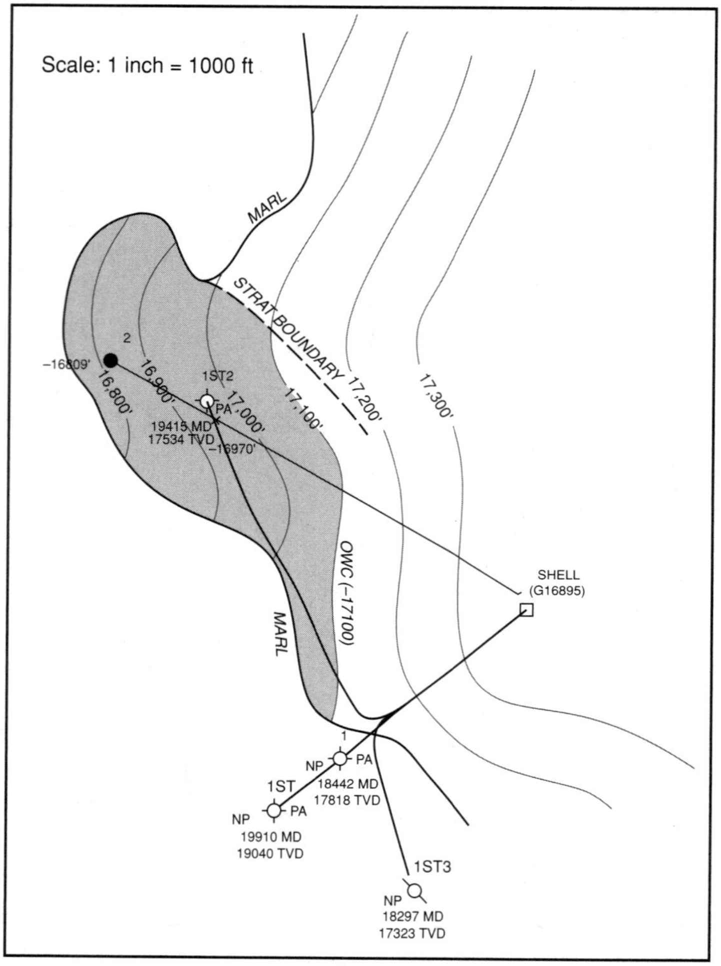

Figure 9.10 Structure map of S10B reservoir.

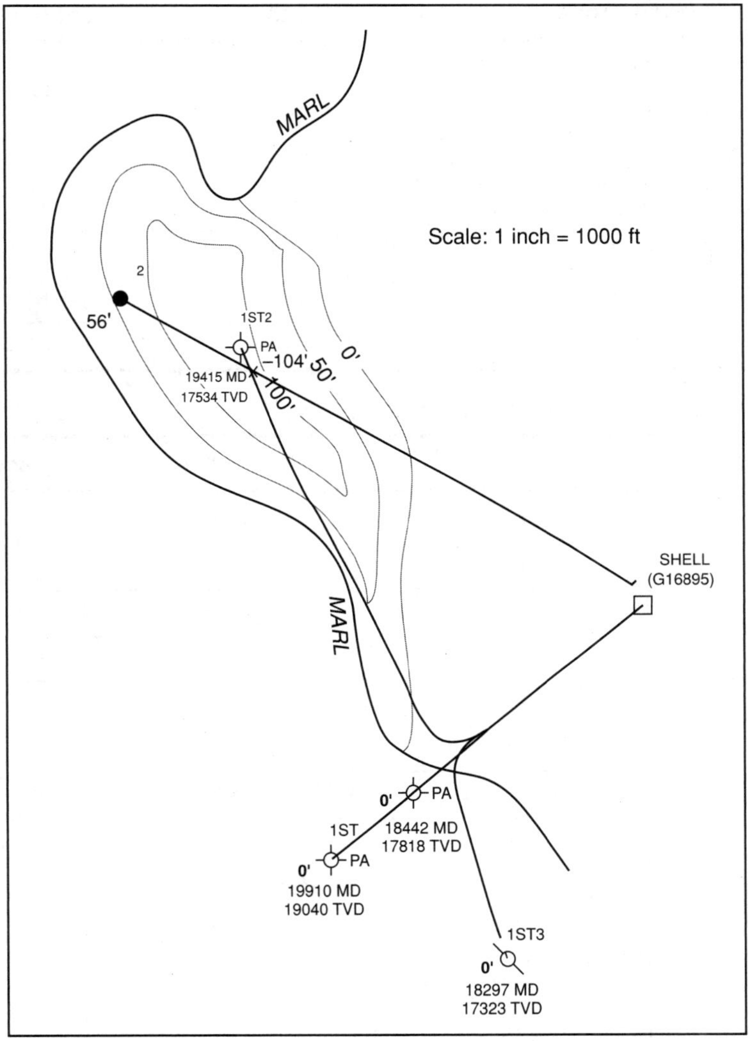

Figure 9.11 Net pay isopach map for S10B reservoir.

Table 9.6 Rock/Fluid Properties for S10A2 and S10B Reservoirs

Rock/Fluid Property	Reservoirs	
	S10A2	S10B
Porosity (%)	27.3	28.0
Permeability (md)	165	400
Initial water saturation (%)	21.7	24.6
Total compressibility (1/psi)	20E-06	20E-06
Temperature (°F)	185	188
Gas Gravity (Air = 1)	0.71	0.70
Oil Gravity (°API)	36.5	35.4

Table 9.7 Oil PVT Data for S10A2 and S10B Reservoirs

Pressure (psia)	FVF (RB/STB)	Rs (scf/STB)	Viscosity (cp)
12000	1.843	1805	0.372
11581	1.847	1805	0.363
11161	1.850	1805	0.354
10742	1.855	1805	0.345
10323	1.859	1805	0.336
9904	1.864	1805	0.327
9484	1.869	1805	0.317
9065	1.875	1805	0.308
8646	1.882	1805	0.298
8227	1.889	1805	0.289
7807	1.897	1805	0.280
7388	1.905	1805	0.271
6969	1.915	1805	0.262
6549	1.927	1805	0.254
6130	1.939	1805	0.246
5273	1.777	1464	0.274
4326	1.614	1123	0.315
3247	1.452	781	0.378
1938	1.289	440	0.498
15	1.126	99	0.884

Table 9.8 Gas PVT Data for S10A2 and S10B Reservoirs

Pressure (psia)	Gas FVF (RB/scf)	Viscosity (cp)
12000	0.000446	0.0446
11369	0.000454	0.0434
10738	0.000462	0.0422
10108	0.000471	0.0409
9477	0.000482	0.0395
8846	0.000493	0.0381
8215	0.000507	0.0366
7584	0.000523	0.0351
6954	0.000542	0.0335
6323	0.000566	0.0317
5692	0.000595	0.0299
5061	0.000633	0.0279
4430	0.000685	0.0258
3800	0.000760	0.0236
3169	0.000876	0.0212
2538	0.001072	0.0188
1907	0.001438	0.0166
1276	0.002230	0.0148
646	0.004674	0.0136
15	0.219731	0.0129

Table 9.9 Historical Production Data for S10A2 Reservoir

DATE	Oil Rate (BOPD)	Gas Rate (MCFPD)	Water Rate (BWPD)	Reservoir Pressure (PSIA)
Jan-02	0	0	0	11225
Feb-02	0	0	0	
Mar-02	0	0	0	
Apr-02	0	0	0	
May-02	0	0	0	
Jun-02	0	0	0	
Jul-02	214	239	0	
Aug-02	362	243	0	
Sep-02	54	119	0	
Oct-02	135	326	0	9600
Nov-02	71	155	0	
Dec-02	0	0	0	
Jan-03	0	0	0	
Feb-03	0	0	0	
Mar-03	249	466	0	
Apr-03	570	951	0	
May-03	1104	1975	0	
Jun-03	927	1592	0	
Jul-03	250	389	0	
Aug-03	878	1331	0	
Sep-03	163	266	0	
Oct-03	1135	1214	0	8400
Nov-03	815	1214	0	
Dec-03	494	724	0	
Jan-04	448	952	0	
Feb-04	488	787	0	
Mar-04	193	197	0	

DATE	Oil Rate (BOPD)	Gas Rate (MCFPD)	Water Rate (BWPD)	Reservoir Pressure (PSIA)
Apr-04	181	423	0	
May-04	235	371	0	8400
Jun-04	46	72	0	
Jul-04	194	318	0	
Aug-04	644	1072	0	
Sep-04	349	502	0	
Oct-04	103	172	0	
Nov-04	180	277	0	
Dec-04	337	640	0	
Jan-05	322	628	0	
Feb-05	0	0	0	
Mar-05	0	0	0	

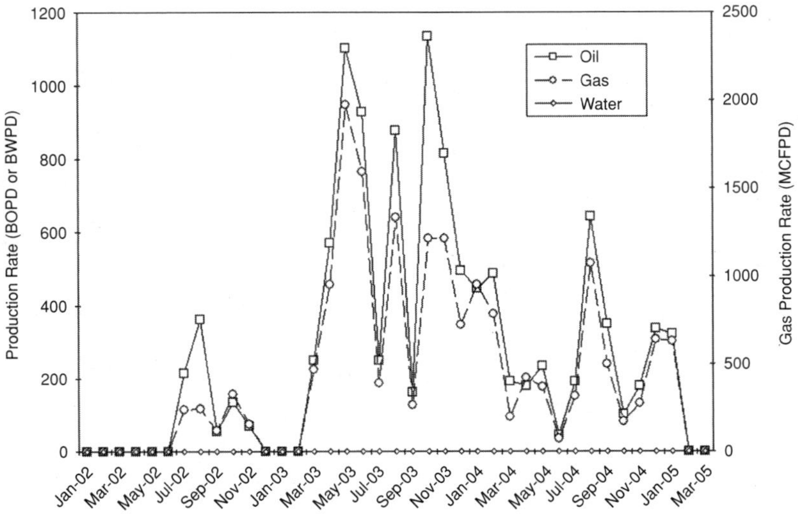

Figure 9.12 Historical production data for S10A2 reservoir.

Table 9.10 Historical Production Data for S10B Reservoir

DATE	Oil Rate (BOPD)	Gas Rate (MCFPD)	Water Rate (BWPD)	Reservoir Pressure (PSIA)
Jan-02	0	0	0	11237
Feb-02	0	0	0	
Mar-02	0	0	0	
Apr-02	0	0	0	
May-02	0	0	0	
Jun-02	0	0	0	
Jul-02	1431	1193	0	
Aug-02	3267	1212	0	
Sep-02	3637	6049	0	
Oct-02	1245	2052	0	9417
Nov-02	4587	7406	0	
Dec-02	5702	9213	0	
Jan-03	6238	10265	0	
Feb-03	6900	10611	0	
Mar-03	8012	13838	0	
Apr-03	6700	11190	0	
May-03	6311	11291	0	
Jun-03	6680	11457	0	
Jul-03	7491	12199	0	
Aug-03	6643	10491	0	
Sep-03	6645	10825	0	
Oct-03	5562	8005	0	8350
Nov-03	5376	8005	0	
Dec-03	5904	8653	0	
Jan-04	4826	10260	0	
Feb-04	3935	6871	0	

DATE	Oil Rate (BOPD)	Gas Rate (MCFPD)	Water Rate (BWPD)	Reservoir Pressure (PSIA)
Mar-04	4512	7337	0	8050
Apr-04	3613	5851	638	
May-04	3544	5591	625	
Jun-04	2986	4647	1538	
Jul-04	1810	2969	1157	
Aug-04	816	1359	1224	
Sep-04	563	809	844	
Oct-04	802	1353	1556	
Nov-04	851	1340	1651	
Dec-04	831	1699	1613	
Jan-05	988	1901	1918	
Feb-05	959	1810	1862	
Mar-05	1046	1875	2030	

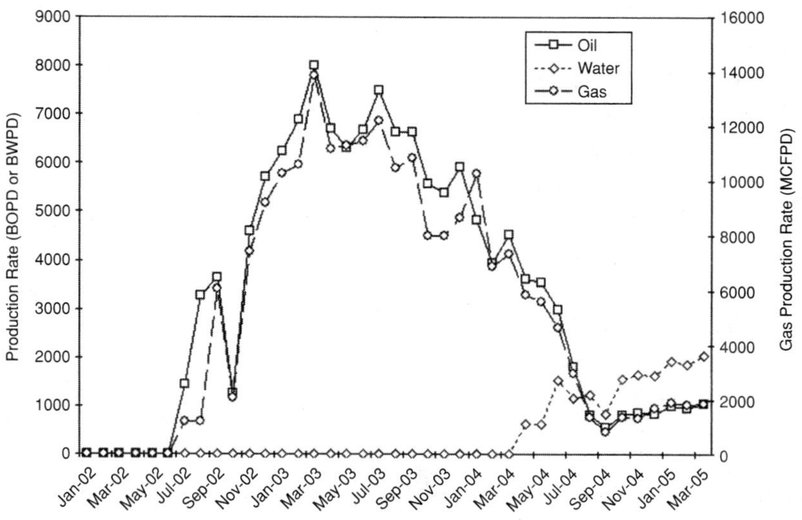

Figure 9.13 Historical production data for S10B reservoir.

Nomenclature

A	reservoir area, acres
B_g	gas formation volume factor, RB/scf
B_o	formation volume factor, RB/STB
B_t	two-phase formation volume factor, RB/STB
B_w	water formation volume factor, RB/STB
c_e	reservoir effective compressibility, psi^{-1}
c_f	formation compressibility, psi^{-1}
c_o	oil compressibility, psi^{-1}
c_w	water compressibility, psi^{-1}
E_{fw}	formation and water expansion term in material balance equation
E_g	gas expansion term in material balance equation
E_o	oil expansion term in material balance equation
F	cumulative reservoir voidage, RB
G	original gas in place, scf
G_{inj}	cumulative gas injection, scf
G_p	cumulative gas produced, scf
h	average net sand thickness, feet
I	drive index
m	ratio of original reservoir gas cap volume to original reservoir oil volume
N	original oil in place, STB
N_p	cumulative oil production, STB
N_r	oil remaining in reservoir, STB
p	pressure, psia
p_b	bubble point pressure, psia
R	oil recovery factor
R_p	cumulative production gas-oil ratio, scf/STB
R_s	solution gas-oil ratio, scf/STB
S_g	gas saturation, fraction
S_o	oil saturation, fraction
S_w	water saturation, fraction
W_e	cumulative water influx, RB
W_{inj}	cumulative water injection, STB
W_p	cumulative water production, STB
ϕ	porosity, fraction

Subscripts

i	initial
g	gas

gc	gas cap
o	oil
oz	oil zone
p	produced
r	remaining or residual
w	water
FWCD	formation and water compressibility drive
GCD	gas cap compressibility drive
OD	oil drive
WD	water drive

Abbreviations

GMBE	General Material Balance Equation
OIP	Oil In Place
OOIP	Original Oil In Place

References

1. Pirson, S.J.: *Elements of Oil Reservoir Engineering*, 2nd Edition, McGraw-Hill, New York (1958).
2. Dykstra, H.: "The Prediction of Oil Recovery by Gravity Drainage," *JPT* (May 1978) 818–830.
3. Wei, M.H., Yu, J.P., Moore, D.M., Ezekwe, N., Querin, M.E., and Williams, L.L.: "Case History of Pressure Maintenance by Crestal Gas Injection in the 26R Gravity Drainage Reservoir," paper SPE 24035 presented at the Western Regional Meeting, Bakersfield, California, March 30–April 1, 1992.
4. Essley, P.L., Hancock, G.L., and Jones, K.E.: "Gravity Drainage Concepts in a Steeply Dipping Reservoir," paper SPE 1029-G presented at the SPE-AIME Petroleum Production and Reservoir Engineering Conference, Tulsa, Oklahoma, March 20–21, 1958.
5. Hall, H.N.: "Analysis of Gravity Drainage." *JPT* (September 1961) 927–936.
6. Li, K., and Horne, R.N.: "Prediction of Oil Production by Gravity Drainage," paper SPE 84184 presented at the SPE Annual Technical Conference and Exhibition in Denver, Colorado, October 5–8, 2003.
7. Havlena, D., and Odeh, A.S.: "The Material Balance as an Equation of a Straight Line," *JPT* (August 1963) 896–900.
8. Poston, S.W., and Chen, H.Y.: "Case History Studies: Abnormal Pressured Gas Reservoirs," paper SPE 18857 presented at the SPE Production Operations Symposium, Oklahoma City, Oklahoma, March 13–14, 1989.
9. Fetkovich, M.J., Reese, D.E., and Whitson, C.H.: "Application of a General Material Balance for High-Pressure Gas Reservoirs," *SPEJ* (March 1998) 3–13.

CHAPTER 10

Fluid Flow in Petroleum Reservoirs

10.1 Introduction

In prior chapters, various methods for the calculation of volumes of hydrocarbons in place in petroleum reservoirs and potential recoveries were presented. These methods do not provide the engineer with techniques for estimating the rates at which the hydrocarbon reserves could be produced. Determination of achievable production rates is a critical element in the economic evaluation of hydrocarbon resources. If the predicted production rates are low due to geologic barriers and/or rock and fluid properties, economic development of the hydrocarbon resource may be delayed or in some cases abandoned if these factors that limit capacity to produce are not resolved in favor of the project. This requirement is especially important in high cost environments, such as deepwater reservoirs, where high production rates are generally required for economic viability of the project. The rate of fluid production at the wells is controlled by the rate of fluid flow in the reservoir. The main factors that control fluid flow in petroleum reservoirs are the permeability of the formation, the viscosity of the fluid, and the pressure difference between the well and the reservoir. (This pressure difference is called the **drawdown**.) The relationship between these factors and fluid flow in reservoirs was established in 1856 by Henry Darcy in an equation that still bears his name. Most equations that represent fluid flow in petroleum reservoirs are based on the Darcy equation. In this chapter, the radial form of the Darcy flow equation is presented for single-phase flow of three main types of fluids. The three main types of fluids that are considered are incompressible fluids, slightly compressible fluids, and compressible fluids. During fluid flow, the state of pressure distribution in the reservoir depends on duration of the flow (time) and distance from the well. Depending on flow time and size of the reservoir, three flow regimes can be distinguished. These are transient flow (or unsteady-state flow), pseudosteady-state flow (or semi-steady state flow), and steady-state flow. The characteristics of each flow regime are presented in this chapter. The continuity equation that governs fluid flow in a porous medium such as petroleum reservoirs is derived. From the continuity equations, diffusivity equations for slightly compressible fluids and compressible fluids are derived. Mathematical solutions to the diffusivity equations for the two types of fluids are presented for transient and pseudosteady-state flow regimes. The theory and application of the principle of superposition are illustrated with examples for the single-well, multi-production rates case; the multi-well reservoir system case; and the single well near a sealing boundary case.

10.2 Fluid Types

The three types of fluids discussed in this section are incompressible fluids, slightly compressible fluids, and compressible fluids. These classifications are based on the response of the fluid types to pressure changes at constant temperature.

10.2.1 Incompressible Fluids

Incompressible fluids are described as fluids whose volume does not change with pressure at constant temperature. However, all fluids are compressible to some degree. Hence, incompressible fluids do not exist in practice. This class of fluids has been defined to simplify the derivation of certain forms of the Darcy equation for radial flow. These equations are widely used to represent fluid flow and are sufficiently accurate for most engineering applications.

10.2.2 Slightly Compressible Fluids

The coefficient of isothermal compressibility, c, of any fluid is expressed as:

$$c = -\frac{1}{V}\left(\frac{\partial V}{\partial p}\right)_T \tag{10.1}$$

For slightly compressible fluids, it is assumed that the compressibility coefficient is small and constant over the range of pressures under consideration. Eq. (10.1) can be integrated between a low base pressure, p_b, where the volume is V_b, and the pressure of interest, p, at volume, V, to yield the expression:

$$\int_{p_b}^{p} -c\,dp = \int_{V_b}^{V} \frac{dV}{V} \tag{10.2}$$

Integrating Eq. (10.2) gives:

$$c(p_b - p) = \ln\left(\frac{V}{V_b}\right) \tag{10.3}$$

Re-arranging Eq. (10.3) gives:

$$V = V_b\left[e^{c(p_b - p)}\right] \tag{10.4}$$

The exponential function, e^x, can be defined by a power series as:

$$e^x = \sum_{n=0}^{\infty} \frac{x^n}{n!} = 1 + x + \frac{x^2}{2!} + \frac{x^3}{3!} + \frac{x^4}{4!} + \cdots \quad \textbf{(10.5)}$$

Applying the expression in Eq. (10.5), for small values of x, Eq. (10.4) can be written as:

$$V = V_b[1 + c(p_b - p)] \quad \textbf{(10.6)}$$

Equation (10.6) is applicable to slightly compressible fluids with small and constant compressibility coefficients. Most liquids can be classified as slightly compressible, especially undersaturated oils.

10.2.3 Compressible Fluids

Generally, all gases can be classified as compressible fluids. As shown in Eq. (4.90), gas compressibility, c_g, can be expressed as:

$$c_g = -\frac{1}{V}\left(\frac{\partial V}{\partial p}\right)_T$$

Note that this equation is equivalent to Eq. (10.1) defined above for slightly compressible fluids. Applying real gas equation of state, it was shown in Eq. (4.95) that gas compressibility for real gases can be expressed as:

$$c_g = \frac{1}{p} - \frac{1}{z}\left(\frac{\partial z}{\partial p}\right)_T$$

By comparing the above equation for gas compressibility to Eq. (10.3) for liquid compressibility, it is apparent that gas compressibility is several times higher than liquid compressibility for the same pressure change.

10.3 Definition of Fluid Flow Regimes

The three types of fluid flow regimes discussed in this section are transient flow, pseudosteady-state flow, and steady-state flow. The flow regimes depend on elapsed time and distance from the wellbore, and have characteristic pressure distributions in the reservoir.

10.3.1 Transient Flow

Suppose you are standing at the banks of a calm, motionless body of water such as a small lake. If you picked up a piece of rock and threw it into this body of water, a radial system of waves will radiate towards you from the point of entry of the rock. Before the waves reach the banks of this body of water, pressure distributions (or pressure transients) caused by your action are changing rapidly and are not constant. They depend only on distance from the point of entry of the rock and the elapsed time since the rock pierced the surface of the water. This is the transient phase of wave motion caused by the rock thrown into the body of water. This analogy can be extended to a circular reservoir with uniform thickness and homogeneous, isotropic properties

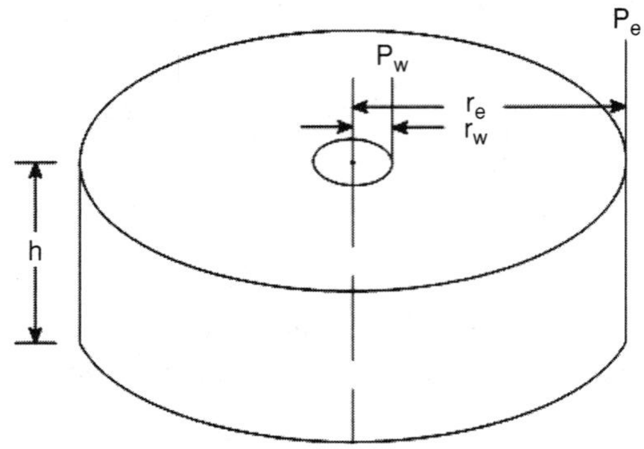

Figure 10.1 Diagram of a circular reservoir with a well at center.

(means constant rock and fluid properties in all directions). Suppose a single well exists at the center of this circular reservoir. The radius of the single wellbore is r_w, and the radius of the outer boundary of the circular reservoir is r_e. A schematic diagram of this simple reservoir is represented as Figure 10.1. If fluid production at a constant rate, q_w, is initiated at this well at time, $t = 0$, pressure transients will radiate from the well and travel towards the outer boundaries of the reservoir. These pressure transients are similar to the waves caused by the rock you threw travelling towards the banks of the lake. The rate of travel of the pressure transient is independent of the production rate at the well but is proportional to a property of the reservoir termed formation diffusivity. The formation diffusivity, η, is defined as:

$$\eta = \frac{k}{\phi \mu c_t} \tag{10.7}$$

In Eq. (10.7), k = formation permeability; ϕ = formation porosity; μ = fluid viscosity; and c_t = total compressibility of the system. As long as the leading front of the pressure transient has not reached the outer boundaries of the reservoir, fluid flow toward the well is in the transient state, and the reservoir acts as if it is infinite in size in response to the well. During the transient flow phase, pressure distribution in the reservoir is not constant and depends on elapsed time and distance from the well. This transient flow regime is illustrated in Figure 10.2 for the

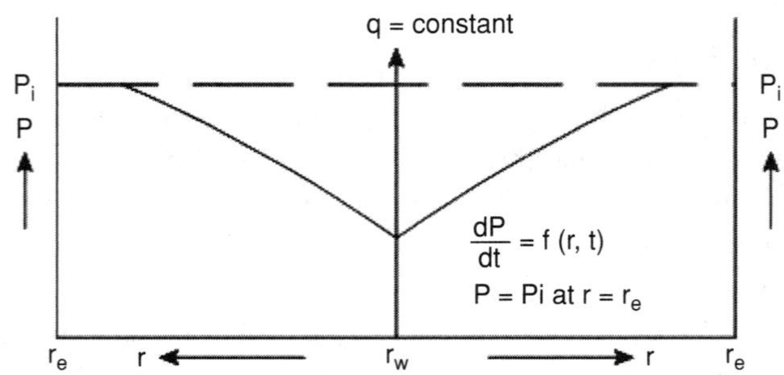

Figure 10.2 Transient flow diagram showing pressure distribution.

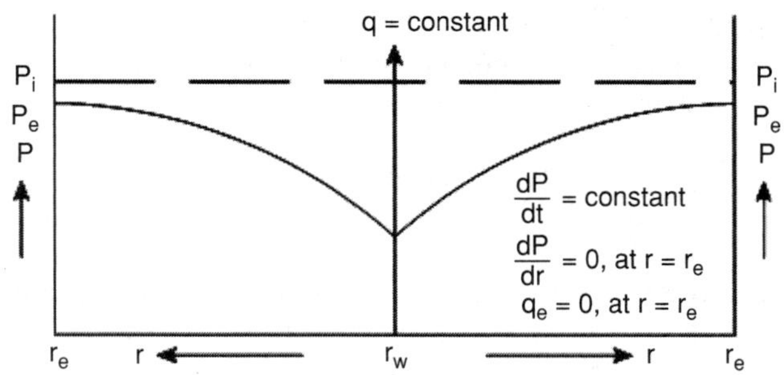

Figure 10.3 Pseudosteady-State flow diagram showing pressure distribution.

circular reservoir in Figure 10.1. Note in Figure 10.2 that reservoir pressure at the outer boundary is still at initial pressure p_i. Also note that transient flow is also called unsteady-state flow.

10.3.2 Pseudosteady-State (PSS) Flow

The pseudosteady-state (PSS) flow regime is sometimes called semi-steady-state flow. Using the illustration in Figure 10.1, PSS flow is achieved in the reservoir after the pressure transient has reached the outer boundaries of the reservoir and the pressure distribution is changing at a constant rate. This is illustrated in Figure 10.3. Note that as shown in Figure 10.3, pressure gradient at the outer boundary is zero. Hence, fluid flow across the outer boundary is also zero. The time, t_{pss}, for the start of PSS flow is given by:

$$t_{pss} = \frac{948\phi\mu c_t r_e^2}{k} \tag{10.8}$$

In Eq. (10.8), t_{pss} = time, hours; ϕ = formation porosity, fraction; μ = fluid viscosity, cp; c_t = total compressibility of system, psi^{-1}; r_e = outer boundary radius, feet; and k = formation permeability, md.

Note that there is a short time period under most flow conditions between the end of the transient flow regime and the establishment of the PSS flow regime. This time period has been called the Late Transient Flow (LTF) regime. The duration of the late transient flow regime is very short and is not considered in most flow systems. For these reasons, it is not covered in this book. However, the time to reach the start of the late transient flow period can be estimated with the following equation:

$$t_{LTF} = \frac{240\phi\mu c_t r_e^2}{k} \tag{10.9}$$

The units of the terms in Eq. (10.9) are the same as in Eq. (10.8). Eqs. (10.8) and (10.9) apply to reservoirs containing slightly compressible fluids such as oil. For gas reservoirs, the start of PSS flow is given by:

$$t_{pss} = \frac{948\phi\mu r_e^2}{k\overline{p}} \tag{10.10}$$

Also, for gas reservoirs the start of the late transient flow period can be estimated with:

$$t_{LTF} = \frac{240\phi\mu r_e^2}{k\overline{p}} \tag{10.11}$$

The average pressure, $\overline{p}$, in Eqs. (10.10) and (10.11) can be estimated as the arithmetic average of initial pressure and flowing well pressure. Other methods for estimating average pressure are presented later in Eqs. (10.93) and (10.94). If the flowing well pressure is not available, average pressure can be assumed to be equal to initial reservoir pressure if the reservoir has not been depleted significantly. In Eqs. (10.10) and (10.11), μ = gas viscosity, cp; ϕ = formation porosity, fraction; r_e = outer boundary radius, feet; k = formation permeability, md; and $\overline{p}$ = average reservoir pressure, psi.

10.3.3 Steady-State (SS) Flow

When the circular reservoir in Figure 10.1 reaches steady-state flow conditions, the pressure distribution at any point in the reservoir will not change with time. This condition is illustrated in Figure 10.4. Note that as shown in Figure 10.4 at steady-state flow conditions, the reservoir pressure at the outer boundary is constant, and the rate of fluid flow into the reservoir at the outer boundary is equal to the flow rate at the well. The SS flow regime is best represented by aquifer influx cases where the rate of water influx into the reservoir is considered equal to the total rate of fluid withdrawal from the reservoir.

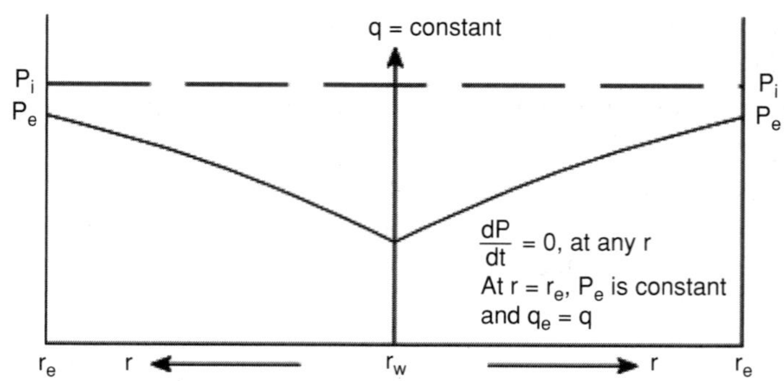

Figure 10.4 Steady-state flow diagram showing pressure distribution.

10.4 Darcy Fluid Flow Equation

In 1856, Henry Darcy demonstrated through a series of experiments that the flow velocity of a homogenous fluid through a porous medium under laminar (non-turbulent) conditions is proportional to the potential gradient. Darcy's law is expressed for linear flow as:

$$\nu = -\frac{k\rho}{\mu}\frac{d\Phi}{dl} \tag{10.12}$$

In Eq. (10.12), k = permeability; ρ = fluid density; μ = fluid viscosity; Φ = potential; and l = length of flow path. If the fluid is assumed to be incompressible, the potential can be represented as:

$$\Phi = \frac{p}{\rho} + gZ \tag{10.13}$$

In Eq. (10.13), p = pressure; Z = elevation above a datum plane; and g = acceleration due to gravity. Substituting Eq. (10.13) into Eq. (10.12) gives:

$$\nu = -\frac{k}{\mu}\left(\frac{dp}{dl} + \rho g\frac{dZ}{dl}\right) \tag{10.14}$$

In terms of Darcy units, Eq. (10.14) becomes:

$$\nu = -\frac{k}{\mu}\left(\frac{dp}{dl} + \frac{\rho g}{1.0133 \times 10^6}\frac{dZ}{dl}\right) \tag{10.15}$$

In Eq. (10.15), ν = apparent fluid velocity, cm/sec; k = permeability, darcy; μ = fluid viscosity, cp; p = pressure, atm; l = length, cm; ρ = fluid density, gm/cm^3; g = acceleration due to gravity, cm/sec^2; and Z = elevation, cm. In field units, Eq. (10.15) converts to:

$$q = -1.127 \times 10^{-3}\frac{kA}{\mu B}\left[\frac{dp}{dl} + 0.4335\gamma \sin\theta\right] \tag{10.16}$$

There are several important terms to note in Eq. (10.16). The volumetric flow rate, q, is in STB/D. The volumetric flow rate is related to the apparent fluid velocity by the expression:

$$q = \frac{\nu A}{B} \tag{10.17}$$

In Eqs. (10.16) and (10.17), ν = apparent fluid velocity, res bbl/day/ft^2; A = total cross-sectional area, ft^2; and B = formation volume factor, RB/STB. Also in Eq. (10.16), k = permeability, md; μ = fluid viscosity, cp; p = pressure, psia; l = flow path length, ft; γ = fluid specific gravity (water = 1); and θ = dip angle of the reservoir or formation measured counterclockwise from the horizontal to the positive flow path. The negative sign in Eq. (10.16) accounts for the sign convention that flow is considered positive in the positive direction of the flow path length, l, and pressure decreases in the direction of flow. The constant 1.127×10^{-3} is the conversion term from Darcy units to field units. The gravity term $(0.4335\gamma \sin\theta)$ in Eq. (10.16) is small in comparison to the pressure gradient (dp/dl) in most petroleum engineering applications of the linear form of the Darcy flow equation. For this reason, it is generally ignored. By ignoring the gravity term, Eq. (10.16) becomes:

$$q = -1.127 \times 10^{-3}\frac{kA}{\mu B}\left[\frac{dp}{dl}\right] \tag{10.18}$$

Assuming all the terms in Eq. (10.18) are constant except pressure, p, and length, l, of the linear flow system and integrating between its limits, Eq. (10.18) becomes:

$$q\int_0^L dl = -1.127 \times 10^{-3}\frac{kA}{\mu B}\int_{p_1}^{p_2} dp \quad \text{where} \quad p_1 > p_2 \tag{10.19}$$

Integration of Eq. (10.19) and re-arrangement of the terms gives:

$$q = 1.127 \times 10^{-3}\frac{kA(p_1 - p_2)}{\mu BL} \tag{10.20}$$

Equation (10.20) is the steady-state, linear flow equation for incompressible fluids.

10.5 Radial Forms of the Darcy Equation

In this section, radial forms of the Darcy equation for incompressible and compressible fluids are presented for pseudosteady-state and steady-state flows. Also methods for averaging permeabilities for parallel and serial beds are derived.

10.5.1 Steady-State Flow, Incompressible Fluids

Starting from Eq. (10.16) and assuming the gravity term is negligible, the Darcy equation can be represented in radial form in field units as:

$$\frac{qB}{2\pi rh} = 1.127 \times 10^{-3}\frac{k}{\mu}\frac{dp}{dr} \tag{10.21}$$

Note that in Eq. (10.21), the negative sign was dropped because pressure is decreasing in the same direction as decreasing radius (see Figure 10.1). Separating the variables and integrating between the wellbore radius, r_w, and the external reservoir boundary, r_e, within the corresponding pressure limits, p_w and p_e, respectively, gives:

$$q\int_{r_w}^{r_e}\frac{dr}{r} = \frac{1.127 \times 10^{-3} \times 2\pi kh}{\mu B}\int_{p_w}^{p_e} dp \tag{10.22}$$

After integration and simplification, Eq. (10.22) reduces to:

$$q = \frac{0.00708kh(p_e - p_w)}{\mu B \ln\left(\frac{r_e}{r_w}\right)} \tag{10.23}$$

Equation (10.23) is the radial flow equation for an incompressible fluid under steady-state conditions. In Eq. (10.23), q = well production rate, STB/D; k = average permeability of the drainage area of the well, md; h = average reservoir thickness, feet; p_e = pressure at external reservoir boundary or drainage area, psia; p_w = pressure at the wellbore, psia; μ = fluid viscosity, cp; B = fluid formation volume factor, RB/STB; r_e = external radius, feet; and r_w = wellbore radius, feet. Eq. (10.23) is generally used to calculate flow rates in reservoirs where the conditions are assumed to be close to steady-state. These are mainly reservoirs where pressure is maintained by water or gas injection or reservoirs with active water influx from aquifers. Note that p_e can be estimated as the average static pressure of the reservoir or drainage area of the well.

10.5.2 Average Permeability of Parallel Beds

Consider the sketch in Figure 10.5, which represents a formation consisting of sand beds of varying thickness and permeability producing into a common wellbore. If it is assumed that the fluids flowing in the sand beds have the same properties and the same drawdown was applied across all the sand beds, then the total flow rate of the well is the sum of the flow rate from each sand bed. This statement can be expressed as:

$$q_t = q_1 + q_2 + q_3 + \cdots\cdots + q_n \tag{10.24}$$

if there are n sand beds in the formation.

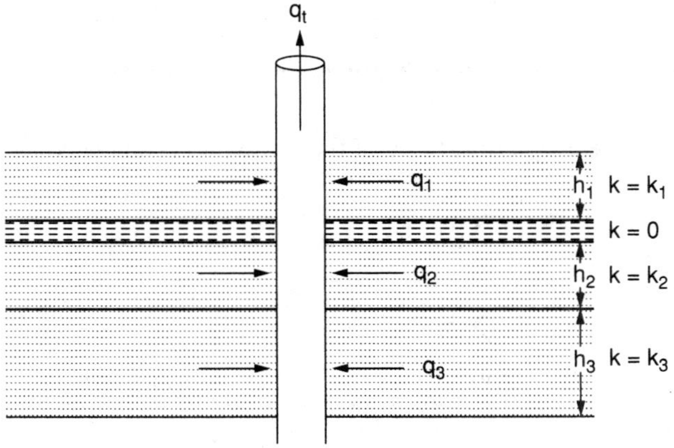

Figure 10.5 Radial flow in parallel beds of different permeabilities.

Applying Eq. (10.23) to Eq. (10.24) gives:

$$\frac{0.00708 k_{avg} h_t (p_e - p_w)}{\mu B \ln(r_e/r_w)} = \frac{0.00708 k_1 h_1 (p_e - p_w)}{\mu B \ln(r_e/r_w)} + \frac{0.00708 k_2 h_2 (p_e - p_w)}{\mu B \ln(r_e/r_w)} + \cdots + \frac{0.00708 k_n h_n (p_e - p_w)}{\mu B \ln(r_e/r_w)} \tag{10.25}$$

Cancelling like terms in Eq. (10.25) yields:

$$k_{avg} h_t = k_1 h_1 + k_2 h_2 + \cdots\cdots + k_n h_n \tag{10.26}$$

$$k_{avg} = \frac{\sum_i^n k_i h_i}{\sum_i^n h_i}$$

The product (kh) is a measure of the productive capacity of a sand bed or an interval. Assuming the intervals contain fluids with similar properties, the (kh) product is frequently used in a qualitative sense to assess the relative productivity of intervals in a well or a reservoir.

Example 10.1 Calculation of Average Permeability for Parallel Beds

Problem

Calculate the average permeability of the reservoir in Figure 10.5 if the permeability and thickness of the three beds (intervals) are as follows:

Interval	Permeability (md)	Thickness (Feet)
1	500	15
2	350	17
3	200	22

Solution

Using Eq. (10.26), the average permeability of the reservoir is calculated as:

$$k_{avg} = \frac{\sum_i^n k_i h_i}{\sum_i^n h_i}$$

$$= \frac{500 \times 15 + 350 \times 17 + 200 \times 22}{15 + 17 + 22}$$

$$= 330.56 \text{ md}$$

10.5.3 Average Permeability of Serial Concentric Segments

Consider the diagram in Figure 10.6, which shows radial flow in serial, concentric segments with different permeabilities but the same thickness. The flow through each segment is the same but the pressure drop across each segment is different. The total pressure drop across all the segments is equal to the sum of the pressure drop across each segment. This statement is represented in the form of an equation as:

$$(p_e - p_w) = (p_e - p_2) + (p_2 - p_1) + (p_1 - p_w) \quad \textbf{(10.27)}$$

Substituting Eq. (10.27) with Eq. (10.23) gives:

$$\frac{\mu B \ln(r_e/r_w)q}{0.00708k_{avg}h} = \frac{\mu B \ln(r_e/r_2)q}{0.00708k_e h} + \frac{\mu B \ln(r_2/r_1)q}{0.00708k_2 h} + \frac{\mu B \ln(r_1/r_w)q}{0.00708k_1 h} \quad \textbf{(10.28)}$$

Note as shown in Figure 10.6, $k_e = k_3$ and $r_e = r_3$.

Eliminating similar terms in Eq. (10.28) results in:

$$\frac{\ln(r_e/r_w)}{k_{avg}} = \frac{\ln(r_e/r_2)}{k_e} + \frac{\ln(r_2/r_1)}{k_2} + \frac{\ln(r_1/r_w)}{k_1} \quad \textbf{(10.29)}$$

Equation (10.29) can be re-arranged so that:

$$k_{avg} = \frac{\ln(r_e/r_w)}{\frac{1}{k_e}\ln\left(\frac{r_e}{r_2}\right) + \frac{1}{k_2}\ln\left(\frac{r_2}{r_1}\right) + \frac{1}{k_1}\ln\left(\frac{r_1}{r_w}\right)} \quad \textbf{(10.30)}$$

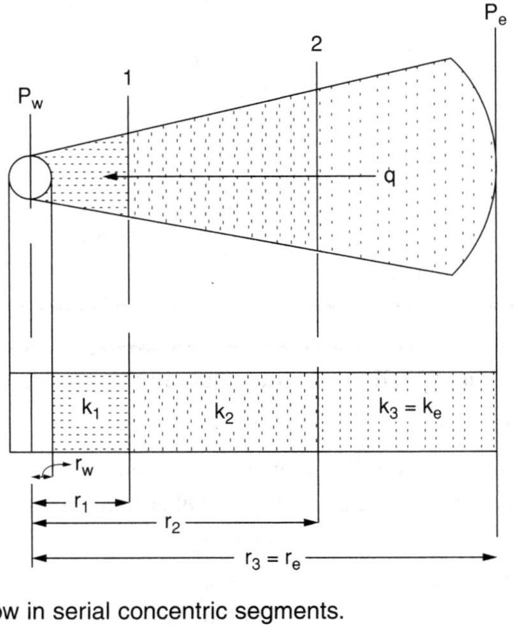

Figure 10.6 Radial flow in serial concentric segments.

A general form of Eq. (10.30) for n segments in series is:

$$k_{avg} = \frac{\ln\left(\frac{r_e}{r_w}\right)}{\sum_{i=1}^{n} \frac{1}{k_i} \ln\left(\frac{r_i}{r_{i-1}}\right)} \tag{10.31}$$

Example 10.2 Calculation of Average Permeability for Concentric Segments in Series

Problem

Suppose the drainage area of a well consists of three segments with different permeabilities. A damaged zone extending 10 feet from the wellbore has a reduced permeability of 25 md. This is followed by another segment extending 300 feet from the wellbore with a permeability of 350 md. Beyond this segment, the permeability of the formation is 250 md. The drainage radius of the well is 1000 feet and the wellbore radius is 0.5 feet. Calculate the average permeability of the drainage area.

Solution

From Eq. (10.31), the average permeability of the drainage area is:

$$\begin{aligned} k_{avg} &= \frac{\ln\left(\frac{r_e}{r_w}\right)}{\sum_{i=1}^{n} \frac{1}{k_i} \ln\left(\frac{r_i}{r_{i-1}}\right)} \\ &= \frac{\ln(r_e/r_w)}{\frac{1}{k_e}\ln\left(\frac{r_e}{r_2}\right) + \frac{1}{k_2}\ln\left(\frac{r_2}{r_1}\right) + \frac{1}{k_1}\ln\left(\frac{r_1}{r_w}\right)} \\ &= \frac{\ln\left(\frac{1000}{0.5}\right)}{\frac{1}{250}\ln\left(\frac{1000}{300}\right) + \frac{1}{350}\ln\left(\frac{300}{10}\right) + \frac{1}{25}\ln\left(\frac{10}{0.5}\right)} \\ &= 56.57 \text{ md} \end{aligned}$$

Example 10.3 Calculation of Well Production Rate in Steady-State Flow

Problem

Calculate the oil production rate of a well in reservoir whose pressure is maintained at 3500 psia by water injection. The flowing well pressure is 500 psia. Other fluid and reservoir data are as given:

Wellbore radius, r_w	0.5 ft
External radius of reservoir, r_e	2500 ft
Average permeability, k	300 md

Average reservoir thickness, h	25 ft
Oil viscosity, μ_o	4.5 cp
Oil FVF, B_o	1.25 RB/STB

Solution

Note that since reservoir pressure was maintained at 3500 psia, fluid properties (oil viscosity and oil FVF) should be evaluated at this pressure and not at the average pressure of 2000 psia between the well and the reservoir. Most of the pressure drop in a flowing well occurs close to the wellbore. Hence, the volumetric average pressure of the reservoir is closer to 3500 psia than 2000 psia. Substituting Eq. (10.23) with the given data, the oil production rate of the well is:

$$q = \frac{0.00708 \times 300 \times 25 \times (3500 - 500)}{4.5 \times 1.25 \times \ln\left(\frac{2500}{0.5}\right)}$$

$$= 3325 \text{ STB/D}$$

10.5.4 Pseudosteady State, Incompressible Fluids

The radial form of the Darcy flow equation can be expressed as:

$$q = \frac{2\pi rkh}{\mu}\frac{dp}{dr} \tag{10.32}$$

As discussed previously for pseudosteady-state flow, the flow rate is zero at the external boundary of the reservoir and increases to the constant flow rate at the wellbore (Figure 10.3). Since the compressibility of the fluid is constant, the flow rate at any radius, r, in the reservoir is proportional to the reservoir volume between the external radius, r_e, and radius, r. The flow rate, q, at radius, r, can be related to the flow rate, q_w, at the wellbore with radius, r_w, by the expression:

$$\frac{q}{q_w} = \frac{\left(r_e^2 - r^2\right)}{\left(r_e^2 - r_w^2\right)} \tag{10.33}$$

Substituting Eq. (10.33) into Eq. (10.32) gives:

$$q_w\left(\frac{r_e^2 - r^2}{r_e^2 - r_w^2}\right) = \frac{2\pi rkh}{\mu}\frac{dp}{dr} \tag{10.34}$$

The term, r_w^2, is very small in comparison to the term, r_e^2. Neglecting the term, r_w^2, Eq. (10.34) becomes:

$$q_w\left(1 - \frac{r^2}{r_e^2}\right) = \frac{2\pi rkh}{\mu}\frac{dp}{dr} \tag{10.35}$$

Separating the variables in Eq. (10.35) and showing the limits of integration gives:

$$q_w \int_{r_w}^{r_e} \frac{dr}{r} - \frac{q_w}{r_e^2} \int_{r_w}^{r_e} r dr = \frac{2\pi kh}{\mu} \int_{p_w}^{p_e} dp \tag{10.36}$$

Integrating between the limits as shown in Eq. (10.36) gives:

$$q_w \ln \frac{r_e}{r_w} - \frac{q_w}{r_e^2}\left(\frac{r_e^2}{2} - \frac{r_w^2}{2}\right) = \frac{2\pi kh(p_e - p_w)}{\mu} \tag{10.37}$$

Again, the term, r_w^2 is small compared with r_e^2. Neglecting r_w^2, Eq. (10.37) becomes:

$$q_w\left(\ln \frac{r_e}{r_w} - \frac{1}{2}\right) = \frac{2\pi kh(p_e - p_w)}{\mu} \tag{10.38}$$

From Eq. (10.38), the flow rate at the wellbore is given by:

$$q_w = \frac{2\pi kh(p_e - p_w)}{\mu\left[\ln\left(\frac{r_e}{r_w}\right) - \frac{1}{2}\right]} \tag{10.39}$$

In field units, Eq. (10.39) is converted to:

$$q_w = \frac{0.00708kh(p_e - p_w)}{\mu B\left[\ln\left(\frac{r_e}{r_w}\right) - \frac{1}{2}\right]} \tag{10.40}$$

The units of the variables in Eq. (10.40) are the same as the units of the variables in Eq. (10.23). Note that Eq. (10.40) is very similar to Eq. (10.23) except for the constant term, $(-1/2)$, that occurs in the denominator of Eq. (10.40). Because of the presence of this term in the denominator, it is obvious that well production rates calculated with Eq. (10.40) will be slightly higher that the rates calculated with Eq. (10.23) using the same data. Eq. (10.40) is applicable to depletion-drive reservoirs producing by fluid and rock expansion above the bubblepoint at constant compressibility of the fluid.

Example 10.4 Calculation of Well Production Rate in Pseudosteady-State Flow

Problem

Calculate the oil production rate of a well in reservoir whose pressure at the reservoir boundary is estimated to be 3500 psia. The flowing well pressure is 500 psia. Other fluid and reservoir data are as given:

Wellbore radius, r_w	0.5 ft
External radius of reservoir, r_e	2500 ft
Average permeability, k	300 md

Average reservoir thickness, h	25 ft
Oil viscosity, μ_o	4.0 cp
Oil FVF, B_o	1.26 RB/STB
Bubblepoint pressure, p_b	1800 psia

Solution

Note that fluid properties in this example were evaluated at an arithmetic average reservoir pressure of 2000 psia. This approach is a quick approximation since reservoir pressure varies from the wellbore to the boundary of the drainage area. Substituting Eq. (10.40) with the data provided, the well production rate is calculated as:

$$q = \frac{0.00708 \times 300 \times 25 \times (3500 - 500)}{4.0 \times 1.26 \times \left[\ln\left(\frac{2500}{0.5}\right) - \frac{1}{2}\right]}$$

$$= 3{,}942 \text{ STB/D}$$

10.5.5 Steady-State Flow, Compressible Fluids

Gas is the reservoir fluid that is highly compressible at reservoir conditions. Gas flow rate is typically measured in units of standard cubic feet per day (scf/D). This can be converted into flow rate at reservoir conditions using the gas formation volume factor. Gas formation volume factor was derived in Eq. (4.75) as:

$$B_g = \frac{zTp_{sc}}{T_{sc}p} \quad \text{in} \quad \frac{\text{ft}^3}{\text{scf}}$$

$$= \frac{zTp_{sc}}{5.615T_{sc}p} \quad \text{in} \quad \frac{\text{RB}}{\text{scf}}$$

Thus, gas flow rate q in scf/D is converted to RB/D by:

$$qB_g = q\left(\frac{zTp_{sc}}{5.615T_{sc}p}\right) \tag{10.41}$$

Substituting Eq. (10.41) into Eq. (10.21) gives:

$$\frac{q}{2\pi rh}\left(\frac{zTp_{sc}}{5.615T_{sc}p}\right) = 1.127 \times 10^{-3}\frac{k}{\mu}\frac{dp}{dr} \tag{10.42}$$

Re-arranging Eq. (10.42) gives:

$$\frac{q(z\mu)Tp_{sc}}{0.03976T_{sc}pkhr} = \frac{dp}{dr} \tag{10.43}$$

Assuming the product $(z\mu)$ is constant and integrating between the limits r_e and r_w (Figure 10.1) gives:

$$\frac{q(z\mu)Tp_{sc}}{0.03976T_{sc}kh}\int_{r_w}^{r_e}\frac{dr}{r} = \int_{p_w}^{p_e} pdp$$
$$\frac{q(z\mu)Tp_{sc}}{0.03976T_{sc}kh}\ln\left(\frac{r_e}{r_w}\right) = \frac{1}{2}\left(p_e^2 - p_w^2\right) \quad (10.44)$$

Re-arranging Eq. (10.44) gives:

$$q = \frac{0.01988T_{sc}kh\left(p_e^2 - p_w^2\right)}{(z\mu)Tp_{sc}\ln\left(\dfrac{r_e}{r_w}\right)} \quad (10.45)$$

In Eq. (10.45), q = gas flow rate, scf/D; T_{sc} = standard temperature, °R; k = permeability, md; h = formation thickness, feet; p_e = external boundary pressure, psia; p_w = wellbore flowing pressure, psia; z = compressibility factor; μ = gas viscosity, cp; T = reservoir temperature, °R; p_{sc} = standard pressure, psia; r_e = external radius, feet; and r_w = wellbore radius, feet. Note that the assumption that the product $(z\mu)$ is constant is generally valid for pressures up to 2000 psia. Consequently, Eq. (10.45) is only applicable for this pressure range.

10.6 Derivation of the Continuity Equation in Radial Form

Consider the volume element of thickness, Δr, in Figure 10.7 located at a distance, r, from a well at the center of the circular system. Suppose there is radial fluid flow through the element towards the well at the center. The reservoir system is assumed to have constant thickness, h, and

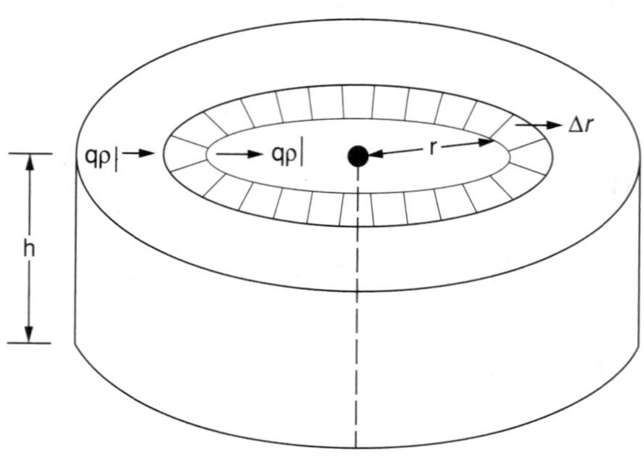

Figure 10.7 Volume element diagram for derivation of the radial continuity equation.

constant rock and fluid properties. Over a time period, Δt, the principle of mass conservation can be applied over the volume element such that:

$$\begin{aligned}&\text{Mass flowing into volume element in } \Delta t - \text{Mass flowing out of volume element in } \Delta t \\ &= \text{Rate of mass accumulation in volume element in } \Delta t\end{aligned} \tag{10.46}$$

$$\text{Mass rate IN} = 2\pi(r + \Delta r)h\rho\nu|_{r+\Delta r} \tag{10.47}$$

$$\text{Mass rate OUT} = 2\pi rh\rho\nu|_{r} \tag{10.48}$$

$$\text{Rate of Accumulation} = \frac{2\pi r\Delta rh[\{\phi\rho\}_{t+\Delta t} - \{\phi\rho\}_{t}]}{\Delta t} \tag{10.49}$$

Combining Eqs. (10.47), (10.48) and (10.49) into Eq. (10.46) gives:

$$2\pi(r + \Delta r)h\rho\nu|_{r+\Delta r} - 2\pi rh\rho\nu|_{r} = \frac{2\pi r\Delta rh[\{\phi\rho\}_{t+\Delta t} - \{\phi\rho\}_{t}]}{\Delta t} \tag{10.50}$$

Dividing both sides of Eq. (10.50) by $2\pi r\Delta rh$ and taking limits as Δr and Δt approach zero gives:

$$\frac{\partial}{\partial r}(\rho\nu) + \frac{1}{r}(\rho\nu) = \frac{\partial}{\partial t}(\phi\rho) \tag{10.51}$$

Re-arranging Eq. (10.51) gives:

$$\frac{1}{r}\left[\frac{\partial(r\rho\nu)}{\partial r}\right] = \frac{\partial(\phi\rho)}{\partial t} \tag{10.52}$$

Equation (10.52) is the continuity equation which represents the conservation of mass for flow in radial systems. In Eq. (10.52), ρ = fluid density; ν = fluid flow velocity; and ϕ = porosity of the system.

10.7 Derivation of Radial Diffusivity Equation for Slightly Compressible Fluids

In deriving the diffusivity equation for slightly compressible fluids, the following assumptions are made:

1. Compressibility of the fluid is small and constant over the pressure range of interest.
2. Viscosity of the fluid is constant and independent of pressure.
3. Permeability of the medium is constant and isotropic.

The above assumptions are necessary to simplify the derivation of the diffusivity equation for slightly compressible fluids.

The Darcy flow equation expressed in radial form in terms of fluid velocity is represented as:

$$\nu = 2.637 \times 10^{-4} \frac{k}{\mu} \frac{\partial p}{\partial r} \tag{10.53}$$

Note that in Eq. (10.53) the negative sign was not included because the pressure gradient necessary for flow to the well decreases as radius, r, decreases. In Eq. (10.53), ν = volumetric fluid velocity, ft^3/hr-ft^2; k = permeability, md; μ = fluid viscosity, cp; p = pressure, psia; and r = radius, ft.

Substituting Eq. (10.53) into Eq. (10.52) and re-arranging gives:

$$\frac{1}{r} \frac{\partial}{\partial r}\left(r\rho \frac{\partial p}{\partial r} \right) = \frac{1}{2.637 \times 10^{-4}} \frac{\mu}{k} \frac{\partial(\phi\rho)}{\partial t} \tag{10.54}$$

Applying the chain rule to the right-side and left-side of Eq. (10.54) yields:

$$\frac{\rho}{r} \frac{\partial}{\partial r}\left(r \frac{\partial p}{\partial r} \right) + \frac{\partial p}{\partial r} \frac{\partial \rho}{\partial p} \frac{\partial p}{\partial r} = \frac{1}{2.637 \times 10^{-4}} \frac{\mu}{k} \phi\rho \left(\frac{1}{\phi} \frac{\partial \phi}{\partial p} \frac{\partial p}{\partial t} + \frac{1}{\rho} \frac{\partial \rho}{\partial p} \frac{\partial p}{\partial t} \right) \tag{10.55}$$

From Eq. (10.1), fluid compressibility is expressed as:

$$\begin{aligned} c &= -\frac{1}{V}\left(\frac{\partial V}{\partial p} \right)_T \\ &= \frac{1}{\rho}\left(\frac{\partial \rho}{\partial p} \right)_T \end{aligned}$$

For an isothermal system, the preceding equation can be re-arranged so that:

$$\frac{\partial \rho}{\partial p} = c\rho \tag{10.56}$$

Note that the compressibility, c, in Eq. (10.56) is assumed to be small and constant. Pore volume compressibility, c_f, is defined by the following equation:

$$c_f = \frac{1}{\phi} \frac{\partial \phi}{\partial p} \tag{10.57}$$

The total compressibility, c_t, of the system is the sum of the fluid compressibility and formation compressibility:

$$c_t = c + c_f \tag{10.58}$$

Substituting Eqs. (10.56), (10.57) and (10.58) into Eq. (10.55) and dividing by ρ gives:

$$\frac{1}{r} \frac{\partial}{\partial r}\left(r \frac{\partial p}{\partial r} \right) + c\left(\frac{\partial p}{\partial r} \right)^2 = \frac{\phi\mu c_t}{2.637 \times 10^{-4} k} \frac{\partial p}{\partial t} \tag{10.59}$$

The pressure gradient is assumed to be small, so the term $(\partial p/\partial r)^2$ can be ignored. Hence, Eq. (10.59) reduces to:

$$\frac{1}{r}\frac{\partial}{\partial r}\left(r\frac{\partial p}{\partial r}\right) = \frac{\phi\mu c_t}{2.637 \times 10^{-4}k}\frac{\partial p}{\partial t} \tag{10.60}$$

Eq. (10.60) is the radial diffusivity equation for liquid flow in a porous medium. The coefficient $\left(\frac{\phi\mu c_t}{2.637 \times 10^{-4}k}\right)$ is the reciprocal of the formation diffusivity, η, as defined in Eq. (10.7). The units of η in Eq. (10.60) are feet^2 per hour.

10.8 Solutions of the Radial Diffusivity Equation for Slightly Compressible Fluids

There are two main solutions to the diffusivity equation for slightly compressible fluids that have wide applications in reservoir engineering. These are called the constant terminal rate solution and the constant terminal pressure solution. These designations are based on the initial and boundary conditions imposed on the diffusivity equation to achieve solutions. Note that these solutions apply to a bounded, cylindrical (circular) reservoir with a well located at its center (Figure 10.1). This model of the reservoir is modified as necessary for special cases.

10.8.1 Constant Terminal Rate Solution

For all the cases under the constant terminal rate solution, the following initial and boundary conditions apply:

1. At time, $t = 0$, the reservoir is at uniform pressure, $p = p_i$, for all r.
2. At time, $t > 0$, a single well of radius, r_w, at the center of the cylinder (reservoir) produces at a constant rate, q.

10.8.1.1 Case 1: Line Source Well, Infinite Reservoir

This is a special case of the constant terminal rate solution. This case assumes that the well located at the center of the cylinder has zero radius and the reservoir acts as if it is infinite in size. Hence, in addition to the conditions stated for the constant terminal rate solution, the following two conditions apply to this special case:

1. The well has zero radius, $r_w = 0$.
2. The reservoir is infinite in size (i.e. $p \rightarrow p_i$ as $r \rightarrow \infty$ for all t).

A solution to the diffusivity equation for a line source well in an infinite reservoir with the stated initial and boundary conditions was given by Matthews and Russell[1] as:

$$p(r, t) = p_i - \frac{q\mu}{2\pi kh}\left[-\frac{1}{2}Ei\left(-\frac{\phi\mu c_t r^2}{4kt}\right)\right] \tag{10.61}$$

The exponential integral, Ei, function is defined as:

$$-Ei(-x) = \int_x^{\infty} \frac{e^{-u}}{u}\,du \tag{10.62}$$

For $x < 0.02$,

$$Ei(-x) = \ln(1.781x) \tag{10.63}$$

Other values of the exponential integral can be read from a chart in Appendix 10A.
In field units, Eq. (10.61) can be written as:

$$p(r,t) = p_i - \frac{70.6qB\mu}{kh}\left[-Ei\left(-\frac{948\phi\mu c_t r^2}{kt}\right)\right] \tag{10.64}$$

In Eq. (10.64), p = pressure, psia; q = flow rate, STB/D; B = formation volume factor, RB/STB; μ = fluid viscosity, cp; k = permeability, md; h = formation thickness, ft; ϕ = porosity, fraction; c_t = total compressibility, psi^{-1}; r = radius, ft; and t = time, hr. Note that Eq. (10.64) can be used to calculate pressures at any distance, r, from the wellbore if:

$$t > \frac{3.79 \times 10^5 \phi\mu c_t r_w^2}{k} \tag{10.65}$$

Also, Eq. (10.64) can be used to calculate pressure at any distance, r, from the wellbore if the reservoir is assumed to be infinite-acting. The reservoir is assumed to be infinite-acting if:

$$t < \frac{948\phi\mu c_t r_e^2}{k} \tag{10.66}$$

Example 10.5 Calculation of Reservoir Pressure Assuming Line Source Well

Problem

A single well located in a deepwater reservoir is producing oil at a constant rate of 5500 STB/D. Calculate the pressure at a location 500 feet away from the well after 5 days, and 10 days of production. The drainage boundary of the well is located 1000 feet away. Other rock and fluid properties of the reservoir are as follows:

Initial pressure, p_i	19,250 psia
Formation thickness, h	345 ft
Formation permeability, k	7 md
Formation porosity, ϕ	0.19
Total compressibility, c_t	4×10^{-6} psi^{-1}
Oil viscosity, μ	4.5 cp
Oil FVF, B_o	1.1 RB/STB
Reservoir boundary, r_e	1000 ft
Wellbore radius, r_w	0.5 ft

Solution

Step 1: Check that the conditions set by Eqs. (10.65) and (10.66) are satisfied.
Before using Eq. (10.64), it is important to check that the conditions set by Eqs. (10.65) and (10.66) are met. Eq. (10.65) is used to check that the well has been on production for sufficient length of time so that the accuracy of Eq. (10.64) is not affected. Equation (10.66) is used to check that the reservoir is still infinite-acting.
The length of time the well is on production must be greater than the time calculated from Eq. (10.65):

$$t > \frac{3.79 \times 10^5 \phi \mu c_t r_w^2}{k} > \frac{3.79 \times 10^5 \times 0.19 \times 4.5 \times 4 \times 10^{-6} \times (0.5)^2}{7}$$

$$> 0.05 \text{ hrs}$$

Typically, the time required to exceed this limit as is the case in this example is very small. Since the calculations are performed after 5 and 10 days of production, the application of Eq. (10.64) is valid.
Next, it is important to check that the reservoir is infinite-acting after 10 days of production. This time of production must be less than the time calculated from Eq. (10.66):

$$t < \frac{948 \phi \mu c_t r_e^2}{k} < \frac{948 \times 0.19 \times 4.5 \times 4 \times 10^{-6} \times (1000)^2}{7}$$

$$< 463 \text{ hrs or } 19.3 \text{ days}$$

The reservoir will cease to be infinite-acting after about 19.3 days of production. Hence, after 10 days of production, the reservoir is still infinite-acting.

Step 2: Calculate reservoir pressure after 5 days, and 10 days of production at 500 feet from well.
Substituting Eq. (10.64) with the data provided gives:

$$p(r,t) = p_i - \frac{70.6 qB\mu}{kh}\left[-Ei\left(-\frac{948 \phi \mu c_t r^2}{kt}\right)\right]$$

$$= 19250 - \frac{70.6 \times 5500 \times 1.1 \times 4.5}{7 \times 345}$$

$$\left[-Ei\left(\frac{-948 \times 0.19 \times 4.5 \times 4 \times 10^{-6} \times (500)^2}{7 \times 24 \times t}\right)\right]$$

$$= 19250 - 795.89\left[-Ei\left(\frac{-4.8246}{t}\right)\right]$$

At 5 days of production, the $-Ei(-x)$ integral is:

$$-Ei\left(\frac{-4.8246}{t}\right) = -Ei\left(\frac{-4.8246}{5}\right) = -Ei(-0.96)$$

From the chart in Appendix 10A, $-Ei(-0.96) = 0.24$

$$\begin{aligned} p(r,t) &= 19250 - 795.8944 \times 0.24 \\ &= 19{,}059 \text{ psia} \end{aligned}$$

At 10 days of production, the $-Ei(-x)$ integral is:

$$-Ei\left(\frac{-4.8246}{t}\right) = -Ei\left(\frac{-4.8246}{10}\right) = -Ei(-0.48)$$

From Appendix 10A, $-Ei(-0.48) = 0.585$

$$\begin{aligned} p(r,t) &= 19250 - 795.8944 \times 0.585 \\ &= 18{,}784 \text{ psia} \end{aligned}$$

10.8.1.2 Case 2: Bounded Cylindrical Reservoir, No Flow at Outer Boundary

This case is a variation of the constant terminal rate solution. In addition to the initial and boundary conditions stated earlier for the constant terminal rate solution, the following conditions apply to this case:

1. The well has a radius, r_w.
2. There is no flow across the outer boundary. Hence, at $r = r_e$, $q = 0$.
3. Since there is no flow across the outer boundary, the pressure gradient at the outer boundary is zero. Thus, at $r = r_e$, $\frac{\partial p}{\partial r} = 0$.

The solution to the diffusivity equation at constant terminal rate under the stated initial and boundary conditions was reported in the classical work of van Everdingen and Hurst.[2] The solution is presented in terms of dimensionless variables of pressure, time, and radius. The dimensionless pressure variable, p_D, is defined as:

$$p_D = \frac{0.007082kh(p_i - p)}{qB\mu} \tag{10.67}$$

In Eq. (10.67), k = permeability, md; h = formation thickness, ft; p_i = initial pressure, psia: p = pressure at radius, r, and time, t; q = flow rate, STB/D; B = FVF, RB/STB; and μ = fluid viscosity, cp.

The dimensionless time variable, t_D, is defined as:

$$t_D = \frac{0.0002637kt}{\phi\mu c_t r_w^2} \tag{10.68}$$

In Eq. (10.68), the same units apply as for Eq. (10.67). The units for time, t, are in hours.

The outer boundary radius as a dimensionless variable, r_{eD}, is defined as:

$$r_{eD} = \frac{r_e}{r_w} \tag{10.69}$$

The dimensionless radius variable, r_D, is defined as:

$$r_D = \frac{r}{r_w} \tag{10.70}$$

Note that at the wellbore, $r_D = 1$.

An important solution by van Everdingen and Hurst[2] is for pressure at the wellbore. This is expressed as:

$$(p_D)_{r_D=1} = f(t_D, r_{eD}) \tag{10.71}$$

The tabulations of p_D as functions of t_D at fixed r_{eD} are provided in Appendix 10C for finite systems. Before the pressure transient caused by production at the wellbore reaches the outer boundary, the reservoir is acting as if it is infinite in size. For this condition, values of p_D as functions of t_D are provided in Appendix 10B. In using Appendix 10B to obtain values of p_D, two possible extensions of the tabulations are worthy of note. For $t_D < 0.01$, p_D can be estimated with the expression:

$$p_D \cong 2\sqrt{\frac{t_D}{\pi}} \tag{10.72}$$

For $100 < t_D < 0.25r_{eD}^2$, p_D can be estimated from the expression:

$$p_D \cong 0.5(\ln t_D + 0.80907) \tag{10.73}$$

Note that the reservoir is infinite-acting at a given r_{eD} if t_D values are smaller than the values listed in Appendix 10C. If this situation occurs, the tabulations in Appendix 10B should be used instead.

Example 10.6 Calculation of Wellbore Pressure for the Well in Example 10.5

Problem

For the well in Example 10.5, calculate flowing wellbore pressure after 5 days, and 10 days of production. The reservoir properties and conditions are as listed in Example 10.5.

Solution

Step 1: Calculate t_D.

From Eq. (10.68),

$$\begin{aligned} t_D &= \frac{0.0002637kt}{\phi\mu c_t r_w^2} \\ &= \frac{0.0002637 \times 7 \times 24 \times t}{0.19 \times 4.5 \times 4 \times 10^{-6} \times (0.5)^2} \\ &= 51{,}814.74t \end{aligned}$$

At t = 5 days, $t_D = 259{,}074$
At t = 10 days, $t_D = 518{,}147$

Step 2: Calculate r_{eD}.
From Eq. (10.69),

$$r_{eD} = \frac{r_e}{r_w} = \frac{1000}{0.5} = 2000.$$

Step 3: Calculate p_D.
Since $100 < t_D < 0.25r_{eD}^2$, use Eq. (10.73) to estimate p_D:

$$p_D \cong 0.5(\ln t_D + 0.80907)$$

At $t_D = 259{,}074$, $p_D = 6.6370$
At $t_D = 518{,}147$, $p_D = 6.9835$

Step 4: Calculate p_{wf}.
From Eq. (10.67),

$$\begin{aligned} p_{wf} &= p_i - \frac{141.2qB\mu}{kh} \times p_D \\ &= 19{,}250 - \frac{141.2 \times 5500 \times 1.1 \times 4.5}{7 \times 345} \times p_D \\ &= 19{,}250 - 1591.7888 \times p_D \end{aligned}$$

At t = 5 days, $p_D = 6.6370$, $p_{wf} = 8685$ psia
At t = 10 days, $p_D = 6.9835$, $p_{wf} = 8134$ psia

10.8.1.3 Case 3: Bounded Cylindrical Reservoir, Constant Pressure at Outer Boundary

This is another variation of the constant terminal rate solution. In addition to the initial and boundary conditions stated earlier for the constant terminal rate solution, the following conditions apply to this case:

1. The well has a radius, r_w.
2. At outer boundary where $r = r_e$, $p = p_i$.

The dimensionless variables, p_D, t_D, r_{eD}, and r_D as defined in Case 2 apply to Case 3. The solution of the diffusivity equation for these conditions by van Everdingen and Hurst[2] in terms of p_D at $r_D = 1$ as a function of t_D for fixed r_{eD} are tabulated in Appendix 10D. The reservoir is infinite-acting at values of t_D less than those listed in Appendix 10D. In such cases, p_D should be obtained from Appendix 10B. Case 3 conditions are applicable to waterflood reservoirs where the pressure at reservoir boundary or drainage volume of the well can be assumed to be constant.

10.8.2 Constant Terminal Pressure Solution

For the constant terminal pressure solution, the following initial and boundary conditions apply:

1. At time, $t = 0$, the reservoir is at uniform pressure, $p = p_i$, for all r.
2. At time, $t > 0$, a single well of radius, r_w, at the center of the cylinder (reservoir) produces at a constant pressure, p_{wf}.
3. There is no flow across the outer boundary. Hence, at $r = r_e$, $q = 0$.
4. Since there is no flow across the outer boundary, the pressure gradient at the outer boundary is zero. Thus, at $r = r_e$, $\partial p/\partial r = 0$.

van Everdingen and Hurst[2] published a solution of the diffusivity equation for slightly compressible fluids for the constant terminal pressure conditions as stated above. The solution provides the means for calculating the cumulative volume of fluid that will flow across any specified inner radius over a specified time interval in response to a specified drop in pressure between an outer boundary and the specified inner radius. The solution is expressed in terms of dimensionless variables, Q_D, t_D, and r_{eD}. Q_D is the dimensionless cumulative flow across the inner radius and is expressed as:

$$Q_D = \frac{QB}{1.119h\phi c_t r^2(p_i - p_{\text{at radius } r})} \tag{10.74}$$

In Eq. (10.74), Q = cumulative flow across radius, r, STB; B = FVF, RB/STB; h = formation thickness, ft.; ϕ = porosity, fraction; c_t = total compressibility, psi^{-1}; r = inner radius, ft.; p_i = initial pressure, psia; and $p_{\text{at radius r}}$ = pressure at radius, r, psia. Note that for a well producing at the center of a circular reservoir, pressure at radius, r, is the same as the flowing bottomhole pressure, p_{wf}.

The remaining dimensionless variables, t_D and r_{eD}, are as defined previously in Eqs. (10.68) and (10.69), respectively. Tabulations of Q_D as functions of t_D for infinite systems are presented in Appendix 10E. Further tabulations of Q_D as functions of t_D for specific r_{eD} are provided in Appendix 10F. For $r_{eD} = \infty$ (essentially $r_{eD} > 20$) and $t_D \geq 200$, Q_D can be estimated with a correlation developed by Edwardson et al.[3] as:

$$Q_D = \frac{(-4.29881 + 2.02566t_D)}{\ln t_D} \tag{10.75}$$

The constant terminal pressure method can be used to calculate cumulative production at a well for a specified pressure drop. This is illustrated in Example 10.7. However, this method is used primarily to calculate cumulative water influx from an aquifer into a reservoir. In this case, the inner radius is the reservoir boundary and the outer boundary is the external boundary of the aquifer. The cumulative flow, Q, is the same as the cumulative water influx, W_e, as designated in Chapters 8 and 9. The use of Eq. (10.74) for calculating cumulative water influx into a reservoir is presented in more detail under application of the superposition principle later in this chapter.

Example 10.7 Calculation of Cumulative Oil Production for the Well in Example 10.5

Problem

Suppose the well in Example 10.5 was produced at a constant flowing bottom-hole pressure of 13,500 psia starting from initial reservoir pressure of 19,250 psia. Calculate cumulative oil production after 5 days and 10 days of production. The reservoir properties and conditions are listed in Example 10.5.

Solution

Step 1: Calculate t_D.
From Eq. (10.68),

$$\begin{aligned} t_D &= \frac{0.0002637kt}{\phi\mu c_t r_w^2} \\ &= \frac{0.0002637 \times 7 \times 24 \times t}{0.19 \times 4.5 \times 4 \times 10^{-6} \times (0.5)^2} \\ &= 51{,}814.74t \end{aligned}$$

At t = 5 days, $t_D = 259{,}074$
At t = 10 days, $t_D = 518{,}147$

Step 2: Calculate r_{eD}.
From Eq. (10.69),

$$\begin{aligned} r_{eD} &= \frac{r_e}{r_w} \\ &= \frac{1000}{0.5} = 2000. \end{aligned}$$

Step 3: Calculate Q_D.
Since $r_{eD} > 20$ and $t_D > 200$, use Eq. (10.75):

$$Q_D = \frac{(-4.29881 + 2.02566t_D)}{\ln t_D}$$

For $t_D = 259{,}074$, $Q_D = 42{,}101.65$
For $t_D = 518{,}147$, $Q_D = 79{,}767.76$

Step 4: Calculate cumulative production, Q.
From Eq. (10.74),

$$Q = \frac{1.119h\phi c_t\left(r_w^2\right)(p_i - p_{wf}) \times Q_D}{B}$$

$$= \frac{1.119 \times 345 \times 0.19 \times 4 \times 10^{-6} \times (0.5)^2 \times (19250 - 13500) \times Q_D}{1.1}$$

$$= 0.3834 \times Q_D$$

At $t = 5$ days, cumulative production:

$$Q = 0.3834 \times Q_D$$
$$= 0.3834 \times 42{,}101.65 = 16{,}142 \text{ STB}.$$

At $t = 10$ days, cumulative production:

$$Q = 0.3834 \times Q_D$$
$$= 0.3834 \times 79{,}767.76 = 30{,}583 \text{ STB}.$$

10.9 Derivation of the Radial Diffusivity Equation for Compressible Fluids

In this derivation, compressible fluids are assumed to be gases. From Eq. (4.12), the real gas equation of state is:

$$\rho_g = \frac{pM_g}{zRT} \tag{10.76}$$

In terms of gas density, ρ_g, Eq. (10.54) can be written as:

$$\frac{1}{r}\frac{\partial}{\partial r}\left(r\rho_g\frac{\partial p}{\partial r}\right) = \frac{1}{2.637 \times 10^{-4}}\frac{\mu}{k}\frac{\partial(\phi\rho_g)}{\partial t} \tag{10.77}$$

Substituting Eq. (10.76) into Eq. (10.77) gives:

$$\frac{1}{r}\frac{\partial}{\partial r}\left[r\left(\frac{pM_g}{zRT}\right)\frac{\partial p}{\partial r}\right] = \frac{1}{2.637 \times 10^{-4}}\frac{\mu}{k}\frac{\partial\left[\phi\left(\frac{pM_g}{zRT}\right)\right]}{\partial t} \tag{10.78}$$

Eliminating M_g/RT in Eq. (10.78) gives:

$$\frac{1}{r}\frac{\partial}{\partial r}\left[r\left(\frac{p}{z}\right)\frac{\partial p}{\partial r}\right] = \frac{1}{2.637 \times 10^{-4}}\frac{\mu}{k}\frac{\partial\left[\phi\left(\frac{p}{z}\right)\right]}{\partial t} \tag{10.79}$$

Applying the product rule to right-side of Eq. (10.79) gives:

$$\begin{aligned}\frac{\partial}{\partial t}\left[\phi\left(\frac{p}{z}\right)\right] &= \phi\frac{\partial}{\partial t}\left(\frac{p}{z}\right) + \left(\frac{p}{z}\right)\frac{\partial \phi}{\partial t} \\ &= \phi\frac{\partial}{\partial p}\left(\frac{p}{z}\right)\frac{\partial p}{\partial t} + \left(\frac{p}{z}\right)\frac{\partial \phi}{\partial p}\frac{\partial p}{\partial t} \\ &= \phi\left(\frac{p}{z}\right)\frac{\partial p}{\partial t}\left[\left(\frac{z}{p}\right)\frac{\partial}{\partial p}\left(\frac{p}{z}\right) + \frac{1}{\phi}\frac{\partial \phi}{\partial p}\right]\end{aligned} \tag{10.80}$$

Gas compressibility was derived in Eq. (4.95) as:

$$c_g = \frac{1}{p} - \frac{1}{z}\left(\frac{\partial z}{\partial p}\right)_T$$

Expansion of one of the terms in Eq. (10.80) gives:

$$\begin{aligned}\left(\frac{z}{p}\right)\frac{\partial}{\partial p}\left(\frac{p}{z}\right) &= \left(\frac{z}{p}\right)\left(\frac{1}{z} - \frac{p}{z^2}\frac{\partial z}{\partial p}\right) \\ &= \frac{1}{p} - \frac{1}{z}\frac{\partial z}{\partial p} = c_g\end{aligned} \tag{10.81}$$

Pore volume compressibility was defined in Eq. (10.57) as:

$$c_f = \frac{1}{\phi}\frac{\partial \phi}{\partial p}$$

Replacing the terms in Eq. (10.80) with the expressions for c_g and c_f from Eqs. (10.81) and (10.57), respectively, gives:

$$\begin{aligned}\frac{\partial}{\partial t}\left[\phi\left(\frac{p}{z}\right)\right] &= \phi\left(\frac{p}{z}\right)\frac{\partial p}{\partial t}(c_g + c_f) \\ &= \phi c_t\left(\frac{p}{z}\right)\frac{\partial p}{\partial t}\end{aligned} \tag{10.82}$$

Substituting Eq. (10.82) into Eq. (10.79) gives:

$$\frac{1}{r}\frac{\partial}{\partial r}\left[r\left(\frac{p}{z}\right)\frac{\partial p}{\partial r}\right] = \frac{\phi \mu c_t}{2.637 \times 10^{-4} k}\left(\frac{p}{z}\right)\frac{\partial p}{\partial t} \tag{10.83}$$

Eq. (10.83) is the radial diffusivity equation in field units for compressible fluids such as gases.

10.10 Transformation of the Gas Diffusivity Equation with Real Gas Pseudo-Pressure Concept

Al-Hussainy et al.[4] introduced the concept of real gas pseudo-pressure to transform the gas diffusivity equation shown as Eq. (10.83) into a form that is suitable for solution and applicable to gas reservoir engineering problems. The real gas pseudo-pressure approach has two major advantages over previous methods used to linearize the gas diffusivity equation. These advantages are:

1. Real gas flow equations based on real gas pseudo-pressure approach account for pressure dependent gas properties such as gas viscosity and gas compressibility factor.
2. No assumption of small pressure gradients in the entire flow system is necessary. This means that the second degree pressure gradient terms $(\partial p/\partial r)^2$ are not ignored as was the case in the derivation of the diffusivity equation for liquids (see Eq. (10.59)).

Al-Hussainy et al.[4] defined the real gas pseudo-pressure, p_p, as:

$$p_p = 2\int_{p_b}^{p} \frac{p}{\mu(p)z(p)}dp \tag{10.84}$$

In Eq. (10.84), gas viscosity, μ, and gas compressibility factor, z, are shown as functions of pressure, and p_b is low reference base pressure, usually close to atmospheric pressure. Writing Eq. (10.84) in differential form gives:

$$\frac{\partial p_p}{\partial p} = \frac{2p}{\mu z} \tag{10.85}$$

Applying the chain rule to Eq. (10.85) yields the following equations:

$$\frac{\partial p_p}{\partial r} = \frac{\partial p_p}{\partial p}\frac{\partial p}{\partial r} \tag{10.86}$$

$$\frac{\partial p_p}{\partial t} = \frac{\partial p_p}{\partial p}\frac{\partial p}{\partial t} \tag{10.87}$$

Substituting for $\partial p_p/\partial p$ in Eqs. (10.86) and (10.87) with Eq. (10.85) give:

$$\frac{\partial p_p}{\partial r} = \frac{2p}{\mu z}\frac{\partial p}{\partial r} \tag{10.88}$$

$$\frac{\partial p_p}{\partial t} = \frac{2p}{\mu z}\frac{\partial p}{\partial t} \tag{10.89}$$

Substituting Eqs. (10.88) and (10.89) into Eq. (10.83) gives:

$$\frac{\partial^2 p_p}{\partial r^2} + \frac{1}{r}\frac{\partial p_p}{\partial r} = \frac{\phi \mu c_t}{2.637 \times 10^{-4}k}\frac{\partial p_p}{\partial t} \tag{10.90}$$

Equation (10.90) is the gas diffusivity equation in terms of real gas pseudo-pressure. This equation is still non-linear since gas viscosity and gas compressibility are functions of pressure. Al-Hussainy and Ramey[5] achieved an approximate solution to Eq. (10.90) which can be used for gas flow problems. For a radial infinite-acting reservoir with a well producing at a constant rate at the center, the flowing pressure at the well in terms gas pseudo-pressure is given by:

$$p_{pwf} = p_{pi} - \frac{1637qT}{kh}[\log t_D + 0.3513] \tag{10.91}$$

In Eq. (10.91), q = gas flow rate, Mscf/D, at standard conditions of 60°F and 14.7 psia; T = reservoir temperature, °R; k = permeability, md; h = formation thickness, ft; and t_D = dimensionless time. The dimensionless time, t_D, was defined in Eq. (10.68) as:

$$t_D = \frac{0.0002637kt}{\phi\mu c_t r_w^2}$$

Note that to calculate the dimensionless time, t_D, from the above equation, gas viscosity and total compressibility should be evaluated at the initial pressure, p_i.

For pseudosteady-state flow in a radial gas reservoir with closed outer boundary and a well producing at the center at constant rate, the gas flow rate can be calculated with an approximate equation from Al-Hussainy and Ramey[5] as:

$$q = \frac{1.987 \times 10^{-3} khT_{sc}}{Tp_{sc}} \left[\frac{\overline{p}_p - p_{pwf}}{\ln \frac{0.472r_e}{r_w}} \right] \tag{10.92}$$

The units in Eq. (10.92) are the same as in Eq. (10.91). The average pressure, $\overline{p}$, can be evaluated rigorously from reservoir isobaric maps as the volume-weighted average pressure thus:

$$\overline{p} = \frac{\sum_{j=1}^{n} \overline{p}_j V_j}{\sum_{j=1}^{n} V_j} \tag{10.93}$$

In Eq. (10.93), $\overline{p}_j$ is the arithmetic average of pressure data in the jth drainage volume, V_j. The average pressure, $\overline{p}$, can also be calculated from a less rigorous method suggested by Carter:[6]

$$\overline{p} = \sqrt{\frac{(p_i^2 + p_{wf}^2)}{2}} \tag{10.94}$$

Example 10.8 Calculation of the Flowing Bottomhole Pressure of a Gas Well

Problem

Calculate the flowing bottom-hole pressure of a gas well on production for 35 days at a constant rate of 250 Mscf/D. The reservoir and fluid properties data are given below:

Initial pressure, p_i	3000 psia
Reservoir temperature, T	121°F
Formation thickness, h	125 ft
Formation permeability, k	4 md
Formation porosity, ϕ	0.24
Total compressibility, c_t	1.8×10^{-4} psi^{-1}
Gas viscosity, μ	0.021 cp
Reservoir boundary, r_e	3100 ft
Wellbore radius, r_w	0.5 ft

The gas viscosity and compressibility data for the reservoir are provided in Table 10.1.

Solution

Step 1: Check that the reservoir is infinite-acting after 35 days of production.
Using Eq. (10.11), time for start of late transient flow is:

$$t_{LTF} = \frac{240\mu\phi r_e^2}{k\overline{p}} = \frac{240 \times 0.021 \times 0.24 \times (3100)^2}{4 \times 3000}$$
$$= 968.69 \text{ hours or } 40.36 \text{ days}$$

Since the reservoir was on production at this rate for 35 days and start of late transient flow was calculated as 40.4 days, the reservoir was still infinite-acting. Hence, Eq. (10.91) can be used to calculate flowing well pressure. Note that in estimating the time for start of late transient flow, it was assumed that average reservoir pressure was equal to initial reservoir pressure.

Table 10.1 Gas Properties for Example 10.8

Pressure, p (psia)	Gas Compressibility factor, z	Gas Viscosity, μ_g (cp)
500	0.9380	0.01200
1000	0.8800	0.01300
1500	0.8230	0.01413
2000	0.7630	0.01600
2500	0.7018	0.01859
3000	0.6843	0.02086
3500	0.6818	0.02285

Table 10.2 Calculation of Real-Gas Pseudo-Pressure in Example 10.8

Pressure, p (psia)	Gas Compressibility factor, z	Gas Viscosity, μ_g (cp)	$2p/(\mu z)$ (psia/cp)	Δp (psia)	p_p (psia^2/cp)
500	0.9380	0.01200	88,842	500	22,210,377
1000	0.8800	0.01300	174,825	500	88,127,047
1500	0.8230	0.01413	257,976	500	196,327,334
2000	0.7630	0.01600	327,654	500	342,734,826
2500	0.7018	0.01859	383,246	500	520,459,742
3000	0.6843	0.02086	420,330	500	721,353,662
3500	0.6818	0.02285	449,319	500	938,765,931

Step 2: Calculate real gas pseudo-pressure, p_p.
From Eq. (10.84),

$$p_p = 2\int_{p_b}^{p} \frac{p}{\mu(p)z(p)}dp$$

The real gas pseudo-pressure is calculated as shown in Table 10.2 using the trapezoidal rule for integration.

Step 3: Calculate dimensionless time, t_D:
From Eq. (10.68):

$$\begin{aligned} t_D &= \frac{0.0002637kt}{\phi\mu c_t r_w^2} \\ &= \frac{0.0002637 \times 4 \times 24 \times 35}{0.24 \times 0.021 \times 1.8 \times 10^{-4} \times (0.5)^2} \\ &= 3.907 \times 10^6 \end{aligned}$$

Step 4: Calculate flowing well pressure, p_{wf}:
From Eq. (10.91),

$$\begin{aligned} p_{pwf} &= p_{pi} - \frac{1637qT}{kh}[\log t_D + 0.3513] \\ &= 721{,}353{,}662 - \frac{1637 \times 250 \times 581}{4 \times 125}[\log (3.907 \times 10^6) + 0.3513] \\ &= 718{,}051{,}878 \end{aligned}$$

Note that p_{pi} was obtained from Table 10.2. Also, by interpolation in Table 10.2, the flowing bottom-hole pressure at $p_{pwf} = 718{,}051{,}878$ was transformed to be:

$$p_{wf} = 2991.8 \text{ psia}$$

10.11 The Superposition Principle

The superposition principle states that a new solution (or equation) to a fluid flow problem can be derived by the addition of solutions (or equations) to the original linear differential equation that represents the fluid flow problem. Thus, the superposition principle can be used in conjunction with the constant terminal rate solutions or the constant terminal pressure solutions to develop solutions to fluid flow problems on petroleum reservoirs.

10.11.1 Applications of Constant Terminal Rate Solutions with Superposition Principle

For the purposes of the discussions that follow, the exponential integral solution (Eq. 10.64) to the diffusivity equation for slightly compressible fluids is used. Remember that this particular solution assumes that the reservoir is infinite-acting. This assumption is also made here. However, other constant terminal rate solutions could be used instead of the exponential integral solution in the application of the superposition principle. In the application of the superposition principle, three special cases are presented for illustrations. These are:

1. Single well, multi-production rates case.
2. Multi-well reservoir system case.
3. Single well near a boundary (Image well) case.

10.11.1.1 Single Well, Multi-Production Rates Case

Suppose there is a single well producing at a constant rate, q_1, at time, $t = 0$. At time, $t = t_1$, the production rate was increased to q_2. The well was produced at this rate until time, $t = t_2$, when its rate was reduced to q_3. The well was then kept at q_3 for the total time, $t = t$. These flow conditions are illustrated in Figure 10.8. Assume there is no near-wellbore damage in this well. Applying the superposition principle, the total pressure drop in this well at time, t, is the sum of the pressure drop caused by the production rate at q_1 over time, t, and the pressure drop caused

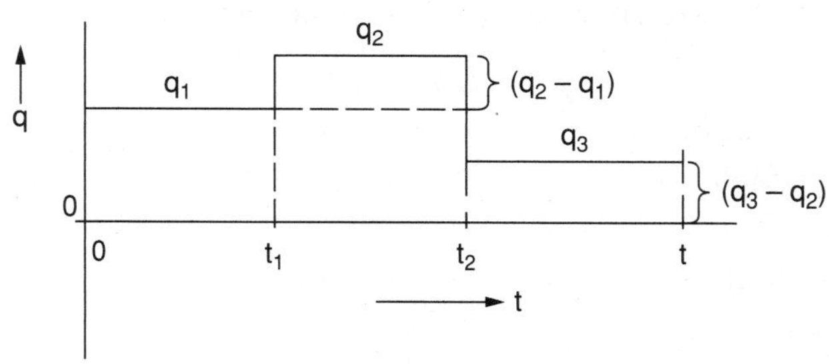

Figure 10.8 Diagram for a single well with multi-production rates.

by the production rate at $(q_2 - q_1)$ over time, $(t - t_1)$, and the pressure drop caused by the production rate at $(q_3 - q_2)$ over time, $(t - t_2)$. This is represented in the form of an equation as:

$$\Delta p_t = \Delta p_{q_1@t} + \Delta p_{(q_2-q_1)@(t-t_1)} + \Delta p_{(q_3-q_2)@(t-t_2)} \tag{10.95}$$

From Eq. (10.64),

$$\Delta p_{q_1@t} = (p_i - p_{wf})_{q_1@t} = \frac{70.6 q_1 B\mu}{kh}\left[-Ei\left(-\frac{948\phi\mu c_t r^2}{kt}\right)\right] \tag{10.96}$$

Similarly,

$$\Delta p_{(q_2-q_1)@(t-t_1)} = (p_i - p_{wf})_{(q_2-q_1)@(t-t_1)} = \frac{70.6(q_2 - q_1)B\mu}{kh}\left[-Ei\left(-\frac{948\phi\mu c_t r^2}{k(t - t_1)}\right)\right] \tag{10.97}$$

$$\Delta p_{(q_3-q_2)@(t-t_2)} = (p_i - p_{wf})_{(q_3-q_2)@(t-t_2)} = \frac{70.6(q_3 - q_2)B\mu}{kh}\left[-Ei\left(-\frac{948\phi\mu c_t r^2}{k(t - t_2)}\right)\right] \tag{10.98}$$

Substituting Eqs. (10.96), (10.97), and (10.98) into Eq. (10.95) gives:

$$\begin{aligned}\Delta p_t = {} & \frac{70.6 q_1 B\mu}{kh}\left[-Ei\left(-\frac{948\phi\mu c_t r^2}{kt}\right)\right] \\ & + \frac{70.6(q_2 - q_1)B\mu}{kh}\left[-Ei\left(-\frac{948\phi\mu c_t r^2}{k(t - t_1)}\right)\right] \\ & + \frac{70.6(q_3 - q_2)B\mu}{kh}\left[-Ei\left(-\frac{948\phi\mu c_t r^2}{k(t - t_2)}\right)\right]\end{aligned} \tag{10.99}$$

Example 10.9 Calculation of Total Pressure Drop due to Multi-Production Rates

Problem

Calculate the total pressure drop at an observation well located 1500 feet from a producing well in a virgin reservoir. The single production well was maintained at a constant production rate of 300 STB/D for 10 days; then the rate was increased to a constant rate of 500 STB/D for an additional 25 days. The initial reservoir pressure was 2500 psia. Assume the reservoir was infinite-acting during this test. Other reservoir and fluid properties are as follows:

Initial pressure, p_i	2500 psia
Formation thickness, h	15 ft
Formation permeability, k	300 md
Formation porosity, ϕ	0.24
Total compressibility, c_t	6×10^{-6} psi^{-1}
Oil viscosity, μ	4.2 cp
Oil FVF, B_o	1.2 RB/STB
Wellbore radius, r_w	0.5 ft

Solution

Step 1: Calculate pressure drop caused by initial production at 300 STB/D.
For this event, $q_1 = 300$ STB/D and total time $t = 35 \times 24 = 840$ hrs. Substituting Eq. (10.96), the pressure drop is:

$$\Delta p_{q_1@t} = (p_i - p_{wf})_{q_1@t} = \frac{70.6 q_1 B\mu}{kh}\left[-Ei\left(-\frac{948\phi\mu c_t r^2}{kt}\right)\right]$$
$$= \frac{70.6 \times 300 \times 1.2 \times 4.2}{300 \times 15}\left[-Ei\left(-\frac{948 \times 0.24 \times 4.2 \times 6 \times 10^{-6} \times (1500)^2}{300 \times 840}\right)\right]$$
$$= 23.7216[-Ei(-0.0512)]$$

From the chart in Appendix 10A, $-Ei(-0.0512) = 2.5$.
Thus, $\Delta p_{q_1@t} = 23.7216 \times 2.5 = 59.3$ psia.

Step 2: Calculate the pressure drop caused by production increase to 500 STB/D after 10 days.
The production increment, $q_2 - q_1 = 500 - 300 = 200$STB/D and the duration is $t - t_1 = (35 - 10) \times 24 = 600$ hrs. Substituting Eq. (10.97), the pressure drop is:

$$\Delta p_{(q_2-q_1)@(t-t_1)} = (p_i - p_{wf})_{(q_2-q_1)@(t-t_1)} = \frac{70.6(q_2 - q_1)B\mu}{kh}\left[-Ei\left(-\frac{948\phi\mu c_t r^2}{k(t - t_1)}\right)\right]$$
$$= \frac{70.6 \times 200 \times 1.2 \times 4.2}{300 \times 15}\left[-Ei\left(-\frac{948 \times 0.24 \times 4.2 \times 6 \times 10^{-6} \times (1500)^2}{300 \times 600}\right)\right]$$
$$= 15.8144 \times [-Ei(-0.0717)]$$

From the chart in Appendix 10A, $-Ei(-0.0717) = 2.2$.
Thus, $\Delta p_{(q_2-q_1)@(t-t_1)} = 15.8144 \times 2.2 = 34.8$ psia.

Step 3: Calculate the total pressure drop at the observation well.
The total pressure drop is:

$$\Delta p_t = \Delta p_{q_1@t} + \Delta p_{(q_2-q_1)@(t-t_1)} = 59.3 + 34.8 \text{ psia}.$$
$$= 94.1 \textit{ psia}$$

10.11.1.2 Multi-Well Reservoir System Case

Consider three wells (A, B, C) in a reservoir (Figure 10.9) producing at constant rates, q_A, q_B, and q_C, respectively, from time, $t = 0$, and located at distances, r_{OA}, r_{OB}, and r_{OC}, from an observation well at location, O. By the superposition principle, the total pressure drop at the observation well located at O after time, t, is the sum of the pressure drops caused by Wells A, B, and C, due to production at rates, q_A, q_B, and q_C. In the form of an equation, this can be expressed as:

$$\Delta p_t = \Delta p_A + \Delta p_B + \Delta p_C \quad \textbf{(10.100)}$$

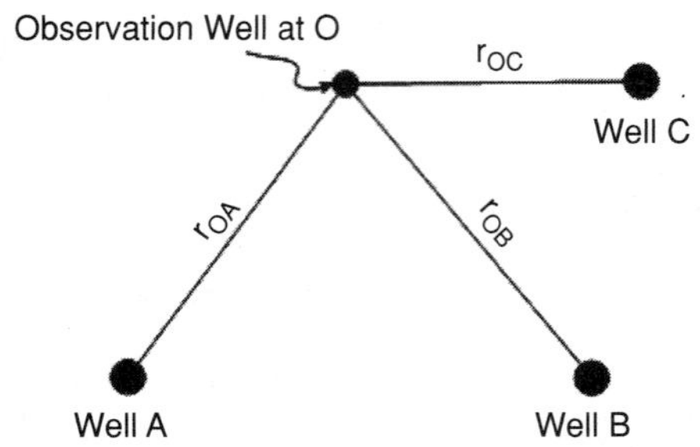

Figure 10.9 Diagram for a multi-well reservoir system.

Applying Eq. (10.64) and using the notation in Figure 10.9, Eq. (10.100) becomes:

$$\begin{aligned}\Delta p_t &= \frac{70.6q_A B\mu}{kh}\left[-Ei\left(-\frac{948\phi\mu c_t r_{OA}^2}{kt}\right)\right]\\ &+ \frac{70.6q_B B\mu}{kh}\left[-Ei\left(-\frac{948\phi\mu c_t r_{OB}^2}{kt}\right)\right]\\ &+ \frac{70.6q_C B\mu}{kh}\left[-Ei\left(-\frac{948\phi\mu c_t r_{OC}^2}{kt}\right)\right]\end{aligned} \tag{10.101}$$

Example 10.10 Calculation of Total Pressure Drop in a Multi-Well Reservoir System

Problem

A reservoir at initial pressure of 2500 psia has two wells (Well A and Well B) on production for 35 days from initial production. Well A produced at a constant rate of 500 STB/D and is located 750 feet from an observation well. Well B produced at a constant rate of 300 STB/D and is located 1000 feet from the same observation well. If the reservoir is infinite in size, calculate the total pressure drop measured at the observation well at the end of 35 days. This reservoir has the same rock and fluid properties as the reservoir in Example 10.9.

Solution

Step 1: Calculate the pressure drop caused by Well A.
From Eq. (10.101):

$$\begin{aligned}\Delta p_A &= \frac{70.6q_A B\mu}{kh}\left[-Ei\left(-\frac{948\phi\mu c_t r_{OA}^2}{kt}\right)\right]\\ &= \frac{70.6\times 500\times 1.2\times 4.2}{300\times 15}\left[-Ei\left(-\frac{948\times 0.24\times 4.2\times 6\times 10^{-6}\times (750)^2}{300\times 35\times 24}\right)\right]\\ &= 39.5360\times[-Ei(-0.0128)]\end{aligned}$$

Since $x < 0.02$, Eq. (10.63) is used for calculation of the exponential integral. From Eq. (10.63):

$$Ei(-x) = \ln(1.781x) = \ln(1.781 \times 0.0128) = -3.7811$$

Hence,

$$\begin{aligned}\Delta p_A &= 39.5360 \times [-(-3.7811)] \\ &= 149.49 \text{ psia}\end{aligned}$$

Step 2: Calculate the pressure drop caused by Well B.
From Eq. (10.101):

$$\begin{aligned}\Delta p_B &= \frac{70.6 q_B B\mu}{kh}\left[-Ei\left(-\frac{948\phi\mu c_t r_{OB}^2}{kt}\right)\right] \\ &= \frac{70.6 \times 300 \times 1.2 \times 4.2}{300 \times 15}\left[-Ei\left(-\frac{948 \times 0.24 \times 4.2 \times 6 \times 10^{-6} \times (1000)^2}{300 \times 35 \times 24}\right)\right] \\ &= 23.7216 \times [-Ei(-0.0228)]\end{aligned}$$

From the chart in Appendix 10A, $-Ei(-0.0228) = 3.1$.
Hence,

$$\begin{aligned}\Delta p_B &= 23.7216 \times 3.1 \\ &= 73.53 \text{ psia}\end{aligned}$$

Step 3: Calculate the total pressure drop at the observation well.
The total pressure drop at the observation well is:

$$\begin{aligned}\Delta p_t &= \Delta p_A + \Delta p_B = 149.49 + 73.53 \\ &= 223 \text{ psia}\end{aligned}$$

10.11.1.3 Single Well Near A Sealing Boundary Case

Suppose Well A producing at a constant rate, q_A, is located at a distance, r, from a sealing fault as shown in Figure 10.10. There is no flow across the sealing fault (i.e., it is a closed boundary). The pressure drop at observation Well O at a distance, r_{OA}, from Well A, is the sum of the pressure drop caused by Well A and an image Well I located behind the sealing fault at an equal distance, r, from the fault. If the distance from the image well to the observation well is r_{OI}, then the total pressure drop at the observation well, assuming the reservoir is infinite-acting is:

$$\begin{aligned}\Delta p_{t@well\,O} &= \Delta p_A + \Delta p_I \\ &= \frac{70.6 q_A B\mu}{kh}\left[-Ei\left(-\frac{948\phi\mu c_t r_{OA}^2}{kt}\right)\right] \\ &+ \frac{70.6 q_A B\mu}{kh}\left[-Ei\left(-\frac{948\phi\mu c_t r_{OI}^2}{kt}\right)\right]\end{aligned} \quad (10.102)$$

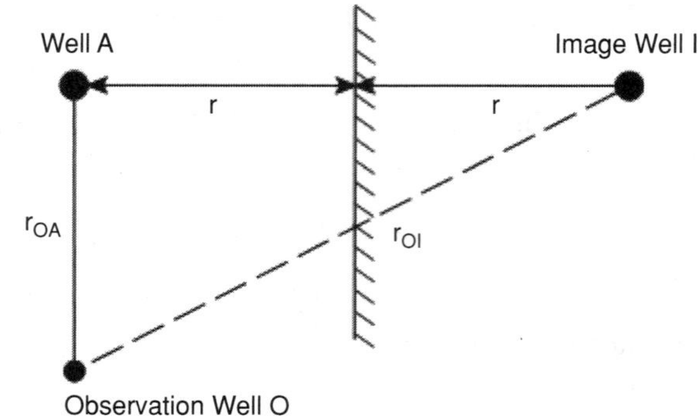

Figure 10.10 Diagram of a well near a sealing boundary.

10.11.2 Applications of Constant Terminal Pressure Solution with Superposition Principle

The constant terminal pressure solution coupled with the superposition principle is used primarily to calculate cumulative water influx into a reservoir from an aquifer. Consider a radial hydrocarbon reservoir surrounded by a radial aquifer as shown in Figure 10.11. Applying the terminology that was used to define the initial and boundary conditions of the constant terminal pressure solution, the hydrocarbon reservoir becomes the well with constant pressure at its boundary, r_R, which separates it from the aquifer. The extent of the aquifer, designated as r_a, is the outer boundary of the system. For this reservoir-aquifer system, the dimensionless radius, r_{aD}, is defined as:

$$r_{aD} = \frac{r_a}{r_R} \tag{10.103}$$

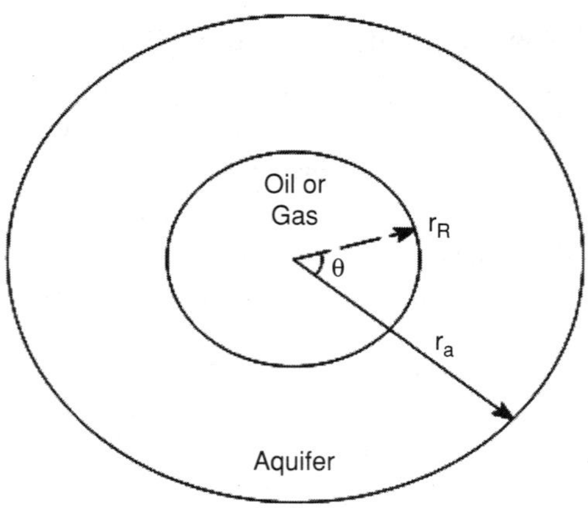

Figure 10.11 Diagram of a reservoir with a radial aquifer.

The dimensionless time, t_D, is defined as:

$$t_D = \frac{0.0002637kt}{\phi\mu c_t r_R^2} \tag{10.104}$$

Rewriting Eq. (10.74) in terms of the reservoir-aquifer system, the cumulative water influx, W_e, in reservoir barrels (RB) is given by:

$$W_e = 1.119h\phi c_t r_R^2 \Delta p Q_D \tag{10.105}$$

In Eqs. (10.103), (10.104) and (10.105), r = radius, ft; k = permeability, md; t = time, hours; ϕ = porosity, fraction; μ = fluid viscosity, cp; c_t = total compressibility, psi^{-1}; h = formation thickness, ft; Δp = pressure drop, psi; and Q_D = dimensionless cumulative water influx evaluated as a function of r_{aD} and t_D. Note that all the fluid and rock properties in Eqs. (10.104) and (10.105) are based on the aquifer.

In hydrocarbon reservoirs, pressure change occurs continuously over time. Sudden or step pressure change as required by Eq. (10.105) does not occur under normal conditions. For a reservoir with a continuous pressure history as shown in Figure 10.12, the superposition principle is used to transform the pressure history into a series of stair-step pressure changes which is kept constant over a specific time interval. For instance, to calculate the cumulative water influx at the end of four years for the reservoir with the pressure history shown in Figure 10.12, the first

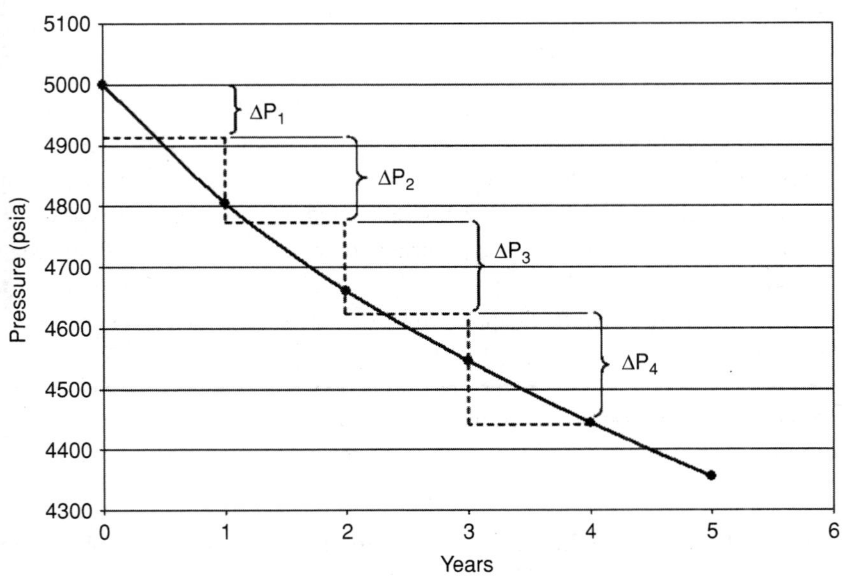

Figure 10.12 Typical pressure history profile at the reservoir/aquifer boundary.

pressure drop, Δp_1, is active for four years, the next pressure drop, Δp_2, is active for three years, the next pressure drop, Δp_3, is active for two years, and the final pressure drop, Δp_4, is active for one year. This concept can be translated into an equation for cumulative water influx after n time periods on the basis of Eq. (10.105) as:

$$W_{en} = 1.119h\phi c_t r_R^2 \sum_{i=1}^{n} \Delta p_i Q_{D@t_{D(n-i+1)}} \quad \textbf{(10.106)}$$

Equation (10.106) can be written in a more concise form as:

$$W_{en} = \beta \sum_{i=1}^{n} \Delta p_i Q_{D@t_{D(n-i+1)}} \quad \text{where} \quad \beta = 1.119h\phi c_t r_R^2 \quad \textbf{(10.107)}$$

In Eqs. (10.106) and (10.107), W_{en} = cumulative water influx at end of n time periods, RB; Δp_i =pressure change at i time period, psi; Q_D = dimensionless cumulative water influx evaluated as a function of r_{aD} and t_D from Appendices 10E and 10F; and $t_{D(n-i+1)}$ = dimensionless time at $(n - i + 1)$ time period. If the angle of contact between the reservoir and the aquifer is less than 360°, then β is calculated as:

$$\beta = 1.119h\phi c_t r_R^2 \left(\frac{\theta}{360} \right) \quad \textbf{(10.108)}$$

In Eq. (10.108), θ is the angle of contact between the reservoir and aquifer in degrees (Figure 10.11).

The step-wise pressure change as illustrated in Figure 10.12 has been simplified for demonstration purposes. The pressure change for each time period can be calculated with the expression:

$$\Delta p_i = \frac{1}{2}(p_{i-2} - p_i), \quad \text{for} \quad i = 1,2,\ldots,n \quad \textbf{(10.109)}$$

For Eq. (10.109), $p_{\leq 0} = p_0 = p_i$, where p_i is the initial pressure. The application of Eq. (10.109) is illustrated in Example 10.11.

The application of the constant terminal pressure solution with superposition principle as developed here is actually the van Everdingen-Hurst Method for calculating cumulative water influx into a reservoir. It was considered appropriate to present this method in this chapter after the introduction of the principle of superposition. It is an alternative method to the Carter-Tracy and Fetkovich methods presented in Chapter 9 for calculating cumulative water influx.

Example 10.11 Calculation of Cumulative Water Influx Using the van Everdingen-Hurst Method

Problem

Calculate the cumulative water influx after five years for the reservoir with pressure history at the reservoir/aquifer boundary shown in Table 10.3 and Figure 10.13. Assume the aquifer is infinite-acting. Other properties of the aquifer are as follows:

Formation thickness, h	65 ft
Formation permeability, k	150 md
Formation porosity, ϕ	0.20
Total compressibility, c_t	7×10^{-6} psi^{-1}
Water viscosity, μ	1.1 cp
Reservoir radius, r_R	3000 ft
Angle of contact, θ	360°

Table 10.3 Pressure History for Example 10.11

Time, *t* (days)	Pressure, *p* (psia)
0	5000
365	4804
730	4662
1095	4547
1460	4443
1825	4355

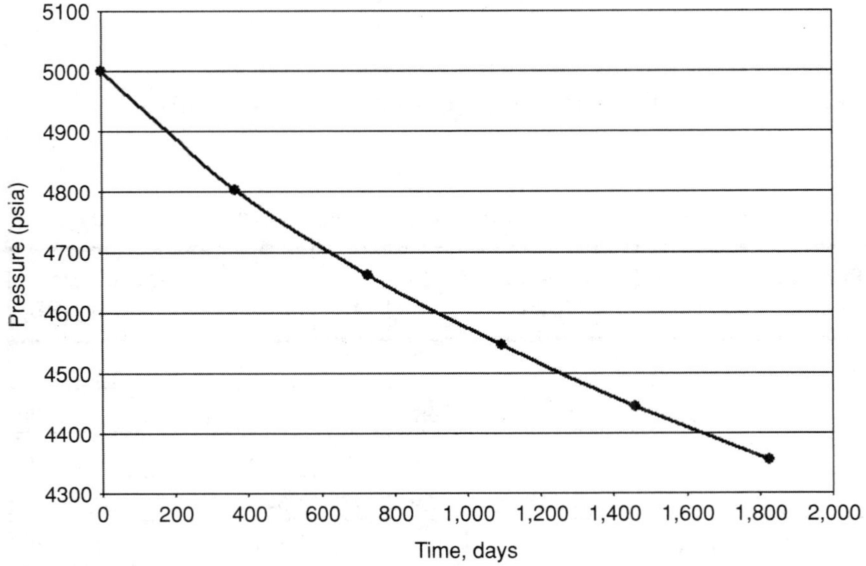

Figure 10.13 Pressure history at the reservoir/aquifer boundary for Example 10.11.

Solution

Step 1: Calculate the pressure change for each time period.

From Eq. (10.109), the pressure change at each time period is given by:

$$\Delta p_i = \frac{1}{2}(p_{i-2} - p_i), \quad \text{for} \quad i = 1, 2, \ldots, n. \text{ Note that for } p_{\le 0} = p_0 = p_i$$

$$\text{For } i = 1,\ \Delta p_1 = \frac{1}{2}(p_0 - p_1) = \frac{1}{2}(5000 - 4804) = 98 \text{ psia.}$$

$$\text{For } i = 2,\ \Delta p_2 = \frac{1}{2}(p_0 - p_2) = \frac{1}{2}(5000 - 4662) = 169 \text{ psia.}$$

$$\text{For } i = 3,\ \Delta p_3 = \frac{1}{2}(p_1 - p_3) = \frac{1}{2}(4804 - 4547) = 128.5 \text{ psia.}$$

$$\text{For } i = 4,\ \Delta p_4 = \frac{1}{2}(p_2 - p_4) = \frac{1}{2}(4662 - 4443) = 109.5 \text{ psia.}$$

$$\text{For } i = 5,\ \Delta p_5 = \frac{1}{2}(p_3 - p_5) = \frac{1}{2}(4547 - 4355) = 96 \text{ psia.}$$

Step 2: Calculate dimensionless time for each time period.

From Eq. (10.104), the dimensionless time is calculated as:

$$t_D = \frac{0.0002637kt}{\phi \mu c_t r_R^2} = \frac{0.0002637 \times 150 \times 24 \times t}{0.20 \times 1.1 \times 7 \times 10^{-6} \times (3000)^2}$$
$$= 0.0685t$$

For $i = 1, t = 365$ days, $t_D = 0.0685 \times 365 = 25.00$

Values of t_D for subsequent time periods are shown in Table 10.4.

Table 10.4 Calculation of Cumulative Water Influx in Example 10.11

Time Periods, *n*	Time, *t* (days)	Pressure, *p* (psia)	Δp	t_D	Q_D	Cum.Water Influx, W_e (RB)
0	0	5000	0	0	0.00	0
1	365	4804	98	25	14.55	1,306,780
2	730	4662	169	50	24.82	4,482,690
3	1095	4547	129	75	34.14	8,630,533
4	1460	4443	109	100	43.00	13,537,388
5	1825	4355	96	125	51.41	19,072,834

Step 3: Calculate dimensionless cumulative water influx term, Q_D.
From Appendix 10E, obtain values of Q_D for infinite aquifer at each t_D. Values of Q_D are listed in Table 10.4 at each t_D.

Step 4: Calculate cumulative water influx at each time period.
From Eq. (10.108),

$$\beta = 1.119 h\phi c_t r_R^2\left(\frac{\theta}{360}\right) = 1.119 \times 65 \times 0.2 \times 7 \times 10^{-6} \times (3000)^2 \times \frac{360}{360}$$
$$= 916.46 \text{ rb/psi}$$

From Eq. (10.107),

$$W_{en} = \beta \sum_{i=1}^{n} \Delta p_i Q_{D@t_{D(n-i+1)}}$$

For example at $n = 1$,

$$W_{e1} = \beta \sum_{i=1}^{1} \Delta p_1 Q_{D@t_{D(1-1+1)}} = \beta \times \Delta p_1 \times Q_{D@t_{D1}}$$
$$= 916.46 \times 98 \times 14.55$$
$$= 1{,}306{,}780 \text{ rbbls.}$$

Another example at $n = 3$,

$$W_{e3} = \beta \sum_{i=1}^{3} \Delta p_i Q_{D@t_{D(3-i+1)}}$$
$$= \beta[\Delta p_1 Q_{D@t_{D3}} + \Delta p_2 Q_{D@t_{D2}} + \Delta p_3 Q_{D@t_{D1}}]$$
$$= 916.46[98 \times 34.14 + 169 \times 24.82 + 129 \times 14.55]$$
$$= 8{,}630{,}533 \text{ rbbls.}$$

Similarly, at $n = 5$

$$W_{e6} = \beta \sum_{i=1}^{5} \Delta p_i Q_{D@t_{D(5-i+1)}}$$
$$= \beta[\Delta p_1 Q_{D@t_{D5}} + \Delta p_2 Q_{D@t_{D4}} + \Delta p_3 Q_{D@t_{D3}} + \Delta p_4 Q_{D@t_{D2}} + \Delta p_5 Q_{D@t_{D1}}]$$
$$= 916.46[98 \times 51.41 + 169 \times 43 + 129 \times 34.14 + 109 \times 24.82 + 96 \times 14.55]$$
$$= 19{,}072{,}834 \text{ rbbls.}$$

The cumulative water influxes at time periods, $n = 2$ and $n = 4$, are shown in Table 10.4.

10.12 Well Productivity Index

The productivity index (PI) of a well is the ratio of its production rate and the pressure drawdown taken at the midpoint of the producing interval. PI is expressed in equation form as:

$$J = \frac{q}{\overline{p} - p_{wf}} \tag{10.110}$$

In Eq. (10.110), J = productivity index, STB/psia-day; q = production rate, STB/D; $\overline{p}$ = average reservoir pressure, psia; and p_{wf} = flowing bottom-hole pressure measured at midpoint of the producing interval, psia. The average reservoir pressure can be estimated quickly as equal to the arithmetic average of all current, static bottom-hole pressures corrected to a datum or determined more rigorously with Eq. (10.93). Productivity index should be calculated after production at the well has stabilized. Productivity index is typically used as a measure to monitor the condition of the well. If the production rate of a well is falling, the productivity index will also be in decline. Factors that affect the productivity index of a well are:

1. Near-wellbore reduction of permeability due to fines migration, evolution of free gas due to pressure drop, and trapping of condensates for gas reservoirs.
2. Non-Darcy flow near the wellbore due to high pressure drawdown.
3. Increase in oil viscosity at pressures below bubblepoint for undersaturated reservoirs.
4. Reduction in reservoir pressure for depletion drive reservoirs.
5. Increase in multiphase flow.

Generally, in water drive reservoirs, productivity index will decline at water breakthrough if calculation of productivity was previously based only on oil production rate. To ensure that the productivity of such well has not declined, water production rate should be included in Eq. (10.110). In reservoir simulation, productivity indices of wells are sometimes modified as necessary to history-match the reported well production data.

10.13 Well Injectivity Index

For injection wells, injectivity index is used in place of productivity index. Injectivity index (I) is expressed as:

$$I = \frac{q_{inj}}{p_{winj} - \overline{p}} \tag{10.111}$$

In Eq. (10.111), I = injectivity index, STB/psia-day; q_{inj} = injection rate, STB/D; p_{winj} = injection pressure at midpoint of interval, psia; and $\overline{p}$ = average reservoir pressure, psia. Injectivity index is a measure of the ability to inject fluid into a reservoir. A reduction of injectivity index with time is an indicator of near-wellbore reduction in permeability due to formation damage.

Nomenclature

A	cross-sectional area, ft^2
B	formation volume factor, RB/STB
B_g	gas formation volume factor, RB/scf
c	isothermal compressibility, psi^{-1}
c_f	formation compressibility, psi^{-1}
c_g	isothermal gas compressibility, psi^{-1}
c_t	total compressibility, psi^{-1}
g	acceleration due to gravity, ft/sec^2
h	reservoir or formation thickness, ft
I	well injectivity index, STB/psi
J	well productivity index, STB/psi
k	permeability, md
l	length of flow path or linear reservoir
M_g	molecular weight of gas, lbm/lbmole
p_p	real gas pseudo-pressure, $psia^2/cp$
p	pressure, psia
$\overline{p}$	average pressure, psia
p_b	base low pressure, psia
p_D	dimensionless pressure
p_{wf}	well bottom-hole flowing pressure, psia
Δp	pressure change, psi
q	volume flow rate, STB/D
Q	cumulative flow or production, STB or MMscf
Q_D	dimensionless cumulative flow or production
r	radius or distance, ft
r_{aD}	dimensionless aquifer radius or distance
r_e	reservoir radius or distance, ft
r_{eD}	dimensionless reservoir radius or outer boundary
r_D	dimensionless radius or distance
r_w	wellbore radius, ft
R	gas constant, 10.732 psia-ft^2/lbmol$-$°R
t	time, hours or days
t_D	dimensionless time
T	temperature, °R
V	volume, ft^2 or RB
V_b	volume at base low pressure, p_b
v	apparent fluid velocity, ft/sec

W_e	cumulative water influx, RB
z	gas compressibility factor
Z	elevation above a datum plane, ft
β	constant in equation in Eq. (10.107)
γ	fluid specific gravity, water = 1
η	formation diffusivity, ft^2/hr
ϕ	porosity, fraction
μ	fluid viscosity, cp
ρ	fluid density, lbm/ft^3
θ	dip angle of formation or contact angle between reservoir and aquifer, degrees
Φ	potential

Subscripts

a	aquifer
avg	average
g	gas
i	initial conditions
inj	injection
LTF	late transient flow
pss	pseudo steady state
R	reservoir
sc	standard conditions
w	well

References

1. Matthews, C.S., and Russell, D.G.: *Pressure Buildup and Flow Tests in Wells*, Monograph Volume 1, Society of Petroleum Engineering of AIME, Dallas, Texas (1967).
2. van Everdingen, A.F., and Hurst, W.: "The Application of the Laplace Transformation to Flow Problems in Reservoirs," *Trans.*, AIME (1949) 186, 305–324.
3. Edwardson, M.J., Girner, H.M., Parkison, H.R., Williams, C.D., and Matthews, C.S.: "Calculation of Formation Temperature Disturbances Caused by Mud Circulation," *JPT* (April 1962) 416–426.
4. Al-Hussainy, R., Ramey, H.J., Jr., and Crawford, P.R.: "The Flow of Real Gases Through Porous Media," *JPT* (May 1966) 624–636.
5. Al-Hussainy, R., and Ramey, H.J., Jr.: "Application of Real Gas Flow Theory to Well Testing and Deliverability Forecasting," *JPT* (May 1966) 637–642.
6. Carter, R.D.: "Solutions of Unsteady-State Radial Gas Flow," *JPT* (May 1962) 549–554.

General Reading

- Swift, G.W., and Kiel, O.G.: "The Prediction of Gas Well Performance Including the Effect of Non-Darcy Flow," *JPT* (July 1962) 791–798.
- Russell, D.G., Goodrich, J.H., and Bruskotter, J.F.: "Methods for Predicting Gas Well Performance," *JPT* (January 1966) 99–108.
- Bruce, G.H., Peaceman, D.W., Rachford, Jr., H.H., and Rice, J.D.: "Calculations of Unsteady-State Gas Flow Through Porous Media," *Trans.*, AIME (1953) 198, 79–92.
- van Everdingen, A.F.: "The Skin Effect and its Influence on the Productive Capacity of a Well," *Trans.*, AIME (1953) 198, 171–176.

Chart for Exponential Integral

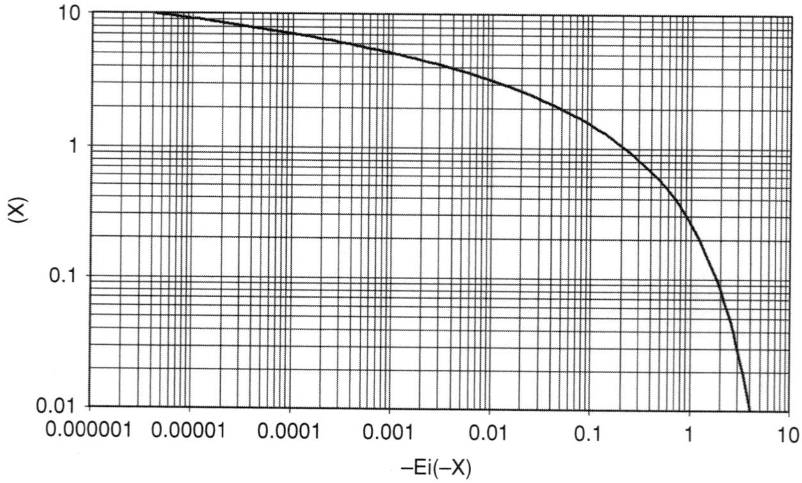

APPENDIX 10B

Tabulation of p_D vs t_D for Radial Flow, Infinite Reservoirs with Constant Terminal Rate at Inner Boundary*

t_D	p_D	t_D	p_D
0.01	0.112	15.0	1.829
0.05	0.229	20.0	1.960
0.10	0.315	25.0	2.067
0.15	0.376	30.0	2.147
0.20	0.424	40.0	2.282
0.25	0.459	50.0	2.388
0.30	0.503	60.0	2.476
0.40	0.564	70.0	2.550
0.50	0.616	80.0	2.615
0.60	0.659	90.0	2.672
0.70	0.702	100.0	2.723
0.80	0.735	150.0	2.921
0.90	0.772	200.0	3.064
1.00	0.802	250.0	3.173
1.50	0.927	300.0	3.263
2.00	1.020	400.0	3.406
2.50	1.101	500.0	3.516

(*continued*)

*From van Everdingen and Hurst

t_D	p_D	t_D	p_D
3.00	1.169	600.0	3.608
4.00	1.275	700.0	3.684
5.00	1.362	800.0	3.750
6.00	1.436	900.0	3.809
7.00	1.500	1000.0	3.860
8.00	1.556		
9.00	1.604		
10.00	1.651		

Tabulation of p_D vs t_D for Radial Flow, Finite Reservoirs with Closed Outer Boundary and Constant Terminal Rate at Inner Boundary*

*From van Everdingen and Hurst

$r_{eD} = 1.5$		$r_{eD} = 2.0$		$r_{eD} = 2.5$		$r_{eD} = 3.0$		$r_{eD} = 3.5$		$r_{eD} = 4.0$		$r_{eD} = 4.5$	
t_D	p_D	t_D	p_D	t_D	p_D	t_D	p_D	t_D	p_D	t_D	p_D	t_D	p_D
0.06	0.251	0.22	0.443	0.40	0.565	0.52	0.627	1.0	0.802	1.5	0.927	2.0	1.023
0.08	0.288	0.24	0.459	0.42	0.576	0.54	0.636	1.1	0.830	1.6	0.948	2.1	1.040
0.10	0.322	0.26	0.476	0.44	0.587	0.56	0.645	1.2	0.857	1.7	0.968	2.2	1.056
0.12	0.355	0.28	0.492	0.46	0.598	0.60	0.662	1.3	0.882	1.8	0.988	2.3	1.072
0.14	0.387	0.30	0.507	0.48	0.608	0.65	0.683	1.4	0.906	1.9	1.007	2.4	1.087
0.16	0.420	0.32	0.522	0.50	0.618	0.70	0.703	1.5	0.929	2.0	1.025	2.5	1.102
0.18	0.452	0.34	0.536	0.52	0.628	0.75	0.721	1.6	0.951	2.2	1.059	2.6	1.116
0.20	0.484	0.36	0.551	0.54	0.638	0.80	0.740	1.7	0.973	2.4	1.092	2.7	1.130
0.22	0.516	0.38	0.565	0.56	0.647	0.85	0.758	1.8	0.994	2.6	1.123	2.8	1.144
0.24	0.548	0.40	0.579	0.58	0.657	0.90	0.776	1.9	1.014	2.8	1.154	2.9	1.158
0.26	0.580	0.42	0.593	0.60	0.666	0.95	0.791	2.0	1.034	3.0	1.184	3.0	1.171
0.28	0.612	0.44	0.607	0.65	0.688	1.0	0.806	2.25	1.083	3.5	1.255	3.2	1.197
0.30	0.644	0.46	0.621	0.70	0.710	1.2	0.865	2.50	1.130	4.0	1.324	3.4	1.222
0.35	0.724	0.48	0.634	0.75	0.731	1.4	0.920	2.75	1.176	4.5	1.392	3.6	1.246
0.40	0.804	0.50	0.648	0.80	0.752	1.6	0.973	3.0	1.221	5.0	1.460	3.8	1.269
0.45	0.884	0.60	0.715	0.85	0.772	2.0	1.076	4.0	1.401	5.5	1.527	4.0	1.292
0.50	0.964	0.70	0.782	0.90	0.792	3.0	1.328	5.0	1.579	6.0	1.594	4.5	1.349
0.55	1.044	0.80	0.849	0.95	0.812	4.0	1.578	6.0	1.757	6.5	1.660	5.0	1.403
0.60	1.124	0.90	0.915	1.0	0.832	5.0	1.828			7.0	1.727	5.5	1.457
		1.0	0.982	2.0	1.215					8.0	1.861	6.0	1.510

		2.0	1.649	3.0	1.596					9.0	1.994	7.0	1.615
		3.0	2.316	4.0	1.977					10.0	2.127	8.0	1.719
		5.0	3.649	5.0	2.358							9.0	1.823
												10.0	1.927
												11.0	2.031
												12.0	2.135
												13.0	2.239
												14.0	2.343
												15.0	2.447

$r_{eD} = 5.0$		$r_{eD} = 6.0$		$r_{eD} = 7.0$		$r_{eD} = 8.0$		$r_{eD} = 9.0$		$r_{eD} = 10$	
t_D	p_D	t_D	p_D	t_D	p_D	t_D	p_D	t_D	p_D	t_D	p_D
3.0	1.167	4.0	1.275	6.0	1.436	8.0	1.556	10.0	1.651	12.0	1.732
3.1	1.180	4.5	1.322	6.5	1.470	8.5	1.582	10.5	1.673	12.5	1.750
3.2	1.192	5.0	1.364	7.0	1.501	9.0	1.607	11.0	1.693	13.0	1.768
3.3	1.204	5.5	1.404	7.5	1.531	9.5	1.631	11.5	1.713	13.5	1.784
3.4	1.215	6.0	1.441	8.0	1.559	10.0	1.653	12.0	1.732	14.0	1.801
3.5	1.227	6.5	1.477	8.5	1.586	10.5	1.675	12.5	1.750	14.5	1.817
3.6	1.238	7.0	1.511	9.0	1.613	11.0	1.697	13.0	1.768	15.0	1.832
3.7	1.249	7.5	1.544	9.5	1.638	11.5	1.717	13.5	1.786	15.5	1.847
3.8	1.259	8.0	1.576	10.0	1.663	12.0	1.737	14.0	1.803	16.0	1.862
3.9	1.270	8.5	1.607	11.0	1.711	12.5	1.757	14.5	1.819	17.0	1.890
4.0	1.281	9.0	1.638	12.0	1.757	13.0	1.776	15.0	1.835	18.0	1.917
4.2	1.301	9.5	1.668	13.0	1.801	13.5	1.795	15.5	1.851	19.0	1.943
4.4	1.321	10.0	1.698	14.0	1.845	14.0	1.813	16.0	1.867	20.0	1.968
4.6	1.340	11.0	1.757	15.0	1.888	14.5	1.831	17.0	1.897	22.0	2.017
4.8	1.360	12.0	1.815	16.0	1.931	15.0	1.849	18.0	1.926	24.0	2.063
5.0	1.378	13.0	1.873	17.0	1.974	17.0	1.919	19.0	1.955	26.0	2.108
5.5	1.424	14.0	1.931	18.0	2.016	19.0	1.986	20.0	1.983	28.0	2.151
6.0	1.469	15.0	1.988	19.0	2.058	21.0	2.051	22.0	2.037	30.0	2.194
6.5	1.513	16.0	2.045	20.0	2.100	23.0	2.116	24.0	2.090	32.0	2.236
7.0	1.556	17.0	2.103	22.0	2.184	25.0	2.180	26.0	2.142	34.0	2.278

7.5	1.598	18.0	2.160	24.0	2.267	30.0	2.340	28.0	2.193	36.0	2.319
8.0	1.641	19.0	2.217	26.0	2.351	35.0	2.499	30.0	2.244	38.0	2.360
9.0	1.725	20.0	2.274	28.0	2.434	40.0	2.658	34.0	2.345	40.0	2.401
10.0	1.808	25.0	2.560	30.0	2.517	45.0	2.817	38.0	2.446	50.0	2.604
11.0	1.892	30.0	2.846					40.0	2.496	60.0	2.806
12.0	1.975							45.0	2.621	70.0	3.008
13.0	2.059							50.0	2.746		
14.0	2.142										
15.0	2.225										

APPENDIX 10D

Tabulation of p_D vs t_D for Radial Flow, Finite Reservoirs with Constant Pressure Outer Boundary and Constant Terminal Rate at Inner Boundary*

*From van Everdingen and Hurst

$r_{eD} = 1.5$		$r_{eD} = 2.0$		$r_{eD} = 2.5$		$r_{eD} = 3.0$		$r_{eD} = 3.5$		$r_{eD} = 4.0$		$r_{eD} = 6.0$	
t_D	p_D	t_D	p_D	t_D	p_D	t_D	p_D	t_D	p_D	t_D	p_D	t_D	p_D
0.050	0.230	0.20	0.424	0.30	0.502	0.50	0.617	0.50	0.620	1.0	0.802	4.0	1.275
0.055	0.240	0.22	0.441	0.35	0.535	0.55	0.640	0.60	0.665	1.2	0.857	4.5	1.320
0.060	0.249	0.24	0.457	0.40	0.564	0.60	0.662	0.70	0.705	1.4	0.905	5.0	1.361
0.070	0.266	0.26	0.472	0.45	0.591	0.70	0.702	0.80	0.741	1.6	0.947	5.5	1.398
0.080	0.282	0.28	0.485	0.50	0.616	0.80	0.738	0.90	0.774	1.8	0.986	6.0	1.432
0.090	0.292	0.30	0.498	0.55	0.638	0.90	0.770	1.0	0.804	2.0	1.020	6.5	1.462
0.10	0.307	0.35	0.527	0.60	0.659	1.0	0.799	1.2	0.858	2.2	1.052	7.0	1.490
0.12	0.328	0.40	0.552	0.70	0.696	1.2	0.850	1.4	0.904	2.4	1.080	7.5	1.516
0.14	0.344	0.45	0.573	0.80	0.728	1.4	0.892	1.6	0.945	2.6	1.106	8.0	1.539
0.16	0.356	0.50	0.591	0.90	0.755	1.6	0.927	1.8	0.981	2.8	1.130	8.5	1.561
0.18	0.367	0.55	0.606	1.0	0.778	1.8	0.955	2.0	1.013	3.0	1.152	9.0	1.580
0.20	0.375	0.60	0.619	1.2	0.815	2.0	0.980	2.2	1.041	3.4	1.190	10.0	1.615
0.22	0.381	0.65	0.630	1.4	0.842	2.2	1.000	2.4	1.065	3.8	1.222	12.0	1.667
0.24	0.386	0.70	0.639	1.6	0.861	2.4	1.016	2.6	1.087	4.5	1.266	14.0	1.704
0.26	0.390	0.75	0.647	1.8	0.876	2.6	1.030	2.8	1.106	5.0	1.290	16.0	1.730
0.28	0.393	0.80	0.654	2.0	0.887	2.8	1.042	3.0	1.123	5.5	1.309	18.0	1.749
0.30	0.396	0.85	0.660	2.2	0.895	3.0	1.051	3.5	1.158	6.0	1.325	20.0	1.762
0.35	0.400	0.90	0.665	2.4	0.900	3.5	1.069	4.0	1.183	7.0	1.347	22.0	1.771
0.40	0.402	0.95	0.669	2.6	0.905	4.0	1.080	5.0	1.215	8.0	1.361	24.0	1.777
0.45	0.404	1.0	0.673	2.8	0.908	4.5	1.087	6.0	1.232	9.0	1.370	26.0	1.781

(continued)

$r_{eD} = 1.5$		$r_{eD} = 2.0$		$r_{eD} = 2.5$		$r_{eD} = 3.0$		$r_{eD} = 3.5$		$r_{eD} = 4.0$		$r_{eD} = 6.0$	
t_D	p_D	t_D	p_D	t_D	p_D	t_D	p_D	t_D	p_D	t_D	p_D	t_D	p_D
0.50	0.405	1.2	0.682	3.0	0.910	5.0	1.091	7.0	1.242	10.0	1.376	28.0	1.784
0.60	0.405	1.4	0.688	3.5	0.913	5.5	1.094	8.0	1.247	12.0	1.382	30.0	1.787
0.70	0.405	1.6	0.690	4.0	0.915	6.0	1.096	9.0	1.250	14.0	1.385	35.0	1.789
0.80	0.405	1.8	0.692	4.5	0.916	6.5	1.097	10.0	1.251	16.0	1.386	40.0	1.791
		2.0	0.692	5.0	0.916	7.0	1.097	12.0	1.252	18.0	1.386	50.0	1.792
		2.5	0.693	5.5	0.916	8.0	1.098	14.0	1.253				
		3.0	0.693	6.0	0.916	10.0	1.099	16.0	1.253				

$r_{eD} = 8$		$r_{eD} = 10$		$r_{eD} = 15$		$r_{eD} = 20$		$r_{eD} = 25$		$r_{eD} = 30$		$r_{eD} = 40$	
t_D	p_D	t_D	p_D	t_D	p_D	t_D	p_D	t_D	p_D	t_D	p_D	t_D	p_D
7.0	1.499	10.0	1.651	20.0	1.960	30.0	2.148	50.0	2.389	70.0	2.551	120.0	2.813
7.5	1.527	12.0	1.730	22.0	2.003	35.0	2.219	55.0	2,434	80.0	2.615	140.0	2.888
8.0	1.554	14.0	1.798	24.0	2.043	40.0	2.282	60.0	2.476	90.0	2.672	160.0	2.953
8.5	1.580	16.0	1.856	26.0	2.080	45.0	2.338	65.0	2.514	100.0	2.723	180.0	3.011
9.0	1.604	18.0	1.907	28.0	2.114	50.0	2.388	70.0	2.550	120.0	2.812	200.0	3.063
9.5	1.627	20.0	1.952	30.0	2.146	60.0	2.475	75.0	2.583	140.0	2.886	220.0	3.109
10.0	1.648	25.0	2.043	35.0	2.218	70.0	2.547	80.0	2.614	160.0	2.950	240.0	3.152
12.0	1.724	30.0	2.111	40.0	2.279	80.0	2.609	85.0	2.643	165.0	2.965	260.0	3.191
14.0	1.786	35.0	2.160	45.0	2.332	90.0	2.658	90.0	2.671	170.0	2.979	280.0	3.226
16.0	1.837	40.0	2.197	50.0	2.379	100.0	2.707	95.0	2.697	175.0	2.992	300.0	3.259
18.0	1.879	45.0	2.224	60.0	2.455	105.0	2.728	100.0	2.721	180.0	3.006	350.0	3.331
20.0	1.914	50.0	2.245	70.0	2.513	110.0	2.747	120.0	2.807	200.0	3.054	400.0	3.391
22.0	1.943	55.0	2.260	80.0	2.558	115.0	2.764	140.0	2.878	250.0	3.150	450.0	3.440
24.0	1.967	60.0	2.271	90.0	2.592	120.0	2.781	160.0	2.936	300.0	3.219	500.0	3.482
26.0	1.986	65.0	2.279	100.0	2.619	125.0	2.796	180.0	2.984	350.0	3.269	550.0	3.516
28.0	2.002	70.0	2.285	120.0	2.655	130.0	2.810	200.0	3.024	400.0	3.306	600.0	3.545
30.0	2.016	75.0	2.290	140.0	2.677	135.0	2.823	220.0	3.057	450.0	3.332	650.0	3.568
35.0	2.040	80.0	2.293	160.0	2.689	140.0	2.835	240.0	3.085	500.0	3.351	700.0	3.588
40.0	2.055	90.0	2.297	180.0	2.697	145.0	2.846	260.0	3.107	600.0	3.375	800.0	3.619
45.0	2.064	100.0	2.300	200.0	2.701	150.0	2.857	280.0	3.126	700.0	3.387	900.0	3.640
50.0	2.070	110.0	2.301	220.0	2.704	160.0	2.876	300.0	3.142	800.0	3.394	1000.0	3.655
60.0	2.076	120.0	2.302	240.0	2.706	180.0	2.906	350.0	3.171	900.0	3.397	1200.0	3.672
70.0	2.078	130.0	2.302	260.0	2.707	200.0	2.929	400.0	3.189	1000.0	3.399	1400.0	3.681
80.0	2.079	140.0	2.302	280.0	2.707	240.0	2.958	450.0	3.200	1200.0	3.401	1600.0	3.685
		160.0	2.303	300.0	2.708	280.0	2.975	500.0	3.207	1400.0	3.401	1800.0	3.687
						300.0	2.980	600.0	3.214			2000.0	3.688
						400.0	2.992	700.0	3.217			2500.0	3.689
						500.0	2.995	800.0	3.218				
								900.0	3.219				

$r_{eD} = 50$		$r_{eD} = 60$		$r_{eD} = 70$		$r_{eD} = 80$		$r_{eD} = 90$		$r_{eD} = 100$		$r_{eD} = 200$	
t_D	p_D	t_D	p_D	t_D	p_D	t_D	p_D	t_D	p_D	t_D	p_D	t_D	p_D
200.0	3.064	300.0	3.257	500.0	3.512	600.0	3.603	800.0	3.747	1000.0	3.859	1500.0	4.061
220.0	3.111	400.0	3.401	600.0	3.603	700.0	3.680	900.0	3.803	1200.0	3.949	2000.0	4.205
240.0	3.154	500.0	3.512	700.0	3.680	800.0	3.747	1000.0	3.858	1400.0	4.026	2500.0	4.317
260.0	3.193	600.0	3.602	800.0	3.746	900.0	3.805	1200.0	3.949	1600.0	4.092	3000.0	4.408
280.0	3.229	700.0	3.676	900.0	3.803	1000.0	3.857	1300.0	3.988	1800.0	4.150	3500.0	4.485
300.0	3.263	800.0	3.739	1000.0	3.854	1200.0	3.946	1400.0	4.025	2000.0	4.200	4000.0	4.552
350.0	3.339	900.0	3.792	1200.0	3.937	1400.0	4.019	1500.0	4.058	2500.0	4.303	5000.0	4.663
400.0	3.405	1000.0	3.832	1400.0	4.003	1600.0	4.080	1800.0	4.144	3000.0	4.379	6000.0	4.754
450.0	3.461	1200.0	3.908	1600.0	4.054	1800.0	4.130	2000.0	4.192	3500.0	4.434	7000.0	4.829
500.0	3.512	1400.0	3.959	1800.0	4.095	2000.0	4.171	2500.0	4.285	4000.0	4.478	8000.0	4.894
550.0	3.556	1600.0	3.996	2000.0	4.127	2500.0	4.248	3000.0	4.349	4500.0	4.510	9000.0	4.949
600.0	3.595	1800.0	4.023	2500.0	4.181	3000.0	4.297	3500.0	4.394	5000.0	4.534	10000.0	4.996
650.0	3.630	2000.0	4.043	3000.0	4.211	3500.0	4.328	4000.0	4.426	5500.0	4.552	12000.0	5.072
700.0	3.661	2500.0	4.071	3500.0	4.228	4000.0	4.347	4500.0	4.448	6000.0	4.565	14000.0	5.129
750.0	3.688	3000.0	4.084	4000.0	4.237	4500.0	4.360	5000.0	4.464	6500.0	4.579	16000.0	5.171
800.0	3.713	3500.0	4.090	4500.0	4.242	5000.0	4.368	6000.0	4.482	7000.0	4.583	18000.0	5.203
850.0	3.735	4000.0	4.092	5000.0	4.245	6000.0	4.376	7000.0	4.491	7500.0	4.588	20000.0	5.227
900.0	3.754	4500.0	4.093	5500.0	4.247	7000.0	4.380	8000.0	4.496	8000.0	4.593	25000.0	5.264
950.0	3.771	5000.0	4.094	6000.0	4.247	8000.0	4.381	9000.0	4.498	9000.0	4.598	30000.0	5.282
1000.0	3.787	5500.0	4.094	6500.0	4.248	9000.0	4.382	10000.0	4.499	10000.0	4.601	35000.0	5.290
1200.0	3.833			7000.0	4.248	10000.0	4.382	11000.0	4.499	12500.0	4.604	40000.0	5.294
1400.0	3.862			7500.0	4.248	11000.0	4.382	12000.0	4.500	15000.0	4.605		

$r_{eD} = 300$		$r_{eD} = 400$		$r_{eD} = 500$		$r_{eD} = 600$		$r_{eD} = 700$		$r_{eD} = 800$		$r_{eD} = 900$	
t_D	p_D	t_D	p_D	t_D	p_D	t_D	p_D	t_D	p_D	t_D	p_D	t_D	p_D
6.0×10^3	4.754	1.5×10^4	5.212	2.0×10^4	5.356	4.0×10^4	5.703	5.0×10^4	5.814	7.0×10^4	5.983	8.0×10^4	6.049
8.0×10^3	4.808	2.0×10^4	5.356	2.5×10^4	5.468	4.5×10^4	5.762	6.0×10^4	5.905	8.0×10^4	6.049	9.0×10^4	6.108
10.0×10^3	5.010	3.0×10^4	5.556	3.0×10^4	5.559	5.0×10^4	5.814	7.0×10^4	5.982	9.0×10^4	6.108	10.0×10^4	6.161
12.0×10^3	5.101	4.0×10^4	5.689	3.5×10^4	5.636	6.0×10^4	5.904	8.0×10^4	6.048	10.0×10^4	6.160	12.0×10^4	6.251
14.0×10^3	5.177	5.0×10^4	5.781	4.0×10^4	5.702	7.0×10^4	5.979	9.0×10^4	6.105	12.0×10^4	6.249	14.0×10^4	6.327
16.0×10^3	5.242	6.0×10^4	5.845	4.5×10^4	5.759	8.0×10^4	6.041	10.0×10^4	6.156	14.0×10^4	6.322	16.0×10^4	6.392
18.0×10^3	5.299	7.0×10^4	5.889	5.0×10^4	5.810	9.0×10^4	6.094	12.0×10^4	6.239	16.0×10^4	6.382	18.0×10^4	6.447
20.0×10^3	5.348	8.0×10^4	5.920	6.0×10^4	5.894	10.0×10^4	6.139	14.0×10^4	6.305	18.0×10^4	6.432	20.0×10^4	6.494
24.0×10^3	5.429	9.0×10^4	5.942	7.0×10^4	5.960	12.0×10^4	6.210	16.0×10^4	6.357	20.0×10^4	6.474	25.0×10^4	6.587
28.0×10^3	5.491	10.0×10^4	5.957	8.0×10^4	6.013	14.0×10^4	6.262	18.0×10^4	6.398	25.0×10^4	6.551	30.0×10^4	6.652
30.0×10^3	5.517	11.0×10^4	5.967	9.0×10^4	6.055	16.0×10^4	6.299	20.0×10^4	6.430	30.0×10^4	6.599	40.0×10^4	6.729
40.0×10^3	5.606	12.0×10^4	5.975	10.0×10^4	6.088	18.0×10^4	6.326	25.0×10^4	6.484	35.0×10^4	6.630	45.0×10^4	6.751
50.0×10^3	5.652	12.5×10^4	5.977	12.0×10^4	6.135	20.0×10^4	6.345	30.0×10^4	6.514	40.0×10^4	6.650	50.0×10^4	6.766
60.0×10^3	5.676	13.0×10^4	5.980	14.0×10^4	6.164	25.0×10^4	6.374	35.0×10^4	6.530	45.0×10^4	6.663	55.0×10^4	6.777
70.0×10^3	5.690	14.0×10^4	5.983	16.0×10^4	6.183	30.0×10^4	6.387	40.0×10^4	6.540	50.0×10^4	6.671	60.0×10^4	6.785
80.0×10^3	5.696	16.0×10^4	5.988	18.0×10^4	6.195	35.0×10^4	6.392	45.0×10^4	6.545	55.0×10^4	6.676	70.0×10^4	6.794
90.0×10^3	5.700	18.0×10^4	5.990	20.0×10^4	6.202	40.0×10^4	6.395	50.0×10^4	6.548	60.0×10^4	6.679	80.0×10^4	6.798
100.0×10^3	5.702	20.0×10^4	5.991	25.0×10^4	6.211	50.0×10^4	6.397	60.0×10^4	6.550	70.0×10^4	6.682	90.0×10^4	6.800
120.0×10^3	5.703	24.0×10^4	5.991	30.0×10^4	6.213	60.0×10^4	6.397	70.0×10^4	6.551	80.0×10^4	6.684	100.0×10^4	6.801
140.0×10^3	5.704	26.0×10^4	5.991	35.0×10^4	6.214			80.0×10^4	6.551	100.0×10^4	6.684		
150.0×10^3	5.704			40.0×10^4	6.214								

$r_{eD} = 1000$		$r_{eD} = 1200$		$r_{eD} = 1400$		$r_{eD} = 1600$		$r_{eD} = 1800$		$r_{eD} = 2000$		$r_{eD} = 2200$	
t_D	p_D	t_D	p_D	t_D	p_D	t_D	p_D	t_D	p_D	t_D	p_D	t_D	p_D
1.0×10^5	6.161	2.0×10^5	6.507	2.0×10^5	6.507	2.5×10^5	6.619	3.0×10^5	6.710	4.0×10^5	6.854	5.0×10^5	6.966
1.2×10^5	6.252	3.0×10^5	6.704	2.5×10^5	6.619	3.0×10^5	6.710	4.0×10^5	6.854	5.0×10^5	6.966	5.5×10^5	7.013
1.4×10^5	6.329	4.0×10^5	6.833	3.0×10^5	6.709	3.5×10^5	6.787	5.0×10^5	6.965	6.0×10^5	7.056	6.0×10^5	7.057
1.6×10^5	6.395	5.0×10^5	6.918	3.5×10^5	6.785	4.0×10^5	6.853	6.0×10^5	7.054	7.0×10^5	7.132	6.5×10^5	7.097
1.8×10^5	6.452	6.0×10^5	6.975	4.0×10^5	6.849	5.0×10^5	6.962	7.0×10^5	7.120	8.0×10^5	7.196	7.0×10^5	7.133
2.0×10^5	6.503	7.0×10^5	7.013	5.0×10^5	6.950	6.0×10^5	7.046	8.0×10^5	7.188	9.0×10^5	7.251	7.5×10^5	7.167
2.5×10^5	6.605	8.0×10^5	7.038	6.0×10^5	7.026	7.0×10^5	7.114	9.0×10^5	7.238	10.0×10^5	7.298	8.0×10^5	7.199
3.0×10^5	6.681	9.0×10^5	7.056	7.0×10^5	7.082	8.0×10^5	7.167	10.0×10^5	7.280	12.0×10^5	7.374	8.5×10^5	7.229
3.5×10^5	6.738	10.0×10^5	7.067	8.0×10^5	7.123	9.0×10^5	7.210	15.0×10^5	7.407	14.0×10^5	7.431	9.0×10^5	7.256
4.0×10^5	6.781	12.0×10^5	7.080	9.0×10^5	7.154	10.0×10^5	7.244	20.0×10^5	7.459	16.0×10^5	7.474	10.0×10^5	7.307
4.5×10^5	6.813	14.0×10^5	7.085	10.0×10^5	7.177	15.0×10^5	7.334	30.0×10^5	7.489	18.0×10^5	7.506	12.0×10^5	7.390
5.0×10^5	6.837	16.0×10^5	7.088	15.0×10^5	7.229	20.0×10^5	7.364	40.0×10^5	7.495	20.0×10^5	7.530	16.0×10^5	7.507
5.5×10^5	6.854	18.0×10^5	7.089	20.0×10^5	7.241	25.0×10^5	7.373	50.0×10^5	7.495	25.0×10^5	7.566	20.0×10^5	7.579
6.0×10^5	6.868	19.0×10^5	7.089	25.0×10^5	7.243	30.0×10^5	7.376	51.0×10^5	7.495	30.0×10^5	7.584	25.0×10^5	7.631
7.0×10^5	6.885	20.0×10^5	7.090	30.0×10^5	7.244	35.0×10^5	7.377	52.0×10^5	7.495	35.0×10^5	7.593	30.0×10^5	7.661
8.0×10^5	6.895	21.0×10^5	7.090	31.0×10^5	7.244	40.0×10^5	7.378	53.0×10^5	7.495	40.0×10^5	7.597	35.0×10^5	7.677
9.0×10^5	6.901	22.0×10^5	7.090	32.0×10^5	7.244	42.0×10^5	7.378	54.0×10^5	7.495	50.0×10^5	7.600	40.0×10^5	7.686
10.0×10^5	6.904	23.0×10^5	7.090	33.0×10^5	7.244	44.0×10^5	7.378	56.0×10^5	7.495	60.0×10^5	7.601	50.0×10^5	7.693
12.0×10^5	6.907	24.0×10^5	7.090							64.0×10^5	7.601	60.0×10^5	7.695
14.0×10^5	6.907											70.0×10^5	7.696
16.0×10^5	6.908											80.0×10^5	7.696

r_{eD} = 2400		r_{eD} = 2600		r_{eD} = 2800		r_{eD} = 3000	
t_D	p_D	t_D	p_D	t_D	p_D	t_D	p_D
6.0×10^5	7.057	7.0×10^5	7.134	8.0×10^5	7.201	1.0×10^6	7.312
7.0×10^5	7.134	8.0×10^5	7.201	9.0×10^5	7.260	1.2×10^6	7.403
8.0×10^5	7.200	9.0×10^5	7.259	10.0×10^5	7.312	1.4×10^6	7.480
9.0×10^5	7.259	10.0×10^5	7.312	12.0×10^5	7.403	1.6×10^6	7.545
10.0×10^5	7.310	12.0×10^5	7.401	16.0×10^5	7.542	1.8×10^6	7.602
12.0×10^5	7.398	14.0×10^5	7.475	20.0×10^5	7.644	2.0×10^6	7.651
16.0×10^5	7.526	16.0×10^5	7.536	24.0×10^5	7.719	2.4×10^6	7.732
20.0×10^5	7.611	18.0×10^5	7.588	28.0×10^5	7.775	2.8×10^6	7.794
24.0×10^5	7.668	20.0×10^5	7.631	30.0×10^5	7.797	3.0×10^6	7.820
28.0×10^5	7.706	24.0×10^5	7.699	35.0×10^5	7.840	3.5×10^6	7.871
30.0×10^5	7.720	28.0×10^5	7.746	40.0×10^5	7.870	4.0×10^6	7.908
35.0×10^5	7.745	30.0×10^5	7.765	50.0×10^5	7.905	4.5×10^6	7.935
40.0×10^5	7.760	35.0×10^5	7.799	60.0×10^5	7.922	5.0×10^6	7.955
50.0×10^5	7.775	40.0×10^5	7.821	70.0×10^5	7.930	6.0×10^6	7.979
60.0×10^5	7.780	50.0×10^5	7.845	80.0×10^5	7.934	7.0×10^6	7.992
70.0×10^5	7.782	60.0×10^5	7.856	90.0×10^5	7.936	8.0×10^6	7.999
80.0×10^5	7.783	70.0×10^5	7.860	100.0×10^5	7.937	9.0×10^6	8.002
90.0×10^5	7.783	80.0×10^5	7.862	120.0×10^5	7.937	10.0×10^6	8.004
95.0×10^5	7.783	90.0×10^5	7.863	130.0×10^5	7.937	12.0×10^6	8.006
		100.0×10^5	7.863			15.0×10^6	8.006

APPENDIX 10E

Tabulation of Q_D vs t_D for Radial Flow, Infinite Reservoirs with Constant Terminal Pressure at Inner Boundary*

*From van Everdingen and Hurst

t_D	Q_D	t_D	Q_D	t_D	Q_D
0.01	0.112	15.0	9.965	3000.0	759.00
0.05	0.278	20.0	12.29	4000.0	975.70
0.10	0.404	25.0	14.55	5000.0	1188.00
0.15	0.520	30.0	16.81	6000.0	1395.00
0.20	0.606	40.0	20.88	7000.0	1599.00
0.25	0.689	50.0	24.82	8000.0	1800.00
0.30	0.758	60.0	28.60	9000.0	1999.00
0.40	0.898	70.0	32.28	10000.0	2196.00
0.50	1.020	80.0	35.99	15000.0	3146.00
0.60	1.140	90.0	39.42	20000.0	4079.00
0.70	1.251	100.0	43.01	25000.0	4994.00
0.80	1.359	150.0	59.80	30000.0	5891.00
0.90	1.469	200.0	75.86	40000.0	7634.00
1.00	1.570	250.0	91.20	50000.0	9342.00
1.50	2.032	300.0	105.80	60000.0	11030.0
2.00	2.442	400.0	134.80	70000.0	12690.0
2.50	2.838	500.0	162.40	80000.0	14330.0
3.00	3.209	600.0	189.70	90000.0	15950.0
4.00	3.879	700.0	216.00	100000.0	17560.0
5.00	4.541	800.0	242.30	150000.0	25380.0
6.00	5.148	900.0	267.70	200000.0	33080.0
7.00	5.749	1000.0	293.10	250000.0	40660.0
8.00	6.314	1500.0	413.60	300000.0	48170.0
9.00	6.861	2000.0	531.50	400000.0	62670.0
10.00	7.417	2500.0	646.60	500000.0	76990.0

t_D	Q_D	t_D	Q_D	t_D	Q_D
6.0×10^5	9.113×10^4	9.0×10^7	9.911×10^6	2.0×10^{10}	1.697×10^9
7.0×10^5	1.051×10^5	1.0×10^8	10.95×10^6	2.5×10^{10}	2.103×10^9
8.0×10^5	1.189×10^5	1.5×10^8	1.604×10^7	3.0×10^{10}	2.505×10^9
9.0×10^5	1.326×10^5	2.0×10^8	2.108×10^7	4.0×10^{10}	3.299×10^9
1.0×10^6	1.462×10^5	2.5×10^8	2.607×10^7	5.0×10^{10}	4.087×10^9
1.5×10^6	2.126×10^5	3.0×10^8	3.100×10^7	6.0×10^{10}	4.868×10^9
2.0×10^6	2.781×10^5	4.0×10^8	4.071×10^7	7.0×10^{10}	5.643×10^9
2.5×10^6	3.427×10^5	5.0×10^8	5.032×10^7	8.0×10^{10}	6.414×10^9
3.0×10^6	4.064×10^5	6.0×10^8	5.984×10^7	9.0×10^{10}	7.183×10^9
4.0×10^6	5.313×10^5	7.0×10^8	6.928×10^7	1.0×10^{11}	7.948×10^9
5.0×10^6	6.544×10^5	8.0×10^8	7.865×10^7	1.5×10^{11}	1.170×10^{10}
6.0×10^6	7.761×10^5	9.0×10^8	8.797×10^7	2.0×10^{11}	1.550×10^{10}
7.0×10^6	8.965×10^5	1.0×10^9	9.725×10^7	2.5×10^{11}	1.920×10^{10}
8.0×10^6	1.016×10^6	1.5×10^9	1.429×10^8	3.0×10^{11}	2.290×10^{10}
9.0×10^6	1.134×10^6	2.0×10^9	1.880×10^8	4.0×10^{11}	3.020×10^{10}
1.0×10^7	1.252×10^6	2.5×10^9	2.328×10^8	5.0×10^{11}	3.750×10^{10}
1.5×10^7	1.828×10^6	3.0×10^9	2.771×10^8	6.0×10^{11}	4.470×10^{10}
2.0×10^7	2.398×10^6	4.0×10^9	3.645×10^8	7.0×10^{11}	5.190×10^{10}
2.5×10^7	2.961×10^6	5.0×10^9	4.510×10^8	8.0×10^{11}	5.890×10^{10}
3.0×10^7	3.517×10^6	6.0×10^9	5.368×10^8	9.0×10^{11}	6.580×10^{10}
4.0×10^7	4.610×10^6	7.0×10^9	6.220×10^8	1.0×10^{12}	7.280×10^{10}
5.0×10^7	5.689×10^6	8.0×10^9	7.066×10^8	1.5×10^{12}	1.080×10^{11}
6.0×10^7	6.758×10^6	9.0×10^9	7.909×10^8	2.0×10^{12}	1.420×10^{11}
7.0×10^7	7.816×10^6	1.0×10^{10}	8.747×10^8		
8.0×10^7	8.866×10^6	1.5×10^{10}	1.288×10^9		

APPENDIX 10F

Tabulation of Q_D vs t_D for Radial Flow, Finite Reservoirs with Closed Outer Boundary and Constant Terminal Pressure at Inner Boundary*

*From van Everdingen and Hurst

$r_{eD} = 1.5$		$r_{eD} = 2.0$		$r_{eD} = 2.5$		$r_{eD} = 3.0$		$r_{eD} = 3.5$		$r_{eD} = 4.0$		$r_{eD} = 4.5$	
t_D	Q_D	t_D	Q_D	t_D	Q_D	t_D	Q_D	t_D	Q_D	t_D	Q_D	t_D	Q_D
0.050	0.276	0.050	0.278	0.10	0.408	0.30	0.755	1.00	1.571	2.00	2.442	2.5	2.835
0.060	0.304	0.075	0.345	0.15	0.509	0.40	0.895	1.20	1.761	2.20	2.598	3.0	3.196
0.070	0.330	0.10	0.404	0.20	0.599	0.50	1.023	1.40	1.940	2.40	2.748	3.5	3.537
0.080	0.354	0.125	0.458	0.25	0.681	0.60	1.143	1.60	2.111	2.60	2.893	4.0	3.859
0.090	0.375	0.150	0.507	0.30	0.758	0.70	1.256	1.80	2.273	2.80	3.034	4.5	4.165
0.10	0.395	0.175	0.553	0.35	0.829	0.80	1.363	2.00	2.427	3.00	3.170	5.0	4.454
0.11	0.414	0.200	0.597	0.40	0.897	0.90	1.465	2.20	2.574	3.25	3.334	5.5	4.727
0.12	0.431	0.225	0.638	0.45	0.962	1.00	1.563	2.40	2.715	3.50	3.493	6.0	4.986
0.13	0.446	0.250	0.678	0.50	1.024	1.25	1.791	2.60	2.849	3.75	3.645	6.5	5.231
0.14	0.461	0.275	0.715	0.55	1.088	1.50	1.997	2.80	2.976	4.00	3.792	7.0	5.464
0.15	0.474	0.300	0.751	0.60	1.140	1.75	2.184	3.00	3.098	4.25	3.932	7.5	5.684
0.16	0.486	0.325	0.785	0.65	1.195	2.00	2.353	3.25	3.242	4.50	4.068	8.0	5.892
0.17	0.497	0.350	0.817	0.70	1.248	2.25	2.507	3.50	3.379	4.75	4.198	8.5	6.089
0.18	0.507	0.375	0.848	0.75	1.229	2.50	2.646	3.75	3.507	5.00	4.323	9.0	6.276
0.19	0.517	0.400	0.877	0.80	1.348	2.75	2.772	4.00	3.628	5.50	4.560	9.5	6.453
0.20	0.525	0.425	0.905	0.85	1.395	3.00	2.886	4.25	3.742	6.00	4.779	10.0	6.621
0.21	0.533	0.450	0.932	0.90	1.440	3.25	2.990	4.50	3.850	6.50	4.982	11.0	6.930
0.22	0.541	0.475	0.958	0.95	1.484	3.50	3.084	4.75	3.951	7.00	5.169	12.0	7.208
0.23	0.548	0.500	0.983	1.0	1.526	3.75	3.170	5.00	4.047	7.50	5.343	13.0	7.457
0.24	0.554	0.550	1.028	1.1	1.605	4.00	3.247	5.50	4.222	8.00	5.504	14.0	7.680
0.25	0.559	0.600	1.070	1.2	1.679	4.25	3.317	6.00	4.378	8.50	5.653	15.0	7.880

0.26	0.565	0.650	1.108	1.3	1.747	4.50	3.381	6.50	4.516	9.00	5.790	16.0	8.060
0.28	0.574	0.700	1.143	1.4	1.811	4.75	3.439	7.00	4.639	9.50	5.917	18.0	8.365
0.30	0.582	0.750	1.174	1.5	1.870	5.00	3.491	7.50	4.749	10.00	6.035	20.0	8.611
0.32	0.588	0.800	1.203	1.6	1.924	5.50	3.581	8.00	4.846	11.00	6.246	22.0	8.809
0.34	0.594	0.900	1.253	1.7	1.975	6.00	3.656	8.50	4.932	12.00	6.425	24.0	8.968
0.34	0.599	1.00	1.295	1.8	2.022	6.50	3.717	9.00	5.009	13.00	6.580	26.0	9.097
0.36	0.603	1.1	1.330	2.0	2.106	7.00	3.767	9.50	5.078	14.00	6.712	28.0	9.200
0.38	0.606	1.2	1.358	2.2	2.178	7.50	3.809	10.00	5.138	15.00	6.825	30.0	9.283
0.40	0.613	1.3	1.382	2.4	2.241	8.00	3.843	11.00	5.241	16.00	6.922	34.0	9.404
0.50	0.617	1.4	1.402	2.6	2.294	9.00	3.894	12.00	5.321	17.00	7.004	38.0	9.481
0.60	0.621	1.6	1.432	2.8	2.340	10.00	3.928	13.00	5.385	18.00	7.076	42.0	9.532
0.70	0.623	1.7	1.444	3.0	2.380	11.00	3.951	14.00	5.435	20.00	7.189	46.0	9.565
0.80	0.624	1.8	1.453	3.4	2.444	12.00	3.967	15.00	5.476	22.00	7.272	50.0	9.586
		2.0	1.468	3.8	2.491	14.00	3.985	16.00	5.506	24.00	7.332	60.0	9.612
		2.5	1.487	4.2	2.525	16.00	3.993	17.00	5.531	26.00	7.377	70.0	9.621
		3.0	1.495	4.6	2.551	18.00	3.997	18.00	5.551	30.00	7.434	80.0	9.623
		4.0	1.499	5.0	2.570	20.00	3.999	20.00	5.579	34.00	7.464	90.0	9.624
		5.0	1.500	6.0	2.599	22.00	3.999	25.00	5.611	38.00	7.481	100.0	9.625
				7.0	2.613	24.00	4.000	30.00	5.621	42.00	7.490		
				8.0	2.619			35.00	5.624	46.00	7.494		
				9.0	2.622			40.00	5.625	50.00	7.497		
				10.0	2.624								

$r_{eD} = 5.0$		$r_{eD} = 6.0$		$r_{eD} = 7.0$		$r_{eD} = 8.0$		$r_{eD} = 9.0$		$r_{eD} = 10.0$	
t_D	Q_D	t_D	Q_D	t_D	Q_D	t_D	Q_D	t_D	Q_D	t_D	Q_D
3.0	3.195	6.0	5.148	9.00	6.861	9	6.861	10	7.417	15	9.965
3.5	3.542	6.5	5.440	9.50	7.127	10	7.398	15	9.945	20	12.32
4.0	3.875	7.0	5.724	10.0	7.389	11	7.920	20	12.26	22	13.22
4.5	4.193	7.5	6.002	11.0	7.902	12	8.431	22	13.13	24	14.09
5.0	4.499	8.0	6.273	12.0	8.397	13	8.930	24	13.98	26	14.95
5.5	4.792	8.5	6.537	13.0	8.876	14	9.418	26	14.79	28	15.78
6.0	5.074	9.0	6.795	14.0	9.341	15	9.895	28	15.59	30	16.59
6.5	5.345	9.5	7.047	15.0	9.791	16	10.361	30	16.35	32	17.38
7.0	5.605	10.0	7.293	16.0	10.23	17	10.82	32	17.10	34	18.16
7.5	5.854	10.5	7.533	17.0	10.65	18	11.26	34	17.82	36	18.91
8.0	6.094	11.0	7.767	18.0	11.06	19	11.70	36	18.52	38	19.65
8.5	6.325	12.0	8.220	19.0	11.46	20	12.13	38	19.19	40	20.37
9.0	6.547	13.0	8.651	20.0	11.85	22	12.95	40	19.85	42	21.07
9.5	6.760	14.0	9.063	22.0	12.58	24	13.74	42	20.48	44	21.76
10.0	6.965	15.0	9.456	24.0	13.27	26	14.50	44	21.09	46	22.42
11.0	7.350	16.0	9.829	26.0	13.92	28	15.23	46	21.69	48	23.07
12.0	7.706	17.0	10.19	28.0	14.53	30	15.92	42	22.26	50	23.71
13.0	8.035	18.0	10.53	30.0	15.11	34	17.22	50	22.82	52	24.33
14.0	8.339	19.0	10.85	35.0	16.39	38	18.41	52	23.36	54	24.94
15.0	8.620	20.0	11.16	40.0	17.49	40	18.97	54	23.89	56	25.53
16.0	8.879	22.0	11.74	45.0	18.43	45	20.26	56	24.39	58	26.11

18.0	9.338	24.0	12.26	50.0	19.24	50	21.42	52	24.88	60	26.67
20.0	9.731	25.0	12.50	60.0	20.51	55	22.46	60	25.36	65	28.02
22.0	10.07	31.0	13.74	70.0	21.45	60	23.40	65	26.48	70	29.29
24.0	10.35	35.0	14.40	80.0	22.13	70	24.98	70	27.52	75	30.49
26.0	10.59	39.0	14.93	90.0	22.63	80	26.26	75	28.48	80	31.61
28.0	10.80	51.0	16.05	100.0	23.00	90	27.28	80	29.36	85	32.67
30.0	10.98	60.0	16.56	120.0	23.47	100	28.11	85	30.18	90	33.66
34.0	11.26	70.0	16.91	140.0	23.71	120	29.31	90	30.93	95	34.60
38.0	11.46	80.0	17.14	160.0	23.85	140	30.08	95	31.63	100	35.48
42.0	11.61	90.0	17.27	180.0	23.92	160	30.58	100	32.27	120	38.51
46.0	11.71	100.0	17.36	200.0	23.96	180	30.91	120	34.39	140	40.89
50.0	11.79	110.0	17.41	500.0	24.00	200	31.12	140	35.92	160	42.57
60.0	11.91	120.0	17.45			240	31.34	160	37.04	180	44.21
70.0	11.96	130.0	17.46			280	31.43	180	37.85	200	45.36
80.0	11.98	140.0	17.48			320	31.47	200	38.44	240	46.95
90.0	11.99	150.0	17.49			360	31.49	240	39.17	280	47.94
100.0	12.00	160.0	17.49			400	31.50	280	39.56	320	48.54
120.0	12.00	180.0	17.50			500	31.50	320	39.77	360	48.91
		200.0	17.50					360	39.88	400	49.14
		220.0	17.50					400	39.94	440	49.28
								440	39.97	480	49.36
								480	39.98		

CHAPTER 11

Well Test Analysis: Straightline Methods

11.1 Introduction

Well testing by activation of pressure transients in the wellbore is routinely conducted to collect data which can be analyzed for well and reservoir properties. During drilling operations, well testing is conducted with drillstem tests (DST) or with openhole formation tester tools. For producing wells, testing is typically performed with either flow (drawdown) or shut-in (buildup) tests depending on well conditions, and reservoir properties to be calculated from the test data. For injection wells, test data can be obtained by changing injection rates (injection tests) or by shutting-down injection (falloff tests). Well test data are analyzed with methods that have been developed over time. In this chapter, these methods are reviewed in context of their historical development and significance. The first methods for well test analysis that gained wide acceptance in the petroleum industry are the straightline methods, which were presented by Miller et al.[1] and Horner.[2] Straightline methods (sometimes called conventional methods) are based on straight lines drawn over middle time region data on semi-log plots of well test data, and are used primarily to calculate reservoir permeability, skin factor, average reservoir pressure, and estimates of reservoir size or the drainage area of the well. Straightline methods are still in use in well test analysis although as supplementary to other more advanced methods. In 1970, Ramey[3] introduced the use of type curves for well test analysis. Type curves are theoretical solutions to reservoir flow equations under specified initial and boundary conditions. The application of type curves for well test analysis was advanced by Gringarten et al.[4] with the introduction of a unified set of type curves and integrated methodologies for well test analysis that incorporated straightline methods. Further revolutionary advancement in the application of type curves for well test analysis occurred in 1983 with the introduction of derivative type curves by Bourdet et al.[5,6,7] The combined applications of log-log type curves by Gringarten et al.[4] and derivative type curves from Bourdet et al.[5,6,7] for well test analysis have enabled identifications of reservoir behaviors and geologic features which were previously undetectable from well test data analysis using straightline methods. Another major advancement in well test analysis was initiated with the introduction of stable deconvolution methods by von Schroeter et al.[8] Deconvolution is the process of converting variable rate pressure data into a constant rate pressure data with duration equal to the duration of the test data. Deconvolution methods provide more data for analyses which aid in the identification of geologic features that would have been remained "hidden" with limited test data. A detailed historical account of the development of well test analysis methods has been documented by Gringarten.[9] In this book, well test analysis methods are presented in the following order starting with straightline methods in this chapter, log-log type curves and derivative type curves in Chapter 12, fractured wells and fractured reservoirs in Chapter 13, and deconvolution methods in Chapter 14. The presentation in this chapter covers slightly compressible fluids (liquids) and compressible fluids (gases). Examples are presented to illustrate the application of straightline solution methods. This chapter begins with reviews of basic concepts commonly encountered in well test analysis.

11.2 Basic Concepts in Well Test Analysis

Basic concepts commonly encountered in the acquisition and analyses of pressure transient tests data are reviewed in this section. The concepts that are presented include: radius of investigation, skin and skin factor, flow efficiency and damage ratio, effective wellbore radius, drawdown well tests, buildup well tests, and wellbore storage. These concepts are important in understanding the fundamental principles of pressure transient test analysis discussed in this chapter, and also in Chapters 12, 13, and 14.

11.2.1 Radius of Investigation

When a transient is introduced in a well by changing its production rate or wellbore pressure, the transient will travel away from the wellbore towards the outer boundaries of the reservoir at a rate which depends only on the diffusivity constant of the reservoir. In Chapter 10, the reservoir diffusivity constant, η, was defined in Eq. (10.7) as:

$$\eta = \frac{k}{\phi \mu c_t}$$

In Eq. (10.7) as shown here, k = permeability; ϕ = porosity; μ = fluid viscosity; and c_t = total compressibility. In field units, the reservoir diffusivity constant is given by:

$$\eta = \frac{0.00633k}{\phi \mu c_t} \tag{11.1}$$

In Eq. (11.1), η = diffusivity constant, ft^2/day; k = permeability, md; ϕ = porosity, fraction; μ = fluid viscosity, cp; and c_t = total compressibility, psi^{-1}. In well testing, the distance a transient has travelled after it was initiated is termed its radius of investigation. It is a measure of the extent of the reservoir through which the transient has travelled since it was initiated. The radius of investigation, r_i, for an oil reservoir is calculated with the expression:

$$r_i = \sqrt{\frac{kt}{948 \phi \mu c_t}} \tag{11.2}$$

In Eq. (11.2), r_i = radius of investigation, feet; k = permeability, md; t = time, hrs; ϕ = porosity, fraction; μ = fluid viscosity, cp; and c_t = total compressibility, psi^{-1}. For a gas reservoir, the radius of investigation is estimated with the expression:

$$r_i = \sqrt{\frac{kt\overline{p}}{948\phi\mu}} \tag{11.3}$$

Equation (11.3) has the same units as Eq. (11.2), and $\overline{p}$ is average reservoir pressure in psia. Note that Eq. (11.3) is the same as Eq. (10.10) in Chapter 10. Note also that in Eqs. (11.2) and (11.3), t is total elapsed time since initiation of the transient. The radius of investigation is calculated in most commercial well test analysis reports.

Example 11.1 Calculation of Radius of Investigation for an Oil Reservoir

Problem

Calculate the radius of investigation of an oil well producing at a constant rate for 20 hours from initial conditions in a reservoir with properties as follows:

Formation permeability, k	200 md
Oil viscosity, μ	2.5 cp
Formation porosity, ϕ	0.28
Total formation compressibility, c_t	5×10^{-6} psi^{-1}

Solution

By substituting Eq. (11.2) with the given reservoir properties, the radius of investigation is calculated as:

$$\begin{aligned} r_i &= \sqrt{\frac{kt}{948\phi\mu c_t}} \\ &= \sqrt{\frac{200 \times 20}{948 \times 0.28 \times 2.5 \times 5 \times 10^{-6}}} \\ &= 1097.97 \text{ feet.} \end{aligned}$$

11.2.2 Skin and Skin Factor

The use of the term "skin" to describe a zone close to the wellbore with altered permeability that is different from the permeability of the formation was introduced by van Everdingen.[10] This concept is illustrated in Figure 11.1, showing a wellbore with a zone of altered permeability close to the borehole. The permeability of the zone close to the wellbore can be reduced due to

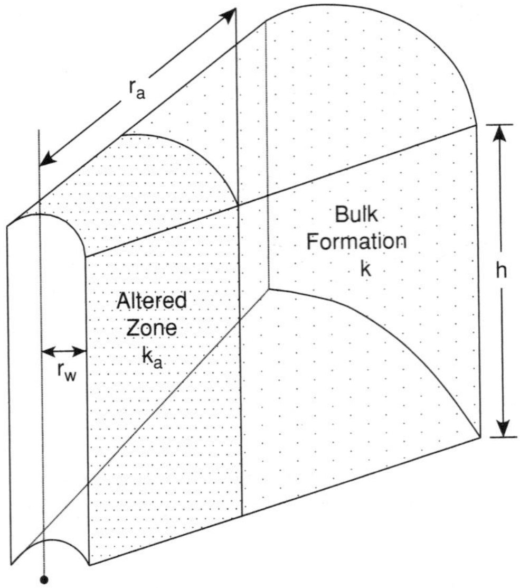

Figure 11.1 Wellbore diagram showing zone with altered permeability.

damage from mud filtrate invasion during drilling or completion operations. Near wellbore damage can also result in producing wells from migration and deposition of fine particles from the reservoir or deposition of waxy residue from the produced oil. On the other hand, the permeability in the altered zone could be improved to be higher than formation permeability through acid stimulation and/or hydraulic fracturing. Generally, the primary objective of most stimulative remedial programs is to remove near wellbore damage, thereby increasing permeability in this zone.

The presence of formation damage near the wellbore increases the pressure drawdown required for a given production rate. This is illustrated in Figure 11.2. As shown in Figure 11.2, the pressure drop, $(p_a - p_{wf})$, across the damaged zone (from r_a to r_w) is greater than the pressure drop, $(p_a - p'_{wf})$, that would have existed for a given production rate if the well was not damaged. Assuming steady-state radial flow, the pressure drop across the damaged zone can be calculated with Eq. (10.23) as:

$$p_a - p_{wf} = \frac{141.2qB\mu}{k_a h} \ln \frac{r_a}{r_w} \tag{11.4}$$

In Eq. (11.4), p_a is the pressure at the boundary of the damaged zone; p_{wf} is the flowing wellbore pressure; k_a is the permeability of the damaged zone; r_a is the radius of the damaged zone; and r_w is the radius of the wellbore. If the zone was undamaged, the pressure drop across the undamaged zone would have been calculated as:

$$p_a - p'_{wf} = \frac{141.2qB\mu}{kh} \ln \frac{r_a}{r_w} \tag{11.5}$$

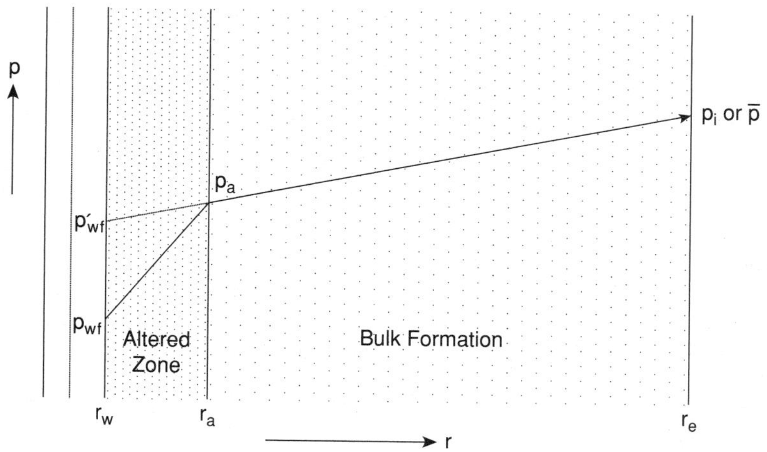

Figure 11.2 Pressure drawdown due to near wellbore formation damage.

In Eq. (11.5), p'_{wf} is the flowing wellbore pressure if the zone was not damaged, and k is the permeability of the undamaged formation. Subtracting Eq. (11.5) from Eq. (11.4) gives:

$$\begin{aligned} p'_{wf} - p_{wf} &= \frac{141.2qB\mu}{h}\left(\frac{1}{k_a} - \frac{1}{k}\right)\ln\frac{r_a}{r_w} \\ &= \frac{141.2qB\mu}{kh}\left(\frac{k}{k_a} - 1\right)\ln\frac{r_a}{r_w} \end{aligned} \tag{11.6}$$

The pressure difference, $(p'_{wf} - p_{wf})$, is equal to the additional pressure drop (Δp_s) caused by the presence of damage or skin in the near wellbore zone. Substituting in Eq. (11.6) gives:

$$\begin{aligned} p'_{wf} - p_{wf} &= \Delta p_s \\ &= \frac{141.2qB\mu}{kh}\left(\frac{k}{k_a} - 1\right)\ln\frac{r_a}{r_w} \end{aligned} \tag{11.7}$$

The zone of altered permeability near the wellbore is often represented in terms of a skin factor. The skin factor, s, is defined as:

$$s = \left(\frac{k}{k_a} - 1\right)\ln\frac{r_a}{r_w} \tag{11.8}$$

With this definition of skin factor in Eq. (11.8), Eq. (11.7) can be written as:

$$\begin{aligned} \Delta p_s &= \frac{141.2qB\mu}{kh} \times s \\ &= \frac{70.6qB\mu}{kh} \times 2s \end{aligned} \tag{11.9}$$

As can be deduced from Eq. (11.8), a positive value for skin factor denotes that the near wellbore zone is damaged, while a negative value for skin factor indicates that the near wellbore zone is stimulated. A zero value for skin factor indicates that the near wellbore zone is neither damaged nor stimulated. It is evident from Eq. (11.8) that skin factors representing damage in wellbores can be very high because the damaged zone permeability is in the denominator. For this reason, the magnitude of skin factors should be used qualitatively rather than quantitatively to assess damage in the near wellbore zone. (For the remainder of this chapter, the near wellbore zone is termed the wellbore for discussions of damage or stimulation.)

11.2.3 Flow Efficiency and Damage Ratio

Flow efficiency and damage ratio are two parameters that are used to measure the condition of a wellbore. Flow efficiency is the productivity of a damaged wellbore relative to a wellbore that is not damaged. Thus, flow efficiency, E, is calculated as:

$$E = \frac{\overline{p} - p_{wf} - \Delta p_s}{\overline{p} - p_{wf}} \tag{11.10}$$

In Eq. (11.10), $\overline{p}$ = average reservoir pressure; p_{wf} = flowing wellbore pressure; Δp_s = pressure drop due to skin. As is evident from Eq. (11.10), if flow efficiency is less than one, the well appears to be damaged, while flow efficiency greater than one represents a stimulated wellbore. Note that average reservoir pressure can be estimated as the arithmetic average of initial reservoir pressure and flowing bottomhole pressure. However, it can be calculated more rigorously with either Eqs. (10.93) or (10.94).

Damage ratio is the reciprocal of flow efficiency. From Eq. (11.10), damage ratio, DR, is represented as:

$$\begin{aligned} DR &= \frac{1}{E} \\ &= \frac{\overline{p} - p_{wf}}{\overline{p} - p_{wf} - \Delta p_s} \end{aligned} \tag{11.11}$$

In contrast to flow efficiency, a damage ratio that is greater than one indicates a damaged wellbore, whereas a damage ratio that is less than one indicates a stimulated wellbore.

11.2.4 Effective Wellbore Radius

Effective wellbore radius is a measure that can be used to denote the level of damage or stimulation existing in a wellbore. Effective wellbore radius, r_{wa}, is defined as the radius of an undamaged well that exhibits the same pressure drawdown as a damaged or stimulated well. From Eqs. (11.7) and (11.8), additional pressure drop across the damaged zone is given as:

$$\Delta p_s = \frac{141.2 qB\mu}{kh} \times s \tag{11.12}$$

For a reservoir exhibiting homogenous behavior, the same additional pressure drop would be achieved if the wellbore radius was r_{wa} instead of r_w such that:

$$\Delta p_s = \frac{141.2 q B \mu}{kh} \ln \frac{r_w}{r_{wa}} \tag{11.13}$$

From Eqs. (11.12) and (11.13), it follows that:

$$s = \ln \frac{r_w}{r_{wa}} \tag{11.14}$$

Re-arranging Eq. (11.14), effective wellbore radius is given by:

$$r_{wa} = r_w e^{-s} \tag{11.15}$$

From Eq. (11.15), it can be deduced that the effective wellbore radius of a damaged well is small, while that of a stimulated well is relatively large. For instance, the effective wellbore radius of a vertically fractured well is one-half of the fracture length:

$$r_{wa} = \frac{1}{2} L_f \tag{11.16}$$

In Eq. (11.16), L_f is the length of one wing of the fracture assuming the fracture has two wings of equal length.

Example 11.2 Calculation of Skin Factor and Effective Wellbore Radius

Problem

Calculate the skin factor and effective wellbore radius of a well with the following properties:

Formation permeability, k	250 md
Damaged zone permeability, k_a	50 md
Wellbore radius, r_w	0.5 ft
Damaged zone radius, r_a	3.0 ft

Solution

From Eq. (11.8), skin factor is calculated using the given properties of the wellbore as:

$$s = \left(\frac{k}{k_a} - 1\right) \ln \frac{r_a}{r_w} = \left(\frac{250}{50} - 1\right) \ln \left(\frac{3}{0.5}\right)$$
$$= 7.17$$

Using Eq. (11.15), effective wellbore radius is calculated as:

$$r_{wa} = r_w e^{-s} = 0.5 e^{-7.17}$$
$$= 3.85 \times 10^{-4} \text{ ft.}$$

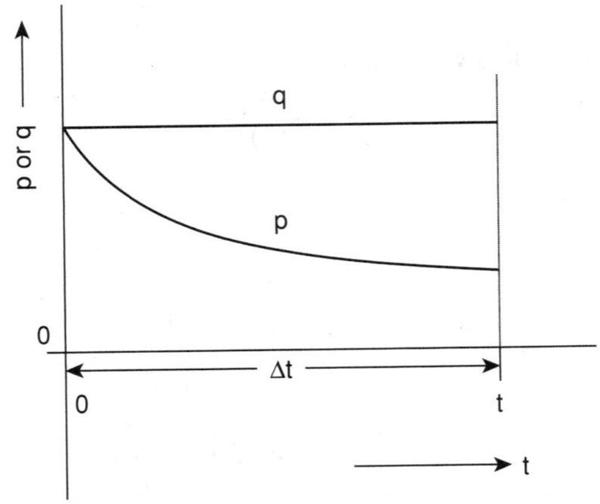

Figure 11.3 Constant single rate drawdown test.

11.2.5 Drawdown Well Tests

One type of flow test that can be conducted on a well is called a drawdown test. In a drawdown test, a well at zero production in a reservoir or drainage volume at initial or stabilized static pressure is placed on production for a specified length of time. There are two main types of drawdown tests, namely, constant single rate tests and multiple rate (or multi-rate) tests. In constant single rate tests, the well is produced at a constant, single rate for the duration of the test. This is shown in Figure 11.3. In multi-rate tests, the production rates of the well are maintained constant at multiple rates over various time periods as illustrated in Figure 11.4. Drawdown tests are affected by following operational problems:

1. It is difficult, sometimes impossible, to maintain the well at a constant production rate over the duration of the test either for single rate or multi-rate tests.
2. Except for virgin reservoirs, the static pressure within the drainage volume of the well may not have been stabilized to a specific average pressure.
3. Early test data may be distorted by unloading of fluids in the wellbore at initiation of production.

As shown in Figure 11.5, a typical plot of drawdown pressure data versus time on a semilog scale can be divided into three time regions: early time region, middle time region, and late time region.

Early Time Region (ETR). In ETR, well test data is influenced by unloading of fluids in the wellbore (wellbore storage effects). The pressure transient is also passing through the near wellbore zone of altered permeability (damaged or stimulated) shown in Figure 11.1. Well test data collected during ETR can be used to calculate wellbore storage coefficient and wellbore skin factor.

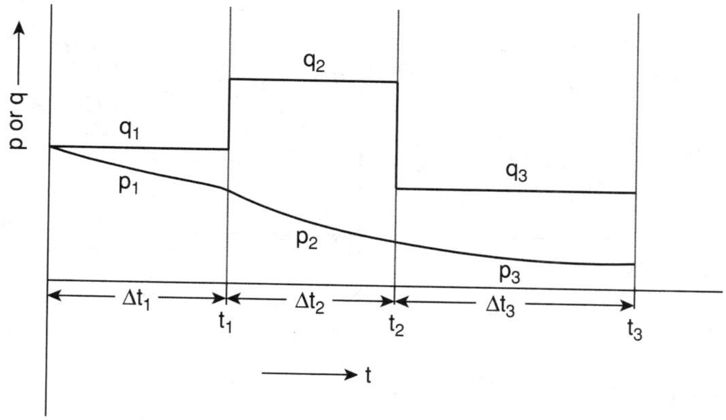

Figure 11.4 Multiple rates drawdown test.

Middle Time Region (MTR). The pressure transient is presumed to have entered MTR after unloading of wellbore fluids have ceased (termination of wellbore storage effects), and the pressure transients have entered zones in the formation with virgin permeability. Well test data collected during MTR can be used to calculate formation permeability and skin factor, as shown later in the chapter.

Late Time Region (LTR). Late time region is reached when the pressure transient is affected by the reservoir or drainage volume boundaries or other severe heterogeneities (such as sealing faults) that may be present in the formation. This is characterized by a change in the slope of the pressure drawdown curve as shown in Figure 11.5.

11.2.6 Buildup Well Tests

A typical buildup well test is shown in Figure 11.6. A buildup test is characterized by production at a constant rate over a time period denoted as, t_p, followed by shut-in either at the wellhead or

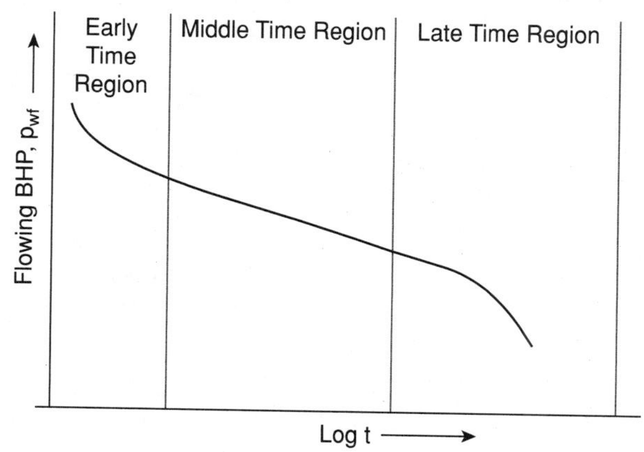

Figure 11.5 Drawdown test time regions.

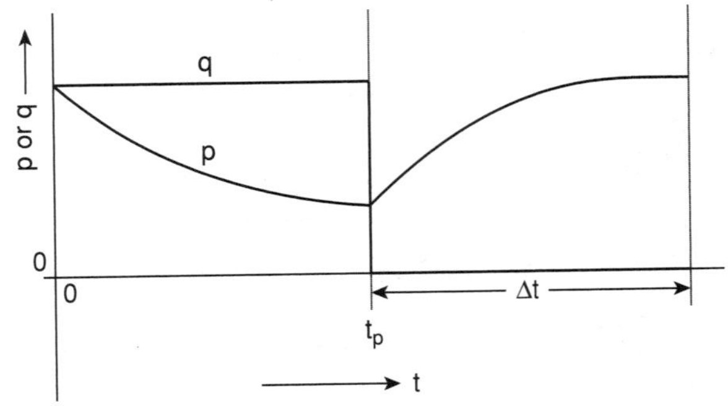

Figure 11.6 Typical buildup test.

downhole with tools over another time period denoted as, Δt. Analyses of buildup test data are affected by the following well conditions:

1. Difficulties of maintaining production at constant rate prior to shut-in.
2. For old producing wells or wells with extended production history, the challenge of calculating production time, t_p, prior to shut-in. The production time before shut-in, t_p, is needed as input for the preparation of the Horner plot.[2]
3. The distortion of early well test data by flows that occur (called afterflow) after the well has been shut in.

A plot of shut-in pressure versus time on a semi-log scale of buildup test data can be divided into time regions similar to a drawdown test as shown in Figure 11.7. The time regions are also named as in drawdown tests as: early time region, middle time region, and late time region.

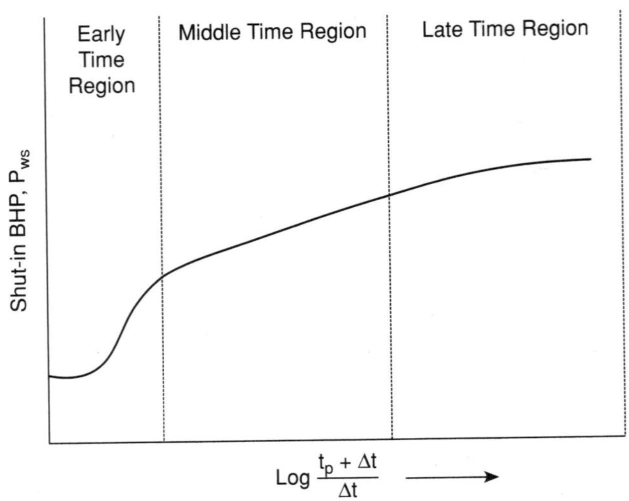

Figure 11.7 Buildup test time regions.

Early Time Region (ETR). Buildup data in this time region is affected by afterflow and the altered permeability zone close to the wellbore (Figure 11.1). These data can be used to calculate wellbore storage coefficient and skin factor.

Middle Time Region (MTR). MTR is reached when afterflow ends (end of wellbore storage) and the pressure transient has entered the areas of the formation with virgin permeability. The slope of the straight line drawn through MTR data on a semi-log plot similar to Figure 11.7 is used to calculate formation permeability, skin factor, and an estimate of average reservoir pressure. This application is demonstrated later in the chapter with the Horner Plot.[2]

Late Time Region (LTR). LTR data in buildup tests are affected by reservoir boundaries and other reservoir heterogeneities, such as a sealing fault. The data from LTR can be used to calculate the size of the drainage volume of the well.

11.2.7 Wellbore Storage

In drawdown flow tests, unloading of wellbore fluids at start of production affects early time test data. Similarly, in buildup tests, continued flow of reservoir fluids into the wellbore after the well has been shut in (called afterflow) affects early time test data. Well unloading and afterflow distortions of early time test data are collectively described as wellbore storage effects. Unloading of wellbore fluids in drawdown tests ceases when the rate of fluid production at the wellhead equals the rate of fluid entry from the reservoir. Afterflow in buildup tests ends when hydrostatic pressure of the fluid flowing into the wellbore equals reservoir pressure at the sandface. Wellbore storage coefficient, C, is a parameter used to quantify wellbore storage. It is defined as change in volume of wellbore fluids per unit change in wellbore pressure. This is represented as:

$$C = \frac{\Delta V}{\Delta p} \tag{11.17}$$

In Eq. (11.17), C is wellbore storage coefficient, bbl/psi; ΔV is in barrels; and Δp is in psi. For a well with a rising or falling liquid level, wellbore storage coefficient is calculated as:

$$\begin{aligned} C &= \frac{144 A_{wb}}{\rho_{wb}}, \frac{\text{ft}^3}{\text{psi}} \\ &= \frac{25.646 A_{wb}}{\rho_{wb}}, \frac{\text{bbl}}{\text{psi}} \end{aligned} \tag{11.18}$$

In Eq. (11.18), A_{wb} is the area of the well bore, ft^2; and ρ_{wb} is the density of the fluid in the wellbore, Ibm/ft^3. For a wellbore filled with a single-phase fluid, wellbore storage coefficient is given by:

$$C = V_{wb} c_{wb} \tag{11.19}$$

For Eq. (11.19), V_{wb} = volume of fluid in wellbore, bbl; and c_{wb} = compressibility of fluid in the wellbore, psi^{-1}. In dimensionless form, wellbore storage coefficient is defined as:

$$C_D = \frac{0.894 C}{\phi c_t h r_w^2} \tag{11.20}$$

In Eq. (11.20), C_D = dimensionless wellbore storage coefficient; C = wellbore storage coefficient, bbl/psi; ϕ = porosity, fraction; c_t = total compressibility, psi^{-1}; h = interval thickness, ft.; and r_w = wellbore radius, ft.

Wellbore storage effects can be reduced in drawdown tests by measuring flow rates downhole close to the producing interval rather at the wellhead. This could increase the cost of conducting drawdown tests. In buildup tests, wellbore storage effects can be minimized by using downhole shut-in tools. This is common practice in conducting buildup tests if there are no obstructions in the wellbore or no substantial risk of losing the tool in the wellbore. The use of downhole shut-in tools reduces the time and cost of buildup tests, especially on wells with high production rates.

11.3 Line Source Well, Infinite Reservoir Solution of the Diffusivity Equation with Skin Factor

In Chapter 10, the constant terminal rate solution of the diffusivity equation for slightly compressible liquids assuming a line source well in an infinite reservoir was presented. The solution presented as Eq. (10.64) in field units is:

$$p(r,t) = p_i - \frac{70.6qB\mu}{kh}\left[-Ei\left(-\frac{948\phi\mu c_t r^2}{kt}\right)\right]$$

In the above equation, p = pressure, psia; q = flow rate, STB/D; B = formation volume factor, RB/STB; μ = fluid viscosity, cp; k = permeability, md; h = formation thickness, ft; ϕ = porosity, fraction; c_t = total compressibility, psi^{-1}; r = radius, ft; and t = time, hr.

It is important to observe that Eq. (10.64) does not include the pressure drop, Δp_s caused by the zone of altered permeability close to the wellbore as shown in Figure 11.1. If the pressure drop, Δp_s, caused by skin is included for calculation of wellbore pressure, p_w, in a well, Eq. (10.64) can be modified to be:

$$p_{(w,t)} = p_i - \frac{70.6qB\mu}{kh}\left[-Ei\left(-\frac{948\phi\mu c_t r_w^2}{kt}\right)\right] - \Delta p_s \quad \textbf{(11.21)}$$

Combining Eqs. (11.9) and (11.21) gives:

$$p_{(w,t)} = p_i - \frac{70.6qB\mu}{kh}\left[-Ei\left(-\frac{948\phi\mu c_t r_w^2}{kt}\right) + 2s\right] \quad \textbf{(11.22)}$$

Introducing the logarithmic approximation to the exponential integral for $x < 0.02$ gives the expression:

$$Ei(-x) \approx \ln(1.781x) \quad \textbf{(11.23)}$$

Assuming the logarithmic approximation to the exponential integral applies, using Eq. (11.23) in Eq. (11.22) yields:

$$p_{(w,t)} = p_i + \frac{70.6qB\mu}{kh}\left[\ln\frac{1688\phi\mu c_t r_w^2}{kt} - 2s\right] \quad \textbf{(11.24)}$$

Converting from natural logarithm to base-10 logarithm, Eq. (11.24) becomes:

$$p_{(w,t)} = p_i - \frac{162.6qB\mu}{kh}\left[\log\frac{kt}{\phi\mu c_t r_w^2} - 3.23 + 0.8691s\right] \tag{11.25}$$

Equation (11.25) is the basic equation used in the development of equations for drawdown and buildup analyses. Consequently, note that the assumptions made to achieve the line source solution of the diffusivity equation for slightly compressible fluids for an infinite reservoir apply to drawdown and buildup analyses.

Example 11.3 Calculation of Total Pressure Drop of a Well in a Multi-Well Reservoir System

Problem

A reservoir at initial pressure of 2500 psia has three wells (Wells A, B, C) as shown in Figure 11.8 on production for 45 days from initial conditions. Well A was produced at a constant rate of 600 STB/D. Well B was produced at a constant rate of 350 STB/D and is located 1000 feet from Well A. Well C was produced at a constant rate of 150 STB/D and is located 750 feet from Well A. Assuming the reservoir is infinite in size, calculate the total pressure drop in Well A at the end of 45 days. Rock and fluid properties for the reservoir, including parameters for Well A, are as follows:

Initial pressure, p_i	2500 psia
Formation thickness, h	15 ft
Formation permeability, k	300 md
Formation porosity, ϕ	0.24
Total compressibility, c_t	6×10^{-6} psi^{-1}
Oil viscosity, μ_o	4.2 cp
Oil FVF, B_o	1.2 RB/STB
Wellbore radius, r_w	0.5 ft
Skin factor, Well A	4.5

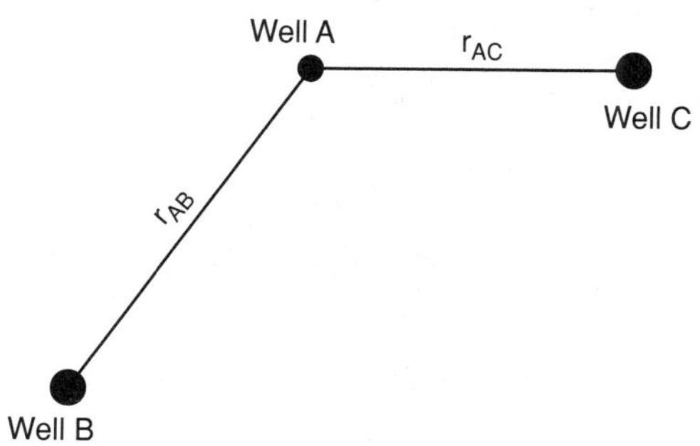

Figure 11.8 Diagram of well locations for Example 11.3.

Solution

This example is similar to Example 10.10 in Chapter 10. It is presented in this chapter to illustrate the application of the superposition principle for calculation of pressure in a well with a near wellbore zone of altered permeability. This application of the superposition principle was deferred to this chapter because the concepts of skin and skin factor were not discussed in Chapter 10. To calculate pressure in a well, it is important to include the pressure drop due to presence of skin near the wellbore. This means that Eq. (11.22) should be used in the formulation of superposition equations. For calculations of pressures away from the wellbore, use Eq. (10.64).

Step 1: Calculate the pressure drop caused by Well A.
From Eq. (11.22),

$$
\begin{aligned}
\Delta p_A &= \frac{70.6 q_A B \mu}{kh}\left[-Ei\left(-\frac{948 \phi \mu c_t r_w^2}{kt}\right) + 2s\right] \\
&= \frac{70.6 \times 600 \times 1.2 \times 4.2}{300 \times 15}\left[-Ei\left(\frac{948 \times 0.24 \times 4.2 \times 6 \times 10^{-6} \times (0.5^2)}{300 \times 45 \times 24}\right) + (2 \times 4.5)\right] \\
&= 47.4432 \times [-Ei(-4.424 \times 10^{-9}) + 9]
\end{aligned}
$$

Since $x < 0.2$, use the logarithmic approximation for calculation of the exponential integral.

$$Ei(-x) = \ln(1.781x) = \ln(1.781 \times 4.424 \times 10^{-9}) = -18.659$$

Thus,

$$
\begin{aligned}
\Delta p_A &= 47.4432 \times [-(-18.659) + 9] \\
&= 1312.2337 \text{ psia}
\end{aligned}
$$

Note that Eq. (11.24) could have been used instead to calculate Δp_A.

Step 2: Calculate the pressure drop caused by Well B.
From Eq. (10.64),

$$
\begin{aligned}
\Delta p_B &= \frac{70.6 q_B B \mu}{kh}\left[-Ei\left(-\frac{948 \phi \mu c_t r_{AB}^2}{kt}\right)\right] \\
&= \frac{70.6 \times 350 \times 1.2 \times 4.2}{300 \times 15}\left[-Ei\left(-\frac{948 \times 0.24 \times 4.2 \times 6 \times 10^{-6} \times (1000)^2}{300 \times 45 \times 24}\right)\right] \\
&= 27.6752 \times [-Ei(-\,0.017696)]
\end{aligned}
$$

Since $x < 0.02, Ei(-x) = \ln(1.781x) = \ln(1.781 \times 0.017696) = -3.4572$
Hence,

$$
\begin{aligned}
\Delta p_B &= 27.6752 \times [-(-3.4572)] \\
&= 95.6799 \text{ psia.}
\end{aligned}
$$

Step 3: Calculate the pressure drop caused by Well C.
From Eq. (10.64),

$$\Delta p_C = \frac{70.6 q_C B\mu}{kh}\left[-Ei\left(-\frac{948\phi\mu c_t r_{AC}^2}{kt}\right)\right]$$
$$= \frac{70.6 \times 150 \times 1.2 \times 4.2}{300 \times 15}\left[-Ei\left(-\frac{948 \times 0.24 \times 4.2 \times 6 \times 10^{-6} \times (750)^2}{300 \times 45 \times 24}\right)\right]$$
$$= 11.8608 \times [-Ei(-0.009954)]$$

Since $x < 0.02, Ei(-x) = \ln(1.781x) = \ln(1.781 \times 0.009954) = -4.0326$

Hence,

$$\Delta p_C = 11.8608 \times [-(-4.0326)]$$
$$= 47.8299 \text{ psia.}$$

The total pressure drop at Well A is:

$$\Delta p_t = \Delta p_A + \Delta p_B + \Delta p_C$$
$$= 1312.23 + 95.68 + 47.83 \text{ psia.}$$
$$= 1455.74 \text{ psia.}$$

11.4 Well Test Analyses with Straightline Methods

In keeping with historical perspective, it is instructive to begin presentation of well test analysis methods with straightline methods. The earliest techniques used to analyze well test data were based on straightline methods. These methods gained wide acceptance in the petroleum industry with publications of the procedures by Miller et al.[1] and Horner.[2] Straightline methods are based on the ability to identify the presence of a straight line on a plot of pressure data versus prescribed function of elapsed time within the dominant flow regime. For instance, the slope of the straight line representing radial flow regime (middle time region) in a drawdown test can be used to calculate reservoir permeability, reservoir pressure, and skin factor, as shown later in this chapter. Straightline methods are applicable to slightly compressible fluids (liquids) and compressible fluids (gases). The application of the method to slightly compressible fluids is presented first, followed by application to compressible fluids.

11.4.1 Slightly Compressible Fluids

In Chapter 10, the derivation of the diffusivity equation for slightly compressible fluids was presented. The constant terminal rate solution to the diffusivity equation assuming line source well in an infinite reservoir was also presented. The key initial and boundary conditions for the solution are repeated in this section as follows:

1. The well has zero radius, $r_w = 0$.
2. The reservoir is infinite in size (i.e., $p \rightarrow p_i$ as $r \rightarrow \infty$ for all t).

3. At time, $t = 0$, the reservoir is at uniform pressure, $p = p_i$, for all r.
4. At time, $t > 0$, a single well of radius, r_w, at the center of the cylinder (reservoir) produces at a constant rate, q.

It is important to remember the above initial and boundary conditions (especially items 2, 3, and 4) in the application of the solution of the diffusivity equation to drawdown and buildup analyses.

11.4.1.1 Constant Rate Drawdown Tests

Consider a single well in a reservoir producing a single phase fluid at constant rate from initial conditions. The flowing bottomhole pressure of the well, p_{wf}, can be calculated with Eq. (11.25) as:

$$p_{wf} = p_i - \frac{162.6qB\mu}{kh}\left[\log\frac{kt}{\phi\mu c_t r_w^2} - 3.23 + 0.8691s\right] \tag{11.26}$$

Equation (11.26) can be re-arranged as:

$$p_{wf} = p_i - \frac{162.6qB\mu}{kh}\left[\log t + \log\frac{k}{\phi\mu c_t r_w^2} - 3.23 + 0.8691s\right] \tag{11.27}$$

Equation (11.27) is in the form of the equation for a straight line: $y = b + mx$. Assigning terms in Eq. (11.27) to representative terms in the straight line equation gives:

$$y \equiv p_{wf} \tag{11.28}$$

$$x \equiv \log t \tag{11.29}$$

$$m \equiv \left|\frac{162.6qB\mu}{kh}\right| \tag{11.30}$$

$$b \equiv p_i - \frac{162.6qB\mu}{kh}\left[\log\frac{k}{\phi\mu c_t r_w^2} - 3.23 + 0.8691s\right] \tag{11.31}$$

A plot of p_{wf} vs. log t produces a curve similar to Figure 11.5. The part of this curve in the ETR is affected by wellbore storage and the altered permeability zone close to the wellbore. The slope of the straight line representing the MTR is equal to m. Note that by convention the absolute value of m is used as the slope of the line. The slope of the MTR line is simply determined as the absolute value of the difference between two pressure points, (p_{wf2} and p_{wf1}), exactly one log cycle apart. Reservoir permeability, k, or its productive capacity, kh, can be calculated from Eq. (11.30) given the slope of the MTR line as m. Generally, it is frequently difficult to determine precisely the end of wellbore storage effects or end of ETR. A rule of thumb generally used in analyses of drawdown tests is that MTR begins approximately one-and-one-half log cycles from the apparent end of ETR.

By substituting Eq. (11.30) into Eq. (11.26) and rearranging terms, skin factor, s, can be represented as:

$$s = 1.151\left[\frac{p_i - p_{wf}}{m} - \log\frac{kt}{\phi\mu c_t r_w^2} + 3.23\right] \tag{11.32}$$

Using Eq. (11.32), the skin factor, s, can be calculated by selecting any point on the straight line representing the MTR on the plot of p_{wf} vs. log t. By convention however, p_{wf} is determined at $t = 1$ hour sometimes by extrapolation of the MTR straight line. With this approach, Eq. (11.32) is re-written as:

$$s = 1.151\left[\frac{p_i - p_{1hr}}{m} - \log\frac{k}{\phi\mu c_t r_w^2} + 3.23\right] \tag{11.33}$$

It is important to reiterate some of the assumptions (constant flow rate, infinite-acting reservoir, and uniform initial pressure) made in achieving the exponential integral solution to the diffusivity equation, which is the basis for the equations used in the analyses of constant rate drawdown flow tests. In practice, some of these assumptions are difficult to achieve or maintain for the duration of the test. Consequently, analyses of drawdown tests are affected because these conditions are rarely achieved in an actual test.

Example 11.4 Calculation of Permeability and Skin Factor from Analyses of a Single, Constant Rate Drawdown Test

Problem

The drawdown data of a discovery well in an oil reservoir are shown in Table 11.1. The well was produced at a single, constant rate of 550 STB/D for a total time of 725 hrs. Other reservoir and fluid properties data are as follows:

Formation thickness, h	20 ft
Formation porosity, ϕ	0.23
Initial water saturation, S_{wi}	0.26
Total compressibility, c_t	6×10^{-5} psi^{-1}
Oil viscosity, μ_o	0.42 cp
Oil FVF, B_o	1.45 RB/STB
Wellbore radius, r_w	0.5 ft

Solution

Step 1: Plot p_{wf} vs. log t.

From the data in Table 11.1, make a plot of p_{wf} vs. t on a semilog graph. A plot of p_{wf} vs. t is shown as Figure 11.9. This can be done easily on an electronic spreadsheet or with any graphing software.

Step 2: Determine the slope, m, of the MTR straight line.

By examining Figure 11.9, it is evident that wellbore storage effects ended approximately after one (1) hour of flow. Applying the one-and-half cycle rule of thumb, the flow test was definitely in the MTR after 50 hours of flow. Boundary effects appear to start after 200 hours of flow. Thus, the MTR straight line was drawn between flow times

Table 11.1 Drawdown Flow Test Data for Example 11.4

Time, *t* (hrs.)	P_{wf} (psia)
0.00	5,665
0.12	5,645
0.22	5,615
0.42	5,590
0.62	5,577
1.10	5,566
2.05	5,548
4.20	5,525
8.40	5,493
10.20	5,481
15.50	5,462
24.50	5,438
36.20	5,417
48.40	5,405
72.50	5,385
97.00	5,368
122.00	5,359
146.00	5,350
196.00	5,335
244.00	5,320
300.00	5,305
370.00	5,293
485.00	5,265
608.00	5,237
725.00	5,209

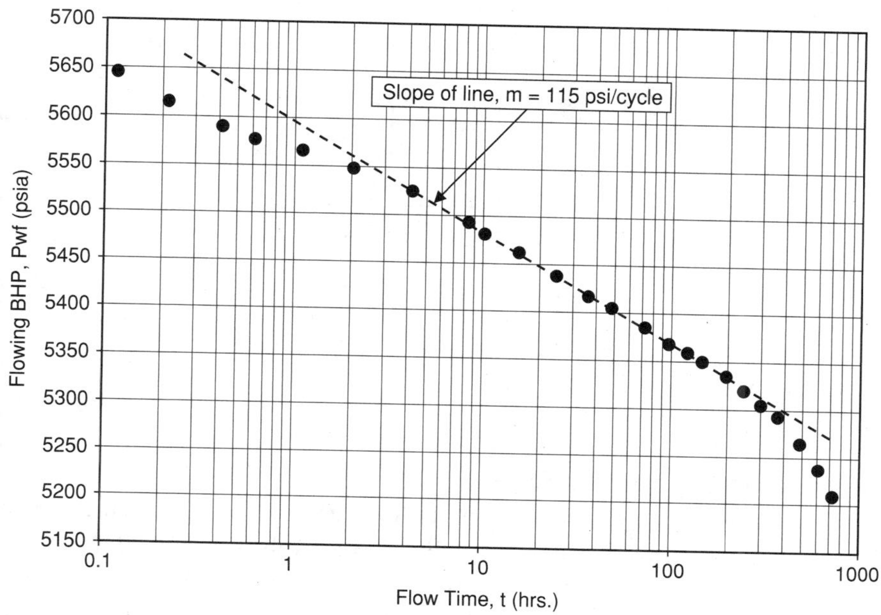

Figure 11.9 Plot of p_{wf} vs. t for Example 11.4.

of 50 hrs and 200 hrs. The slope of the MTR straight line was determined one cycle apart between $t = 10$ hrs. and $t = 100$ hrs. as:

$$\begin{aligned} m &= |p_{wf@10hrs.} - p_{wf@100hrs.}| \\ &= 5486 - 5371 \\ &= 115 \text{ psi/cycle} \end{aligned}$$

Step 3: Calculate formation permeability, k, from m.
Using Eq. (11.30):

$$\begin{aligned} k &= \frac{162.6qB\mu}{mh} \\ &= \frac{162.6 \times 550 \times 1.45 \times 0.42}{115 \times 20} \\ &= 23.68 \text{ md.} \end{aligned}$$

Step 4: Calculate skin factor, s.
From Figure 11.9, $p_{wf@1hr.} = 5595$ psia on the extrapolated MTR straight line. Using Eq. (11.33):

$$\begin{aligned} s &= 1.151\left[\frac{p_i - p_{1hr}}{m} - \log\frac{k}{\phi\mu c_t r_w^2} + 3.23\right] \\ &= 1.151\left[\frac{5665 - 5595}{115} - \log\frac{23.68}{0.23 \times 0.42 \times 6 \times 10^{-5} \times (0.5)^2} + 3.23\right] \\ &= -3.88 \end{aligned}$$

The negative skin indicates that this well may have a stimulated zone near the wellbore.

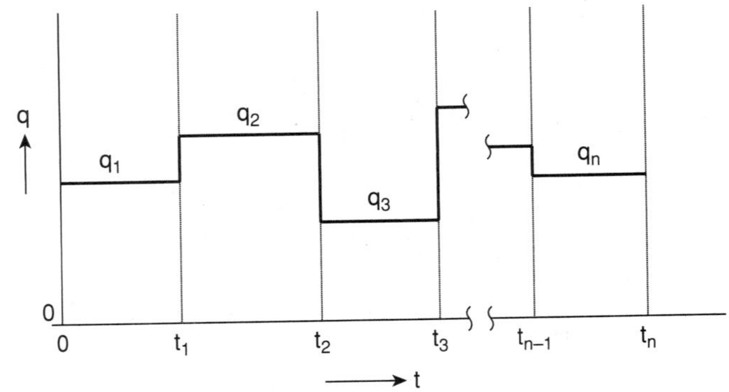

Figure 11.10 Multiple rate flow test.

11.4.1.2 Multi-Rate Drawdown Tests

It is difficult to maintain a single, constant production rate on a well during the duration of a drawdown test. For this reason, it is more desirable to test a well at multiple flow rates of varying durations. This type of test at multiple flow rates is represented in Figure 11.10. As shown in Figure 11.10, flow at constant rate, q_1, starts at time, $t = 0$; increases to q_2 at t_1; changes to q_3 at t_2; and finally changes to q_n at t_{n-1}. Note that the flow rate is constant at each flow period. An equation can be derived to represent the pressure at the well at any time period using the principle of superposition. To simplify the application of superposition principle, Eq. (11.26) can be written in the following form:

$$p_i - p_{wf} = m_m q[\log(t) + b_m] \tag{11.34}$$

where

$$m_m = \frac{162.6B\mu}{kh} \tag{11.35}$$

and

$$b_m = \log\left(\frac{k}{\phi\mu c_t r_w^2}\right) - 3.23 + 0.8691s \tag{11.36}$$

Applying the superposition principle on the flow rates shown in Figure 11.10 for n rates gives:

$$\begin{aligned} p_i - p_{wf} &= m_m q_1[\log(t) + b_m] + m_m(q_2 - q_1)[\log(t - t_1) + b_m] \\ &\quad + m_m(q_3 - q_2)[\log(t - t_2) + b_m] + \cdots\cdots \\ &\quad \cdots\cdot + m_m(q_n - q_{n-1})[\log(t - t_{n-1}) + b_m] \end{aligned} \tag{11.37}$$

Equation (11.37) can be written as:

$$\frac{p_i - p_{wf}}{q_n} = m_m \sum_{j=1}^{n}\left[\frac{(q_j - q_{j-1})}{q_n}\log(t_n - t_{j-1})\right] + m_m b_m \quad \text{for} \quad q_n \neq 0 \tag{11.38}$$

Substituting for b_m in Eq. (11.38) yields:

$$\frac{p_i - p_{wf}}{q_n} = m_m \sum_{j=1}^{n} \left[\frac{(q_j - q_{j-1})}{q_n} \log(t_n - t_{j-1}) \right] + m_m \left[\log\left(\frac{k}{\phi \mu c_t r_w^2} \right) - 3.23 + 0.8691s \right] \tag{11.39}$$

Equation (11.39) can be adapted for the analyses of multi-rate tests as illustrated with Example 11.5. Note that because this equation is based on the line source solution for an infinite-acting reservoir, the condition of a reservoir or drainage volume that is infinite in size applies for the entire duration of the test for the flow test to be valid. Also, note that Eq. (11.39) is in the form of the equation for a straight line. If Y_F is used to represent values on the y-axis, and X_F for values on the x-axis, then Eq. (11.39) can be written as:

$$Y_F \equiv \frac{p_i - p_{wf}}{q_n} \tag{11.40}$$

$$X_F \equiv \sum_{j=1}^{n} \left[\frac{(q_j - q_{j-1})}{q_n} \log(t_n - t_{j-1}) \right] \tag{11.41}$$

$$slope, m_m = \frac{162.6B\mu}{kh} \tag{11.42}$$

$$Y_{intercept} \equiv b_I = m_m \left[\log\left(\frac{k}{\phi \mu c_t r_w^2} \right) - 3.23 + 0.8691s \right] \tag{11.43}$$

A Cartesian plot of Y_F vs. X_F results in a straight line with slope, m_m, and intercept, b_I, if the reservoir is infinite-acting. This type of plot is sometimes called rate-normalized plot or multi-rate superposition plot.[11] An expression for calculating skin factor can be derived by rearranging Eq. (11.43) to get:

$$s = 1.151 \left[\frac{b_I}{m_m} - \log\left(\frac{k}{\phi \mu c_t r_w^2} \right) + 3.23 \right] \tag{11.44}$$

Example 11.5 Calculation of Permeability and Skin Factor from a Multi-Rate Drawdown Test

Problem

A multi-rate test was conducted on an oil well at production rates and flowing bottom-hole pressures shown in Table 11.2. Other reservoir properties and well data are listed as follows:

Initial reservoir pressure, p_i	2906 psia
Formation thickness, h	40 ft
Formation porosity, ϕ	0.24
Total compressibility, c_t	5×10^{-5} psi^{-1}
Oil viscosity, μ_o	0.60 cp
Oil FVF, B_o	1.27 RB/STB
Wellbore radius, r_w	0.5 ft

Table 11.2 Multi-Rate Test Data and Plotting Functions for Example 11.5 (from Tiab et al.[12] © 1999 SPE, Reproduced with permission)

n	t (hrs.)	q_n (STB/D)	P_{wf} (psi)	Y_F (psi/STB-D)	X_F
1	0	0	2906	—	—
1	1.00	1580	2023	0.5589	0.0000
1	1.50	1580	1968	0.5937	0.1761
1	1.89	1580	1941	0.6108	0.2765
1	2.40	1580	—	—	—
2	3.00	1490	1892	0.6805	0.5193
2	3.45	1490	1882	0.6872	0.5690
2	3.98	1490	1873	0.6933	0.6241
2	4.50	1490	1867	0.6973	0.6732
2	4.80	1490	—	—	—
3	5.50	1440	1853	0.7313	0.7870
3	6.05	1440	1843	0.7382	0.8193
3	6.55	1440	1834	0.7444	0.8485
3	7.00	1440	1830	0.7472	0.8739
3	7.20	1440	—	—	—
4	7.50	1370	1827	0.7876	0.9737
4	8.95	1370	1821	0.7920	1.0091
4	9.60	1370	—	—	—
5	10.00	1300	1815	0.8392	1.1242
5	12.00	1300	1797	0.8531	1.1535
6	14.40	1260	—	—	—
7	15.00	1190	1775	0.9504	1.3374
7	18.00	1190	1771	0.9538	1.3553
7	19.20	1190	—	—	—
8	20.00	1160	1772	0.9776	1.4225
8	21.60	1160	—	—	—
9	24.00	1137	1756	1.0114	1.4851
10	28.80	1106	—	—	—

n	t (hrs.)	q_n (STB/D)	P_{wf} (psi)	Y_F (psi/STB-D)	X_F
11	30.00	1080	1751	1.0694	1.6067
11	33.60	1080	—	—	—
12	36.00	1000	—	—	—
13	36.20	983	1756	1.1699	1.7883
13	48.00	983	1743	1.1831	1.7995

Solution

Step 1: Calculate the Y_F function.

From Eq. (11.40), $Y_F = \dfrac{p_i - p_{wf}}{q_n}$

As example, at $t = 1.89$ hrs, $p_i = 2906$ psi; $p_{wf} = 1941$ psi; $q_n = 1580$ STB/D

$$Y_F = \frac{2906 - 1941}{1580} = 0.6108 \text{ psi/STB-D}$$

Similarly, at $t = 20$ hrs, $p_{wf} = 1772$ psi; $q_n = 1160$ STB/D

$$Y_F = \frac{2906 - 1772}{1160} = 0.9776 \text{ psi/STB-D}$$

The remaining values for the Y_F function are shown in Table 11.2 for other test times.

Step 2: Calculate the X_F function.

From Eq. (11.41), $X_F = \sum_{j=1}^{n}\left[\dfrac{(q_j - q_{j-1})}{q_n}\log(t_n - t_{j-1})\right]$

As example, at $t = 1.89$ hrs,

$$X_F = \left(\frac{1580 - 0}{1580}\right)\log(1.89 - 0)$$
$$= 0.2765$$

At $t = 20$ hrs,

$$X_F = \frac{1}{1160} \times \begin{bmatrix} (1580 - 0)\log(20 - 0) + (1490 - 1580)\log(20 - 2.4) \\ + (1440 - 1490)\log(20 - 4.8) + (1370 - 1440)\log(20 - 7.2) \\ + (1300 - 1370)\log(20 - 9.6) + (1260 - 1300)\log(20 - 12) \\ + (1190 - 1260)\log(20 - 14.4) + (1160 - 1190)\log(20 - 19.2) \end{bmatrix}$$
$$= 1.4225$$

Values for X_F are shown in Table 11.2 for the remaining data points. Calculation of X_F (superposition time) can be tedious if done manually. However, the calculation can be performed quite readily on an electronic spreadsheet.

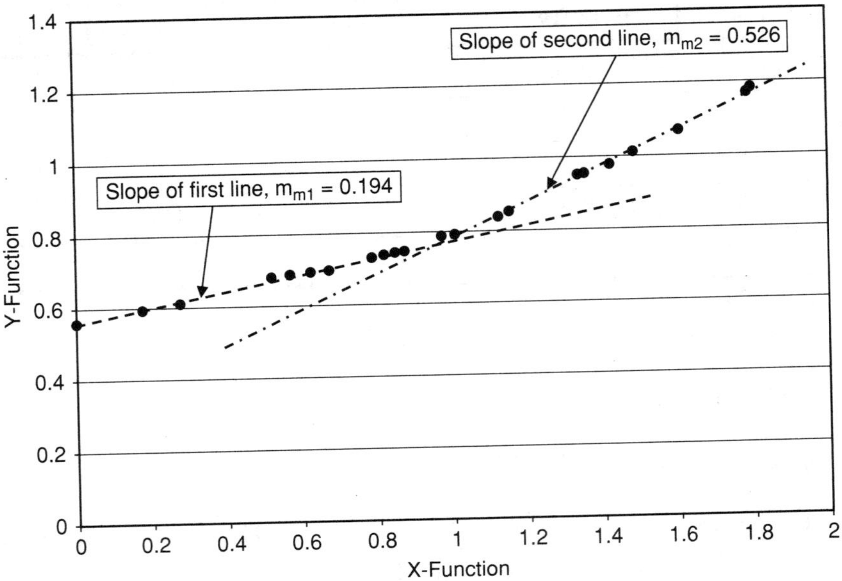

Figure 11.11 Plot of Y_F vs. X_F for Example 11.5.

Step 3: Plot X_F vs. Y_F on a Cartesian graph.
A plot of X_F vs. Y_F based on the data in Table 11.2 is shown in Figure 11.11.

Step 4: Determine the slope, m_m.
Note that Figure 11.11 has two straight lines. The slope of the first straight line is determined thus:

$$m_m = \frac{Y_{F2} - Y_{F1}}{X_{F2} - X_{F2}}$$

Using the values for Y_F and X_F at $t = 6.55$ hrs and $t = 3.00$ hrs give:

$$m_m = \frac{0.7444 - 0.6805}{0.8485 - 0.5193} = 0.194 \text{ psi/(STB/D-cycle)}$$

Similarly the slope of the second line is determined to be 0.526 psi/(STB/D-cycle). It is apparent that this second straight line represents late time region data that may have been affected by boundary effects. The slope of this line is not used in the calculations of permeability and skin factor since it is influenced by boundary effects.

Step 5: Calculate formation permeability.
From Eq. (11.42),

$$k = \frac{162.6B\mu}{m_m h}$$

$$= \frac{162.6 \times 1.27 \times 0.6}{0.194 \times 40} = 15.97 \text{ md.}$$

Step 6: Calculate skin factor.
Using Eq. (11.44),

$$s = 1.151\left[\frac{b_I}{m_m} - \log\left(\frac{k}{\phi\mu c_t r_w^2}\right) + 3.23\right]$$

From Figure 11.11, the intercept on the y-axis is $b_I = 0.5589$. Substituting gives:

$$s = 1.151\left[\frac{0.5589}{0.194} - \log\left(\frac{15.97}{0.24 \times 0.6 \times 5 \times 10^{-5} \times (0.5)^2}\right) + 3.23\right]$$
$$= -0.96$$

11.4.1.3 Two-Rate Drawdown Tests

Russell[13] proposed the two-rate test as a substitute for buildup tests, especially in wells where long shut-in periods may result in severe economic losses. Other benefits of the two-rate test are elimination of the effects of afterflow and phase redistribution in buildup tests. The equations representing the two-rate test can be developed directly from Eq. (11.39). If q_1 represents the production rate of the first test of duration, t, as shown in Figure 11.12, and q_2 is the production rate of the second test which starts at t_1, substitution of respective terms in Eq. (11.39) gives:

$$p_i - p_{wf} = \frac{162.6 q_2 B\mu}{kh}\left[\left(\frac{q_1}{q_2}\right)\log(t) + \left(\frac{q_2 - q_1}{q_2}\right)\log(t - t_1)\right] + \frac{162.6 q_2 B\mu}{kh}\left[\log\left(\frac{k}{\phi\mu c_t r_w^2}\right) - 3.23 + 0.8691 s\right] \quad \textbf{(11.45)}$$

By setting $\Delta t = t - t_1$ and re-arranging, Eq. (11.45) becomes:

$$p_{wf} = p_i - \frac{162.6 q_1 B\mu}{kh}\left[\log\left(\frac{t_1 + \Delta t}{\Delta t}\right) + \left(\frac{q_2}{q_1}\right)\log(\Delta t)\right] - \frac{162.6 q_2 B\mu}{kh}\left[\log\left(\frac{k}{\phi\mu c_t r_w^2}\right) - 3.23 + 0.8691 s\right] \quad \textbf{(11.46)}$$

Equation (11.46) is in the form of a straight line equation such that the following assignments can be made for y and x axes:

$$Y_F \equiv p_{wf} \quad \textbf{(11.47)}$$

$$X_F \equiv \left[\log\left(\frac{t_1 + \Delta t}{\Delta t}\right) + \left(\frac{q_2}{q_1}\right)\log(\Delta t)\right] \quad \textbf{(11.48)}$$

$$slope, m = -\frac{162.6 q_1 B\mu}{kh} \quad \textbf{(11.49)}$$

$$Y_{intercept} = p_i + m\left(\frac{q_2}{q_1}\right)\left[\log\left(\frac{k}{\phi\mu c_t r_w^2}\right) - 3.23 + 0.8691 s\right] \quad \textbf{(11.50)}$$

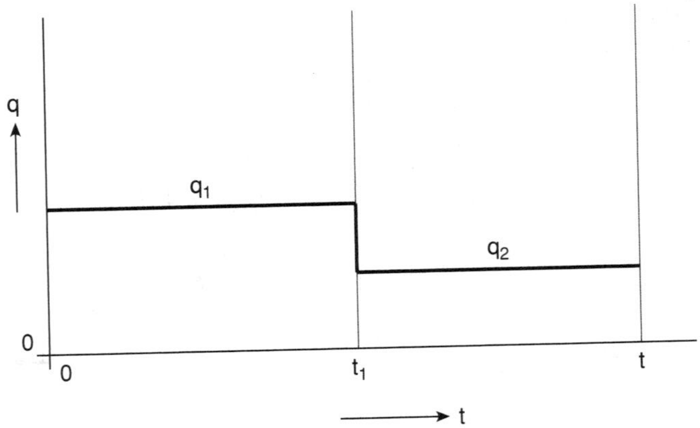

Figure 11.12 Two-rate flow test.

As shown by Russell,[13] the skin factor for a two-rate test is calculated from the equation:

$$s = 1.151\left[\frac{q_1}{(q_1 - q_2)}\left(\frac{p_{wf1} - p_{1hr}}{m}\right) - \log\left(\frac{k}{\phi\mu c_t r_w^2}\right) + 3.23\right] \tag{11.51}$$

Note that in Eq. (11.51), p_{1hr} is the pressure at $\Delta t = 1$ hr on the straight line drawn on the plot or extrapolation of the straight line. p_{wf1} is the flowing pressure at $\Delta t = 0$ or the time of the rate change to q_2. The two-rate test is valid only if the rate of the second test q_2 is maintained constant for the duration of the test. This requirement is difficult to achieve in most tests. Odeh and Jones[14] proposed a method that can handle variable flow rate in the second flow period. Note also that the two-rate flow test is accurate when the reservoir is assumed to be infinite-acting for the entire duration of the test.

Example 11.6 Calculation of Permeability and Skin Factor from Two-Rate Drawdown Test

Problem

A two-rate drawdown test was conducted on a low permeability oil reservoir. The reservoir was produced at the first rate of 220 STB/D. The cumulative production at the first rate is 50,985 STB. The production rate of the well was then reduced to 143 STB/D. The flowing bottom-hole pressure of the well at this second rate versus time is shown in Table 11.3. Using data given for this reservoir, calculate formation permeability and skin factor of the well.

Formation thickness, h	26 ft
Formation porosity, ϕ	0.09
Total compressibility, c_t	24×10^{-6} psi^{-1}
Oil viscosity, μ_o	0.76 cp
Oil FVF, B_o	1.24 RB/STB
Wellbore radius, r_w	0.25 ft
Flowing BHP at $\Delta t = 0$	2925 psia

Table 11.3 Two-Rate Test Data and Plotting Function for Example 11.6

Time, Δt Hrs.	P_{wf} (psia)	X_F
1.00	3075	3.7453
1.50	3082	3.6837
2.00	3088	3.6400
2.50	3100	3.6061
3.00	3110	3.5785
4.00	3125	3.5348
5.00	3138	3.5010
6.00	3148	3.4733
7.00	3155	3.4500
8.00	3162	3.4298
9.00	3169	3.4119
10.00	3175	3.3960
15.00	3197	3.3348
20.00	3212	3.2914
25.00	3225	3.2579

Solution

Step 1: Calculate production time, t_1, at first rate.

$$t_1 = \frac{24N_p}{q_1}$$
$$= \frac{24 \times 50{,}985}{220} = 5562 \text{ hrs.}$$

Step 2: Calculate plotting function X_F.
From Eq. (11.48),

$$X_F = \left[\log\left(\frac{t_1 + \Delta t}{\Delta t}\right) + \left(\frac{q_2}{q_1}\right)\log(\Delta t)\right]$$

As an example, at $\Delta t = 1.50$, $t_1 = 5{,}562$, $q_1 = 220$ STB/D, $q_2 = 143$ STB/D:

$$X_F = \left[\log\left(\frac{5562 + 1.50}{1.50}\right) + \left(\frac{143}{220}\right)\log(1.5)\right]$$
$$= 3.6837$$

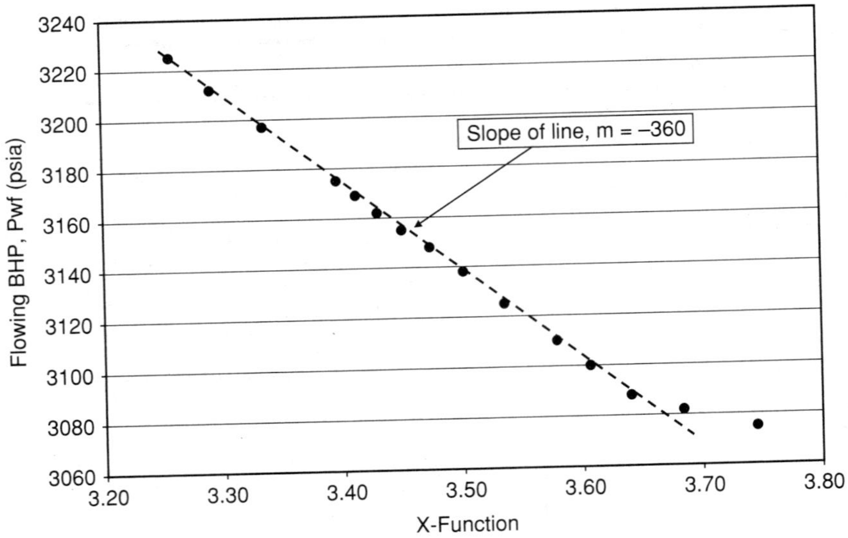

Figure 11.13 Two-rate flow test.

Step 3: Plot p_{wf} vs. X_F.

A plot of p_{wf} vs. X_F based on the data in Table 11.3 is shown in Figure 11.13. From the plot, the slope of the line, m, is:

$$m = \frac{p_{wf2} - p_{wf1}}{X_{F2} - X_{F1}}$$

Using data points at $\Delta t = 4$ hrs and $\Delta t = 15$ hrs,

$$m = \frac{3{,}197 - 3{,}125}{3.3348 - 3.5348} = -360 \text{ psi/(STB/D-cycle)}$$

Step 4: Calculate formation permeability, k.

From Eq. (11.49),

$$k = -\frac{162.6 q_1 B\mu}{mh}$$

$$= -\frac{162.6 \times 220 \times 1.24 \times 0.76}{(-360) \times 26}$$

$$= 3.60 \text{ md.}$$

Step 5: Calculate skin factor, s.

From Eq. (11.51),

$$s = 1.151\left[\frac{q_1}{(q_1 - q_2)}\left(\frac{p_{wf1} - p_{1hr}}{m}\right) - \log\left(\frac{k}{\phi\mu c_t r_w^2}\right) + 3.23\right]$$

Substituting with $p_{wf1} = 2925$ psia and $p_{1hr} = 3075$ psia gives:

$$s = 1.151\left[\frac{220}{220 - 143}\left(\frac{2925 - 3075}{-360}\right) - \log\frac{3.6}{0.09 \times 0.76 \times 24 \times 10^{-6} \times (0.25)^2} + 3.23\right]$$
$$= -3.60$$

The negative skin factor calculated for this example indicates that the near wellbore region may be stimulated.

11.4.1.4 Buildup Tests

Most engineers have encountered buildup tests rather than drawdown tests. In a typical buildup test, a producing well is shut in and its bottomhole pressure (BHP) is measured as a function of time. The production profile and pressure response of a typical buildup test was shown in Figure 11.6. The ideal buildup test represented in Figure 11.6 assumes that the production rate before the well was shut in is constant. This is rarely the case since it is extremely difficult to maintain the production rate of a well constant for an extended period of time as was noted for drawdown tests. The equations for analysis of buildup tests are derived based on the assumption of constant production rate before the well was shut in. This is followed with equations for buildup tests with variable production rates before shut-in. Modification of the buildup analysis for wells with variable production rates or long production history before shut-in introduced by Horner[2] is then presented.

11.4.1.5 Buildup Tests with Constant Production Rate Before Shut-In

For this case, the basic assumption is that the well is producing at a constant rate before it was shut in. The equation representing pressure drop in a well, assuming the logarithmic approximation to the line source solution of the radial diffusivity equation for slightly compressible fluids applies, is obtained by re-arranging Eq. (11.25) as:

$$p_i - p_{(w,t)} = \frac{162.6qB\mu}{kh}\left[\log\frac{kt}{\phi\mu c_t r_w^2} - 3.23 + 0.8691s\right] \tag{11.52}$$

Referring to Figure 11.6, suppose a well was produced at a constant rate, q, for a time period, t_p, before it was shut in for a duration represented as Δt. Applying the principle of superposition in time, the production at rate, q, is presumed to have occurred for the total duration of the test, $t_p + \Delta t$, and the shut-in at a rate equal to $(-q)$ has a duration equal to the shut-in time, Δt. The total pressure drop at the wellbore, $(p_i - p_{ws})$, is the sum of the pressure drop caused by production at rate, q, over a time period of $t_p + \Delta t$ and the pressure drop caused by production at rate $(-q)$ over a time period of Δt. Expressing this statement in equation form gives:

$$p_i - p_{ws} = \Delta p_{@q(t_p + \Delta t)} + \Delta p_{@-q(\Delta t)} \tag{11.53}$$

In Eq. (11.53), p_{ws} is the shut-in BHP measured over time, Δt. Substituting Eq. (11.52) into Eq. (11.53) gives:

$$p_i - p_{ws} = \frac{162.6qB\mu}{kh}\left[\log\frac{k(t_p + \Delta t)}{\phi\mu c_t r_w^2} - 3.23 + 0.8691s\right] + \frac{162.6(-q)B\mu}{kh}\left[\log\frac{k\Delta t}{\phi\mu c_t r_w^2} - 3.23 + 0.8691s\right] \quad (11.54)$$

Grouping and cancelling like terms in Eq. (11.54) yields:

$$p_i - p_{ws} = \frac{162.6qB\mu}{kh}\left[\log\frac{(t_p + \Delta t)}{\Delta t}\right] \quad (11.55)$$

Rearranging Eq. (11.55) gives:

$$p_{ws} = p_i - \frac{162.6qB\mu}{kh}\left[\log\frac{(t_p + \Delta t)}{\Delta t}\right] \quad (11.56)$$

Eq. (11.56) is the basic equation for buildup analysis assuming a constant production rate before shut-in. Eq. (11.56) is in the form of an equation for a straight line, $y = mx + b$ where:

$$y \equiv p_{ws} \quad (11.57)$$

$$x \equiv \log\frac{(t_p + \Delta t)}{\Delta t} \quad (11.58)$$

$$m = \left|\frac{162.6qB\mu}{kh}\right| \quad (11.59)$$

$$b \equiv p_i \quad (11.60)$$

As the above equations suggest, a plot of p_{ws} vs. $\log\frac{t_p + \Delta t}{\Delta t}$ gives a straight line with slope, m, and intercept equal to p_i at infinite shut-in time. This plot is widely known as the Horner[2] plot and the ratio, $(t_p + \Delta t)/\Delta t$, is called the Horner time ratio. The slope, m, is determined from the MTR straight line when the pressure transient is in the infinite-acting radial flow (IARF) regime. On a semilog plot, this slope is the pressure difference between two points one cycle apart on the MTR straight line. The permeability of the formation is calculated from the slope of the MTR line as shown in Eq. (11.59).

Skin factor for the well can also be calculated from buildup tests. Starting with Eq. (11.25), the equation for BHP at one hour of shut-in time is:

$$p_{ws@1hr} = p_{wf} + \frac{162.6qB\mu}{kh}\left[\log\frac{k}{\phi\mu c_t r_w^2} - 3.23 + 0.8691s\right] \quad (11.61)$$

Re-arranging Eq. (11.61) and solving for skin factor, s, gives:

$$s = 1.151\left[\frac{p_{1hr} - p_{wf}}{m} - \log\frac{k}{\phi\mu c_t r_w^2} + 3.23\right] \quad (11.62)$$

Note that in Eq. (11.62), p_{1hr} is the shut-in pressure at $\Delta t = 1$ hr on the MTR straight line or extrapolation of the MTR straight line representing IARF regime.

Theoretically, if the reservoir is shut in for a very long time such that shut-in time approaches infinity ($\Delta t \rightarrow \infty$), then the Horner time ratio approaches unity as: $[\{(t_p + \Delta t)/\Delta t\} \rightarrow 1]$. Under this condition, the shut-in pressure approaches initial reservoir pressure, p_i, for a new reservoir or extrapolated pressure, p^*, for a partially depleted reservoir. The average reservoir pressure, $\overline{p}$, can be determined from the extrapolated pressure, p^*, with a method developed by Matthews et al.[15] Using correlations by Matthews et al.,[15] the average reservoir pressure is calculated as:

$$\overline{p} = p^* - \frac{m \times f_{pf}}{2.303} \tag{11.63}$$

In Eq. (11.63), m = slope of MTR line; and f_{pf} = pressure function for the drainage area of the well. Matthews et al.[15] published pressure functions for well locations in various shapes of the drainage area of the well. These pressure functions have not been reproduced in this book because they are rarely used by engineers, since it is difficult to determine the appropriate shape of the drainage area of the well. However, the pressure functions can be obtained from Matthews et al.[15] if needed. Another method for determining average reservoir pressure was proposed by Muskat[16] and later modified by Larson.[17] A robust method that does require prior knowledge of the reservoir and its fluid properties was proposed by Crump and Hite.[18] These methods and other similar methods[19–22] in the petroleum literature provide alternative means of rigorously calculating average reservoir pressure from buildup data. However, for many applications, many engineers assume that the extrapolated pressure, p^*, is a good estimate of the average reservoir pressure of the drainage area of the well. This approximation may be permissible depending on the application of the results from analysis of the buildup test but it can be improved with published methods[15–22] as suggested earlier.

Example 11.7 Calculation of Permeability, Reservoir Pressure, and Skin Factor from Buildup Tests

Problem

The discovery well in an oil reservoir was shut in for a buildup test after producing 20,000 STB at a constant rate of 480 STB/D. The shut-in pressures measured versus time are shown in Table 11.4. Calculate the effective permeability of the formation, average reservoir pressure, and the skin factor of the well using the following reservoir and well parameters:

Formation thickness, h	35 ft
Formation porosity, ϕ	0.18
Total compressibility, c_t	2.6×10^{-5} psi^{-1}
Oil viscosity, μ_o	1.5 cp
Oil FVF, B_o	2.2 RB/STB
Wellbore radius, r_w	0.50 ft
Flowing BHP at $\Delta t = 0$	2832 psia

Table 11.4 Buildup Test Data for Example 11.7

Δt (hrs.)	P_{ws} (psia)	$(t_p + \Delta t)/\Delta t$
0.00	2,832	—
0.10	2,842	10,001
0.35	2,867	2,858
0.80	2,912	1,251
1.40	2,972	715
2.00	3,032	501
3.00	3,092	334
3.50	3,112	287
4.00	3,132	251
5.00	3,152	201
6.00	3,172	168
7.00	3,182	144
8.00	3,197	126
9.00	3,207	112
10.00	3,216	101
15.00	3,249	68
20.00	3,275	51

Solution

Step 1: Calculate the total production time, t_p.

The total production time before shut-in is:

$$t_p = \frac{24N_p}{q}$$

$$= \frac{24 \times 20{,}000}{480} = 1000 \text{ hrs}.$$

Step 2: Calculate Horner time ratio.

The Horner time ratio is $\dfrac{(t_p + \Delta t)}{\Delta t}$. For example, at shut-in time, $\Delta t = 0.35$ hrs., Horner time ratio is:

$$\frac{(t_p + \Delta t)}{\Delta t} = \frac{(1{,}000 + 0.35)}{0.35} = 2858.1$$

Horner time ratios for other shut-in times are shown in Table 11.4.

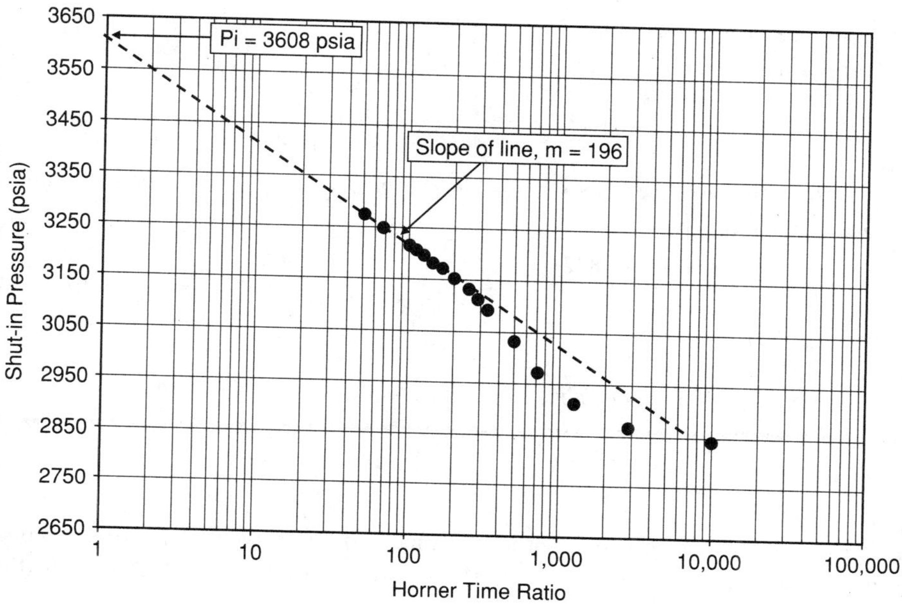

Figure 11.14 Horner plot for Example 11.7.

Step 3: Plot p_{ws} vs. $\log\left(\frac{t_p + \Delta t}{\Delta t}\right)$ and determine the slope of MTR straight line, m.

A plot of p_{ws} vs. $\log\left(\frac{t_p + \Delta t}{\Delta t}\right)$ is shown in Figure 11.14. By examining Figure 11.14, wellbore storage ended after $\Delta t = 3.5$ hrs. The slope of the MTR straight line representing infinite-acting radial flow (IARF) regime is determined to be 196 psi per cycle. That is, $m = 196$ psi per cycle.

Step 4: Calculate effective formation permeability.

From Eq. (11.59),

$$k = \frac{162.6qB\mu}{mh}$$

$$= \frac{162.6 \times 480 \times 2.2 \times 1.5}{196 \times 35}$$

$$= 37.54 \text{ md}.$$

Step 5: Calculate average reservoir pressure.

By extrapolation of the MTR straight line in Figure 11.14,

$$p_i = \overline{p} = p^* = 3{,}608 \text{ psia}.$$

Note that it has been assumed that initial reservoir pressure is equal to average reservoir pressure which is approximated as equal to extrapolated pressure. This approximation is not accurate but is permissible for many applications. Use the methods suggested by Matthews et al.,[15] Muskat,[16] Larson,[17] Crump and Hite,[18] and other authors[19–22] to determine average reservoir pressure if more accurate value is required.

Step 6: Calculate skin factor for the well.
From Eq. (11.62),

$$s = 1.151\left[\frac{p_{1hr} - p_{wf}}{m} - \log\frac{k}{\phi\mu c_t r_w^2} + 3.23\right]$$

By extrapolation of the MTR straight line in Figure 11.14, $p_{1hr} = 3019$ psia. Substituting above equation gives:

$$s = 1.151\left[\frac{3{,}019 - 2{,}832}{196} - \log\left(\frac{37.54}{0.18 \times 1.5 \times 2.6 \times 10^{-5} \times (0.5)^2}\right) + 3.23\right]$$
$$= -3.62$$

The negative skin factor indicates that this well may be stimulated.

11.4.1.6 MDH Plot

Miller, Dyes, and Hutchinson[1](MDH) proposed a method for analyzing buildup tests conducted on old wells with long production period. The theoretical basis of the MDH method can be derived from Eq. (11.56) which can be written as:

$$p_{ws} = p_i - m\left[\log\frac{(t_p + \Delta t)}{\Delta t}\right] \quad \textbf{(11.64)}$$

Expanding Eq. (11.64) gives:

$$p_{ws} = p_i - m\log(t_p + \Delta t) + m\log(\Delta t) \quad \textbf{(11.65)}$$

If $t_p \gg \Delta t$, then $\log(t_p + \Delta t) \simeq \log(t_p) =$ constant. Eq. (11.65) can then be written as:

$$p_{ws} = \kappa + m\log\Delta t, \text{ where } \kappa = p_i - m\log(t_p) = \text{constant}. \quad \textbf{(11.66)}$$

A plot of p_{ws} vs. $\log\Delta t$ yields a straight line in the MTR region representing IARF regime. The slope of this line on the MDH plot is equal to the slope of the line on the Horner plot over the same time-function on the MTR straight line.

The MDH plot can be used to determine effective formation permeability and skin factors for wells with long production history but can not be used to obtain extrapolated pressure, p^*. The MDH plot should be used as the last resort if it is difficult to estimate the length of time the well has been on production before the buildup test.

Example 11.8 Calculation of Permeability and Skin Factor from Buildup Tests in Example 11.7 Using the MDH Plot

Problem

Suppose the cumulative production of the well in Example 11.7 is not known. Calculate permeability of the formation and skin factor of the well using the buildup test data in Table 11.4. Reservoir and fluid properties data are as given in Example 11.7.

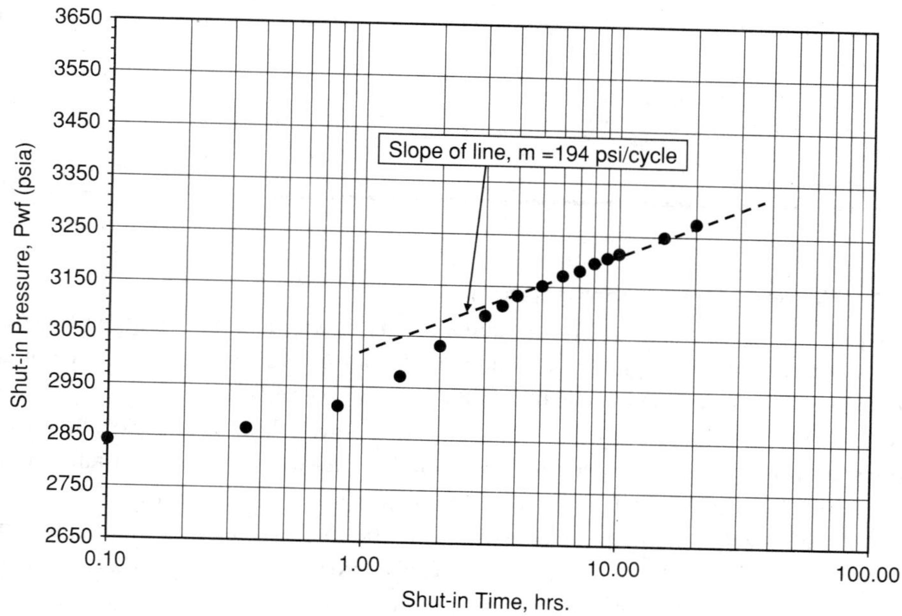

Figure 11.15 MDH plot for Example 11.8.

Solution

Step 1: Plot p_{ws} vs. log Δt.
A plot of p_{ws} vs. log Δt is shown in Figure 11.15.

Step 2: Determine the slope of the MTR straight line.
From Figure 11.15, the slope of the MTR line, $m = 194$ psi/cycle. Note that this is very close to the slope of the MTR line of the Horner plot in Example 11.7.

Step 3: Calculate formation permeability, k.
Using Eq. (11.59),

$$k = \frac{162.6qB\mu}{mh}$$

$$= \frac{162.6 \times 480 \times 2.2 \times 1.5}{194 \times 35}$$

$$= 37.93 \text{ md.}$$

Step 4: Calculate skin factor, s, for the well.
From Eq. (11.62),

$$s = 1.151\left[\frac{p_{1hr} - p_{wf}}{m} - \log\frac{k}{\phi\mu c_t r_w^2} + 3.23\right]$$

By extrapolation of the MTR straight line in Figure 11.15, $p_{1hr} = 3018$ psia. Substituting above equation gives:

$$s = 1.151\left[\frac{3{,}018 - 2{,}832}{194} - \log\left(\frac{37.93}{0.18 \times 1.5 \times 2.6 \times 10^{-5} \times (0.5)^2}\right) + 3.23\right]$$
$$= -3.62$$

11.4.1.7 Horner's Time Approximation

In developing the equations for analyzing buildup tests, it was assumed that the well was produced at a constant rate throughout its production history before shut-in. In practice, it is very difficult to maintain the production rate of a well relatively constant over a reasonable time period. The production rates of most wells vary due to changing downhole and surface conditions. These conditions are usually difficult to control over time in order to maintain a constant production rate at the well. As a result, most wells have variable production rates over time. The production histories of wells with variable production rates can be modeled accurately by application of superposition principle as shown in Eq. (11.39). This approach can be very tedious, especially if the well has a long production history with many variable rates. To simplify estimation of total producing time in such situations, Horner[2] introduced an approximation which is widely used in place of superposition time. Horner's pseudo-producing time, t_{pH}, is defined as:

$$t_{pH} = \frac{24N_p}{q_{last}} \quad \textbf{(11.67)}$$

In Eq. (11.67), N_p = cumulative oil production before shut-in, STB; q_{last} = production rate of the well just before shut-in, STB/D. In practice, q_{last} should be maintained constant and stabilized as long as possible before starting the buildup phase of the test. Note that by definition, Horner's pseudo-producing time, t_{pH}, is different from t_p used in Eq. (11.56). This t_p assumes a constant producing rate over the entire history of the well. However, t_{pH} can be used to replace t_p and q_{last} to replace q in Eq. (11.56) to yield:

$$p_{ws} = p_i - \frac{162.6q_{last}B\mu}{kh}\left[\log\frac{(t_{pH} + \Delta t)}{\Delta t}\right] \quad \textbf{(11.68)}$$

A plot of p_{ws} vs. $\log\frac{t_{pH} + \Delta t}{\Delta t}$ gives a straight line with slope, m, and intercept equal to p_i, at infinite shut-in time. This is the same as the Horner plot. The slope of the MTR straight line in the Horner plot, m, is given as:

$$m = \frac{162.6q_{last}B\mu}{kh} \quad \textbf{(11.69)}$$

All other equations derived previously for the ideal case Horner plot apply to this case with the Horner pseudo-producing time.

Example 11.9 Calculation of Permeability, Reservoir Pressure, and Skin Factor from Buildup Tests Using Horner Pseudo-Producing Time Approximation

Problem

The discovery well in an oil reservoir was tested at production rates and duration shown in Table 11.5. The well was then shut in for a buildup test. The pressures versus time data recorded during the shut-in period are shown in Table 11.6. If the flowing bottomhole pressure of the well before it was shut in was 3535 psia, calculate the formation permeability, initial reservoir pressure, and skin factor of the well from the buildup test data. Other reservoir and fluid properties data are given as follows:

Formation thickness, h	50 ft
Formation porosity, ϕ	0.20
Total compressibility, c_t	2.4×10^{-5} psi^{-1}
Oil viscosity, μ_o	0.75 cp
Oil FVF, B_o	1.25 RB/STB
Wellbore radius, r_w	0.50 ft
Flowing BHP at $\Delta t = 0$	3535 psia

Solution

Step 1: Calculate Horner pseudo-producing time, t_{pH}.
Using Eq. (11.67),

$$\begin{aligned} t_{pH} &= \frac{24N_p}{q_{last}} \\ &= \frac{24(450 \times 7 + 600 \times 4 + 500 \times 10)}{500} \\ &= 506.4 \text{ hrs}. \end{aligned}$$

Step 2: Calculate Horner time ratio.
The Horner time ratio is $(t_{pH} + \Delta t)/\Delta t$. For example, at shut-in time, $\Delta t = 0.50$ hrs, Horner time ratio is:

$$\frac{(t_{pH} + \Delta t)}{\Delta t} = \frac{(506.4 + 0.50)}{0.50} = 1013.8$$

Horner time ratios for other shut-in times are shown in Table 11.6.

Table 11.5 Production History of the Well in Example 11.9

Production Rate, STB/D	Duration, days
450	7
600	4
500	10

Table 11.6 Buildup Test Data for Example 11.9

Δt (hrs.)	P_{ws} (psia)	$(t_{pH} + \Delta t)/\Delta t$
0.25	3,605	2026.60
0.50	3,632	1013.80
1.00	3,795	507.40
1.50	3,960	338.60
2.00	4,230	254.20
3.00	4,320	169.80
4.00	4,370	127.60
7.00	4,390	73.34
10.00	4,395	51.64
15.00	4,410	34.76
24.00	4,422	22.10

Step 3: Plot p_{ws} vs. $\log\left(\frac{t_{pH} + \Delta t}{\Delta t}\right)$ and determine the slope of MTR straight line, m.

A plot of p_{ws} vs. $\log\left(\frac{t_{pH} + \Delta t}{\Delta t}\right)$ is shown in Figure 11.16. By examining Figure 11.16, wellbore storage ended after $\Delta t = 3.0$ hrs. The slope of the MTR straight line representing infinite-acting radial flow (IARF) regime is determined to be 71.4 psi per cycle. That is, $m = 71.4$ psi per cycle.

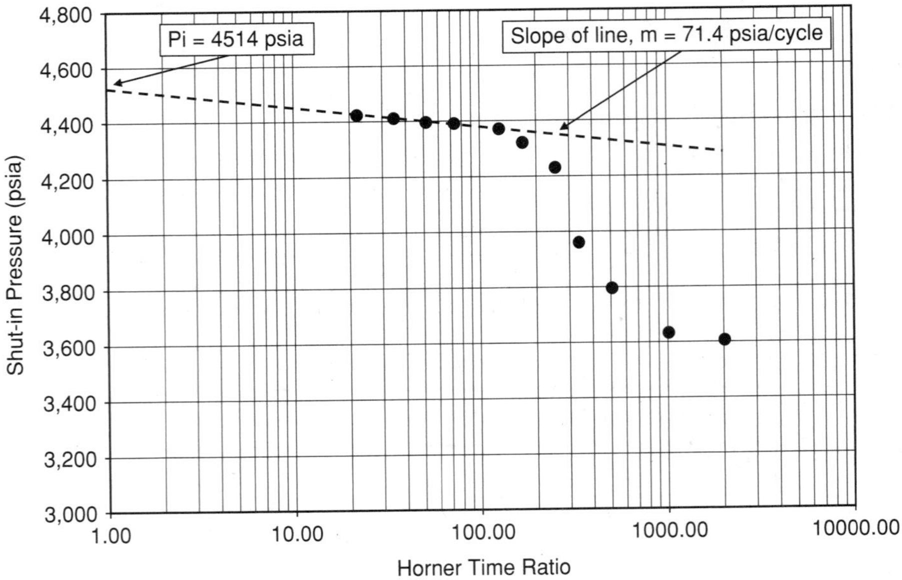

Figure 11.16 Plot of p_{ws} vs. $\{(t_{pH} + \Delta t)/\Delta t\}$ for Example 11.9.

Step 4: Calculate effective formation permeability, k.
From Eq. (11.59),

$$k = \frac{162.6qB\mu}{mh} = \frac{162.6 \times 500 \times 1.25 \times 0.75}{71.4 \times 50} = 21.35 \text{ md.}$$

Step 5: Calculate average reservoir pressure, $\overline{p}$.
By extrapolation of the MTR straight line in Figure 11.16,

$$p_i = \overline{p} = p^* = 4514 \text{ psia.}$$

Step 6: Calculate skin factor, s, for the well.
From Eq. (11.62),

$$s = 1.151\left[\frac{p_{1hr} - p_{wf}}{m} - \log\frac{k}{\phi\mu c_t r_w^2} + 3.23\right]$$

By extrapolation of the MTR straight line in Figure 11.16, $p_{1hr} = 4329$ psia. Substituting above equation gives:

$$s = 1.151\left[\frac{4329 - 3535}{71.4} - \log\left(\frac{21.35}{0.20 \times 0.75 \times 2.4 \times 10^{-5} \times (0.5)^2}\right) + 3.23\right]$$
$$= 8.03$$

The positive skin factor calculated for the well indicates that it may be damaged in the near wellbore region.

11.4.2 Compressible Fluids

In this section, the application of straightline methods for analyses of drawdown and buildup tests in gas reservoirs is presented. It is worthwhile to note that notional forms of equations developed for liquid reservoirs are applicable to gas reservoirs with slight or no modifications by using transforming pressure functions which account for the variation of gas properties with pressure. In Chapter 10, the concept of pseudo-pressure was introduced and used to linearize to some extent the gas diffusivity equation. In this section, two additional concepts are introduced that will assist in the manipulation of the equations that are used to represent gas flow equations for drawdown and buildup tests.

11.4.2.1 Real Gas Pseudo-Pressure and Pseudo-Time

Real gas pseudo-pressure, p_p, as defined by Al-Hussainy et al.[23] is:

$$p_p = 2\int_{p_b}^{p} \frac{p}{\mu(p)z(p)}dp \tag{11.70}$$

Note that Eq. (11.70) is the same as Eq. (10.84). An analogous function called real gas pseudo-time, t_A, was defined by Agarwal[24] as:

$$t_A = \int_0^t \frac{1}{\mu(p)c_t(p)} dt \tag{11.71}$$

The combined use of these pseudo-variables provides an "effective" linearization of the gas diffusivity equations, especially in applications to type-curve analyses. The use of type-curves in well test analysis is discussed in Chapter 12.

Al-Hussainy and Ramey[25] achieved an approximate solution to the gas diffusivity equation expressed in terms of pseudo-pressure as:

$$p_{pwf} = p_{pi} - \frac{5.792 \times 10^4 q_g p_{sc} T}{khT_{sc}} \left[\log \frac{kt}{\phi \mu c_t r_w^2} - 3.23 + 0.8691 s' \right] \tag{11.72}$$

In Eq. (11.72), s' is the apparent skin factor which is defined as:

$$s' = s + Dq_g \tag{11.73}$$

In Eq. (11.73), s = skin factor due to formation damage or stimulation; D = non-Darcy flow coefficient; and q_g = gas flow rate.

If standard conditions are set at T_{sc} = 60°F and p_{sc} = 14.7 psia, Eq. (11.72) becomes:

$$p_{pwf} = p_{pi} - \frac{1637 q_g T}{kh} \left[\log \frac{kt}{\phi (\mu c_t)_i r_w^2} - 3.23 + 0.8691 s' \right] \tag{11.74}$$

11.4.2.2 Normalized Gas Pseudo-Pressure and Gas Pseudo-Time

Meunier et al.[26] proposed normalization of gas pseudo-pressure and pseudo-time at any arbitrary reference pressure to produce pseudo-variables whose magnitudes are close to those of untransformed pressure and time, and have the same units. For gas pseudo-pressure, the normalized pseudo-pressure is expressed as:

$$\begin{aligned} p_{pn} &= \frac{1}{2}\left(\frac{\mu z}{p}\right)_r p_p \\ &= \left(\frac{\mu z}{p}\right)_r \int_{p_b}^{p} \frac{p}{\mu(p)z(p)} dp \end{aligned} \tag{11.75}$$

For the gas pseudo-time, the normalized pseudo-time is expressed as:

$$\begin{aligned} t_{An} &= (\mu c_t)_r \int_0^t \frac{1}{\mu(p)c_t(p)} dt \\ &= (\mu c_t)_r t_A \end{aligned} \tag{11.76}$$

The choice of the reference pressure for the normalization of the pseudo-variables is completely arbitrary. Typically, the initial reservoir pressure or the average reservoir pressure is selected if known. If the average reservoir pressure is the reference pressure, then Eqs. (11.75) and (11.76) can be written as:

$$p_{pn} = \left(\frac{\overline{\mu}\,\overline{z}}{\overline{p}}\right)_{\overline{p}} \int_{p}^{p} \frac{p}{\mu(p)z(p)} dp \tag{11.77}$$

and

$$t_{An} = (\overline{\mu}\,\overline{c}_t)_{\overline{p}} \int_0^t \frac{dt}{\mu c_t} \tag{11.78}$$

Using Eq. (11.77), Eq. (11.74) can be written in terms of normalized pseudo-variables as:

$$p_{pnwf} = p_{pni} - \frac{162.6 q_g \overline{B}\,\overline{\mu}}{kh}\left[\log \frac{kt}{\phi \overline{\mu c_t} r_w^2} - 3.23 + 0.8691 s'\right] \tag{11.79}$$

11.4.2.3 Pressure and Pressure-Squared as Variables in Gas Flow Equations

In Eqs. (11.74) and (11.79), gas flow equations were expressed as functions of pseudo-pressures and normalized pseudo-pressures, respectively. As explained earlier, this transformation was necessary because gas properties such as viscosity and compressibility factor vary with pressure. However, under certain conditions, some assumptions can be made with respect to the variation of gas viscosity and gas compressibility factor with pressure. The first assumption that can be made is that at pressures above 3500 psia, the gas property, $p/(\mu z)$, is constant. With this assumption, it can be written that at any pressure above the 3500 psia threshold that:

$$\frac{p}{\mu z} = \frac{\overline{p}}{\overline{\mu z}}, \text{ at the average reservoir pressure, } \overline{p} \tag{11.80}$$

If the condition represented in Eq. (11.80) applies, then it can be derived from Eq. (11.70) that:

$$p_p = \frac{2\overline{p}p}{\overline{\mu z}} \tag{11.81}$$

Substituting Eq. (11.81) into Eq. (11.74) and applying some conversions gives:

$$p_{wf} = p_i - \frac{162.6 q_g \overline{B}\,\overline{\mu}}{kh}\left[\log \frac{kt}{\phi \overline{\mu}\,\overline{c}_t r_w^2} - 3.23 + 0.8691 s'\right] \tag{11.82}$$

Note that with the stated assumption, Eq. (11.82) is similar in form to Eq. (11.79) developed for normalized pseudo-pressures.

The second assumption that is applicable at low pressures, ($p < 2000$ psia), is that the gas property, (μz), is constant. With this assumption, it can be written that:

$$(\mu z) = (\overline{\mu}\,\overline{z}), \text{ at the average reservoir pressure, } \overline{p} \tag{11.83}$$

If the condition expressed in Eq. (11.83) applies, then it can be derived from Eq. (11.70) that:

$$p_p = \frac{p^2}{\overline{\mu}\,\overline{z}} \tag{11.84}$$

Substituting Eq. (11.84) into Eq. (11.74) gives:

$$p_{wf}^2 = p_i^2 - \frac{1637 q_g \overline{\mu}\,\overline{z} T}{kh}\left[\log \frac{kt}{\phi \overline{\mu}\,\overline{c}_t r_w^2} - 3.23 + 0.8691 s'\right] \tag{11.85}$$

11.4.2.4 Constant Rate Drawdown Tests for Gas Wells

In the previous section, several equations were derived for gas flow in porous media. It is important to note the various assumptions that were used to derive each equation. The most important assumption that applies to all the gas flow equations is that the reservoir is radial and infinite-acting. This is the underlying assumption that applies to the solution of the radial diffusivity equation for compressible fluids.

In this section, the gas flow equations derived in the previous section are applied in the analysis of gas drawdown tests at constant rate. When drawdown tests at constant rate for oil wells were discussed, it was pointed out that it is very difficult to maintain the production rate of an oil well constant for the entire duration of a test. It is even more difficult to maintain the production rate of a gas well constant for the duration of a test. Typically, gas wells are produced at conditions that attempt to achieve constant pressures either at the wellhead or bottomhole. These conditions will cause the production rate to vary with time. In the next section, methods for analyzing gas wells with varying production rates are presented. However, it is instructive to review methods for analyzing gas wells at constant production rates. These methods are applicable if the production rate of the gas well is assumed to be approximately constant.

In terms of pseudo-pressure, the expressions for analyses of drawdown tests on gas wells at constant rate can be deduced from Eq. (11.74). On the basis of Eq. (11.74), a plot of p_{pwf} vs. log t gives a straight line. The slope of the straight line, m, is:

$$m = \left|\frac{1637 q_g T}{kh}\right| \tag{11.86}$$

It can be derived from Eq. (11.74) that the apparent skin is given by:

$$s' = 1.151\left[\frac{p_{pi} - p_{p,1hr}}{m} - \log\left(\frac{k}{\phi(\mu c_t)_i r_w^2}\right) + 3.23\right] \tag{11.87}$$

In terms of normalized pseudo-pressure, the expressions for analysis of drawdown tests on gas wells at constant rate can be deduced from Eq. (11.79). From Eq. (11.79), a plot of p_{pnwf} vs. log t gives a straight line. The slope of the straight line, m, is:

$$m = \left|\frac{162.6 q_g \overline{B}\,\overline{\mu}}{kh}\right| \tag{11.88}$$

It can be derived from Eq. (11.79) that the apparent skin is given by:

$$s' = 1.151\left[\frac{p_{pni} - p_{pn,1hr}}{m} - \log\left(\frac{k}{\phi(\mu c_t)_i r_w^2}\right) + 3.23\right] \quad \textbf{(11.89)}$$

In terms of pressure, the expressions for analyses of drawdown tests on gas wells at constant rate can be deduced from Eq. (11.82). From Eq. (11.82), a plot of $p_{wf}\, vs \log t$ gives a straight line. The slope of the straight line, m, is:

$$m = \left|\frac{162.6 q_g \overline{B}\,\overline{\mu}}{kh}\right| \quad \textbf{(11.90)}$$

It can be derived from Eq. (11.82) that the apparent skin is given by:

$$s' = 1.151\left[\frac{p_i - p_{1hr}}{m} - \log\left(\frac{k}{\phi(\mu c_t)_i r_w^2}\right) + 3.23\right] \quad \textbf{(11.91)}$$

Similarly in terms of pressure-squared, the expressions for analyses of drawdown tests on gas wells at constant rate can be deduced from Eq. (11.85). From Eq. (11.85), a plot of p_{wf}^2 vs. log t gives a straight line. The slope of the straight line, m, is:

$$m = \left|\frac{1{,}637 q_g \overline{\mu}\,\overline{z} T}{kh}\right| \quad \textbf{(11.92)}$$

It can be derived from Eq. (11.85) that the apparent skin is given by:

$$s' = 1.151\left[\frac{p_i^2 - p_{1hr}^2}{m} - \log\left(\frac{k}{\phi(\mu c_t)_i r_w^2}\right) + 3.23\right] \quad \textbf{(11.93)}$$

Note that by convention, the absolute values of m are used as the slopes of the straight line in Eqs. (11.86), (11.88), (11.90), and (11.92). It is important to observe that in all preceding equations for gas flow, reservoir pressure was assumed to be at initial pressure, p_i. Under this condition, reservoir properties such as gas viscosity, gas formation volume factor, etc., should be evaluated at initial reservoir pressure. In many applications, average reservoir pressure, $\overline{p}$, at the start of the flow test is used. In such cases, reservoir properties should be evaluated at the average reservoir pressure.

Example 11.10 Calculation of Permeability and Skin Factor from Analyses of Single, Constant Rate Drawdown Test Using Pseudo-Pressure, Normalized Pseudo-Pressure, Pressure, and Pressure-Squared Methods

Problem

The discovery well in a dry gas reservoir was produced at a constant rate of 6000 Mcf/D. The flowing BHP measured at various time intervals are shown in Table 11.7. Initial reservoir pressure

Table 11.7 Drawdown Flow Test Data for Example 11.10

Flowing Time t (hrs.)	Flowing BHP P_{wf} (psia)
0.0	4,275
0.5	4,260
0.9	4,245
1.3	4,230
1.7	4,180
2.1	4,130
2.4	4,087
2.7	4,047
3.2	4,010
4.0	3,991
4.8	3,975
6.1	3,950
8.0	3,934
10.1	3,914
12.7	3,895
15.2	3,885
20.0	3,867
25.0	3,858
30.0	3,850
35.0	3,844
40.0	3,840
50.0	3,831
60.0	3,824
70.0	3,819
80.0	3,815
90.0	3,810
100.0	3,805
110.0	3,802
120.0	3,798
130.0	3,793
140.0	3,791

before the test was measured at 4275 psia. Reservoir temperature is 215°F. Assuming the reservoir gas has no non-hydrocarbon impurities, gas properties are summarized as follows:

Gas gravity, γ_g	0.70
Critical Pressure, p_c	660 psia
Critical Temperature, T_c	–78.5°F
Gas FVF, B_{gi}	0.7553 RB/Mcf
Gas compressibility, c_g	1.71×10^{-4} psi^{-1}
Total compressibility, c_t	1.71×10^{-4} psi^{-1}

Reservoir and well data are as follows:

Porosity, ϕ	0.15
Well radius, r_w	0.29 ft
Formation thickness, h	24 ft

Solution

Step 1: Construct a table of gas properties consisting of gas compressibility factor and gas viscosity versus pressure using the procedure shown in Example 4.4. The calculated gas properties are shown in Table 11.8.

Step 2: Calculate pseudo-pressure from data obtained in Step 1 using Eq. (11.70) as illustrated in Example 10.8. The calculated pseudo-pressures are shown in Table 11.9.

Step 3: Calculate normalized pseudo-pressure from Eq. (11.75). The calculated normalized pseudo-pressures are also shown in Table 11.9.

Step 4: Pseudo-pressure plot: Plot p_{pwf} vs. log t.
A plot of p_{pwf} vs. log t from the data in Table 11.9 is shown in Figure 11.17. From the plot, the slope of the MTR straight line is determined as $m = 32.5 \times 10^6$ psia^2/cp-cycle. From Eq. (11.86), formation permeability, k, is calculated as:

$$\begin{aligned} k &= \left| \frac{1637 q_g T}{mh} \right| \\ &= \frac{1637 \times 6000 \times 675}{32.5 \times 10^6 \times 24} \\ &= 8.50 \text{ md} \end{aligned}$$

The apparent skin factor, s', from Eq. (11.87) is:

$$\begin{aligned} s' &= 1.151\left[\frac{p_{pi} - p_{p,1hr}}{m} - \log\left(\frac{k}{\phi(\mu c_t)_i r_w^2}\right) + 3.23\right] \\ &= 1.151\left[\frac{(191.9 - 85.5) \times 10^6}{32.5 \times 10^6} - \log\left(\frac{8.50}{0.15 \times 0.0248 \times 1.71 \times 10^{-4} \times (0.29)^2}\right) + 3.23\right] \\ &= -1.9 \end{aligned}$$

Table 11.8 Calculated Gas Properties for Example 11.10

Pressure (psia)	z-Factor z	Gas Viscosity μ_g (cp)
3750	0.91	0.0234
3775	0.91	0.0235
3800	0.91	0.0236
3825	0.91	0.0237
3850	0.92	0.0237
3875	0.92	0.0238
3900	0.92	0.0239
3925	0.92	0.0241
3950	0.92	0.0241
3975	0.92	0.0242
4000	0.92	0.0243
4025	0.93	0.0243
4050	0.93	0.0244
4075	0.93	0.0245
4100	0.93	0.0246
4125	0.93	0.0247
4150	0.93	0.0247
4175	0.94	0.0247
4200	0.94	0.0247
4225	0.95	0.0247
4250	0.95	0.0248
4275	0.95	0.0248
4300	0.96	0.0248

Step 5: Normalized pseudo-pressure plot: Plot p_{pnwf} vs. log t.

A plot of p_{pnwf} vs. log t from the data in Table 11.9 is shown in Figure 11.18. From the plot, the slope of the MTR straight line is determined as $m = 85$ psia/cycle. From Eq. (11.88), formation permeability, k, is calculated as:

$$k = \left| \frac{162.6 q_g \overline{B} \overline{\mu}}{mh} \right|$$

$$= \frac{162.6 \times 6000 \times 0.7553 \times .0248}{85 \times 24}$$

$$= 8.96 \text{ md}$$

Table 11.9 Plot Data for Example 11.10

t (hrs.)	P_{wf} (psia)	P^2_{wf} (psia)^2	P_{pwf} (psia^2/cp)	P_{pnwf} (psia)
0.0	4,275	18,275,625	191,862,124	528.69
0.5	4,260	18,147,600	186,434,492	513.73
0.9	4,245	18,020,025	181,013,842	498.79
1.3	4,230	17,892,900	175,607,153	483.89
1.7	4,180	17,472,400	157,566,955	434.18
2.1	4,130	17,056,900	139,546,662	384.53
2.4	4,087	16,703,569	124,121,092	342.02
2.7	4,047	16,378,209	109,821,590	302.62
3.2	4,010	16,080,100	96,621,360	266.24
4.0	3,991	15,928,081	89,833,919	247.54
4.8	3,975	15,800,625	84,114,523	231.78
6.1	3,950	15,602,500	75,197,224	207.21
8.0	3,934	15,476,356	69,514,380	191.55
10.1	3,914	15,319,396	62,419,441	172.00
12.7	3,895	15,171,025	55,686,205	153.45
15.2	3,885	15,093,225	52,142,781	143.68
20.0	3,867	14,953,689	45,770,988	126.12
25.0	3,858	14,884,164	42,589,073	117.36
30.0	3,850	14,822,500	39,760,705	109.56
35.0	3,844	14,776,336	37,637,140	103.71
40.0	3,840	14,745,600	36,221,431	99.81
50.0	3,831	14,676,561	33,036,084	91.03
60.0	3,824	14,622,976	30,558,224	84.20
70.0	3,819	14,584,761	28,786,744	79.32
80.0	3,815	14,554,225	27,369,560	75.42
90.0	3,810	14,516,100	25,598,080	70.54
100.0	3,805	14,478,025	23,826,600	65.66
110.0	3,802	14,455,204	22,763,712	62.73
120.0	3,798	14,424,804	21,348,185	58.83
130.0	3,793	14,386,849	19,580,849	53.96
140.0	3,791	14,371,681	18,873,914	52.01

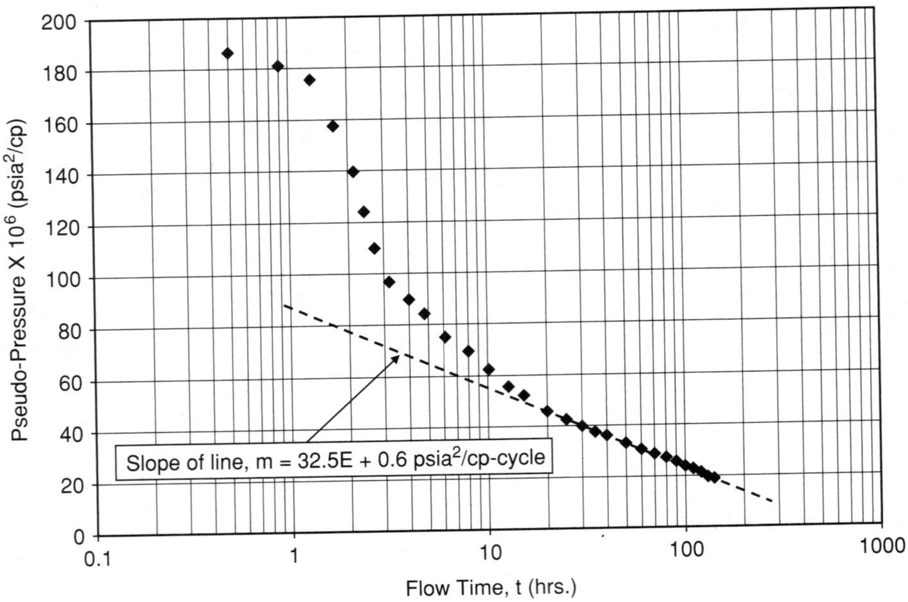

Figure 11.17 Plot of P_{pwf} vs. t for Example 11.10.

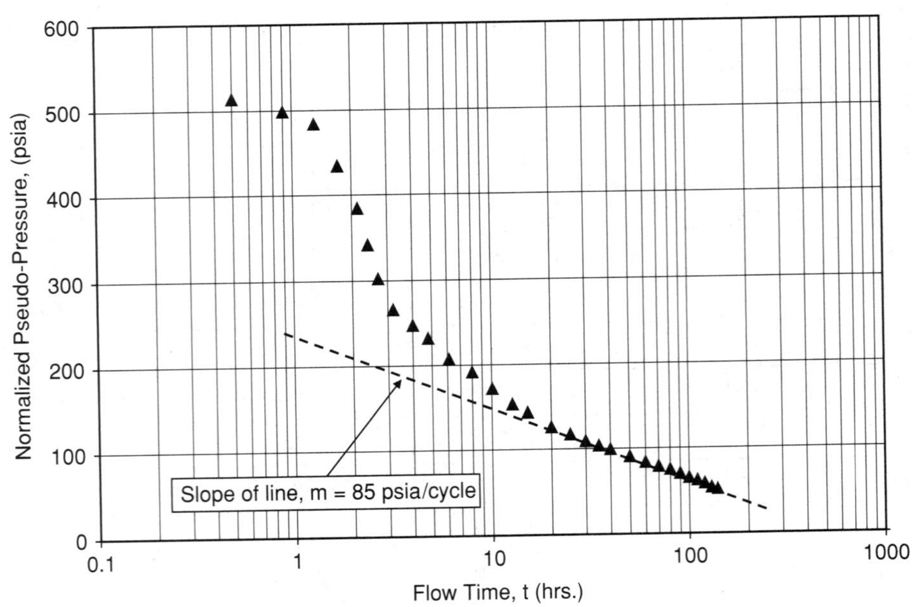

Figure 11.18 Plot of P_{pnwf} vs. t for Example 11.10.

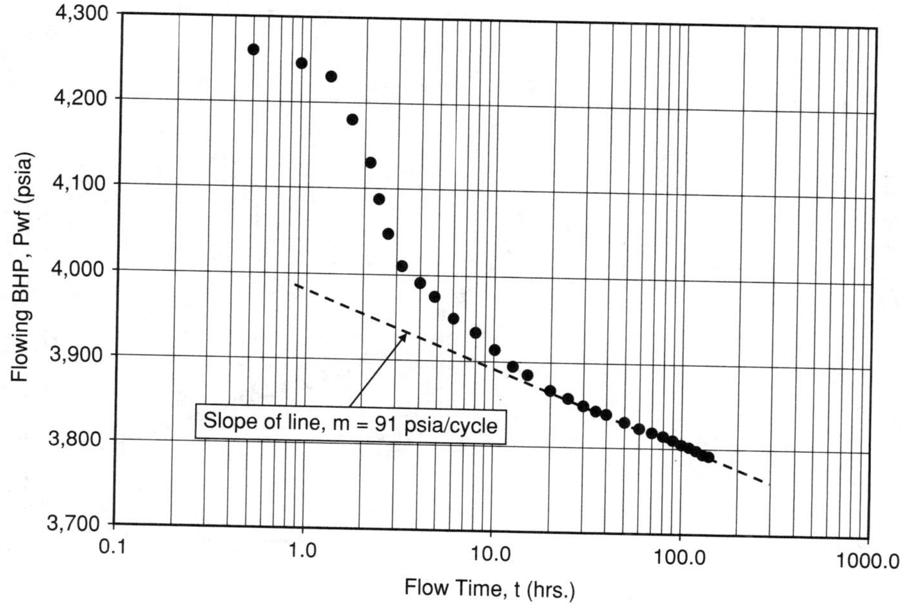

Figure 11.19 Plot of P_{pwf} vs. t for Example 11.10.

The apparent skin factor, s', from Eq. (11.87) is:

$$s' = 1.151\left[\frac{p_{pni} - p_{pn,1hr}}{m} - \log\left(\frac{k}{\phi(\mu c_t)_i r_w^2}\right) + 3.23\right]$$

$$= 1.151\left[\frac{528.69 - 235}{85} - \log\left(\frac{8.96}{0.15 \times 0.0248 \times 1.71 \times 10^{-4} \times (0.29)^2}\right) + 3.23\right]$$

$$= -1.77$$

Step 6: Pressure plot method: Plot p_{wf} vs. log t.
A plot of p_{wf} vs. log t from the data in Table 11.9 is shown in Figure 11.19. From the plot, the slope of the MTR straight line is determined as $m = 91$ psia/cycle. From Eq. (11.90), formation permeability, k, is calculated as:

$$k = \left|\frac{162.6 q_g \overline{B}\,\overline{\mu}}{mh}\right|$$

$$= \frac{162.6 \times 6000 \times 0.7553 \times .0248}{91 \times 24}$$

$$= 8.37 \text{ md}.$$

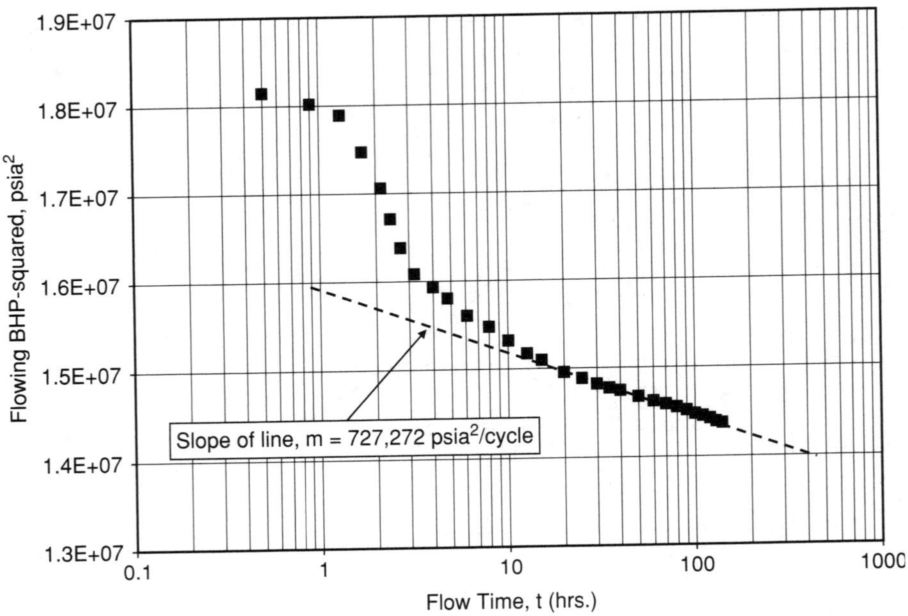

Figure 11.20 Plot of P_{wf}^2 vs. t for Example 11.10.

The apparent skin factor, s', from Eq. (11.91) is:

$$s' = 1.151\left[\frac{p_i - p_{1hr}}{m} - \log\left(\frac{k}{\phi(\mu c_t)_i r_w^2}\right) + 3.23\right]$$

$$= 1.151\left[\frac{4275 - 3986}{91} - \log\left(\frac{8.37}{0.15 \times 0.0248 \times 1.71 \times 10^{-4} \times (0.29)^2}\right) + 3.23\right]$$

$$= -2.06$$

Step 7: Pressure-squared plot: Plot p_{wf}^2 vs. log t.

A plot of p_{wf}^2 vs. log t from the data in Table 11.9 is shown in Figure 11.20. From the plot, the slope of the MTR straight line is determined as m = 727,272.73 psia^2/cycle. From Eq. (11.92), formation permeability, k, is calculated as:

$$k = \left|\frac{1637 q_g \overline{\mu}\,\overline{z} T}{mh}\right|$$

$$= \frac{1637 \times 6000 \times .0248 \times 0.95 \times 675}{727{,}272.73 \times 24}$$

$$= 8.95 \text{ md}$$

Table 11.10 Summary of Results from Example 11.10

Method	Effective Gas Permeability, *k* (md)	Skin Factor, *s*
p_{wf} vs. log t	8.37	–2.06
p_{wf}^2 vs. log t	8.95	–2.00
p_{pwf} vs. log t	8.50	–2.02
p_{pnwf} vs. log t	8.96	–1.77

The apparent skin factor, s', from Eq. (11.93 is:

$$s' = 1.151\left[\frac{p_i^2 - p_{1hr}^2}{m} - \log\left(\frac{k}{\phi(\mu c_t)_i r_w^2}\right) + 3.23\right]$$

$$= 1.151\left[\frac{(4275)^2 - (3988.62)^2}{727{,}272.73} - \log\left(\frac{8.95}{0.15 \times 0.0248 \times 1.71 \times 10^{-4} \times (0.29)^2}\right) + 3.23\right]$$

$$= -2.00$$

A summary of the results from the four methods is shown in Table 11.10. The results in Table 11.10 indicate that there are insignificant differences between the four methods for this example problem. However, prior work reported by Al-Hussainy and Ramey,[25] Meunier et al.,[26] Aziz et al.,[27] Reynolds et al.,[28] and Wattenbarger and Ramey[29] showed that use of pseudo-pressures or normalized pseudo-pressure provides more accurate analyses of gas drawdown test data. Note also for the four methods, semilog analyses were based on actual flow time instead of pseudo-time. The application of pseudo-pressure or normalized pseudo-pressure versus actual flow time is the recommended practice for analysis of gas well drawdown data.

11.4.2.5 Multi-Rate Drawdown Tests for Gas Wells

Multi-rate drawdown tests in gas wells are similar to the same type of tests in oil wells. As was the case for oil wells, these multi-rate tests are conducted in gas wells to overcome the difficulty of maintaining the rate constant in a single rate test. A typical multi-rate test is shown in Figure 11.10. As shown in Figure 11.10, flow at constant rate, q_1, starts at time, $t = 0$; increases to rate, q_2, at time, t_1; changes to rate, q_3, at time, t_2; and finally changes to rate, q_n, at time, t_{n-1}. Note that the flow rate is maintained constant at each flow period. The gas flow equation written in the form of normalized pseudo-pressures (Eq. (11.79) is used in the derivation of a general equation for calculation of the pressure at the well at any time period by applying the principle of superposition. To simplify the application of the superposition principle, Eq. 11.79 can be written in the following form:

$$p_{pni} - p_{pnwf} = m'_m q_g[\log(t) + b'_m] \tag{11.94}$$

where

$$m'_m = \frac{162.6\overline{B}\overline{\mu}}{kh} \tag{11.95}$$

and

$$b'_m = \log\left(\frac{k}{\phi\overline{\mu}\overline{c}_t r_w^2}\right) - 3.23 + 0.8691s' \tag{11.96}$$

Applying the superposition principle on the flow rates shown in Figure 11.10 for n rates gives:

$$\begin{aligned} p_{pni} - p_{pnwf} &= m'_m q_1[\log(t) + b'_m] + m'_m(q_2 - q_1)[\log(t - t_1) + b'_m] \\ &+ m'_m(q_3 - q_2)[\log(t - t_2) + b'_m] + \cdots\cdots \\ &\cdots\cdots + m'_m(q_n - q_{n-1})[\log(t - t_{n-1}) + b'_m] \end{aligned} \tag{11.97}$$

Eq. (11.97) can be written as:

$$\frac{p_{pni} - p_{pnwf}}{q_n} = m'_m \sum_{j=1}^{n}\left[\frac{(q_j - q_{j-1})}{q_n}\log(t_n - t_{j-1})\right] + m'_m b'_m \quad \text{for } q_n \neq 0 \tag{11.98}$$

Substituting for b'_m in Eq. (11.98) gives:

$$\begin{aligned} \frac{p_{pni} - p_{pnwf}}{q_n} &= m'_m \sum_{j=1}^{n}\left[\frac{(q_j - q_{j-1})}{q_n}\log(t_n - t_{j-1})\right] \\ &+ m'_m\left[\log\left(\frac{k}{\phi\mu c_t r_w^2}\right) - 3.23 + 0.8691s'\right] \end{aligned} \tag{11.99}$$

Eq. (11.99) is similar in form to Eq. (11.39) derived for multi-rate tests in oil wells. Note that Eq. (11.99) is valid only for an infinite-acting reservoir. Eq. (11.99) is in the form of the equation for a straight line. If Y_F is used to represent values on the y-axis and X_F for values on the x-axis, then Eq. (11.99) can be written as:

$$Y_F \equiv \frac{p_{pni} - p_{pnwf}}{q_n} \tag{11.100}$$

$$X_F \equiv \sum_{j=1}^{n}\left[\frac{(q_j - q_{j-1})}{q_n}\log(t_n - t_{j-1})\right] \tag{11.101}$$

$$slope, m'_m = \frac{162.6\overline{B}\overline{\mu}}{kh} \tag{11.102}$$

$$Y_{intercept} = b'_I = m'_m\left[\log\left(\frac{k}{\phi\overline{\mu}\overline{c}_t r_w^2}\right) - 3.23 + 0.8691s'\right] \tag{11.103}$$

A Cartesian plot of Y_F vs. X_F results in a straight line with slope, m'_m, and intercept, b'_I, if the reservoir is infinite-acting. This type of plot is sometimes called the rate-normalized plot or multi-rate superposition plot.[11] An expression for calculating skin factor, s', can be derived by re-arranging Eq. (11.103) to get:

$$s' = 1.151\left[\frac{b'_I}{m'_m} - \log\left(\frac{k}{\phi\overline{\mu}\overline{c}_t r_w^2}\right) + 3.23\right] \tag{11.104}$$

The procedures for applying the above equations to multi-rate gas flow tests are the same as were illustrated in Example 11.5 for multi-rate tests in oil wells.

11.4.2.6 Non-Darcy Flow Coefficient

In Eq. (11.73), the apparent skin factor, s', in gas flow equations was represented as $s' = s + Dq_g$, where s is the skin factor due formation damage or stimulation, D is the non-Darcy flow coefficient, and q_g is the gas flow rate. Non-Darcy flow coefficient is used to account for pressure loss due to non-Darcy (or turbulent) flow near the wellbore caused by high gas flow velocities. The effects of non-Darcy flow depend on gas flow rate as shown in the above expression for apparent skin factor. The apparent skin factor can be separated into its components parts by conducting two flow tests of equal duration (called isochronal tests) at different flow rates. This yields two equations which can be solved simultaneously for the skin factor and non-Darcy flow coefficient. This method is illustrated in Example 11.11. If only one flow-rate test is available, the non-Darcy coefficient can be estimated with the expression:[30]

$$D = \frac{2.715 \times 10^{-15} \beta k M p_{sc}}{h T_{sc} r_w \mu_{@p_{wf}}} \tag{11.105}$$

In Eq. (11.105, β = turbulence parameter; k = gas permeability, md; M = gas molecular weight, lbm/lb-mole; p_{sc} = pressure at standard conditions, psia; h = formation thickness, feet; T_{sc} = temperature at standard conditions, °R; r_w = wellbore radius, feet; and $\mu_{@p_{wf}}$ = gas viscosity, cp, evaluated at flowing BHP. The turbulence parameter, β, can be determined experimentally or estimated from correlations[31] as:

$$\ln \beta = -1.201 \ln k + 23.83 \tag{11.106}$$

The non-Darcy coefficient was assumed to be constant for the preceding presentation. However, studies performed by Wattenbarger et al.,[29] Firoozabadi et al.,[31] and Swift and Kiel[32] show that non-Darcy coefficient varies with gas flow rates. Rigorous procedures for calculating the variation of non-Darcy coefficients with rate have been published by Fligelman et al.[30] These procedures can be used in analyzing gas flow tests if warranted by flow conditions.

Example 11.11 Calculation of Non-Darcy Flow Coefficient from Isochronal Tests

Problem

An isochronal flow test was conducted on a well in a dry gas reservoir at two flow rates of 2000 Mcf/D and 4000 Mcf/D. The measured flowing BHP at the two flow rates are shown in Table 11.11. The properties of the reservoir gas are shown in Table 11.12. Calculate the formation gas permeability, skin factor, and non-Darcy flow coefficient of the well. Assume non-Darcy flow coefficient for the tests is constant. Other reservoir data are provided as follows:

Initial reservoir pressure, p_i	3000 psia
Reservoir temperature, T	121°F
Formation thickness, h	15 ft
Formation porosity, ϕ	0.20
Total compressibility, c_t	4.5×10^{-4} psi^{-1}
Wellbore radius, r_w	0.50 ft

Table 11.11 Isochronal Flow Test Data for Example 11.11

	Flow #1 q_1 = 2000 Mcf/D	Flow #2 q_2 = 4000 Mcf/D
Time, t (hrs.)	p_{wf} (psia)	p_{wf} (psia)
0.25	2,576	1,170
0.40	2,555	1,085
0.60	2,533	1,019
0.80	2,524	971
1.00	2,515	925
1.50	2,495	815
2.00	2,479	755
3.00	2,456	650
4.00	2,445	570
5.00	2,434	495
6.00	2,425	435
7.00	2,416	385

Solution

Step 1: Using gas properties in Table 11.12, calculate pseudo-pressures for the reservoir pressure range using Eq. (11.70). The results are shown in Table 11.13.

Step 2: Calculate pseudo-pressures for the two flow tests using Table 11.13. The results are shown in Table 11.14.

Table 11.12 Gas Properties for Example 11.11

Pressure, p (psia)	Gas Compressibility factor, z	Gas Viscosity, μ_g (cp)
500	0.9380	0.01200
1,000	0.8800	0.01300
1,500	0.8230	0.01413
2,000	0.7630	0.01600
2,500	0.7018	0.01859
3,000	0.6843	0.02086
3,500	0.6818	0.02285

Table 11.13 Calculation of Real Gas Pseudo-Pressure in Example 11.11

Pressure, p (psia)	Gas Compressibility factor, z	Gas Viscosity, μ_g (cp)	$2p/(\mu z)$	Δp	p_p
500	0.9380	0.01200	88,842	500	22,210,377
1,000	0.8800	0.01300	174,825	500	88,127,047
1,500	0.8230	0.01413	257,976	500	196,327,334
2,000	0.7630	0.01600	327,654	500	342,734,826
2,500	0.7018	0.01859	383,246	500	520,459,742
3,000	0.6843	0.02086	420,330	500	721,353,662
3,500	0.6818	0.02285	449,319	500	938,765,931

Step 3: For each flow rate, plot p_{pwf} vs. log t. The plot for the flow rate at 2000 Mcf/D is shown in Figure 11.21, and the plot for the 4000 Mcf/D is shown in Figure 11.22.

Step 4: Determine slope, m, of the MTR straight line for each plot. Calculate gas permeability at each flow rate.

From Figure 11.21, $m = 40 \times 10^6$ psia^2/cp-cycle for Flow #1 at 2000 Mcf/D.

Table 11.14 Pseudo-Pressures for Flow Test Data for Example 11.11

Time, t (hrs.)	Flow #1 q_1 = 2000 Mcf/D		Flow #2 q_2 = 4000 Mcf/D	
	p_{wf} (psia)	p_p (psia)^2/cp	p_{wf} (psia)	p_p (psia)^2/cp
0.25	2,576	544,393,512	1,170	130,336,695
0.40	2,555	536,431,402	1,085	113,051,234
0.60	2,533	528,147,966	1,019	100,238,733
0.80	2,524	524,776,341	971	91,255,016
1.00	2,515	521,414,618	925	82,909,917
1.50	2,495	513,979,573	815	64,003,115
2.00	2,479	508,066,744	755	54,313,768
3.00	2,456	499,621,880	650	38,416,368
4.00	2,445	495,605,892	570	27,208,683
5.00	2,434	491,604,696	495	17,412,033
6.00	2,425	488,341,992	435	10,069,810
7.00	2,416	485,089,190	385	4,287,467

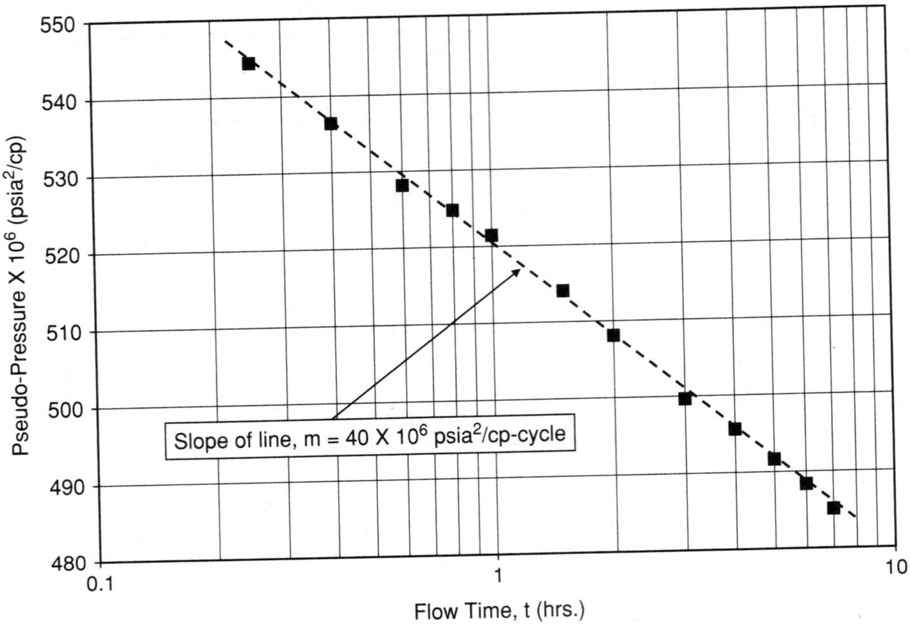

Figure 11.21 Plot of P_{pwf} vs. t for Flow #1 (2000 Mcf/D) in Example 11.11.

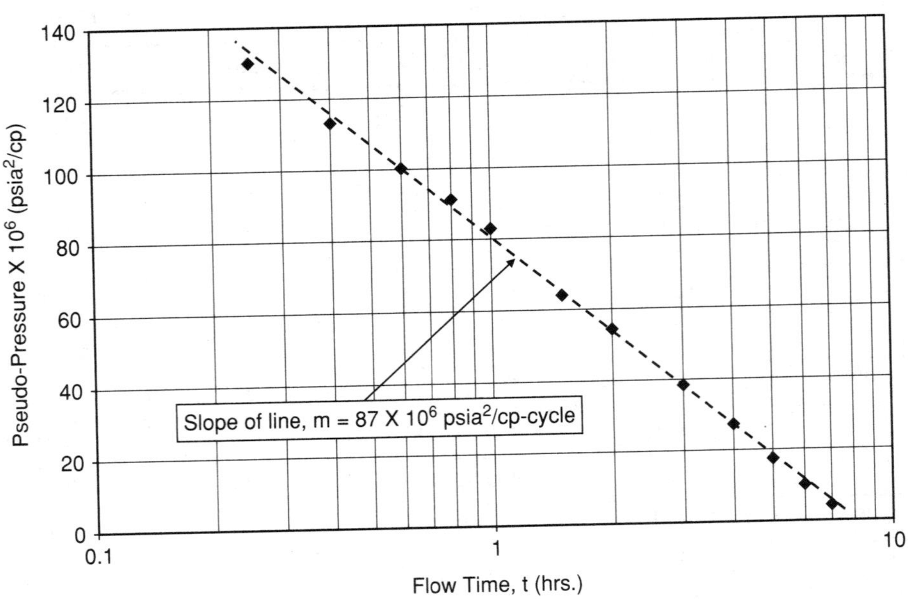

Figure 11.22 Plot of P_{pwf} vs. t for Flow #2 (4000 Mcf/D) in Example 11.11.

From Figure 11.22, $m = 87 \times 10^6$ psia^2/cp-cycle for Flow #2 at 4000 Mcf/D.
For Flow #1, using Eq. (11.86):

$$k = \frac{1637 q_g T}{mh} = \frac{1637 \times 2000 \times 581}{40 \times 10^6 \times 15}$$
$$= 3.17 \text{ md}.$$

For Flow #2, using Eq. (11.86):

$$k = \frac{1637 q_g T}{mh} = \frac{1637 \times 4000 \times 581}{87 \times 10^6 \times 15}$$
$$= 2.91 \text{ md}.$$

Step 5: Calculate apparent skin factor, s'.
For Flow #1, applying Eq. (11.87),

$$s' = 1.151\left[\frac{p_{pi} - p_{p,1hr}}{m} - \log\left(\frac{k}{\phi(\mu c_t)_i r_w^2}\right) + 3.23\right]$$
$$= 1.151\left[\frac{(721.4 - 520.0) \times 10^6}{40 \times 10^6} - \log\left(\frac{3.17}{0.20 \times 0.02086 \times 4.5 \times 10^{-4} \times (0.50)^2}\right) + 3.23\right]$$
$$= 1.6522$$

For Flow #2, applying Eq. (11.87),

$$s' = 1.151\left[\frac{p_{pi} - p_{p,1hr}}{m} - \log\left(\frac{k}{\phi(\mu c_t)_i r_w^2}\right) + 3.23\right]$$
$$= 1.151\left[\frac{(721.4 - 80.0) \times 10^6}{87 \times 10^6} - \log\left(\frac{2.91}{0.20 \times 0.02086 \times 4.5 \times 10^{-4} \times (0.50)^2}\right) + 3.23\right]$$
$$= 4.3853$$

Step 6: Determine skin factor, s, and non-Darcy coefficient, D.
Using Eq. (11.73), the equations for the two flow rates are:

Flow #1: $s + q_1 D = 1.6522$ or $s + 2000D = 1.6522$
Flow #2: $s + q_2 D = 4.3853$ or $s + 4000D = 4.3853$
Solving the two equations simultaneously gives: $s = -1.0809$, and $D = 1.37 \times 10^{-3}$ (Mcf/D)$^{-1}$

From the results of this test, it appears that the well is not damaged and all pressure loss near the wellbore is caused by turbulent or non-Darcy flow.

11.4.2.7 Buildup Tests for Gas Wells with Constant Production Rate Before Shut-In

The equations for analysis of buildup tests in gas wells with constant production rate before shut-in are similar in form to the equations developed previously for oil wells. The equations for buildup analyses in gas wells are developed initially in terms of pseudo-pressure. Other forms of the equations in terms of normalized pseudo-pressure, pressure, and pressure-squared are presented later.

From Eq. (11.74), the gas flow equation in terms of pseudo-pressure can be re-arranged as:

$$p_{pi} - p_{pwf} = \frac{1637 q_g T}{kh}\left[\log\frac{kt}{\phi(\mu c_t)_i r_w^2} - 3.23 + 0.8691 s'\right] \tag{11.107}$$

As shown in Figure 11.6, if a gas well was produced at constant rate, q_g, for time period, t_p, before it was shut in for the duration designated as Δt, the total pressure drop that occurred in the wellbore can be represented by application of the superposition principle in time as:

$$p_{pi} - p_{pws} = \Delta p_{p@q_g(t_p+\Delta t)} + \Delta p_{p@-q_g(\Delta t)} \tag{11.108}$$

In Eq. (11.108), p_{pws} is the shut-in pseudo-pressure measured over time, Δt. Substituting Eq. (11.107) into Eq. (11.108) yields:

$$\begin{aligned} p_{pi} - p_{pws} = {} & \frac{1637 q_g T}{kh}\left[\log\frac{k(t_p + \Delta t)}{\phi(\mu c_t)_i r_w^2} - 3.23 + 0.8691 s'\right] \\ & + \frac{1637(-q_g) T}{kh}\left[\log\frac{k(\Delta t)}{\phi(\mu c_t)_i r_w^2} - 3.23 + 0.8691 s'\right] \end{aligned} \tag{11.109}$$

Grouping and eliminating similar terms in Eq. (11.109) and re-arranging gives:

$$p_{pws} = p_{pi} - \frac{1637 q_g T}{kh}\left[\log\frac{(t_p + \Delta t)}{\Delta t}\right] \tag{11.110}$$

Eq. (11.110) is in the form of an equation for a straight line, $y = mx + b$ where:

$$y \equiv p_{pws} \tag{11.111}$$

$$x \equiv \log\frac{(t_p + \Delta t)}{\Delta t} \tag{11.112}$$

$$m = \left|\frac{1637 q_g T}{kh}\right| \tag{11.113}$$

$$b = p_{pi} \tag{11.114}$$

As shown by the above equations, a plot of p_{pws} vs. $\log\frac{t_p + \Delta t}{\Delta t}$ yields a straight line with slope, m, and intercept, p_{pi}, at infinite shut-in time.

The equation for calculating the apparent skin factor, s', for the well can be derived from Eq. (11.107) as:

$$s' = 1.151\left[\frac{p_{p,1\text{hr}} - p_{pwf}}{m} - \log\left(\frac{k}{\phi(\mu c_t)_i r_w^2}\right) + 3.23\right] \tag{11.115}$$

The analysis of the buildup test for a gas well is identical to that of an oil well by using the preceding equations and applying the procedures illustrated in Example 11.7.

The equations for analyses of buildup test in gas wells can be developed in terms of normalized pseudo-pressures. Starting with Eq. (11.79), the gas flow equation in terms of normalized pseudo-pressure can be re-arranged as:

$$p_{pni} - p_{pnwf} = \frac{162.6 q_g \overline{B}\,\overline{\mu}}{kh}\left[\log\frac{kt}{\phi\overline{\mu}\,\overline{c}_t r_w^2} - 3.23 + 0.8691 s'\right] \tag{11.116}$$

Applying the same procedures used previously for derivation of buildup equations in terms of pseudo-pressure, it can be shown that the corresponding equation in terms normalized pseudo-pressure is:

$$p_{pnws} = p_{pni} - \frac{162.6 q_g \overline{B}\,\overline{\mu}}{kh}\left[\log\frac{(t_p + \Delta t)}{\Delta t}\right] \tag{11.117}$$

A plot of p_{pnws} vs. $\log\left(\frac{t_p + \Delta t}{\Delta t}\right)$ yields a straight line with slope, m, and intercept, p_{pni}, at infinite shut-in time. The slope of the straight line, m, is represented as:

$$m = \left|\frac{162.6 q_g \overline{B}\,\overline{\mu}}{kh}\right| \tag{11.118}$$

The apparent skin factor, s', of the well is calculated from the expression:

$$s' = 1.151\left[\frac{p_{pn,1\mathrm{hr}} - p_{pnwf}}{m} - \log\left(\frac{k}{\phi(\mu c_t)_i r_w^2}\right) + 3.23\right] \tag{11.119}$$

Analysis of buildup data from gas wells in terms of pseudo-pressure and normalized pseudo-pressure have been presented by plotting p_{pws} or p_{pnws} vs. $\log\left(\frac{t_p + \Delta t}{\Delta t}\right)$. The ratio, $(t_p + \Delta t)/\Delta t$, is the standard Horner time ratio. Wattenbarger and Ramey[29] showed that buildup data on gas wells can be analyzed accurately by plotting p_{pws} or p_{pnws} vs. $\log\left(\frac{t_p + \Delta t}{\Delta t}\right)$ if production rate before shut-in is low. Production rate before shut-in is considered low if the dimensionless flow rate, $q_D \simeq 0.01$. If the production rate before shut-in is high, the semilog slope, m, based on the standard Horner time ratio could be approximately 10% higher than the "correct" slope. Production rate before shut-in is considered high if the dimensionless flow rate, $q_D \simeq 0.1$. According to Wattenbarger and Ramey,[29] dimensionless flow rate is defined as:

$$q_D = \frac{q_g p_{sc} T}{1.987 \times 10^{-5} kh T_{sc} p_{pi}} \tag{11.120}$$

If $p_{sc} = 14.7$ psia and $T_{sc} = 520°\mathrm{R}$ ($T_{sc} = 60°\mathrm{F}$), Eq. (11.120) becomes:

$$q_D = \frac{1422.7 q_g T}{kh p_{pi}} \tag{11.121}$$

Reynolds et al.[28] recommended application of two types of Horner time ratios that could be used to achieve accurate analysis of gas well buildup data. The first ratio is based on normalized shut-in time and represented as:

$$\text{Horner Time Ratio} = \frac{(t_p + \Delta t)/(\mu c_t)_i}{\Delta t/(\mu c_t)} \tag{11.122}$$

In Eq. (11.122), $(\mu c_t)_i$ is evaluated at initial pressure, p_i, or average pressure, $\overline{p}$. The term in the denominator, (μc_t) is evaluated at the shut-in pressure, p_{ws} that is measured at shut-in time, Δt. The second ratio is based on shut-in pseudo-time and expressed as:

$$\text{Horner Time Ratio} = \frac{(t_p + \Delta t)/(\mu c_t)_i}{\Delta t_A} \tag{11.123}$$

Eqs. (11.122) and (11.123) yield almost identical results, although it is easier in terms of computation to apply Eq. (11.122).

Other equations for the analyses of gas well buildup data can be expressed in terms of pressure or pressure-squared from Eqs. (11.82) and (11.85), respectively. In terms of pressure, a plot of p_{ws} vs. $\log\left(\frac{t_p + \Delta t}{\Delta t}\right)$ yields a straight line with slope, m. The slope of the straight line is represented as:

$$m = \left|\frac{162.6 q_g \overline{B}\,\overline{\mu}}{kh}\right| \tag{11.124}$$

The apparent skin factor, s', is determined from the equation:

$$s' = 1.151\left[\frac{p_{1\text{hr}} - p_{wf}}{m} - \log\left(\frac{k}{\phi(\mu c_t)_i r_w^2}\right) + 3.23\right] \tag{11.125}$$

In terms of pressure-squared, a plot of p_{ws}^2 vs. $\log\left(\frac{t_p + \Delta t}{\Delta t}\right)$ gives a straight line with slope, m. The slope of the straight line is expressed as:

$$m = \left|\frac{1637 q_g T \overline{\mu}\,\overline{z}}{kh}\right| \tag{11.126}$$

The apparent skin factor, s', for the well is calculated from the equation:

$$s' = 1.151\left[\frac{p_{1\text{hr}}^2 - p_{wf}^2}{m} - \log\left(\frac{k}{\phi(\mu c_t)_i r_w^2}\right) + 3.23\right] \tag{11.127}$$

Analyses of gas well buildup data using the pressure or the pressure-squared technique is not recommended. Studies performed by Al-Hussainy and Ramey,[25] Meunier et al.,[26] Aziz et al.,[27] Reynolds et al.,[28] and Wattenbarger and Ramey[29] showed that use of the pseudo-pressure or normalized pseudo-pressure techniques yield more accurate analyses of gas buildup data in comparison to either pressure or pressure-squared methods. The application of pseudo-pressure or normalized pseudo-pressure is the practice recommended for analyses of gas well buildup data, irrespective of the gas flow rate before shut-in.

11.4.2.8 Horner's Time Approximation

The equations developed previously for the analysis of buildup tests in gas wells assumed that the wells were produced at constant rates before shut-in. As was discussed previously for oil wells, it is even more difficult to maintain the production rate of a gas well constant over extended time intervals. Following the approach developed for oil wells by Horner,[2] an approximation of total producing time for gas wells, also termed Horner's pseudo-producing time, t_{pH}, is defined as:

$$t_{pH} = \frac{24G_p}{q_{g,last}} \tag{11.128}$$

In Eq. (11.128), G_p = cumulative gas production before shut-in, Mscf; and $q_{g,last}$ = gas production rate just before shut-in, Mcf/D. In practice, q_{last} should be maintained constant and stabilized as long as possible before starting the buildup phase of the test. Note that by definition, Horner's pseudo-producing time, t_{pH}, is different from t_p used in previous equations such as Eq. (11.110). This t_p assumes a constant producing rate over the entire history of the well. However, t_{pH} can be used to replace t_p and q_{last} in place of q in Eq. (11.110) to yield:

$$p_{pws} = p_{pi} - \frac{1637 q_{g,last} T}{kh}\left[\log\frac{(t_{pH} + \Delta t)}{\Delta t}\right] \tag{11.129}$$

A plot of p_{pws} vs. $\log\frac{t_{pH} + \Delta t}{\Delta t}$ gives a straight line with slope, m, and intercept equal to p_{pi} at infinite shut-in time. This is the same as the Horner plot. The slope of the Horner plot, m, is given as:

$$m = \left|\frac{1637 q_{g,last} T}{kh}\right| \tag{11.130}$$

All other equations derived previously for the ideal Horner plot can be modified accordingly as shown for the pseudo-pressure case with the Horner pseudo-producing time.

Example 11.12 Calculation of Permeability, Reservoir Pressure, and Skin Factor from Gas Well Buildup Tests Using Horner Pseudo-Producing Time Approximation

Problem

A buildup test was conducted on a discovery well in a dry gas reservoir. The production history of the well before the test is shown in Table 11.15. The data taken during the buildup test are

Table 11.15 Production History of Gas Well in Example 11.12

Flow Time (Days)	Production Rate (Mcf/D)
0.5	3500
0.5	0
0.5	4500
0.5	0
0.5	9600
0.5	0
3.0	9000

shown in Table 11.16 as shut-in pressures versus time. The gas PVT data of the reservoir are the same as in Example 11.11. Other reservoir data are presented as follows:

Reservoir temperature, T	121 °F
Formation thickness, h	30 ft
Formation porosity, ϕ	0.20
Total compressibility, c_t	4.5×10^{-4} psi^{-1}
Wellbore radius, r_w	0.50 ft

Table 11.16 Buildup Data for Example 11.12

Δt (hrs.)	p_{ws}(psia)
0	2130
0.25	2540
0.5	2580
1	2630
2	2680
4	2740
8	2798
12	2830
24	2880
36	2909
48	2925
72	2948
96	2960

Solution

Step 1: Calculate Horner's pseudo-time, t_{pH}.
Using Eq. (11.128) and production history in Table 11.15:

$$t_{pH} = \frac{24G_p}{q_{g,\,last}}$$
$$= \frac{(3500 \times 0.5 + 0 \times 0.5 + 4500 \times 0.5 + 0 \times 0.5 + 9600 \times 0.5 + 0 \times 0.5 + 9000 \times 3) \times 24}{9000}$$
$$= 95.47 \text{ hrs}.$$

Step 2: Plot p_{pws} vs. $\log\left(\frac{t_{pH} + \Delta t}{\Delta t}\right)$ from data in Table 11.17. Determine slope of MTR line.

From Figure 11.23, slope $m = 85.72 \times 10^6 \text{psia}^2/\text{cp-cycle}$.

Step 3: Calculate formation permeability, k, and apparent skin factor, s'.
Using Eq. (11.130), formation permeability, k, is:

$$k = \left|\frac{1637 q_{g,last} T}{mh}\right|$$
$$= \frac{1637 \times 9000 \times 581}{85.72 \times 10^6 \times 30}$$
$$= 3.33 \text{ md}.$$

Table 11.17 Plot Data for Figure 11.23 of Example 11.12

$\frac{(t_{pH} + \Delta t)}{\Delta t}$	$p_{pws} \times 10^6$ (psia)^2/cp
382.88	530.78
191.94	545.92
96.47	565.12
48.74	584.62
24.87	608.43
12.93	631.86
8.96	644.97
4.98	665.69
3.65	677.86
2.99	684.61
2.33	694.37
1.99	699.49

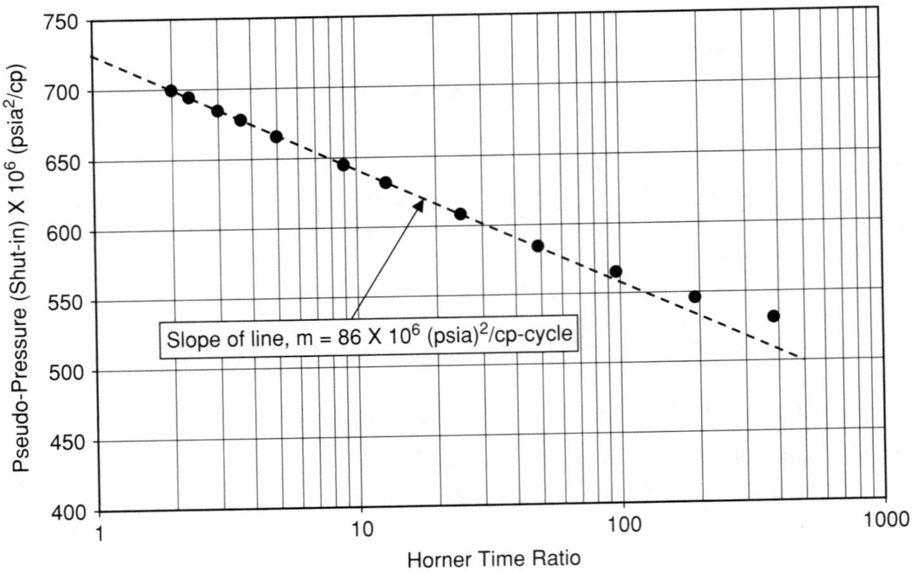

Figure 11.23 Plot of p_{pws} vs. $\{(t_{pH} + \Delta t)/\Delta t\}$ for Example 11.12.

From Eq. (11.115), apparent skin factor, s', is calculated as:

$$s' = 1.151\left[\frac{p_{p,1\text{hr}} - p_{pwf}}{m} - \log\left(\frac{k}{\phi(\mu c_t)_i r_w^2}\right) + 3.23\right]$$

$$= 1.151\left[\frac{(559.52 - 386.88) \times 10^6}{85.72 \times 10^6} - \log\left(\frac{3.33}{0.20 \times 0.02086 \times 4.5 \times 10^{-4} \times (0.5)^2}\right) + 3.23\right]$$

$$= -1.85$$

Step 4: Calculate initial reservoir pressure, p_i.

From Figure 11.23, by extrapolation of the MTR straight line to Horner time ratio =1, the intercept is . This translates to initial reservoir pressure, $p_i = 2993$ psia.

Alternative Method: Using Normalized Horner Time Ratio

Step 1: Same as in previous method.

Step 2: Plot p_{pws} vs. $\log\dfrac{(t_{pH} + \Delta t)/(\mu c_t)_i}{\Delta t/(\mu c_t)}$ from data in Table 11.18. Determine slope of MTR line.

From Figure 11.24, slope, $m = 82.43 \times 10^6 \text{psia}^2/\text{cp-cycle}$.

Step 3: Calculate formation permeability, k, and apparent skin factor, s'.

Using Eq. (11.130), formation permeability, k, is:

$$k = \left|\frac{1637 q_{g,last} T}{mh}\right|$$

$$= \frac{1637 \times 9000 \times 581}{82.43 \times 10^6 \times 30}$$

$$= 3.46 \text{ md.}$$

Table 11.18 Plot Data for Figure 11.24 of Example 11.12

$\frac{(t_{pH} + \Delta t)/(\mu c_t)_i}{\Delta t/(\mu c_t)}$	$p_{pws} \times 10^6$ (psia)^2/cp
405.86	530.78
202.18	545.92
100.85	565.12
50.59	584.62
25.61	608.43
13.22	631.86
9.12	644.97
5.04	665.69
3.69	677.86
3.01	684.61
2.34	694.37
2.00	699.49

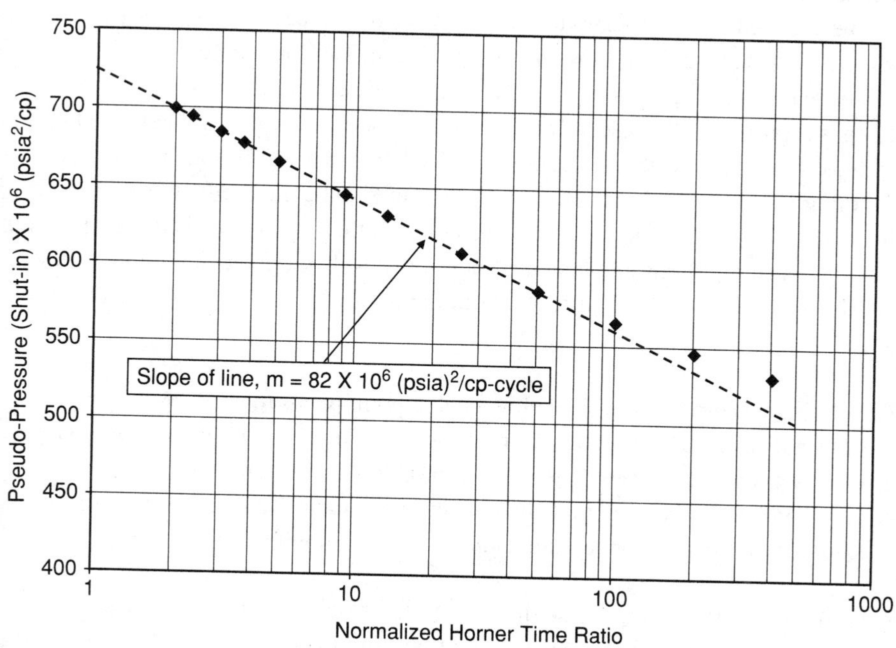

Figure 11.24 Plot of p_{pws} vs. $\{(t_{pH} + \Delta t)/(\mu c_t)_i\}/(\Delta t/\mu c_t)$ for Example 11.12.

From Eq. (11.115), apparent skin factor, s' is calculated as:

$$s' = 1.151\left[\frac{p_{p,1hr} - p_{pwf}}{m} - \log\left(\frac{k}{\phi(\mu c_t)_i r_w^2}\right) + 3.23\right]$$
$$= 1.151\left[\frac{(559.46 - 386.88) \times 10^6}{82.43 \times 10^6} - \log\left(\frac{3.46}{0.20 \times 0.02086 \times 4.5 \times 10^{-4} \times (0.5)^2}\right) + 3.23\right]$$
$$= -1.78$$

Step 4: Calculate initial reservoir pressure, p_i.
From Figure 11.24, by extrapolation of the MTR straight line to Horner time ratio =1, the intercept is $p_{pi} = 718 \times 10^6 (\text{psia})^2/\text{cp}$. This translates to initial reservoir pressure, $p_i = 2993$ psia.

For this example, using the standard Horner time ratio or the normalized Horner time ratio gave almost identical results. However, the normalized Horner time ratio was shown by Reynolds et al.[28] to give more accurate results. The application of normalized Horner time ratio based on Eq. (11.122) or Eq. (11.123) is recommended.

11.5 Special Topics in Well Test Analyses

11.5.1 Multiphase Flow

All the equations developed previously in this chapter for analyses of drawdown and buildup tests have been based on the assumption that only a single phase fluid (oil or gas) was flowing in the reservoir. This could be the case if, for an oil reservoir, the pressure conditions during the test were above the bubble point pressure. This single phase condition also applies to a gas reservoir if dry gas is present in the reservoir or for a condensate gas reservoir where the flowing pressure conditions are outside the retrograde condensation region. In either situation described above for oil or gas reservoirs, connate water is assumed to be immobile in the reservoirs. When more than one fluid phase is flowing in the reservoir at the same time, then multiphase interactions occur between the fluids and the equations developed assuming single-phase flow are rendered invalid, and are no longer applicable. Multiphase flow occurs in reservoirs under following conditions:

1. Solution gas drive reservoirs below bubble point pressure with gas and/or water saturations above critical saturations. That is, gas and/or water are mobile in the reservoir.
2. Oil reservoirs above bubble point with mobile water saturations.
3. Gas condensate reservoirs with condensate and/or water saturations above critical saturations. That is, condensate and/or water are mobile in the reservoir.

If a reservoir is under multiphase flow, the flow capacity of each fluid phase is reduced due to relative permeability effects since the flow of any fluid phase impedes the flow of the other fluid phases. As discussed in Chapter 2, the relative permeability of a fluid is defined as the ratio of its effective permeability to the absolute permeability of the porous medium. This can be expressed for any fluid as:

$$k_{ri} = \frac{k_i}{k} \tag{11.131}$$

In Eq. (11.131), k_{ri} = relative permeability of the fluid, i; k_i = effective permeability of the fluid, i; and k = absolute permeability of the porous medium. For instance, if the fluid is oil, the relative permeability of oil is given by $k_{ro} = k_o/k$. As discussed in Chapter 2, relative permeability is a function of fluid saturation. Hence, effective permeability also depends on saturation. In a well test, the distribution of fluid saturations changes with time. Thus, effective permeability, which affects flow, also changes with time. Furthermore, mobility of a fluid in a porous medium is defined as the ratio of its effective permeability to its viscosity. In equation form, this expressed as:

$$\lambda_i = \frac{k_i}{\mu_i} \tag{11.132}$$

In Eq. (11.132), λ_i = mobility of the fluid, i; and μ_i = viscosity of the fluid, i. If the fluid is oil, the mobility of oil is expressed as $\lambda_o = k_o/\mu_o$. A fluid with high mobility flows faster than a fluid with less mobility. Since fluid mobilities depend on effective permeability which varies with saturation and time, it is important to consider multiphase flow in well test analyses.

Multiphase flow well test analysis is presented on the basis of an approximate method developed by Perrine[33] and later verified theoretically by Martin.[34] This approach uses the concepts of total mobility, λ_t, and total compressibility, c_t. Total mobility is defined as:

$$\lambda_t = \frac{k_o}{\mu_o} + \frac{k_w}{\mu_w} + \frac{k_g}{\mu_g} \tag{11.133}$$

Total compressibility is defined as:

$$c_t = c_o S_o + c_w S_w + c_g S_g + c_f \tag{11.134}$$

where

$$c_o = \left(-\frac{1}{B_o}\frac{dB_o}{dp}\right) + \frac{B_g}{1000 B_o}\frac{dR_s}{dp} \tag{11.135}$$

and

$$c_w = \left(-\frac{1}{B_w}\frac{dB_w}{dp}\right) + \frac{B_g}{1000 B_w}\frac{dR_{sw}}{dp} \tag{11.136}$$

In Eqs. (11.135) and (11.136), the units of B_g are RB/Mcf. The total flow rate, q_t, under reservoir conditions is defined thus:

$$(qB)_t = q_o B_o + q_w B_w + \left(q_g - \frac{q_o R_s}{1000} - \frac{q_w R_{sw}}{1000}\right) B_g \tag{11.137}$$

Application of the Perrine's method to multiphase well test analyses makes the following assumptions:

1. Pressure gradients in the drainage area of the well are small.
2. Saturation gradients in the same area are small.
3. Saturation changes during the test are negligible.

Assuming the above conditions are applicable, well test analyses equations and methods developed earlier for single phase fluid systems are applicable to multiphase fluid systems. Thus, to analyze test data from a well producing two or three phases at the same time, prepare a plot of the test data as was described earlier for a single phase system. For instance, for a buildup test, prepare a plot of p_{pws} vs. $\log\dfrac{(t_p + \Delta t)/(\mu c_t)_i}{\Delta t/(\mu c_t)}$ as shown in Example 11.12. From the plot, determine the slope, m, of the MTR straight line. From the slope, calculate the effective permeabilities of each fluid phase as follows:

For oil:

$$k_o = \frac{162.6 q_o B_o \mu_o}{mh} \tag{11.138}$$

For water:

$$k_w = \frac{162.6 q_w B_w \mu_w}{mh} \tag{11.139}$$

For gas:

$$k_g = \frac{162.6[q_g - (q_o R_s/1000) - (q_w R_{sw}/1000)] B_g \mu_g}{mh} \tag{11.140}$$

The total mobility, λ_t, is calculated from:

$$\lambda_t = \frac{162.6(qB)_t}{mh} \tag{11.141}$$

The absolute permeability, k, can be calculated from total mobility as:

$$k = \frac{\lambda_t}{\left[\dfrac{k_{ro}}{\mu_o} + \dfrac{k_{rw}}{\mu_w} + \dfrac{k_{rg}}{\mu_g}\right]} \tag{11.142}$$

The skin factor, s, is calculated in terms of total mobility from:

$$s = 1.151\left[\frac{p_{1\text{hr}} - p_{wf}}{m} - \log\frac{\lambda_t}{\phi c_t r_w^2} + 3.23\right] \tag{11.143}$$

Perrine's method for analysis of multiphase flow tests is simple and straight-forward but approximate. There are other methods that have been proposed by Raghavan[35,36] and Al-Khalifah et al.[37] that claim to be improvements over Perrine's method. These other methods can be useful depending on the state of the reservoir. However, Perrine's method is widely used in most applications of well test analysis.

11.5.2 Wellbore Storage Effects

The concept of wellbore storage was discussed earlier in this chapter as the unloading of wellbore fluids at the start of drawdown tests or the continued flow of reservoir fluids into the wellbore (called afterflow) after a well has been shut in at the start of buildup tests. The main effects of wellbore storage on well test data are the distortion of early time data and delay of the start of middle time region data. For wells with high wellbore storage coefficient, most of the MTR data could be lost due to wellbore storage effects. A critical step in the application of straightline methods in well test analysis is determination of the time at which wellbore storage effects have ended. This time marks the beginning of MTR data which in turn establishes the slope of the MTR straight line. Since this step is subject to the interpretation of the data by the analyst, it is reasonable to expect that different analysts will pick different points as the start of the MTR depending on the quality of the well test data being analyzed. This was a persistent problem in the application of straightline methods until the introduction of type curves in the analysis of well test data. In Chapter 12, the application of type curves as tools to assist in the identification of end of wellbore storage effects are presented. The joint application of straightline methods and type curves improved the interpretation of well test data.[9] It should be noted that in all the examples shown in this chapter, the start of MTR straight line appear to have been "magically" determined without further discussion on the procedures that could be used to select the end of wellbore storage. The procedures for determination of the end of wellbore storage are discussed in Chapter 12 in conjunction with the application of type curves for well test analysis.

11.5.3 Wellbore Phase Redistribution Effects

Another wellbore phenomenon which is closely tied to wellbore storage is wellbore phase redistribution effects. Wellbore phase redistribution effects occur in wells with multiphase flow of gas and liquid. When such a well is shut in for buildup tests, gravity effects cause the gas to rise towards the top of the wellbore and the liquid to fall towards the bottom of the wellbore.[38–40] Since the liquid phase is almost incompressible, and the gas has no room to expand in such a closed system, a high-pressure gas column develops at the top of the wellbore with a column of liquid below it. In some cases, the sum of the pressure due to the gas column and the pressure due to the liquid column will exceed the formation pressure at the sandface and cause liquid to flow back into the formation until equilibrium is eventually established. At early times, this phenomenon shows up in buildup data as an anomalous pressure hump as shown in Figure 11.25. In Figure 11.25, pressure curve A represents normal wellbore storage which affects only ETR portion of the buildup test. Pressure curve B represents a buildup test in which ETR and MTR are completely distorted by phase redistribution in the wellbore. If the pressure hump caused by phase redistribution is large, it will delay and sometime distort the MTR data completely. This makes it difficult to determine properties such as formation permeability and skin factor from the buildup data.

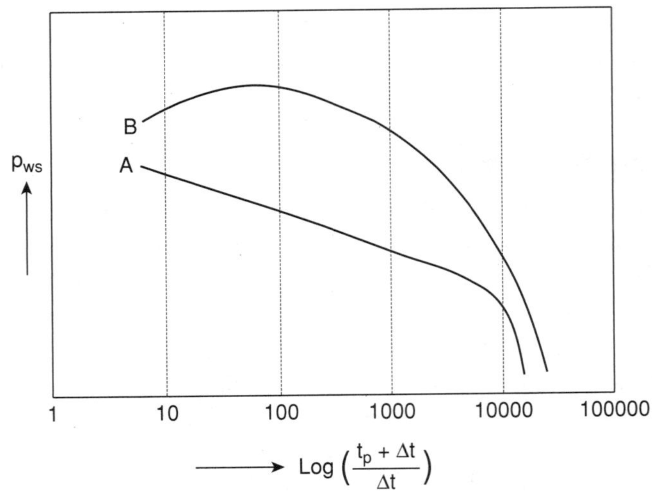

Figure 11.25 Pressure hump caused by phase redistribution in the wellbore.

11.5.4 Boundary Effects

The flow equations that form the bases of drawdown and buildup analyses assume an infinite-acting reservoir. This assumption does not apply in some tests. Depending on the size of the drainage area of the well, the presence of a boundary can influence the pressure response in the wellbore. The boundary can be a no-flow boundary (or closed boundary) created by a sealing fault or impermeable barriers. The boundary can also be a constant pressure boundary imposed by an aquifer, a large gas cap, or fluid injectors. These boundary effects generally terminate the MTR straight line and generate characteristic shapes of the pressure response curves. One of the major advances in well test analyses is the use of derivative type curves to identify specific types of boundary effects. These characteristic type curves are presented in Chapter 12.

11.5.5 Multilayered Reservoirs

It is necessary to restate that all equations developed previously in this chapter for the analyses of drawdown and buildup tests are based on the flow of a single phase fluid in a **single layer** with homogeneous properties. Most reservoirs consist of more than one layer and in many cases the rock and fluid properties of the layers are very different. In this section, some observations on the analyses of well test data from multilayered reservoirs are presented. The discussions on multilayered reservoirs proceed by defining two main reservoir systems commonly used in describing these reservoirs. These are commingled reservoirs and crossflow reservoirs. Commingled reservoirs are multilayered reservoirs in which fluid flow between layers takes place only through the wellbore (Figure 11.26). Essentially, the vertical permeabilities across layer boundaries are zero for commingled reservoirs. Crossflow reservoirs are multilayered reservoirs in which fluid flow can occur between layers within the reservoir (Figure 11.27). In crossflow

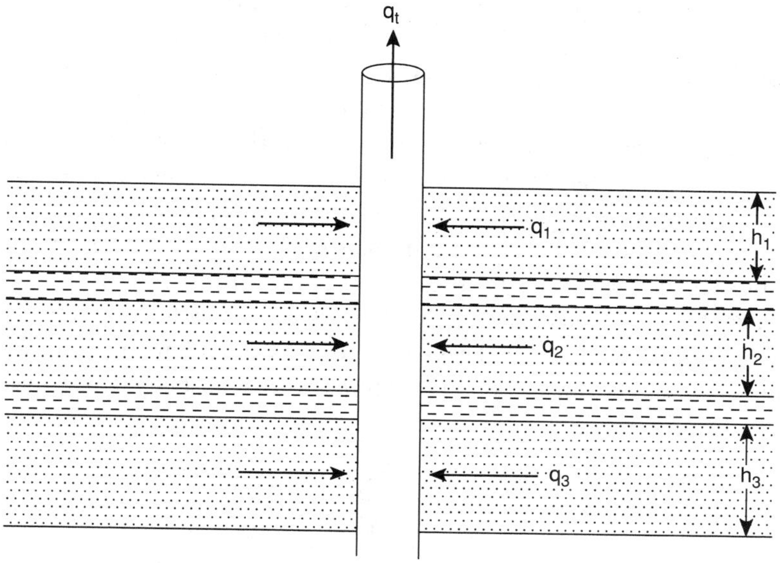

Figure 11.26 Multilayered commingled reservoirs.

reservoirs, interlayer communication or fluid transfer between the layers exist inside the reservoir and also through the wellbore. Note that the term crossflow as used here is different from common industry usage of crossflow as fluid transfer between layers only through the wellbore. This is sometimes described as backflow.

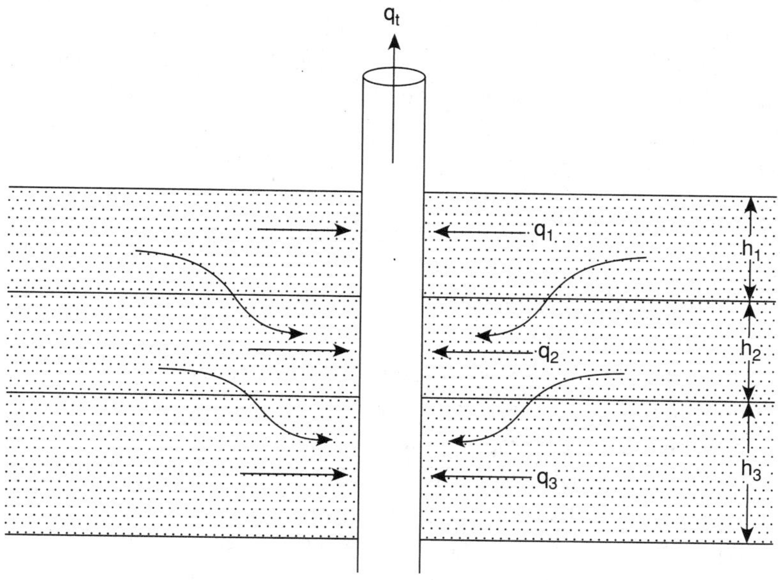

Figure 11.27 Multilayered crossflow reservoirs.

The pressure response of commingled and crossflow reservoirs can be examined by comparing the transmissivities (kh/μ) and storativities ($\phi c_t h$) of the composite layers. If the transmissivities and storativities of the composite layers are nearly the same, the reservoirs will exhibit homogeneous behavior. On the other hand, if these properties are remarkably different, the reservoirs will exhibit heterogeneous behavior. The derivative type curve from a test on a reservoir that shows homogeneous behavior is identical to that of single layer reservoir. For a reservoir that shows heterogeneous behavior, the derivative type curve has a characteristic shape. The application of derivative type curves as diagnostic plots for well test analysis of heterogeneous reservoirs are presented in Chapter 12.

Numerous studies[41–45] on commingled reservoirs concluded that conventional straightline methods can be used to determine thickness-averaged formation flow capacity, $\overline{k}h$, and equivalent total skin factor, s_t. The thickness-averaged formation permeability, $\overline{k}$ is defined as:

$$\overline{k} = \frac{\sum_{j=1}^{n} k_j h_j}{\sum_{j=1}^{n} h_j} \tag{11.144}$$

Tariq[46] showed that the equivalent total skin factor of well producing a commingled reservoir with different skin factor for each layer can be approximated by the following expression:

$$s_t = \sum_{j=1}^{n} \frac{k_j h_j s_j}{\overline{k}h} \tag{11.145}$$

Published studies[47–51] on crossflow reservoirs show that early-time well response of such reservoirs are identical to that of comparably located commingled reservoirs. Russell and Prats[47] reported that a multilayered crossflow reservoir will ultimately behave as a single layer reservoir with equivalent pore volume and drainage area. Furthermore, the total flow capacity is equal to the sum of the flow capacities of the composite layers (same as Eq. (11.144)) for commingled reservoirs. The equivalent total skin factor for crossflow reservoirs can be estimated as reported by Prijambodo et al.[50] as:

$$s_t = \sum \frac{s_j q_j}{q_t} \tag{11.146}$$

In Eq. (11.146), s_j and q_j are the skin factor and flow rate for layer j, respectively, and q_t is the total flow rate.

The important decision to note on the testing of multilayered reservoir is whether to measure the total flow rate of all the layers or the flow rate of each layer. If total flow rate is measured, then only average reservoir properties for the multilayer system can be determined from the well test. Testing individual layers will yield determination of individual layer properties such as layer permeability and skin factor. Many studies[52–54] have been published on methods to test and analyze data from individual layers in a multilayer system. These methods are generally expensive but could be justified if the results will lead to improved and accelerated depletion of multilayered reservoirs.

Nomenclature

A_{wb}	area of wellbore, feet^2
B	formation volume factor, RB/STB
c	compressibility, psi^{-1}
c_f	formation compressibility, psi^{-1}
c_t	total compressibility, psi^{-1}
c_{wb}	compressibility of wellbore fluid, psi^{-1}
C	wellbore storage coefficient, bbl/psi
C_D	dimensionless wellbore storage coefficient
D	non-Darcy flow coefficient, D/Mcf
E	flow efficiency
f_{pf}	pressure function of the drainage area as defined by Matthews et al.[15]
G_p	cumulative gas production, Mcf
h	reservoir or formation thickness, feet
k	permeability, md
$\bar{k}$	thickness-averaged formation permeability, md
k_a	permeability of the altered zone, md
k_i	effective permeability of fluid, i
k_{ri}	relative permeability of fluid, i
L_f	length of fracture wing, feet
m	slope of MTR straight line
M_g	molecular weight of gas, lbm/lb-mole
N_p	cumulative oil production, STB
Δp	pressure change
Δp_s	pressure change due to skin, psia
p^*	extrapolated pressure in buildup test, psia
p	pressure, psia
$\bar{p}$	average pressure, psia
p_a	pressure at boundary of damaged zone, psia
p_b	low base pressure or reference pressure
p_p	real gas pseudo-pressure, psia^2/cp
p_{pn}	normalized gas pseudo-pressure, psia
p_{pnwf}	normalized flowing bottomhole pseudo-pressure, psia
p_{pnws}	normalized shut-in bottomhole pseudo-pressure, psia
p'_{wf}	flowing bottom-hole pressure of undamaged zone
p_{wf}	well bottom-hole flowing pressure
p_{pws}	shut-in bottomhole pseudo-pressure, psia^2/cp
p_{ws}	shut-in bottomhole pressure, psia
q	volume flow rate, STB/D

q_D	dimensionless flow rate
q_{last}	production rate before shut-in, STB/D
$q_{g,last}$	gas production rate before shut-in, Mcf/D
r	radius or distance, feet
r_a	radius of damaged zone, feet
r_i	radius of investigation, feet
r_w	wellbore radius, feet
r_{wa}	effective wellbore radius, feet
R_s	solution gas-oil ratio, scf/STB
R_{sw}	solution gas-water ratio, scf/STB
s	skin factor
s'	apparent skin factor
S	fluid saturation
Δt	shut-in time in buildup test, hours
t	time, hours
t_A	real gas pseudo-time, hr-psi/cp
t_{An}	normalized real gas pseudo-time, hr
t_p	total producing time before shut-in, hours
t_{pH}	Horner's pseudo-producing time, hours
T	temperature, °F or °R
V	volume, feet^3 or barrels
V_{wb}	volume of wellbore, barrels
z	gas compressibility factor
γ_g	gas gravity, air $= 1$
η	formation diffusivity, ft^2/D
ϕ	porosity
μ	fluid viscosity, cp
ρ_{wb}	density of fluid in wellbore, lbm/ft^3
β	turbulence parameter
λ	mobility, md/cp

Subscripts

g	gas
f	formation
i	initial conditions
o	oil
sc	standard conditions
t	total
w	water

Abbreviations

BHP	Bottom Hole Pressure
DR	Damage Ratio
ETR	Early Time Region
FVF	Formation Volume Factor
IARF	Infinite Acting Radial Flow
LTR	Late Time Region
MTR	Middle Time Region

References

1. Miller, C.C., Dyes, A.B., and Hutchinson, C.A.: "The Estimation of Permeability and Reservoir Pressure From Bottom Hole Pressure Build-up Characteristics," *Trans.* AIME (1950) 189, 91–104.
2. Horner, D.R.: "Pressure Buildup in Wells," Pressure Analysis Methods, Reprint Series, SPE, Richardson, Texas (1967) 9, 25–43.
3. Ramey, H.J.: "Short-Time Well Test Data Interpretation in the Presence of Skin Effect and Wellbore Storage," *JPT* (January 1970) 97–104.
4. Gringarten, A.C., Bourdet, D.P., Landel, P.A., and Kniazeff, V.J.: "A Comparison Between Different Skin and Wellbore Storage Type-Curves for Early-Time Transient Analysis," Paper SPE 8205 presented at the SPE Annual Technical Conference and Exhibition, Las Vegas, Nevada, September 23–26, 1979.
5. Bourdet, D., Whittle, T.M., Douglas, A.A., and Pirard, Y.M.: "A New Set of Type Curves Simplifies Well Test Analysis," *World Oil* (May 1983) 95–106.
6. Bourdet, D., Ayoub, J.A., Whittle, T.M., Pirard, Y.M., and Kniazeff, V.: "Interpreting Well Tests in Fractured Reservoirs," *World Oil* (October 1983) 77–87.
7. Bourdet, D., Ayoub, J.A., and Pirard, Y.M.: "Use of Pressure Derivative in Well-Test Interpretation," *SPEFE* (June 1989) 293–302.
8. von Schroeter, T., Hollaender, F., and Gringarten, A.C.: "Deconvolution of Well-Test Data as a Nonlinear Total Least-Squares Problem," *SPEJ* (December 2004) 375–390.
9. Gringarten, A.C.: "From Straight Lines to Deconvolution: The Evolution of the State of the Art in Well Test Analysis," *SPEREE* (February 2008) 41–62.
10. van Everdingen, A.F.: "The Skin Effect and its Influence on the Productive Capacity of a Well," *Trans.* AIME (1953) 198, 171–176.
11. Horne, R.N.: *Modern Well Test Analysis—A Computer-Aided Approach*, 2nd Edition, Petroway, Inc., California (2008).
12. Tiab, D., Ispas, I.N., Mongi, A., and Berkat, A.: "Interpretation of Multirate Tests by the Pressure Derivative—1. Oil Reservoirs," Paper SPE 53935 presented at the 1999 SPE Latin

American and Caribbean Petroleum Engineering Conference, Caracas, Venezuela, April 21–23, 1999.

13. Russell, D.G.: "Determination of Formation Characteristics From Two-Rate Flow Tests," *JPT* (December 1963) 1347–1355.

14. Odeh, A.S., and Jones, L.G.: "Two-Rate Flow Test, Variable-Rate Case—Application to Gas-Lift and Pumping Wells," *JPT* (January 1974) 93–99.

15. Matthews, C.S., Brons, F., and Hazebroek, P.: A Method for Determination of Average Pressure in a Bounded Reservoir," *Trans.* AIME (1954) 201, 182–191.

16. Muskat, M.: "Use of Data on the Build-up of Bottom-Hole Pressures," *Trans.* AIME (1937) 123, 44–48.

17. Larson, V.C.: "Understanding the Muskat Method of Analyzing Pressure Buildup Curves," *J. Cdn. Pet. Tech.* (Fall 1963) 2, 136.

18. Crump, J.G., and Hite, R.H.: "A New Method for Estimating Average Reservoir Pressure: The Muskat Plot Revisited," *SPEREE* (April 2008) 298–306.

19. Dietz, D.N.: "Determination of Average Reservoir Pressure from Buildup Surveys," *JPT* (1965) 955–959.

20. Mead, H.N.: "A Practical Approach to Transient Pressure Behavior," Paper SPE 9901 presented at the SPE California Regional Meeting, Bakersfield, California, March 25–27, 1981.

21. Kabir, C.S., and Hasan, A.R.: "Estimating Average Reservoir Pressure Using the Hyperbola Approach: New Algorithm and Field Examples," Paper SPE 36255 presented at the SPE Abu Dhabi International Petroleum Exhibition and Conference, Abu Dhabi, United Arab Emirates, October 13–16, 1996.

22. Chacon, A., Djebrouni, A., and Tiab, D.: "Determining the Average Reservoir Pressure from Vertical and Horizontal Well Test Analysis Using the Tiab's Direct Synthesis Technique," Paper SPE 88619 presented at the SPE Asia Pacific Oil and Gas Conference and Exhibition, Perth, Australia, October 18–20, 2004.

23. Al-Hussainy, R., Ramey, H.J., Jr., and Crawford, P.R.: "The Flow of Real Gases Through Porous Media," *JPT* (May 1966) 624–636.

24. Agarwal, Ram G.: "Real Gas Pseudo-Time—A New Function for Pressure Buildup Analysis of MHF Gas Wells," Paper SPE 8279 presented at the 54th Annual Fall Technical Conference and Exhibition of SPE, Las Vegas, Nevada, September 23–26, 1979.

25. Al-Hussainy, R., and Ramey, H.J., Jr.: "Application of Real Gas Flow Theory to Well Testing and Deliverability Forecasting," *JPT* (May 1966) 637–642.

26. Meunier, D.F., Kabir, C.S., and Wittmann, M.J.: "Gas Well Test Analysis: Use of Normalized Pseudovariables" *SPEFE* (December 1987) 629–636.

27. Aziz, K., Mattar, L., Ko, S., and Brar, G.S.: "Use of Pressure, Pressure-Squared or Pseudo-Pressure in the Analysis of Transient Pressure Drawdown Data from Gas Wells," *J. Cdn. Pet. Tech.* (April–June 1976) 58–65.

28. Reynolds, A.C., Bratvold, R.B., and Ding, W.: "Semilog Analysis of Gas Well Drawdown and Buildup Data," *SPEFE* (December 1987) 657–670.

29. Wattenbarger, R.A., and Ramey, H.J.: "Gas Well Testing with Turbulence, Damage and Wellbore Storage," *JPT* (August 1968) 877–887.

30. Fligelman, H., Cinco-Ley, H., Ramey Jr., H.J., Braester, C., and Couri, F.: "Pressure-Drawdown Test Analysis of a Gas Well-Application of New Correlations," *SPEFE* (September 1989) 406–412.

31. Firoozabadi, A., and Katz, D.L.: "An Analysis of High Velocity Gas Flow Through Porous Media," *JPT* (February 1979) 211–216.

32. Swift, G.W., and Kiel, O.G.: "The Prediction of Gas Well Performance Including the Effect of Non-Darcy Flow," *JPT* (July 1962) 791–798.

33. Perrine, R.L.: "Analysis of Pressure Buildup Curves," *Drill. and Prod. Prac.,* API (1956), 482–509.

34. Martin, J.C.: "Simplified Equations of Flow in Gas Drive Reservoirs and the Theoretical Foundation of Multiphase Pressure Buildup Analyses," *Trans.* AIME (1959) 216, 321–323.

35. Raghavan, R.: "Well Test Analysis: Wells Producing by Solution Gas Drive," *SPEJ* (August 1976) 196–208.

36. Raghavan, R.: "Well Test Analysis for Multiphase Flow," *SPEFE* (December 1989) 585–594.

37. Al-Khalifah, A-J.A., Aziz, K., and Horne, R.N.: "A New Approach to Multiphase Well Test Analysis," paper SPE 16473 presented at the 62nd Annual SPE Technical Conference and Exhibition, Dallas, Texas, September 27–30, 1987.

38. Stegemeier, G.L., and Mathews, C.S.: "A Study of Anomalous Pressure Build-up Behavior," *Trans.* AIME (1958) 213, 44–50.

39. Fair, W.B.: "Pressure Buildup Analysis with Wellbore Phase Redistribution," *SPEJ* (April 1981) 259–270.

40. Hegeman, P.S., Hallford, D.L., and Joseph, J.A.: "Well Test Analysis with Changing Wellbore Storage," *SPEFE* (September 1993) 201–207.

41. Lefkovits, H.C., Hazebroek, P., Allen, E.E., and Matthews, C.S.: "A Study of the Behavior of Bounded Reservoirs Composed of Stratified Layers," *SPEJ* (March 1961) 43–58.

42. Cobb, W.M., Ramey, H.J. Jr., and Miller, F.G.: "Well-Test Analysis for Wells Producing Commingled Zones," *JPT* (January 1972) 27–37.

43. Raghavan, R., Topaloglu, H.N., Cobb, W.M., and Ramey, H.J. Jr.: "Well Test Analysis for Wells Producing From Two Commingled Zones of Unequal Thickness," *JPT* (September 1974) 1035–1043

44. Earlougher, R.C. Jr., Kersch, K.M., and Kunzman, W.J.: "Some Characteristics of Pressure Buildup Behavior in Bounded Multiple-Layer Reservoirs Without Crossflow," *JPT* (October 1974) 1178–1186.

45. Raghavan, R.: "Behavior of Wells Completed in Multiple Producing Zones," *SPEFE* (June 1989) 219–230.

46. Tariq, S.M.: "A Study of the Behavior of Layered Reservoirs with Wellbore Storage and Skin Effect," PhD dissertation, Stanford University, Stanford, California (September 1977).

47. Russell, D.G., and Prats, M.: "Performance of Layered Reservoirs with Crossflow-Single-Compressible-Fluid Case," *SPEJ* (March 1962) 53–67.

48. Katz, M.L., and Tek, M.R.: "A Theoretical Study of Pressure Distribution and Fluid Flux in Bounded Stratified Porous Systems with Crossflow," *SPEJ* (March 1962) 68–82.

49. Pendergrass, J.D., and Berry, V.J. Jr.: "Pressure Transient Performance of a Multilayered Reservoir with Crossflow," *SPEJ* (December 1962) 347–354.

50. Prijambodo, R., Raghavan, R., and Reynolds, A.C.: "Well Test Analysis for Wells Producing Layered Reservoirs with Crossflow," *SPEJ* (June 1985) 380–396.

51. Bourdet, D.: "Pressure Behavior of Layered Reservoirs with Crossflow," paper SPE 13628 presented at the SPE California Regional Meeting, Bakersfield, California, March 27–29, 1985.

52. Dogru, A.H., and Seinfeld, J.H.: "Design of Well Tests to Determine the Properties of Stratified Reservoirs," paper SPE 7694 presented at the SPE Symposium on Reservoir Simulation, Denver, Colorado, February 1–2, 1979.

53. Kuchuk, F., Karakas, M., and Ayestaran, L.: "Well Testing and Analysis Techniques for Layered Reservoirs," *SPEFE* (August 1986) 342–354.

54. Ehlig-Economides, C.A., and Joseph, J.: "A New Test for Determination of Individual Layer Properties in a Multilayered Reservoir," *SPEFE* (September 1987) 261–283.

General Reading

- Earlougher, R.C.: *Advances in Well Test Analysis*, SPE Monograph Volume 5, SPE, Richardson, Texas (1977).
- Lee, J., Rollins, J.B., and Spivey, J.P.: *Pressure Transient Testing*, SPE Textbook Series Vol. 9, SPE, Richardson, Texas (2003).

CHAPTER 12

Well Test Analysis: Type Curves

12.1 Introduction

In Chapter 11, well test analysis with straightline methods were presented. In that chapter, the difficulties associated with correctly identifying the time region over which straightline methods are applicable were discussed. In some cases, engineers analyzing the same well test data could identify different times for the start of the infinite-acting radial flow (IARF) regime, which is crucial to the application of straightline methods for calculation of well and reservoir properties. Well test analysis was improved tremendously by combined application of straightline methods and type curves on the same well test data. In this chapter, representative type curves, such as Ramey type curve, Gringarten type curve, and Bourdet derivative type curve are presented. The chapter concludes with a manual illustration of the application of type curves in well test analysis.

12.2 What Are Type Curves?

Type curves are graphic plots of theoretical solutions to flow equations under specific initial and boundary conditions of the interpretation model representing a reservoir-well system.[1] The most common forms of type curves are presented in terms of dimensionless pressure versus dimensionless time. The use of type curves for well test analysis was introduced by Ramey[2] in 1970. Ramey type curves[3] for a single well with skin and wellbore storage in an infinite system is shown in Figure 12.1. Numerous type curves for various well and reservoir boundary conditions were subsequently published by other authors. The best known of these type curves are the McKinley[4] type curves for wellbore storage effects, type curves for a well with infinite conductivity vertical fracture by Gringarten et al.,[5] and type curves for a well with a finite conductivity vertical fracture by Cinco-Ley et al.[6] The application of type curves for well test analysis was advanced by Gringarten et al.[7] with the introduction of a unified set of type curves shown in

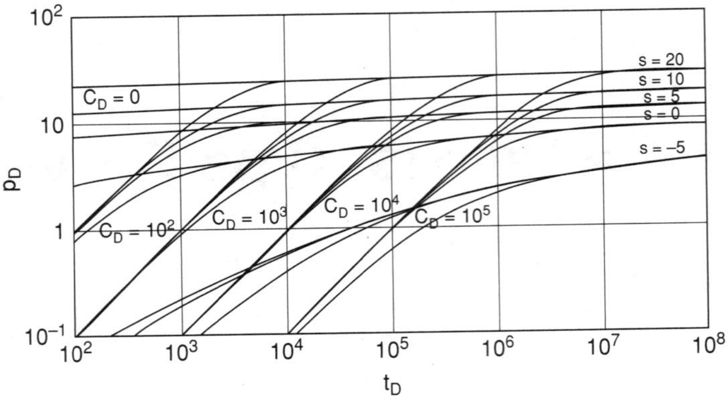

Figure 12.1 Type curves for a single well with skin and wellbore storage in an infinite reservoir (from Agarwal et al.[3] © 1970 SPE, Reproduced with permission).

Figure 12.2. The most important advancement that revolutionalized the application of type curves for well test analysis began in 1983 with the introduction of derivative type curves by Bourdet et al.[8–10] The Bourdet derivative type curves are shown in Figure 12.3. The combined application of Gringarten type curves and Bourdet derivative type curves (Figure 12.4) in a single plot for well test analysis has enabled identification of reservoir behaviors and geologic features which were previously undetectable from well test data using only straightline methods.[11]

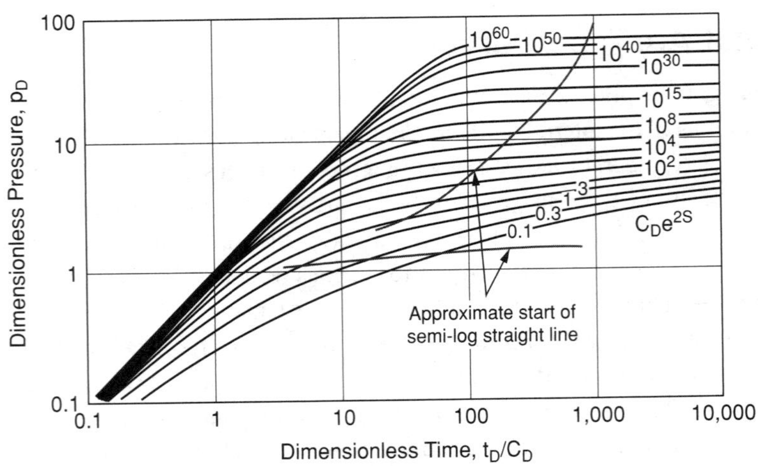

Figure 12.2 Type curves for a single well with skin and wellbore storage in an infinite reservoir (from Gringarten et al.[7] © 1979 SPE, Reproduced with permission).

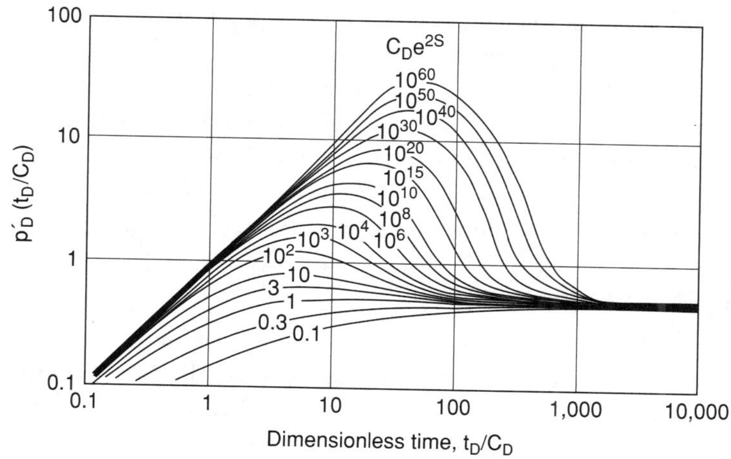

Figure 12.3 Bourdet derivative type curves (from Bourdet et al.[8]).

12.3 Gringarten Type Curves

The Gringarten[7] type curves were derived from solutions to the diffusivity equation of a slightly compressible fluid produced from a well with skin and wellbore storage in an infinite-acting reservoir exhibiting homogeneous behavior. In Gringarten type curves (Figure 12.2), dimensionless pressure, p_D, is plotted against dimensionless variables, t_D/C_D, with C_De^{2s} as the correlating parameter. The range of C_De^{2s} as the correlating parameter includes damaged and stimulated wells. Also shown in Figure 12.2 are two lines denoting the loci of approximate start

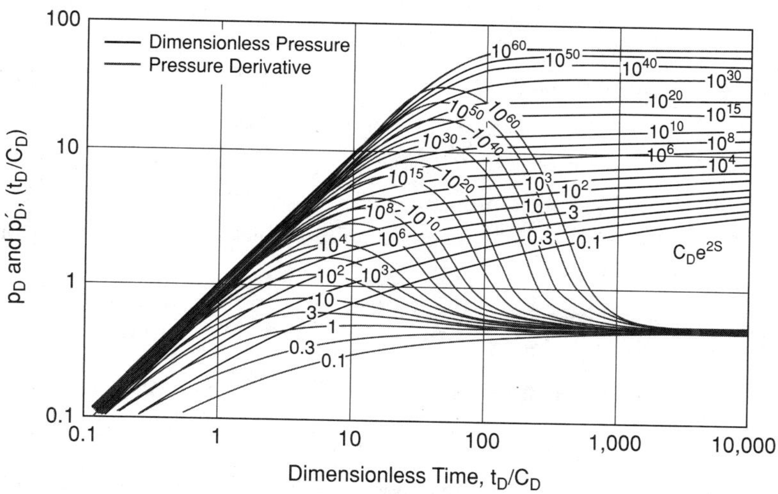

Figure 12.4 Combined Gringarten type curves and Bourdet derivative type curves.

of radial flow for an infinite-acting reservoir. This is the approximate location of the middle time region (MTR) or more appropriately the location of the infinite-acting radial flow (IARF) regime. The MTR or IARF regime on the Gringarten type curves corresponds to the MTR straight line region identified on semi-log plots using straightline methods as shown in Chapter 11. Consequently, Gringarten type curves can be used with straightline methods to determine more accurately the start of IARF flow regime. This diagnostic application of Gringarten type curves was further enhanced by the introduction of derivative type curves as discussed later.

12.3.1 Unit-Slope Line

At early times, fluid flow is dominated by wellbore storage. During this flow regime, the type curves merge into a straight line whose slope is equal to unity. This line is termed the unit-slope line. Note that some of the Gringarten type curves with low values of C_De^{2s} have slopes which are less than one. This indicates stimulation, either by acidization or hydraulic fracturing. All points on the unit slope line imply that:

$$p_D = \frac{t_D}{C_D} \tag{12.1}$$

In Eq. (12.1), p_D is the dimensionless pressure defined as:

$$p_D = \frac{kh(p_i - p)}{141.2qB\mu} \tag{12.2}$$

For Eq. (12.2), p = pressure, psia; k = permeability, md; h = formation thickness, ft; q = flow rate, STB/D; B = formation volume factor, RB/STB; and μ = viscosity, cp.
Also in Eq. (12.1), t_D is the dimensionless time defined as:

$$t_D = \frac{0.0002637kt}{\phi\mu c_t r_w^2} \tag{12.3}$$

In Eq. (12.3), k = permeability, md; t = time, hrs.; ϕ = porosity, fraction; μ = viscosity, cp; c_t = total compressibility, psi^{-1}; and r_w = wellbore radius, ft.
Finally in Eq. (12.1), C_D is the dimensionless wellbore storage coefficient defined as:

$$C_D = \frac{0.894C}{\phi c_t h r_w^2} \tag{12.4}$$

In Eq. (12.4), the units of all terms are as previously defined for Eqs. (12.2) and (12.3). The term, C, is the wellbore storage coefficient in bbl/psi which can be calculated for a rising or falling fluid level in the wellbore as:

$$C = \frac{25.65A_{wb}}{\rho_{wb}} \tag{12.5}$$

If the wellbore is filled with a single phase fluid, then C can be calculated as:

$$C = V_{wb}c_{wb} \tag{12.6}$$

In Eqs. (12.5) and (12.6), A_{wb} = wellbore area, ft^2; ρ_{wb} = density of wellbore fluid, lbm/ft^3; V_{wb} = volume of fluid in wellbore, ft^3; and c_{wb} = compressibility of wellbore fluid, psi^{-1}.

From Eq. (12.1):

$$C_D = \frac{t_D}{p_D} \tag{12.7}$$

Substituting Eq. (12.7) with Eqs. (12.2) and (12.3), it can be shown that for any point on the unit-slope line:

$$C_D = \frac{0.03723qB}{\phi c_t h r_w^2}\left(\frac{\Delta t}{\Delta p}\right)_{USL} \tag{12.8}$$

Furthermore, substituting Eq. (12.4) into Eq. (12.8) gives:

$$C = 0.04165qB\left(\frac{\Delta t}{\Delta p}\right)_{USL} \tag{12.9}$$

Equations (12.8) and (12.9) can be applied by making a log-log plot of Δp vs. Δt for any test data. If a unit-slope line exists for the test data, any point on the line can be used to determine $(\Delta t/\Delta p)_{USL}$.

12.4 Bourdet Derivative Type Curves

The Bourdet[8–10] derivative type curves were developed from pressure derivatives of the analytical solutions of the same flow equations used in the generation of the Gringarten type curves. Taking derivatives of Eq. (12.1) gives:

$$\frac{dp_D}{d(t_D/C_D)} = p'_D = 1 \tag{12.10}$$

Also, differentiating dimensionless pressure with respect to logarithm of dimensionless time yields:

$$\begin{aligned}\frac{dp_D}{[d\ln(t_D/C_D)]} &= (t_D/C_D)\{dp_D/[d(t_D/C_D)]\} \\ &= (t_D/C_D)p'_D\end{aligned} \tag{12.11}$$

Combining Eqs. (12.10) and (12.11) gives:

$$\left(\frac{t_D}{C_D}\right)p'_D = \frac{t_D}{C_D} \tag{12.12}$$

As can be determined from Eq. (12.12), a plot of $(t_D/C_D)p'_D$ vs. (t_D/C_D) gives a line with slope equal to unity within the early time region (ETR). Thus, at early times when wellbore storage is dominant, the Bourdet derivative type curves coincides with the unit-slope lines of the Gringarten type curves for damaged wellbores. Stimulated wellbores have low values of the correlating parameter, C_De^{2s}, which indicate stimulation, either by acidization or hydraulic fracturing.

The line-source solution of the diffusivity equation for slightly compressible fluids (Eq. 11.24) can be transformed in terms of dimensionless variables into:

$$p_D = 0.5[\ln t_D + 0.80907 + 2s] \tag{12.13}$$

By introducing the term, $\ln C_D$, into Eq. (12.13), it can be derived that:

$$p_D = 0.5\left[\ln (t_D/C_D) + 0.80907 + \ln\left(C_D e^{2s}\right)\right] \tag{12.14}$$

Taking derivative of Eq. (12.14) with respect to (t_D/C_D) gives:

$$\frac{dp_D}{d(t_D/C_D)} = p'_D = \frac{0.5}{(t_D/C_D)} \tag{12.15}$$

Thus, when infinite-acting radial flow (IARF) has been reached, the Bourdet derivative type curves become horizontal at constant value of $p'_D(t_D/C_D) = 0.5$. Thus for Bourdet derivative type curves, there are two asymptotes defined by the unit-slope line and the horizontal line at a constant value of 0.5. Between these two asymptotes, the Bourdet derivative type curves exhibit characteristic shapes which depend on values of the correlating parameter, $C_D e^{2s}$. The Bourdet derivative type curves are shown in Figure 12.3 as a log-log plot of $p'_D(t_D/C_D)$ vs. (t_D/C_D) with $C_D e^{2s}$ as the correlating parameter. In practice, the Bourdet derivative type curves are combined with Gringarten type curves in a single plot as shown in Figure 12.4. When both type curves are used together in type-curve matching of test data, it helps to eliminate most of the non-uniqueness of a type-curve match that may occur if only the Gringarten type curves are used. Further, the Bourdet derivative type curves have been used to identify characteristic response shapes representing various conditions of well models, reservoir models, and reservoir boundaries. These have led to significant advances in the interpretation of well test data. With the aid of derivative type curves, it is possible to identify fracture response in wells, double porosity behavior in naturally fractured reservoirs, and boundary effects in reservoirs. A selection of some of the characteristic shapes of derivative type curves for various well, reservoir, and boundary models are shown in Appendix 12A.

12.5 Agarwal Equivalent Time

The Gringarten type curves and Bourdet derivative type curves are based on analytical solutions to the diffusivity equations of slightly compressible fluids at constant production rates. Essentially, these are pressure drawdown type curves. As shown in Figure 12.5, drawdown and buildup type curves are not identical because buildup type curves are affected by the duration of the producing time, t_p, before shut-in. If the duration of the producing time, t_p, before shut-in is relatively short, the deviation between drawdown and buildup type curves tends to be more pronounced. This could lead to erroneous interpretation of buildup tests with drawdown type curves. To reduce the effects of producing time before shut-in and allow the use of drawdown type curves

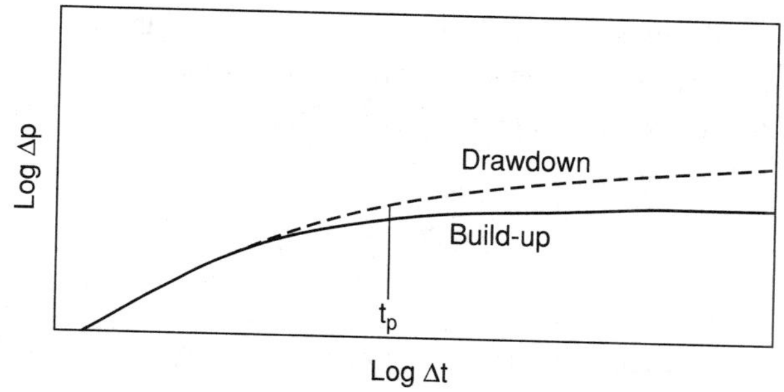

Figure 12.5 Drawdown and buildup type curves comparison.

for analysis of buildup tests, Agarwal[12] proposed an equivalent drawdown time which is called Agarwal equivalent time. Agarwal[12] equivalent time, Δt_e, is defined as:

$$\Delta t_e = \frac{t_p \Delta t}{t_p + \Delta t} \tag{12.16}$$

Agarwal equivalent time is strictly applicable to infinite-acting radial flow reservoirs with no wellbore storage. However, it has been shown to be applicable to wells with wellbore storage and fracture effects.[12]

12.6 Type-Curve Matching

Type-curve matching is the methodical process of superimposing a plot of the test data on a set of type curves on the same scale and using the characteristic shapes of the type curves to identify an interpretation model for the test data. The process of matching a plot of the test data to a type curve can be done manually as demonstrated in Example 12.1 or by using graph fitting techniques based on regression analysis as implemented in most commercial software for well test analysis. The identification of an interpretation model (or type curve) for the test data during type-curve matching was enhanced considerably with the introduction of derivative type curves. With application of derivative curves, an interpretation model can be selected that encompasses the entire flow regimes of the test data. This means that by selecting an interpretation model that fits the test data, the interpretation model also includes a well model that represents flow regimes at the early times (ETR), a reservoir model for the flow regime at the intermediate times (MTR), and a boundary model for the late times (LTR). Thus, the test data can be analyzed at various flow regimes if the flow regimes are present. Note that selection of an interpretation model during type-curve matching is a non-unique solution because it is possible to find another interpretation model that can match the test data. In well test analysis using type-curve matching, it is important to include a validation step that checks the results calculated from type curves against other methods such as straightline methods.[7,11] The validation step may incorporate data from other external sources such as geological, petrophysical, and seismic data.

Once a match has been achieved with a type curve, well and reservoir properties corresponding to the test data can be calculated from the dimensionless parameters of the type curve. From the unit-slope line, it was shown previously in Eq. (12.8) that dimensionless wellbore storage coefficient can be calculated as:

$$C_D = \frac{0.03723qB}{\phi c_t h r_w^2}\left(\frac{\Delta t}{\Delta p}\right)_{USL}$$

Similarly, wellbore storage coefficient can be determined from Eq. (12.9) as:

$$C = 0.04165qB\left(\frac{\Delta t}{\Delta p}\right)_{USL}$$

Starting with Eq. (12.3), it can be derived that at the Time Match Point (TMP) along the x-axis, the dimensionless wellbore storage coefficient can be calculated from:

$$C_D = \frac{0.0002637k}{\phi\mu c_t r_w^2}\left[\frac{\Delta t}{t_D/C_D}\right]_{TMP} \qquad (12.17)$$

Note that it is common practice to check the values for C_D calculated from Eqs. (12.8) and (12.17) for consistency.

By rearranging Eq. (12.2), formation permeability, k, at the Pressure Match Point (PMP) along the y-axis can be calculated from the expression:

$$k = \frac{141.2qB\mu}{h}\left[\frac{p_D}{\Delta p}\right]_{PMP} \qquad (12.18)$$

As part of the validation process, the formation permeability calculated from Eq. (12.18) should be compared to the formation permeability calculated from straightline methods for the same test data.

From the correlating parameter, $C_D e^{2s}$, for the type curve, the skin factor for the test data can be calculated from the relationship:

$$s = 0.5\ln\left(\frac{C_D e^{2s}}{C_D}\right) \qquad (12.19)$$

Also, as part of the validation process, the skin factor calculated from Eq. (12.19) should be compared to the skin factor calculated from straightline methods.

12.7 Procedures for Manual Application of Type-Curve Matching in Well Test Analysis

The procedures presented here apply only to manual use of type-curve matching in analysis of well test data. Since the advent of computer-aided well test analysis,[13–15] type-curve matching is generally accomplished in all commercial software through automatic fitting of the test data using linear or non-linear regression techniques. Most engineers may never have to manually plot the test data and match them against a set of type curves. However, these procedures are presented here to enable engineers to gain thorough understanding of the processes that have been automated in computer-aided well test analysis. The procedures presented here are based on the use of Gringarten-Bourdet combined type curves as shown in Figure 12.4. The procedures for manual type-curve matching are as follows:

1. Prepare the test data for analysis by tabulating pressure change versus time. For drawdown test data, tabulate pressure change as $\Delta p = (p_i \text{ or } \overline{p}) - p_{wf}$ versus flowing time, Δt. For buildup test data, tabulate pressure change as $\Delta p = (p_{ws} - p_{wf@\Delta t=0})$ versus equivalent shut-in time, Δt_e, calculated from Eq. (12.16). Plot Δp vs. Δt or Δt_e.
2. If a unit slope line is present on the plot of the test data, select any point $(\Delta p, \Delta t \text{ or } \Delta t_e)_{USL}$ on the unit slope line. Calculate dimensionless wellbore storage coefficient from Eq. (12.8) as:

$$C_D = \frac{0.03723qB}{\phi c_t h r_w^2}\left(\frac{\Delta t \text{ or } \Delta t_e}{\Delta p}\right)_{USL}$$

From Eq. (12.9), calculate wellbore storage coefficient as:

$$C = 0.04165qB\left(\frac{\Delta t \text{ or } \Delta t_e}{\Delta p}\right)_{USL}$$

3. For the derivative curve, calculate derivative of pressure with respect to natural logarithm of time. For drawdown test data, calculate the derivative as:

$$-\frac{dp_{wf}}{d(\ln t)} = -t\left[\frac{dp_{wf}}{dt}\right] = t\Delta p' \qquad \textbf{(12.20)}$$

For buildup test data, calculate the derivative as:

$$\frac{dp_{ws}}{d(\ln \Delta t_e)} = \Delta t_e\left[\frac{dp_{ws}}{d(\Delta t_e)}\right] = \Delta t_e \Delta p' \qquad \textbf{(12.21)}$$

Plot $t\Delta p'$(or $\Delta t_e \Delta p'$) versus Δt (or Δt_e) on the same graph plotted in Step 1. An algorithm for calculation of the derivatives is provided in Appendix 12C.
4. Select a set of Gringarten-Bourdet type curves on the same scale as the graph plotted in Steps 1 and 2. Note that it is important that the two graphs are on the same scale.
5. Superimpose the plot of the test data on the Gringarten-Bourdet type curves. Achieve a match in the vertical direction by aligning the IARF stabilization line of the test data with similar line on the type curve indicated by $p'_D(t_D/C_D) = 0.5$.
6. If a unit slope line is present on the plot of the test data, achieve a match in the horizontal direction by aligning the unit-slope line of the test data with the unit slope line of the type curve.

7. From the pressure match point $(p_D, \Delta p)_{PMP}$ in Step 5, calculate formation permeability from Eq. (12.18) as:

$$k = \frac{141.2qB\mu}{h}\left[\frac{p_D}{\Delta p}\right]_{PMP}$$

8. From the time match point in Step 6, use the match point $(t_D/C_D, \Delta t \text{ or } \Delta t_e)_{TMP}$ to calculate dimensionless wellbore storage coefficient from Eq. (12.17) as:

$$C_D = \frac{0.0002637k}{\phi\mu c_t r_w^2}\left[\frac{\Delta t \text{ or } \Delta t_e}{t_D/C_D}\right]_{TMP}$$

Compare the values of C_D calculated from Steps 2 and 8. If inconsistent, repeat match and re-calculate.

9. Using the correlating parameter, C_De^{2s}, of the match type curve and value for C_D from Step 8, calculate skin factor for the test data from Eq. (12.19) as:

$$s = 0.5 \ln\left(\frac{C_De^{2s}}{C_D}\right)$$

10. Use straightline methods presented in Chapter 11 to prepare a plot of p_{wf} vs. $\log t$ for drawdown data, and p_{ws} vs. $\log(t_p + \Delta t/\Delta t)$ for buildup data. Determine formation permeability from the time region defined by the stabilization line identified with the derivative type curve in Step 4. Also, calculate skin factor from this MTR line. Compare results from type curve matching and straightline methods. If results are inconsistent, repeat the entire process until reasonable consistency is achieved.

12.8 Stages of the Type-Curve Matching Procedures

The procedures presented earlier for type-curve matching can be grouped into three stages as suggested by Gringarten.[11] These stages are:

1. Identification of interpretation model (Steps 1, 2, 3, 4, 5, 6).
2. Calculation from interpretation model parameters (Steps 7, 8, 9).
3. Validation of the interpretation model (Step 10).

12.8.1 Identification of the Interpretation Model

The key step in well test analysis is the identification of the interpretation model based on the characteristic shapes of Bourdet derivative type curves. A sample of these characteristic shapes is shown in Appendix 12A. In ETR, the characteristic shape may indicate a damaged well with wellbore storage and skin or a fractured well with no wellbore storage. During the MTR flow regime, the characteristic shape of the derivative curve may indicate double porosity reservoir or double permeability reservoir. In the LTR, reservoir boundary effects may indicate a sealing fault or a constant pressure boundary. Each of these characteristic shapes may indicate an interpretation model that could assist in the complete analysis of the test data. However, it is important to note that none of these characteristic shapes is unique to a particular well, reservoir, or boundary model. A characteristic shape of the derivative type curve may potentially represent two or more well, reservoir, or boundary responses that are similar in appearance. For instance, the characteristic shape of a double porosity (fractured) reservoir closely resembles the response of a double permeability (multilayered) reservoir. Consequently, it is often necessary to incorporate external data from geological, geophysical, and petrophysical sources to assist in the identification of an interpretation model.

12.8.2 Calculation from Interpretation Model Parameters

After an interpretation model has been selected and fitted to the test data either by type-curve matching or regression analysis, the reservoir parameters can be calculated from the interpretation model parameters using equations developed in this chapter and Chapter 13. The results from the analysis are independent of calculation methods as long as the calculations are based on the same interpretation model.[11] Consequently, calculation methods such as straightline methods and type-curve matching should yield the same result as long as they are based on the same interpretation model.

12.8.3 Validation of the Interpretation Model Results

It is important to validate the results calculated with the interpretation model. This could be done by comparing the results calculated from type-curve matching for instance, to results calculated from straightline methods based on the same interpretation model. If the results are consistent, then the interpretation model should be accepted as being representative of the test data at least for that test period. However, if the results are grossly in disagreement, the entire process of analyzing the test data should be repeated.

Example 12.1 Calculation of Reservoir Parameters from Buildup Test Using the Gringarten-Bourdet Type Curves

Problem

A well in a reservoir above its bubble point pressure was producing oil at a constant rate of 185 BOPD before it was shut-in for a buildup test. The buildup test data are given in Appendix 12B as Table 12B.1 showing elapsed shut-in time, Δt, and shut-in pressure, p_{ws}. Other reservoir and well data are given below:

Formation thickness, h	114 ft
Formation porosity, ϕ	0.28
Total compressibility, c_t	4.1×10^{-6} psi^{-1}
Oil viscosity, μ_o	2.2 cp

Oil FVF, B_o	1.1 RB/STB
Wellbore radius, r_w	0.50 ft
Flowing BHP, p_{wf} at $\Delta t = 0$	2820 psia
Producing time before shut-in, t_p	540 hrs

Calculate dimensionless wellbore storage, C_D; wellbore storage coefficient, C; formation permeability, k; and skin factor, s. Use the technique of type curve matching with the combined Gringarten-Bourdet type curve. Compare the results from type curves to the results obtained from using straightline methods based on the Horner plot.

Solution

Step 1: Prepare the test data for analysis.
Calculate Agarwal equivalent shut-in time, Δt_e, and pressure change, $\Delta p = (p_{ws} - p_{wf@\Delta t=0})$. The calculated data are shown in Appendix 12B as Table 12B.2. Plot Δp vs. Δt_e on a log-log scale as shown in Figure 12.6.

Step 2: Calculate pressure derivative with respect to natural logarithm of equivalent time.
Calculate pressure derivatives with the algorithm shown in Appendix 12C. The calculated derivatives are shown in Table 12B.2. Plot $\Delta p'$ vs. Δt_e on a log-log scale as shown in Figure 12.6.

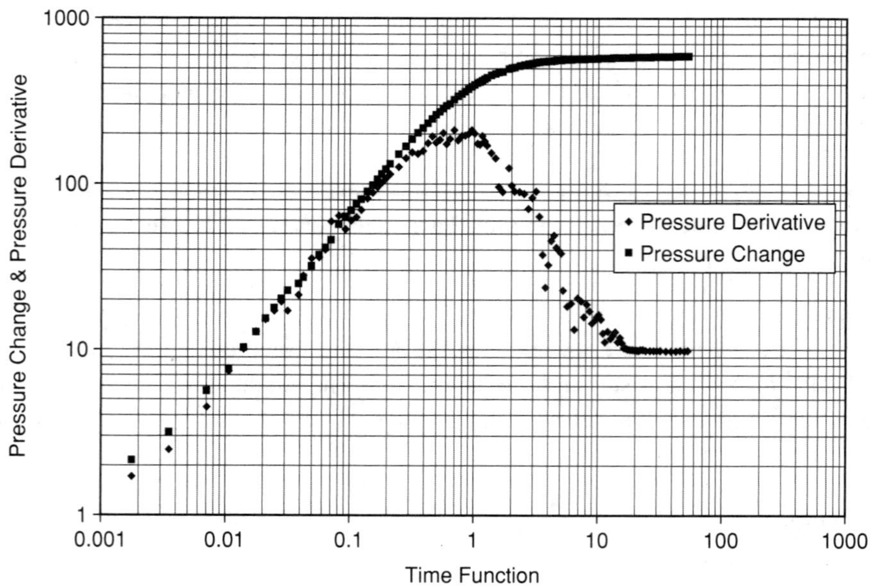

Figure 12.6 Plot of Δp & $\Delta p'$ vs. Δt_e for Example 12.1.

Step 3: Calculate C_D and C from the unit-slope line.

Select any point on the unit-slope line from Figure 12.6. One such point on the unit-slope line is $(\Delta t_e/\Delta p)_{USL} = (0.06374/41.13)$. Substituting Eq. (12.8) gives:

$$\begin{aligned} C_D &= \frac{0.03723qB}{\phi c_t h r_w^2}\left(\frac{\Delta t_e}{\Delta p}\right)_{USL} \\ &= \frac{0.03723 \times 185 \times 1.1}{0.28 \times 4.1 \times 10^{-6} \times 114 \times (0.5)^2}\left[\frac{0.06374}{41.13}\right] \\ &= 358.86 \end{aligned}$$

From Eq. (12.9):

$$\begin{aligned} C &= 0.04165qB\left(\frac{\Delta t_e}{\Delta p}\right)_{USL} \\ &= 0.04165 \times 185 \times 1.1 \times \left(\frac{0.06374}{41.13}\right) \\ &= 0.013 \text{ RB/psi} \end{aligned}$$

Step 4: Perform type-curve matching using Figure 12.6 and Gringarten-Bourdet type curve.

From Figure 12.7, the pressure match point (PMP) is $\left(\frac{p_D}{\Delta p}\right)_{PMP} = \left(\frac{4.5}{100}\right)$. Substituting Eq. (12.18) gives:

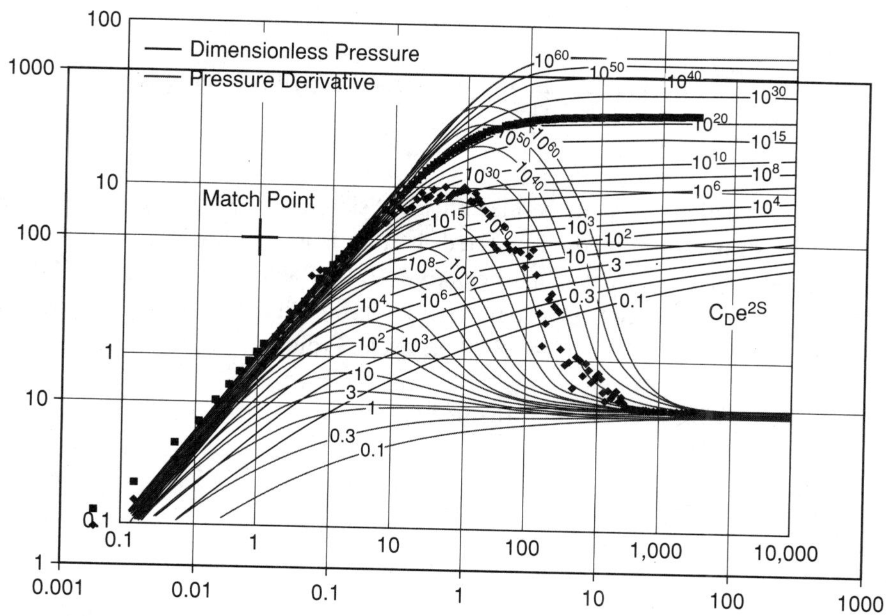

Figure 12.7 Type-curve match for Example 12.1.

$$k = \frac{141.2qB\mu}{h}\left[\frac{p_D}{\Delta p}\right]_{PMP}$$
$$= \frac{141.2 \times 185 \times 1.1 \times 2.2}{114}\left(\frac{4.5}{100}\right)$$
$$= 24.95 \text{ md}.$$

Step 5: Calculate C_D from time match point (TMP).

From Figure 12.7, the time match point (TMP) is $\left(\frac{\Delta t_e}{t_D/C_D}\right)_{TMP} = \left(\frac{0.03}{1}\right)$. Substituting into Eq. (12.17) gives:

$$C_D = \frac{0.0002637k}{\phi\mu c_t r_w^2}\left[\frac{\Delta t_e}{t_D/C_D}\right]_{TMP}$$
$$= \frac{0.0002637 \times 24.95}{0.28 \times 2.2 \times 4.1 \times 10^{-6} \times (0.5)^2}\left(\frac{0.03}{1}\right)$$
$$= 312.61$$

Note the value for dimensionless storage coefficient calculated from TMP by type-curve matching is reasonably close to the value calculated in Step 3 from unit-slope line.

Step 6: Calculate skin factor, s.

From Eq. (12.19):

$$s = 0.5 \ln\left(\frac{C_D e^{2s}}{C_D}\right)$$
$$= 0.5 \ln\left(\frac{10^{21}}{312.61}\right), \text{ where } C_D e^{2s} = 10^{21} \text{ from Figure 12.7}.$$
$$= 22.46$$

Step 7: For validation, calculate permeability and skin factor using straightline methods.

The straightline method applied is the Horner plot. A tabulation of p_{ws} and $(t_p + \Delta t)/\Delta t$ is shown in Appendix 12B as Table 12B.3. Figure 12.8 is a plot of p_{ws} vs. $(t_p + \Delta t/\Delta t)$ on a semilog scale. From Figure 12.8, slope of the IARF line is: $m = 3413.25 - 3384.46 = 28.79$ psi/cycle. From Eq. 11.59,

$$k = \frac{162.6qB\mu}{mh}$$
$$= \frac{162.6 \times 185 \times 1.1 \times 2.2}{28.79 \times 114}$$
$$= 22.18 \text{ md}.$$

Note the formation permeability calculated here is close to the value of 24.95 md calculated from type-curve matching in Step 4.

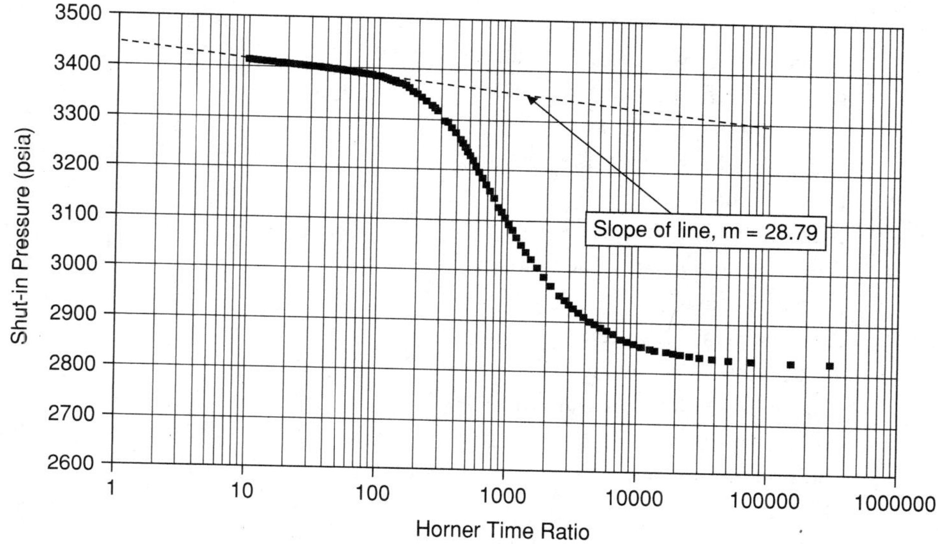

Figure 12.8 Plot of p_{ws} vs. $(t_p + \Delta t/\Delta t)$ for Example 12.1.

From Figure 12.8, $p_{1hr} = 3360$ psia. From Eq. 11.62,

$$s = 1.151\left[\frac{p_{1hr} - p_{wf}}{m} - \log\frac{k}{\phi\mu c_t r_w^2} + 3.23\right]$$

$$= 1.151\left[\frac{3360 - 2820}{28.79} - \log\left(\frac{22.18}{0.28 \times 2.2 \times 4.1 \times 10^{-6} \times (0.5)^2}\right) + 3.23\right]$$

$$= 16.63$$

Again, note the skin factor calculated here is close to the value of 22.46 from type-curve matching. This validation process indicates that the results obtained from analysis of the test data are consistent.

Nomenclature

A_{wb}	area of wellbore, feet
B	formation volume factor, RB/STB
c_f	formation compressibility, psi^{-1}
c_t	total compressibility, psi^{-1}
c_{wb}	compressibility of wellbore fluid, psi^{-1}

C	wellbore storage coefficient, bbl/psi
C_D	dimensionless wellbore storage coefficient
h	reservoir or formation thickness, feet
k	permeability, md
m	slope of MTR straight line
Δp	pressure change
$\Delta p'$	pressure change derivative
p'_D	dimensionless pressure derivative
p	pressure, psia
p_D	dimensionless pressure
p_{wf}	well bottomhole flowing pressure, psia
p_{ws}	shut-in bottomhole pressure, psia
q	volume flow rate, STB/D
r_w	wellbore radius, feet
s	skin factor
Δt	shut-in time in buildup test, hours
Δt_e	Agarwal equivalent time
t	time, hours
t_D	dimensionless time
t_p	total producing time before shut-in, hours
V_{wb}	volume of wellbore, barrels
ϕ	porosity, fraction
μ	fluid viscosity, cp
ρ_{wb}	density of fluid in wellbore, lbm/ft^3

Subscripts

g	gas
f	formation
i	initial conditions
o	oil
t	total
w	water

Abbreviations

BHP	Bottom Hole Pressure
ETR	Early Time Region
FVF	Formation Volume Factor
IARF	Infinite Acting Radial Flow
LTR	Late Time Region

MTR	Middle Time Region
PMP	Pressure Match Point
TMP	Time Match Point
USL	Unit Slope Line

References

1. Gringarten, A.C.: "Type-Curve Analysis: What It Can and Cannot Do," *JPT* (January 1987) 11–13.
2. Ramey, H.J.: "Short-Time Well Test Data Interpretation in the Presence of Skin Effect and Wellbore Storage," *JPT* (January 1970) 97–104.
3. Agarwal, R.G., Al-Hussainy, R., and Ramey, H.J., Jr.: "An Investigation of Wellbore Storage and Skin Effect in Unsteady Liquid Flow: I. Analytical Treatment," *SPEJ* (September 1970) 279–290.
4. McKinley, R.M.: "Wellbore Transmissibility from Afterflow-Dominated Pressure Buildup Data," *JPT* (July 1971) 863–872.
5. Gringarten, A.C., Ramey, H.J., Jr., and Raghavan, R.: "Unsteady-State Pressure Distribution Created by a Well with a Single Infinite-Conductivity Vertical Fracture," *SPEJ* (August 1974) 347–360.
6. Cinco-Ley, H., Samaniego-V., F., and Dominguez-A., N.: "Transient Pressure Behavior for a Well with a Finite-Conductivity Vertical Fracture," *SPEJ* (August 1978) 253–264.
7. Gringarten, A.C., Bourdet, D.P., Landel, P.A., and Kniazeff, V.J.: "A Comparison Between Different Skin and Wellbore Storage Type-Curves for Early-Time Transient Analysis," Paper SPE 8205 presented at the SPE Annual Technical Conference and Exhibition, Las Vegas, Nevada, September 23–26, 1979.
8. Bourdet, D., Whittle, T.M., Douglas, A.A., and Pirard, Y.M.: "A New Set of Type Curves Simplifies Well Test Analysis," *World Oil* (May 1983) 95–106.
9. Bourdet, D., Ayoub, J.A., Whittle, T.M., Pirard, Y.M., and Kniazeff, V.: "Interpreting Well Tests in Fractured Reservoirs," *World Oil* (October 1983) 77–87.
10. Bourdet, D., Ayoub, J.A., and Pirard, Y.M.: "Use of Pressure Derivative in Well-Test Interpretation," *SPEFE* (June 1989) 293–302.
11. Gringarten, A.C.: "From Straight Lines to Deconvolution: The Evolution of the State of the Art in Well Test Analysis," *SPEREE* (February 2008) 41–62.
12. Agarwal, Ram G.: "A New Method to Account for Producing Time Effects When Drawdown Type Curves Are Used to Analyze Pressure Buildup and Other Test Data," paper SPE 9289 presented at the 55th Annual Fall Technical Conference and Exhibition, Dallas, Texas, September 21–24, 1980.

13. Gringarten, A.C.: "Computer-Aided Well Test Analysis," paper SPE 14099 presented at the SPE International Meeting on Petroleum Engineering, Beijing, China, March 17–20, 1986.

14. Ehlig-Economides, C.A., Joseph, J.A., Ambrose, R.W., Jr., and Norwood, C.: "A Modern Approach to Reservoir Testing," *JPT* (December 1990) 1554–1563.

15. Horne, R.N.: *Modern Well Test Analysis-A Computer-Aided Approach*, 2nd Edition, Petroway, Inc., California (2008).

Characteristic Shapes of Pressure and Pressure-Derivative Curves for Selected Well, Reservoir, and Boundary Models

1. Well Models

A. Wellbore Storage and Skin

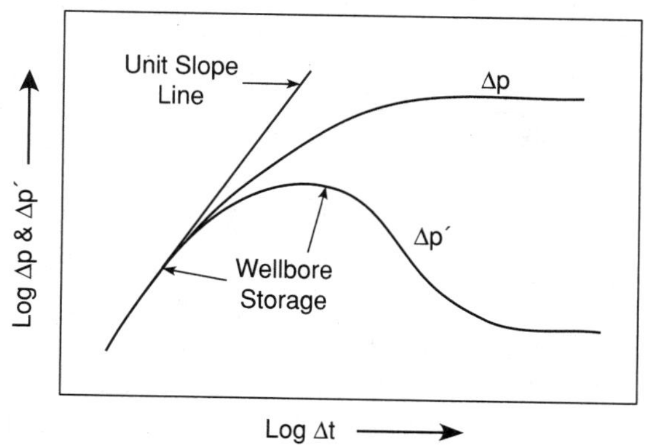

B. Infinite Conductivity Fracture

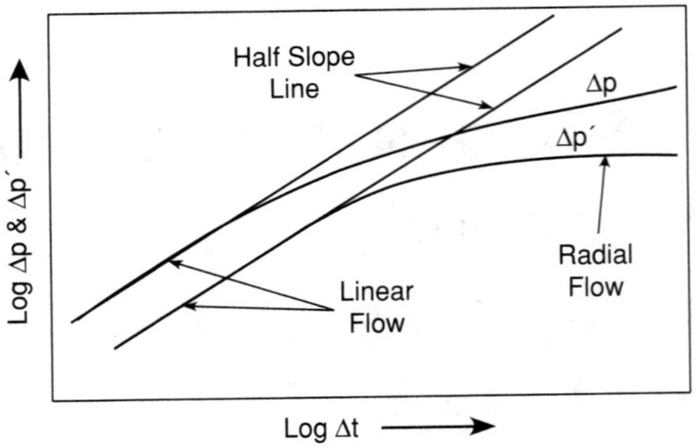

C. Finite Conductivity Fracture

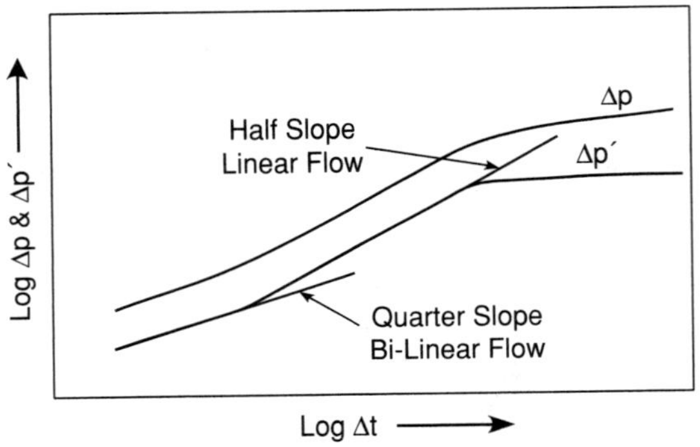

2. Reservoir Models

A. Homogeneous Behavior- Infinite-Acting Radial Flow

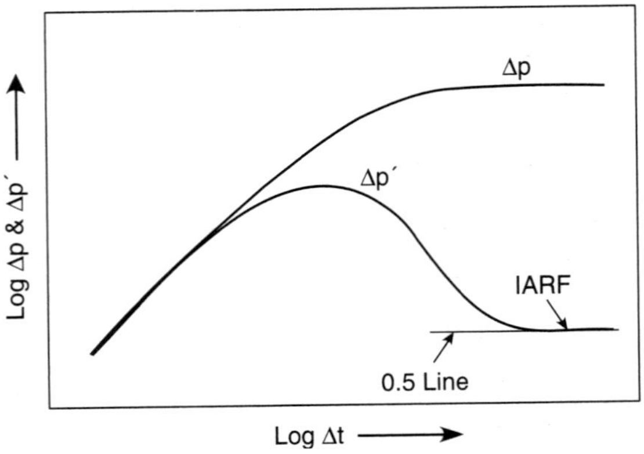

B. Heterogeneous Behavior- Pseudosteady-State Interporosity Flow, Double Porosity Behavior.

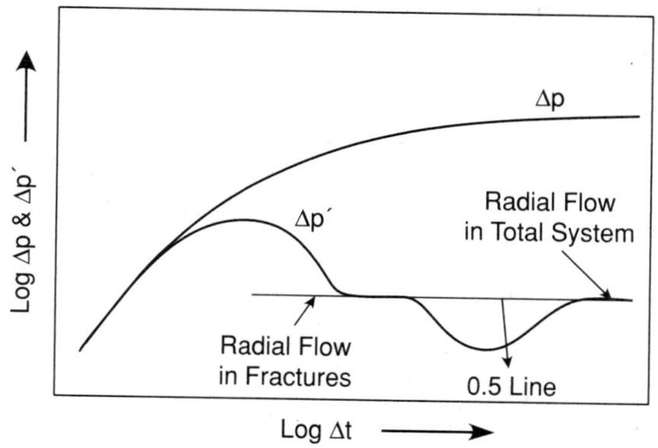

C. Heterogeneous Behavior- Transient Interporosity Flow, Double Porosity Behavior.

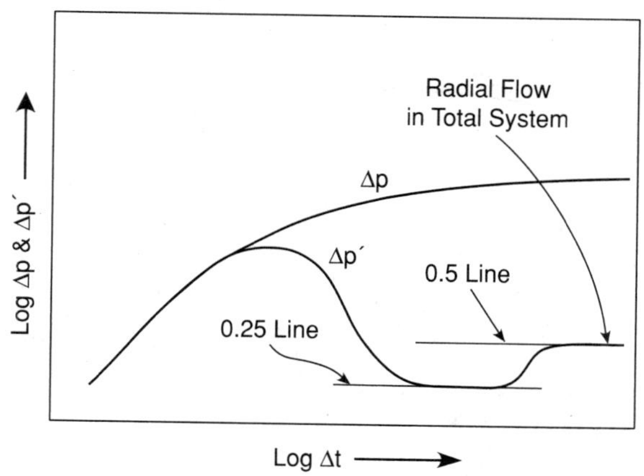

D. Heterogeneous Behavior- Double Permeability Behavior.

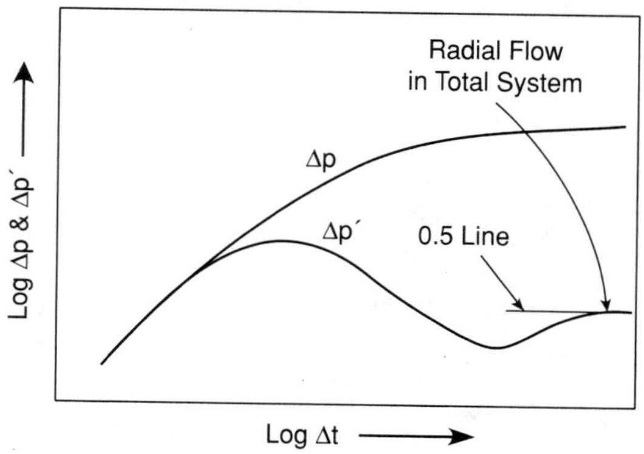

3. Boundary Models

A. Infinite-Acting Reservoir- No Boundary Effects.

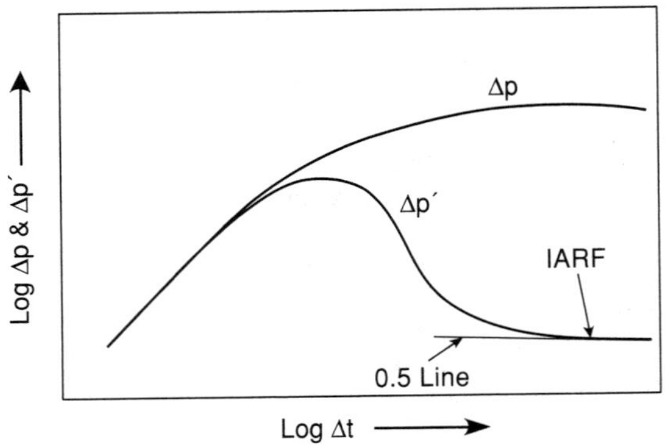

B. Sealing Fault or Single No-Flow Boundary.

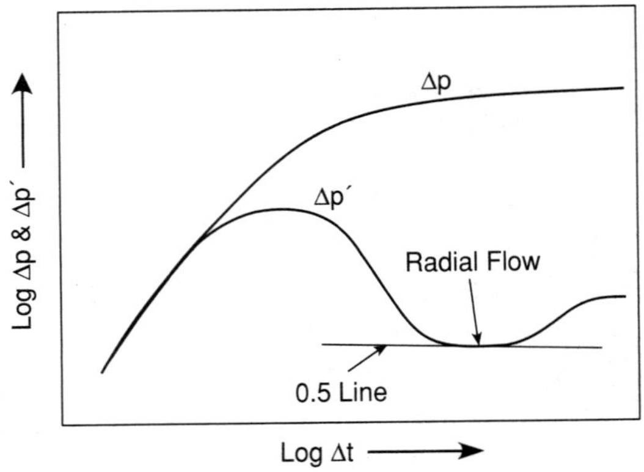

C. Bounded Reservoir or Constant Pressure Boundary Effects.

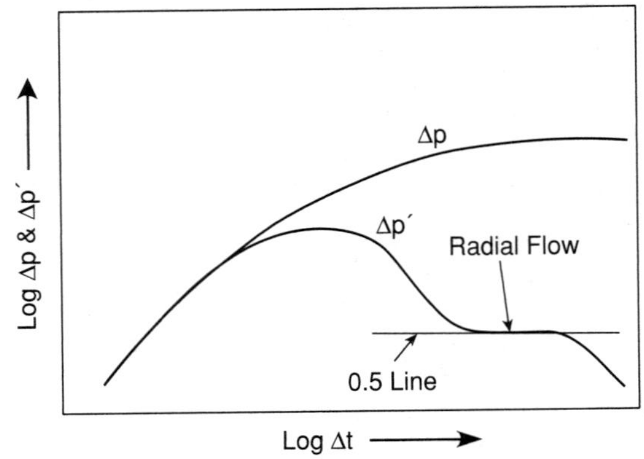

APPENDIX 12B

Buildup Test Data for Example 12.1

Table 12B.1 Buildup Test Data for Example 12.1

Δt (hrs)	p_{ws} (psia)	Δt (hrs)	p_{ws} (psia)	Δt (hrs)	p_{ws} (psia)	Δt (hrs)	p_{ws} (psia)
0.0000	2820.00	0.3188	3006.01	2.5500	3344.09	12.9625	3398.19
0.0018	2822.15	0.3542	3022.46	2.7625	3350.80	13.6000	3398.75
0.0035	2823.18	0.3896	3036.49	2.9750	3355.10	14.2375	3399.31
0.0071	2825.61	0.4250	3051.08	3.1875	3362.38	14.8750	3399.86
0.0106	2827.67	0.4604	3065.68	3.4000	3367.23	15.5125	3400.25
0.0142	2830.28	0.4958	3080.63	3.6125	3370.40	16.1500	3400.80
0.0177	2832.71	0.5313	3091.29	3.8250	3371.72	16.7875	3401.10
0.0213	2835.33	0.5667	3104.95	4.0375	3373.03	17.4250	3401.55
0.0248	2837.76	0.6021	3116.73	4.2500	3375.09	18.0625	3401.83
0.0283	2840.19	0.6375	3125.52	4.4625	3377.52	18.9125	3402.39
0.0319	2842.62	0.6906	3144.23	4.6750	3379.73	19.7625	3402.71
0.0390	2844.86	0.7438	3158.26	4.8875	3381.28	20.6125	3403.21
0.0425	2847.11	0.7969	3170.42	5.1000	3383.15	21.4625	3403.51
0.0496	2851.97	0.8500	3184.07	5.3125	3384.46	22.3125	3403.96
0.0567	2857.19	0.9031	3194.93	5.7375	3384.80	23.1625	3404.24
0.0638	2861.13	0.9563	3207.47	6.1625	3387.00	24.2250	3404.78
0.0708	2865.98	1.0094	3218.31	6.5875	3387.51	25.5000	3405.13
0.0815	2876.72	1.0625	3228.62	7.0125	3388.64	27.6250	3406.22
0.0921	2883.26	1.1156	3235.90	7.4375	3389.96	29.7500	3406.62
0.1027	2889.05	1.1688	3244.89	7.8625	3390.89	31.8750	3407.52
0.1133	2895.59	1.2219	3253.49	8.2875	3391.64	34.0000	3407.90

0.1240	2900.93	1.2750	3260.61	8.7125	3392.76	38.2500	3409.71
0.1381	2909.72	1.3813	3274.46	9.1375	3393.32	42.5000	3410.10
0.1523	2917.76	1.4875	3284.39	9.5625	3394.07	46.7500	3411.42
0.1665	2926.17	1.5938	3294.70	9.9875	3394.63	51.0000	3411.83
0.1806	2934.03	1.7000	3297.68	10.4125	3395.39	55.2500	3412.86
0.1948	2942.06	1.9125	3317.54	10.8375	3395.94	59.5000	3413.25
0.2125	2951.60	2.0188	3323.15	11.2625	3396.56	—	—
0.2479	2970.66	2.1250	3327.83	11.6875	3396.88	—	—
0.2833	2987.87	2.3375	3336.24	12.3250	3397.63	—	—

Table 12B.2 Plot Data for Figure 12.6 in Example 12.1

Δt_e	Δp (psi)	$\Delta t_e \Delta p'$	Δt_e	Δp (psi)	$\Delta t_e \Delta p'$	Δt_e	Δp (psi)	$\Delta t_e \Delta p'$
0.0018	2.15	1.7098	0.7427	338.26	182.7992	8.5742	572.76	17.1367
0.0035	3.18	2.4958	0.7957	350.42	194.7352	8.9855	573.32	14.4583
0.0071	5.61	4.4916	0.8487	364.07	195.1176	9.3961	574.07	14.9275
0.0106	7.67	7.4107	0.9016	374.93	200.1915	9.8061	574.63	15.8172
0.0142	10.28	10.0963	0.9546	387.47	210.0963	10.2155	575.39	16.2879
0.0177	12.71	12.8207	1.0075	398.31	201.1232	10.6243	575.94	15.2523
0.0212	15.33	15.1382	1.0604	408.62	174.7776	11.0324	576.56	12.5318
0.0248	17.76	17.0691	1.1133	415.90	172.1645	11.4399	576.88	11.1707
0.0283	20.19	19.4921	1.1662	424.89	193.8354	12.0500	577.63	12.8853
0.0319	22.62	17.1346	1.2191	433.49	180.5007	12.6586	578.19	11.6352
0.0390	24.86	21.4190	1.2720	440.61	169.6267	13.2659	578.75	12.2291
0.0425	27.11	27.9062	1.3777	454.46	153.1716	13.8718	579.31	12.8085
0.0496	31.97	35.5838	1.4834	464.39	142.3950	14.4762	579.86	11.1516
0.0567	37.19	36.0873	1.5891	474.70	96.3010	15.0793	580.25	11.8288
0.0637	41.13	40.0996	1.6947	477.68	89.7894	15.6810	580.80	11.0220
0.0708	45.98	59.2751	1.9058	497.54	124.5963	16.2813	581.10	10.2644
0.0814	56.71	64.3431	2.0112	503.15	97.7732	16.8803	581.55	10.2107
0.0921	63.26	53.1749	2.1167	507.83	90.6307	17.4779	581.83	10.0441
0.1027	69.05	60.0963	2.3274	516.24	89.6178	18.2725	582.39	10.0083
0.1133	75.59	62.9027	2.5380	524.09	87.2892	19.0648	582.71	9.9814
0.1239	80.93	69.4071	2.7484	530.80	70.8360	19.8546	583.21	9.9659
0.1381	89.72	81.8344	2.9587	535.10	83.0795	20.6421	583.51	9.9290
0.1522	97.76	88.7436	3.1688	542.38	90.3366	21.4271	583.96	9.8879
0.1664	106.17	95.4434	3.3787	547.23	63.7965	22.2098	584.24	9.9722
0.1806	114.03	101.4855	3.5885	550.40	37.4576	23.1849	584.78	9.9844
0.1947	122.06	107.8807	3.7981	551.72	23.8049	24.3501	585.13	9.8985
0.2124	131.60	114.7506	4.0075	553.03	32.6983	26.2806	586.22	9.8435
0.2478	150.66	126.5068	4.2168	555.09	45.4847	28.1966	586.62	9.8889
0.2832	167.87	142.3060	4.4259	557.52	49.0577	30.0984	587.52	9.8851

Δt_e	Δp (psi)	$\Delta t_e \Delta p'$	Δt_e	Δp (psi)	$\Delta t_e \Delta p'$	Δt_e	Δp (psi)	$\Delta t_e \Delta p'$
0.3186	186.01	155.2165	4.6349	559.73	41.3656	31.9861	587.90	9.8516
0.3539	202.46	151.5460	4.8437	561.28	39.7761	35.7198	589.71	9.8177
0.3893	216.49	158.0667	5.0523	563.15	38.2492	39.3991	590.10	9.7815
0.4247	231.08	175.4780	5.2607	564.46	22.7654	43.0251	591.42	9.8236
0.4600	245.68	192.5590	5.6772	564.80	18.3097	46.5990	591.83	9.8396
0.4954	260.63	177.4633	6.0930	567.00	19.0236	50.1218	592.86	9.8026
0.5307	271.29	184.2295	6.5081	567.51	13.2130	53.5947	593.25	9.8448
0.5661	284.95	202.9277	6.9226	568.64	20.5387	—	—	—
0.6014	296.73	173.6664	7.3365	569.96	19.7070	—	—	—
0.6367	305.52	187.2953	7.7497	570.89	15.7560	—	—	—
0.6897	324.23	210.9340	8.1622	571.64	18.7560	—	—	—

Table 12B.3 Plot Data for Figure 12.8 in Example 12.1

$\frac{(t_p + \Delta t)}{\Delta t}$	p_{ws} (psia)	$\frac{(t_p + \Delta t)}{\Delta t}$	p_{ws} (psia)	$\frac{(t_p + \Delta t)}{\Delta t}$	p_{ws} (psia)	$\frac{(t_p + \Delta t)}{\Delta t}$	p_{ws} (psia)
304990.97	2822.15	1525.70	3022.46	196.48	3350.80	40.71	3398.75
152471.10	2823.18	1387.10	3036.49	182.51	3355.10	38.93	3399.31
76236.32	2825.61	1271.59	3051.08	170.41	3362.38	37.30	3399.86
50824.53	2827.67	1173.85	3065.68	159.82	3367.23	35.81	3400.25
38118.57	2830.28	1090.08	3080.63	150.48	3370.40	34.44	3400.80
30495.17	2832.71	1017.47	3091.29	142.18	3371.72	33.17	3401.10
25412.76	2835.33	953.94	3104.95	134.75	3373.03	31.99	3401.55
21782.41	2837.76	897.89	3116.73	128.06	3375.09	30.90	3401.83
19059.84	2840.19	848.06	3125.52	122.01	3377.52	29.55	3402.39
16942.18	2842.62	782.90	3144.23	116.51	3379.73	28.32	3402.71
13861.97	2844.86	727.05	3158.26	111.49	3381.28	27.20	3403.21
12706.88	2847.11	678.65	3170.42	106.88	3383.15	26.16	3403.51
10891.76	2851.97	636.29	3184.07	102.65	3384.46	25.20	3403.96
9530.39	2857.19	598.92	3194.93	95.12	3384.80	24.31	3404.24
8471.59	2861.13	565.71	3207.47	88.63	3387.00	23.29	3404.78
7624.53	2865.98	535.98	3218.31	82.97	3387.51	22.18	3405.13
6630.16	2876.72	509.24	3228.62	78.01	3388.64	20.55	3406.22
5865.27	2883.26	485.03	3235.90	73.61	3389.96	19.15	3406.62
5258.61	2889.05	463.03	3244.89	69.68	3390.89	17.94	3407.52
4765.72	2895.59	442.94	3253.49	66.16	3391.64	16.88	3407.90
4357.31	2900.93	424.53	3260.61	62.98	3392.76	15.12	3409.71
3910.50	2909.72	391.95	3274.46	60.10	3393.32	13.71	3410.10
3546.82	2917.76	364.03	3284.39	57.47	3394.07	12.55	3411.42
3245.06	2926.17	339.82	3294.70	55.07	3394.63	11.59	3411.83
2990.62	2934.03	318.65	3297.68	52.86	3395.39	10.77	3412.86
2773.19	2942.06	283.35	3317.54	50.83	3395.94	10.08	3413.25
2542.18	2951.60	268.49	3323.15	48.95	3396.56	—	—
2179.14	2970.66	255.12	3327.83	47.20	3396.88	—	—
1906.88	2987.87	232.02	3336.24	44.81	3397.63	—	—
1695.12	3006.01	212.76	3344.09	42.66	3398.19	—	—

Calculation of Pressure Derivatives

The calculation of pressure derivatives is based on an algorithm recommended by Bourdet et al.[1] The pressure derivative at a point of interest denoted as I in Figure 12C.1 is determined by using a preceding point denoted as 1 and a succeeding point denoted as 2 to calculated a weighted mean derivative which is placed at point I. Using the notation in Figure 12C.1, the algorithm can be written as:

$$\left(\frac{dp}{dX}\right)_I = \frac{\left[\left(\frac{\Delta p_1}{\Delta X_1}\right)\Delta X_2 + \left(\frac{\Delta p_2}{\Delta X_2}\right)\Delta X_1\right]}{(\Delta X_1 + \Delta X_2)} \tag{12C.1}$$

This algorithm is illustrated by using a subset of the data from Example 12.1 shown in Table 12B.2. This is shown in Table 12C.1. At $\Delta t = 0.1523$, substituting equation (12C.1) gives:

$$\left(\frac{dp}{dX}\right)_{@\Delta t=0.1523} = \frac{\left[\left(\frac{8.04}{0.0976}\right) \times 0.0889 + \left(\frac{8.41}{0.0889}\right) \times 0.0976\right]}{(0.0976 + 0.0889)}$$

$$= 88.7436$$

Bourdet et al.[1] recommends using a smoothing parameter, L, to reduce the effects of noise associated with pressure transient data on calculation of pressure derivatives. The smoothing parameter, L, can be defined as $\Delta(\ln t)$ for drawdown test data and $\Delta(\ln t_e)$ for buildup test data. The range of the smoothing parameter is $0 \leq L \leq 0.5$. For the pressure derivative calculated as shown earlier, no smoothing was applied, hence $L = 0$. The procedure for applying the smoothing parameter is as follows: For values of $L > 0$, subtract L from I and add L to I, which creates a "window" around the point of interest, I. Then use the data points just above and below this "window" to calculate the derivative at I. The implementation of the procedure for calculation

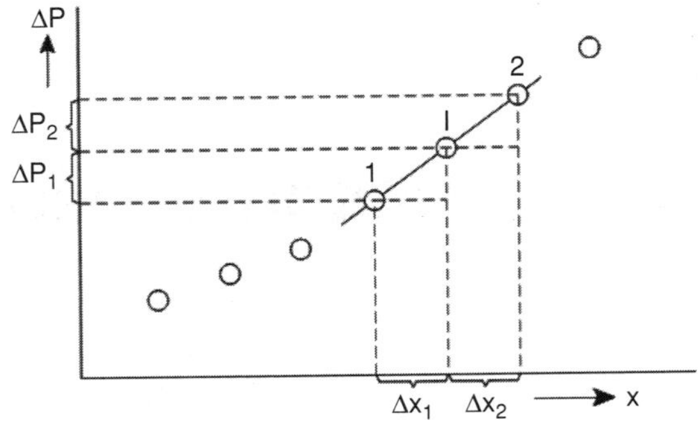

Figure 12C.1 A sketch for calculation of pressure derivative.

of pressure derivatives with smoothing parameter $L > 0$, requires additional programming steps especially for large data sets.

Reference

1. Bourdet, D., Ayoub, J.A., and Pirard, Y.M.: "Use of Pressure Derivative in Well-Test Interpretation," *SPEFE* (June 1989) 293–302.

Table 12C.1 Pressure Derivative Calculation (Subset of Data in Table 12B.2)

Δt (hrs)	Δp (psia)	Δt_e	$\ln \Delta t_e$	ΔX_1	ΔX_2	Δp_1	Δp_2	$\Delta t_e \Delta p'$
0.0921	63.26	0.0921	−2.3852	0.1226	0.1092	6.55	5.78	53.1749
0.1027	69.05	0.1027	−2.2761	0.1092	0.0984	5.78	6.55	60.0963
0.1133	75.59	0.1133	−2.1776	0.0984	0.0896	6.55	5.34	62.9027
0.1240	80.93	0.1239	−2.0880	0.0896	0.1082	5.34	8.79	69.4071
0.1381	89.72	0.1381	−1.9799	0.1082	0.0976	8.79	8.04	81.8344
0.1523	97.76	0.1522	−1.8822	0.0976	0.0889	8.04	8.41	88.7436
0.1665	106.17	0.1664	−1.7933	0.0889	0.0817	8.41	7.86	95.4434
0.1806	114.03	0.1806	−1.7117	0.0817	0.0755	7.86	8.03	101.4855
0.1948	122.06	0.1947	−1.6362	0.0755	0.0870	8.03	9.54	107.8807
0.2125	131.60	0.2124	−1.5492	0.0870	0.1541	9.54	19.07	114.7506
0.2479	150.66	0.2478	−1.3951	0.1541	0.1335	19.07	17.21	126.5068
0.2833	167.87	0.2832	−1.2617	0.1335	0.1177	17.21	18.14	142.3060

CHAPTER 13

Well Test Analysis: Hydraulically Fractured Wells and Naturally Fractured Reservoirs

13.1 Introduction

This chapter presents basic concepts on well test analysis of hydraulically fractured wells and naturally fractured reservoirs. Hydraulically fractured wells and naturally fractured reservoirs form a large part of hydrocarbon production capacity around the world. The well test analysis methods presented for these wells and reservoir types emphasize applications of combined straightline and type-curve matching techniques as previously recommended for unfractured wells and reservoirs. In this chapter, well test analysis for hydraulically fractured wells is presented first followed by well test analysis for naturally fractured reservoirs.

13.2 Hydraulically Fractured Wells

Most wells completed in low permeability sand and shale formations in western regions of the United States of America are hydraulically fractured to achieve commercial production. In fact, many gas wells in this area of United States could be categorized as unproductive without the application of hydraulic fracturing. Also, oil production has advanced into the deepwater regions of the Gulf of Mexico, United States. Deepwater reservoirs are also on production in offshore Brazil, West Africa, and other deepwater regions around the world. Most of the wells completed in these deepwater reservoirs are hydraulically fractured to achieve and maintain production at high commercial rates in consideration of the high cost of operations in the deepwater environment.

Many other oil producing regions in the world are also using hydraulic fracturing to improve well productivity. Consequently, it is important to understand how to analyze pressure transient tests from hydraulically fractured wells to evaluate the effectiveness of the post-fracture operations, predict expected productivity of the well, or evaluate the condition of a fractured well to determine whether additional stimulation is needed.

13.3 Definition of Dimensionless Variables for Fractured Wells

A number of dimensionless variables are defined specifically for fractured wells. The dimensionless fracture conductivity, C_{fD}, is defined as:

$$C_{fD} = \frac{k_f w_f}{kL_f} \tag{13.1}$$

In Eq. (13.1), k_f = fracture permeability, md; w_f = fracture width, ft; k = formation permeability, md; and L_f = fracture half-length, ft.

The dimensionless time, t_{DL_f}, in terms of fracture half-length is defined as:

$$t_{DL_f} = \frac{0.0002637kt}{\phi\mu c_t L_f^2} \tag{13.2}$$

Note that t_{DL_f} is related to the familiar dimensionless time, t_D defined in Chapter 12, Eq. (12.3) by the expression:

$$t_{DL_f} = t_D \frac{r_w^2}{L_f^2} \tag{13.3}$$

In Eqs. (13.2) and (13.3), t = time, hr; ϕ = porosity, fraction; μ = viscosity, cp; c_t = total compressibility, psi^{-1}; and r_w = wellbore radius, ft.

13.4 Flow Regimes in Fractured Wells

Fluid flow in vertically fractured wells can be represented by four flow regimes as shown in Figure 13.1. The four flow regimes are:

1. Fracture linear flow
2. Bilinear flow
3. Formation linear flow
4. Pseudo-radial flow

13.4.1 Fracture Linear Flow

Initial fluid flow in fractured wells is dominated by fracture linear flow regime (Figure 13.1a). During this flow regime, most of the fluid flow towards the wellbore is caused by the expansion

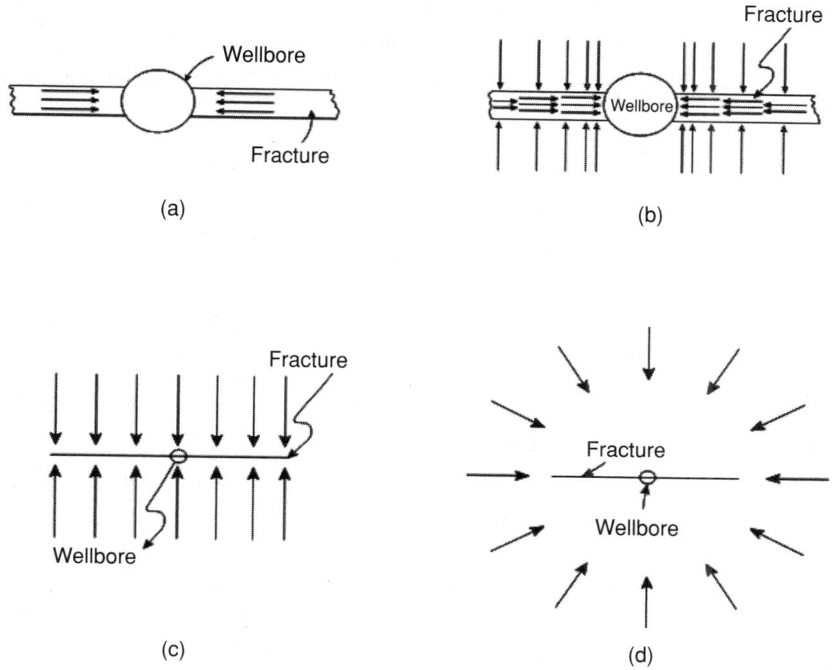

Figure 13.1 Fracture flow regimes (from Cinco-Ley and Samaniego[1] © 1981 SPE, Reproduced with permission).

of the fluid in the fracture. The fluid flow is linear and very short-lived. Fracture linear flow has no practical application in well test analysis because its duration is very short. The end of fracture linear flow can be estimated with the expression:[1]

$$t_{DL_f} = \frac{0.01C_{fD}^2}{\eta_{fD}^2} \tag{13.4}$$

In Eq. (13.4), η_{f_D} is the dimensionless hydraulic diffusivity which is defined as:

$$\eta_{fD} = \frac{k_f \phi c_t}{k \phi_f c_{ft}} \tag{13.5}$$

The new terms in Eq. (13.5) are defined as follows: ϕ_f = fracture porosity, fraction; and c_{ft} = total compressibility of fluid in the fracture, psi^{-1}.

13.4.2 Bilinear Flow

Bilinear flow, if present in a fracture, begins at the end of fracture linear flow. The flow is described as bilinear because two linear flows occur at the same time. One linear flow occurs in the fracture and the other linear flow occurs in the formation (Figure 13.1b). Bilinear flow occurs in fractures with $C_{fD} < 300$. During bilinear flow regime, most of the fluid entering the wellbore comes from the formation until fracture-tip effects begin to affect well behavior. The time for bilinear flow to end depends on dimensionless fracture conductivity. The dimensionless times for bilinear flow to end for various ranges of dimensionless fracture conductivity are given as follows:[1]

$$t_{DL_f} = \frac{0.1}{C_{fD}^2}, \quad \text{for} \quad C_{fD} \geq 3 \tag{13.6}$$

$$t_{DL_f} = 0.0205[C_{fD} - 1.5]^{-1.53} \quad \text{for} \quad 1.6 \leq C_{fD} \leq 3 \tag{13.7}$$

$$t_{DL_f} = \left[\frac{4.55}{\sqrt{C_{fD}}} - 2.5\right]^{-4} \quad \text{for} \quad C_{fD} \leq 1.6 \tag{13.8}$$

13.4.3 Formation Linear Flow

Highly conductive fractures with $C_{fD} \geq 300$ exhibit formation linear flow (Figure 13.1c). Bilinear flow does not occur in these high conductive fractures. Formation linear flow is estimated to begin at:

$$t_{DL_f} \simeq \frac{100}{C_{fD}^2} \tag{13.9}$$

The end of formation linear flow period is given by:

$$t_{DL_f} \simeq 0.016 \tag{13.10}$$

13.4.4 Pseudo-Radial Flow

All fractured wells eventually exhibit pseudo-radial flow (Figure 13.1d) provided the flow period is sufficiently long and the fracture penetrates less than one-third of the drainage radius of the well.[2] Pseudo-radial flow is essentially equivalent to the radial flow used previously to represent fluid flow in unfractured wells. Hence, all equations developed for radial flow also apply to pseudo-radial flow. Pseudo-radial flow is estimated to begin for highly conductive fractures at:[3]

$$t_{DL_f} \simeq 5 \quad \text{for} \quad C_{fD} \geq 300 \tag{13.11}$$

The range of dimensionless time for start of pseudo-radial flow for less conductive fractures is given by: $0.1 \leq C_{fD} \leq 300$ and $1 \leq t_{DL_f} \leq 5$.

13.5 Fractured Well Flow Models

There are three main flow models used to represent fractured wells. These are:

1. Finite conductivity vertical fracture
2. Infinite conductivity vertical fracture
3. Uniform flux vertical fracture

13.5.1 Finite Conductivity Vertical Fracture

Fractures with $C_{fD} < 300$ are considered to be finite conductivity fractures. Finite conductivity fractures can exhibit the four fluid flow regimes which are fracture linear flow, bilinear flow, formation linear flow, and pseudo-radial flow with intervening transition periods. As shown earlier, the duration of fracture linear flow is so short that it has no practical use in well test analysis. Generally, finite conductivity fractures transits from bilinear flow to pseudo-radial flow with the intermediate formation linear flow not fully developed.

13.5.2 Infinite Conductivity Vertical Fracture

Fractures with $C_{fD} \geq 300$ are considered to be infinite conductivity fractures. Infinite conductivity fractures do not exhibit bilinear flow regime. The flow regimes that occur in infinite conductivity fractures are formation linear flow followed by pseudo-radial flow after a transition period. The main characteristic of infinite conductivity fractures is that pressure drop along the length of the fracture is negligible.

13.5.3 Uniform Flux Vertical Fracture

Uniform flux fracture model assumes that the flow into the fracture is uniform along its length.[4] Uniform flux vertical fracture is similar to infinite conductivity vertical fracture in terms of flow behavior. As shown in Figure 13.2, there is little difference between the two fracture models. Uniform flux vertical fractures also exhibit only formation linear and pseudo-radial flow regimes.

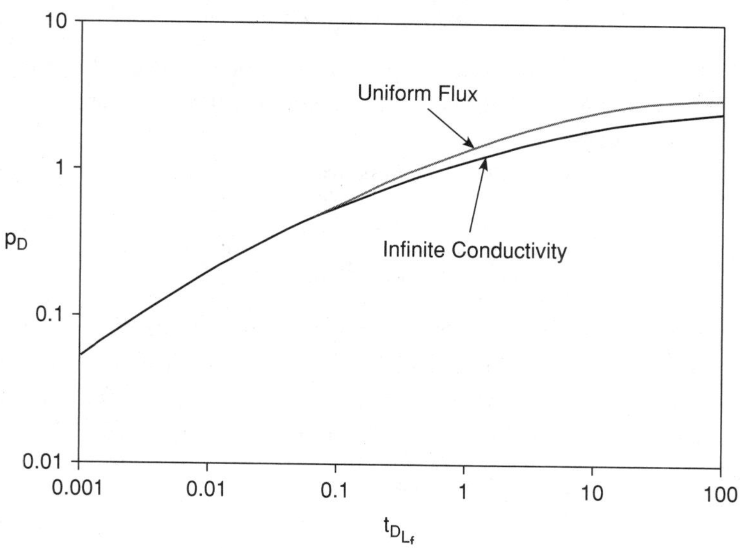

Figure 13.2 Infinite conductivity vertical fracture versus uniform flux vertical fracture.

13.6 Fractured Well Test Analysis: Straightline Methods

Well test analysis in fractured wells is used to evaluate the effectiveness of hydraulic fracturing by determining fracture parameters, such as fracture half-length, fracture conductivity, and permeability of the formation. The flow regimes that exist during a test are determined by the conductivity of the fracture and the duration of the test. The procedures for fractured well test analysis are presented according to the following flow regimes: bilinear flow, formation linear flow, and pseudo-radial flow.

13.6.1 Bilinear Flow

Bilinear flow occurs in vertical fractures with $C_{fD} < 300$ at early flow times. The duration of bilinear flow for various ranges of C_{fD} are given in Eqs. (13.6), (13.7), and (13.8). The dimensionless wellbore pressure for bilinear flow is expressed as:

$$p_D = \frac{2.45}{\sqrt{C_{fD}}} t_{DL_f}^{1/4} \tag{13.12}$$

Taking derivatives with respect to dimensionless time, t_{DL_f}, Eq. (13.12) becomes:

$$t_{DL_f} \frac{dp_D}{dt_{DL_f}} = \frac{0.6125}{\sqrt{C_{fD}}} t_{DL_f}^{1/4} \tag{13.13}$$

For an oil well, Eq. (13.12) can be transformed into:

$$\Delta p = \frac{44.1 qB\mu}{h(k_f w_f)^{1/2}(\phi\mu c_t k)^{1/4}} t^{1/4} \tag{13.14}$$

In Eq. (13.14), Δp = pressure change, psia; q = oil rate, STB/D; B = formation volume factor of oil, RB/STB; and h = net formation thickness, ft. Other terms are as defined in previous equations in this chapter.

Similarly, for a gas well, Eq. (13.12) can be transformed into:

$$\Delta p_p = \frac{444.75 q_g T}{h(k_f w_f)^{1/2}(\phi\mu c_t k)^{1/4}} t^{1/4} \tag{13.15}$$

In Eq. (13.15), Δp_p = real gas pseudo-pressure change, psia^2/cp; q_g = gas rate, Mcf/D; and T = temperature, °R. It can be deduced from Eqs. (13.14) and (13.15) that a plot of Δp or Δp_p vs. $\sqrt[4]{t}$ on Cartesian coordinates will given a straight line with a slope, m_{bf}, that passes through the origin (Figure 13.3). From this plot, the fracture conductivity can be calculated as follows:

For an oil well:

$$k_f w_f = \left(\frac{44.1 qB\mu}{m_{bf} h}\right)^2 \left(\frac{1}{\phi\mu c_t k}\right)^{1/2} \tag{13.16}$$

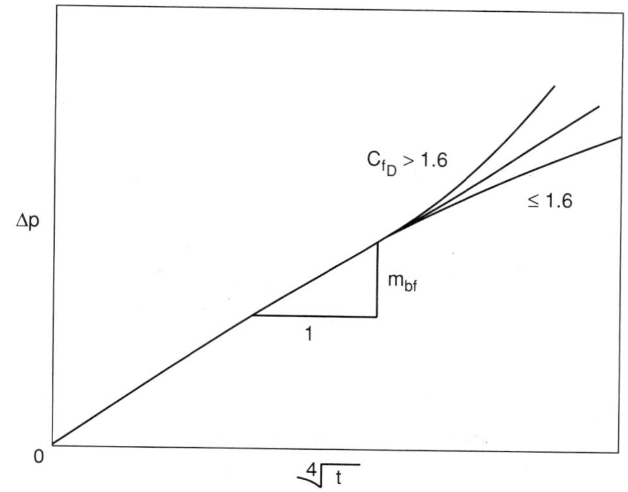

Figure 13.3 Plot of Δp versus $\sqrt[4]{t}$ for bilinear flow (from Cinco-Ley and Samaniego[1] © 1981 SPE, Reproduced with permission).

For a gas well:

$$k_f w_f = \left(\frac{444.75 q_g T}{m_{bf} h}\right)^2 \left(\frac{1}{\phi \mu c_t k}\right)^{1/2} \tag{13.17}$$

Note that in Eqs. (13.16) and (13.17), other formation properties such as porosity, ϕ, fluid viscosity, μ, and total compressibility, c_t, are required. The formation permeability, k, can be determined from prefracture well test data.

As Eqs. (13.14) and (13.15) imply, bilinear flow is evident in a log-log plot of Δp vs. t which yields a straight line with a quarter-slope as shown in Figure 13.4. This is a diagnostic plot that can be used to demonstrate the presence of bilinear flow within the test period.

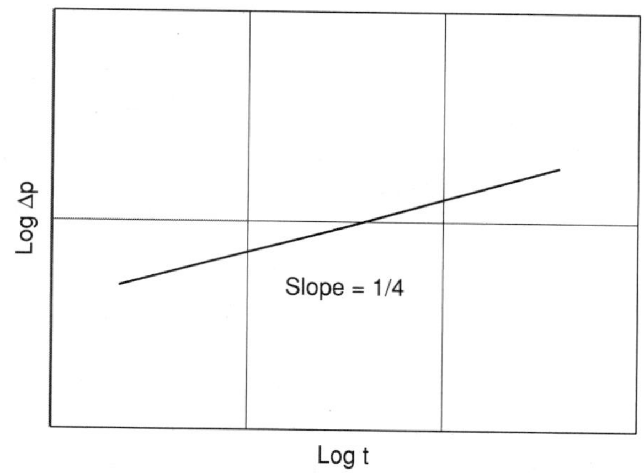

Figure 13.4 Plot of log Δp versus log t for bilinear flow (from Cinco-Ley and Samaniego[1] © 1981 SPE, Reproduced with permission).

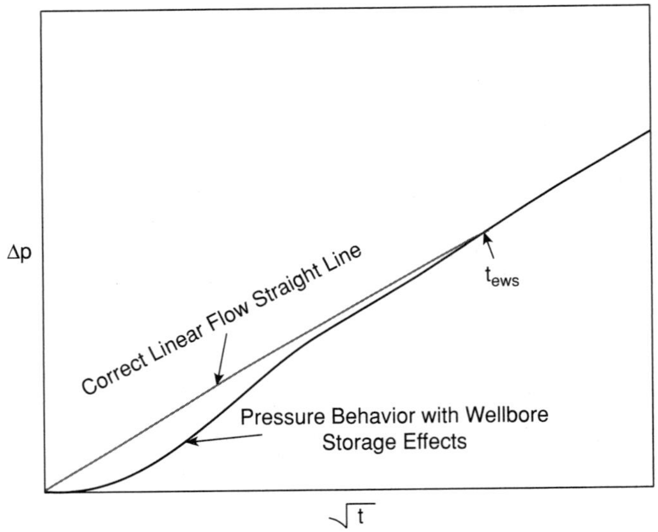

Figure 13.5 Effects of Wellbore Storage on Δp versus $\sqrt[4]{t}$ plot (from Cinco-Ley and Samaniego[1] © 1981 SPE, Reproduced with permission).

Bilinear flow regime may be distorted by effects of wellbore storage as shown in Figure 13.5 or flow restriction within the fracture which causes additional pressure drop as shown in Figure 13.6.

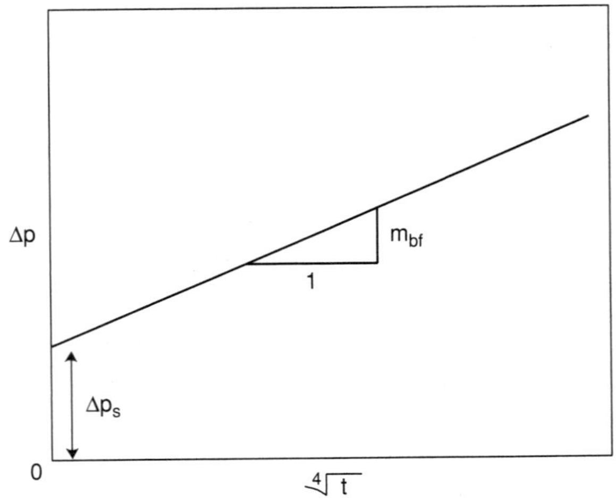

Figure 13.6 Additional pressure drop Δp_s due to fracture restriction on Δp versus $\sqrt[4]{t}$ plot (from Cinco-Ley and Samaniego[1] © 1981 SPE, Reproduced with permission).

13.6.2 Procedure for Application of Straightline Methods on Well Test Data During Bilinear Flow Regime

1. Plot p_{wf} vs. $\sqrt[4]{t}$ for a constant rate flow test or p_{ws} vs. $\sqrt[4]{t}$ for a buildup test. For gas wells, use appropriate pressure transformation functions, such as real gas pseudo-pressure, as shown in Chapter 11.
2. Determine the slope, m_{bf}, from the straight line region of the plot.
3. Calculate fracture conductivity from Eqs. (13.16) or (13.17).

Note that it is important to validate the results obtained with straightline methods with the results obtained from application of type-curve matching. This is discussed further under type-curve matching of fractured well tests.

13.6.3 Formation Linear Flow

Formation linear flow occurs in highly conductive fractures with $C_{fD} \geq 300$. The duration of formation linear flow is estimated with Eqs. (13.9) and (13.10). The dimensionless wellbore pressure change for formation linear flow is expressed as:

$$p_D = \sqrt{\pi t_{DL_f}} \tag{13.18}$$

By logarithmic expansion, Eq. (13.18) gives:

$$\log p_D = \frac{1}{2}\log(t_{DL_f}) + \frac{1}{2}\log(\pi) \tag{13.19}$$

Taking derivative of p_D in Eq. (13.18) with respect to t_{DL_f} yields:

$$t_{DL_f}\frac{dp_D}{dt_{DL_f}} = \frac{1}{2}\sqrt{\pi t_{DL_f}} \tag{13.20}$$

Again, applying logarithmic expansion to Eq. (13.20) gives:

$$\log\left(t_{DL_f}\frac{dp_D}{dt_{DL_f}}\right) = \frac{1}{2}\log(t_{DL_f}) + \frac{1}{2}\log(\pi) \tag{13.21}$$

From Eq. (13.19), it is evident that a log-log plot of Δp vs. t will yield a half-slope straight line (Figure 13.7). Similarly from Eq. (13.21), a log-log plot of the pressure derivative function versus time will also yield a half-slope straight line.

For an oil well, Eq. (13.18) can be expanded to give:

$$\Delta p = \frac{4.064qB}{hL_f}\sqrt{\frac{\mu t}{k\phi c_t}} \tag{13.22}$$

Similarly for a gas well, Eq. (13.18) can be expanded to give:

$$\Delta p_p = \frac{40.925q_gT}{hL_f}\sqrt{\frac{t}{k\phi\mu c_t}} \tag{13.23}$$

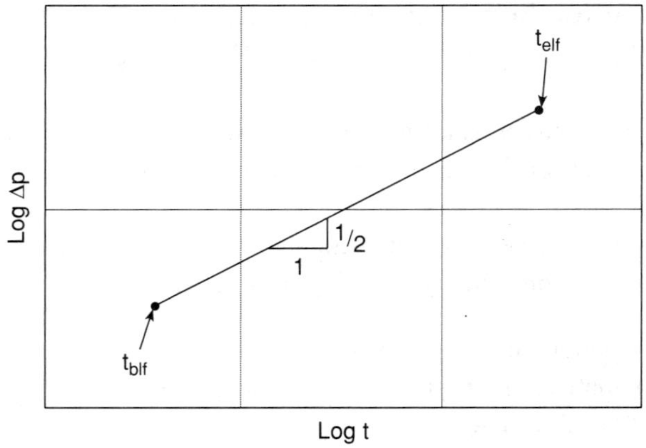

Figure 13.7 Plot of log Δp versus log t for linear flow (from Cinco-Ley and Samaniego[1] © 1981 SPE, Reproduced with permission).

From Eqs. (13.22) and (13.23), it is evident that a plot of Δp or Δp_p vs. $\sqrt{t}$ on Cartesian coordinates will give a straight line with a slope, m_{Lf}, that passes through the origin (Figure 13.8). The fracture half-length, L_f, can be calculated from the slope, m_{Lf}, as follows:

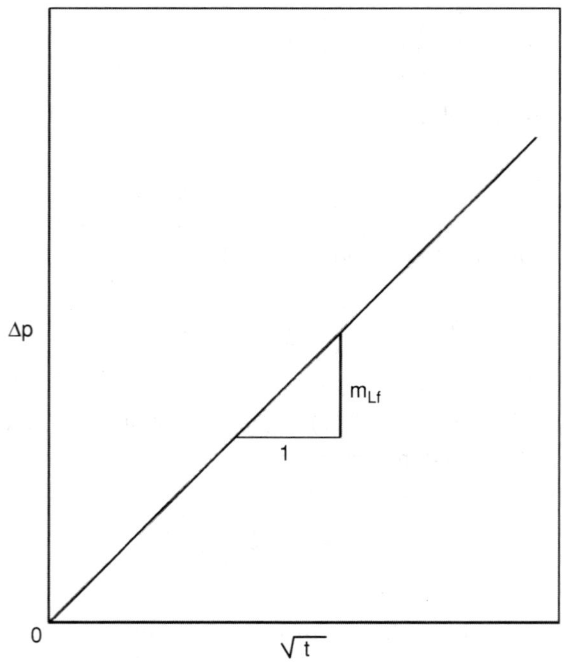

Figure 13.8 Plot of Δp versus $\sqrt{t}$ for linear flow (from Cinco-Ley and Samaniego[1] © 1981 SPE, Reproduced with permission).

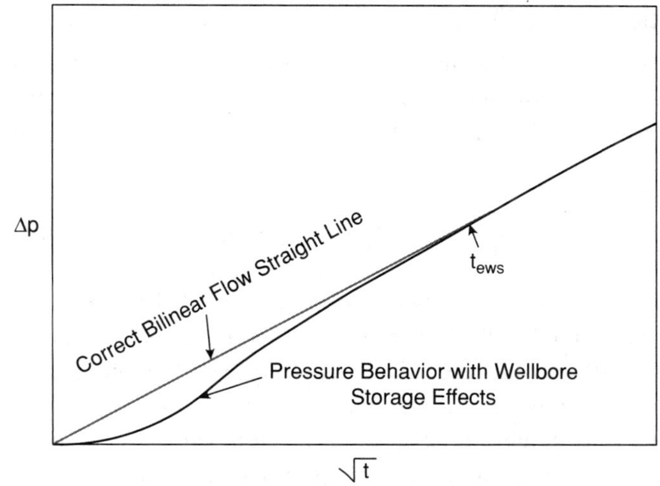

Figure 13.9 Effects of wellbore storage on Δp versus $\sqrt{t}$ plot (from Cinco-Ley and Samaniego[1] © 1981 SPE, Reproduced with permission).

For an oil well:

$$L_f = \frac{4.064qB}{m_{Lf}h}\sqrt{\frac{\mu}{k\phi c_t}} \tag{13.24}$$

For a gas well:

$$L_f = \frac{40.925q_gT}{m_{Lf}h}\sqrt{\frac{1}{k\phi\mu c_t}} \tag{13.25}$$

Formation linear flow regime can be distorted by wellbore storage effects, as shown in Figure 13.9, or flow restrictions within the fracture that will cause additional pressure drop close to the wellbore.

13.6.4 Procedure for Application of Straightline Methods on Well Test Data During Formation Linear Flow Regime

1. Plot p_{wf} vs. $\sqrt{t}$ for a constant rate flow test or p_{ws} vs. $\sqrt{\Delta t}$ for a buildup test. For gas wells, use appropriate pressure transformation functions such as real gas pseudo-pressure, as shown in Chapter 11.
2. Determine the slope, m_{Lf}, from the straightline region of the plot.
3. Calculate fracture half-length, L_f, from Eqs. (13.24) or (13.25).

Note that it is important to validate the results obtained with straightline methods with the results obtained from application of type-curve matching. This is discussed further under type-curve matching of fractured well tests.

13.6.5 Pseudo-Radial Flow

Finite and infinite conductivity fractures eventually exhibit pseudo-radial flow if the duration of the test is sufficiently long. Infinite conductivity fractures with $C_{fD} \geq 300$ will reach pseudo-radial flow when $t_{DL_f} \simeq 5$. Pseudo-radial flow for finite conductivity fractures is estimated to begin at:[2]

$$t_{DL_f} \simeq 5\exp\left[-0.5(C_{fD})^{-0.6}\right] \quad \text{for} \quad 0.1 \leq C_{fD} < 300. \tag{13.26}$$

Pseudo-radial flow is theoretically equivalent to the radial flow used previously to represent fluid flow in unfractured wells. Hence, all equations developed for radial flow and well test analysis in Chapters 10 and 11, respectively, apply to pseudo-radial flow in fractured wells. Consequently, a semilog plot of p_{wf} vs. t for a drawdown test or p_{ws} vs. $(t_p + \Delta t)/\Delta t$ for a buildup test will yield a straight line. For a drawdown test in an oil well, the slope, m, of the straightline can be used to calculate the following formation properties as shown in Chapter 11:

$$k = \left|\frac{162.6qB\mu}{mh}\right| \tag{13.27}$$

$$s = 1.151\left[\frac{\overline{p} - p_{1\text{hr}}}{m} - \log\left(\frac{k}{\phi\mu c_t r_w^2}\right) + 3.23\right] \tag{13.28}$$

For a buildup test in an oil well:

$$k = \left|\frac{162.6qB\mu}{mh}\right| \tag{13.29}$$

$$s = 1.151\left[\frac{p_{1\text{hr}} - p_{wf}}{m} - \log\left(\frac{k}{\phi\mu c_t r_w^2}\right) + 3.23\right] \tag{13.30}$$

For a drawdown test in a gas well using real gas pseudo-pressure, p_p:

$$k = \left|\frac{1{,}637q_gT}{mh}\right| \tag{13.31}$$

$$s' = 1.151\left[\frac{\overline{p}_p - p_{p,1\text{hr}}}{m} - \log\left(\frac{k}{\phi\overline{\mu c}_t r_w^2}\right) + 3.23\right] \tag{13.32}$$

For a buildup test in a gas well using real gas pseudo-pressure, p_p:

$$k = \left|\frac{1{,}637q_gT}{mh}\right| \tag{13.33}$$

$$s' = 1.151\left[\frac{p_{p,1\text{hr}} - p_{p,wf}}{m} - \log\left(\frac{k}{\phi\overline{\mu c}_t r_w^2}\right) + 3.23\right] \tag{13.34}$$

In Eqs. (13.32) and (13.34), $s' = s + Dq_g$, where s' = apparent skin factor; s = skin factor due to damage or stimulation; D = non-Darcy flow coefficient; and q_g = gas flow rate, Mcf/D. Gas properties are evaluated either at average reservoir pressure or initial pressure (Chapters 10 and 11).

The effective wellbore radius, r_{wa}, is related to skin factor, s, by:

$$r_{wa} = r_w e^{-s} \tag{13.35}$$

The half-length of a vertical fracture, L_f, can be estimated from effective wellbore radius as:

$$L_f = 2r_{wa} \tag{13.36}$$

Note that Eqs. (13.27) to (13.36) were previously derived and presented in Chapter 11.

13.6.6 Procedure for Application of Straightline Methods on Well Test Data During Pseudo-Radial Flow Regime

1. Plot p_{wf} vs. t for a constant rate drawdown test or p_{ws} vs. $(t_p + \Delta t)/\Delta t$ for a buildup test on a semilog scale. For gas wells, use appropriate pressure transformation functions, such as real gas pseudo-pressure, as shown in Chapter 11.
2. Determine the slope, m, from the straight line region of the plot.
3. Calculate formation properties from Eqs. (13.27) to (13.34) depending on the type of test and the type of fluid in the reservoir.

Note that it is important to validate the results obtained with straightline methods with the results obtained from application of type-curve matching. This is discussed further under type-curve matching of fractured well tests.

13.7 Fractured Well Test Analysis: Type-Curve Matching

In Chapter 12, analysis of well test data by type-curve matching was presented as consisting of three main stages:

1. Identification of interpretation model.
2. Calculations from interpretation model parameters.
3. Validation of the interpretation model.

These stages are also applicable to analysis of test data from fractured wells by type-curve matching. Fractured well type curves have similar shapes (Figures 13.10 and 13.11) and tend to result in a match that is non-unique. Thus, a systematic application of the above three stages in analyzing fractured well test data in an iterative mode leads ultimately to achievement of results that are consistent with current knowledge of the well and the reservoir.

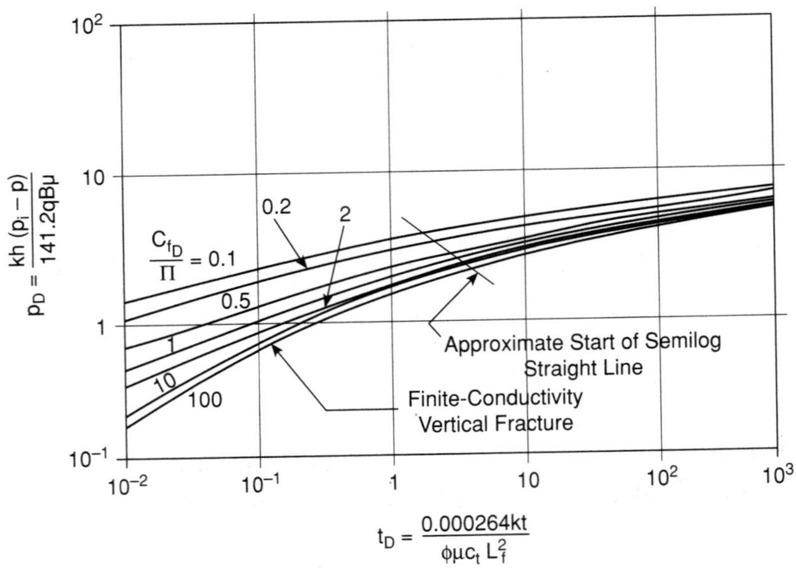

Figure 13.10 Type curve for finite conductivity vertical fracture (from Cinco-Ley et al.[3] © 1978 SPE, Reproduced with permission).

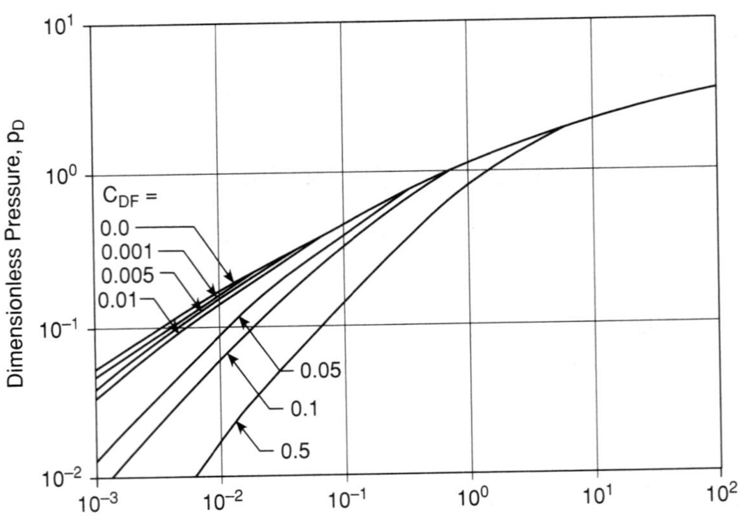

Figure 13.11 Type curve for an infinite conductivity vertical fracture with wellbore storage (from Cinco-Ley[2] © 1982 SPE, Reproduced with permission).

13.7.1 Identification of the Interpretation Model

Earlier in this chapter, three flow regimes that could be present in testing a fractured well were discussed. These flow regimes were identified as bilinear flow, formation linear flow, and pseudo-radial flow. The fourth flow regime, which is fracture linear flow, has such a very short duration at early times that it has little or no application in well test analysis. The presence of bilinear flow, formation linear flow, and pseudo-radial flow can be identified with type curves. Note that all these flow regimes may not be present in a test either because of the duration of the test or the conductivity of the fracture. Typically, finite conductivity fractures ($C_{fD} < 300$) will exhibit bilinear flow followed by transition period before reaching pseudo-radial flow if the duration of the test is sufficiently long. The formation linear flow regime is not fully developed before the emergence of pseudo-radial flow in most finite conductivity fractures. On the other hand, infinite conductivity fractures ($C_{fD} \geq 300$) exhibit formation linear flow before transiting to pseudo-radial flow depending on the duration of the test. Bilinear flow regime is typically absent in tests of infinite conductive fractures.

In analyzing fractured well test data, type curves could be used initially to identify the presence of these flow regimes. For instance, the presence of bilinear flow is indicated by the presence of the quarter-slope line in a log-log plot of Δp vs. t in a drawdown test or Δp vs. Δt_e for a buildup test. The quarter-slope line should match the quarter-slope line on a type curve. Similarly, the presence of formation linear flow is indicated by the presence of half-slope line in a log-log plot of Δp vs. t in a drawdown test or Δp vs. Δt_e for a buildup test. The half-slope line should match the half-slope line on a type curve. Note that derivative type curves could also be plotted. In that case, on the same log-log plot, plot $t\Delta p'$ vs. t for a drawdown test or $\Delta t_e \Delta p'$ vs. Δt_e for a buildup test. Typically, for fractured wells, the pressure curve and the derivative curve have the same slope as demonstrated in Eqs. (13.13) and (13.21). The derivative plot is useful in the identification of attainment of pseudo-radial flow.[5] Note that the presence of wellbore storage and fracture damage could distort the early flow regimes.[6,7]

13.7.2 Calculation from Interpretation Model Parameters

After achieving a match with a type curve, fracture and reservoir parameters can be calculated from the match points. From the pressure match point (PMP), the formation permeability, k, can be calculated from the following equations as:

$$k = \frac{141.2qB\mu}{h}\left(\frac{p_D}{\Delta p}\right)_{PMP} \quad \text{for an oil well} \tag{13.37}$$

$$k = \frac{1422q_gT}{h}\left(\frac{p_D}{\Delta p_p}\right)_{PMP} \quad \text{for a gas well} \tag{13.38}$$

The units for the terms in Eqs. (13.37) and (13.38) are as follows: k = permeability, md; q = oil rate, STB/D; B = FVF, RB/STB; μ = oil viscosity, cp; h = formation thickness, ft;

q_g = gas flow rate, Mcf/D; and T = temperature, °R. Eq. (13.38) is based on real gas pseudo-pressure, p_p. From the time match point (TMP), the fracture half-length, L_f, can be calculated from the following equations as:

$$L_f = \sqrt{\left(\frac{0.0002637k}{\phi \mu c_t}\right)\left(\frac{\Delta t \text{ or } \Delta t_e}{t_{DL_f}}\right)_{TMP}} \quad \text{for an oil well} \tag{13.39}$$

$$L_f = \sqrt{\left(\frac{0.0002637k}{\phi \overline{\mu c_t}}\right)\left(\frac{\Delta t \text{ or } \Delta t_e}{t_{DL_f}}\right)_{TMP}} \quad \text{for a gas well} \tag{13.40}$$

In Eqs. (13.39) and (13.40), L_f = fracture half-length, ft; ϕ = porosity, fraction; and c_t = total compressibility, psi^{-1}. Gas properties are evaluated either at average reservoir pressure or initial pressure (Chapters 10 and 11).

From the match of the dimensionless fracture conductivity, C_{fD}, the fracture conductivity can be calculated as:

$$w_f k_f = C_{fD} \times kL_f \tag{13.41}$$

13.7.3 Validation of the Interpretation Model Results

The validation step involves using straightline methods to calculate fracture and reservoir properties as previously discussed and comparing them to the results obtained from type-curve matching. If the results are reasonably close, then it can be presumed that the interpretation model is representative of the test data, and that the results are inherently consistent. If the results are widely divergent, then the analysis should begin again from type-curve matching by choosing another interpretation model, and repeating the entire process. This process is highly iterative and should be continued until reasonable agreement is achieved with the results from type-curve matching and straightline methods.

For instance, if the interpretation model had indicated bilinear flow, use the procedures presented earlier to calculate fracture conductivity. Compare it to the fracture conductivity calculated from type-curve matching. If the results are in reasonable agreement, then the analysis is terminated. If not, repeat the entire process starting from selection of another interpretation model.

Similarly, if the interpretation model had indicated formation linear flow, use the procedures enumerated earlier to calculate the fracture half-length. Compare it to the fracture half-length from type-curve matching. If the results are in reasonable agreement, then the analysis is terminated. If not, repeat the entire process starting from selection of another interpretation model.

If the duration of the test is sufficiently long to reach pseudo-radial flow, use the procedures enumerated above to calculate formation permeability. Compare it to the formation permeability from type-curve matching. If the results are in reasonable agreement, then the analysis is terminated. If not, repeat the entire process starting from selection of another interpretation model. From pseudo-radial flow analysis, other well parameters such as skin factor and fracture half-length can be estimated as shown in the procedures for analysis of pseudo-radial flow using straightline methods.

13.7.4 Procedure for Analysis of Well Test from Hydraulically Fractured Wells

The procedure for analysis of well test from hydraulically fractured wells is summarized in the following steps:

1. Plot $\log \Delta p$ vs. $\log t$ for a drawdown test or $\log \Delta p$ vs. $\log \Delta t_e$ for a buildup test on the same scale size as the type curves graph.
2. Select either finite conductivity or infinite conductivity type curves. Most wells exhibit finite conductivity behavior. For this reason, make the first pass with finite conductivity type curves.
3. Match your plot to one of the type curves. Identify potential flow regimes on the type curve. Select a match point. Determine pressure and time match points.
4. Calculate reservoir and well parameters from the match points.
5. Make appropriate plots of p_{wf} or p_{ws} vs. $\sqrt[4]{t}$ for bilinear flow; p_{wf} or p_{ws} vs. $\sqrt{t}$ for formation linear flow; p_{wf} vs. $\log t$ for drawdown test or $[p_{ws}$ vs. $\log\{(t_p + \Delta t)/\Delta t\}]$ for a buildup test if pseudo-radial is indicated.
6. Calculate reservoir and well parameters from straightline methods of Step 5.
7. Compare results from Step 4 to results from Step 6.
8. If results are in reasonable agreement, terminate the analysis. If not, repeat Steps 2 to 7.

Note that the above procedure is for manual analysis of fractured well tests. Most modern well test analysis is computer-aided including non-linear regression matching of the test data to an interpretation model. For this reason, the well test analyst may never have to conduct manual type-curve matching of the test data. As pointed out previously, this procedure was included to illustrate the steps actually involved in computer-aided well test analysis. The goal for this approach is to provide the reader with thorough understanding of the processes embedded in computer-aided analysis of well tests.

Example 13.1 Calculation of Well and Reservoir Parameters from a Drawdown Test Data Using Finite Conductivity Type Curves

Problem

A drawdown test was conducted on a hydraulically fractured well. The test data are shown in Table 13.1. Calculate fracture half-length and reservoir properties from the test data. Well and other reservoir properties are given as follows:

Table 13.1 Test and Plot Data for Example 13.1

t (hrs.)	p_{wf} (psia)	$\sqrt{t}$ (hrs.)$^{0.5}$	Δp (psia)	t (hrs.)	p_{wf} (psia)	$\sqrt{t}$ (hrs.)$^{0.5}$	Δp (psia)
0.3	2488.64	0.548	61.36	60.0	2182.91	7.746	367.09
0.5	2477.09	0.707	72.91	65.0	2172.85	8.062	377.15
1.0	2457.86	1.000	92.14	70.0	2163.30	8.367	386.70
1.5	2444.35	1.225	105.65	75.0	2154.18	8.660	395.82
2.0	2433.57	1.414	116.43	80.0	2145.46	8.944	404.54
2.5	2424.46	1.581	125.54	85.0	2137.10	9.220	412.90
3.0	2416.49	1.732	133.51	90.0	2129.05	9.487	420.95
3.5	2409.35	1.871	140.65	95.0	2121.30	9.747	428.70
4.0	2402.87	2.000	147.13	100.0	2113.81	10.000	436.19
4.5	2396.90	2.121	153.10	105.0	2106.56	10.247	443.44
5.0	2391.36	2.236	158.64	110.0	2099.54	10.488	450.46
5.5	2386.17	2.345	163.83	115.0	2092.73	10.724	457.27
6.0	2381.28	2.449	168.72	120.0	2086.11	10.954	463.89
6.5	2376.66	2.550	173.34	130.0	2078.51	11.402	471.49
7.0	2372.27	2.646	177.73	140.0	2070.91	11.832	479.09
7.5	2368.08	2.739	181.92	150.0	2063.31	12.247	486.69
8.0	2364.07	2.828	185.93	160.0	2055.71	12.649	494.29
8.5	2360.23	2.915	189.77	170.0	2048.11	13.038	501.89
9.0	2356.53	3.000	193.47	180.0	2040.51	13.416	509.49
9.5	2352.97	3.082	197.03	190.0	2032.91	13.784	517.09
10.0	2349.53	3.162	200.47	200.0	2025.31	14.142	524.69
11.0	2342.97	3.317	207.03	210.0	2020.11	14.491	529.89
12.0	2336.80	3.464	213.20	220.0	2014.91	14.832	535.09
13.0	2330.96	3.606	219.04	230.0	2009.71	15.166	540.29
14.0	2325.41	3.742	224.59	240.0	2004.51	15.492	545.49
15.0	2320.12	3.873	229.88	250.0	1999.31	15.811	550.69
16.0	2315.05	4.000	234.95	260.0	1994.11	16.125	555.89
17.0	2310.19	4.123	239.81	270.0	1988.91	16.432	561.09
18.0	2305.52	4.243	244.48	280.0	1983.71	16.733	566.29

t (hrs.)	p_{wf} (psia)	$\sqrt{t}$ (hrs.)$^{0.5}$	Δp (psia)	t (hrs.)	p_{wf} (psia)	$\sqrt{t}$ (hrs.)$^{0.5}$	Δp (psia)
19.0	2301.02	4.359	248.98	290.0	1978.51	17.029	571.49
20.0	2296.67	4.472	253.33	300.0	1973.31	17.321	576.69
22.0	2288.38	4.690	261.62	310.0	1968.71	17.607	581.29
24.0	2280.58	4.899	269.42	320.0	1964.11	17.889	585.89
26.0	2273.20	5.099	276.80	330.0	1959.51	18.166	590.49
28.0	2266.19	5.292	283.81	340.0	1954.91	18.439	595.09
30.0	2259.50	5.477	290.50	350.0	1950.31	18.708	599.69
35.0	2243.98	5.916	306.02	360.0	1945.71	18.974	604.29
40.0	2229.87	6.325	320.13	370.0	1941.11	19.235	608.89
45.0	2216.88	6.708	333.12	380.0	1936.51	19.494	613.49
50.0	2204.82	7.071	345.18	390.0	1931.91	19.748	618.09
55.0	2193.53	7.416	356.47	400.0	1927.31	20.000	622.69

Formation thickness, h	60 ft
Formation porosity, ϕ	0.20
Total compressibility, c_t	20×10^{-6} psi^{-1}
Oil viscosity, μ_o	2.0 cp
Oil FVF, B_o	1.5 RB/STB
Wellbore radius, r_w	0.25 ft.
Production rate, q_o	200 STB/D
Initial average pressure, $\overline{p}$	2550 psia

Solution

Step 1: Plot $\log \Delta p$ vs. $\log t$.

The data in Table 13.1 were used to make a log-log plot of Δp vs. t as shown in Figure 13.12.

Step 2: Type-curve matching.

The type curve shown as Figure 13.10 was used in the type-curve matching procedure. Figure 13.12 was superimposed on Figure 13.10. The matched (superimposed) plot of Figures 13.10 and 13.12 is shown as Figure 13.13. The matched points are as follows:

$$\text{Pressure match point (PMP)}, (p_D/\Delta p)_{PMP} = (0.6/100)$$

$$\text{Time match point (TMP)}, (\Delta t/t_{DL_f})_{TMP} = (100/3.0)$$

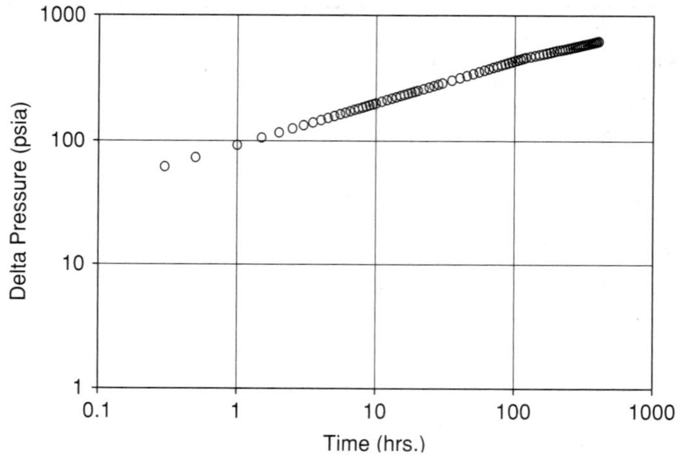

Figure 13.12 Log-log plot of Δp vs. t for Example 13.1.

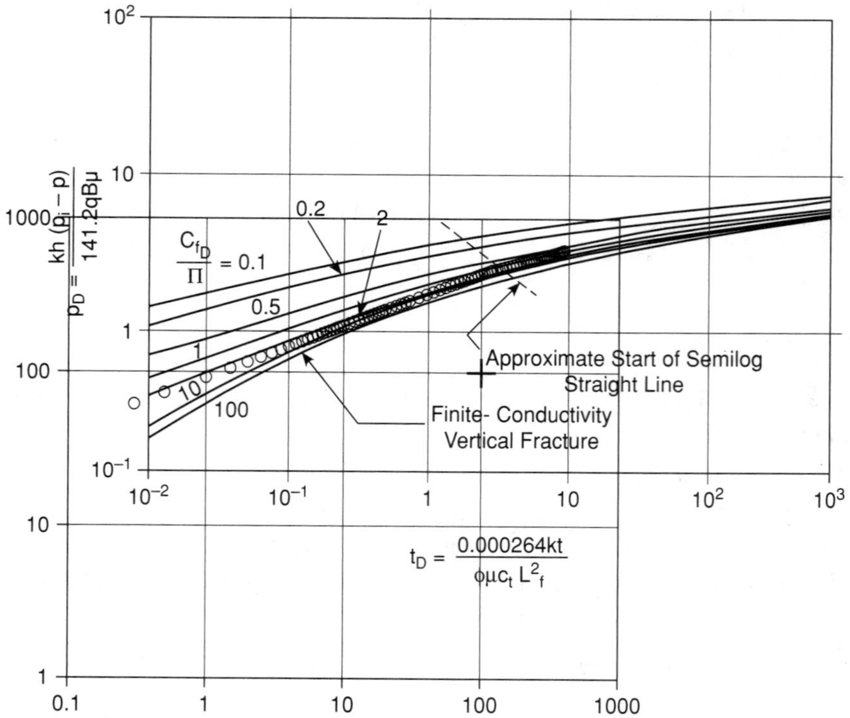

Figure 13.13 Type-curve match for Example 13.1.

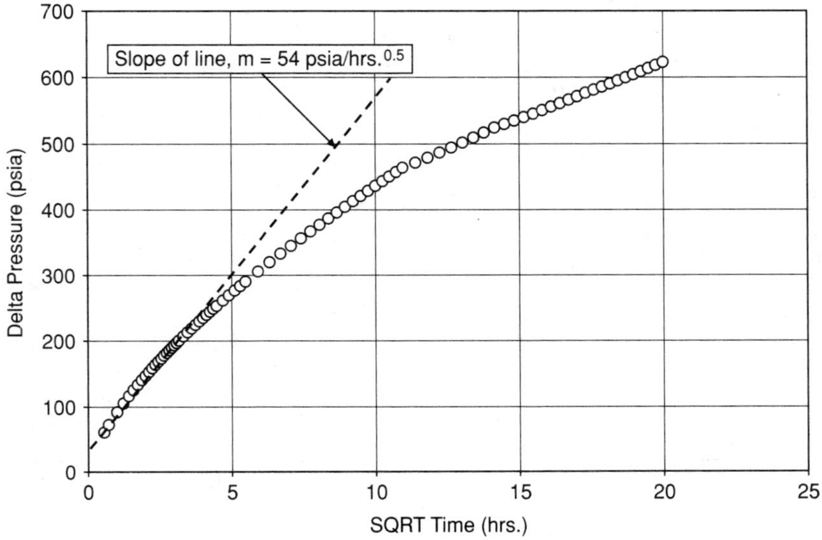

Figure 13.14 Plot of Δp vs. $\sqrt{t}$ for Example 13.1.

Using Eq. (13.37),

$$k = \frac{141.2 qB\mu}{h}\left(\frac{p_D}{\Delta p}\right)_{PMP} = \frac{141.2 \times 200 \times 1.5 \times 2}{60} \times \left(\frac{0.6}{100}\right) = 8.47 \text{ md}$$

From Eq. (13.39),

$$L_f = \sqrt{\left(\frac{0.0002637k}{\phi\mu c_t}\right)\left(\frac{\Delta t}{t_{DL_f}}\right)_{TMP}} = \sqrt{\frac{0.0002637 \times 8.47}{0.2 \times 2 \times 20 \times 10^{-6}} \times \left(\frac{100}{3}\right)} = 96.47 \text{ ft.}$$

From Figure 13.13, the matched type curve is designated as: $C_{fD} = 10\pi$. From Eq. (13.41), fracture conductivity is calculated as:

$$w_f k_f = C_{fD} \times kL_f = 10\pi \times 8.47 \times 96.47 = 25{,}670 \text{ md-ft}$$

Step 3: Plot Δp vs. $\sqrt{t}$.

The data in Table 13.1 were used to make a plot of Δp vs. $\sqrt{t}$ as shown in Figure 13.14. The slope of the straight line is:

$$m_{Lf} = 54 \text{ psia/hrs}^{0.5}$$

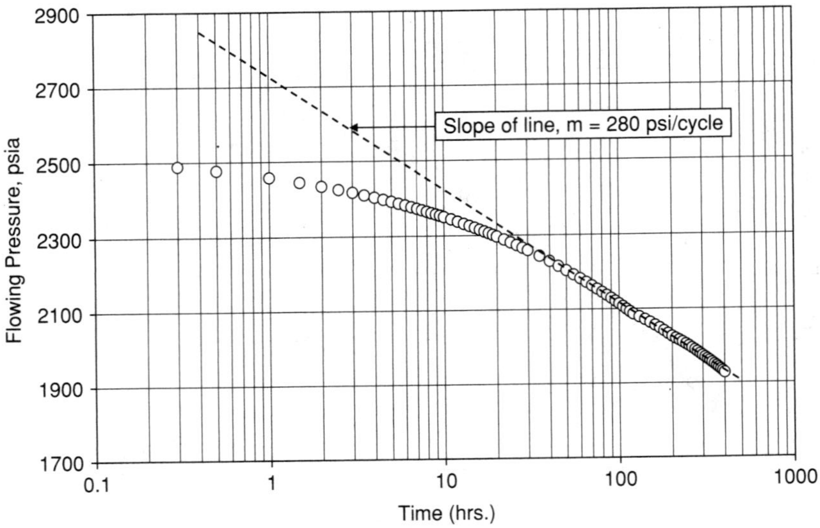

Figure 13.15 Semilog plot of p_{wf} vs. t for Example 13.1.

Applying Eq. (13.24),

$$L_f = \frac{4.064qB}{m_{Lf}h}\sqrt{\frac{\mu}{k\phi c_t}}$$
$$= \frac{4.064 \times 200 \times 1.5}{54 \times 60}\sqrt{\frac{2}{8.47 \times 0.2 \times 20 \times 10^{-6}}}$$
$$= 91.43 \text{ ft.}$$

Step 4: Plot p_{wf} vs. log t.
The data in Table 13.1 were used to prepare a semilog plot of p_{wf} vs. t as shown in Figure 13.15. The slope of the semilog straight line is: $m = 280$ psi/cycle.
Applying Eq. (13.27),

$$k = \frac{162.6qB\mu}{mh}$$
$$= \frac{162.6 \times 200 \times 1.5 \times 2}{280 \times 60} = 5.81 \text{ md}$$

Also, from Figure 13.15, $p_{1hr} = 2738$ psia.
Applying Eq. (13.28),

$$s = 1.151\left[\frac{\bar{p} - p_{1\text{hr}}}{m} - \log\left(\frac{k}{\phi\mu c_t r_w^2}\right) + 3.23\right]$$
$$= 1.151\left[\frac{2550 - 2738}{280} - \log\left(\frac{5.81}{0.2 \times 2 \times 20 \times 10^{-6} \times (0.25)^2}\right) + 3.23\right]$$
$$= -5.19$$

Applying Eq. (13.35),

$$\begin{aligned} r_{wa} &= r_w e^{-s} \\ &= (0.25)e^{-(-5.19)} = 44.7 \text{ ft}. \end{aligned}$$

Further by applying Eq. (13.36),

$$\begin{aligned} L_f &= 2r_{wa} \\ &= 2 \times 44.7 = 89.4 \text{ ft}. \end{aligned}$$

There are reasonable agreements between well and reservoir parameters calculated from type-curve matching and straightline methods. For instance, the formation permeabilities calculated from the two methods are reasonably close. This indicates that the results obtained from analyzing the well test data with the two methods are reasonably consistent.

13.8 Naturally Fractured Reservoirs

Many reservoirs in different petroleum producing formations across the world are naturally fractured. In a majority of these cases, the productivity of these reservoirs depends on the presence of these fractures. The fractures act as conduits that convey fluids from the matrix to the wellbore. Production rates of wells in naturally fractured reservoirs are relatively high, although these wells may have short lives due to channeling of gas and/or water through the fracture system to the wellbore. The pressure response of naturally fractured reservoirs during a well test is different from that of unfractured reservoirs. In this section, well test analysis is presented as a tool that can be applied in the diagnosis of naturally fractured reservoirs.

13.9 Naturally Fractured Reservoir Models

Naturally fractured reservoirs (NFRs) are composed of random distributions of fractures, vugs, and matrices, as shown in Figure 13.16. The distribution of fractures in the reservoir can be massive, localized, oriented, or clustered along a fault. The nature of the fractures and their distribution has considerable influence on the pressure response of a well in a test. In well test analysis, naturally fractured reservoirs can be represented with different models based on the type and distribution of the fractures within the system. These representative models are:[8]

1. Homogeneous reservoir
2. Multiple region or composite reservoir
3. Anisotropic reservoir
4. Single fracture reservoir
5. Double-porosity reservoir

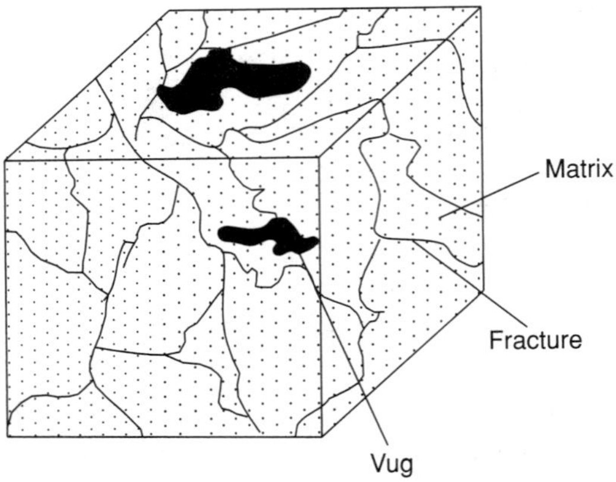

Figure 13.16 Naturally fractured rock (from Cinco-Ley[8] © 1996 SPE, Reproduced with permission).

13.9.1 Homogeneous Reservoir Model

Homogeneous reservoir model can be used to represent massively fractured formations with small matrix blocks (Figure 13.17a) or fractured formations in which most of the fluids are stored within the fractures (Figure 13.17b). In these types of NFRs, the fractures and the rock matrix behave as a single indistinguishable system. Such NFRs exhibit homogeneous type of behavior in flow tests. Their pressure responses in well tests are similar to those of other homogeneous-acting reservoirs previously described in Chapters 11 and 12. Consequently, all well test analysis techniques composed of straightline methods and type-curve matching developed earlier in Chapters 11 and 12 are equally applicable to these types of NFRs. For instance, a plot of

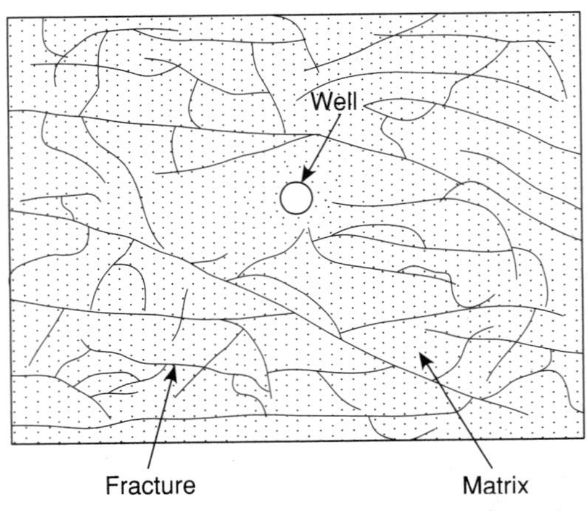

Figure 13.17a Massively fractured reservoir (from Cinco-Ley[8] © 1996 SPE, Reproduced with permission).

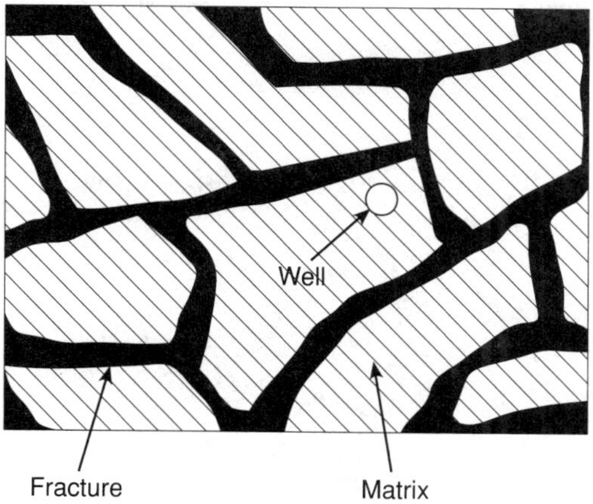

Figure 13.17b Naturally fractured reservoir with low porosity/permeability matrix (from Cinco-Ley[8] © 1996 SPE, Reproduced with permission).

p_{wf} vs. log t for a drawdown test or a Horner plot, p_{ws} vs. $\log\{(t_p + \Delta t)/\Delta t\}$, for a buildup test yields a straightline with slope, m, from which the flow capacity, kh, of the entire system (fracture plus matrix) can be calculated.

13.9.2 Multiple Region or Composite Reservoir Model

In Figure 13.18, a fractured reservoir composed of two regions is shown. The rock formation in one region is fractured, while the rest of the reservoir is not fractured. The fractured region will most likely exhibit higher transmissibility than the unfractured region. This NFR type can be represented with a composite radial model. Wells completed in the fractured region of the reservoir will tend to exhibit higher production rates than wells completed in the unfractured region. The flow capacity of the composite system depends on the flow capacity of the fractured and unfractured regions. If a test is conducted on a well located in the fractured region, the pressure

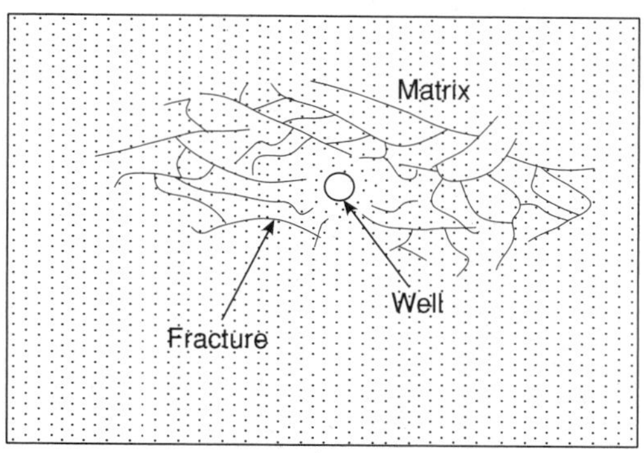

Figure 13.18 Regionally fractured reservoir (from Cinco-Ley[8] © 1996 SPE, Reproduced with permission).

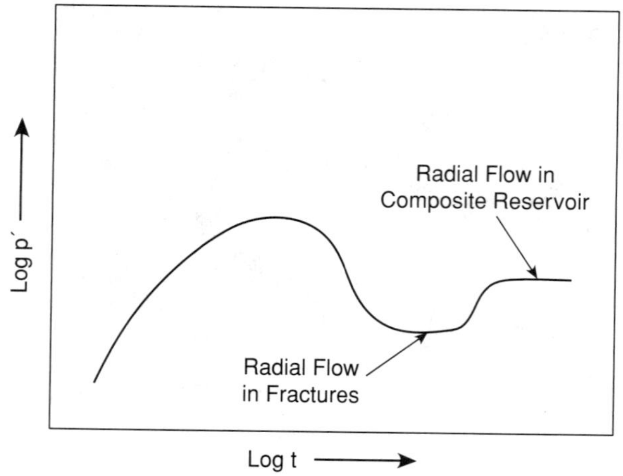

Figure 13.19a Derivative plot for a composite reservoir model (from Cinco-Ley[8] © 1996 SPE, Reproduced with permission).

response will be influenced initially by nearby fractures, and later by the unfractured region if the duration of the test is sufficiently long. Figure 13.19a is schematic log-log plot of the pressure derivative. The first horizontal line in Figure 13.19a represents radial flow from the fracture system after wellbore storage effects have ended. After a transition period, the second horizontal line represents radial flow from the fractured and unfractured regions (composite system) of the reservoir. Figure 13.19b is a schematic semilog plot of the pressure response. The first straight line in Figure 13.19b represents the fracture system and the second straight line represents the composite reservoir. The flow capacities of the fractured region and the composite reservoir can be calculated from the slopes of these straight lines (m_1 and m_2), respectively.

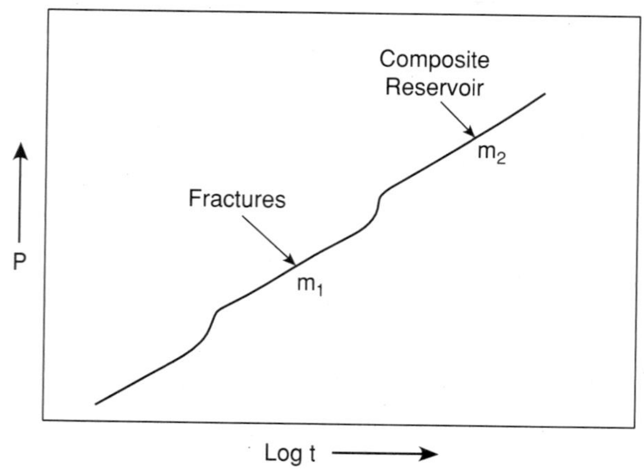

Figure 13.19b Semilog plot for a composite reservoir model (from Cinco-Ley[8] © 1996 SPE, Reproduced with permission).

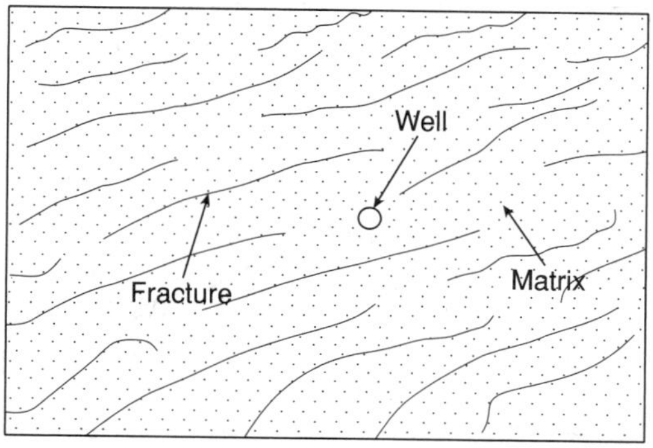

Figure 13.20 Naturally fractured reservoir with directional fractures (from Cinco-Ley[8] © 1996 SPE, Reproduced with permission).

13.9.3 Anisotropic Reservoir Model

The fractures in some NFRs are aligned in a particular direction as shown in Figure 13.20. In these reservoirs, the composite permeability in the direction of the fractures is higher than the composite permeability in the directional normal to the fractures. Consequently, these reservoirs exhibit anisotropic behavior. The reservoir exhibits a maximum permeability, $k_{\max}$, and a minimum permeability, k_{min}, as shown in Figure 13.21. In NFR exhibiting anisotropic behavior, interference tests are the best method for determining $k_{\max}$ and k_{min} plus the orientation of the principal axis of permeability.[8] The application of interference testing in wells is not covered in this book.

13.9.4 Single Fracture Model

In Figure 13.22, a well located near a high conductivity fracture system is shown. This fracture system may consist of a permeable fault through which fluids can flow to the well from other parts of the reservoir. Wells located in this region of the reservoir are capable of producing at

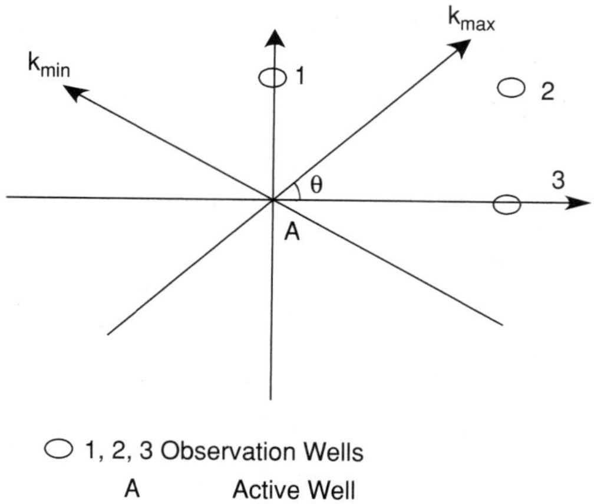

Figure 13.21 Orientation of fracture permeability for anisotropic NFR (from Cinco-Ley[8] © 1996 SPE, Reproduced with permission).

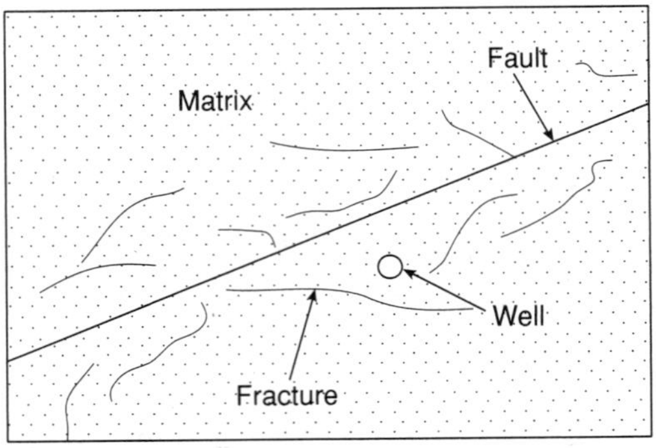

Figure 13.22 Naturally fractured reservoir with highly conductive fault (from Cinco-Ley[8] © 1996 SPE, Reproduced with permission).

very high, sustained rates. They are also subject to premature water breakthrough if the permeable fault traverses the aquifer. Several reservoir and well parameters can be calculated from analysis of a well test in this reservoir. Figure 13.23 is a log-log pressure derivative plot of a well test for such reservoir. As shown in Figure 13.23, there is a radial flow regime representing the entire system. After a transition period, the response to the permeable fault appears to indicate a constant pressure boundary. This is then followed by bilinear flow representing the flow through the fracture system. Several specialized plots (Figure 13.24) are useful for analysis of flow tests on the well. These include a semilog plot to calculate the flow capacity, kh, of the total system and a plot of p vs. $\sqrt[4]{t}$ for the bilinear flow regime to calculate fracture conductivity, $w_f k_f$.

13.9.5 Double Porosity Model

NFRs have traditionally been described as consisting of a matrix system and a fracture system as shown in Figure 13.25. Reservoirs' fluids flow from the matrix system to the fracture system and the fracture system delivers the fluids to the wellbore. The capacity of the matrix system to

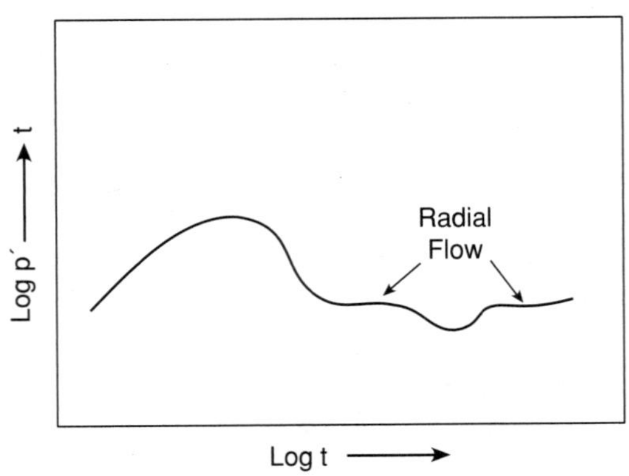

Figure 13.23 Derivative plot for NFR with highly conductive fault (from Cinco-Ley[8] © 1996 SPE, Reproduced with permission).

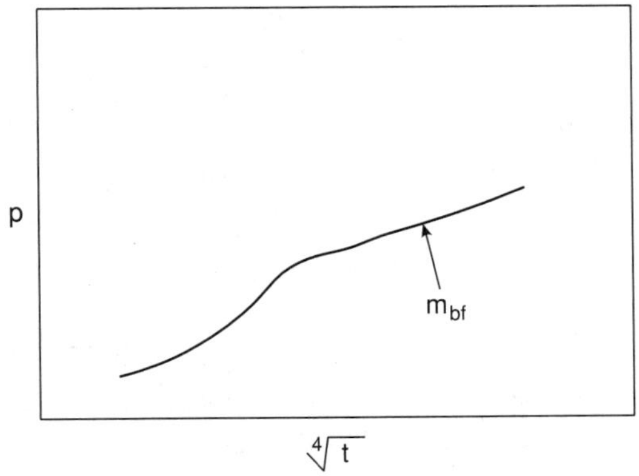

Figure 13.24 Specialized plot for a well near a highly conductive fault (from Cinco-Ley[8] © 1996 SPE, Reproduced with permission).

store and transport fluids is measured by its porosity, ϕ_{ma}, and permeability, k_{ma}, respectively. Similarly, the storage and transport capacity of the fracture system is measured by its porosity, ϕ_f, and permeability, k_f, respectively.

Since reservoir fluids are stored in both the matrix and fractures, the double porosity model is used to represent the total system. Two flow conditions are used to describe the flow of fluids from the matrix to the fractures. These are pseudosteady-state flow[9,10] and transient flow[11–13] conditions. It is not readily apparent which flow condition is prevalent in a flow test. Some tests appear to indicate pseudosteady-state interporosity flow behavior while others indicate transient interporosity flow behavior.[14] In the analysis of a flow test, pseudosteady-state flow should be assumed on a trial basis. If analysis of the test data does not conform to this model, then transient flow behavior can be assumed as an alternative model.

The term interporosity flow describes the exchange of fluids between the fracture and the matrix systems. A measure of fluid transfer between the matrix and the fracture systems is called the interporosity flow coefficient. Interporosity flow coefficient, λ, is defined as:[14]

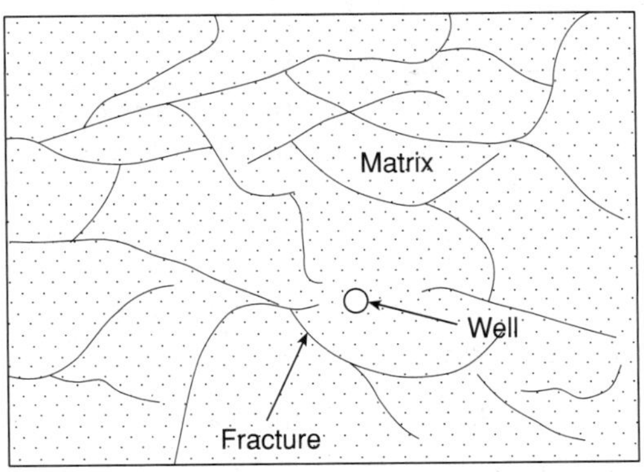

Figure 13.25 NFR with fracture and matrix permeability (from Cinco-Ley[8] © 1996 SPE, Reproduced with permission).

$$\lambda = \alpha r_w^2 \frac{k_{ma}}{k_f} \tag{13.42}$$

In Eq. (13.42), k_{ma} = matrix permeability; k_f = fracture permeability; and α = a parameter characteristic of the geometry of the system. The parameter, α, is defined as:

$$\alpha = \frac{4n(n + 2)}{l^2} \tag{13.43}$$

In Eq. (13.43), n = the number of normal sets of planes limiting the least permeable medium; and l = characteristic dimension of the matrix block.[14] Figure 13.26 shows an idealized matrix block model of a fractured reservoir. For this representative model, $n = 3$. Figure 13.27 shows a slab model of a fractured reservoir. For this model, $n = 1$. The range of values for λ is from 10^{-9} to 10^{-4}. A high value for λ indicates fast interaction between the fracture system and the matrix system.[8]

The term storativity represents the storage and expansion capacity of the fluid in the particular system. The storativity of the fractures relative to the storativity of the total system is called the storativity ratio. It is a measure of the relative storage and expansion capacity of the fractures in the entire system. Storativity ratio, ω, is defined as:

$$\omega = \frac{(\phi V c_t)_f}{(\phi V c_t)_f + (\phi V c_t)_{ma}} \tag{13.44}$$

In Eq. (13.44), V = ratio of the total volume of medium to the bulk volume of the total system. The subscripts, f and ma, denote fracture and matrix, respectively. The range of typical values for ω is from 0.001 to 0.5.[8]

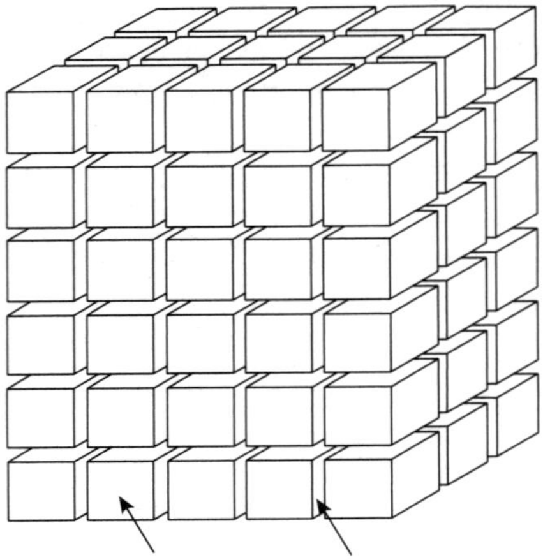

Figure 13.26 Idealized matrix block model for NFRs (from Warren and Root[10] © 1963 SPE, Reproduced with permission).

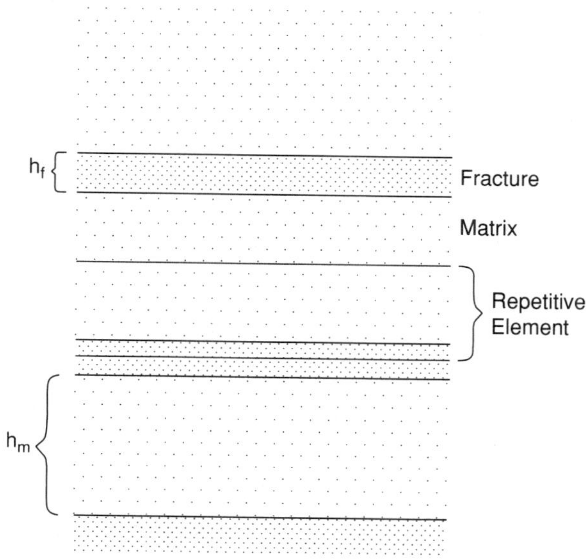

Figure 13.27 Slab model for NFRs (from Serra et al.[13] © 1983 SPE, Reproduced with permission).

13.10 Well Test Analysis in Naturally Fractured Reservoirs Based on Double Porosity Model

Many NFRs exhibit pseudosteady-state flow from the matrix to the fractures.[14] Pseudosteady-state flow implies that the rate of pressure change in the matrix is constant. Initially, fluid flow to the wellbore is from the fracture system. This followed by a transition period during which fluids begin to flow from the matrix. Finally, an equilibrium condition is reached at which fluids flow from the matrix to the fractures, and then from the fractures to the wellbore. A semilog plot of Δp vs. t representing this behavior is shown in Figure 13.28. In Figure 13.28, Line 1 represents flow in the fracture system exhibiting radial, homogeneous behavior, and Line 2 represents radial, homogeneous flow in the total system (fracture plus matrix). Line 1 is parallel to Line 2. For this reason, both lines have the same slope. From the slope, the flow capacity, $(kh)_{f+ma}$, of the total system can be calculated. The flow capacity, $(kh)_{f+ma}$, is essentially equal to the flow capacity of the fracture, $(kh)_f$, since the permeability of the fracture is much greater than the permeability of the matrix (i.e., $k_f \gg k_m$).

Some naturally fractured reservoirs exhibit transient flow from the matrix to the fractures.[13] Transient flow conditions imply the rate of pressure change varies with time and location in the matrix. As previously described for pseudosteady-state flow, flow originates in the fractures under transient conditions. This followed by transition period before stabilized flow goes from the matrix to the fractures, and finally from the fractures to the wellbore. A semilog plot of Δp vs. t representing this transient behavior is shown in Figure 13.29. In Figure 13.29, Line 1 represents flow in the fracture system exhibiting radial, homogeneous behavior, and Line 3 represents radial, homogeneous flow in the total system (fracture plus matrix). A third straight

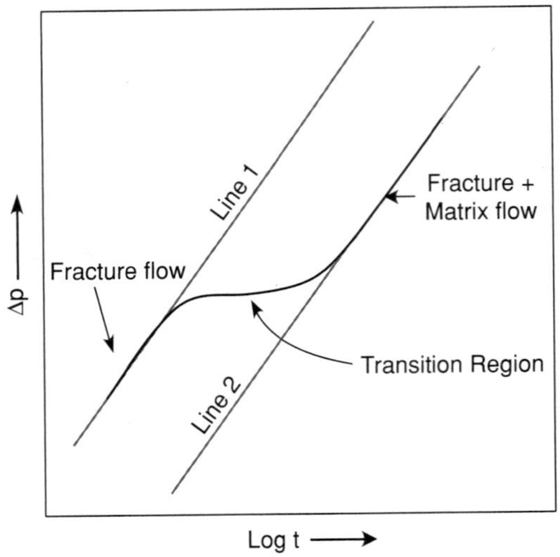

Figure 13.28 Semilog pressure response of a NFR assuming pseudosteady-state flow.

line, Line 2, represents transitional flow regime from the matrix to the fractures. Lines 1 and 3 have the same slope. The slope of Line 2 is equal to half the slope of either Lines 1 or 3.

13.11 Well Test Analysis in NFRs: Straightline Methods

Fluid flow in the matrix of NFRs is described under the double porosity model either by assuming pseudosteady-state conditions or transient state conditions. Most flow tests in NFRs appear to exhibit pseudosteady-state behavior.[14] Consequently, the application of straightline methods in the analysis of well tests in NFRs is based primarily on the assumption of pseudosteady-state

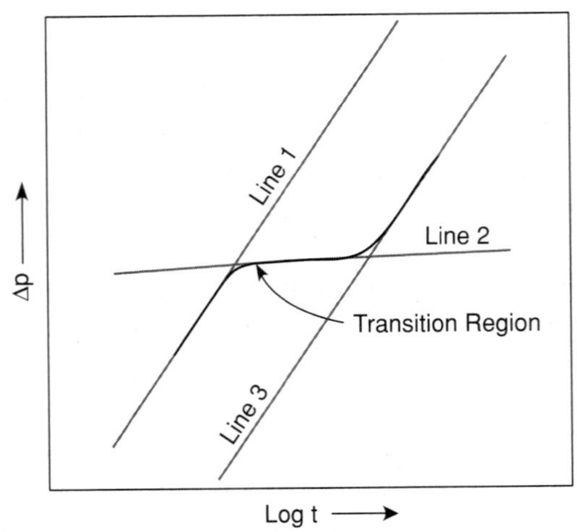

Figure 13.29 Semilog pressure response of a NFR assuming transient flow.

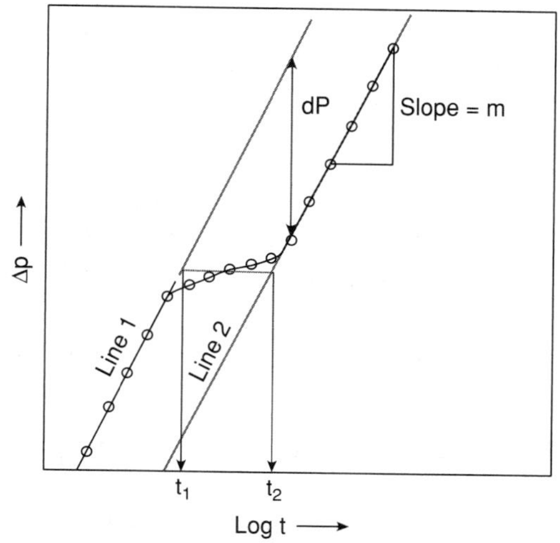

Figure 13.30 Semilog plot of Δp vs. t for a NFR flow test.

flow conditions. However, the equations presented here are applicable if transient flow conditions are assumed.

A semilog plot of pressure change versus time (Δp vs. $\log t$) is shown in Figure 13.30, assuming pseudosteady-state flow. Lines 1 and 2 are two parallel lines representing homogeneous flow in the fractures and the total system (fractures plus matrix), respectively. The slope, m, of either straight line can be used to calculate the permeability-thickness product, kh, of the fracture or the total system, thus:

$$(kh)_f \simeq (kh)_{total} = \frac{162.6 q B \mu}{m} \tag{13.45}$$

In Eq. (13.45), $(kh)_{total} = [(kh)_f + (kh)_m]$. Since fracture permeability, k_f, is much greater than matrix permeability, k_m, (*i.e.*, $k_f \gg k_m$), then $(kh)_f \simeq (kh)_{total}$ as indicated in Eq. (13.45).

The storativity ratio, ω, can be calculated from the vertical pressure displacement, δp, between Lines 1 and 2 as shown in Figure 13.30. Thus,

$$\omega = 10^{-\left(\frac{\delta p}{m}\right)} \tag{13.46}$$

As shown in Figure 13.30, a straight line drawn through the middle of the transition curve will intersect Line 1 at t_1 and Line 2 at t_2. The interporosity flow coefficient, λ, can be calculated using either t_1 or t_2 from the following equations:[14]

$$\lambda = \frac{(\phi V c_t)_f \mu r_w^2}{\gamma k_f t_1} = \frac{(\phi V c_t)_{f+m} \mu r_w^2}{\gamma k_f t_2}, \quad \text{for drawdown tests} \tag{13.47}$$

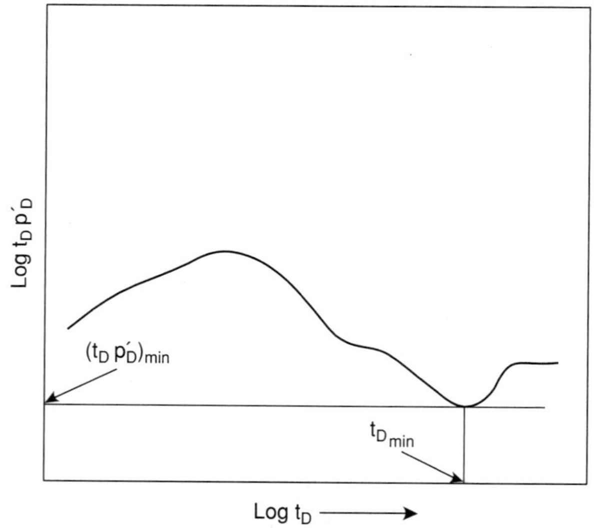

Figure 13.31 NFR derivative plot.

$$\lambda = \frac{(\phi V c_t)_f \mu r_w^2 (t_p + \Delta t_1)}{\gamma k_f t_p \Delta t_1} = \frac{(\phi V c_t)_{f+m} \mu r_w^2 (t_p + \Delta t_2)}{\gamma k_f t_p \Delta t_2}, \quad \text{for buildup tests} \tag{13.48}$$

In Eqs. (13.47) and (13.48), γ is the exponential of Euler's constant, which is approximately 1.781.

It is clearly evident from Figure 13.30 that these methods used earlier to calculate the storativity ratio, ω, and the interporosity flow coefficient, λ, depend on the presence of the two semilog straight lines, Line 1 and Line 2. Line 1, which is based on early flow in the fractures, is usually distorted by wellbore storage and may not be observed. Consequently, parameters needed for Eqs. (13.46) to (13.48) may not be obtainable from a semilog plot similar to Figure 13.30.

An alternative method for calculating the storativity ratio and interporosity flow coefficient is based on the characteristic minimum point that is typically present on the pressure derivative plots of flow tests from NFRs, as shown in Figure 13.31. The dimensionless time value, $t_{D\,min}$, at which this minimum point occurs is given as:[15]

$$t_{D\,min} = \frac{\omega}{\lambda}\left[\ln\left(\frac{1}{\omega}\right)\right] \tag{13.49}$$

This minimum point also corresponds to the dimensionless pressure derivative value derived as:[16]

$$\left[t_D \frac{\partial p_D}{\partial t_D}\right]_{min} = 0.5\left[1 + \omega^{\{1/(1-\omega)\}} - \omega^{\{\omega/(1-\omega)\}}\right] \tag{13.50}$$

Equation (13.50) can be solved for ω by using either successive substitution methods or Newton-Raphson numerical analysis techniques. An initial estimate for ω can be obtained from the expression:[16]

$$\log \omega = \frac{\left[0.01765 + \log\left(t_D \frac{\partial p}{\partial t_D}\right)_{\min}\right]}{0.94903} \tag{13.51}$$

The interporosity flow coefficient is calculated from Eq. (13.49) by substituting the value of the storativity ratio calculated from Eq. (13.50).

The skin factor, s, is calculated by extrapolating the second semilog straight line (Line 2, Figure 13.30) to $p_{1\text{hr}}$. The skin factor is calculated from the conventional expression:

$$s = 1.151\left[\frac{\Delta p_{1\text{hr}}}{m} - \log\left(\frac{k}{\phi \mu c_t r_w^2}\right) + 3.23\right] \tag{13.52}$$

Note that in Eq. (13.52), $\Delta p_{1\text{hr}} = (p_i - p_{1\text{hr}})$ for a drawdown test or $\Delta p_{1\text{hr}} = (p_{1\text{hr}} - p_{wf@\Delta t=0})$ for a buildup test.

13.12 Well Test Analysis in NFRs: Type Curves

Type curves, in conjunction with straightline methods, are useful in well test analysis of NFRs, especially in cases where early time flows in the fractures are distorted by wellbore storage. A typical analysis using type curves is illustrated in Figure 13.32, which shows the pressure behavior of a well

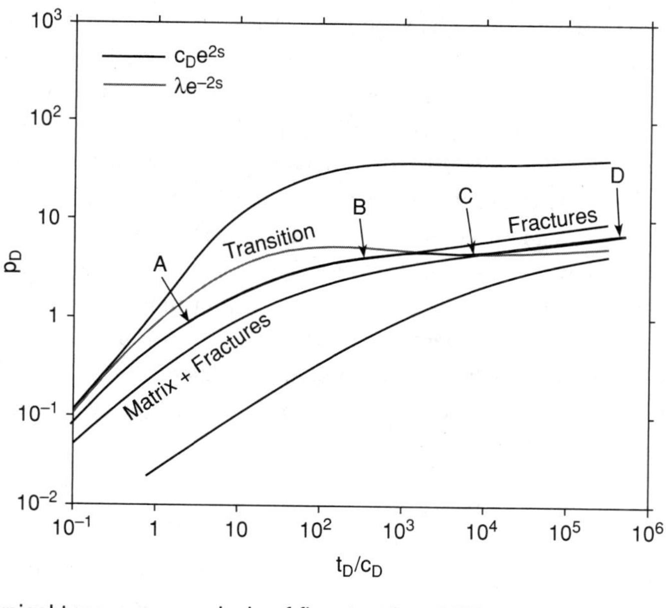

Figure 13.32 Typical type curve analysis of flow test in a NFR (from Gringarten[14] © 1984 SPE, Reproduced with permission).

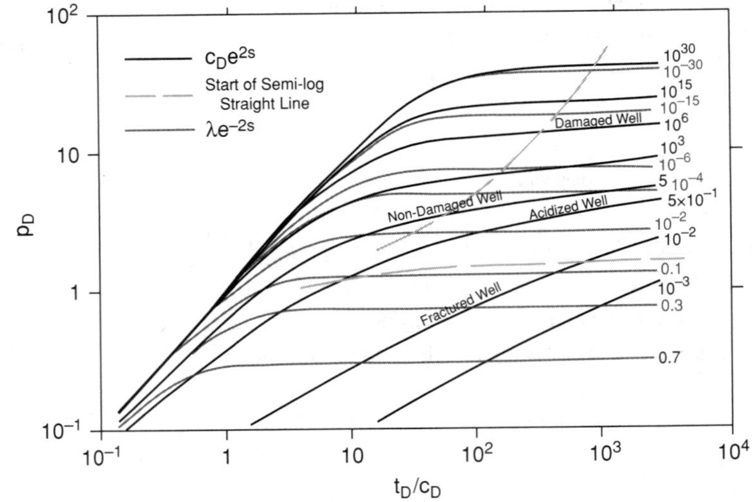

Figure 13.33 Type curve for NFRs under pseudosteady-state flow (from Gringarten[14] © 1984 SPE, Reproduced with permission).

test in a double-porosity reservoir. During early time flow from the fractures, the pressure response follows one of the homogeneous curves with $C_De^{2s} = \left(C_De^{2s}\right)_f$. This is shown as the heavy dark line between points A and B in Figure 13.32. When interporosity flow begins from the matrix to the fractures (termed the transition period), the pressure response curve departs from the C_De^{2s} curve and follow the transition curve characterized by the parameter λ^{e-2s}. This is represented by the segment between points B and C in Figure 13.32. At later times, when both the fractures and matrix are contributing to production, the pressure curve leaves the transition curve and follows a new C_De^{2s} that represents the entire system. This is shown as the segment between points C and D in Figure 13.32.

The process of determining reservoir and fracture parameters by type-curve matching is the same as discussed previously in Chapter 12. The process begins by making a log-log plot of pressure change versus time on the same scale as the type curve. By examining the plotted data, it is possible to identify different flow periods such as fracture flow, transition flow, and total system flow. This plot is then superimposed on a set of type curves such as shown in Figure 13.33 for pseudosteady-state flow, and Figure 13.34 for transient flow. By shifting the two plots in the manner described in Chapter 12 for type-curve matching, a match point is obtained. From the pressure match point (PMP), the permeability-thickness product for the total system (fracture + matrix), $(kh)_{f+ma}$, is calculated as:

$$(kh)_{f\,+\,ma} = 141.2qB\mu\left[\frac{p_D}{\Delta p}\right]_{PMP} \tag{13.53}$$

The permeability of the fracture system is much higher than the permeability of the matrix (i.e., $k_f \gg k_{ma}$). For this reason, $(kh)_{f+ma}$ is essentially equal to $(kh)_f$.

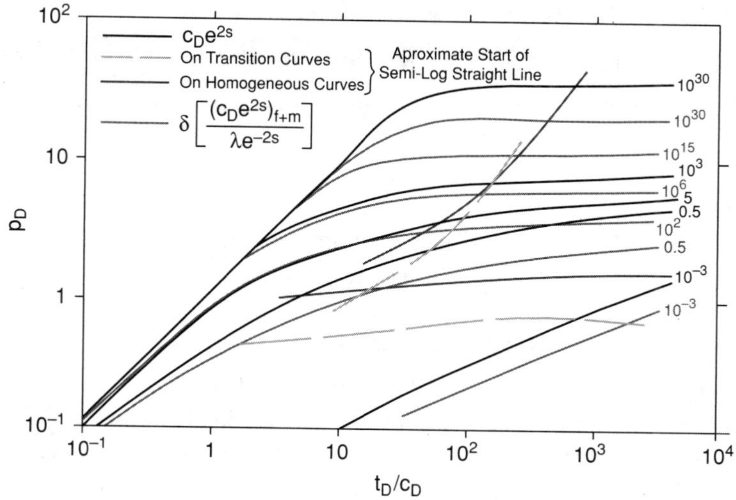

Figure 13.34 Type curve for NFRs under transient flow (from Gringarten[14] © 1984 SPE, Reproduced with permission).

From the time match point (TMP), the wellbore storage coefficient, C_D, can be calculated as:

$$C_D = \frac{0.0002637k}{\phi\mu c_t r_w^2}\left(\frac{\Delta t}{t_D/C_D}\right)_{TMP} \tag{13.54}$$

From the fit of the early data that characterize the fracture system, determine the value of $\left(C_De^{2s}\right)_f$ from the match. Also, determine the value of λe^{-2s} that characterizes the transition region. Finally, determine the value of $\left(C_De^{2s}\right)_{f+m}$ that characterizes flow of the total system in the later flow period. The storativity ratio, ω, is calculated from these matched values as:

$$\omega = \frac{\left(C_De^{2s}\right)_{f+m}}{\left(C_De^{2s}\right)_f} \tag{13.55}$$

The skin factor, s, is calculated as:

$$s = 0.5\ln\left[\frac{\left(C_De^{2s}\right)_{f+m}}{C_D}\right] \tag{13.56}$$

The interporosity flow coefficient, λ, is calculated from:

$$\lambda = \left(\lambda e^{-2s}\right)e^{2s} \tag{13.57}$$

13.13 Procedure for Analysis of Well Test from NFRs Assuming Double Porosity Behavior

The procedure for analysis of well test data from NFRs can be classified as consisting of three main processes:

1. Identification of flow periods
2. Calculation of fracture and reservoir parameters from type curves
3. Validation of results with straightline methods

13.13.1 Identification of Flow Periods

As discussed earlier, there are potentially three flow periods in the test data from NFRs. These are fracture flow at early times, transition flow period marking the onset of fluid flow from the matrix, and stabilized flow from both fracture and matrix at later times. In many cases, fracture flow and transition flow periods may be distorted or masked by wellbore storage. The presence of these flow regimes can be identified on a plot of pressure change and pressure-derivative versus time. The pressure-derivative plot yields the characteristic dip below the homogeneous behavior level which is indicative of double porosity behavior.

1. Plot $\log\Delta p$ vs. $\log t$ for flow test or $\log \Delta p$ vs. $\log \Delta t_e$ for a buildup test. Plot $t\Delta p'$(or $\Delta t_e \Delta p'$) versus t (or Δt_e) on the same graph. Use the same scale size as the type curves graph to be used later for type-curve matching.
2. Identify any flow periods that may be present. Note if double porosity behavior is present.

13.13.2 Calculation of Fracture and Reservoir Parameters from Type Curves

Fracture and reservoir parameters can be calculated by matching the plot from Step 1 to the type curves in Figure 13.33 or Figure 13.34. Since most test data from NFRs appear to indicate pseudosteady-state flow conditions, use Figure 13.33 for the first trial.

1. Match the flow test data plot of $\log\Delta p$ vs. $\log t$ for flow test or $\log \Delta p$ vs. $\log \Delta t_e$ for a buildup test against type curves of Figure 13.33.
2. Determine pressure and time match points (PMP and TMP, respectively). Also determine $\left(C_D e^{2s}\right)_f$ for the fracture flow period, λe^{-2s} for the transition flow period, and $\left(C_D e^{2s}\right)_{f+m}$ for the stabilized flow from the combined system (fracture plus matrix).
3. Calculate fracture and reservoir parameters using Eqs. (13.53) to (13.57).

13.13.3 Validation of Results with Straightline Methods

The validation step involves using straightline methods to calculate fracture and reservoir parameters, and comparing them to the results obtained with type-curve matching. If the results are reasonably close, then it can be presumed that the type curves selected from type-curve matching are representative of the test data and that the results are inherently consistent. If the results are widely divergent, then the analysis should begin again from type-curve matching by choosing another set of type curves and repeating the entire process. This process is highly iterative and should be continued until reasonable agreement is achieved with the results from type-curve matching and straightline methods.

1. Plot p_{wf} vs. t for a drawdow test or p_{ws} vs. $(t_p + \Delta t)/\Delta t$ for a buildup test on a semilog scale.
2. Identify flow periods present in plot using the knowledge gained from type-curve matching.
3. If the flow periods illustrated in Figure 13.30 are present, use Eqs. (13.45) to (13.48) to calculate fracture and reservoir parameters.
4. Use the derivative plot to obtain $t_{D\,\min}$. Calculate ω and λ from Eqs. (13.49) and (13.50).
5. Calculate skin factor from Eq. (13.52).
6. Compare the results obtained from straightline methods to the results obtained from type-curve analysis.
7. If the results are in reasonable agreement, then the analysis is internally consistent. If not, repeat process by using another interpretation model.

The discussions on analysis of test data from NFRs have been conducted on the basis of slightly compressible fluids such as oil reservoirs. For gas reservoirs, it is necessary to use appropriate pressure transformation functions such as real gas pseudo-pressure, as demonstrated in Chapter 11.

Example 13.2 Calculation of Fracture and Reservoir Parameters from the Test Data of a Well in a Naturally Fractured Reservoir (NFR)

Problem

A drawdown test was conducted on a well in a naturally fractured reservoir. The test data are shown in Table 13.2. Calculate fracture and reservoir parameters from the test data. Well and other reservoir properties are given as follows:

Formation thickness, h	30 ft
Formation porosity, ϕ_m	0.12
Total compressibility, c_t	14×10^{-6} psi^{-1}
Oil viscosity, μ_o	1.2 cp
Oil FVF, B_o	1.3 RB/STB
Wellbore radius, r_w	0.5 ft
Production rate, q_o	2950 STB/D
Initial average pressure, $\overline{p}$	4480 psia

Table 13.2 Test and Plot Data for Example 13.2

Time, *t* (hrs.)	p_{wf} (psia)	Δp (psia)	$t\Delta p'$
0.00	4480.00	0.00	—
0.05	4462.00	18.00	128.87
0.06	4440.00	40.00	136.28
0.08	4403.00	77.00	147.34
0.09	4379.20	100.80	141.03
0.12	4338.50	141.50	135.82
0.17	4296.50	183.50	137.56
0.20	4270.00	210.00	138.68
0.25	4240.00	240.00	145.16
0.33	4196.00	284.00	144.70
0.40	4170.00	310.00	141.94
0.51	4135.00	345.00	149.17
0.57	4117.00	363.00	139.44
0.65	4100.00	380.00	146.64
0.74	4080.00	400.00	120.92
0.92	4070.00	410.00	99.51
1.00	4060.00	420.00	101.18
1.10	4052.40	427.60	76.42
1.40	4036.00	444.00	67.38
1.65	4025.00	455.00	67.00
1.93	4014.49	465.51	69.02
2.30	4002.00	478.00	72.41
2.93	3984.07	495.93	78.05
3.93	3959.73	520.27	85.48
4.93	3939.90	540.10	89.70
5.93	3923.00	557.00	95.16
6.93	3907.69	572.31	115.62
7.93	3890.08	589.92	134.18
8.50	3880.64	599.36	137.50
10.50	3850.62	629.38	138.42

Time, t (hrs.)	p_{wf} (psia)	Δp (psia)	$t\Delta p'$
11.60	3837.00	643.00	144.72
13.40	3814.45	665.55	140.49
15.00	3800.00	680.00	144.40
16.30	3787.00	693.00	146.00
17.40	3778.00	702.00	144.80
18.35	3770.00	710.00	146.74
19.20	3763.50	716.50	144.14
20.50	3754.00	726.00	143.71
21.50	3747.20	732.80	144.00
22.50	3740.60	739.40	143.85
23.50	3734.40	745.60	143.76
25.00	3725.40	754.60	148.33

Solution

Step 1: Plot $\log \Delta p$ vs. $\log t$.

Using the data in Table 13.2 and plotted in Figure 13.35, make a log-log plot of Δp vs. t as shown in Figure 13.36. Also, plot the derivative curve of $t\Delta p'$ vs. t on the same graph using the data in Table 13.2. The derivative curve clearly shows evidence of double porosity behavior which is indicated by the characteristic dip below the homogeneous

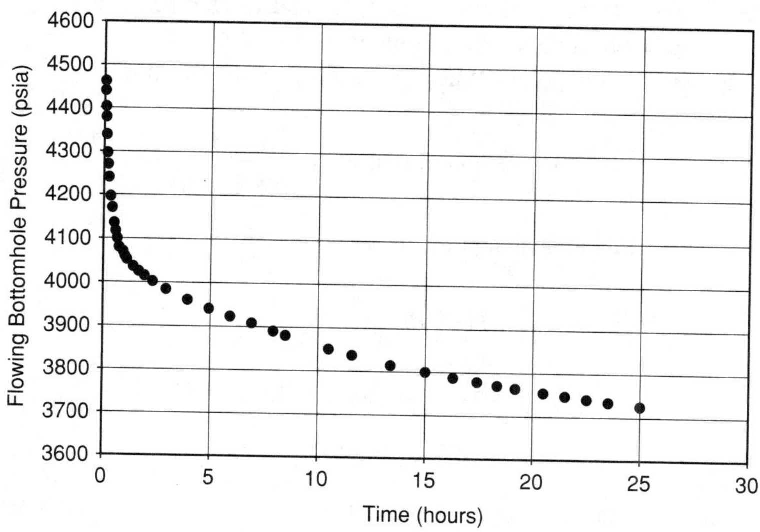

Figure 13.35 Plot of flow test data for Example 13.2.

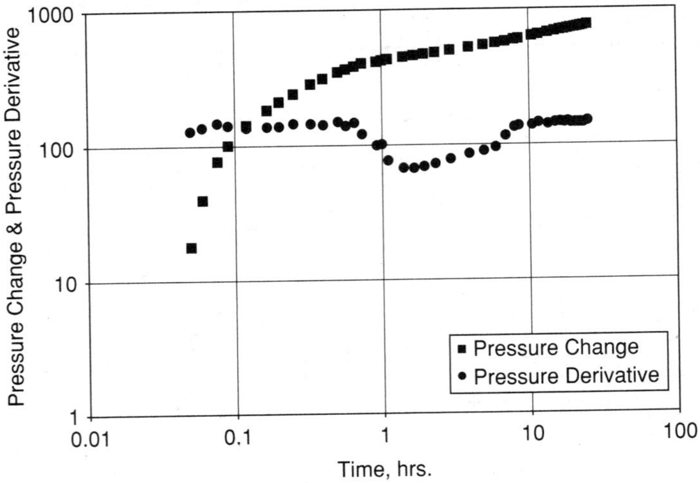

Figure 13.36 Log-Log plot of Δp & $t\Delta p'$ vs. t for Example 13.2.

behavior level. Also, the derivative curve in Figure 13.36 shows pseudosteady-state flow from the combined system of fracture and matrix which is represented by the stabilized leveling of the derivative curve after the dip.

Step 2: Type-curve matching.

The type curve for NFRs under pseudosteady-state flow shown as Figure 13.33 was used in the type-curve matching procedure. Figure 13.36 was matched against Figure 13.33. The matched plot of Figures 13.33 and 13.36 together is shown as Figure 13.37. The matched data points are as follows:

Pressure match point (PMP), $(p_D/\Delta p)_{PMP} = (0.35/100)$

$$\text{Time match point (TMP),} \quad \left(\frac{\Delta t}{t_D/C_D}\right)_{TMP} = \left(\frac{1}{10}\right)$$

For the pressure match point, applying Eq. (13.53),

$$\begin{aligned}(kh)_{f+m} &= 141.2qB\mu\left[\frac{p_D}{\Delta p}\right]_{PMP} \\ &= 141.2 \times 2950 \times 1.3 \times 1.2 \times \left(\frac{0.35}{100}\right) \\ &= 2274.31 \text{ md-ft}.\end{aligned}$$

Since the matrix permeability is relatively small in comparison to fracture permeability, $(kh)_{f+ma}$ is essentially equal to $(kh)_f$.

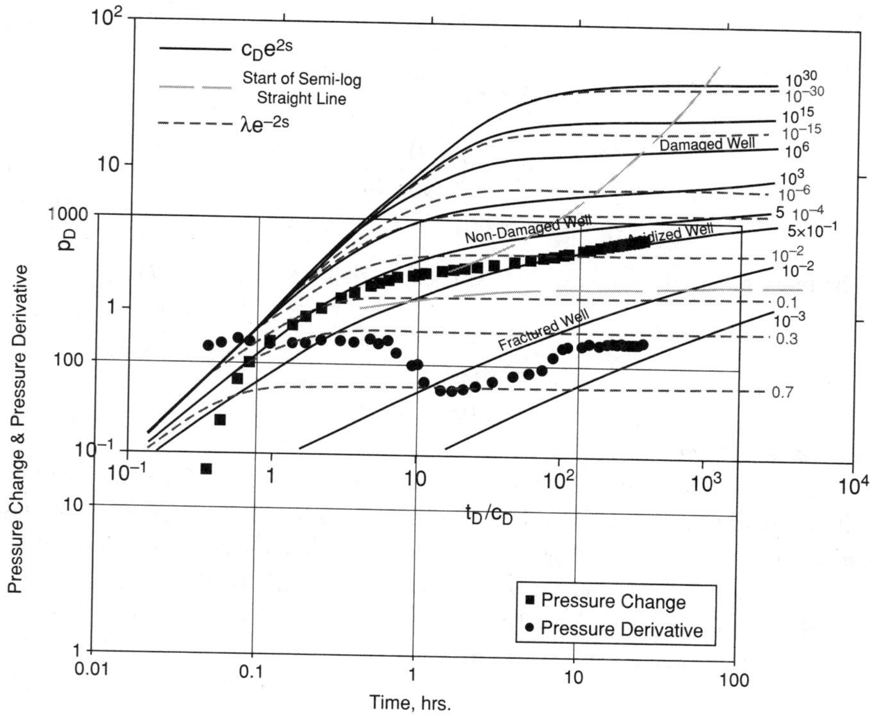

Figure 13.37 Type-curve match for Example 13.2.

For the time match point, applying Eq. (13.54),

$$C_D = \frac{0.0002637k}{\phi\mu c_t r_w^2}\left(\frac{\Delta t}{t_D/C_D}\right)_{TMP}$$

$$= \frac{0.0002637 \times 75.81}{0.12 \times 1.2 \times 14 \times 10^{-6} \times (0.5)^2}\left(\frac{1}{10}\right)$$

$$= 3966.49$$

From Figure 13.37, type-curve match parameters were estimated as follows:

$$\left(C_De^{2s}\right)_{f+ma} = 0.4\,;\ \left(C_De^{2s}\right)_f = 2.0 \text{ and } \left(\lambda e^{-2s}\right) = 0.06$$

Applying Eq. (13.55), the storativity ratio, ω, is calculated to be:

$$\omega = \frac{\left(C_De^{2s}\right)_{f+m}}{\left(C_De^{2s}\right)_f}$$

$$= \frac{0.4}{2} = 0.2$$

The skin factor, s, is calculated from Eq. (13.56) as:

$$s = 0.5 \ln \left[\frac{\left(C_D e^{2s}\right)_{f+m}}{C_D} \right]$$
$$= 0.5 \ln \left(\frac{0.4}{3966.49} \right)$$
$$= -4.601$$

The interporosity flow coefficient, λ, is calculated from Eq. (13.57) to be:

$$\lambda = (\lambda e^{-2s}) e^{2s}$$
$$= (0.06) e^{2(-4.601)}$$
$$= 6.05 \times 10^{-6}$$

Step 3: Plot p_{wf} vs. $\log t$.
Prepare a semilog plot of p_{wf} vs. t from the data in Table 13.2. The semilog plot is shown as Figure 13.38. From Figure 13.38, it is obvious that fracture flow lasted for about one hour. This is followed by a transition period to matrix flow of duration of one to nine hours. Pseudosteady-state flow from both matrix and fractures started approximately after ten hours of flow. These flow periods conform to the flow periods identifiable on the derivative curve of Figure 13.36.

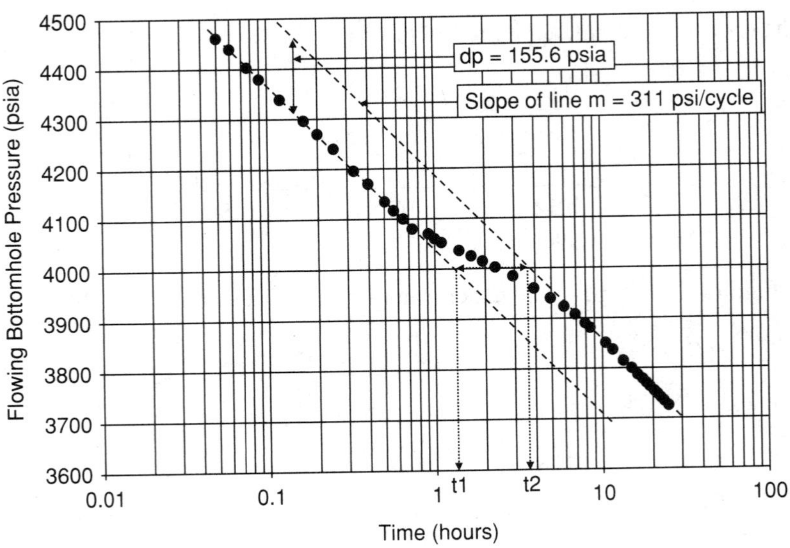

Figure 13.38 Semilog plot of p_{wf} vs. t for Example 13.2.

The slope of the semilog straightline in Figure 13.38 is $m = 311$ psi/cycle. Applying Eq. (13.45), the total flow capacity of the formation is:

$$(kh)_{total} = \frac{162.6qB\mu}{m}$$
$$= \frac{162.6 \times 2950 \times 1.3 \times 1.2}{311}$$
$$= 2406.06 \text{ md-ft}$$

From Figure 13.38, the vertical pressure displacement, δp, is 155.56 psia. Applying Eq. (13.46), the storativity ratio, ω, is calculated to be:

$$\omega = 10^{-\left(\frac{\delta p}{m}\right)}$$
$$= 10^{-\left(\frac{155.6}{311}\right)} = 0.32$$

A straight line through the middle of the transition curve intersects the two parallel straight lines as shown in Figure 13.38 at $t_1 = 1.7$ hrs and $t_2 = 3.7$ hrs, respectively. From Eq. (13.47), the interporosity flow coefficient, λ, at t_2 is calculated as:

$$\lambda = \frac{(\phi V c_t)_{f+m}\mu r_w^2}{\gamma k_f t_2}$$
$$= \frac{0.12 \times 14 \times 10^{-6} \times 1.2 \times (0.5)^2}{1.78 \times 80.2 \times 3.7}$$
$$= 9.54 \times 10^{-10}$$

Note that V_{f+m} is essentially equal to 1.0. Also, $k_f = (kh)_f/h = 2406.1/30 = 80.2$ md. Re-arranging Eq. (13.44) yields:

$$(\phi V c_t)_f = \frac{\omega}{1-\omega}(\phi V c_t)_{ma}$$
$$= \frac{0.32}{1-0.32}\left(0.12 \times 14 \times 10^{-6}\right)$$
$$= 7.9059 \times 10^{-7} \text{ psi}^{-1}$$

Applying Eq. (13.47), the interporosity flow coefficient, λ, at t_1 is calculated to be:

$$\lambda = \frac{(\phi V c_t)_f \mu r_w^2}{\gamma k_f t_1}$$
$$= \frac{7.9059 \times 10^{-7} \times 1.2 \times (0.5)^2}{1.78 \times 80.2 \times 1.7}$$
$$= 9.77 \times 10^{-10}$$

From Figure 13.38, $p_{1hr} = 4190$ psia. The skin factor, s, is given by Eq. (13.52) as:

$$s = 1.151\left[\frac{\Delta p_{1hr}}{m} - \log\left(\frac{k}{\phi\mu c_t r_w^2}\right) + 3.23\right]$$
$$= 1.151\left[\frac{4480 - 4190}{311} - \log\frac{80.2}{0.12 \times 1.2 \times 14 \times 10^{-6} \times (0.5)^2} + 3.23\right]$$
$$= -4.65$$

For this example, the flow capacities calculated from type-curve matching and straightline methods are reasonably in close agreement. The same close agreement was achieved for the skin factors calculated from both methods. However, there are poor agreement between the storativity ratios and interporosity flow coefficients calculated from type-curve matching and straightline methods. The storativity ratio and interporosity flow coefficient calculated from type-curve matching should at best be considered as gross estimates because of the difficulties associated with estimating the correct values for the type-curve parameters $\left(C_De^{2s}\right)_{f+ma}$, $\left(C_De^{2s}\right)_f$, and λe^{-2s}. Under these circumstances, the storativity ratio and interporosity flow coefficient calculated from straightline methods should be considered to be more representative of the actual values.

Nomenclature

B	formation volume factor, RB/STB
c_{ft}	total compressibility of fluid in fracture, psi^{-1}
c_t	total compressibility, psi^{-1}
C_D	dimensionless wellbore storage coefficient
C_{Df}	dimensionless fracture storage coefficient, $C_{Df} = 0.894C/\left(\phi c_t h L_f^2\right)$
C_{fD}	dimensionless fracture conductivity
h	reservoir or formation thickness, feet
k	permeability, md
k_f	fracture permeability, md
k_{ma}	matrix permeability, md
L_f	fracture half-length, feet
m	slope of straight line during pseudo-radial flow
m_{bf}	slope of straight line during bilinear flow
m_{Lf}	slope of straight line during formation linear flow
Δp	pressure change, psia
Δp_s	pressure drop due to skin, psia
$\Delta p'$	pressure change derivative
p'_D	dimensionless pressure derivative
p	pressure, psia
p_D	dimensionless pressure
p_p	real gas pseudo-pressure, psia^2/cp

p_{wf} well bottom-hole flowing pressure, psia
p_{ws} shut-in bottomhole pressure, psia
q volume flow rate, STB/D
q_g gas volume flow rate, Mcf/D
r_w wellbore radius, feet
r_{wa} effective wellbore radius, feet
s skin factor
s' apparent skin factor.
Δt shut-in time in buildup test, hours
Δt_e Agarwal equivalent time
t time, hours
t_{blf} time, begining of linear flow
t_{elf} time, end of linear flow
t_{ews} time, end of wellbore storage
t_D dimensionless time
t_{DL_f} dimensionless time defined in terms of L_f
t_p total producing time before shut-in, hours
T temperature, °R
V ratio of the total volume of medium to the bulk volume of the total system
w_f fracture width, feet
ϕ porosity, fraction
ϕ_f fracture porosity, fraction
μ fluid viscosity, cp
λ interporosity flow coefficient
α characteristic parameter of geometry of system
ω storativity ratio
η_{fD} dimensionless hydraulic diffusivity
γ exponential Euler's constant approximately 1.781

Subscripts

g gas
f fracture
i initial conditions
o oil
ma matrix
t total
w water

Abbreviations

PMP Pressure Match Point
TMP Time Match Point

References

1. Cinco-Ley, H., and Samaniego-V, F.: "Transient Pressure Analysis for Fractured Wells," *JPT* (September 1981) 1749–1766.
2. Cinco-Ley, H.: "Evaluation of Hydraulic Fracturing By Transient Pressure Analysis Methods," paper SPE 10043 presented at the International Petroleum Exhibition and Technical Symposium, Beijing, China, March 18–26, 1982.
3. Cinco-Ley, H., Samaniego-V, F., and Dominguez-A, N.: "Transient Pressure Behavior for a Well with a Finite-Conductivity Vertical Fracture," *SPEJ* (August 1978) 253–264.
4. Gringarten, A.C., Ramey, H.J., Jr., and Raghavan, R.: "Applied Pressure Analysis for Fractured Wells," *JPT* (July 1975) 887–892.
5. Wong, D.W., Harrington, A.G., and Cinco-Ley, H.: "Application of the Pressure-Derivative Function in the Pressure-Transient Testing of Fractured Wells," *SPEFE* (October 1986) 470–480.
6. Cinco-Ley, H., and Samaniego-V., F.: "Effect of Wellbore Storage and Damage on the Transient Pressure Behavior of Vertically Fractured Wells," paper SPE 6752 presented at the 52nd Annual Technical Conference and Exhibition, Denver, Colorado, October 9–12, 1977.
7. Cinco-Ley, H., and Samaniego-V., F.: "Transient Pressure Analysis: Finite Conductivity Fracture Case Versus Damaged Fracture Case," paper SPE 10179 presented at the 56th Annual Fall Technical Conference and Exhibition, San Antonio, Texas, October 5–7, 1981.
8. Cinco-Ley, H.: "Well-Test Analysis for Naturally Fractured Reservoirs," *JPT* (January 1996) 51–54.
9. Barenblatt, G.E., Zheltov, I.P., and Kochina, I.N.: "Basic Concepts in the Theory of Homogeneous Liquids in Fissured Rocks," *J. Appl. Math. Mech.* (1960) 24, 852–864.
10. Warren, J.E., and Root, P.J.: "The Behavior of Naturally Fractured Reservoirs," *SPEJ* (September 1963) 245–255.
11. de Swaan-O., A.: "Analytical Solutions for Determining Naturally Fractured Reservoir Properties by Well Testing," *SPEJ* (June 1976) 117–122.
12. Najurieta, H.L.: "A Theory for Pressure Transient Analysis in Naturally Fractured Reservoirs," *JPT* (July 1980) 1241–1250.
13. Serra, K., Reynolds, A.C., and Raghavan, R.: "New Pressure Transient Analysis Methods for Naturally Fractured Reservoirs," *JPT* (December 1983) 2271–2283.
14. Gringarten, A.C.: "Interpretation of Tests in Fissured and Multilayered Reservoirs with Double-Porosity Behavior: Theory and Practice," *JPT* (April 1984) 549–564.
15. Tiab, D., Restrepo, D.P., and Igbokoyi, A.: "Fracture Porosity of Naturally Fractured Reservoirs," paper SPE 104056 presented at the First International Oil Conference and Exhibition, Cancun, Mexico, August 31–September 2, 2006.
16. Horne, R.N.: *Modern Well Test Analysis—A Computer-Aided Approach*, 2nd Edition, Petroway, Inc., California (2008).

General Reading

- Bourdet, D., Ayoub, J.A., and Pirard, Y.M.: "Use of Pressure Derivative in Well-Test Interpretation," *SPEFE* (June 1989) 293–302.
- Tiab, D., and Puthigai, S.K.: "Pressure-Derivative Type Curves for Vertically Fractured Wells," *SPEFE* (March 1988) 156–158.
- Bourdet, D., and Gringarten, A.C.: "Determination of Fissure Volume and Block Size in Fractured Reservoirs by Type-Curve Analysis," paper SPE 9293 presented at the 55th Annual Fall Technical Conference and Exhibition, Dallas, Texas, September 21–24, 1980.
- Gringarten, A.C.: "How to Recognize 'Double-Porosity' Systems from Well Tests," *JPT* (June 1987) 631–633.
- de Swaan, A.: "Analysis of Well Tests in Multiple Fractured Reservoirs—Field Case Applications," paper SPE 59015 presented at the 2000 SPE International Petroleum Conference and Exhibition, Villarhermosa, Mexico, February 1–3, 2000.
- Cinco-Ley, H.: "Pressure Transient Analysis of Wells with Finite Conductivity Vertical Fractures in Double Porosity Reservoirs," paper SPE 18172 presented at the 63rd Annual Technical Conference and Exhibition, Houston, Texas, October 2–5, 1988.
- Okoye, C.U., Oshinuga, A.D., and Ghalambor, A.: "Pressure Transient Behavior in Multilayer Vertically Fractured Gas Reservoir with Finite Conductivity," paper SPE 18960 presented at the SPE Joint Rocky Mountain Region/Low Permeability Reservoirs Symposium and Exhibition, Denver, Colorado, March 6–8, 1989.
- Tiab, D.: "Multi-Rate Testing for Vertical Wells in Naturally Fractured Reservoirs," paper SPE 88558 presented at the SPE Asia Pacific Oil and Gas Conference and Exhibition, Perth, Australia, October 18–20, 2004.
- Gringarten, A.C.: "From Straight Lines to Deconvolution: The Evolution of the State of the Art in Well Test Analysis," *SPEREE* (February 2008) 41–62.
- Lee, J., Rollins, J.B., and Spivey, J.P.: *Pressure Transient Testing,* SPE Textbook Series Vol. 9, SPE, Richardson, Texas (2003).

CHAPTER 14

Well Test Analysis: Deconvolution Concepts

14.1 Introduction

In the preceding chapters on well test analysis, applications of conventional methods on the analyses of well test data were presented. These conventional methods include straightline methods discussed in Chapter 11 and the application of type curves discussed in Chapter 12. In Chapter 13, analysis of well test data from hydraulically fractured wells and naturally fractured reservoirs using both straightline methods and type curves were presented. In this chapter, deconvolution is introduced as another tool which can be used in the analysis of well test data. It is important to note that deconvolution is not a new interpretation method or a replacement for conventional methods in the analysis of well test data. It is rather a process that is used to convert variable-rate pressure data into constant-rate pressure data, thereby providing more pressure data for the determination of an interpretation model. Deconvolution is rapidly gaining wide application, especially on extensive well test data collected from permanent downhole gauges installed in deepwater wells, and other wells where constant pressure monitoring is desired. In this book, basic concepts of deconvolution are presented with examples and guidelines for application of deconvolution method.

14.2 What Is Deconvolution?

Deconvolution is the process of transforming variable-rate pressure data into a constant-rate drawdown pressure response data with duration equal to the total duration of the test. A corresponding pressure derivative normalized to a unit rate is also obtained during the deconvolution process.

The process of deconvolving pressure-rate data is not simply an advanced numerical analysis technique, but requires considerable skill and knowledge of the deconvolution algorithm.

In the application of conventional methods (semilog and derivative type plots), well test analysis is generally limited to transient pressure behavior for a specific test period. This could be either a drawdown or a buildup test period. The rate history of the well is typically accounted for with a superposition-time transform which is meant to remove the effects of variability present in the production rates before the test period. The superposition-time transforms do not remove all the effects due to variability of production rates prior to either drawdown or buildup tests. These residual superposition effects sometimes distort and complicate analysis of well test data with conventional methods. Furthermore, analysis of well test data based on individual test periods using conventional methods are sometimes inconclusive if the test was terminated before characteristic curve shapes evident in conventional plots (semilog or log-log derivative plots) have been reached.

The deconvolution process, on the other hand, incorporates all pressure-rate data obtained during a well test sequence. This includes all the variable rate flow periods and shut-in periods of varying durations. Consequently, the deconvolution process removes the effects of rate variations from the pressure-rate data and yields a constant-rate drawdown pressure response which could be more representative of the well/reservoir system. The deconvolved pressure response typically has a much longer time span and may reveal characteristic features of the reservoir which will not be evident, if only conventional methods had been used in the test analysis. It is important to note that the deconvolution process helps in the identification of an interpretation model for the test data rather than the direct calculation of reservoir parameters. However, once an acceptable interpretation model has been identified through application of deconvolution, reservoir parameters can be estimated based on the interpretation model.

Deconvolution techniques have been applied to variable-rate test data since 1959 with limited success. These techniques can be classified into two methods, namely:[1] Time-domain methods and Spectral Methods. Time-domain methods use various interpolation schemes to evaluate the convolution integral directly in the time domain.[1] Spectral methods are based on Laplace transform techniques.[1] The detailed mathematical application of either of these methods is not covered in this book. However, it is important to observe that application of robust and stable deconvolution techniques were not available until von Schroeter et al.[1–3] introduced a new time-domain based method in 2001. This method is based on a regularized, nonlinear total least squares (TLS) formulation of the deconvolution problem. The von Schroeter et al.[3] deconvolution method was evaluated by Levitan,[4] Levitan et al.,[5] and Onur et al.,[6] and found to be relatively stable under certain guidelines.

14.3 The Pressure-Rate Deconvolution Model

The pressure-rate deconvolution model used by von Schroeter et al.,[1–3] Levitan,[4] and Levitan et al.[5] is based on the following convolution integral for a single well during a variable well test:

$$p(t) = p_0 - \int_0^t q(\tau)\frac{dp_u(t-\tau)}{dt}d\tau \qquad \textbf{(14.1)}$$

In Eq. (14.1), $q(t)$ is the measured well rate, $p(t)$ is the measured bottomhole pressure, and p_0 is the initial reservoir pressure. Also in Eq. (14.1), $p_u(t)$ is the drawdown pressure response of the well/reservoir system at a constant unit rate production, assuming that the reservoir is in equilibrium and pressure is also uniform throughout the reservoir at start of production. Eq. (14.1) is valid only for linear systems (that is, single phase flow of slightly compressible fluid in a porous medium). Extension of Eq. (14.1) to multiwell test data has been reported by Levitan.[7]

Deconvolution is the process of reconstructing the constant-rate drawdown pressure response, $p_u(t)$, together with the initial pressure, p_0, from the pressure, $p(t)$, and rate, $q(t)$, data measured in a variable-rate well test.[4] The deconvolution problem is the same as solving Eq. (14.1) for $p_u(t)$ and p_0 given the measured data of $p(t)$ and $q(t)$. Thus, the deconvolution problem as defined is an inverse problem. The solution of an inverse problem is non-unique.[8] This means that different interpretation models can produce the same constant-rate drawdown pressure response, $p_u(t)$, from the same input of pressure, $p(t)$, and rate, $q(t)$, data measured in a variable-rate well test. It is important to remember this point when evaluating the pressure response data from the deconvolution process.

The deconvolution problem is also ill-conditioned. This means that slight variations in the input data (measured pressure and rate data) can cause large changes in the output data (deconvolved pressure response and initial pressure). The ill-conditioned form of the deconvolution problem combined with errors typically present in measurements of pressure and rate data make the solution of the deconvolution problem very unstable. As reported by von Schroeter et al.,[3] several deconvolution algorithms proposed in the literature were not sufficiently stable in the solution of the deconvolution problem. The algorithm proposed by von Schroeter et al.[1–3] and evaluated by Levitan,[4] Levitan et al.,[5] and Onur et al.[6] have been found to be stable under certain guidelines.

14.3.1 The von Schroeter et al.[3] Deconvolution Algorithm

The von Schroeter algorithm is summarized under three novel concepts as proposed in the work by Levitan:[4]

1. Implicit Constraint Encoding. For this approach, Eq. (14.1) is solved in terms of the function:

$$z(\sigma) = \ln\left[\frac{dp_u(t)}{d\ln t}\right] = \ln\left[\frac{dp_u(\sigma)}{d\sigma}\right] \tag{14.2}$$

In Eq. (14.2), $\sigma = \ln(t)$. In terms of the function, $z(\sigma)$, Eq. (14.2) becomes:

$$p(t) = p_0 - \int_{-\infty}^{\ln t} q(t - e^{\sigma})\, e^{z(\sigma)} d\sigma \tag{14.3}$$

The selection of $z(\sigma)$ as the new solution variable ensures that the term, $dp_u(t)/d\ln t$, is positive. This condition should be satisfied by the constant-unit-rate response of the system. The use of the variable, $z(\sigma)$ makes Eq. (14.3) non-linear.

2. Regularization Based on Curvature. The deconvolution problem is ill-conditioned and very sensitive to errors in measured pressure and rate data. Regularization based on the curvature of $z(\sigma)$ is imposed to attain some degree of smoothness of the solution and improve the conditioning of the deconvolution problem.
3. Total Least Squares Error Model. The deconvolution problem is formulated as unconstrained nonlinear minimization with the objective function defined to include errors in the pressure and rate data, residuals of Eq. (14.3), and regularization curvature constraints.

A detailed presentation of the mathematical formulation and implementation of the von Schroeter deconvolution algorithm is not presented in this book. Additional details on the implementation of the algorithm were documented by von Schroeter et al.,[3] and Levitan.[4]

14.4 Application of Deconvolution to Pressure-Rate Data

The convolution integral as written in Eq. (14.1) is valid for a single well producing a single-phase fluid in a variable-rate well test. It is also assumed that wellbore storage and/or skin factor are constant for the entire duration of the well test. This is the model on which the von Schroeter deconvolution algorithm is based. Consequently, for well tests in which these conditions are not met, the deconvolved constant-rate drawdown pressure response may yield erroneous results. In extended well tests, it is possible for wellbore storage and skin factor to change over the duration of the test. If the reservoir pressure is close to saturation pressure of the fluids within the reservoir, multiphase flow will occur when reservoir pressure falls sufficiently below the saturation pressure. These constraining conditions should be observed when subjecting pressure-rate data to the deconvolution process.

Levitan[4] proposed a novel procedure for checking the consistency of the variable-rate test data and ensuring the successful application of the von Schroeter deconvolution algorithm. Inconsistent test data sets are defined as data sets that do not conform to the conditions stipulated for the deconvolution model of Eq. (14.1) as stated earlier. This includes data sets obtained under conditions in which wellbore storage and/or skin factor changed or multiphase flow occurred during the duration of the test. For an inconsistent data set, Levitan[4] applied a variant of the von Schroeter deconvolution algorithm on pressure-rate data from individual flow periods within the test sequence while specifying initial pressure as input. Typically, pressure buildup (PBU) data from the individual flow periods are preferable because PBU data generally have fewer errors associated with them. Levitan[4] found that for a consistent data set, the constant-rate drawdown pressure response deconvolved from PBU data of individual flow periods appear to be identical or show similar trends. This procedure could be used to check variable-rate test data for consistency. Levitan et al.[5] listed three factors that could affect pressure response obtained from applying the von Schroeter deconvolution algorithm on well test data. These are:

1. Accuracy of initial reservoir pressure.
2. Consistency of test data.
3. Effects of local events on deconvolved pressure response.

The deconvolved pressure response is very sensitive to initial reservoir pressure. For this reason, it is crucial to ensure that the source of initial reservoir pressure is reliable and accurate. An inconsistent data set will produce deconvolved system response that is completely misleading. The approach devised by Levitan et al.[5] should be used to check test data for consistency. It was also found by Levitan et al.[5] that localized events such as mismatch of pressure and rate data distort the constant-rate drawdown pressure response at late times.

14.5 Examples on the Application of the von Schroeter Deconvolution Algorithm to Real Well Test Data

The two examples presented in this book are based on the work of Levitan et al.[5] These examples are intended simply to illustrate the application of the von Schroeter algorithm and present the results that could be obtained from deconvolution of well test data.

Example 14.1

The production history of a new gas well is shown in Figure 14.1. The total duration of the variable-rate test is 210 hours. Included in the test history are two pressure buildups (PBU) labeled PBU 1

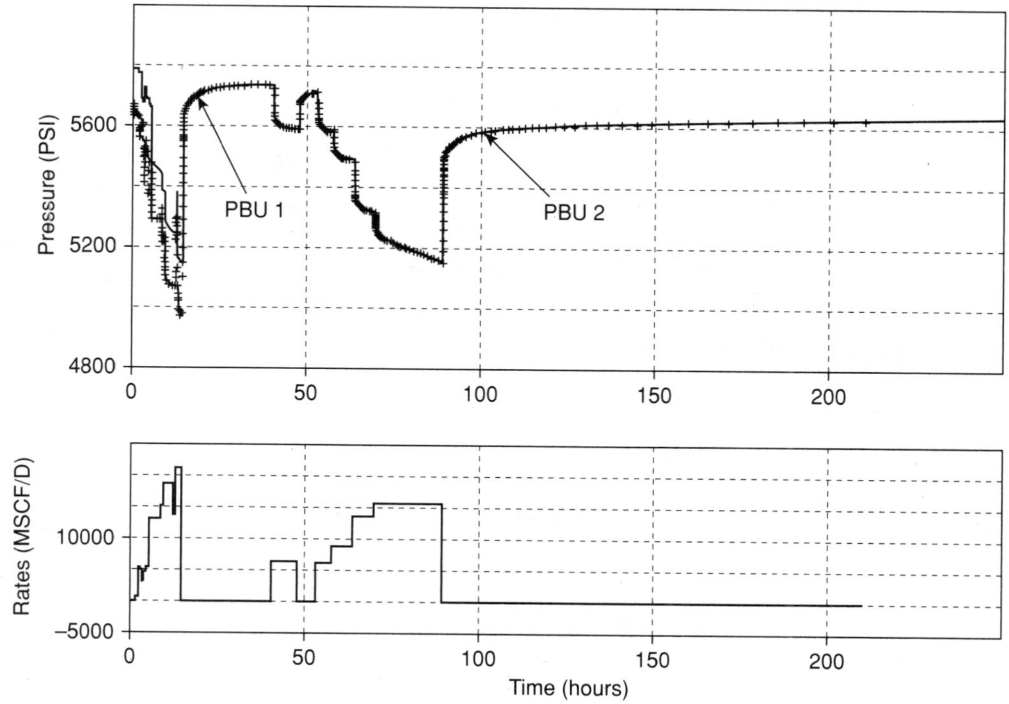

Figure 14.1 Pressure-rate history for Example 14.1 (from Levitan et al.[5] © 2006 SPE, Reproduced with permission).

and PBU 2 in Figure 14.1. PBU 2 is the longer buildup with duration of 121 hours. The pressure and pressure derivative of the two PBUs (PBU 1 and PBU 2) shown in Figure 14.2 indicate similar trends. However, the pressure derivative curves do not indicate definitive boundary effects. This appears to be in contradiction with the pressure history shown in Figure 14.1, which shows pressure depletion during PBU 2, indicative of a closed reservoir compartment.[5] Figure 14.3 shows the pressure and pressure derivative of the deconvolved constant-rate drawdown responses derived from PBU 1 and PBU 2 data. The slopes of both pressure derivatives at late times approach unity. This is indicative of a closed reservoir system. Figure 14.4 compares the pressure derivative for PBU 2 obtained by differentiating PBU data with respect to superposition time and the pressure derivative obtained by differentiating the deconvolved pressure response with respect to the logarithm of time. Note the obvious differences between the two curves. The pressure derivative obtained from the deconvolved pressure response is more representative of the interpretation model assumed to be a partially closed system. By assuming a reservoir model that is closed on three sides with the fourth side open for limited pressure support, the pressure buildup data for PBU 2 is matched as shown in Figure 14.5. This model also matches most of the sequence shown in Figure 14.1.

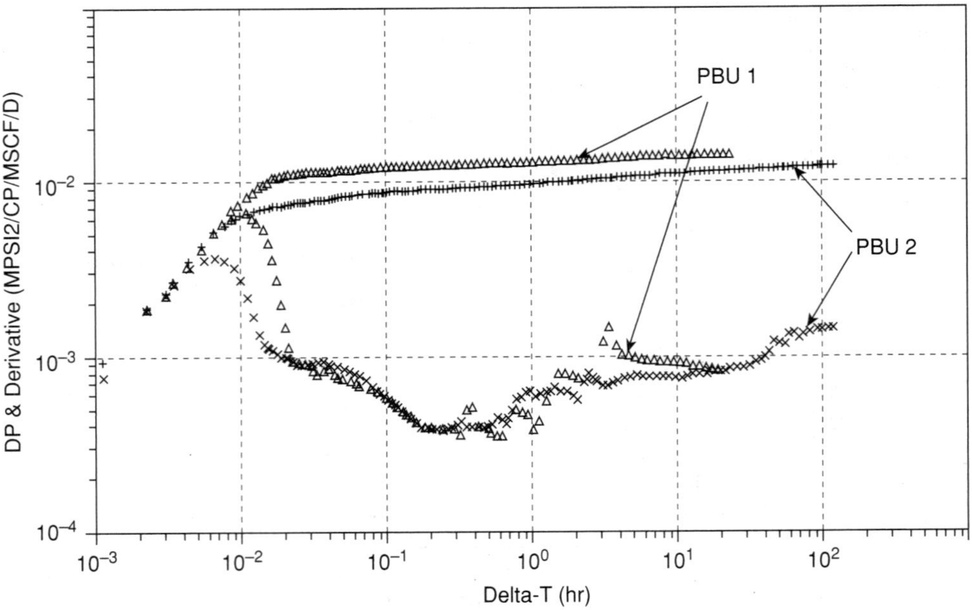

Figure 14.2 Comparison of derivative plots of two PBUs for Example 14.1 (from Levitan et al.[5] © 2006 SPE, Reproduced with permission).

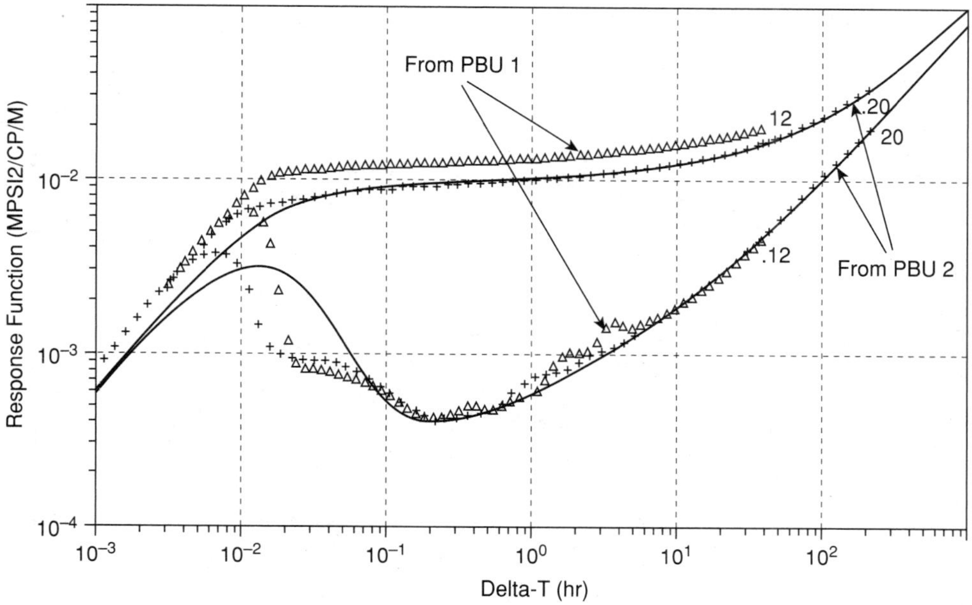

Figure 14.3 Comparison of deconvolved responses for PBU 1 and PBU 2 for Example 14.1 (from Levitan et al.[5] © 2006 SPE, Reproduced with permission).

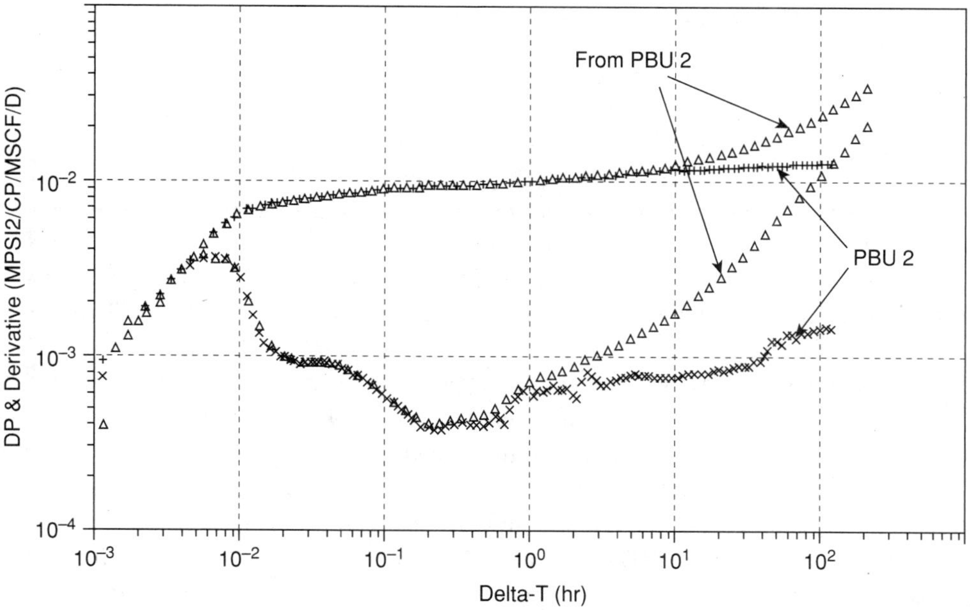

Figure 14.4 Comparison of deconvolved responses for PBU 2 data (triangles) with derivative plot of PBU 2 data (crosses) for Example 14.1 (from Levitan et al.[5] © 2006 SPE, Reproduced with permission).

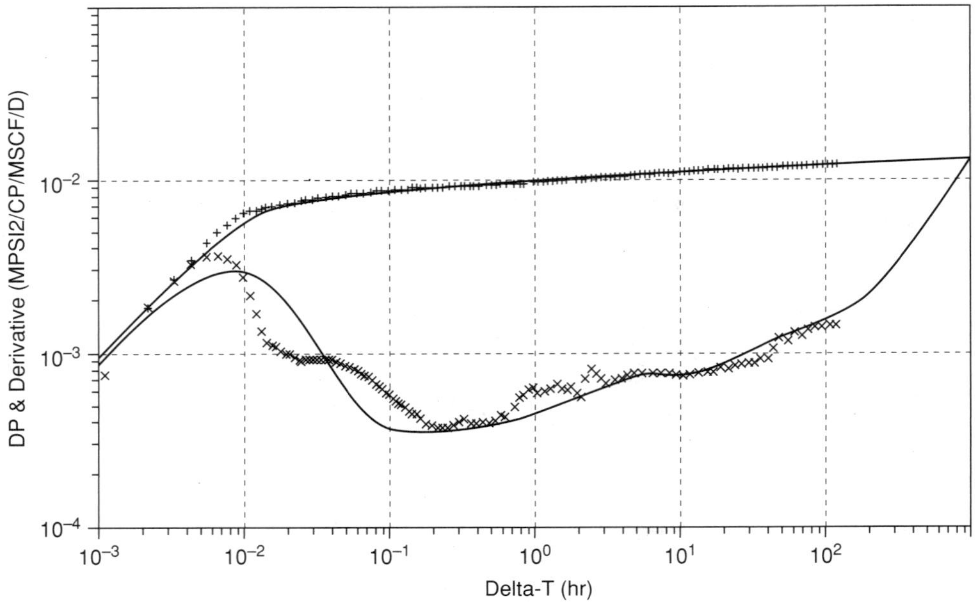

Figure 14.5 Model match of PBU 2 data assuming partially closed reservoir model for Example 14.1 (from Levitan et al.[5] © 2006 SPE, Reproduced with permission).

Example 14.2

The rate and permanent gauge pressure data from a single horizontal well in a gas reservoir is shown in Figure 14.6. The time span of the production history is 12,800 hours. Two pressure buildups (PBU) labeled PBU 1 and PBU 2 with durations of 167 hours and 170 hours, respectively, were selected for analysis. The pressure and pressure derivative plots for PBU 1 and PBU 2 are shown in Figure 14.7. The pressure derivative curves for PBU 1 and PBU 2 show increasing trends at late times. Figure 14.8 shows the pressure and pressure derivative plots for PBU 1 and PBU 2 obtained from constant-rate drawdown pressure response of the two PBUs over time intervals of 5400 and 11,900 hours, respectively. The pressure derivative plots of the two PBUs indicate unit slope at late times. This is indicative of boundary effects. Consequently, a reservoir model of a horizontal well in an elongated rectangular reservoir matched PBU 2 data as shown in Figure 14.8 by the solid curve.

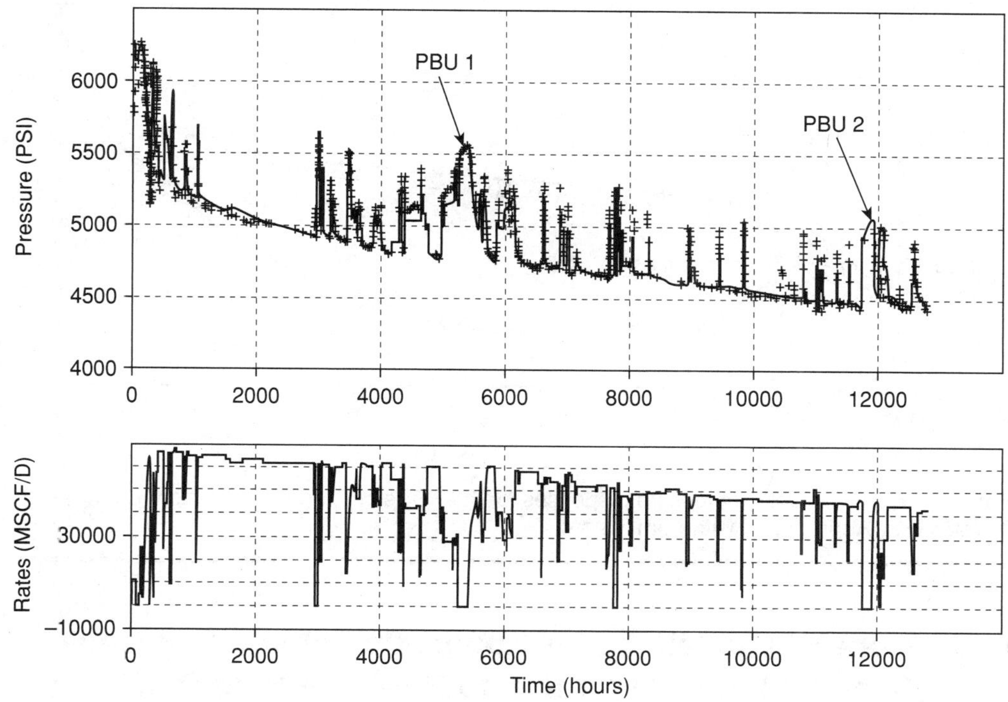

Figure 14.6 Pressure-rate history for Example 14.2 (from Levitan et al.[5] © 2006 SPE, Reproduced with permission).

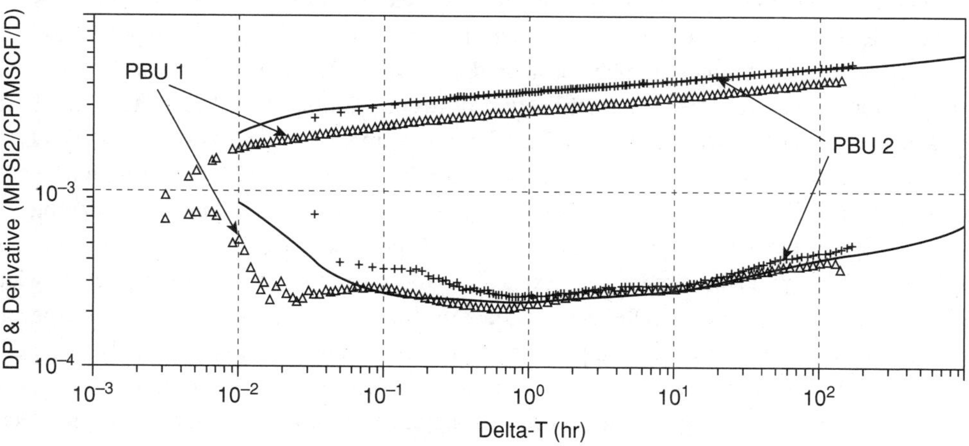

Figure 14.7 Comparison of derivative plots of two PBUs for Example 14.2 (from Levitan et al.[5] © 2006 SPE, Reproduced with permission).

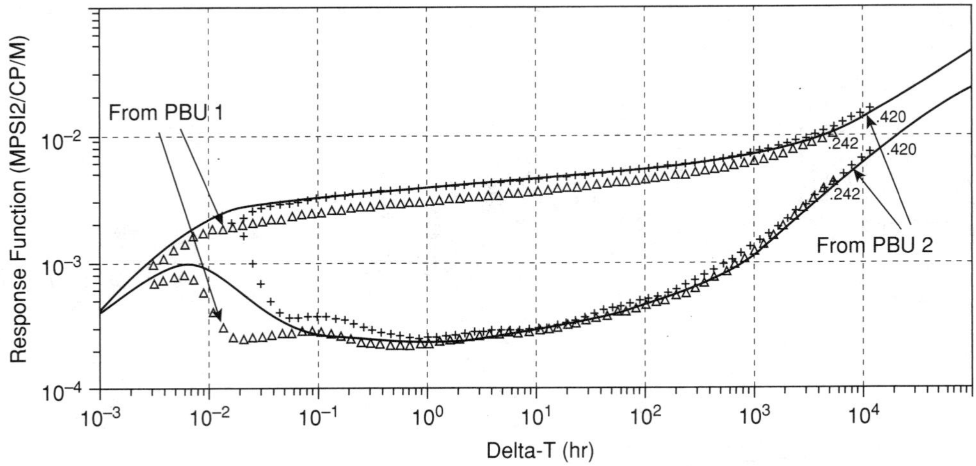

Figure 14.8 Deconvolved pressure responses for the PBUs in Example 14.2 (from Levitan et al.[5] © 2006 SPE, Reproduced with permission)

14.6 General Guidelines for Application of von Schroeter Deconvolution Algorithm to Pressure-Rate Data from Well Tests

It is necessary to reiterate that deconvolution is not a new method for interpretation of well test data. It is rather a process for converting variable-rate pressure data into constant-rate pressure data, thereby providing more pressure data for the determination of an interpretation model. Furthermore, deconvolution is not a replacement for conventional methods or type curves in the analysis of well test data. It should be considered as another tool in the toolkit of the analyst for the evaluation of well test data. The application of deconvolution to well test data should yield results that are consistent with results obtained from other methods (such as straightline methods, type curve analysis, etc.), and in conformity with known geological interpretations of the reservoir. The application of deconvolution algorithms to well test data requires considerable skill and knowledge of the technology. As more robust and stable deconvolution algorithms are made available in commercial software for well test analysis, deconvolution methods will become widely used because of its capacity to include all the data recorded in a test sequence. This is especially important in the analysis of data from permanent downhole gauges that are routinely installed in wells located in challenging environments, such as found in deepwater reservoirs.

The following are suggested guidelines for application of the von Schroeter-type deconvolution algorithm to variable-rate well test data:[5]

1. The von Schroeter deconvolution algorithm is based on linear systems. This means that this algorithm is applicable to systems with single-phase flow in the reservoir. Reservoir

pressure should be above bubble point pressure for oil reservoirs or dewpoint pressure for gas reservoirs when applying this algorithm to test data.

2. The reservoir should be at equilibrium and at a uniform pressure at the start of the well test.
3. A reliable estimate of the initial reservoir pressure at the start of the test sequence will aid in the deconvolution of the test data.
4. The test data should be evaluated for consistency. This means that the test data should comply with the assumptions of constant wellbore storage and/or skin factor implied in the convolution integral of Eq. (14.1). This can be done by selecting individual pressure buildup data from the test sequence for deconvolution and comparing the results.
5. Check the pressure-rate data for local events such as mismatched pressure-rate data. This type of error may distort the deconvolution pressure transform at late times and create non-existent artifacts in the transformed data.
6. Compare the interpretation model generated from the deconvolved pressure response to other interpretation models based on derivative plots, type curves, or geologic data. All the interpretation models generated from all these methods should be inherently consistent. For instance, if the deconvolved pressure response is indicating a closed reservoir system, this should be confirmed with other interpretations based on derivative plots, type curves, or known geologic data. If the interpretation is inconsistent, the evaluation process should be re-examined in its entirety.
7. Perform a thorough check on the quality of the test data. Poor quality data will generally yield misleading results when subjected to deconvolution.

References

1. von Schroeter, T., Hollaender, F., and Gringarten, A.C.: "Deconvolution of Well-Test Data as a Nonlinear Total Least-Squares Problem," *SPEJ* (December 2004) 375–390.
2. von Schroeter, T., Hollaender, F., and Gringarten, A.C.: "Analysis of Well Test Data From Permanent Downhole Gauges by Deconvolution," paper SPE 77688 presented at the 2002 SPE Annual Technical Conference and Exhibition, San Antonio, Texas, September 29–October 2, 2002.
3. von Schroeter, T., Hollaender, F., and Gringarten, A.C.: "Deconvolution of Well-Test Data as a Nonlinear Total Least-Squares Problem," paper SPE 71574 presented at the 2001 SPE Annual Technical Conference and Exhibition, New Orleans, Louisiana, September 30–October 3, 2001.
4. Levitan, M.M.: "Practical Application of Pressure/Rate Deconvolution to Analysis of Real Well Tests," *SPEREE* (April 2005) 113–121.
5. Levitan, M.M., Crawford, G.E., and Hardwick, A.: "Practical Considerations for Pressure-Rate Deconvolution of Well-Test Data," *SPEJ* (March 2006) 35–46.

6. Onur, M., Cinar, M., Ilk, D., Valko, P.P., and Blasingame, T.A.: "An Investigation of Recent Deconvolution Methods for Well-Test Data Analysis," *SPEJ* (June 2008) 226–247.
7. Levitan, M.M.: "Deconvolution of Multiwell Test Data," *SPEJ* (December 2007) 420–428.
8. Gringarten, A.C.: "From Straight Lines to Deconvolution: The Evolution of the State of the Art in Well Test Analysis," *SPEREE* (February 2008) 41–62.

General Reading

- Kuchuk, F.J., Carter, R.G., and Ayestaran, L.: "Deconvolution of Wellbore Pressure and Flow Rate," *SPEFE* (March 1990) 53–59.
- Baygun, B., Kuchuk, F.J., and Arikan, O.: "Deconvolution Under Normalized Autocorrelation Constraints," *SPEJ* (September 1997) 246–253.
- Gringarten, A.C., von Schroeter, T., Rolfsvaag, T., and Bruner, J.: "Use of Downhole Permanent Gauge Data to Diagnose Production Problems in a North Sea Horizontal Well," paper SPE 84470 presented at the SPE Annual Technical Conference and Exhibition, Denver, Colorado, October 5–8, 2003.
- Du, K.: "The Determination of Tested Drainage Area and Reservoir Characterization from Entire Well-Test History by Deconvolution and Conventional Pressure-Transient Analysis Techniques," paper SPE 115720 presented at the 2008 SPE Annual Technical Conference and Exhibition, Denver, Colorado, September 21–24, 2008.
- Whittle, T., and Gringarten, A.: "The Determination of Minimum Tested Volume from the Deconvolution of Well Test Pressure Transients," paper SPE 116575 presented at the 2008 SPE Annual Technical Conference and Exhibition, Denver, Colorado, September 21–24, 2008.

CHAPTER 15

Immiscible Fluid Displacement

15.1 Introduction

Most of the oil and gas recovered from reservoirs is displaced immiscibly by water and/or gas. The displacement could be in the form of solution gas drive, gas cap expansion, water influx from aquifers or injection of water and/or gas. Solution-gas drive, gas cap expansion, and water influx from aquifers are essentially natural processes that supply energy to the reservoir for hydrocarbon recovery. Gas and water injection are designed and installed to artificially supply energy to the reservoir and thereby improve hydrocarbon recovery. It is important to understand the fundamental processes that occur when reservoir fluids are displaced immiscibly by gas or water. The displacement process is affected by the wettability of the rock, and the mobility ratio between the displaced and the displacing fluids. The total efficiency of the displacement process is measured in terms of the effectiveness of water or gas in displacing the reservoir fluids, and the proportion of the reservoir actually contacted by the displacing fluids. In this chapter, basic concepts in immiscible fluid displacement are presented. These are then followed with the presentation of the fractional flow equation, the Buckley-Leverett equation, and the Welge method for estimating average water saturation in a water displacement process. These equations are presented to familiarize the engineer with some of the classical developments in the analysis of immiscible displacement processes before the advent and widespread application of reservoir simulation techniques. This approach is intended to enable the engineer to become conversant with some of the terms generally used in the industry to analyze and discuss the results from reservoir simulation when applied to immiscible displacement processes.

15.2 Basic Concepts in Immiscible Fluid Displacement

In this section, basic concepts in the immiscible fluid displacement process are presented. These concepts are useful in discussion and understanding of immiscible displacement processes. Some of the concepts are also used in Chapters 16 and 17 in discussions of secondary and enhanced hydrocarbon recovery processes. The concepts that are presented in this section are: rock wettability, capillary pressure, relative permeability, mobility and mobility ratio, fluid displacement efficiency, volumetric displacement efficiency, and total recovery efficiency.

15.2.1 Rock Wettability

Rock wettability is the tendency of either the water phase or the oil phase to preferentially maintain contact with the rock surface in a multiphase fluid system. Thus, the surface of a water-wet rock, preferentially maintains contact with water, while the surface of an oil-wet rock will preferentially maintain contact with oil in an oil-water system. The most common method of determining rock wettability is by measurement of the contact angle, θ, between the rock surface and the fluid system.[1] For an oil-water system in contact with a solid surface, the contact angle, θ, is the angle between the fluid-solid interface measured through the water phase. The rock surface is considered to be water-wet when $\theta < 90°$, and oil-wet when $\theta > 90°$ (Figure 15.1). When $\theta \approx 90°$, the rock surface is considered to be intermediate- or neutral-wet. There are other methods for determining rock

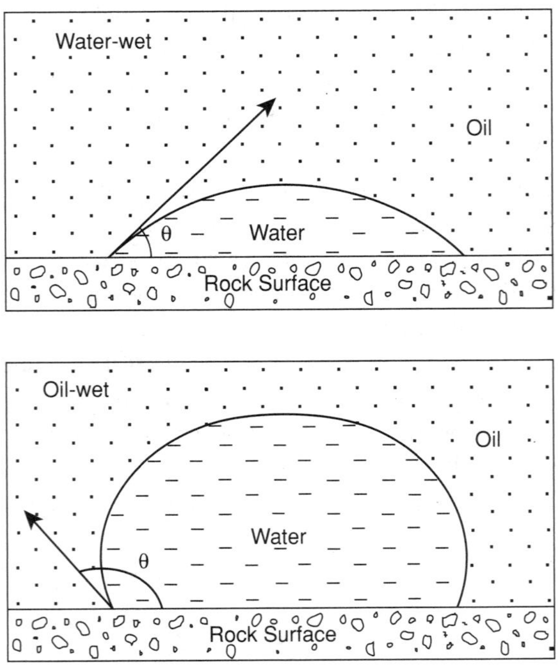

Figure 15.1 Water-wet rock and oil-wet rock.

wettability based on displacement studies. These are the Amott[2] wettability test and the United States Bureau of Mines (USBM)[3] wettability test. The Amott and USBM methods are relatively fast, with tests completed in a few days, in comparison to the contact angle method, which may require several months for completion. Additional methods for determining rock wettability are: Imbibition rate test,[4] Hysteresis of relative permeability curve,[5] and Nuclear Magnetic Relaxation (NMR).[6]

Rock wettability is classified into uniform or homogeneous wetting, and non-uniform or heterogeneous wetting. Uniform wetting, in contrast to non-uniform wetting, assumes that the entire reservoir rock has a single form of wettability. Non-uniform wettability is characterized by distinct zones within the reservoir rock system that are preferentially water-wet or oil-wet. The distribution and extent of wetting heterogeneity is determined by the variation of the reservoir rock minerals. Heterogeneous wettability consists of wetting states that have been described as mixed wettability, fractional wettability, "Dalmatian" wetting, and speckled wetting.[7]

The composition of the reservoir oil affects the wettability of the rock. The wetting state of reservoir rock is affected by the presence of polar compounds such as asphaltenes, film forming components, and high molecular weight paraffins. Other factors that may affect rock wettability include the type of minerals present in the rock, the reservoir rock type (quartz, silica, calcite, etc.), and salinity of the connate water.

Rock wettability is usually measured from core plugs or whole length cores. The wettability of the core samples can be altered during preparation of the core samples before testing. Furthermore, in-situ rock wettability may be different from laboratory measured rock wettability due to the following reasons:

1. Flash vaporization of reservoir fluids from cores due to depressurization as the cores are retrieved from the reservoir.
2. Invasion of mud filtrate.
3. Asphaltenes deposition or wax precipitation in the cores due to pressure and temperature changes.
4. Drying of cores during storage.

It is important to ensure that the wettability of the core samples is not altered during handling. In cases where the wettability of the core may have been altered, procedures should be undertaken to restore it as close as possible to its native state before initiating the wettability tests.

15.2.2 Capillary Pressure

Capillary pressure, p_c, is commonly defined as the difference in the pressure of the non-wetting phase and the pressure of the wetting phase. This is represented as:

$$p_c = p_{nw} - p_w \tag{15.1}$$

In Eq. (15.1), p_{nw} = pressure in the non-wetting phase; and p_w = pressure in the wetting phase. As defined in Eq. (15.1), it is evident that capillary pressure is related to the wettability of the rock as discussed in Section 15.2.1. For a water-wet rock in an oil/water system, the capillary pressure derived from Eq. (15.1) is:

$$p_c = p_o - p_w \tag{15.2}$$

Similarly, for an oil-wet rock in an oil/water system, the capillary pressure is:

$$p_c = p_w - p_o \tag{15.3}$$

By convention, the capillary pressure in a water-wet rock as represented in Eq. (15.2) is designated as positive. Hence, capillary pressure in an oil-wet rock is designated as negative.

The phenomenon of capillarity in reservoirs can be discussed in terms of capillary pressure as measured in capillary tubes. For a capillary tube, capillary pressure is determined as:

$$p_c = \frac{2\sigma \cos \theta}{r} \tag{15.4}$$

In Eq. (15.4), p_c = capillary pressure, dynes/cm^2; σ = the interfacial tension between the two immiscible phases, dynes/cm; θ = contact angle, degrees; and r = radius of the capillary tube, cm. As shown by Eq. (15.4), capillary pressure is inversely proportional to the radius of the capillary tube. This means that capillary tubes with smaller radii will result in higher capillary pressures if other terms in Eq. (15.4) remain the same. A reservoir can be considered as a porous medium consisting of pores of different sizes. Since the sizes of these pores are at dimensions of capillary tubes, it is easily deduced by application of Eq. (15.4) that capillary pressure will be higher in pores with smaller radii. Capillary pressure can also be related to height above the free water level in a reservoir. The free water level (FWL) is defined as the depth at which oil-water capillary pressure is zero. Applying these definitions, capillary pressure in an oil/water system can be represented also as:

$$p_c = \Delta \rho g H \tag{15.5}$$

In Eq. (15.5), $\Delta \rho = \rho_w - \rho_o$, where ρ_w is the density of water, and ρ_o is the density of oil; g = acceleration due to gravity; and H = height above the FWL. By applying Eqs. (15.4) and (15.5), it can be deduced that reservoirs with relatively high capillary pressure also have large transition zone. Transition zone is defined as height between the FWL and irreducible connate water saturation. The effects of capillary pressure on the interpretation of FWL and transition zone were presented in Chapter 5 (Section 5.5.3).

The capillary pressure curves for a strongly water-wet rock are shown in Figure 15.2. Curve 1 is the imbibition curve and Curve 2 is the drainage curve. Imbibition is described as the process that results in the increase of the saturation of the wetting phase, while drainage is the process that leads to decrease in the saturation of the wetting phase. Figure 15.2 illustrates the path-dependent hysteresis that exists between the imbibition and drainage capillary pressure curves. The capillary pressure curves of an oil-wet rock are shown in Figure 15.3. Note that the capillary pressures are negative by convention. Also, note the hysteresis between the imbibition and

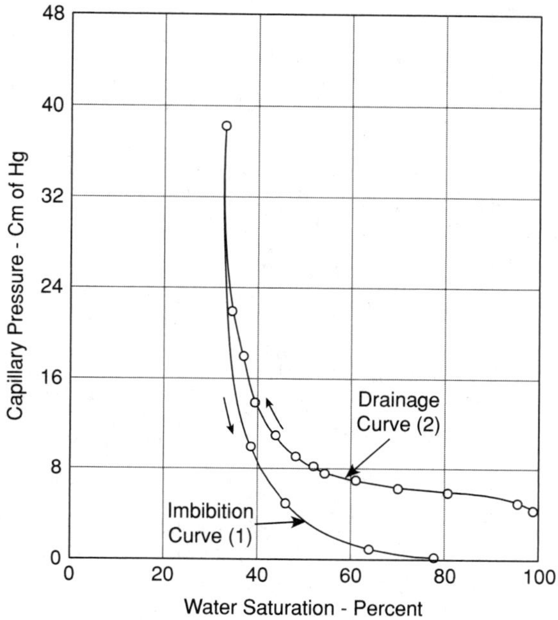

Figure 15.2 Capillary pressure curve for strongly water-wet rock (from Killins et al.[8]).

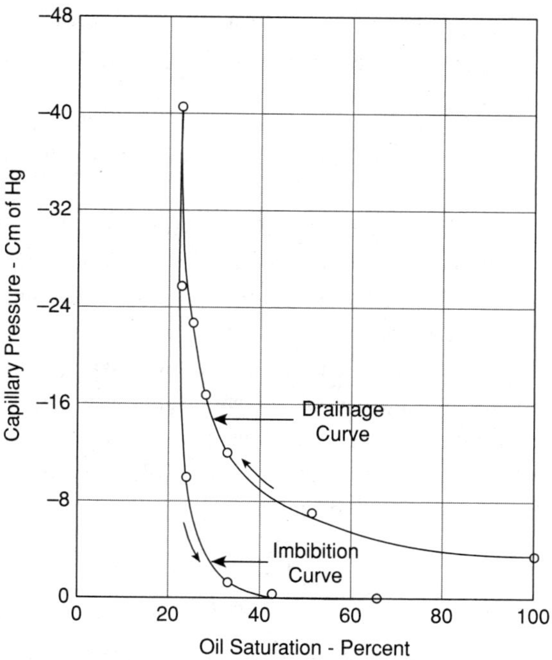

Figure 15.3 Capillary pressure curve for oil-wet rock (from Killins et al.[8]).

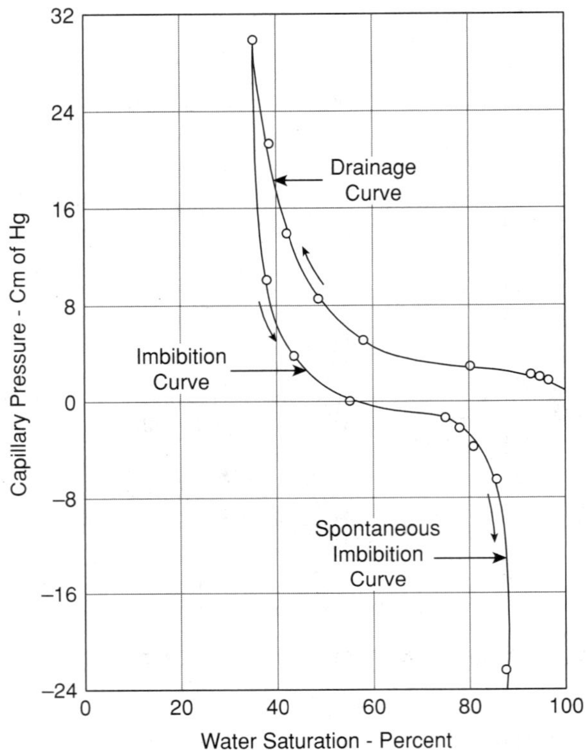

Figure 15.4 Capillary pressure curve for neutral-wet rock (from Killins et al.[8])

drainage capillary pressure curves. The capillary pressure curves for a neutral-wet rock are shown in Figure 15.4.

One of the important applications of capillary pressure curves is the distribution of initial fluid saturations in the reservoir. This application utilizes the dependence of capillary pressure on fluid saturations, the wettability of the reservoir rock, pore distribution, and the sizes of the pores and the pore throat. It is common to use the drainage capillary pressure curve in this application since the reservoir is assumed to be initially filled with water before the migration of oil into it. This essentially assumes a water-wet rock system. This assumption could be in error since it is possible for the wettability of the rock to change with time if the oil contains polar compounds, such as asphaltenes. Capillary pressure data from various core samples can be correlated with the J-function. The J-function is represented as:

$$J(S_w) = \frac{0.217 p_c}{\sigma \cos\theta} \sqrt{\frac{k}{\phi}} \tag{15.6}$$

In Eq. (15.6), $J(S_w)$ = dimensionless function dependent on saturation of the wetting phase; p_c = capillary pressure, psi; σ = interfacial tension, dynes/cm; θ = contact angle, degrees;

k = rock permeability, md; and ϕ = porosity, fraction. The J-function can be used to distribute initial fluid saturations in the reservoir as part of the process of reservoir characterization and building a reservoir model for simulation.

15.2.3 Relative Permeability

Relative permeability of a fluid is the ratio of its effective permeability to the absolute permeability of the porous medium. Basic concepts in the application of relative permeability curves were presented in Chapter 2, Section 2.3. Figures 2.4 and 2.5 show the relative permeability curves for an oil-water system, and gas-oil system, respectively. Note that relative permeability curves are affected by the wettability of the rock. One of the most important factors that affect oil recovery in an immiscible displacement process is the magnitude of the relative permeability of the displacing fluid with respect to the displaced fluid. This is shown by the fractional flow equation developed in this chapter in Section 15.3.

15.2.4 Mobility and Mobility Ratio

The mobility of a fluid, λ, is defined as the ratio of its effective permeability to its viscosity. For a fluid, i, mobility, λ_i, is given by:

$$\lambda_i = \frac{k_i}{\mu_i} \tag{15.7}$$

In Eq. (15.7), k_i = effective permeability of the fluid; and μ_i = viscosity of the fluid. For instance, the mobility of water is $\lambda_w = k_w/\mu_w$. Similarly, the mobility of oil is $\lambda_o = k_o/\mu_o$.

Mobility ratio is the ratio of the mobility of the displacing phase to the mobility of the displaced phase. Mobility ratio, M, is represented as:

$$M = \frac{\lambda_D}{\lambda_d} \tag{15.8}$$

In Eq. (15.8), λ_D = mobility of the displacing phase; and λ_d = mobility of the displaced phase. Mobility ratios are defined with reference to saturations at specific locations of the flood front. In this book, mobility ratio is defined with reference to endpoint saturations assuming piston-like displacement. Using this convention, the mobility ratio for water displacing oil in a waterflood is defined as:

$$M = \frac{(k_w/\mu_w)_{S_{or}}}{(k_o/\mu_o)_{S_{wi}}} \tag{15.9}$$

In terms of relative permeability, Eq. (15.9) can be written as:

$$M = \frac{(k_{rw}/\mu_w)_{S_{or}}}{(k_{ro}/\mu_o)_{S_{wi}}} \tag{15.10}$$

In Eq. (15.10), $(k_{rw})_{S_{or}}$ = water relative permeability at residual oil saturation, S_{or}; and $(k_{ro})_{S_{wi}}$ = oil relative permeability at initial water saturation, S_{wi}.

From Eq. (15.8), it is evident that when $M \leq 1$, the displacement process is considered favorable because the displaced fluid is more mobile than the displacing fluid. When M is much greater than one, the displacement process is considered to be unfavorable. High mobility ratios could lead to viscous fingering, resulting in significant amounts of bypassed hydrocarbon reserves.

15.2.5 Fluid Displacement Efficiency

Fluid displacement efficiency, E_D, is sometimes called microscopic displacement efficiency. It is defined as the volume of oil displaced from the invaded region divided by the volume of oil initially in place in the invaded region. Fluid displacement efficiency is affected by rock wettability, capillary pressure, relative permeability, and mobility ratios of the fluids. If reservoir pressure is maintained at initial conditions and saturation within the invaded region is reduced to residual oil saturation, the displacement efficiency can be estimated with the expression:

$$E_D = 1 - \frac{S_{or}}{S_{oi}} \quad \textbf{(15.11)}$$

In Eq. (15.11), S_{or} = residual oil saturation in the swept (invaded) region; and S_{oi} = initial oil saturation in the swept region.

15.2.6 Volumetric Displacement Efficiency

Volumetric displacement efficiency, E_V, is sometimes termed macroscopic displacement efficiency. It is the fraction of the reservoir volume swept by the displacing fluid. It is composed of two parts, namely, areal sweep efficiency, E_A, and vertical sweep efficiency, E_I. Areal sweep efficiency is the fraction of the reservoir area contacted by the displacing fluid. Vertical sweep efficiency is the fraction of the vertical reservoir cross-section contacted by the displacing fluid. Volumetric displacement efficiency is the product of areal and vertical sweep efficiencies:

$$E_V = E_A \times E_I \quad \textbf{(15.12)}$$

15.2.7 Total Recovery Efficiency

Total recovery efficiency, E_R, is the volume of oil displaced divided by the initial volume of oil in place in the swept portion of the reservoir. Thus, total recovery efficiency is given by:

$$E_R = E_D \times E_V \quad \textbf{(15.13)}$$

15.3 Fractional Flow Equations

The fractional flow equation is used to calculate the flow rate of a fluid as a fraction of the total fluid flow rate when only two fluids are flowing in the reservoir. The flow rate of the fluid at any point in the reservoir depends on its saturation at that point. Since relative permeability of the fluid is dependent on saturation, it follows then that the flow rate of the fluid is dependent on its relative permeability at that point in the reservoir. Fractional flow of a fluid in a reservoir is primarily dependent on its relative permeability, but can be affected by capillary and gravity forces. The fractional flow equations developed in this book are based on the assumption of linear flow.

15.3.1 Fractional Flow Equation for Oil Displaced by Water

The fractional flow equation developed in this section is for water displacing oil in an oil-water reservoir.

Applying Eq. 10.16, the flow rate of water in reservoir barrels is represented as:

$$q_w = -1.127 \times 10^{-3} \frac{k_w A}{\mu_w} \left[\frac{\partial p_w}{\partial l} + 0.00694 \rho_w \sin \alpha \right] \tag{15.14}$$

The oil flow rate is represented as:

$$q_o = -1.127 \times 10^{-3} \frac{k_o A}{\mu_o} \left[\frac{\partial p_o}{\partial l} + 0.00694 \rho_o \sin \alpha \right] \tag{15.15}$$

In Eqs. (15.14) and (15.15), q_w, q_o = flow rates of water and oil, respectively, RB/D; k_w, k_o = effective permeabilities of water and oil, respectively, md; μ_w, μ_o = viscosities of water and oil, respectively, cp; p_w, p_o = phase pressures in water and oil, respectively, psia; A = cross-sectional area for fluid flow, ft^2; l = flow path length, ft; ρ_w, ρ_o = densities of water and oil, respectively, at reservoir conditions, lbm/ft^3; and α = dip angle of the reservoir measured counter clockwise from the horizontal to the positive flow path (Figure 15.5).

The total reservoir liquid flow rate, q_t, is the sum of the water flow rate, q_w, and the oil flow rate, q_o, expressed as:

$$q_t = q_w + q_o \tag{15.16}$$

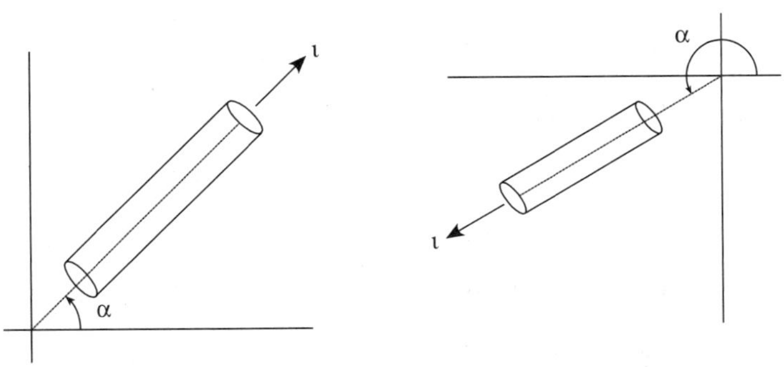

Figure 15.5 Flow path from the horizontal to the positive path.

The fractional flow of water, f_w, as a fraction of the total flow rate is:

$$f_w = \frac{q_w}{q_t} \tag{15.17}$$

The fractional flow of oil, f_o, as fraction of the total flow rate is:

$$f_o = \frac{q_o}{q_t} \tag{15.18}$$

Re-arranging Eqs. (15.17) and (5.18) yields:

$$f_w q_t = q_w \tag{15.19}$$

and

$$(1 - f_w)q_t = q_o \tag{15.20}$$

Substituting Eq. (15.14) into Eq. (15.19) and re-arranging gives the following equations:

$$f_w q_t = -1.127 \times 10^{-3} \frac{k_w A}{\mu_w}\left[\frac{\partial p_w}{\partial l} + 0.00694\rho_w \sin\alpha\right] \tag{15.21}$$

$$-\frac{f_w q_t \mu_w}{k_w A} = 1.127 \times 10^{-3}\left[\frac{\partial p_w}{\partial l} + 0.00694\rho_w \sin\alpha\right] \tag{15.22}$$

Substituting Eq. (15.15) into Eq. (15.20) and re-arranging gives the following equations:

$$(1 - f_w)q_t = -1.127 \times 10^{-3} \frac{k_o A}{\mu_o}\left[\frac{\partial p_o}{\partial l} + 0.00694\rho_o \sin\alpha\right] \tag{15.23}$$

$$-\frac{(1 - f_w)q_t \mu_o}{k_o A} = 1.127 \times 10^{-3}\left[\frac{\partial p_o}{\partial l} + 0.00694\rho_o \sin\alpha\right] \tag{15.24}$$

Subtracting Eq. (15.22) from Eq. (15.24) yields:

$$-\frac{(1 - f_w)q_t \mu_o}{k_o A} + \frac{f_w q_t \mu_w}{k_w A} = 1.127 \times 10^{-3}\left[\frac{\partial p_o}{\partial l} - \frac{\partial p_w}{\partial l} + 0.00694(\rho_o - \rho_w)\sin\alpha\right] \tag{15.25}$$

Applying Eq. (15.2), $p_c = p_o - p_w$ and taking derivatives gives:

$$\frac{\partial p_c}{\partial l} = \frac{\partial p_o}{\partial l} - \frac{\partial p_w}{\partial l} \tag{15.26}$$

Substituting Eq. (15.26) into Eq. (15.25) and simplifying yields:

$$-\frac{q_t \mu_o}{k_o A} + \frac{f_w q_t}{A}\left(\frac{\mu_o}{k_o} + \frac{\mu_w}{k_w}\right) = 1.127 \times 10^{-3}\left[\frac{\partial p_c}{\partial l} + 0.00694(\rho_o - \rho_w)\sin\alpha\right] \tag{15.26}$$

Re-arranging Eq. (15.27) gives:

$$\frac{f_w q_t}{A}\left(\frac{\mu_o}{k_o} + \frac{\mu_w}{k_w}\right) = \frac{q_t \mu_o}{k_o A} + 1.127 \times 10^{-3}\left[\frac{\partial p_c}{\partial l} + 0.00694(\rho_o - \rho_w)\sin\alpha\right] \quad (15.27)$$

Solving for f_w from Eq. (15.28) gives:

$$f_w = \frac{\dfrac{\mu_o}{k_o}}{\left(\dfrac{\mu_o}{k_o} + \dfrac{\mu_w}{k_w}\right)} + \frac{1.127 \times 10^{-3}\dfrac{A}{q_t}\left[\dfrac{\partial p_c}{\partial l} + 0.00694(\rho_o - \rho_w)\sin\alpha\right]}{\left(\dfrac{\mu_o}{k_o} + \dfrac{\mu_w}{k_w}\right)} \quad (15.29)$$

Multiplying the right side of Eq. (15.29) by k_o/μ_o yields:

$$f_w = \frac{1}{\left(1 + \dfrac{\mu_w}{k_w} \times \dfrac{k_o}{\mu_o}\right)} + \frac{1.127 \times 10^{-3}\dfrac{k_o A}{\mu_o q_t}\left[\dfrac{\partial p_c}{\partial l} + 0.00694(\rho_o - \rho_w)\sin\alpha\right]}{\left(1 + \dfrac{\mu_w}{k_w} \times \dfrac{k_o}{\mu_o}\right)} \quad (15.30)$$

Equation (15.30) is the complete form of the fractional flow equation for water flowing linearly in an oil-water reservoir. A quick examination of Eq. (15.30) shows that it has all the factors that affect the flow of water in an oil-water reservoir. These factors are fluid properties (viscosities, densities), rock properties (effective permeabilities, saturations, capillary pressure), total flow rate, and structural inclination of the reservoir (dip angle, α). If it is assumed that fluid properties and total flow rate are constant, then fractional flow of water is only a function of water saturation in the reservoir. A plot of f_w versus S_w shown in Figure 15.6 is commonly called the fractional flow curve. The fractional flow curve can be used to calculate the flow rate of water in the reservoir at any water saturation.

The effect of capillary pressure can be assumed to be negligible. In such cases, Eq. (15.30) reduces to the following form:

$$f_w = \frac{1}{\left(1 + \dfrac{\mu_w}{k_w} \times \dfrac{k_o}{\mu_o}\right)} + \frac{7.821 \times 10^{-6}\dfrac{k_o A}{\mu_o q_t}(\rho_o - \rho_w)\sin\alpha}{\left(1 + \dfrac{\mu_w}{k_w} \times \dfrac{k_o}{\mu_o}\right)} \quad (15.31)$$

Equation (15.30) can be simplified further by assuming that both capillary and gravity effects are negligible to give:

$$f_w = \frac{1}{\left(1 + \dfrac{\mu_w}{k_w} \times \dfrac{k_o}{\mu_o}\right)} \quad (15.32)$$

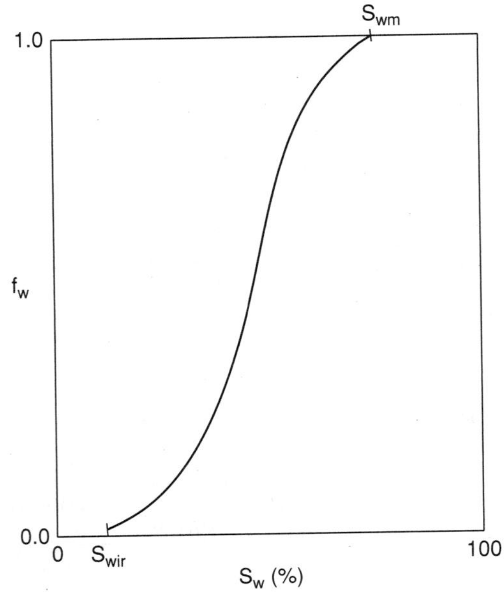

Figure 15.6 Fractional flow curve for water in an oil-water system.

In terms of relative permeabilities, Eq. (15.32) can be written as:

$$f_w = \frac{1}{\left(1 + \frac{\mu_w}{k_{rw}} \times \frac{k_{ro}}{\mu_o}\right)} \tag{15.33}$$

15.3.2 Fractional Flow Equation for Oil Displaced by Gas

For reservoirs in which oil is displaced by gas, a comparable set of equations can be written for fractional flow of gas. The fractional flow equation for gas flow is:

$$f_g = \frac{1}{\left(1 + \frac{\mu_g}{k_g} \times \frac{k_o}{\mu_o}\right)} + \frac{1.127 \times 10^{-3} \frac{k_o A}{\mu_o q_t}\left[\frac{\partial p_c}{\partial l} + 0.00694(\rho_o - \rho_g)\sin\alpha\right]}{\left(1 + \frac{\mu_g}{k_g} \times \frac{k_o}{\mu_o}\right)} \tag{15.34}$$

Equation (15.34) shows that the flow of a gas in a reservoir is affected by fluid properties, rock properties, total flow rate, and structural inclination of the reservoir. If it is assumed that fluid properties and total flow rate are constant, then fractional flow of gas is only a function of gas saturation in the reservoir. A fractional flow curve of f_g versus S_g similar to Figure 15.6 can be generated for gas flow in a reservoir.

If the effects of capillary pressure are assumed to be negligible, Eq. (15.34) reduces to:

$$f_g = \frac{1}{\left(1 + \frac{\mu_g}{k_g} \times \frac{k_o}{\mu_o}\right)} + \frac{7.821 \times 10^{-6} \frac{k_o A}{\mu_o q_t} (\rho_o - \rho_g) \sin\alpha}{\left(1 + \frac{\mu_g}{k_g} \times \frac{k_o}{\mu_o}\right)} \tag{15.35}$$

If the effects of both capillary pressure and gravity are assumed to be negligible, Eq. (15.34) simplifies to the form:

$$f_g = \frac{1}{\left(1 + \frac{\mu_g}{k_g} \times \frac{k_o}{\mu_o}\right)} \tag{15.36}$$

In terms of relative permeability, Eq. (15.36) is written as:

$$f_g = \frac{1}{\left(1 + \frac{\mu_g}{k_{rg}} \times \frac{k_{ro}}{\mu_o}\right)} \tag{15.37}$$

15.4 The Buckley-Leverett Equation

The Buckley-Leverett[9] equation is based on the principle of conservation of mass for linear flow of a fluid (water or gas) through a reservoir at constant total flow rate. To illustrate the derivation of the Buckley-Leverett equation, the case of water displacing oil is used. Note that the same equation can be developed representing the case for gas displacing oil.

Consider a volume element of a linear reservoir model shown in Figure 15.7. Let the thickness of the element be represented as Δx and located at a distance, x, from the inlet face of the

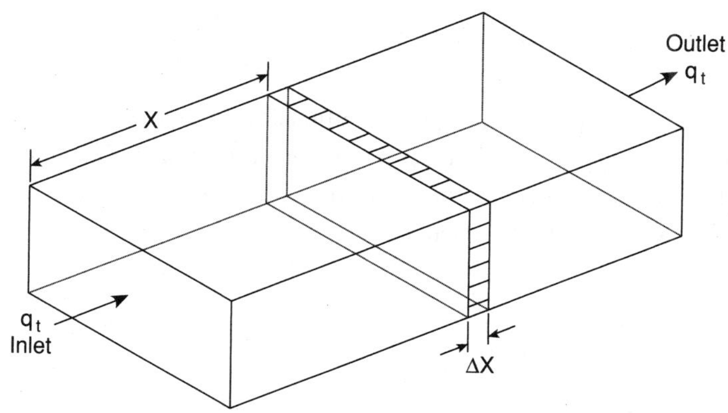

Figure 15.7 Linear reservoir model.

linear model. A volumetric balance in terms of the water phase (assuming density of water is constant) for the element of the reservoir model can be written as:

$$\begin{bmatrix}\text{Vol. of water flowing}\\ \text{into element in time, } \Delta t\end{bmatrix} - \begin{bmatrix}\text{Vol. of water flowing out}\\ \text{of element in time, } \Delta t\end{bmatrix} = \begin{bmatrix}\text{Accumulation of}\\ \text{water in element, } \Delta x\end{bmatrix} \tag{15.38}$$

Equation (15.38) can be expressed algebraically as:

$$[f_w q_t \Delta t]_x - [f_w q_t \Delta t]_{x+\Delta x} = \left[\frac{\phi A \Delta x}{5.615}\Delta S_w\right] \tag{15.39}$$

In Eq. (15.39), f_w = fraction of water in flow stream; q_t = constant total flow rate, RB/D; Δt = time interval, days; ϕ = porosity, fraction; A = cross-sectional area of flow, ft^2; and S_w = water saturation in the element. Re-arranging Eq. (15.39) gives:

$$\left[\frac{\phi A}{5.615 q_t}\frac{\Delta S_w}{\Delta t}\right] = -\frac{[f_w]_{x+\Delta x} - [f_w]_x}{\Delta x} \tag{15.40}$$

Taking limits as $\Delta t \rightarrow 0$ and $\Delta x \rightarrow 0$ yields the continuity equation:

$$\frac{\phi A}{5.615 q_t}\left(\frac{\partial S_w}{\partial t}\right)_x = -\left(\frac{\partial f_w}{\partial x}\right)_t \tag{15.41}$$

As stated in Section 15.3.1, the fractional flow of water is a function of water saturation only if fluid properties and total flow rate are constant. By application of chain rule, $f_w = f_w(S_w)$ can be expressed as:

$$\left(\frac{\partial f_w}{\partial x}\right)_t = \left(\frac{\partial f_w}{\partial S_w}\right)_t \left(\frac{\partial S_w}{\partial x}\right)_t \tag{15.42}$$

Substituting Eq. (15.42) into Eq. (15.41) and re-arranging gives:

$$\left[\frac{\partial S_w}{\partial t}\right]_x = -\frac{5.615 q_t}{\phi A}\left[\left(\frac{\partial f_w}{\partial S_w}\right)_t \left(\frac{\partial S_w}{\partial x}\right)_t\right] \tag{15.43}$$

Equation (15.43) gives water saturation as a function of time at a given location. A more useful equation expressing water saturation as a function of location at a given time can be developed from Eq. (15.43). For any displacement, the distribution of water saturation is a function of both location and time. This is represented as:

$$S_w = S_w(x,t) \tag{15.44}$$

The total derivative of S_w is then:

$$dS_w = \left(\frac{\partial S_w}{\partial x}\right)_t dx + \left(\frac{\partial S_w}{\partial t}\right)_x dt \tag{15.45}$$

Since the focus is on a fixed water saturation, then $dS_w = 0$. And Eq. (15.45) becomes:

$$0 = \left(\frac{\partial S_w}{\partial x}\right)_t dx + \left(\frac{\partial S_w}{\partial t}\right)_x dt \tag{15.46}$$

By re-arrangement, Eq. (15.46) becomes:

$$\left(\frac{dx}{dt}\right)_{S_w} = -\frac{\left(\frac{\partial S_w}{\partial t}\right)_x}{\left(\frac{\partial S_w}{\partial x}\right)_t} \tag{15.47}$$

Substituting Eq. (15.47) into Eq. (15.43) gives:

$$\left(\frac{dx}{dt}\right)_{S_w} = \frac{5.615q_t}{\phi A}\left(\frac{\partial f_w}{\partial S_w}\right)_t \tag{15.48}$$

Since the total flow rate is assumed to be constant, then fractional flow of water is independent of time. Hence,

$$\left(\frac{\partial f_w}{\partial S_w}\right)_t = \frac{df_w}{dS_w} \tag{15.49}$$

Equation (15.48) then becomes:

$$\left(\frac{dx}{dt}\right)_{S_w} = \frac{5.615q_t}{\phi A}\frac{df_w}{dS_w} \tag{15.50}$$

Equation (15.50) is the Buckley-Leverett equation. It is also called the frontal advance equation. Integration of Eq. (15.50) yields a useful form of the Buckley-Leverett equation:

$$x = \frac{5.615q_t t}{\phi A}\left(\frac{df_w}{dS_w}\right)_{S_w} \tag{15.51}$$

In Eq. (15.51), x = distance travelled by a fixed saturation in time t, feet; q_t = total flow rate, RB/D; t = time interval, days; ϕ = porosity, fraction; A = cross-sectional area of flow, ft^2; and $(df_w/dS_w)_{S_w}$ = slope of the fractional flow curve at S_w.

A similar equation can be written for gas displacement of oil as follows:

$$x = \frac{5.615q_t t}{\phi A}\left(\frac{df_g}{dS_g}\right)_{S_g} \tag{15.52}$$

In Eq. (15.52), f_g = fractional flow of gas; and S_g = gas saturation. All previous definitions of terms in Eq. (15.51) apply.

Equation (15.51) can be used to calculate the distribution of water saturation as a function of time in a linear reservoir under water injection or aquifer influx. The distance travelled by a given

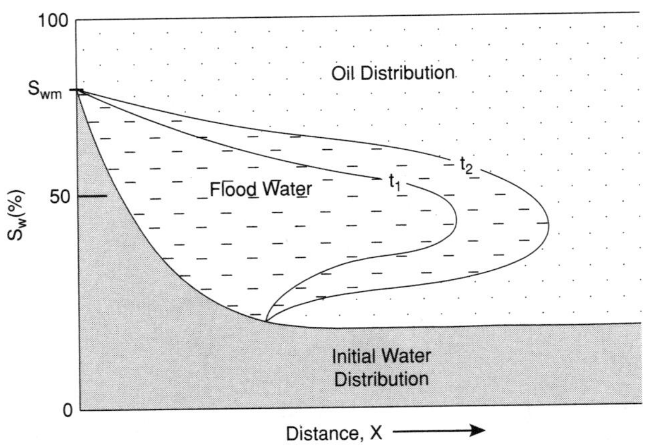

Figure 15.8 Saturation distribution based on frontal advance equation.

saturation in a specified time interval is proportional to the slope of the fractional flow curve at that saturation assuming the total flow rate and reservoir properties are constant. Using this approach, the distribution of water saturation in the reservoir as a function of time can be calculated by determining the slope of the fractional flow curve at that saturation. However, because of the shape of the fractional flow curve (Figure 15.6), it is possible that two slopes of equal value can exist for two different water saturations. Appling Eq. (15.51), this is interpreted to indicate that two different water saturations can exist at the same location in the reservoir at the same time. The appearance of this contradiction in the application of the frontal advance equation is illustrated in Figure 15.8. Buckley and Leverett[9] recognized that a portion of the saturation distribution curve is imaginary and that the real curve is discontinuous at the flood front. The location of the flood front as determined by material balance is represented in Figure 15.9 by a solid

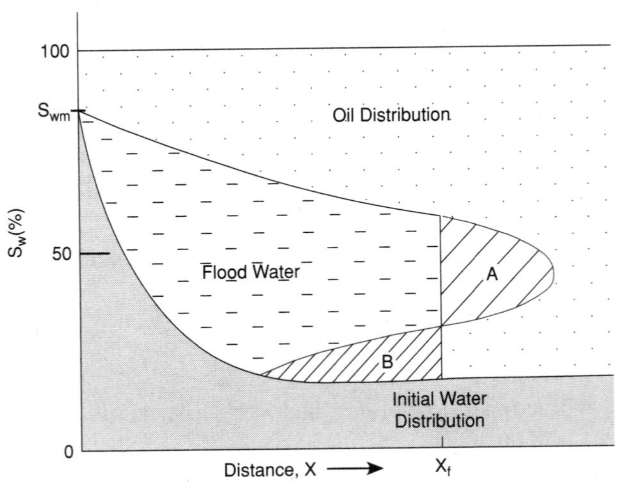

Figure 15.9 Location of the flood front as determined by material balance.

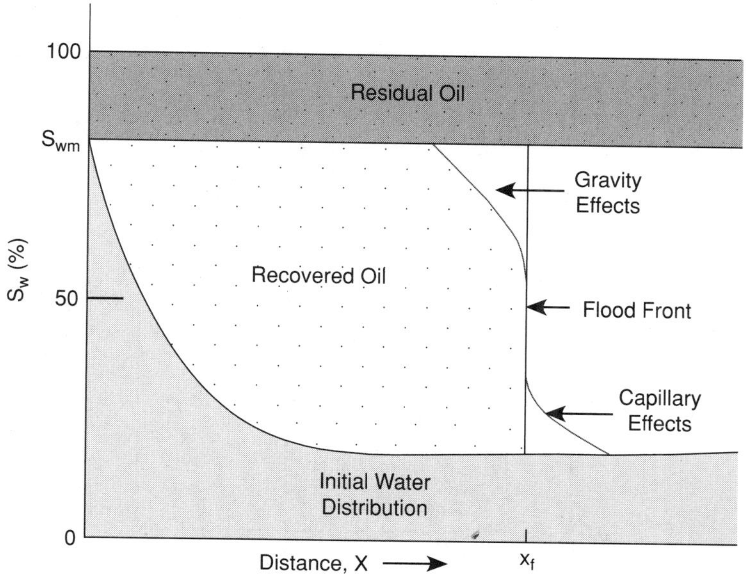

Figure 15.10 Location of the flood front with capillary and gravity effects.

line such that areas A and B are equal. Note the sharp discontinuity of the saturation curve at the flood front as represented in Figure 15.9. This is because capillary and gravity effects were assumed to be negligible. If capillary and gravity effects are considered, the distribution of water saturation at the flood front is more gradual as represented in Figure 15.10.

15.5 The Welge Method

Welge[10] proposed a method for computing oil recovery from gas or water drive that simplified the application of the Buckley-Leverett method. The Welge method is presented with graphical illustrations for the case of water drive in a linear reservoir. The graphical illustrations can be replicated for gas drive by simply replacing f_w and S_w, with f_g and S_g, respectively.

15.5.1 Water Saturation at the Flood Front

Water saturation at the flood front, S_{wf}, can be determined graphically using the Welge method by drawing a straight line from initial water saturation tangent to the fractional flow curve as shown in Figure 15.11. If the initial water saturation is greater than the irreducible water saturation (i.e., $S_{wi} > S_{wir}$), the tangent line is drawn from the initial water saturation as shown in Figure 15.12.

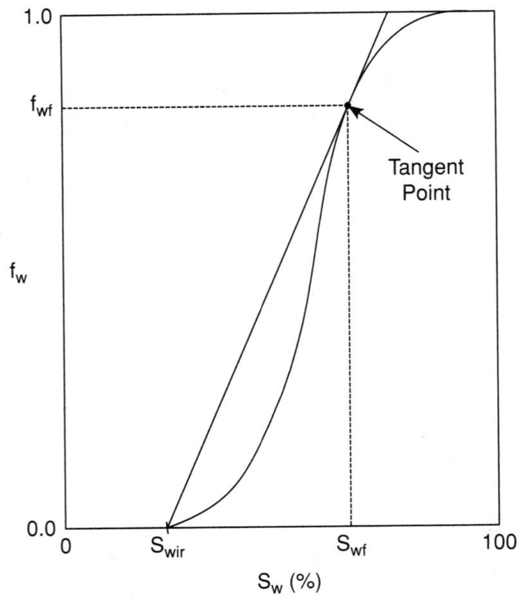

Figure 15.11 Fractional flow curve with application of the Welge method.

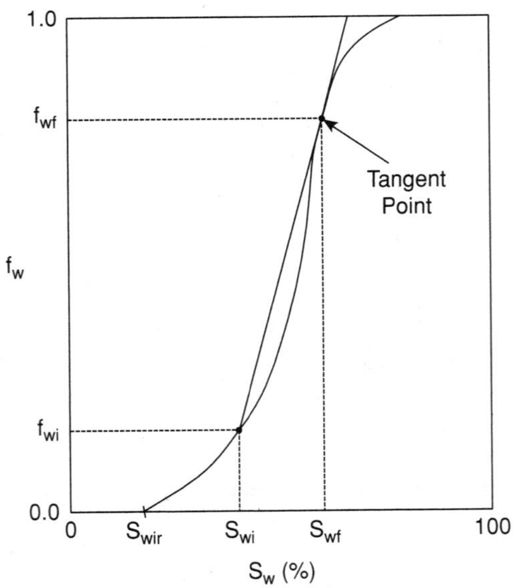

Figure 15.12 Fractional flow curve with application of Welge method for $S_{wi} > S_{wir}$.

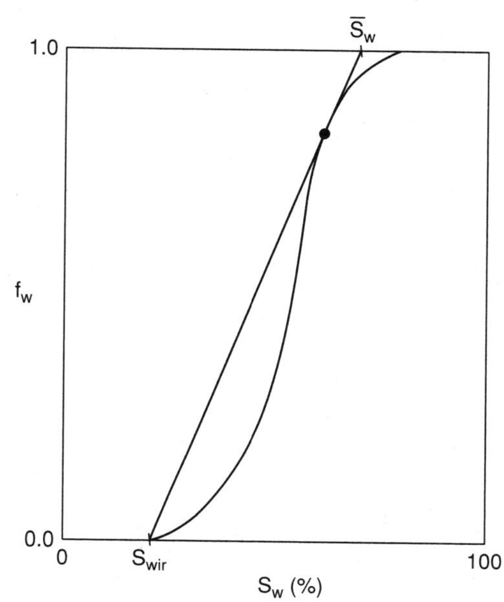

Figure 15.13 Fractional flow curve with application of Welge method for average water saturation.

15.5.2 Average Water Saturation behind the Flood Front

By extending the tangent line drawn to the fractional flow curve as shown in Figures 15.11 or 15.12 to the point where $f_w = 1.0$, the average water saturation, $\overline{S}_w$, behind the flood front can be determined as shown in Figure 15.13. At water breakthrough, $\overline{S}_w = \overline{S}_{wbt}$, where $\overline{S}_{wbt}$ is the average water saturation in the reservoir at water breakthrough.

15.5.3 Average Water Saturation after Water Breakthrough

The average water saturation after water breakthrough is determined as shown in Figure 15.14 by drawing a tangent line to the fractional flow curve at water saturation, S_{w2}, greater than S_{wf} but less than the maximum water saturation, S_{wm} (that is, $S_{wf} < S_{w2} < S_{wm}$). The water saturation, S_{w2}, is the saturation at the outlet end of the linear system after water breakthrough with the corresponding fractional flow of water denoted as f_{w2}. By extending the tangent line to the point where $f_w = 1.0$, the average water saturation, $\overline{S}_{wabt}$, in the system after water breakthrough is determined.

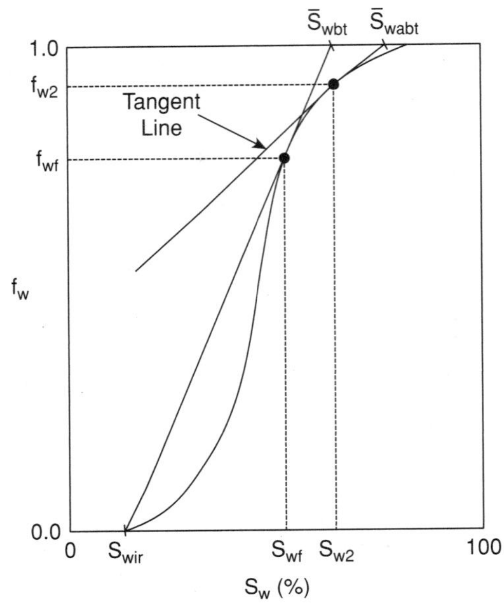

Figure 15.14 Application of Welge method for average water saturation after water breakthrough.

Example 15.1 Calculation of Fractional Flow Curve and Water Saturations Based on the Fractional Flow Curve

Problem

The relative permeability ratio of a linear reservoir as a function of water saturation is given in Table 15.1. Plot the fractional flow curve. Using the fractional flow curve, calculate the following:

1. Average water saturation and fractional flow of water at water breakthrough.
2. Displacement efficiency at water breakthrough.
3. Average water saturation and fractional flow of water after water breakthrough.

Assume capillary pressure and gravity effects are negligible. Other reservoir and fluid properties are as follows:

Porosity, ϕ	0.25
Initial water saturation, S_w	0.20
Viscosity of water, μ_w	0.95 cp
Viscosity of oil, μ_o	1.60 cp

Solution

Step 1: Calculate the fractional flow of water, f_w.

Since capillary and gravity effects are negligible, Eq. (15.33) is used to calculate the fractional flow of water as a function of water saturation. The calculated data are shown in Table 15.2.

Table 15.1 Data for Example 15.1

S_w	$\frac{k_{ro}}{k_{rw}}$
0.20	∞
0.25	110.20
0.30	34.40
0.35	12.20
0.40	5.60
0.45	2.80
0.50	1.40
0.55	0.75
0.60	0.35
0.65	0.12
0.70	0.02
0.75	0.00

Table 15.2 Plot Data for Example 15.1

S_w	f_w
0.20	0.00
0.25	0.02
0.30	0.05
0.35	0.12
0.40	0.23
0.45	0.38
0.50	0.55
0.55	0.69
0.60	0.83
0.65	0.93
0.70	0.99
0.75	1.00

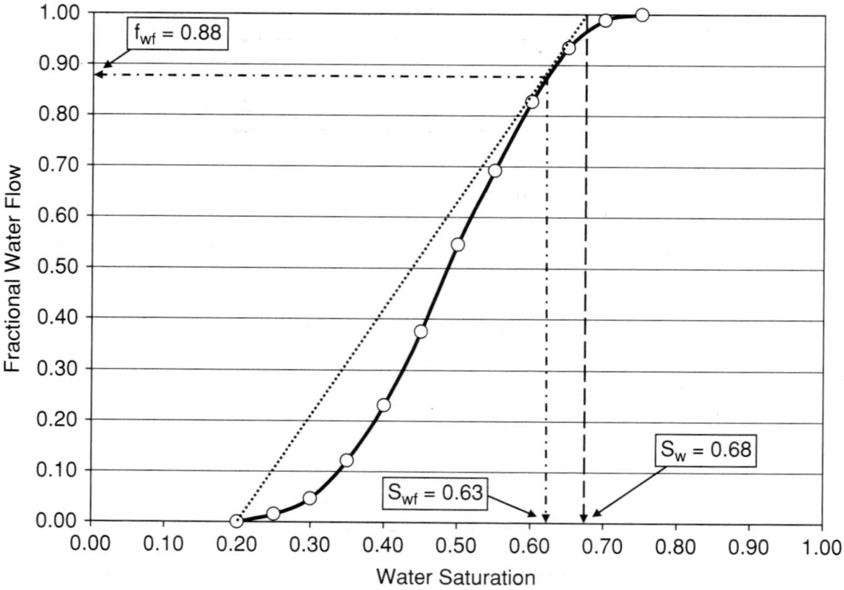

Figure 15.15 Fractional flow curve for Example 15.1.

Step 2: Plot the fraction flow curve.
The data in Table 15.2 are used to plot the fractional flow curve as shown in Figure 15.15.

Step 3: Determine S_{wf}, f_{wf} and $\overline{S}_{wbt}$.
Draw a straight line from the initial water saturation tangent to the fractional flow curve as shown in Figure 15.15. Extend this line to the point where $f_w = 1.0$. Read values for S_{wf}, f_{wf}, and $\overline{S}_{wbt}$ as shown in Figure 15.15. From Figure 15.15, $S_{wf} = 0.63$, $f_{wf} = 0.88$ and $\overline{S}_{wbt} = 0.68$.

Step 4. Calculate displacement efficiency, E_D, at water breakthrough.
By definition, $E_D = \dfrac{S_{wbt} - S_{wi}}{1 - S_{wi}}$.

$$E_D = \frac{0.68 - 0.20}{1 - 0.20} = 0.60$$

Step 5. Determine fractional water flow and water saturations after water breakthrough.
Expand the fractional flow curve as shown in Figure 15.6. Draw tangents at saturations greater than S_{wf}. Read the fractional flow of water and average saturations as shown in Figure 15.16. For instance, at $S_{w2} = 0.68$, $f_{w2} = 0.97$, and $\overline{S}_{wabt} = 0.71$. Note that several average water saturations after water breakthrough can be determined by drawing additional tangents at saturations greater than S_{wf}.

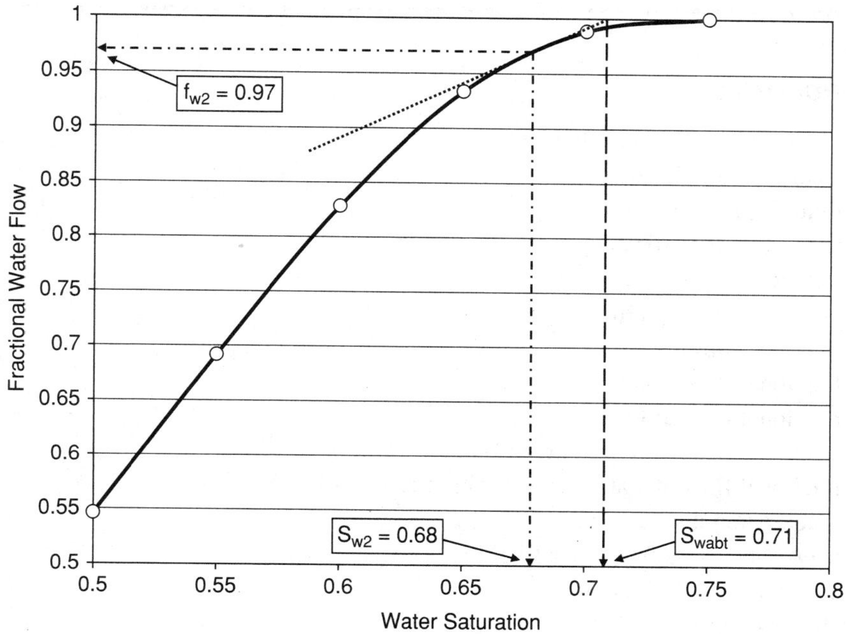

Figure 15.16 Expanded fractional flow curve for Example 15.1.

15.6 Summary

The treatment of immiscible displacement in this chapter is by no means exhaustive. The main objective was to introduce the engineer to basic concepts and some of the historical theoretical developments in immiscible displacements. In the process, several important terms were introduced in discussing the importance of rock wettability, capillary pressure, relative permeability, mobility ratio, and displacement efficiency in immiscible displacements. The fractional flow equation was developed to convey the impact of fluid and rock properties and reservoir geometry on immiscible displacement. In fact, this relatively simple equation can be used to discuss the fractional flow of water or gas in many waterflood or gasflood projects. Finally, the Welge method was introduced to illustrate determination of several terms, such as saturations at the flood front, average saturations behind the flood front, saturations at breakthrough, and saturations after breakthrough. These terms are useful in discussing performance of waterflooding or gasflooding on a wider scale that may involve application of reservoir simulation. For this reason, there was no attempt made in this chapter to present immiscible displacements in more rigorous and quantitative terms.

Nomenclature

A	cross-sectional area to flow, ft^2
E_A	areal sweep efficiency
E_D	fluid displacement efficiency
E_I	vertical sweep efficiency
E_R	total recovery efficiency
E_v	volumetric sweep efficiency
f_g	fractional flow of gas
f_o	fractional flow of oil
f_w	fractional flow of water
f_{wf}	fractional flow of water at the flood front
f_{w2}	fractional flow of water at the outlet after water breakthrough
g	acceleration due to gravity
H	height above the free water level
k	absolute permeability, md
k_g	effective gas permeability, md
k_o	effective oil permeability, md
k_w	effective water permeability, md
k_{rg}	relative gas permeability
k_{ro}	relative oil permeability
k_{rw}	relative water permeability
l	flow path length, ft.
M	mobility ratio
p_c	capillary pressure, psia
p_{nw}	pressure in the non-wetting phase, psia
p_w	pressure in the wetting phase or water, psia
p_o	pressure in the oil phase, psia
q_o	oil flow rate, RB/D
q_t	total flow rate, RB/D
q_w	water flow rate, RB/D
r	radius of the capillary tube, cm
S_{oi}	initial oil saturation, fraction
S_{or}	residual oil saturation, fraction
S_w	water saturation, fraction
S_{w2}	water saturation at the outlet after water breakthrough, fraction
S_{wi}	initial water saturation, fraction
S_{wm}	maximum water saturation, $S_{wm} = 1 - S_{or}$, fraction
S_{wir}	irreducible water saturation, fraction

$\overline{S}_w$	average water saturation, fraction
$\overline{S}_{wabt}$	average water saturation after water breakthrough, fraction
$\overline{S}_{wbt}$	average water saturation at water breakthrough, fraction
S_{wf}	water saturation at the flood front, fraction
t	time, days
α	dip angle of reservoir or formation, degrees
ϕ	porosity, fraction
μ	fluid viscosity, cp
μ_g	gas viscosity, cp
μ_o	oil viscosity, cp
μ_w	water viscosity, cp
ρ	fluid density, lbm/ft^3
ρ_g	gas density, lbm/ft^3
ρ_o	oil density, lbm/ft^3
ρ_w	water density, lbm/ft^3
θ	contact angle, degrees
λ	fluid mobility, md/cp
σ	interfacial tension, dynes/cm

References

1. Treiber, L.E., Archer, D.L., and Owens, W.W.: "A Laboratory Evaluation of the Wettability of Fifty Oil-Producing Reservoirs," *SPEJ* (Dec. 1972) 531–540.
2. Amott, E.: "Observations Relating to the Wettability of Porous Rock," *Trans.* AIME (1959) 216, 156–192.
3. Donaldson, E.C., Thomas, R.D., and Lorenz, P.B.: "Wettability Determination and Its Effect on Recovery Efficiency," *SPEJ* (March 1969) 13–20.
4. Morrow, N.R.: "Wettability and Its Effect on Oil Recovery," *JPT* (Dec. 1990) 1476–1484.
5. Owens, W.W., and Archer, D.L.: "The Effect of Rock Wettability on Oil-Water Relative Permeability Relationships," *JPT* (July 1971) 873–878.
6. Brown, R.J., and Fatt, I.: "Measurements of Fractional Wettability of Oilfield Rocks by Nuclear Magnetic Relaxation Method," *Trans.* AIME (1956) 207, 262–264.
7. Agbalaka, C., Dandekar, A.Y., Patil, S.L., Khataniar, S., and Hemsath, J.R.: "The Effect of Wettability on Oil Recovery: A Review," paper SPE 114496 presented at the 2008 SPE Asia Pacific Oil & Gas Conference and Exhibition, Perth, Australia, October 20–22, 2008.
8. Killins, C.R., Nielson, R.F., and Calhoun, J.C., Jr.: "Capillary Desaturation and Imbibition in Rocks," Prod. Monthly (Feb. 1953) 18, 2, 30–39.

9. Buckley, S.E., and Leverett, M.C.: "Mechanism of Fluid Displacements in Sands," *Trans.* AIME (1942) 107, 107–116.

10. Welge, H.J.: "A Simplified Method for Computing Oil Recovery by Gas or Water Drive," *Trans.* AIME (1952) 195, 91–98.

General Reading

- Willhite, G.P.: *Waterflooding,* SPE Textbook Series, Vol. 3, Richardson, Texas (1986).

CHAPTER 16

Secondary Recovery Methods

16.1 Introduction

Secondary recovery methods are defined as processes that are used to increase hydrocarbon recovery from the reservoir beyond primary recovery. Typical secondary recovery methods are considered to be intervention methods implemented during the primary recovery period to improve projected low hydrocarbon recovery from the primary process. The presentation of secondary recovery methods in this chapter is focused on waterflooding and gasflooding. Waterflooding of hydrocarbon reservoirs is generally an immiscible displacement process since water is virtually immiscible with hydrocarbons even at high pressures. Gasflooding can be an immiscible or miscible displacement process, depending on pressure and temperature of the reservoir, the type of injection gas, the composition of the injection gas, the composition of the reservoir fluid, the nature of reservoir heterogeneities, and other factors. Waterflooding and gasflooding are presented in this chapter as immiscible displacement processes. Gasflooding as a miscible displacement process is presented in Chapter 17 as an enhanced oil recovery process.

In this chapter, secondary recovery methods are presented in descriptive terms under the categories of well patterns, well surveillance, reservoir performance monitoring, and project management. The application of prediction methods such as the Stiles[1] method, Dykstra-Parsons[2] method, Craig-Geffen-Morse[3] method, and other methods for the calculation of recoveries from waterflooding are not presented in this book. Similar methods, such as the Welge[4] method for gasflooding are also not presented. For flood design and calculation of flood performance, these analytical methods have been rendered virtually obsolete by faster and more robust methods based on reservoir simulation. It may be necessary to use these analytical methods to gain a fundamental understanding of immiscible flooding processes, but the same knowledge can be gained by using very simple reservoir models to explore key factors (such as rock and fluid properties, reservoir heterogeneities, well pattern, injection rates, production rates, etc.) that influence hydrocarbon recovery under secondary recovery processes. The application of reservoir simulation for predicting reservoir performance under secondary recovery processes is presented in Chapter 19.

16.2 Waterflooding

Waterflooding is the most common secondary recovery method used worldwide to improve oil recovery from reservoirs. The performance of a waterflood project in terms of oil recovery depends on many factors, such as rock wettability, rock and fluid properties, formation heterogeneities, flood patterns, composition of injection water, water injection rates, fluid production rates, etc. These factors should be considered in the design and installation of a waterflood project.

16.2.1 Waterflood Patterns

Waterflood patterns are based on the arrangement of injectors and producers in a waterflooded reservoir. The patterns are generally designed to improve areal sweep efficiency. It is important to note that waterflood patterns evolved from conversion of onshore oil fields developed with regular well spacing from primary production to secondary production by waterflooding. The criteria for converting these oil fields to waterflooding were to minimize drilling of new wells, maximize water injection, and improve oil recovery. Consequently, the waterflood pattern installed in these oil fields depended on the density of existing wells, reservoir properties, and projected rate of oil recovery. Generally, pattern flooding is not efficient for development of deepwater reservoirs or reservoirs found in similar environments with high development costs. In these high cost environments, the ratio of water injectors to producers, and the location of these wells should be optimized through the use of reservoir simulation models.

Some of the common waterflood patterns are the direct line drive, the staggered line drive, the five-spot, and the nine-spot patterns. The direct line drive pattern is shown in Figure 16.1.

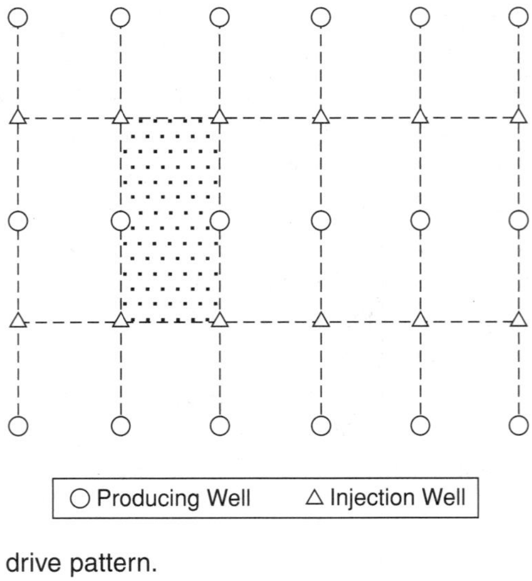

Figure 16.1 Direct line drive pattern.

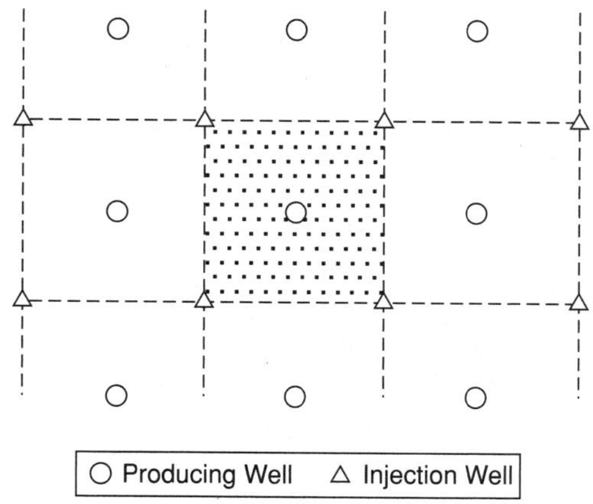

Figure 16.2 Staggered line drive pattern.

For this pattern, the injectors and producers offset each other directly. This could lead to early water breakthrough if the distances between the injectors and producers are small. The impact of early water breakthrough on sweep efficiency for the direct line drive pattern can be reduced somewhat by using the staggered line drive pattern, as shown in Figure 16.2. As seen in Figure 16.2, the alternate rows of injectors and producers are displaced so that the wells do not offset each other directly. Note that the ratio of injectors to producers in both patterns is one. The five-spot pattern is shown in Figure 16.3. The five-spot pattern has been used widely in many field

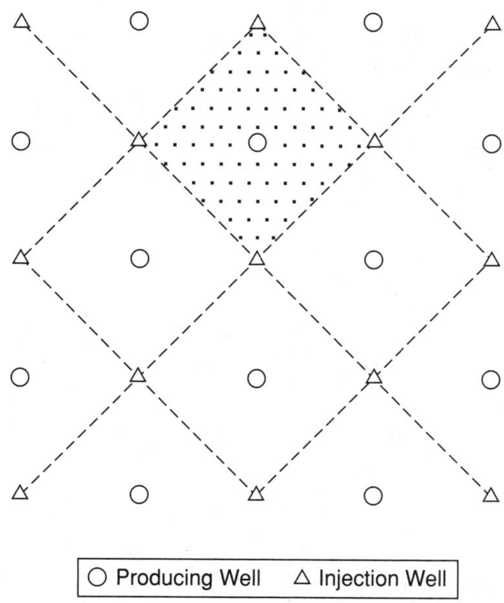

Figure 16.3 Normal five-spot pattern.

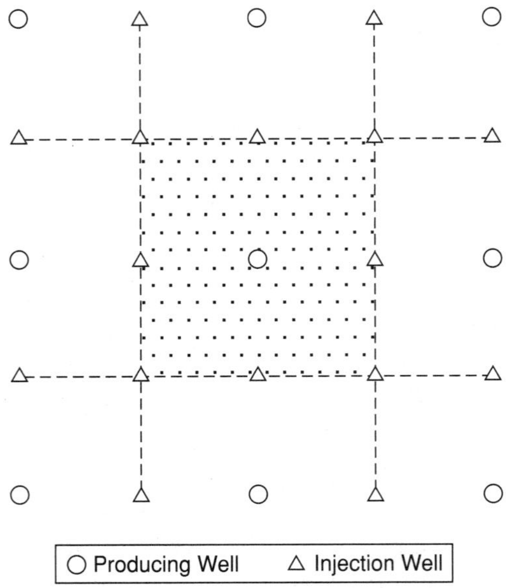

Figure 16.4 Normal nine-spot pattern.

applications and model studies. This pattern generally results in good areal sweep efficiency. The ratio of injectors to producers in a five-spot pattern is also one. The nine-spot pattern is shown in Figure 16.4. This pattern is used in low permeability reservoirs where high water injection volumes are desired. The ratio of injectors to producers in a nine-spot pattern is three. There are numerous variations of these basic flood patterns, such as inverted five-spot and inverted nine-spot patterns. It is not necessary to replicate these other patterns in this book since the engineer can devise flood patterns that are best suited to the development of a particular reservoir with modern reservoir simulation tools.

Peripheral flooding is developed when injectors are located around part or the entire boundary of the reservoir, as shown in Figure 16.5. If the injectors are located along one side of

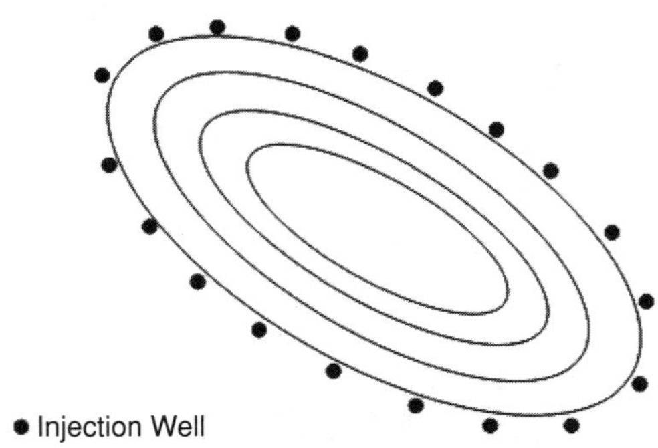

Figure 16.5 Peripheral flood pattern.

the reservoir or the middle of the reservoir, it is called a line flood. Peripheral flooding is generally used in reservoirs where the objectives are to maintain reservoir pressure, minimize water production, and improve oil recovery.

16.2.2 Waterflood Design

The steps recommended in this book for the design of a waterflood are based entirely on the application of numerical reservoir simulation models. This approach reflects the experience of the author that numerical reservoir modeling of a waterflood is superior to all the earlier prediction methods (Stiles method, Dykstra-Parsons method, Craig-Geffen-Morse method, etc.) for waterflood design. Each of these earlier prediction methods made key assumptions (such as linear flow, layered system, no crossflow, piston-like displacement, constant sweep efficiency, etc.) which limit their applicability in the design of a waterflood. Furthermore, none of these methods can incorporate the impact of reservoir heterogeneities, such as fracturing, faults, permeability barriers, spatial variations of fluid and rock properties, etc. in a waterflood design. Numerical reservoir simulation of the waterflooding process can be applied without any of these assumptions, while incorporating any geologic features that may impact the performance of the waterflood. Numerical reservoir modeling also allows the waterflood design to be subjected to more rigorous tests on uncertainty and sensitivity of key variables that could impact the performance of the waterflood.

16.2.3 Recommended Steps in Waterflood Design

1. Construct a geologic model of the reservoir or project area.
   - **a.** Identify and include all faults and other structural features that may affect fluid flow in a geologic frame work model of the reservoir.
   - **b.** Identify and include all reservoir heterogeneities, such as permeability barriers, reservoir unconformities, etc. in the geologic model.
   - **c.** Perform characterization of the geologic model to include areal and vertical variation of reservoir properties such as facies, net pay, porosity, permeability, and saturations. Additional details on geologic modeling and reservoir characterization are presented in Chapter 18.

2. Analyze rock/fluid properties data.
   - **a.** Determine mineralogy of reservoir rocks.
   - **b.** Conduct studies on compatibility of injection water with reservoir rocks.
   - **c.** Determine PVT properties of reservoir fluids, including saturation pressures and oil viscosity.

3. Construct reservoir flow model with data from Steps 1 and 2. (Note: Reservoir simulation concepts are presented in Chapter 19.)
   - **a.** If the reservoir had prior production history, history-match reservoir model to obtain the current depleted state of the reservoir before the start of waterflooding.

   b. At the completion of step 3a, determine gas cap size if a gas cap is present and extent of aquifer influx if reservoir has an active aquifer.
   c. Compare pressure distribution in model after step 3a to actual pressure data. Identify state of reservoir depletion.
   d. Explore distribution of fluid saturation in the model after step 3a to identify potential undepleted areas of the reservoir that would be targets for waterflooding.

4. Run prediction cases.
   a. Run a base prediction case assuming continuation of current depletion strategy.
   b. Run several predictive cases assuming the reservoir is waterflooded with different numbers and locations of water injectors and producers.
   c. Compare results from step 4a and step 4b. If the waterflood cases indicate substantial improvement in total oil recovery, then proceed to Step 5.
5. Optimize waterflood design.
   a. Choose the cases from step 4b with the "best" reservoir performance and optimize the numbers and locations of injectors and producers.
   b. Optimize injection and production rates for each case.
   c. Rank the cases by incorporating project economics.
6. Perform sensitivity analyses.
   a. Select two or three cases from step 5 and perform sensitivity analyses on key reservoir and operational variables of the waterflood design.
   b. Repeat economic analyses of the entire project based on results from step 6a.
7. Conduct a pilot waterflood project.
   a. For large waterflood projects, it is recommended that a pilot project should be installed. The pilot waterflood project would provide data, such as water injection rates, that could be used to adjust the waterflood design and improve predictions before implementation of the project over the entire reservoir.

16.2.4 Waterflood Management

In many cases, proper management of a waterflood project is as important as designing the waterflood. A well-designed waterflood can become an engineering and economic failure if several important surveillance practices are not implemented in the management of the project. Monitoring and surveillance practices are presented for injection wells, injection water, production wells, production water, facilities, and the reservoir.

A. Injection Wells. One of the key measures of a successful waterflood project is achieving and maintaining the desired levels of water injection rates. Water injection wells may exhibit reduced injectivity over time due to scale buildup and plugging of the formation from solid deposits. It is important to test the injection wells periodically to ensure that injectivity had not declined substantially. The most common test that can be used to gauge well injectivity is the fall-off test. Pressure fall-off tests on injectors are similar and are conducted in the same manner as buildup tests on producers. Data from pressure fall-off tests are analyzed with the same techniques demonstrated in Chapter 11 for buildup tests. Some caution should be exercised in analyzing pressure fall-off test data since multiple fluid boundaries may exist around the injector and create complications in the analysis.[5–7]

The condition of an injection well can be monitored periodically by preparing a Hall[8] plot on the performance of the injector. The Hall plot can be prepared from data routinely collected as part of monitoring the performance of the injector. These data include cumulative water injection and injection pressure versus time. The Hall method is based on the radial form of the Darcy flow equation expressed as:

$$q_w = \frac{0.00708 k_w h (p_{inj} - \overline{p})}{\mu_w \ln\left(\dfrac{r_e}{r_w}\right)} \tag{16.1}$$

In Eq. (16.1), q_w = water injection rate, BWPD; k_w = effective permeability to water, md; h = formation thickness, ft; p_{inj} = bottom-hole injection pressure, psia; $\overline{p}$ = average reservoir pressure, psia; μ_w = water viscosity, cp; r_e = external radius, ft; and r_w = wellbore radius, ft. The cumulative water injection, W_i, can be expressed as:

$$W_i = \int_0^t q_w dt \tag{16.2}$$

Substituting Eq. (16.1) into Eq. (16.2) and re-arranging yields:

$$\int_0^t (p_{inj} - \overline{p})\, dt = \left(\frac{141.2 \mu_w \ln\left(\dfrac{r_e}{r_w}\right)}{k_w h}\right) W_i \tag{16.3}$$

In integrated form, Eq. (16.3) can be expressed as:

$$\sum \Delta(p_{inj} - \overline{p}) \times \Delta t = \left(\frac{141.2 \mu_w \ln\left(\dfrac{r_e}{r_w}\right)}{k_w h}\right) W_i \tag{16.4}$$

If it is assumed that all terms on the right hand side of Eq. (16.4) are constant except W_i, then Eq. (16.4) can be written in an equivalent form:

$$\sum \Delta(p_{inj} - \overline{p}) \times \Delta t = m_H W_i, \quad \text{where} \quad m_H = \left(\frac{141.2 \mu_w \ln(r_e/r_w)}{k_w h}\right) \tag{16.5}$$

Equation (16.5) implies that a plot of $\sum \Delta(p_{inj} - \overline{p}) \times \Delta t$ versus W_i should yield a straight line if m_H is constant. The parameter, m_H, would not be constant if permeability is reduced due to plugging or increased due to stimulation or fracturing. A characteristic Hall plot is shown in Figure 16.6. Note the changes in the shape of the curve due to effects of various injection conditions as indicated in Figure 16.6.

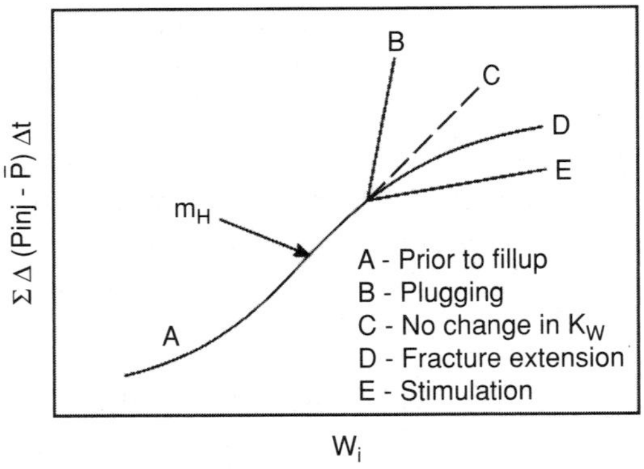

Figure 16.6 Characteristic Hall plot.

To maximize water injection, it is sometimes necessary to determine the maximum injection pressure at which the formation would be fractured. This could be done using the step rate test.[9] A step rate test is a sequential measurement of stabilized water injection rates at injection pressures recorded at specified time intervals. A plot of injection pressures versus injection rates would indicate a discontinuity if the fracture pressure was reached, as shown in Figure 16.7. The point of intersection of straight lines drawn on data points on either side of the discontinuity represents the fracture pressure. After establishing the fracture pressure, water injection is maximized by maintaining injection pressure as high as possible without exceeding the indicated fracture pressure.

It is good waterflood management practice to run injection profile logs on water injection wells to determine the intervals of the formation that is taking injected water. It is recommended that injection profile logs should be run at least once a year on water injectors. Injection profile logs are useful in determining the formation intervals that are being waterflooded, and could be used to plan future remedial work to improve distribution of water injection into the formation.

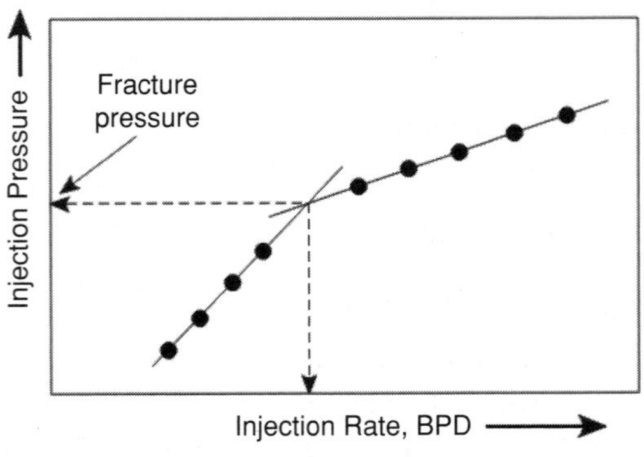

Figure 16.7 Step-rate test plot.

Example 16.1 Preparation of the Hall Plot from Well Injection Data

Problem

Prepare a Hall plot from the data for an injection well shown in Table 16.1. Determine the condition of the well from the Hall plot. Estimate permeability of the well. Other well and reservoir data are as follows:

Formation thickness, h	50 ft
Reservoir radius, r_e	1000 ft
Well radius, r_w	0.5 ft
Water viscosity, μ_w	1.15 cp

Solution

Step 1: Prepare data for Hall plot.
Applying Eq. (16.5), use the data from Table 16.1 to prepare the data for Hall plot as shown in Table 16.2.

Step 2: Make Hall plot.
Use the data in Table 16.2 to make a Hall plot as shown in Figure 16.8. Note that the y-axis is represented as Y-Function in Figure 16.8 where the Y-Function $= \Sigma(\Delta p \times \Delta t)$.

Step 3: Evaluate the Hall plot.

Table 16.1 Injection Well Data for Example 16.1

N	Δt (days)	Injection Pressure, p_{inj} (psia)	Average Pressure, $\bar{p}$ (psia)	Injection volume, (bbls)
1	25	4260	3205	25,000
2	20	4295	3215	19,800
3	30	4285	3210	29,250
4	22	4279	3211	22,220
5	28	4295	3225	27,500
6	30	4305	3230	25,500
7	31	4322	3232	24,025
8	29	4328	3233	22,765
9	24	4343	3227	19,200
10	27	4338	3229	21,465
11	30	4353	3234	23,400
12	21	4360	3236	16,695

Table 16.2 Plot Data for Example 16.1

N	Injection Pressure, p_{inj} (psia)	Average Pressure, $\bar{p}$ (psia)	Δp (psia)	Δt (days)	$\Sigma(\Delta p \times \Delta t)$	Injection volume, (bbls)	Cumulative Injection, W_i (bbls)
1	4260	3205	1055	25	26,375	25,000	25,000
2	4295	3215	1080	20	47,975	19,800	44,800
3	4285	3210	1075	30	80,225	29,250	74,050
4	4279	3211	1068	22	103,721	22,220	96,270
5	4295	3225	1070	28	133,681	27,500	123,770
6	4305	3230	1075	30	165,931	25,500	149,270
7	4322	3232	1090	31	199,721	24,025	173,295
8	4328	3233	1095	29	231,476	22,765	196,060
9	4343	3227	1116	24	258,260	19,200	215,260
10	4338	3229	1109	27	288,203	21,465	236,725
11	4353	3234	1119	30	321,773	23,400	260,125
12	4360	3236	1124	21	345,377	16,695	276,820

Review the Hall plot diagnostically using the characteristic shapes shown in Figure 16.6. For this example, there is no change in the slope of the plot after fill-up of the area around the injector. For this reason, the injector does not appear to be damaged.

Step 4: Estimate the permeability of the formation around the water injector.
From Figure 16.8, the slope of the line, $m_H = 1.4152$
From Eq. (16.5),

$$m_H = \frac{141.2\mu_w \ln \dfrac{r_e}{r_w}}{k_w h}$$

Therefore,

$$k_w = \frac{141.2\mu_w \ln \dfrac{r_e}{r_w}}{m_H h}$$

$$= \frac{141.2 \times 1.15 \times \ln \dfrac{1000}{0.5}}{1.4152 \times 50}$$

$$= 17.44 \text{ md}$$

This is the average effective permeability to water within the area taking water from the injector.

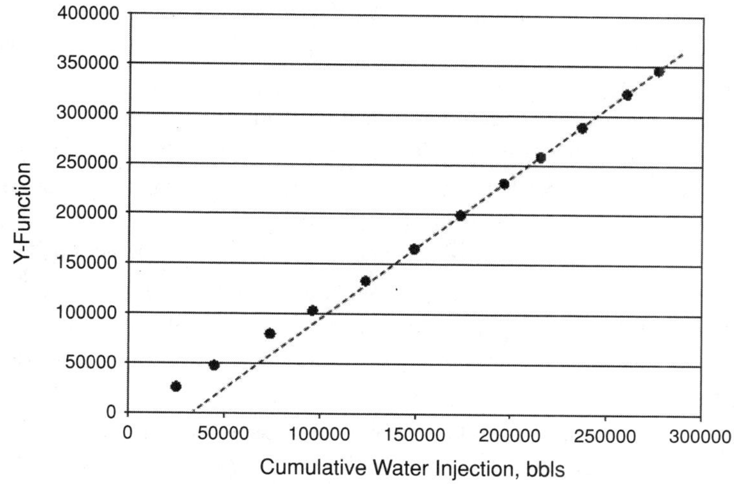

Figure 16.8 Hall plot for Example 16.1.

B. Injection Water. The source and quality of injection water is another key component in the management of a waterflood project. The source of injection water should be sufficient to supply as much water as needed during the early stages of the waterflood project. Later on, as the project matures, produced water could be recycled and used to make up the required volumes of injection water. The quality of injection water should be determined by performing complete water analysis on all sources of water. The water analysis process should evaluate the compatibility of source water with reservoir formation water and swelling clays or other minerals in the reservoir rock, the type of injection facility (open or closed), and the extent of water treatment required before the water can be used safely for injection. Extensive water treatment should be avoided because it could be expensive, and adversely impact the economics of the project. Water analysis should be performed regularly throughout the life of the waterflood project to monitor the quality of injection water.

Injection water, if not treated, could cause corrosion, solid precipitation, and promote growth of bacteria. The major cause of corrosion in injection water is dissolved gases in the form of oxygen, carbon dioxide, and hydrogen sulfide. Oxygen in injection water is very corrosive, and may cause precipitation of ferrous hydroxide, if the injection water contains iron. If oxygen is present in injection water, it should be removed by stripping or other processes. Solid precipitation can be avoided by minimizing the presence of iron, manganese, and barium in the injection water. Water containing barium is likely to precipitate barium sulfate, if injected into a reservoir containing sulfate. Barium sulfate can cause severe plugging of production wells and production facilities.[10,11] Bacteria growth should be controlled by treating injection water with chlorine or other bactericides. The most important bacteria that should be controlled are the iron bacteria and sulfate reducing bacteria. The iron bacteria cause precipitation of iron and manganese hydroxides which can plug the formation. The sulfate reducing bacteria react with sulfate ion to create iron sulfide which can also plug the formation. Hydrogen sulfide is also generated as a by-product of the reaction.

Hydrogen sulfide is a very corrosive and toxic substance that can pose serious problems on the operation of the waterflood project. The generation of hydrogen sulfide by sulfate reducing bacteria should be controlled and eliminated.

C. Production Wells. Production wells must be monitored regularly in a waterflood project just as injection wells. They are equally important for achieving engineering and economic success on the waterflood project. Each production well should be tested frequently (depending on the availability of test equipment) to measure fluid production rates. The measured production rates should be plotted on various graphs to monitor the response of the well to water injection. Typical graphs commonly used are plots of fluid production rates versus time, percent oil cut versus cumulative oil produced, or water-oil ratio versus cumulative oil produced. There are numerous other types of plots that can be used to monitor well performance. These plot types can be tailored to suit the particular needs of the project.

If the production well appears to be damaged, a pressure buildup test should be conducted on the well to determine the extent of damage. Methods for analysis of pressure buildup tests were presented in Chapters 11, 12, and 13. A common cause of reduction in well productivity is deposition of scale on the sandface and in the production string. Some types of scale can be removed by simulation of the wells with acid.

Profile (fluid entry) logs should be run periodically on production wells. These profile logs are extremely important in monitoring the intervals that are contributing to production. Historical reviews of profile production logs can be used to determine intervals that have been flooded, and intervals that were not affected by water injection. These reviews are useful in reallocation of water injection, and targeting of specific formation intervals for water injection.

D. Produced Water. Water analysis should be performed regularly on produced water as a means of monitoring the progress of the waterflood, and tracking potential problems with solids deposition caused by mixing of source water and formation water. At early stages of the waterflood project, produced water consists mostly of formation water. Later, as production wells are affected by water breakthrough, produced water will contain increasing amounts of injected water. The proportion of injected water in produced water is sometimes used as an indicator in monitoring the location of the floodfront.

In many waterflood projects, some of the produced water is recycled and used as injection water. This reduces the amount of produced water to be disposed, thereby reducing the environmental impact of the project. Produced water recycled for re-injection should be treated to remove harmful substances, and improve its quality, as was the case for the source injection water. In some operations, produced water is also treated to remove harmful substances before disposal.

E. Facilities. The facilities used in a waterflood project should be able to withstand potential threat from massive corrosion caused by the water. Protection of injection and production facilities from corrosion begins with careful water analysis, and selection of equipment with proper metallurgy. It is recommended that major waterflood projects should retain the services of a corrosion prevention department. The corrosion engineers should routinely monitor all facilities used in the project for corrosion, and implement preventative measures to forestall onset of corrosion in these facilities.

16.2.5 Management of Waterflooded Reservoirs

Several elements of proper management of waterflooded reservoirs were presented previously under management of injection wells, injection water, production wells, produced water, and facilities. This approach was taken to emphasize the importance of these key sectors in the successful operation of a waterflood project. In addition to these key sectors, the waterflooded reservoir should be managed as a whole. This means that the reservoir should be monitored regularly to determine the location of the floodfront, evaluate the distributions of reservoir pressure and fluid saturations, and exploit opportunities for re-alignment of injectors and producers to improve sweep efficiency and hydrocarbon recovery.

Successful management of waterflooded reservoirs must include sustained programs for gathering data on the reservoir.[12–16] Pressure data should be obtained regularly from producers, injectors, and observation wells, if available. The pressure data should be analyzed and used to track the movement of injected water in the reservoir. The pressure data are also used to monitor reservoir voidage balance, and ensure that reservoir pressure is maintained for waterflood projects designed primarily for pressure maintenance. In addition to pressure data, fluid saturations can be monitored with carbon/oxygen logs.[17] Injection and production profile logs should be obtained regularly and assembled with the fluid saturations data to develop a comprehensive view of the status of the project.

A reservoir model is typically used to design the waterflood project as discussed in Section 16.2.2. This model should be used to manage the waterflood project. The model should be updated with pressure data, injection and production data, and other data acquired as part of a monitoring program. In some cases, it might be necessary to reconstruct the reservoir model entirely, if new data indicate substantial new knowledge on reservoir geology and architecture. The functional reservoir model should be used to locate infill wells, allocate water injection, re-align wells, and predict future performance of the waterflood project. In general, waterflood projects should adopt the principles stated in Chapter 20 for sound management of petroleum reservoirs.

16.3 Gasflooding

Gasflooding is presented in this chapter as an immiscible displacement process. Immiscible displacement occurs in a displacement process where a distinct interface (or boundary) exists between the displaced fluid and the displacing fluid. This includes displacement processes that are described as near-miscible. The key factors that determine whether a gasflooding process is immiscible or miscible are reservoir pressure, reservoir temperature, composition of injection gas, composition of reservoir fluid, and reservoir heterogeneities such as faults and permeability barriers. The impact of each factor can be determined with laboratory measurements and modeling of the displacement process. Gas miscible displacement processes are presented under enhanced oil recovery methods in Chapter 17.

The injection gas in immiscible gasflooding could be nitrogen, hydrocarbon gas, flue gas, carbon dioxide, or any other gas mixtures. In a typical displacement process, sufficient quantities of the gas are injected into the reservoir to displace the reservoir fluid to the producing wells. Gasflooding is sometimes used instead of waterflooding because the fluid displacement (microscopic) efficiency of gas is superior in some applications. In addition, some of the injection gas (such as hydrocarbon gas and carbon dioxide) can dissolve in the reservoir fluid to reduce its viscosity and density, thereby improving its mobility in the reservoir. On the other hand, the volumetric sweep efficiency (composed of areal and vertical sweep efficiencies) of gas displacement processes is generally low due to gravity over-ride, and viscous fingering caused by higher mobility ratios of gas in comparison to liquids. This effect can be compounded in low permeability reservoirs, or reservoirs with severe heterogeneities, such as faults and permeability barriers. Mobility ratios of gas displacement processes can be reduced by alternately injecting water and gas. This is described as a WAG (water-alternating-gas) displacement process. Volumetric sweep efficiency can be improved by application of WAG displacement process. This has been shown to improve oil recovery based on field results.[18]

16.3.1 Applications of Gasflooding

The primary applications of gas injection into a hydrocarbon reservoir are pressure maintenance, gas cycling, and fluid displacement. In many gas injection projects, these three primary functions are achieved at the same time to some degree.[19] Pressure maintenance by gas injection is routinely implemented in many reservoirs. Reservoirs that are good candidates for pressure maintenance may have experienced steep, rapid pressure decline during primary depletion accompanied with the development of a secondary gas cap, or enlargement of a primary gas cap. For such reservoirs, gas is injected into the gas cap at sufficient volumes to reduce or stop the decline of reservoir pressure. In some cases, gas injection for pressure maintenance is initiated early close to start of production to prevent any decline in reservoir pressure. This practice is used, in some cases, to keep reservoir pressure from falling below bubble point for oil reservoirs, or dew point for gas reservoirs. Gas cycling is used to improve recovery from condensate reservoirs by vaporizing fluids that may have condensed in the reservoir with injected gas. Gas cycling can be used also to improve hydrocarbon recovery from secondary gas caps in oil reservoirs. In many steeply dipping reservoirs, gas is injected at the crest of the structure to assist in displacing oil towards downdip wells.[19] The volumetric sweep efficiency can be high for such gravity stable processes.[20] Gasflooding in horizontal reservoirs can lead to gravity over-ride, where the injected gas rises to the top of the reservoir because it is less dense than oil. Severe cases of gravity over-ride can lead to low volumetric sweep efficiencies leaving large portions of the reservoir unswept by injected gas. The effects of gravity over-ride and viscous fingering (early gas breakthrough) can be mitigated by using mobility control measures such as WAG as part of the injection strategy.

16.3.2 Gasflood Design

This section on gasflood design begins with an overview of two main factors that determine the performance of a gas flood. The first key factor in gasflood design is obtaining a fluid property model that represents the interactions that occur between the reservoir fluids and the injected gas. These interactions can be both physical and chemical in nature. For instance, lean hydrocarbon gas is capable of physically vaporizing light hydrocarbons from reservoir oil.[21] Carbon dioxide can dissolve chemically in reservoir oil to reduce its density and viscosity, and improve its ability to flow.

These changes can be measured in the laboratory by conducting PVT studies, slim-tube tests, and core floods with the injection gas and the reservoir fluids.[22] The data from these studies can be used to modify the parameters of an equation of state (EOS) which can then be used to simulate and reproduce the phase behavior resulting from the interaction between the injected gas and reservoir fluids. A procedure for regression of the parameters of an EOS was presented in Chapter 6.

The second key factor in gasflood design is proper geologic modeling and characterization of the reservoir that is the candidate for gasflooding.[23] Geologic modeling and characterization of a reservoir for gasflooding should begin with the construction of a geologic model that incorporates all available geologic data. The geologic model is the framework that includes all structural, stratigraphic, and lithological data. It is important that the geologic model should include all known faults, fractures, and reservoir heterogeneities, including variation of rock properties such as permeability, porosity and saturations.[24,25] The geologic model should represent all reservoir conditions and properties that could impact the gasflooding process. A general procedure for geologic modeling and reservoir characterization is presented in Chapter 18.

The geologic model is then coupled with the EOS fluid model to create a compositional simulation model for the gasflooding process. Note that the compositional simulation model as recommended here can be used to investigate the performance of the gasflooding process under immiscible as well as miscible conditions.

16.3.3 Recommended Steps in Gasflood Design

1. Perform comprehensive PVT studies.
   a. Obtain PVT data on the reservoir fluids.
   b. Measure PVT data on mixtures of reservoir fluids with the injection gas at different compositions.
   c. Conduct multicontact miscibility tests to establish the conditions at which immiscible displacement can occur.
2. Conduct slim tube and core flood tests.
   a. Establish displacement efficiencies with slim tube tests.
   b. Verify results of slim tube tests with core flood tests if whole core samples of the reservoir are available.
   c. Obtain relative permeability data from the slim tube or the core flood tests.
3. Build a fluid properties predictor from an EOS.
   a. Use the data from step 1 to build a fluid properties predictor with an EOS. The procedures for regression of the parameters of an EOS from PVT data are provided in Chapter 6.
   b. Enhance the predictive capabilities of the EOS with the data from slim tube tests and the core flood tests conducted in step 2.
4. Perform geologic modeling and characterization of the reservoir.
   a. Construct a geologic model of the reservoir using all available data. Include all faults and other geologic structures that can affect fluid flow.

   b. Define reservoir stratigraphic framework. Incorporate lithological data from petrophysical analysis. Detailed analysis should include definition of facies and assignment of rock properties based on facies.
   c. Apply geostatistics to populate reservoir properties.
   d. Generate several realization of the geologic model using uncertainty analysis. (Note: Detailed geologic modeling and reservoir characterization procedures are presented in Chapter 18.)
5. Construct compositional reservoir flow models with fluid properties model from step 3 and geologic models from step 4. (Note: Reservoir simulation is presented in Chapter 19.)
   a. If the reservoir had prior production history, history-match flow models to obtain current state of the reservoir.
   b. Review pressure and saturation distributions in history-matched models. Compare results to actual data.
   c. Select two or more flow models for prediction of gasflood performance.
6. Run prediction cases.
   a. Make prediction runs assuming several gas injection rates.
   b. Make prediction runs by changing location and number of gas injection wells.
   c. Make prediction runs by applying mobility control measures such as the WAG injection strategy.
   d. Compare results and select the cases with the optimum hydrocarbon recoveries.
7. Rank gasflood design.
   a. Rank cases from step 6d after performing economic analysis.
8. Perform sensitivity analyses.
   a. Select two or more cases from step 7 and conduct sensitivity analyses on key reservoir and operational variables of the gasflood project.
   b. Incorporate economic analyses in the sensitivity runs.
   c. Select a prediction case for project implementation.
9. Conduct a pilot gasflood project.
   a. A pilot gasflood project is recommended, especially for large projects. The pilot project is essential for testing the data and the results generated with the reservoir models. It also provides the opportunity to check the hydrocarbon recoveries predicted for the project before implementing the project on the entire reservoir. A pilot project is necessary in major projects as a field test that can be used to avoid disastrous outcomes in terms of technical and economic performance of gasflood projects. More details on applications of pilot floods are presented in Chapter 17.

16.3.4 Gasflood Management

The main requirement for sound management of gasflood projects is extensive surveillance and monitoring of gas injectors, producers, gas processing facilities, and the reservoir.[26] The gas injectors, producers, and processing facilities should be maintained in good condition for the project to be successful. The reservoir should be monitored closely to ensure that the gasflood is achieving recoveries as projected in designing the process. The surveillance and monitoring practices for gas injectors, producers, processing facilities, and the reservoir are presented.

A. Gas Injectors. It is essential to maintain the gas injectors in good conditions so that the required gas volumes are injected. This is particularly important for pressure maintenance projects. Gas injectivity may decline with time due to plugging of perforations from deposition of wax, asphaltene, and hydrates. Loss of gas injectivity can be remediated by stimulating the well with acids, solvents, or other agents for removal of well damage. It is also important to run injection profile logs periodically on gas injectors to evaluate the distribution of the injected gas over the perforated interval. If the profile injection logs indicate unfavorable distribution of injection gas across the perforated interval, then the gas injector should undergo remedial work to direct injected gas to the intended formation intervals. If the injected gas is carbon dioxide or flue gas, the injector should be completed with equipment that can resist corrosion, and the well should be inspected regularly for corrosion. Inspection of well equipment for corrosion also applies to wells used in WAG operations.

B. Producers. Producers in gasflood projects should be monitored regularly to prevent excessive production of injected gas which could result from early gas breakthrough due to channeling, or gravity override.[27] If injected gas is allowed to flow to producers without contacting unswept portions of the reservoir, the volumetric sweep efficiency will be reduced resulting in poor hydrocarbon recovery. Gas overrides or channeling is particularly detrimental to gas cycling or gas displacement projects. This problem can be remedied at the producers by isolating intervals showing high gas production with packers, or by squeezing them with cement. As recommended for injectors, producers should be under surveillance regularly for corrosion, especially if the injection gas is carbon dioxide or flue gas. In most gasfloods containing carbon dioxide or other corrosive injectants, producers are outfitted with subsurface and surface equipment that are resistant to corrosion.

C. Processing Facilities. All facilities used in the injection of gas and processing of produced fluids are grouped under the umbrella of processing facilities. These are gas injection facilities, production separators and stock tanks, flow lines, gas plants (for processing, separation and recycling of injection gas), and storage tanks. These facilities must be fitted to resist corrosion, and inspected regularly to avoid the initiation of corrosion in the equipment, especially if the injection gas contains carbon dioxide or other corrosive agents.

16.3.5 Management of Gasflood Reservoirs

Reservoirs under gasflood require extensive program of monitoring and surveillance. The program depends on the type of the gasflood project. For instance, if the gas injection project is for pressure maintenance, extensive program for gathering pressure data across the reservoir should be implemented. In addition, gas production should be monitored and controlled to limit excessive production of injected gas. If gas cycling is the objective of the project, then the surveillance program should ensure that the injected gas is controlled to sweep a large portion of the reservoir. This will require monitoring of the intervals taking gas at the injectors, and measuring the liquid content of produced gas at the producers. The surveillance and monitoring program should be adapted to meet the stated objectives of the gasflood project.

The progression of a gasflood project is rapid because of the high mobility of gas. Consequently, it is highly recommended that the gasflood project should be managed with a reservoir model.[28] The reservoir model should be the same model used in designing the project or an updated version of it. The reservoir model should be updated and refined with data gathered from the monitoring and surveillance programs. If actual performance of the gasflood is substantially different from model predictions, the entire project should be re-evaluated. The reservoir model should be used to adjust injection and production volumes, and predict future performance of the gasflood project. In general, gasflood projects should adopt the principles stated in Chapter 20 for sound management of petroleum reservoirs.

Nomenclature

h	formation thickness, ft
k_w	water effective permeability, md
$\overline{p}$	average reservoir pressure, psia
p_{inj}	injection pressure, psia
q_w	water injection rate, STB/D
r_e	reservoir radius, ft
r_w	wellbore radius, ft
t	time, days
W_i	cumulative water injection, bbls
μ_w	viscosity of water, cp

Abbreviations

EOS	equation of state
PVT	pressure-volume-temperature
WAG	water-alternating-gas

References

1. Stiles, W.E.: "Use of Permeability Distribution in Waterflood Calculations," *Trans.* AIME (1949) 186, 9–13.
2. Dykstra, H., and Parsons, R.L.: "The Prediction of Oil Recovery by Waterflood," Secondary Recovery of Oil in the United States, 2nd Ed., API (1950) 160–174.
3. Craig, F.F., Jr., Geffen, T.M., and Morse, R.A.: "Oil Recovery Performance of Pattern Gas or Water Injection Operations from Model Tests," *Trans.* AIME (1955) 204, 7–15.
4. Welge, H.J.: "A Simplified Method for Computing Oil Recovery by Gas or Water Drive," *Trans.* AIME (1952) 195, 91–98.

5. Earlougher, R.C., Jr.: *Advances in Well Test Analysis,* Monograph Series, SPE, Richardson, Texas (1977).
6. Hazebroek, P., Rainbow, H., and Mathews, C.S.: "Pressure Falloff in Water Injection Wells," *Trans.* AIME (1958) 213, 250–260.
7. Merrill, L.S., Kazemi, H., and Gogarty, W.B.: "Pressure Falloff Analysis in Reservoirs with Fluid Banks," *JPT* (July 1974) 809–818.
8. Hall, H.N.: "How to Analyze Injection Well Performance," *World Oil* (October 1963) 128–130.
9. Felsenthal. M.: "Step Rate Tests Determine Safe Injection Pressures in Floods," *The Oil and Gas Journal* (October 28, 1974) 49–54.
10. Vetter, O.J.: "Oilfield Scale—Can We Handle It?," *JPT* (December 1976) 1402–1408.
11. Vetter, O.J., Kandarpa, V. and Harauka, A.: "Prediction of Scale Problems Due to Injection of Incompatible Waters," *JPT* (February, 1982) 273–284.
12. Stiles, L.H.: "Optimizing Waterflood Recovery in a Mature Waterflood, The Fullerton Clearfork Unit," paper SPE 6198 presented at the 51st SPE Annual Technical Conference and Exhibition, New Orleans, October 3–6, 1976.
13. Ghauri, W.K.: "Production Technology Experience in a Large Carbonate Waterflood, Denver Unit, Wasson San Andres Field," *JPT* (September 1980) 1493–1502.
14. Langston, E.P., Shirer, J.A., and Nelson, D.E.: "Innovative Reservoir Management—Key to Highly Successful Jay/LEC Waterflood," *JPT* (May 1981) 783–791.
15. Ezekwe, N., Smith, S., Wilson, M., Humphrey, M., and Murphy, J.: "Performance of a Peripheral Waterflood Project in the Main Body 'B' Reservoirs (Stevens) at the Elk Hills Oil Field, California," paper SPE 21759 presented at the SPE Western Regional Meeting, Long Beach, California, March 20–21, 1991.
16. Kenawy, F.A., Bibars, O.A., Hanafy, H.H., and Rezk, A.S.: "Waterflood Strategy, Challenges and Innovations," paper SPE 53136 presented at the 1999 Middle East Oil Show, Bahrain, February 20–23, 1999.
17. Weiland, J., Mikulencak, D., Fox, P., Frisch, G., and Azari, M.: "Waterflood Surveillance in the Mars Field Deepwater GOM: Mississippi Canyon Block 807," paper SPE 115365 presented at the 2008 Annual Technical Conference and Exhibition, Denver, Colorado, September 21–24, 2008.
18. Christensen, J.R., Stenby, E.H., and Skauge, A.: "Review of WAG Field Experience," *SPEREE* (April 2001) 97–106.
19. Wei, M.H., Yu, J.P., Moore, D.M., Ezekwe, N., Querin, M.E., and Williams, L.L.: "Case History of Pressure Maintenance by Crestal Gas Injection in the 26R Gravity Drainage Reservoir," paper SPE 24035 presented at the SPE Western Regional Meeting, Bakersfield, California, March 30–April 1, 1992.
20. Naylor, P., and Frorup, M.: "Gravity-Stable Nitrogen Displacement of Oil," paper SPE 19641 presented at the 64th Annual Technical Conference and Exhibition of SPE, San Antonio, Texas, October 8–11, 1989.

21. Shyeh, J.J., and Stadler, M.P.: "Effect of Injectant Composition and Pressure on Displacement of Oil by Enriched Hydrocarbon Gases," *SPERE* (May 1995) 109–115.

22. Egemann, P., Lombard, J.-M., Modavi, A., and Kalam, M.Z.: "Gas Process Displacement Efficiency Comparisons on a Carbonate Reservoir," *SPEREE* (December 2006) 621–629.

23. Thawer, R., Alhendi, A., Al Mazroui, Y., Boyd, D., Masuzawa, T., Sugawara, Y., Hollis, C., and Lowden, B.: "Controls on Vertical and Horizontal Flow in a Carbonate Reservoir that Impact Gasflooding and Waterflooding," paper SPE 87237 presented at the 9th Abu Dhabi International Petroleum Exhibition and Conference, October 15–18, 2000.

24. Wylie, P.L., and Mohanty, K.K.: "Effect of Wettability on Oil Recovery by Near-Miscible Gas Injection," *SPEREE* (December 1999) 558–563.

25. Wylie, P. and Mohanty, K.K.: "Effect of Water Saturation on Oil Recovery by Near-Miscible Gas Injection," *SPERE* (November 1997) 264–268.

26. Young, R.E., Fairfield, W.H., and Dykstra, H.: "Performance of a High-Pressure Gas Injection Project, Swanson River Field, Alaska," *JPT* (February 1977) 99–104.

27. Ruhovets, N., and Wyatt, D.F.: "Quantitative Monitoring of Gasflooding in Oil-Bearing Reservoirs by Use of a Pulsed Neutron Tool," *SPEFE* (December 1995) 255–258.

28. Lawrence, J.J., Teletzke, G.F., Hutfliz, J.M., and Wilkinson, J.R.: "Reservoir Simulation of Gas Injection Processes," paper SPE 81459 presented at the SPE 13th Middle East Oil Show & Conference, Bahrain, April 5–8, 2003.

General Reading

- Willhite, G.P.: *Waterflooding,* SPE Textbook Series, Vol. 3, Richardson, Texas (1986).
- Jackson, R.: "Why Waterfloods Fail," *World Oil* (March 1968) 65.
- Shirer, J.A., Ainsworth, W.J., and White, R.W.: "Selection of a Waterflood Pattern for the Jay-Little Escambia Creek Fields," paper SPE 4978 presented at the 49th Annual Fall Meeting of SPE, Houston, Texas, October 6–9, 1974.
- Kumar, M., Hoang, V., Satik, C., and Rojas, D.: "High-Mobility-Ratio-Waterflood Performance Prediction: Challenges and New Insights," *SPEREE* (February 2008) 186–196.
- Brice, B.W., and Renouf, G.: "Increasing Oil Recovery from Heavy Oil Waterfloods," paper SPE 117327 presented at the 2008 SPE International Thermal Operations and Heavy Oil Symposium, Calgary, Alberta, Canada, October 20–23, 2008.
- Rossen, W.R., and Shen, C.: "Gravity Segregation in Gas-Injection IOR," paper SPE 107262 presented at the SPE Europe/EAGE Annual Conference and Exhibition, London, United Kingdom, June 11–14, 2007.
- Burger, J.E. and Mohanty, K.K.: "Mass Transfer From Bypassed Zones During Gas Injection," *SPERE* (May 1997) 124–130.

CHAPTER 17

Enhanced Oil Recovery

17.1 Introduction

Enhanced oil recovery (EOR) is sometimes called tertiary oil recovery or improved oil recovery. In this book, EOR processes are defined to include **all** processes that increase oil recovery beyond primary or secondary recovery processes. This definition was broadened to include EOR processes that have been used to improve oil recovery beyond the primary stage of reservoir depletion, such as steam flooding, or at the end of secondary recovery processes that involve waterflooding. This definition of EOR processes might appear to be in conflict with the inclusion of immiscible gasflooding as a secondary recovery process in Chapter 16. Immiscible gasflooding was presented in Chapter 16 as a matter of convenience since the immiscible process of waterflooding was also presented in that chapter. A case can be made that immiscible gasflooding is an EOR process. Such classification is also acceptable.

A typical EOR process involves injecting a fluid other than water into the reservoir. These injectants usually have economic values either as marketable fluids or the cost incurred to procure and process them for injection. Consequently, EOR processes are very sensitive to oil prices. The price of oil on a sustainable basis must exceed the cost of the injectant plus operating costs by a sizeable margin for an EOR process to be considered economical. For this reason, an EOR process must be **efficient** in terms of cost per barrel of oil recovered and also **effective** in substantially increasing the volume of oil recovered beyond the current recovery process. Economic evaluation is the key important step in the selection of an EOR process and is emphasized throughout the selection process.

EOR processes are important as technologies that could help meet the growing demand for oil in the world. It is estimated that roughly 65% of the original oil in place (OOIP) remains in the reservoir after primary and secondary recoveries. This remaining oil can be recovered by applying suitable EOR processes. The potential for EOR processes is clearly substantial and is responsible for the growth of EOR projects in all oil producing regions of the world.[1] In this chapter, the three main classes (gas injection, chemical, and thermal) of EOR processes are presented with sub-processes that are grouped under each class. This is followed with a summary of criteria for selection of EOR processes based on reservoir properties. Each EOR process is presented later with more details in terms of the mechanism that facilitates oil recovery from the reservoir, and the reservoir conditions that are suitable for its application. Finally, the steps to be followed in the evaluation, implementation, and management of an EOR project are presented.

17.2 EOR Processes

EOR processes can be classified under three main groups. These are Miscible Gas Injection Processes, Chemical Processes, and Thermal Processes. Under these main groups are specific types of applications of the EOR processes which are named as follows:

A. Miscible Gas Injection Processes
  1. Nitrogen injection
  2. Hydrocarbon (HC) gas injection
  3. Carbon dioxide (CO_2) injection
  4. Sour gas, flue gas, etc. injection

B. Chemical Processes
  1. Polymer flooding
  2. Polymer/surfactant flooding
  3. Alkali-Surfactant-Polymer (ASP) flooding
  4. Microbial

C. Thermal Processes
  1. In-Situ Combustion (ISC) or High Pressure Air Injection (HPAI)
  2. Steam/Hot water injection
  3. Steam Assisted Gravity Drainage (SAGD)

The total number of EOR projects in the United States of America from 1978 to 2008 is shown in Figure 17.1.[1,2] In this figure, the total number of gas injection, chemical, and thermal projects in the United States for the same period are also shown. From 1978 to 1986, the number of thermal and chemical projects rose each year, but has been on decline since 1988. In fact, there appears to be no active commercial chemical project in the United States since 2006.[2] Gas injection projects have been on a relatively steady growth in the United States since 1978, as shown in Figure 17.1. Gas injection projects have surpassed both thermal and chemical projects as EOR processes in the United States since 2002.[2] It is projected that gas injection as an EOR process will continue to grow in the United States in future years because it can be applied to a wide range of reservoirs with favorable economic outcomes. The oil production associated with the three EOR processes, including the total EOR production in the United States from 1980 to 2008, are shown in Figure 17.2.[1,2] Thermal processes produced more oil in the United States than

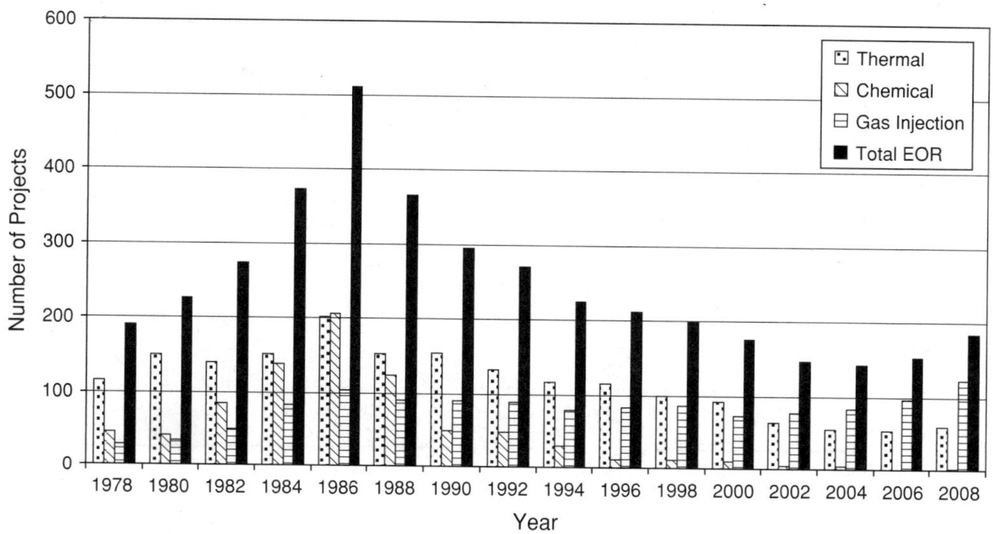

Figure 17.1 Number of EOR projects in the United States from 1978 to 2008.

any other EOR process from 1980 to 2004. However from 2006, gas injection projects produced more oil than thermal projects, as shown in Figure 17.2. In the United States, the contribution to oil production from chemical projects is very limited. The percentage contribution of EOR projects to total oil production in the United States since 1980 is shown in Figure 17.3.[1,3] As shown

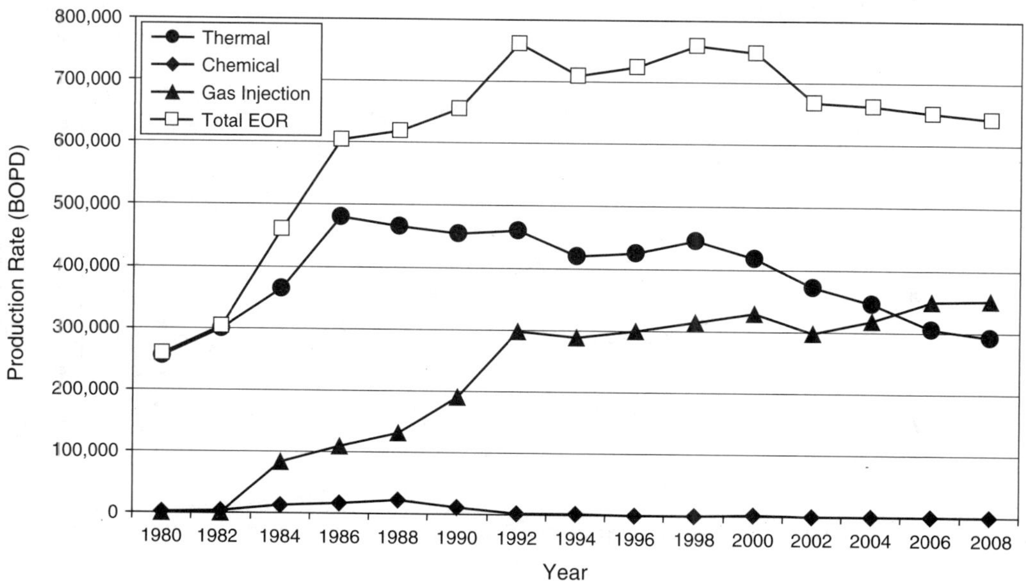

Figure 17.2 Oil production from EOR projects in United States from 1978 to 2008.

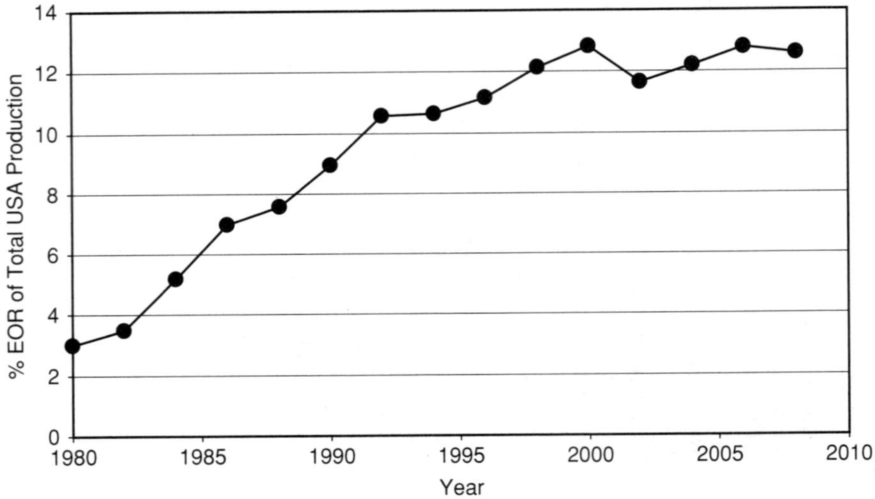

Figure 17.3 Percentage EOR oil Production in United States from 1980 to 2010.

in Figure 17.3, the percentage of total United States oil production due to EOR projects has increased since 1980, and appears to have stabilized at around 12.4% since 1998. Oil prices have substantial influence on the initiation and execution of EOR projects. Figure 17.4 shows the variation of average oil prices with the number of active EOR projects in the United States from 1978 to 2008.[1,4] All EOR projects increased between 1978 and 1986 when there was an increase

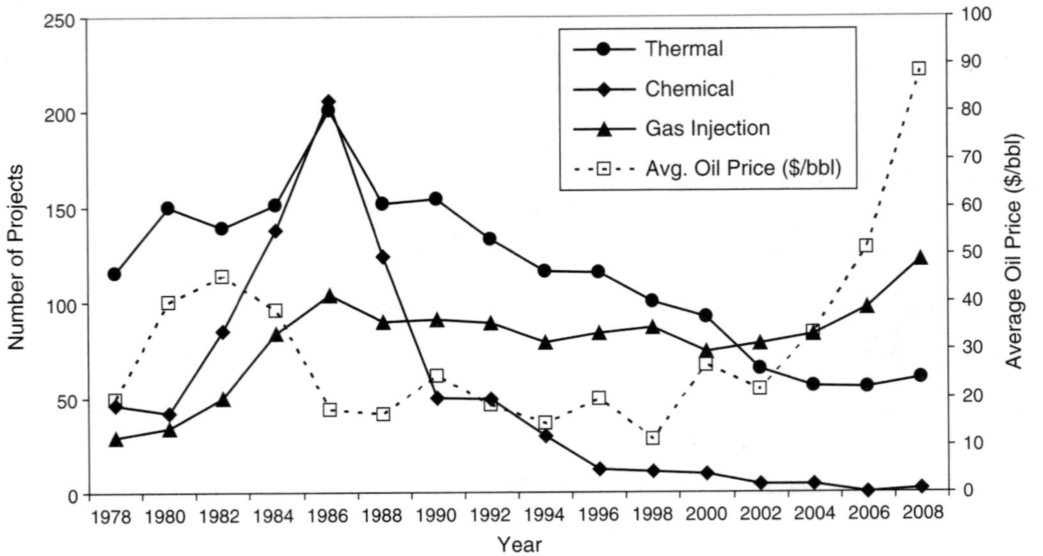

Figure 17.4 EOR projects versus oil prices in the United States from 1978 to 2008.

in average oil prices in the United States. The numbers of EOR projects were flat or decreasing between 1988 and 2000 when average oil prices were relatively stable. Since 2000, gas injection projects have increased in the United States with increase in average oil price. During this period, the number of thermal projects has stabilized while the number of chemical projects has declined almost to zero.

There is significant oil production from EOR projects in other countries.[5] Canada has substantial oil production from gas injection and thermal projects. Indonesia has the single largest EOR project in the world, with production at 190,000 BOPD in 2008 from steam injection in the Duri field. China has numerous EOR projects consisting mainly of chemical and thermal processes. Other countries that have many EOR projects include Brazil, Oman, Mexico, Trinidad, and Venezuela.[5]

17.3 EOR Screening Criteria

The application of EOR processes are both reservoir-specific and reservoir fluid-specific. This literally means that each EOR process must be specifically evaluated before it can be applied to a reservoir. The evaluation process is typically extensive and may include laboratory work, geologic and reservoir modeling, economic analyses, and in many cases field trial in the form of a pilot test. The selection criteria presented here are meant to serve as the first-pass screening procedures that compare the candidate reservoir with other reservoirs that have been produced with an EOR process. They should not replace the rigorous evaluation procedure that each EOR process must undergo before it is actually implemented in the field.

17.3.1 EOR Screening Criteria for Miscible Gas Injection Processes

The EOR screening criteria for miscible gas injection processes are presented in Table 17.1. The injection gases are nitrogen/flue gas, hydrocarbon gas, and carbon dioxide. The screening criteria

Table 17.1 EOR Screening Criteria for Miscible Gas Injection Processes

Gas Injection Process (miscible)	Fluid Properties			Reservoir Properties				
	Gravity (°API)	Viscosity (cp)	Temp. (°F)	Porosity (%)	Perm. (md)	Oil Sat. at start (% PV)	Lithology	Depth (ft)
Nitrogen/ flue gas	>30	<0.5	>250	>10	>30	>50	Carbonate or Sandstone	>7000
Hydrocarbon Gas (HC)	21–57	0.1–1.3	136–290	4–26	10–5000	30–98	Carbonate or Sandstone	4000–14500
Carbon Dioxide (CO_2)	28–44	0.4–3.0	100–250	4–26	2–500	25–90	Carbonate or Sandstone	2000–12000

Table 17.2 EOR Screening Criteria for Chemical Flooding Processes

Chemical Flooding Process	Fluid Properties			Reservoir Properties				
	Gravity (°API)	Viscosity (cp)	Temp. (°F)	Porosity (%)	Perm. (md)	Oil Sat. at start (% PV)	Lithology	Depth (ft)
Polymer/Surfactant Flooding	14–34	5–80	80–160	20–30	170–900	60–75	Sandstone	1300–4600
Alkali/Surfactant/Polymer (ASP) Flooding	14–34	5–80	80–160	20–30	170–900	60–75	Sandstone	1300–4600
Polymer Flooding	14–34	5–80	80–160	20–30	170–900	60–75	Sandstone	1300–4600
Microbial (MEOR)*	—	5–50	<176	≥20	>50	—	Sandstone or Carbonate	<7700

*Includes water salinity <150,000 ppm.

for nitrogen/flue gas and carbon dioxide are based on all the successful field projects reported in the 2008 Worldwide EOR Survey.[5] The criteria for hydrocarbon gas injection are based on the 2008 Worldwide EOR Survey[5] and a survey of North Sea EOR projects reported by Awan et al.[6] The screening criteria for miscible gas injection shown in Table 17.1 indicate that most light oil reservoirs meet the criteria for any of these injection gases. The most important factors that determine the choice of injection gas appear to be availability (gas supply) and cost.

17.3.2 EOR Screening Criteria for Chemical Flooding Processes

The EOR screening criteria for chemical flooding processes are shown in Table 17.2. Note that the same screening criteria are recommended for polymer/surfactant flooding, alkali/surfactant/polymer flooding, and polymer flooding. This is as a result of the screening criteria being based on successful field projects reported in the 2008 Worldwide EOR Survey.[5] Few projects were reported for polymer/surfactant flooding and alkali/surfactant/polymer flooding in the survey. In any case, chemical flooding using polymers, surfactants, and alkalis require extensive laboratory testing and design of the chemicals to suit the properties of the reservoir rock. Hence, the screening criteria for these chemicals serve basically as guides. The criteria for microbial EOR are based on the criteria reported by Lazar,[7] and Bryant and Lindsey.[8] Again, as with all chemical EOR processes, microbial EOR (MEOR) requires extensive laboratory and field testing.

Table 17.3 EOR Screening Criteria for Thermal Flooding Processes

Thermal Process	Fluid Properties			Reservoir Properties				
	Gravity (°API)	Viscosity (cp)	Temp. (°F)	Porosity (%)	Perm. (md)	Oil Sat. at start (% PV)	Lithology	Depth (ft)
ISC or HPAI	19–33	2–660	110–230	17–32	10–1265	50–94	Sandstone or Carbonate	400–8300
Steam	8–30	50–500,000	45–290	15–65	100–10000	44–90	Sandstone* or Tripolite	200–3600

*Includes unconsolidated sandstone.

17.3.3 EOR Screening Criteria for Thermal Processes

The EOR screening criteria for thermal processes are shown in Table 17.3. The criteria for thermal processes are based on successful field projects reported in the 2008 Worldwide EOR Survey.[5] The criteria for steam flooding also apply to hot water injection and the application of steam flooding described as steam-assisted gravity drainage (SAGD). Note that steam flooding has been applied successfully to a wide range of heavy oil sandstone reservoirs.

17.4 Miscible Gas Injection Processes

A miscible displacement process is defined as a fluid displacement where there is **no** interface (or boundary) between the displacing fluid and the displaced fluid. In this chapter, all EOR gas injection processes that are presented assume that reservoir conditions exist that miscibility is achieved between the injected gas and the reservoir fluid. This is in contrast to the immiscible gasflooding processes presented in Chapter 16. However, since the main distinction between miscible and immiscible gasflooding depends largely on the amount of hydrocarbon fluids recovered, most of the material (screening criteria, project design and implementation, etc.) presented for miscible gasflooding are also applicable to immiscible gasflooding processes.

17.4.1 Basic Concepts on Miscibility for Gas Displacement Processes

The concept of miscibility in a gas displacement process can be demonstrated with a ternary-phase diagram. A ternary-phase diagram for the methane/*n*-butane/decane[9] system at 2500 psia and 160°F is shown in Figure 17.5. As shown in Figure 17.5, this system has a two-phase region which is enclosed by a bubble point line and a dew point line that meet at the plait (or critical) point. At 2500 psia and 160°F, mixtures of this system with compositions within the two-phase envelope are immiscible, while mixtures of the system with composition outside the two-phase envelope are miscible. The relative size of the two-phase envelope in Figure 17.5 depends on

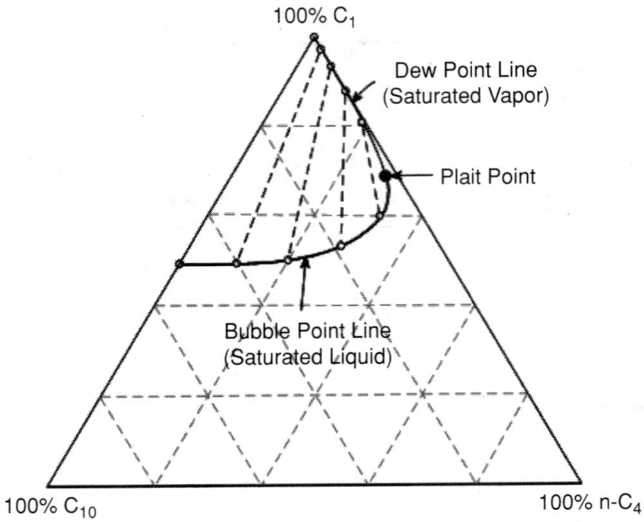

Figure 17.5 Ternary-phase diagram of methane/*n*-butane/decane system at 2500 psia and 160°F (from Hutchinson and Braun[9]).

temperature and pressure of the system. For instance, at constant temperature of 160°F, increasing the pressure of the system from 2500 to 3250 psia will shrink the two-phase envelope, as shown in Figure 17.6. This also means that at higher pressures more compositions of the methane/*n*-butane/decane system are miscible at 160°F. For simplicity, complex hydrocarbon systems can be represented with a pseudoternary diagram, as shown in Figure 17.7. Note that in

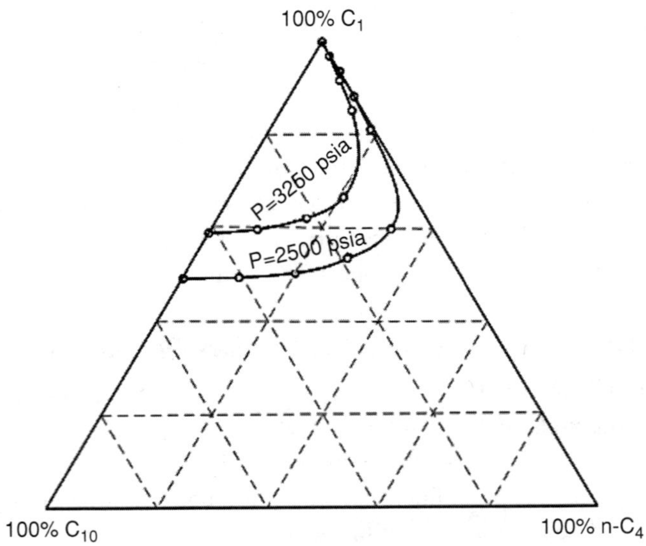

Figure 17.6 Effect of pressure on the two-phase envelope of methane/*n*-butane/decane system at 160°F (from Hutchinson and Braun[9]).

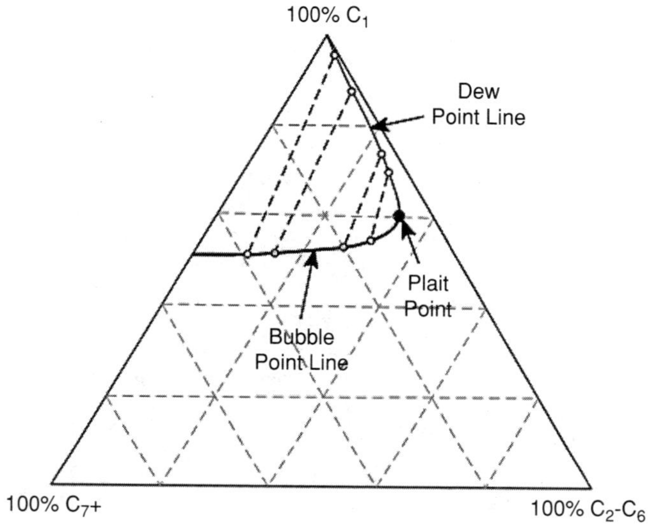

Figure 17.7 Pseudoternary phase diagram for a complex hydrocarbon system (from Hutchinson and Braun[9]).

a pseudoternary diagram, pseudocomponents are used to represent the multicomponent hydrocarbon system, as shown in Figure 17.7. As was the case for simple ternary diagrams, a two-phase region can exist in a multicomponent system formed by a bubble point line and a dew point line that meet at the plait (or critical) point. Consequently, compositions of the system that fall within the two-phase envelope are considered to be immiscible while compositions outside the two-phase envelope can be miscible. These concepts are used in the descriptions of miscible conditions in gas displacement processes.

17.4.2 First-Contact Miscibility (FCM)

The phenomenon of first-contact miscibility (FCM) occurs when the injected fluid is completely miscible with the reservoir fluid under the existing reservoir pressure and temperature. Under FCM, the injected fluid completely mixes with the reservoir fluid to create a transition fluid that miscibly displaces the reservoir fluid. FCM can be considered as an instantaneous process of complete mixing of the injected fluid with the reservoir fluid. The factors that determine FCM are reservoir temperature and pressure, composition of the injected fluid, and composition of the reservoir fluid. Suppose Figure 17.8 is the pseudoternary diagram for a CO_2-hydrocarbon mixture. Line AB is the critical tie line drawn through the critical point of the phase envelope. As a simple illustration of FCM, pure CO_2 will achieve FCM with reservoir fluid in the shaded region of Figure 17.8.

17.4.3 Multiple-Contact Miscibility (MCM)

As the term suggests, multiple-contact miscibility (MCM) is achieved as the result of repeated contacts in the reservoir between the reservoir oil and the injected fluid, and also fluids generated

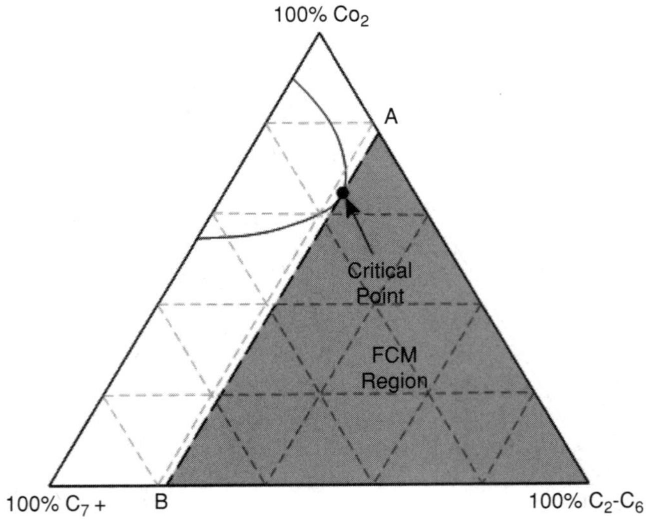

Figure 17.8 Pseudoternary diagram for CO_2-hydrocarbon system.

in-situ by interactions between the injected fluid and the reservoir oil.[10,11] There are two main processes by which MCM is achieved in the reservoir. These are the vaporizing gas drive and the condensing gas drive. In the vaporizing gas drive, the composition of the injected fluid is progressively modified by vaporization of the intermediate hydrocarbon components (C_2 to C_6) in the reservoir oil until the modified fluid becomes ultimately miscible with the original reservoir oil. In the condensing gas drive, intermediate hydrocarbon components in the injection fluid condense in the reservoir oil to generate a modified fluid that becomes miscible with the injection fluid. Zick[12] and Stalkup[13] reported that some MCM processes involve a combination of the vaporizing gas drive and the condensing gas drive. This third MCM process is termed the combined condensing and vaporizing (CV) gas drives. The type of MCM process that is dominant in a displacement drive depends on the composition of the reservoir oil, the composition of the injection fluid, and the temperature and pressure of the reservoir. These processes are illustrated with pseudoternary phase diagrams in the following sections.

17.4.4 Vaporizing Gas Drive MCM Process

The mechanism for achieving MCM by vaporizing gas drive is best illustrated with a pseudoternary diagram, as shown in Figure 17.9. The pseudoternary diagram is simplified to represent a system composed of methane (C_1)/ethane-hexane (C_2-C_6)/Heptane-plus (C_{7+}). Suppose a lean gas composed entirely of methane is injected into a reservoir containing oil with composition represented at point O in Figure 17.9. The methane gas will vaporize the intermediate components of the oil to form a fluid with composition represented at point 1 in the two-phase region. This fluid

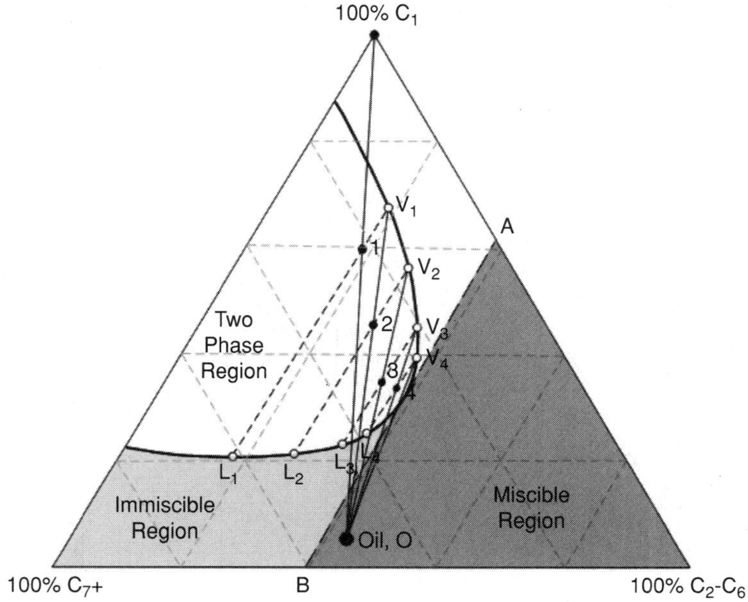

Figure 17.9 MCM by vaporizing gas mechanism.

will consists of a vapor phase shown as V_1, and a liquid phase shown as L_1, on the phase envelope. As the displacement drive progresses, the vapor phase, V_1, advances to contact additional original reservoir oil, and again vaporize more intermediate components to form another fluid with composition represented at point 2 within the two-phase envelope. This fluid separates into a vapor phase, V_2, and a liquid phase, L_2. The vapor phase, V_2, advances to contact more original oil to form a third fluid with composition represented at point 3 within the two-phase envelope. The vapor phase, V_3, from the fluid at point 3 contacts more original oil to form a fluid represented at point 4, whose vapor phase, V_4, achieves miscibility with original oil as shown in Figure 17.9. As described, the vaporizing gas drive MCM process is a mechanism whereby the vapor phase of the displacing fluid is progressively enriched through vaporization of the intermediate components in the original reservoir oil by repeated contacts until a fluid miscible with the original reservoir oil is achieved. Note that reservoir temperature and pressure are assumed to be constant, and local thermodynamic equilibrium is assumed to have been achieved between the intermediate phases during this process.

In Figure 17.9, line AB drawn through the critical point on the two-phase envelope is called the critical tie-line. For the fluid system at fixed conditions represented in Figure 17.9, reservoir oil compositions on the critical tie-line or to the right of the critical tie-line are miscible with injection gas composition to the left of point A on the critical tie-line. Furthermore, Figure 17.9 shows that reservoir oil composition to the left of the critical tie-line will be immiscible with injection gas composition to the left of point A on the critical tie-line. Note that these

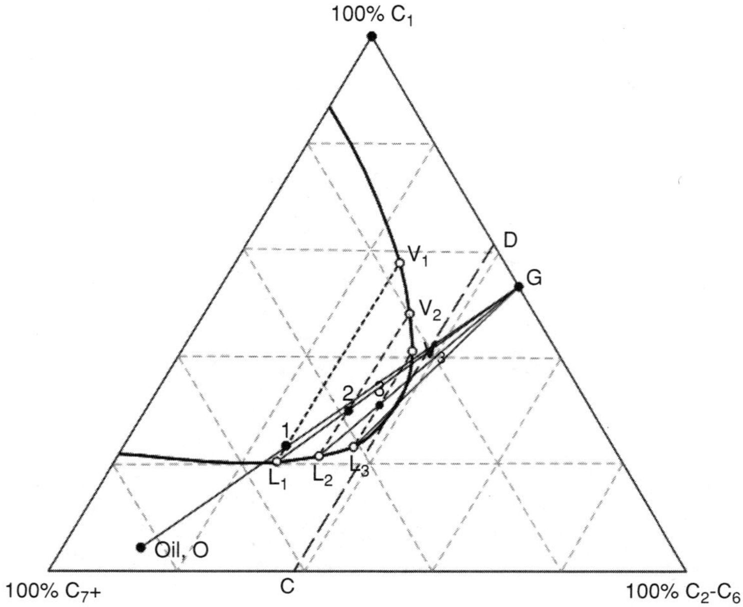

Figure 17.10 MCM by condensing gas mechanism.

descriptive representations on the pseudoternary diagram are applicable only to vaporizing gas drive processes. The critical tie-line can be used also to define the minimum miscibility pressure (MMP) for a fluid system at fixed conditions. The MMP is defined as the minimum pressure at which the reservoir oil composition lies on a critical tie-line extension for a specified fluid system. The MMP represents the minimum pressure at which MCM can be achieved in the reservoir for the fluid system. At pressures less than MMP, the displacement process will be immiscible.

17.4.5 Condensing Gas Drive MCM Process

The mechanism of attaining MCM by condensing gas drive is illustrated with a pseudoternary diagram shown as Figure 17.10. In Figure 17.10, the injected fluid is represented by point G, and the reservoir oil is shown as point O. To initiate the multiple contact process, the intermediate components in the enriched injection gas will condense in the reservoir oil to form a fluid within the two-phase envelope with composition represented at point 1. This fluid is composed of a vapor phase, V_1, and a liquid phase, L_1. As the displacement advances, the liquid phase, L_1, is contacted by more injection gas to form a fluid with composition shown at point 2. This fluid is composed of a vapor phase, V_2, and a liquid phase, L_2. The liquid phase, L_2, is contacted by more injection gas to form a fluid shown at point 3 composed of a vapor phase, V_3, and liquid phase, L_3. Further contact between the liquid, L_3, and the injection gas produces a fluid which is ultimately miscible with the injection gas. The condensing gas drive MCM process can be described as the mechanism of progressive enrichment of the condensed liquid phase with intermediate hydrocarbons in the injection gas resulting from multiple contacts that ultimately produces a fluid that is miscible with the injection gas. Note that this description of the condensing gas drive MCM process assumes that reservoir conditions are fixed, and local thermodynamic equilibrium is attained between the intermediate liquid and vapor phases.

In Figure 17.10, the critical tie-line is shown as line CD. Any injection fluid composition to the left of point D will not achieve miscibility with the oil composition to the left of point C. The oil composition to the left of point C can achieve MCM with injection fluid composition at point D or to the right of point D. For the condensing gas drive, the MMP is defined as the minimum pressure at which the injection fluid composition lies on the extension of a critical tie-line. For the condensing gas drive, the injection fluid composition must lie on the critical tie-line or to the right of the critical tie-line and the oil composition must lie on the opposite (left) side of the critical tie-line as shown in Figure 17.10 for MCM to be achieved. Another important parameter that can be used to define the condensing gas MCM process is the minimum miscibility enrichment (MME). This describes the process of enriching the composition of the injection gas with intermediate hydrocarbons (C_2 to C_6) to achieve miscibility with the reservoir oil. The MME at a given pressure is defined as the minimum enrichment of the injection gas composition, such that the gas composition lies on an extension of a critical tie-line. Thus for a condensing-gas drive process, miscibility can be achieved either by increasing pressure or by enriching the composition of the injection fluid with intermediate hydrocarbons.

17.4.6 Combined Condensing/Vaporizing (CV) Gas Drive MCM Process

Zick[12] reported that a combined condensing/vaporizing gas drive mechanism rather than the traditional condensing gas drive alone is the key mechanism in the displacement of real reservoir oil by enriched fluids. This observation was supported by the work done by Stalkup.[13] The importance of the observation is that MMPs and MMEs estimated by pseudoternary methods are different from MMPs and MMEs observed for combined CV drives. This development should be expected since pseudoternary diagrams are inadequate for representation of a complex fluid system, such as reservoir hydrocarbon systems.

17.5 Methods for Determination of MMP or MME for Gasfloods

The two main techniques used in the determination of MMP or MME can be broadly classified as analytical techniques and experimental methods. Analytical techniques, which are based on numerical computations, can be subdivided further into empirical correlations and compositional simulations. Empirical correlations are based on fitting equations to MMP data obtained from experimental methods, and/or computed from compositional simulations. Compositional simulations methods generally use equations of state to predict MMP or MME from the compositions of the injection gas and the reservoir oil. The three common analytical techniques that use compositional simulations are the tie-line method, the 1-D slimtube simulation, and the mixing cell method. The two most common experimental methods are the slimtube method and the rising bubble apparatus method.

17.5.1 Analytical Techniques for Estimation of MMP or MME

Analytical techniques for the estimation of MMPs or MMEs include all computational methods that do not involve laboratory experiments. Analytical techniques can be subdivided into empirical correlations and compositional simulations. Compositional simulations usually require the use of an equation of state for prediction of phase equilibria, and are comprised of the tie-line method, the 1-D slimtube simulation method, and the multiple mixing-cell method.

17.5.1.1 Empirical Correlations for MMP

Empirical correlations are sets of equations generated from MMP data measured by experimental methods and/or MMP data calculated from compositional simulations. Empirical correlations are used to predict the MMPs for reservoir oils with various types of gas injection. These correlations provide quick estimates of MMPs which can be useful during screening of various gas injection processes for the reservoir. The MMPs calculated from empirical correlations can have large errors and should not replace MMPs obtained from experimental or compositional simulation methods. Empirical correlations should be used to predict MMPs during the early stages of screening the reservoir for various types of miscible gas injection processes. Empirical correlations for prediction of MMPs have been developed on the basis of type and composition of the injection gas. This classification is used in their presentation.

MMP Correlations for Hydrocarbon Injection Gases. A set of equations for prediction of MMPs for hydrocarbon injection gases was presented by Glasø.[14] These equations were derived from the graphical correlations developed by Benham et al.[10] The Glasø correlations state MMP as a function of reservoir temperature, molecular weight of the C_{7+} in the oil, the mole percent of methane in the injection gas, and the molecular weight of the intermediates (C_2 through C_6) in the injection gas. The Glasø correlations for predicting MMP for hydrocarbon gas/oil systems are as follows:

$$MMP_{x=34} = 6329.0 - 25.410y - z(46.745 - 0.185y) + T\left(1.127 \times 10^{-12} y^{5.258} e^{319.8zy^{-1.703}}\right) \tag{17.1}$$

$$MMP_{x=44} = 5503.0 - 19.238y - z(80.913 - 0.273y) + T\left(1.700 \times 10^{-9} y^{3.730} e^{13.567zy^{-1.058}}\right) \tag{17.2}$$

$$MMP_{x=54} = 7437.0 - 25.703y - z(73.515 - 0.214y) + T\left(4.920 \times 10^{-14} y^{5.520} e^{21.706zy^{-1.109}}\right) \tag{17.3}$$

In Eqs. (17.1) to (17.3), MMP = minimum miscibility pressure, psig; x = molecular weight of C_2 through C_6 in the injection gas, lbm/lb-mole; y = corrected molecular weight of

C_{7+} in the stock-tank oil, where $y = \left(2.622/\gamma_{o,C_{7+}}^{-0.846}\right)^{6.588}$ and $\gamma_{o,c_{7+}}$ is the specific gravity of stock-tank oil; z = methane in injection gas, mole percent; and T = reservoir temperature, °F. Glasø[14] reported a deviation of 2 to 8 percent between the MMP calculated from Eqs. (17.1) to (17.3) and experimentally determined MMP data.

Example 17.1 Calculation of MMP from the Glasø Correlations for Hydrocarbon Gas Injection

Problem

Calculate the MMP for a reservoir oil and hydrocarbon gas injection mixture with the following properties:

Molecular weight of C_2 through C_6 in injection gas	32 lbm/lb-mole
Mole percent of methane in injection gas	65.2
Corrected molecular weight of C_{7+} in stock-tank oil	224 lbm/lb-mole
Reservoir temperature	160°F

Solution

Using Eq. (17.1),

$$\begin{aligned} MMP_{x=34} &= 6329.0 - 25.410y - z(46.745 - 0.185y) \\ &\quad + T\left(1.127 \times 10^{-12} y^{5.258} e^{319.8zy^{-1.703}}\right) \\ &= 6329.0 - 25.410 \times 224 - 65.2(46.745 - 0.185 \times 224) \\ &\quad + 160\left(1.127 \times 10^{-12} \times 224^{5.258} e^{319.8 \times 65.2 \times 224^{-1.703}}\right) \\ &= 3557.55 \text{ psig} \end{aligned}$$

Using Eq. (17.2),

$$\begin{aligned} MMP_{x=44} &= 5503.0 - 19.238y - z(80.913 - 0.273y) \\ &\quad + T\left(1.700 \times 10^{-9} y^{3.730} e^{13.567zy^{-1.058}}\right) \\ &= 5503.0 - 19.238 \times 224 - 65.2(80.913 - 0.273 \times 224) \\ &\quad + 160\left(1.70 \times 10^{-9} \times 224^{3.730} e^{13.567 \times 65.2 \times 224^{-1.058}}\right) \\ &= 2749.74 \text{ psig} \end{aligned}$$

By extrapolation, $MMP_{x=32}$ = 3719.26 psig.

The experimental MMP reported in the literature for this system by Glasø[14] is 3700 psig. This represents an error of +0.52%.

MMP Correlations for Nitrogen Injection Gases. Two correlations are presented for prediction of MMP when the injection gas is nitrogen. These correlations were developed by Glasø[14] and Firoozabadi and Aziz.[15] The Glasø correlation for nitrogen/oil systems are based on limited experimental data and should be used with caution. The MMP correlations for nitrogen gas injection by Glasø are expressed as functions of molecular weight of the C_{7+} in the stock-tank oil, mole percent of the intermediates (C_2 through C_6) in the reservoir oil, and reservoir temperature. The Glasø MMP correlations for nitrogen gas injection are as follows:

For molecular weight of C_{7+} >160 and mole percent of intermediates >28, the correlation to use is:

$$MMP = 6364.0 - 12.090M_{C_{7+}} + T\left(1.127 \times 10^{-12}M_{C_{7+}}^{5.258}e^{23,025.0M_{C_{7+}}^{-1.703}} - 20.80\right) \quad \textbf{(17.4)}$$

For molecular weight of C_{7+} <160 and mole percent of intermediates >28, the correlation to use is:

$$MMP = 7695.1 - 12.090M_{C_{7+}} + T\left(1.127 \times 10^{-12}M_{C_{7+}}^{5.258}e^{23,025.0M_{C_{7+}}^{-1.703}} - 39.77\right) \quad \textbf{(17.5)}$$

If the mole percent of intermediates <28, the correlation to use is:

$$\begin{aligned} MMP = {} & 9364.0 - 12.090M_{C_{7+}} + T\left(1.127 \times 10^{-12}M_{C_{7+}}^{5.258}e^{23,025.0M_{C_{7+}}^{-1.703}} - 20.80\right) \\ & - 99.3C_{2-6} \end{aligned} \quad \textbf{(17.6)}$$

In Eqs. (17.4) to (17.6), MMP = minimum miscibility pressure, psig; $M_{C_{7+}}$ = molecular weight of the C_{7+} in the stock-tank oil, lbm/lb-mole; T = reservoir temperature, °F; and C_{2-6} = mole percent of the intermediates (C_2 through C_6) in the reservoir oil.

The Firoozabadi and Aziz[15] correlation for prediction of MMP for nitrogen/oil systems is based on experimental data and compositional simulation using the Peng-Robinson equation of state. This correlation can also be used to predict MMP for lean-gas injection. (A lean gas is composed mostly of methane gas.) There was no accuracy reported for this correlation in comparison with experimental data. This correlation should be used with caution. The Firoozabadi and Aziz[15] correlation for prediction of MMP for nitrogen or lean gas injection is:

$$MMP = 9433 - 188 \times 10^3\left(\frac{C_{2-5}}{M_{C_{7+}}T^{0.25}}\right) + 1430 \times 10^3\left(\frac{C_{2-5}}{M_{C_{7+}}T^{0.25}}\right)^2 \quad \textbf{(17.7)}$$

In Eq. (17.7), MMP = minimum miscibility pressure, psia; $M_{C_{7+}}$ = molecular weight of C_{7+}, lbm/lb-mole; C_{2-5} = mole percent of C_2 through C_5 including CO_2 and H_2S in the reservoir fluid; and T = reservoir temperature, °F.

Example 17.2 Calculation of MMP from the Glasø, and Firoozabadi and Aziz Correlations for Nitrogen Gas Injection

Problem

Calculate the MMP for a reservoir oil and nitrogen gas injection mixture with the following properties:

Mole percent of C_2 through C_5 in reservoir oil	25.17
Molecular weight of C_{7+} in reservoir oil	193.3 lbm/lb-mole
Reservoir temperature	164°F
Bubble point pressure of reservoir oil	4000 psi

Solution

To apply the Glasø correlations to this nitrogen/oil system, Eq. (17.4) was selected because the molecular weight of the C_{7+} fraction was >160 lbm/lb-mole and the mole percent of the intermediates was assumed >25.17. Note that the Glasø correlations are based on intermediate composition of C_2 through C_6, whereas the Firoozabadi and Aziz correlation is based on intermediates composition of C_2 through C_5.
Using Eq. (17.4) of the Glasø correlation:

$$\begin{aligned} MMP &= 6364.0 - 12.090M_{C_{7+}} + T\left(1.127 \times 10^{-12} M_{C_{7+}}^{5.258} e^{23,025.0 M_{C_{7+}}^{-1.703}} - 20.80\right) \\ &= 6364.0 - 12.090 \times 193.3 + 164\left(1.127 \times 10^{-12} \times 193.3^{5.258} \times e^{23,025 \times 193.3^{-1.703}} \right. \\ &\quad \left. - 20.8\right) \\ &= 4295.19 \text{ psig} \end{aligned}$$

Applying the Firoozabadi and Aziz correlation, Eq. (17.7) gives:

$$\begin{aligned} MMP &= 9433 - 188 \times 10^3 \left(\frac{C_{2-5}}{M_{C_{7+}} T^{0.25}}\right) + 1430 \times 10^3 \left(\frac{C_{2-5}}{M_{C_{7+}} T^{0.25}}\right)^2 \\ &= 9433 - 188 \times 10^3 \left(\frac{25.17}{193.3 \times 164^{0.25}}\right) + 1430 \times 10^3 \left(\frac{25.17}{193.3 \times 164^{0.25}}\right)^2 \\ &= 4485.62 \text{ psia} \end{aligned}$$

The reported experimental MMP for this system is 4280 psi. For screening purposes, both correlations gave very good MMP estimates for this nitrogen/reservoir oil system, although the Glasø correlation appears to be superior for this system.

MMP Correlations for Pure Carbon Dioxide Injection Gases. Many correlations have been published in the literature for the prediction of MMP for pure carbon dioxide (CO_2)-oil systems.[16] Three of these correlations were selected on the basis of simplicity, and accuracy as reported by their authors for presentation in this book. The three correlations that are recommended for prediction of MMP for pure CO_2-Oil systems are the Glasø[14] correlation, the Emera and Sarma[16] correlation, and the Yuan et al.[17] correlation. Glasø[14] correlated MMP for pure CO_2-Oil systems as a function of molecular weight of the C_{7+} fraction in the stock-tank oil, the mole percent of the intermediates (C_2 through C_6) in the oil, and reservoir temperature. For mole composition of C_2 through C_6 greater than 18%, the Glasø correlation for pure CO_2-Oil systems is:

$$MMP_{pure} = 810.0 - 3.404 M_{C_{7+}} + T\left(1.700 \times 10^{-9} M_{C_{7+}}^{3.730} e^{786.8 M_{C_{7+}}^{-1.058}}\right) \quad \textbf{(17.8)}$$

For mole composition of C_2 through C_6 less than 18%, the Glasø correlation for pure CO_2-Oil systems is:

$$MMP_{pure} = 2947.9 - 3.404 M_{C_{7+}} + T\left(1.700 \times 10^{-9} M_{C_{7+}}^{3.730} e^{786.8 M_{C_{7+}}^{-1.058}}\right) - 121.2 C_{2-6} \quad \textbf{(17.9)}$$

In Eqs. (17.8) and (17.9), MMP_{pure} = minimum miscibility pressure for pure CO_2 gas injection, psig; $M_{C_{7+}}$ = molecular weight of the C_{7+} in the stock-tank oil, lbm/lb-mole; T = reservoir temperature, °F; and C_{2-6} = mole percent of the intermediates (C_2 through C_6) in the reservoir oil. For a pressure range of 900 to 4400 psig, Glasø[14] reported an average deviation of −4.41% and standard deviation of 11.68% for MMP predicted with these correlations and experimental MMP data reported in the literature.

The Emera[16] correlation for pure CO_2 injection is represented as:

$$MMP_{pure} = 5.0093 \times 10^{-5} \times (1.8T + 32)^{1.164} \times M_{C_{5+}}^{1.2785} \times \left(\frac{C_{C_1+N_2}}{C_{C_{2-4}} + H_2S + CO_2}\right)^{0.1073} \quad \textbf{(17.10)}$$

If the bubble point pressure (p_b) of the oil is less than 50 psi, the Emera[16] correlation for pure CO_2 injection becomes:

$$MMP_{pure} = 5.0093 \times 10^{-5} \times (1.8T + 32)^{1.164} \times M_{C_{5+}}^{1.2785} \quad \textbf{(17.11)}$$

Note that for the Emera correlations, if the MMP is less than p_b, then $MMP_{pure} = p_b$. In Eqs. (17.10) and (17.11), MMP_{pure} = minimum miscibility pressure for pure CO_2 , MPa; $M_{C_{5+}}$ = molecular weight of the C_{5+} in the stock-tank oil, lbm/lb-mole; T = reservoir temperature, °C; $C_{C_1+N_2}$ = mole fraction of the volatiles (C_1 and N_2) in the reservoir oil; and $C_{C_{2-4}+H_2S+CO_2}$ = mole fraction of the intermediates (C_2, C_3, C_4, H_2S, and CO_2) in the reservoir oil. Emera and Sarma[16] reported an average deviation of 5.86% and standard deviation of 7.96% for MMP predicted with their correlations when compared with experimental MMP data available in the literature.

The Yuan et al.[17] correlation is based on analytical MMPs calculated using the tie-line method.[18] The expression of the Yuan correlation for pure CO_2 gas injection is given by:

$$\begin{aligned} MMP_{pure} &= a_1 + a_2 M_{C_{7+}} + a_3 P_{C_{2-6}} + \left(a_4 + a_5 M_{C_{7+}} + a_6 \frac{P_{C_{2-6}}}{M_{C_{7+}}^2}\right) T \\ &\quad + \left(a_7 + a_8 M_{C_{7+}} + a_9 M_{C_{7+}}^2 + a_{10} P_{C_{2-6}}\right) T^2 \end{aligned} \quad \textbf{(17.12)}$$

The coefficients in Eq. (17.12) are: a_1 = −1.4364E + 03; a_2 = 0.6612 E + 01; a_3 = −4.4979 E + 01; a_4 = 0.2139 E + 01; a_5 = 1.1667 E − 01; a_6 = 8.1661 E + 03; a_7 = −1.2258 E − 01; a_8 = 1.2883 E − 03; a_9 = −4.0152 E − 06; a_{10} = −9.2577 E − 04.

In Eq. (17.12), MMP_{pure} = minimum miscibility pressure for pure CO_2 gas injection, psia; $M_{C_{7+}}$ = molecular weight of C_{7+} fraction in reservoir oil, lbm/lb-mole; $P_{C_{2-6}}$ = mole percent of C_2 to C_6 in reservoir oil; and T = reservoir temperature, °F. The range of the molecular weight of the C_{7+} of all the reservoir oils used in the regression of the correlations is 139 to 319 lbm/lb-mole. The mole composition of C_2 to C_6 in the oil has a range of 2.0 to 40.3%, and the reservoir temperature ranged from 71 to 300 °F. Yuan et al.[17] reported an average absolute error of 11.9%, when MMPs computed with the correlation are compared with 41 slimtube MMPs. Also, the Yuan correlation is applicable over a wider temperature range than other existing correlations.[17]

Example 17.3 Calculation of MMP for Pure CO_2 Gas Injection Using the Glasø, Emera, and Yuan Correlations

Problem

Calculate the MMP for pure CO_2 gas injection into a reservoir containing oil with the following properties:

Mole percent of C_1 in reservoir oil	4.4
Mole percent of C_2 through C_6 in reservoir oil	24.0
Molecular weight of C_{5+} in reservoir oil	185 lbm/lb-mole
Molecular weight of C_{7+} in reservoir oil	234 lbm/lb-mole
Reservoir temperature	180°F

Solution

For the Glasø correlation, since mole percent of C_2 to C_6 is >18, using Eq. (17.8) gives:

$$\begin{aligned} MMP_{pure} &= 810.0 - 3.404M_{C_{7+}} + T\left(1.700 \times 10^{-9} M_{C_{7+}}^{3.730} e^{786.8 M_{C_{7+}}^{-1.058}}\right) \\ &= 810.0 - 3.404 \times 234 + 180\left(1.700 \times 10^{-9} \times 234^{3.730} e^{786.8 \times 234^{-1.058}}\right) \\ &= 2451.75 \text{ psig} \end{aligned}$$

For the Emera correlation, substituting the given properties in Eq. (17.10) gives:

$$\begin{aligned} MMP_{pure} &= 5.0093 \times 10^{-5} \times (1.8T + 32)^{1.164} \times M_{C_{5+}}^{1.2785} \times \left(\frac{C_{C_1+N_2}}{C_{C_{2-4}} + H_2S + CO_2}\right)^{0.1073} \\ &= 5.0093 \times 10^{-5} \times (180)^{1.164} \times (185)^{1.2785} \times \left(\frac{4.4}{24}\right)^{0.1073} \\ &= 13.9457 \text{ MPa} \\ &= 2022.63 \text{ psia} \end{aligned}$$

For the Yuan correlation, substituting the given properties in Eq. (17.2) gives:

$$\begin{aligned} MMP_{pure} &= a_1 + a_2 M_{C_{7+}} + a_3 P_{C_{2-6}} + \left(a_4 + a_5 M_{C_{7+}} + a_6 \frac{P_{C_{2-6}}}{M_{C_{7+}}^2}\right)T \\ &\quad + (a_7 + a_8 M_{C_{7+}} + a_9 M_{C_{7+}}^2 + a_{10} P_{C_{2-6}})T^2 \\ &= -1436.4 + 6.612 \times 234 - 44.979 \times 24 \\ &\quad + \left(2.139 + 0.11667 \times 234 + 8166.1 \times \frac{24}{(234)^2}\right) \times 180 \\ &\quad + [-0.12258 + 0.0012883 \times 234 - 4.0152E - 06(234)^2 - 9.257E \\ &\quad - 04 \times 24](180)^2 \\ &= 2927.30 \text{ psia} \end{aligned}$$

The experimental MMP from slimtube measurements for this system was reported by Yuan et al.[17] as 3250 psia. The Yuan correlation prediction of 2927.30 psia is the closest to the experimental result of 3250 psia.

MMP Correlations for Impure Carbon Dioxide Injection Gases. In actual field applications, CO_2 injection gas is not available at 100% purity. CO_2 injection gas contains impurities such as H_2S, SO_X, N_2, O_2, and hydrocarbons (C_1 to C_4). Some of the impurities such as H_2S and SO_X could lower the MMP for the CO_2-Oil system, while impurities such as N_2 and CH_4 could increase the MMP for the system. For these reasons, it is important to account for the effect of impurities present in the CO_2 injection gas on achieving miscibility for the CO_2-Oil system. This impact is reflected in the calculation of MMPs for impure CO_2 injection gas. Numerous correlations have been published for the calculation of MMP for impure CO_2-Oil systems. These include correlations by Yuan et al.,[17] Johnson and Pollin,[18] Alston et al.,[19] Sebastian et al.,[20] Kovarik,[21] Orr and Silva,[22] Enick et al.,[23] Eakin and Mitch,[24] Dong,[25] and Emera and Sarma.[26] Only three of these correlations were selected on the basis of simplicity, and accuracy as reported by their authors for presentation in this book. The recommended correlations for calculation of MMP for impure CO_2 gas injection mixtures are the Sebastian et al.[20] correlation, Yuan et al.[17] correlation, and the Emera and Sarma[26] correlation. The Sebastian correlation was selected because of its simplicity and ability to handle CO_2-rich injection gas containing up to 55 mol% of impurities such as H_2S, N_2, and hydrocarbons (C_1 to C_4). The Yuan correlation was selected because it covers a wider temperature range, especially when applied with the Yuan correlation for pure CO_2. Note however, that the Yuan correlation was derived for impure CO_2 containing only methane as the impurity. The Emera and Sarma correlation was selected because it can be used for gas mixture containing up to 78 mol% of impurities such as H_2S, SO_X, N_2, O_2, and hydrocarbons (C_1 to C_4), and was claimed to be more accurate than all the other correlations by its authors.

The Sebastian correlation is expressed as a function of pseudocritical temperature, $\overline{T}_{cM}$, of the drive gas mixture in the following form:

$$\left(\frac{MMP_{impure}}{MMP_{pure}}\right) = 1.0 - 2.13 \times 10^{-2}\left(\overline{T}_{cM} - 304.2\right) + 2.51 \times 10^{-4}\left(\overline{T}_{cM} - 304.2\right)^2 - 2.35 \times 10^{-7}\left(\overline{T}_{cM} - 304.2\right)^3 \tag{17.13}$$

The pseudocritical temperature, $\overline{T}_{cM}$, is expressed as:

$$\overline{T}_{cM} = \sum_i x_i T_{ci} \tag{17.14}$$

In Eq. (17.13) and (17.14), MMP_{impure} = minimum miscibility pressure for impure CO_2, psi; MMP_{pure} = minimum miscibility pressure for pure CO_2, psi; $\overline{T}_{cM}$ = pseudocritical temperature for impure CO_2 injection gas mixture, °K; x_i = mole fraction of component, i, in the injection gas mixture; and T_{ci} = critical temperature of component, i, in the injection gas mixture, °K. Values of critical temperatures for various gases are given in Table 17.4. Note that the critical temperature for H_2S was reduced from the actual value by Sebastian et al.[20] to improve the accuracy of the correlation. The temperature range of the experimental data used to develop the Sebastian correlation is 100 to 160 °F, and the pressure range is 1000 to 3000 psi. The accuracy of the correlation is an average absolute deviation of 12% and a standard deviation of 17%, when

Table 17.4 Critical Temperature of Gases for the Sebastian et al.[20] Correlation

Component	Critical Temp., T_c (°K)
Hydrogen	33.2
Nitrogen	126.2
Oxygen	154.6
Carbon monoxide	132.9
Methane	190.6
Carbon dioxide	304.2
Ethane	305.2
Hydrogen sulfide	325.0*
Propane	369.8
i-Butane	408.1
n-Butane	425.2
i-Pentane	433.8
n-Pentane	469.6
n-Heptane	507.4

*Reduced from actual to improve correlation fit to data.

the MMPs predicted with the correlation are compared with experimentally measured MMPs as reported by Sebastian et al.[20]

The Yuan correlation, which is based on 119 MMPs calculated by the tie-line method, is given as:

$$\left(\frac{MMP_{impure}}{MMP_{pure}}\right) = 1 + m(P_{CO_2} - 100) \tag{17.15}$$

The term, m, is represented by:

$$m = a_1 + a_2 M_{C_{7+}} + a_3 P_{C_{2-6}} + \left(a_4 + a_5 M_{C_{7+}} + a_6 \frac{P_{C_{2-6}}}{M^2_{C_{7+}}}\right) T + \left(a_7 + a_8 M_{C_{7+}} + a_9 M^2_{C_{7+}} + a_{10} P_{C_{2-6}}\right) T^2 \tag{17.16}$$

The coefficients in Eq. (17.16) are: $a_1 = -6.5996\text{E} - 02$; $a_2 = -1.5246\ \text{E} - 04$; $a_3 = 1.3807\ \text{E} - 03$; $a_4 = 6.2384\ \text{E} - 04$; $a_5 = -6.7725\ \text{E} - 07$; $a_6 = -2.7344\ \text{E} - 02$; $a_7 = -2.6953\ \text{E} - 06$; $a_8 = 1.7279\ \text{E} - 08$; $a_9 = -3.1436\ \text{E} - 11$; $a_{10} = -1.9566\ \text{E} - 08$.

Note that Eq. (17.16) is strictly applicable for CO_2–rich gas injection mixtures containing up to 40 mol% of methane only. In Eqs. (17.15) and (17.16), P_{CO_2} = mole percent of CO_2 in the injection gas. All other terms are as defined previously in Eq. (17.12). The maximum average error reported for this correlation is 9% for the MMP ratio for an injection gas containing 10 to 20 mol% methane.

The Emera correlation, which can be used for CO_2–rich gas injection mixtures containing H_2S, SO_X, N_2, O_2, and hydrocarbons (C_1 to C_4), is expressed as:

$$\left(\frac{p_{r,impure}}{p_{r,pure}}\right) = 3.406 + 5.786 \times \left(\frac{1.8T_{cw} + 32}{1.8T_{c,CO_2} + 32}\right) - 23.0 \times \left(\frac{1.8T_{cw} + 32}{1.8T_{c,CO_2} + 32}\right)^2 + 20.48 \times \left(\frac{1.8T_{cw} + 32}{1.8T_{c,CO_2} + 32}\right)^3 - 5.7 \times \left(\frac{1.8T_{cw} + 32}{1.8T_{c,CO_2} + 32}\right)^4 \quad (17.17)$$

In Eq. (17.17),

$$p_{r,impure} = \frac{MMP_{impure}}{p_{cw}} \quad (17.18)$$

$$p_{cw} = \sum_{i=}^{n} w_i p_{ci} \quad (17.19)$$

$$p_{r,pure} = \frac{MMP_{pure}}{p_{c,CO_2}} \quad (17.20)$$

$$T_{cw} = \sum_{i=1}^{n} w_i T_{ci} MF_i \quad (17.21)$$

Values for the multiplying factors, MF_i, are given in Table 17.5. In Eq. (17.17) to (17.21), T_{cw} = weight averaged pseudocritical temperature adjusted with MF_i, °C; T_{c,CO_2} = pure CO_2 critical temperature, °C; MMP_{impure} = MMP for impure CO_2-Oil, MPa; MMP_{pure} = MMP for pure CO_2-Oil, MPa; w_i = weight fraction of component i, fraction; p_{ci} = critical pressure of component i, MPa; T_{ci} = critical temperature of component i, °C; and MF_i = multiplying factors. The Emera correlation is valid for impure CO_2 containing up to 78 mol% of impurities and up to 20 mol% of N_2. The temperature range of the data used to develop the correlation is 105.4 to

Table 17.5 Multiplication Factors for the Emera and Sarma[26] Correlation

Component	Multiplication Factor, *MF*
Sulfur dioxide	0.30
Hydrogen sulfide	0.59
Carbon dioxide	1.00
Ethane	1.10
Methane	1.60
Nitrogen	1.90
All other gases	1.00

234.0°F. The range of molecular weight of C_{5+} fraction in the oils used to develop and test the correlation is 166.2 to 267.5. The average error reported for the correlation is 4.7% with a standard deviation of 6.3%, when compared with experimental data.

Example 17.4 Calculation of MMP for Impure CO_2 Gas Injection Using the Sebastian, Yuan, and Emera Correlations

Problem

Calculate the MMP for impure CO_2 gas injection into reservoir oil with the following properties:

Mole percent of CO_2 in injection gas mixture	90
Mole percent of CH_4 in injection gas mixture	10
Mole percent of C_2 to C_6 in the oil	23.62
Molecular weight of C_{7+} fraction in the oil	240 lbm/lb-mole
Reservoir temperature	106°F

Solution

Step 1: Calculate MMP for pure CO_2 gas injection.
The Yuan correlation is used to calculate the MMP for pure CO_2 gas injection because it appears to be superior to other correlations for this purpose as shown in Example 17.3. Substituting the properties given for the oil into Eq. (17.12) for the Yuan correlation, $MMP_{pure} = 1890.32$ psia or 13.03 MPa.

Step 2: Calculate MMP for impure CO_2 gas injection using the Sebastian correlation.
Using Eq. (17.14) to calculate the mole average critical temperature for the injection gas mixture:

$$\begin{aligned}\overline{T}_{cM} &= \sum_i x_i T_{ci} \\ &= 0.9 \times 304.2 + 0.1 \times 190.6 = 292.84°\text{K}\end{aligned}$$

Values for T_{ci} were taken from Table 17.4.
Substituting the value for $\overline{T}_{cM}$ in Eq. (17.13) gives:

$$\begin{aligned}\left(\frac{MMP_{impure}}{MMP_{pure}}\right) &= 1.0 - 2.13 \times 10^{-2}(\overline{T}_{cM} - 304.2) + 2.51 \times 10^{-4}(\overline{T}_{cM} - 304.2)^2 \\ &= -2.35 \times 10^{-7}(\overline{T}_{cM} - 304.2)^3 \\ &= 1.0 - 2.13 \times 10^{-2}(292.84 - 304.2) + 2.51 \times 10^{-4}(292.84 - 304.2)^2 \\ &= -2.35 \times 10^{-7}(292.84 - 304.2)^3 \\ &= 1.2747\end{aligned}$$

Using the MMP for pure CO_2 gas injection calculated in Step 1, $MMP_{impure} = 2409.59$ psia.

Step 3: Calculate MMP for impure CO_2 gas injection using the Yuan correlation. Substituting the given properties of the oil into Eq. (17.16) gives:

$$\begin{aligned} m &= a_1 + a_2 M_{C_{7+}} + a_3 P_{C_{2-6}} + \left(a_4 + a_5 M_{C_{7+}} + a_6 \frac{P_{C_{2-6}}}{M^2_{C_{7+}}}\right)T \\ &\quad + \left(a_7 + a_8 M_{C_{7+}} + a_9 M^2_{C_{7+}} + a_{10} P_{C_{2-6}}\right)T^2 \\ &= -0.065996 - 0.00015246 \times 240 + 0.0013807 \times 23.62 + \\ &\quad \left(0.00062384 - 6.7725E - 07 \times 240 - 0.027344 \times \frac{23.62}{(240)^2}\right) \times 106 \\ &\quad + [-2.6953E - 06 + 1.7279E - 08 \times 240 - 3.1436E \\ &\quad - 11 \times (240)^2 - 1.9566E - 08 \times 23.62](106)^2 \\ &= -0.0315 \end{aligned}$$

For the Yuan correlation, Eq. (17.15) gives:

$$\begin{aligned} \left(\frac{MMP_{impure}}{MMP_{pure}}\right) &= 1 + m(P_{CO_2} - 100) \\ &= 1 - 0.0315(90 - 100) \\ &= 1.315 \end{aligned}$$

$MMP_{impure} = 2485.77$ psia.

Step 4: Calculate MMP for impure CO_2 gas injection using the Emera correlation. The critical properties for carbon dioxide and methane were obtained from Table 4.1. Thus, $p_{c,co_2} = 1070$ psia; $T_{c,co_2} = 87.76$°F; $p_{c,c_1} = 667$ psia; and $T_{c,c_1} = -116.66$°F. Using Eq. (17.19) to calculate weight average pseudocritical pressure for the impure CO_2 injection gas gives:

$$\begin{aligned} p_{cw} &= \sum_{i=}^{n} w_i p_{ci} \\ &= 0.9611 \times 7.3774 + 0.03892 \times 4.5988 \\ &= 7.2694 \text{ MPa} \end{aligned}$$

Also, using Eq. (17.21) and Table 17.5 to calculate the weight average pseudocritical temperature for the impure CO_2 injection gas yields:

$$\begin{aligned} T_{cw} &= \sum_{i=1}^{n} w_i T_{ci} MF_i \\ &= 0.9611 \times 30.98 \times 1.0 + 0.0389 \times -82.59 \times 1.6 \\ &= 24.6345°\text{C} \end{aligned}$$

Table 17.6 Comparison of Impure CO_2 MMPs Calculated in Example 17.4

Correlation Type	MMP_{impure}/MMP_{pure}	MMP_{impure} (psia)	Error (%)
Sebastain, et al.[20]	1.2747	2409.6	−4.2
Yuan, et al.[17]	1.3149	2485.6	−1.1
Emera and Sarma[26]	1.2342	2333.0	−7.2
Experimental	1.3300	2514.1	+0.0

From Eq. (17.20) and $MMP_{pure} = 13.03$ MPa from Step 1;

$$p_{r,pure} = MMP_{pure}/p_{c,CO_2} = 13.03/7.38 = 1.77$$

Substituting Eq. (17.17) with $T_{cw} = 24.6284$ °C and $T_{c,co_2} = 30.98$°C yields:

$$\begin{aligned}\left(\frac{p_{r,impure}}{p_{r,pure}}\right) &= 3.406 + 5.786 \times \left(\frac{1.8T_{cw} + 32}{1.8T_{c,CO_2} + 32}\right) - 23.0 \times \left(\frac{1.8T_{cw} + 32}{1.8T_{c,CO_2} + 32}\right)^2 \\ &\quad + 20.48 \times \left(\frac{1.8T_{cw} + 32}{1.8T_{c,CO_2} + 32}\right)^3 - 5.7 \times \left(\frac{1.8T_{cw} + 32}{1.8T_{c,CO_2} + 32}\right)^4 \\ &= 3.406 + 5.786 \times \left(\frac{1.8 \times 24.63 + 32}{1.8 \times 30.98 + 32}\right) - 23.0 \times \left(\frac{1.8 \times 24.63 + 32}{1.8 \times 30.98 + 32}\right)^2 \\ &\quad + 20.48 \times \left(\frac{1.8 \times 24.63 + 32}{1.8 \times 30.98 + 32}\right)^3 - 5.7 \times \left(\frac{1.8 \times 24.63 + 32}{1.8 \times 30.98 + 32}\right)^4 \\ &= 1.2524\end{aligned}$$

$\left(\frac{p_{r,impure}}{p_{r,pure}}\right) = 1.2524$; $p_{r,impure} = 2.2167$ and from Eq. (17.18),
$MMP_{impure} = p_{r,impure} \times p_{cw} = 2.2167 \times 7.2692 = 16.1139$ MPa or 2337.11 psia.
$MMP_{impure}/MMP_{pure} = 1.2342$.

The results from the three correlations are compared in Table 17.6 to actual experimental data reported for the CO_2 gas injection mixture by Sebastian et al.[20] Note that the Yuan correlation predicted the impure MMP that was closest to the experimental value. However, the Yuan correlation can be applied to impure CO_2 gas injection mixtures containing only methane as the impurity. The Sebastian and Emera correlations can be applied to CO_2 mixtures containing other gases, such as H_2S, SO_x, N_2, O_2, and hydrocarbons (C_1 to C_4). For this reason, either the Sebastian correlation or the Emera correlation is recommended for the calculation of MMP for impure CO_2 mixtures, especially if the mixture contains impurities other than methane.

17.5.1.2 Compositional Simulations

Compositional simulations are used to predict MMP or MME from the compositions of the oil and the injection gas. The three main methods for prediction of MMP or MME by compositional simulations are the tie-line method, the 1-D slimtube simulation, and the multiple mixing-cell method. Compositional simulations require the use of an equation of state (EOS) to represent the phase behavior of the oil and the injection gas. The ability of an EOS to adequately represent the phase behavior of a complex mixture of oil and injection gas depends on the characterization of the oil and "tuning" of the EOS to experimental data. Fluid characterization of the oil and tuning of the EOS to experimental data are crucial in the development of an EOS that can be used reliably in compositional simulations to predict MMPs or MMEs.[27,28] In Chapter 6, Section 11, a procedure for characterization of fluid systems and tuning of an EOS was presented. This procedure is recommended for characterizations of oil and gas systems and tuning of the EOS used in compositional simulations to predict MMPs or MMEs.

The Tie-Line Method. The tie-line method, sometimes called the analytical method, is used to calculate MMP by compositional simulation along the displacement path of the oil by the injected fluid. The tie-line method for calculating MMP is fast in comparison with other methods, and the computed MMP is not affected by either physical or numerical dispersion. The tie-line method is based on the model for one-dimensional flow of a multicomponent mixture without the influence of dispersion.[29,30] The analytical theory, as developed by Johns and Orr[31] and Wang and Orr[32] showed that the displacement path is controlled by $n_c - 1$ tie lines. These include the tie line for the oil, the tie line for the injection fluid, and $n_c - 3$ crossover tie lines. In the tie-line method, MMP is calculated as the minimum pressure at which one of the tie lines becomes a critical tie line. A critical tie line is defined as the tie line that has zero length and is tangent to the critical locus. If the oil tie line becomes the critical tie line, the displacement type is vaporizing gas drive. If the gas (for the injection fluid) tie line becomes the critical tie line, the displacement type is condensing gas drive. If the development of MCM is controlled by one of the crossover tie lines, then the displacement type is combined condensing-vaporizing gas drive. Methods for calculation of MMPs using the tie-line method have been developed by Johns and Orr,[31] Wang and Orr,[32] Jessen et al.,[33] and Yuan and Johns.[34]

1-D Slimtube Simulation. MMP can be determined by conducting compositional simulation of the process in a 1-D reservoir model. In 1-D slimtube simulation, oil recovery at 1.2 HCPV (hydrocarbon pore volume) of gas injection is plotted against displacement pressure. MMP is determined at the bend or break-over (also called the "knee") of the plot as shown in Figure 17.11. The MMP determined by 1-D compositional simulation is affected by numerical dispersion from truncation error. The magnitude of numerical dispersion is directly proportional to the number of grid blocks in the model. Thus, the effect of numerical dispersion on MMP can be eliminated by increasing the number of grid blocks in the 1-D simulation model. The dispersion-free MMPs calculated by the tie-line method are close to MMPs determined by the 1-D slimtube simulation for most oil-gas injection systems. MMPs based on the tie-line method are about 100 to 200 psi lower than MMPs from slimtube simulations.[28]

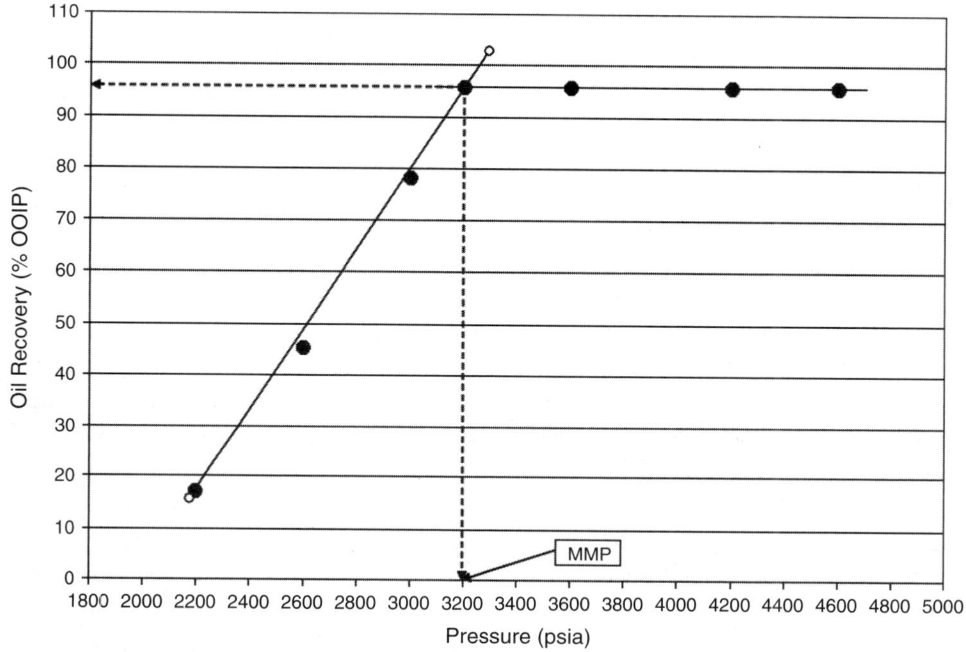

Figure 17.11 Determination of MMP from 1-D slimtube simulation or experiments.

Multiple Mixing-Cell Method. The multiple mixing-cell method can be used to estimate MMP. It is similar to other computational methods because it relies on an EOS to predict the equilibrated compositions of the mixtures from flash calculations. There have been many variations of the multiple mixing-cell method published in the literature.[35] However, the basic model of the method involves analytically mixing injection gas with reservoir oil, separating the equilibrated mixture into vapor and liquid phases, and further analytically mixing fresh oils with either the vapor phase or the liquid phase. To simulate vaporizing gas drive, the equilibrated vapor phase from the previous cell is contacted with fresh oil in the next cell. This process of multiple contacts is repeated for several steps until the resultant vapor phase is miscible with fresh oil. Similarly, for condensing gas drive, the equilibrated liquid phase from the previous cell is contacted with fresh injection gas. Miscibility is achieved when the condensed liquid phase is miscible with injection gas after several steps in the multi-cell mixing process. This procedure can be used to predict reliable MMPs for either purely vaporizing gas drive or condensing gas drive. However, Zick[12] and Stalkup[13] have demonstrated that most real oil displacements are combined condensing-vaporizing gas drive processes. Ahmadi and Johns[35] have published a multiple mixing-cell method that calculates MMPs for condensing-vaporizing gas drives using a modification of the tie-line method. They claim that the accuracy of MMP predicted with their method is comparable to MMP predicted by the tie-line method.

17.5.2 Experimental Methods

Two experimental methods are used in the measurement of MMP or MME. These are slimtube and the rising-bubble apparatus (RBA) methods. The slimtube method is by far more widely used than the RBA method. However, the RBA method can be used to obtain comparable MMP data in much less time than the slim tube method.

17.5.2.1 Slimtube Method

The slimtube equipment consists of a 40–60 feet long, 0.25 inch internal diameter, stainless steel tubing, packed with either fine sands or glass beads of a size in the range of 100–200 mesh. This tubing is spirally coiled to render gravity effects insignificant in the displacement process. Also, viscous fingering is considered negligible because of the small diameter of the tubing. Fluids stored in stainless cylinders are injected into the coiled tubing at a constant rate by an injection pump. Pressure across the coiled tubing is controlled by a backpressure regulator. A sight glass is used to monitor fluids flowing out of the coiled tube. The effluent fluids are measured with simple equipment that range from a wet-test meter to more elaborate equipment that may include a gas chromatograph. Most of the slimtube apparatus is enclosed in an air-bath maintained at any desired constant temperature. A schematic diagram of a typical slimtube apparatus is shown in Figure 17.12. Slimtube measurements are affected by the length and diameter of the tube, the type of packing material, particle size of the packing material, and the rate of displacement.[36–38]

In a typical slimtube experiment, the packed medium in the coiled tube is saturated with the oil that is the subject of the experiment. The apparatus is maintained at the desired temperature

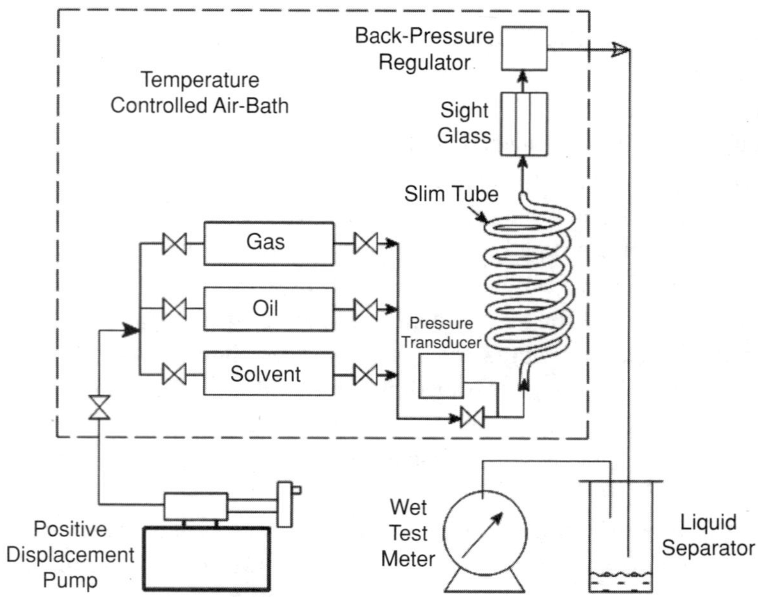

Figure 17.12 Schematic diagram of slimtube apparatus.

and pressure set for the experiment. The displacement fluid is then injected at constant rate until a predetermined pore volume of the fluid has been injected. This is typically set at injection of 1.2 pore volumes of displacement fluid. The oil recovery at this pore volume of injection and pressure is recorded. The experiment is repeated at different pressures with all the other variables (such as injection rate, temperature, etc.) kept constant. A plot of oil recovery versus displacement pressure similar to Figure 17.11 will yield the MMP at the "bend" or break-over of the plot. Slimtube experiments can be used to screen reservoirs for miscible processes, and should be included in the design and evaluation of reservoirs for gasfloods.

17.5.2.2 Rising-Bubble Apparatus (RBA) Method

Since it was introduced in the early 1980s, the rising-bubble apparatus (RBA) has gained acceptance in the petroleum industry as an alternative method for measuring MMP because it is quicker than the slimtube method. The essential parts of the RBA consist of a high pressure visual cell equipped with a camera to record observations during the experiment. Inside the visual cell, there is a flat glass tube into which the reservoir fluid is charged before the experiment is initiated. The glass tube is made flat so that bubbles rising in the oil can be seen more clearly. The internal dimensions of the glass tube are typically 0.04 by 0.2 inches and about 8 inches long. A schematic diagram of the RBA is shown in Figure 17.13. A detailed description and operational

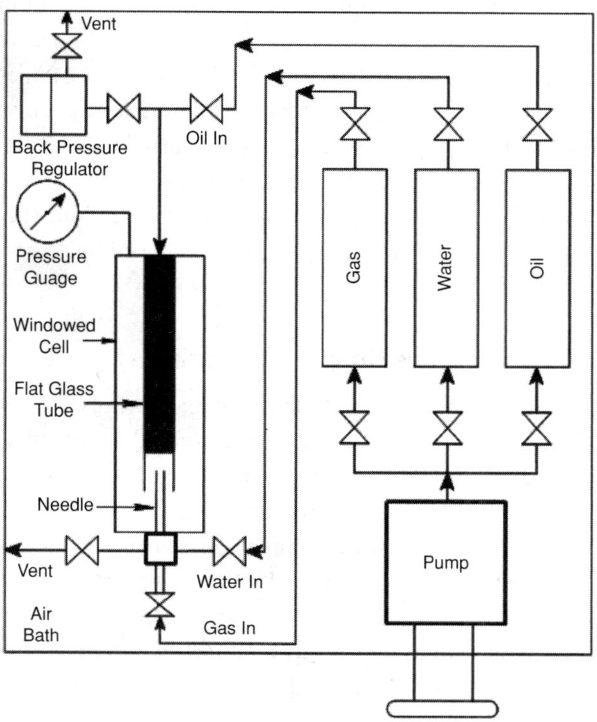

Figure 17.13 Schematic diagram of rising-bubble apparatus.

procedure for the RBA is provided by Novosad et al.,[39] Christiansen and Haines,[40] Elsharkawy et al.,[41] and Thomas et al.[42] Measurements of MMP with RBA are very dependent on monitoring and interpretation of the shapes of bubbles formed as the gas bubbles travel through the oil column in the glass tube. Data reported in the literature by Novosad et al.,[39] Elsharkawy et al.,[41] Thomas et al.,[42] Zhou and Orr,[43] and Bon et al.[44] show that MMPs measured with RBA are comparable in magnitude with those measured with slimtubes. The key advantage of the RBA is that it can be used to determine MMP at approximately one-tenth of the time required for a slimtube experiment.

17.6 Types of Miscible Gasflooding

Miscible displacement can be achieved by injecting different types of gases or gas mixtures, depending on the composition of fluids in the reservoir, reservoir conditions, and injection pressure. The types of gases used for miscible flooding include nitrogen, flue gas, light hydrocarbons (C_1 to C_6), and carbon dioxide. Flue gas is generated from power plants, internal combustion engines, etc. and contains a mixture of gases including a high percentage of nitrogen (>66%), carbon dioxide (9%–12%), and other gases, such as sulfur dioxide and oxygen in smaller amounts. The MMP required for flue gas injection is close to the requirements for nitrogen injection. For this reason, flue gas is grouped with nitrogen in the discussion of miscible gasflooding. Note however that flue gas is very corrosive because it contains carbon dioxide and sulfur dioxide. The manner by which miscibility is achieved depends on the type of injection gas. Generally, nitrogen and flue-gas achieve MCM by the vaporizing gas drive mechanism. Hydrocarbon gas and carbon dioxide can achieve MCM by vaporizing gas drive, condensing gas drive or combined condensing-vaporizing gas drive mechanisms, depending on the compositions of the reservoir fluid and injection gas, and injection pressure. In this section, the nature of miscible gasflooding with the different gases is presented. This is followed by a presentation on different types of gas injection strategies.

17.6.1 Nitrogen/Flue-gas Miscible Gasflooding

Nitrogen/flue-gas is sometimes selected as the injection gas for miscible flooding because it is relatively cheap and abundantly available in comparison with other potential injection gases such as hydrocarbon (HC) gas or carbon dioxide. Nitrogen/flue-gas can achieve miscibility at higher MMP than either HC gas or CO_2 at the same reservoir conditions. This condition limits the application of nitrogen/flue gas to reservoirs with relatively high pressures. The mechanism by which nitrogen/flue-gas achieves miscibility with oil is through vaporization of the light intermediate components present in the oil. This is essentially the mechanism of vaporizing gas drive. For this reason, nitrogen/flue-gas is suitable for light oil reservoirs as candidates for miscible flooding. The criteria for screening reservoirs for miscible flooding with nitrogen were given in Table 17.1. The correlations for predicting MMP with nitrogen as the injection gas were provided as Eqs. (17.4) to (17.7). Evaluation of nitrogen for use as an injection gas should include the cost of removing nitrogen from the produced gas before it can be sold. Note that flue-gas is corrosive and may not be suitable for application in many reservoirs without preventative procedures implemented to combat corrosion in processing equipment.

17.6.2 Hydrocarbon (HC) Miscible Gasflooding

In hydrocarbon (HC) miscible gasflooding, light hydrocarbons (C_1 to C_6) are injected into the reservoir at pressures sufficiently high to achieve miscibility with the reservoir oil. Many pressure maintenance projects with HC gas injection involve some aspects of miscible gasflooding, if reservoir pressures are maintained above MMP. HC gas injection can be classified broadly into three categories. The first category involves the injection of liquefied petroleum gas (LPG), such as propane at pressures sufficiently high to achieve FCM. The second category involves the injection of enriched gas (containing C_2 to C_6 at enriched levels) to achieve MCM by predominantly condensing gas drive mechanism. The third category entails the injection of lean gas (containing mostly methane) at high pressures to achieve MCM by vaporizing the light components (C_2 to C_6) in the oil using predominantly the vaporizing drive mechanism. In most real applications, the drive mechanisms for the second and third categories involve some degree of combination of condensing-vaporizing gas drives. The criteria for screening reservoirs for miscible flooding with HC gas were given in Table 17.1. The correlations for predicting MMP with light hydrocarbons as the injection gas were provided as Eqs. (17.1) to (17.3). In addition to enhanced oil recovery due to miscible conditions, HC gasflood improves oil recovery by swelling the oil, reducing oil viscosity, and physical displacement of the oil from the pore spaces. The application of HC gasflooding is limited by poor sweep efficiencies due to potential for viscous fingering, and the high economic value of hydrocarbon gases.

17.6.3 Carbon Dioxide Gasflooding

Carbon dioxide (CO_2) gas is a very effective agent for miscible displacement of oil. In most cases, CO_2 achieves miscibility with reservoir oil by extracting light to intermediate (C_1 to C_6) components of the oil in a MCM, vaporizing gas drive mechanism. Also, CO_2 improves oil recovery by swelling the oil, reducing its viscosity, and lowering interfacial tension between the oil and the CO_2-enriched fluid. The criteria for screening reservoirs for miscible flooding with CO_2 gas were given in Table 17.1. The correlations for predicting MMP with CO_2 as the injection gas were provided as Eqs. (17.8) to (17.12) for pure CO_2, and Eqs. (17.13) to (17.21) for CO_2 containing impurities. CO_2 can cause extensive corrosion and should be used in a field project after extensive corrosion preventative measures have been implemented. Carbon dioxide can be obtained from naturally occurring CO_2 reservoirs or separated from flue gases for injection. There is a persistent clamor around the world to reduce the amount of carbon dioxide in the atmosphere which is considered as a major contributor to climate change. The amount of carbon dioxide emitted into the atmosphere can be reduced by removing carbon dioxide from flue gas and injecting it into reservoirs to improve oil recovery. This a novel proposal which can contribute to solving the dangers posed to the entire world by climate change, and at the same time contribute to meeting the increasing need around the world for additional oil supply. Oil and gas reservoirs are good candidates for storage or sequestration of carbon dioxide because these reservoirs had geologic seals that originally trapped the hydrocarbons provided the seals had not been breached by field operations.[45,46] It is anticipated that large-scale operations in the future will combine the use of carbon dioxide for oil recovery and sequestration of large volumes of carbon dioxide in the same reservoir and its associated aquifers. A model that estimates the cumulative oil recovery from CO_2 flooding and CO_2 storage potential for a given reservoir has been provided by Wood et al.[47] This model can be used to screen many reservoirs to determine the best candidates for CO_2 flooding and storage.

17.6.4 Types of Miscible Gas Injection Strategies

Mobility control is a major concern in miscible gas injection processes because of the relatively low viscosity of the injected gas. The mobility ratio between injected gas and displaced oil is unfavorable in most applications. This causes viscous fingering which creates low volumetric sweep efficiency. To remedy this condition, most miscible gas injection processes, inject gas and water to improve volumetric sweep of the reservoir. If water and gas are injected in alternating cycles, the process is called water-alternating-gas (WAG) injection strategy. In some application, water and gas are injected simultaneously. This injection strategy is called simultaneous-water-and-gas (SWAG). SWAG injection strategy is considered to be a variant of WAG. WAG injection strategy is used in most miscible flood projects around the world. A WAG ratio of 1:1 (pore volume of water injected versus pore volume of gas injected) is used in most field applications. Note that application of WAG results in loss of injectivity of water and gas over time. Some miscible gasfloods are initiated with the strategy of continuous gas injection (CGI) which can be switched later to WAG or SWAG to achieve better mobility control. The volume of gas injected can be reduced as the flood progresses to reduce cost and improve mobility control. This injection strategy is described as tapering. Tapering is used in most carbon dioxide and hydrocarbon injection projects to reduce the cost of injecting these gases into reservoirs for hydrocarbon recovery.

17.7 Chemical Flooding Processes

Chemical flooding processes are displacement processes that improve oil recovery from reservoirs with the aid of chemicals that are generally called surfactants, polymers, and alkalis. These chemicals are combined in various proportions to create injection systems which are termed polymer/surfactant (micellar) process, and alkali/surfactant/polymer (ASP) process. Only polymers are injected for mobility control and improved oil recovery in polymer flooding. Chemical flooding processes improve oil recovery by lowering the interfacial tension (IFT) between oil and water, solubilization of the oil in the micelles, emulsification of the oil and water, alteration of the wettability of the rock, and enhancement of the mobility of the displacing fluids. For each application of chemical flooding, the chemicals must be formulated and tailored to the properties of the reservoir rock and its fluid system. This usually requires extensive testing of various combinations of the chemicals in laboratory coreflood tests followed with a pilot test in the field.[48] There are few field-wide chemical flooding projects in the world in comparison with either CO_2 or HC gas injections projects. The decline of chemical flooding can be attributed to the high cost of the chemicals and higher risk associated with achieving predicted oil recoveries with field-wide projects. The chemical flooding processes that have been applied in the field are surfactant/polymer floods, ASP floods, and polymer floods. In some fields, microbial flooding has been tested. These chemical flooding processes are discussed in more detail in the following sections.

17.7.1 Polymer/Surfactant Flooding

In a typical polymer/surfactant flood (also called micellar/polymer flood), a slug of micellar solution (consisting of water, surfactant, electrolytes, etc.) is injected into the reservoir. The size of the micellar slug can vary from 3% to 30% of the flood pattern pore volume (PV) depending on the flood design. The micellar slug is followed with a mobility buffer slug made up of polymer and water for mobility control. The entire micellar/polymer slug is then chased with drive water. The polymer solution sometimes contains a biocide to control microorganisms that can degrade the polymer and reduce the viscosity of the solution. The micellar slug is formulated such that a favorable mobility ratio exists between the slug and the oil bank that is formed by the flooding process. This sometimes requires the addition of small amounts of polymers in the composition of the micellar solution. The micellar slug is usually the most expensive solution used in micellar/polymer flooding. The surfactants in the micellar slug are degraded with the advancement of the flood due to adsorption on rock surfaces and trapping of some of the micellar solution as a residual phase. Adsorption of surfactants on rock surfaces can be reduced by preflushing the flood zone with fresh water. Preflushing with fresh water also reduces incompatibilities between the surfactant and formation brine. The application of micellar/polymer floods have declined because of the high cost of micellar solution and limited economic success of the process in field tests.[49]

17.7.2 Alkali/Surfactant/Polymer (ASP) Flooding

Alkali/surfactant/polymer (ASP) flooding is more widely used than micellar/polymer flooding in field applications. ASP formulation utilizes the favorable mechanisms of its key components (alkali, surfactant and polymer) to improve oil recovery.[50] The main functions of the alkaline component are to promote emulsification of the crude oil, reduce IFT, reduce adsorption of the surfactants, and regulate phase behavior of the mixtures. The main function of the surfactants is to reduce IFT between the oil and the injected slugs. The polymers are used for mobility control and improvement of sweep efficiencies. The favorable attributes of ASP formulations have been tested and proven in many laboratory and field tests.[51–53] ASP flooding has been applied in most cases to sandstone reservoirs but laboratory and pilot tests suggest that it can also be applied to carbonate reservoirs.[54,55]

17.7.3 Polymer Flooding

Polymer flooding is the most widely used chemical flooding process for both sandstone and carbonate reservoirs. Polymer flooding can be considered as a waterflooding process in which polymers have been added to the solution to improve mobility control. Thus, polymer flooding is able to achieve higher oil recovery than ordinary waterflooding due to improved volumetric sweep efficiency resulting from better mobility control. Factors affecting the application of polymer flooding are the cost of the polymers, the permeability of the reservoir, and the temperature of the reservoir. The cost of polymers limits the concentration and slug size of the polymers that can be used to achieve higher oil recovery. Low rock permeability can limit the use of more viscous polymers to improve mobility control, and can cause low injectivity. Polymers are not stable at reservoir temperatures higher than 175°F. Many field applications of polymer flooding have been reported as technical and economic success.[56–58]

17.7.4 Microbial Enhanced Oil Recovery (MEOR)

Microbial activities in the reservoir are believed to increase oil recovery by altering the wettability of the rock, reducing interfacial tension, reducing oil viscosity, generation of gases such as carbon dioxide, production of surfactants, etc. These mechanisms, which combine to improve oil recovery by the MEOR process, are complex and not yet fully understood. Consequently, after many years of laboratory and pilot tests, MEOR processes have not been applied on a large scale in the field in comparison with other EOR processes. There are no active MEOR projects in the United States in 2008,[5] and small-scale MEOR projects are in operation in China,[5] and the North Sea.[6] The screening criteria for microbial processes are presented in Table 17.2. Generally, MEOR processes are applicable to reservoirs with injected and connate water salinities <150,000 ppm; oil viscosity in the 5–50 cp range; reservoir temperature <176°F; rock porosity ≤20%; rock permeability ≥50 md.; and reservoir depth <7700 ft. Most MEOR tests have been on sandstone reservoirs, although there have been tests in carbonate reservors.[6] The bacteria source in MEOR processes can be indigenous or exogenous. Indigenous bacteria are already present in the reservoir while exogenous bacteria are cultivated and injected into the reservoir. Microbial processes in the reservoir can be aerobic or anaerobic. For aerobic processes, oxygen must be injected to maintain the required concentration of bacteria. Injection of oxygen into the wellbore may cause severe corrosion problems. Anaerobic processes require injection of large amounts of nutrients in the form of sugars to sustain the growth of bacteria. This will increase the cost of the MEOR process due to the additional costs of purchasing, processing, and injecting the nutrients. Large-scale field utilization of MEOR processes have not occurred in the industry because of limited knowledge of the process, and lack of models to predict potential oil recovery from application of the technology.

17.8 Thermal Processes

Thermal EOR processes are defined to include all processes that supply heat energy to the rocks and fluids contained in a reservoir thereby enhancing the ability of oil (including other fluids) to flow by primarily reducing its viscosity. The oil caused to flow by the supply of thermal energy is produced through nearby wells. The two main thermal processes that are presented in this section are steamflooding and in-situ combustion. Steamflooding has achieved substantial commercial successes around the world and is ranked as one of the top EOR processes. In-situ combustion process, by comparison, has attained limited commercial success, and is actively applied in few reservoirs around the world. The main obstacle to the application of in-situ combustion technology is ability to manage and control the advancement and progression of the combustion process. Steamflooding has been used on highly viscous, shallow reservoirs at depths of 200 to 3600 feet while in-situ combustion can be applied to less viscous, deeper reservoirs at depths reaching 8300 feet (Table 17.3). In principle, in-situ combustion is the only viable thermal process for deeper reservoirs where steam can not be used due to thermodynamic limitations of steam (critical pressure of steam = 3200.1 psia, critical temperature of steam = 705.1°F from Table 4.1).

The number of thermal projects has declined in the United States from 1990 to 2004 but appears to be increasing from 2006 to 2008 (Figure 17.1). In the United States, oil production from thermal processes has declined from a peak production of 480,000 BOPD in 1986 to 293,000 BOPD in 2008 (Figure 17.2). Commercial thermal projects are very sensitive to oil prices. This is illustrated in Figure 17.3, where the number of active thermal projects in the United States is plotted against oil prices. The peak in the number of thermal projects was reached in 1986 in response to the rise in oil prices from 1978 to 1982. It is expected that the number of thermal projects in the United States (also around the world) will continue to increase after 2008 in response to the rise in oil prices from 2002 to 2008.

Thermal EOR processes have significant impact on oil recovery in other regions of the world. Canada has substantial oil reserves deposited as heavy oil or in tar sands which are good candidates for thermal oil recovery processes. China has heavy oil deposits of more than 1.9 billion tons of oil reserves.[59] First commercial development in China started in 1982. Between 1993 and 2004, oil production from thermal processes in China reached the level of 10 to 13 $\times$ 10^6 tons per year.[59] Venezuela has many thermal projects in the Tia Juana field.[5] The Duri steamflood project in Indonesia is the largest thermal project in the world averaging about 190,000 BOPD in 2008.[5]

In this section, more emphasis is placed on steamflooding, since it is by a wide margin the most commercially applied thermal recovery process. The presentation includes various methods of steamflooding, description of steamflood models, and management of steamflood projects. This is followed with a descriptive presentation of in-situ combustion.

17.8.1 Steamflooding Methods

Steamflooding is an established EOR technique that has been applied successfully on many heavy oil reservoirs around the world. The process started in early 1960 with cyclic steam injection in the Tia Juana Field in Venezuela, the Schoonebeek Field in the Netherlands, and the Yorba Linda Field in California.[60,61] The three main methods of steamflooding are cyclic steam stimulation (CSS), steam drive (SD), and steam assisted gravity drainage (SAGD). SAGD can be considered

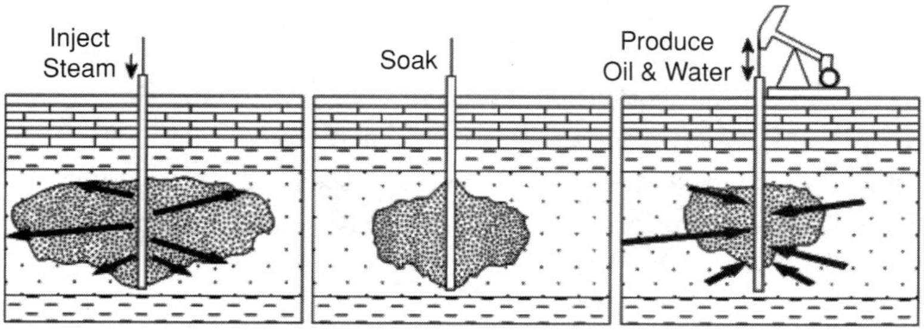

Figure 17.14 Cyclic steam stimulation (CSS) process.

as a modification of SD for heavy oil reservoirs including tar sands. This section concludes with a discussion of water-alternating-steam process (WASP) as a steam injection strategy for mobility control.

17.8.1.1 Cyclic Steam Stimulation (CSS)

Cyclic steam stimulation (CSS) is a simple and cheap method of applying thermal recovery process on a reservoir. It involves injecting steam into a well for several weeks, shutting the well in as long as necessary to allow the steam to heat the oil in the areas around the well, and putting the well back on production to recover the heated oil (Figure 17.14). This process is repeated when the production from the well declines to a low level. The cycle is repeated many times until the ratio of oil produced to steam injected termed oil-steam ratio (OSR) drops to a level that is considered uneconomic. CSS was the first steamflooding technique used in heavy oil reservoirs. It can still be used in a limited pilot test on a well to investigate whether steam injection can enhance oil recovery in a particular field before undertaking more expensive pattern pilot tests that will require more wells and more equipment.

17.8.1.2 Steam Drive (SD)

In steam drives, steam is injected continuously at injectors with the aim of driving oil towards producers. Typically in most projects, steam injection is organized in patterns. For instance, in a normal five-spot pattern (Figure 16.3), the producer is located in the middle of the pattern and surrounded by four steam injectors. Various pattern configurations of injectors and producers have been used in many fields. The selection of the pattern that is best suited for a particular reservoir is an essential part of the design and pilot testing process, before field implementation of the project. Steam drive was initially described as essentially a frontal displacement process similar to waterflooding, whereby steam displaces oil towards a producer. Later, experimental and field evidence showed that the mechanism of oil recovery by steam is dominated by gravity drainage caused by steam rising towards the top of the formation due to gravity override (Figure 17.15). The accumulation of steam at the top of the formation creates a steam zone which continues to expand with more steam injection. This steam zone can propagate and breakthrough at the producers to cause

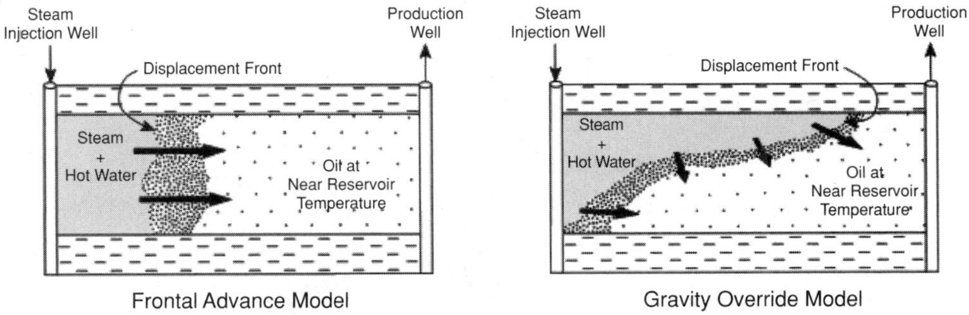

Figure 17.15 Steam drive models.

inefficient utilization of the energy from the steam due to cycling. The main challenges of managing a steam drive project are monitoring the size and growth of the steam zone, monitoring producers for early steam breakthrough, adjusting steam inject rates, and measuring the trend of oil-steam ratio for each pattern.

17.8.1.3 Steam-Assisted Gravity Drainage (SAGD)

The concept of steam-assisted gravity drainage (SAGD) was tested and shown to be successful by the Alberta Oil Sands Technology and Research Authority (AOSTRA) at the Underground Test Facility in Fort McMurray, Alberta, Canada between 1987 and 2004.[62,63] In 2008, more than ten commercial SAGD projects were in operation in the Athabasca area of Canada, and SAGD has quickly become the preferred thermal recovery process for oil sands in the area.[63] The SAGD process is applicable to other heavy oil and oil sands deposits in other countries such as China, Venezuela, and Russia. The SAGD process consists of two horizontal wells about 15 feet apart located close to the bottom of the formation (Figure 17.16). Steam is injected into the top

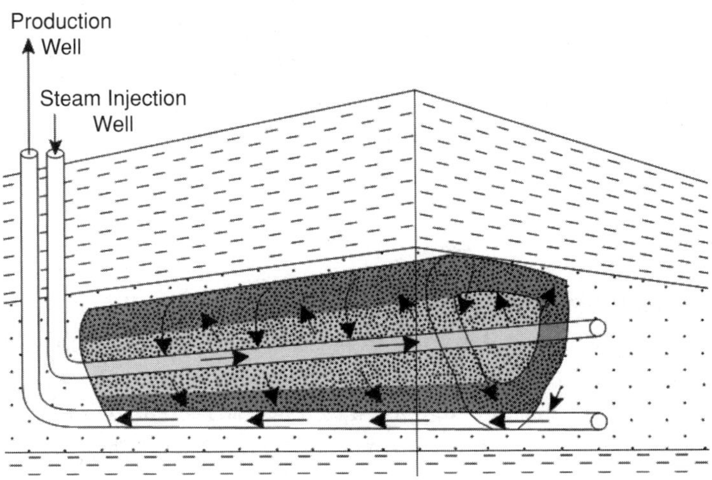

Figure 17.16 SAGD process diagram.

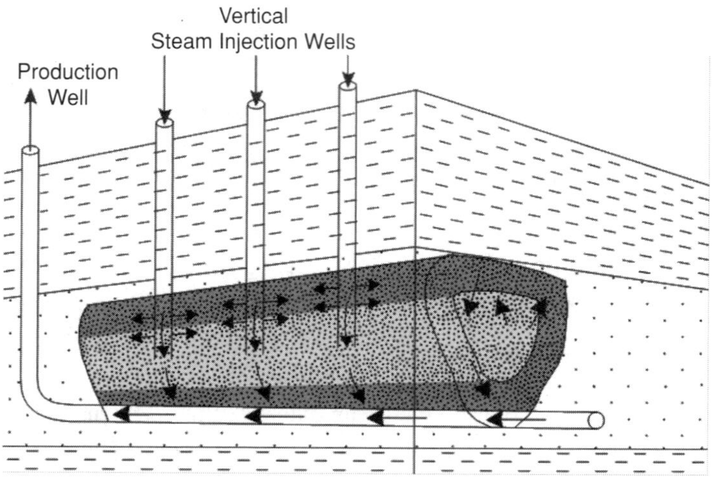

Figure 17.17 XSAGD process diagram.

horizontal well, while the horizontal well below it functions as the producer. The steam creates an expanding steam chamber around the injector as more steam is injected. Within the steam chamber and at its boundaries, as the viscosity of the oil is reduced, its mobility increases causing it to drain under gravity towards the production well. The SAGD process should be applied to reservoirs with formation thickness greater than 50 feet, good vertical permeability, and absence of thief zones.[63] In some SAGD projects, a hydrocarbon solvent (such as methane) is added to the steam to improve oil recovery. This variation of the SAGD process has been described as the expanding solvent-SAGD (ES-SAGD).[64] In another variation, vertical wells are used for steam injection, while the producer remains a horizontal well as shown in Figure 17.17. This configuration of injectors and producer is termed Cross SAGD (XSAGD).[65] Oil recoveries with SAGD in pilot tests can reach 65% of the oil-in-place in the project area in comparison with CSS that achieves oil recovery in the range of 25% to 30%.[66] The heavy oil/bitumen in-place in Canada is estimated to be 1.6 trillion barrels.[66] With this huge resource base, SAGD is positioned to be a very important technology that is bound to be used to meet future energy needs in the world.

17.8.1.4 Water-Alternating-Steam Process (WASP)

Water-alternating-gas (WAG) was described in Section 17.6.4 as an injection strategy used in gasfloods to reduce viscous fingering, improve vertical sweep, and thereby increase oil recovery. This concept is also used in steamflooding by injecting water alternating with steam to reduce fingering of steam, and improve its vertical sweep efficiency. This strategy is called water-alternating-steam process or WASP. The main benefit of WASP over continuous steam injection is reduction or elimination of premature steam breakthrough at producing wells. Early steam breakthrough at producers wastes heat energy, and reduces productivity of the wells.

17.8.2 Steamflood Models

Steamflood models are used to predict the performance of the process in terms of oil production rates, oil-steam ratios, cumulative oil recovery, steam zone size, etc. Consequently, steamflood models are essential tools used in the screening, pilot testing, field implementation, and management of steamflood projects. The two main mathematical steamflood models widely used in the industry are analytical models and numerical simulation models. Analytical steamflood models are based on heat flow and energy balance equations, assuming uniform reservoir properties and shape of the steam zone. Numerical steamflood simulation models are based on simultaneous solutions of material balance and energy balance equations, including possibly an equation of state for component phase equilibrium calculations. Numerical simulation models can incorporate most of the known geologic features (such as structures, faults, permeability barriers, etc.) and spatial variation of rock and fluid properties for the reservoir. Numerical models are more rigorous than analytical models in representing the physical processes of steamflooding, and achieving results that include the impact of geological heterogeneities on the performance of steamfloods. Consequently, numerical simulation models are recommended as the preferred method for prediction of steamflood performance.

17.8.2.1 Analytical Steamflood Models

With the introduction of steamflooding technology in 1960, analytical models were developed to predict the performance of steamfloods. The complexity of published analytical models has grown over many years with the spread of steamflooding. The first steamflood analytical model was published by Marx and Langenheim in 1959.[67] Analytical or semi-analytical models published by other authors in chronological order are as follows: Boberg and Lantz, 1966;[68] Mandl and Volek, 1967;[69] Farouq Ali, 1970;[70] Myhill and Stegemeir, 1978;[71] Gomaa, 1980;[72] Rhee and Doscher, 1980;[73] Jones, 1981;[74] Aydelotte and Pope, 1983;[75] van Lookeren, 1983;[76] Vogel, 1984;[77] Neuman, 1985;[78] Miller and Leung, 1985;[79] and Chandra and Mamora, 2007.[80] This list is not meant to include all analytical steamflood models ever published in the petroleum literature, but to serve to illustrate the historical development work performed on this application. The theoretical development and applications of any of these analytical models are not presented in this book. This is because the utility of these models has been surpassed by numerical steamflood simulation models as pointed out in the previous section.

17.8.2.2 Numerical Steamflood Simulation Models

Early works on the development of numerical simulators for steamfloods were published by Shutler,[81,82] Abdalla and Coats,[83] Weinstein et al.,[84] Coats et al.,[85] Coats,[86,87] and Grabowski et al.[88] The application of numerical simulators on steamfloods was boosted with the publication of the Fourth SPE Comparative Solution Project on steamflood simulators in 1987.[89] Because of the complexities of the physical processes involved in steamflooding, numerical simulation of steamfloods require smaller grid sizes and large computation times, in comparison with black oil simulators. However, recent advances in computer hardware and improvements in formulation of solution algorithms have made it possible to construct very heterogeneous reservoir models for simulation of steam floods, and achieve considerable reduction in computation times. For instance, parallel computing and dynamic gridding are some of the advances in application of numerical steamflood simulators.[90,91] The development of material and energy balance equations representing numerical simulators for steamfloods are not presented in this book. However, the interested reader is encouraged to consult published literature that is referenced in this book and other books that have been published on numerical simulations. In addition, there are several competent thermal numerical simulators available commercially in the petroleum industry which can be used to model any of the described steamflood processes.

17.8.3 Management of Steamflood Projects

Management of a steamflood project offers unique challenges because the heat energy supplied to the reservoir must be utilized efficiently and effectively to maximize oil recovery. The management of steamflood projects can be subdivided into the management of three interrelated areas: surface equipment management, well management, and reservoir management. These areas are presented briefly in the next sections.

17.8.3.1 Surface Equipment Management

The key objective of surface equipment management in steamfloods is to minimize heat losses and optimize the process of generating steam. To minimize heat losses, most projects use insulated tubing and coupling joints coated with Teflon. Many projects have also incorporated cogeneration of steam and power using cogeneration plants as a cost effective means of improving the economics of steamflooding.

17.8.3.2 Well Management

The well system in a typical steamflood project consists of steam injectors, oil producers, and observation wells. Management of steam injectors usually requires profile control so that the steam can reach the targeted intervals. This is achieved by using surfactants or high temperature polymers to reduce mobility of steam and re-direct steam to unswept intervals, thereby improving sweep efficiency. Another practice is the use of multizone completion techniques to isolate and inject steam into separate target zones. Steam injectors can be remediated by cementing existing perforated intervals, and recompleting the well in other intervals that has not been flooded with steam. Similar remedial practice is conducted at the producers by either shutting-in updip producers as steam breaks through their perforated intervals or recompleting them to produce from lower intervals. Producers should be monitored regularly to measure fluid production rates and detect steam breakthrough. One of the important activities in managing steamfloods is the use of observation wells to monitor the temperature and growth of the steam zone. This can be accomplished by running frequent temperature surveys or equipping the observation wells with fiber-optic distributed temperature sensing (DTS) technology.[92,93]

17.8.3.3 Steamflood Reservoir Management Strategy

Steamflooded reservoirs require higher levels of monitoring and surveillance because of the close spacing of injectors and producers, the potential for early steam breakthrough from gravity override, and the need to ensure that the energy supplied by the steam is effectively used in heating the reservoir to produce oil. Heat management in steamfloods have evolved over several years.[94,95] The recommended practice is to manage the steamflooding process with a numerical thermal simulator with the goal of optimal utilization of the heat energy from steam. The simulator should be used to adjust steam injection rates, monitor the steam zone, adjust or modify the flood patterns, and predict future flood performance.

17.8.4 In-Situ Combustion (ISC)/High Pressure Air Injection (HPAI)

In-situ combustion (ISC)/High pressure air injection (HPAI) are EOR processes in which compressed air is injected into heavy and light oil reservoirs under conditions, such that the oxygen in the injected air reacts with a small fraction of the crude oil at high temperatures to create a combustion front. The distinction between ISC and HPAI processes is blurred, although both processes are definitely related. Turta et al.[96] state that an ISC process is associated with high peak temperatures (350°C to 600°C), and may require ignition to initiate it, whereas an HPAI process may not have high peak temperature and does not require ignition. The ISC process appears to be more applicable to heavy oil reservoirs, whereas HPAI is better suited for light oil reservoirs. In this book, both processes are considered to be air injection processes (AIP). A schematic diagram of an air injection displacement process is shown in Figure 17.18. The displacement path consists of an air zone behind the combustion/reaction zone. The combustion/

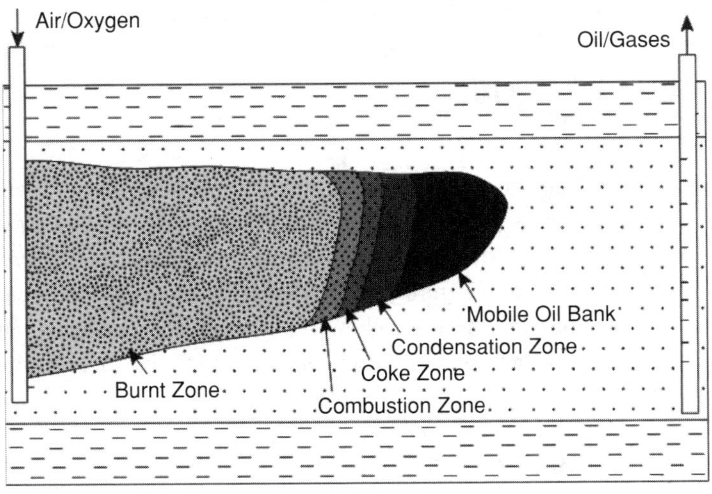

Figure 17.18 Air injection process diagram.

reaction zone is preceded by a coke zone, a condensation zone, and mobilized oil bank. High oil recoveries have been reported from laboratory and field tests of AIP.[97–99] However, there are few air injection projects around the world. In the United States, the remaining few air injection projects are based on relatively small reservoirs located in North and South Dakota.[5] Steamflooding have surpassed air injection as the preferred EOR process even though both processes can satisfy the screening criteria for the same light or heavy oil reservoirs. This is the result of air injection being considered as more risky than steamflooding because the propagation of its combustion/reaction front is more difficult to predict and control.

17.9 Implementation of EOR Projects

Taking an EOR project from initial screening to field implementation is a long, labor-intensive process.[100] It is one of the most elaborate processes in project management and execution that requires the involvement of all the major areas of petroleum engineering. EOR projects require input from reservoir engineers, geologists, petrophysicists, geophysicists, drilling engineers, production engineers, facilities engineers, field personnel, economists, etc. for proper evaluation and implementation of the process. Assembling these professionals in a functional team is more efficient and effective for execution of the project.

The procedure of screening, testing, and implementing an EOR process is divided into stages. The stages are organized sequentially as follows:

1. Process screening and selection.
2. Quick economic evaluation of selected processes.
3. Geologic and reservoir modeling of selected processes.
4. Expanded economic evaluation of processes.
5. Pilot testing, if necessary.
6. Update/upgrade simulation models with pilot test data/results.
7. Detailed economic evaluation of selected process.
8. Field implementation of process.
9. Project management.

These stages should not be followed rigidly as enumerated. The exchange of data and knowledge between the stages as the evaluation progresses should be incorporated as an integral part of the process. The stages for evaluation and implementation of an EOR process have been organized into a flowchart shown as Figure 17.19. Additional suggestions and guidance on work required at each stage are provided in the following sections. These suggestions and guidance are not exhaustive by any measure. Potential users are advised to use them as starting points for more detailed investigation at each stage of the process.

17.9.1 Process Screening and Selection

EOR processes tend to be specific to geologic characteristics of the reservoir and the type of fluid present in the reservoir. These observations are evident by reviewing the screening criteria

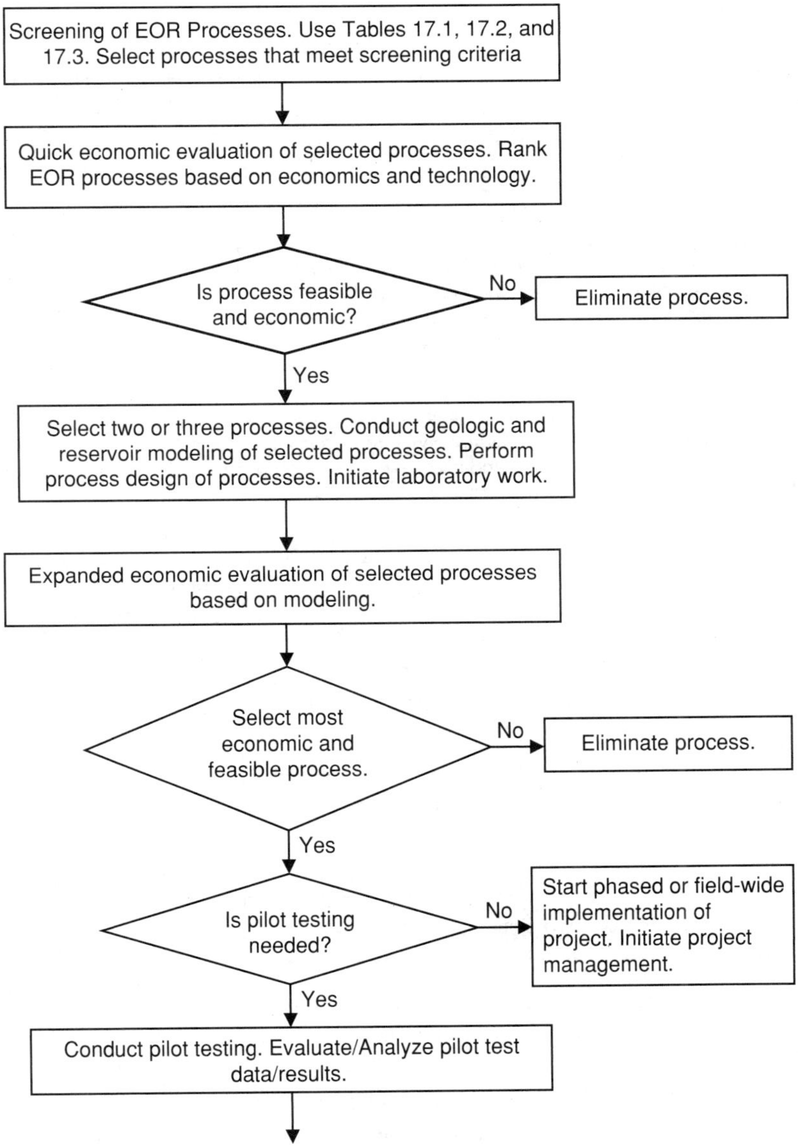

Figure 17.19 Flow chart for EOR process implementation. (*continues*)

presented in Tables 17.1 to 17.3 for various EOR processes. For instance, generally thermal processes apply to heavy oils in shallow reservoirs, while miscible gas injection processes apply to lighter oils in deeper reservoirs. To use the screening criteria presented in Tables 17.1 to 17.3, key fluid and reservoir properties for the candidate reservoir should be averaged and compared

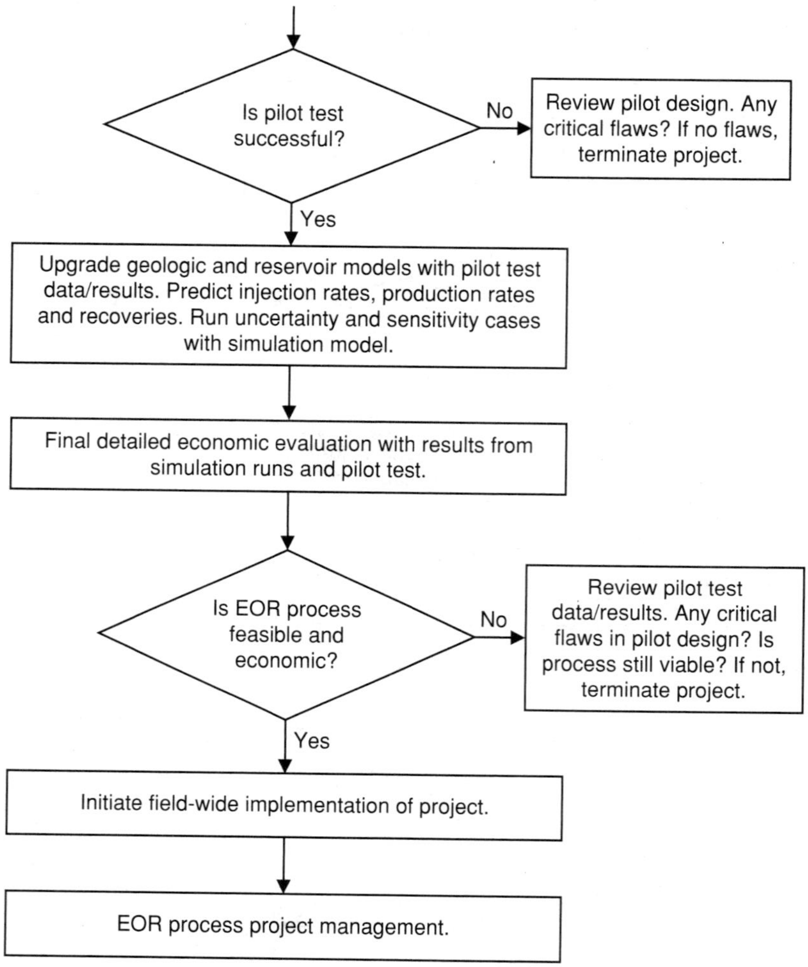

Figure 17.19 Flow chart for EOR process implementation.

with the ranges listed in the tables. If the average rock and fluid properties of the candidate reservoir meet the screening criteria, the process is selected for further evaluation.

As part of the screening process, it is important to assemble all data (including geologic, rock properties, fluid properties, pressure transients, production, injection, static and flowing pressure, production logs, etc.) on the candidate reservoir. These data should be reviewed and evaluated to determine the volume of oil to be targeted with the EOR process, and the approximate distribution of the targeted oil in the reservoir. Other issues to address at this stage are whether they are known geologic features existing in the reservoir, such as high permeability layers or permeability barriers that may cause the potential EOR process to fail. Some questions that can help with the selection process are as follows:

1. Do the rock and fluid properties of the reservoir match the screening criteria for the selected process as listed in Tables 17.1 to 17.3?
2. Are there known geologic features such as faults, high permeability layers, permeability barriers, thief zones, etc. in the reservoir that might be detrimental to the selected process?
3. Is the target oil volume in the reservoir sufficient for this process?
4. Can the selected process recover the targeted oil?
5. What additional data are needed to improve the evaluation of the process?

Addressing these and many other questions will lay the groundwork necessary for evaluation of the EOR processes before proceeding to the next stage of the evaluation process.

17.9.2 Quick Economic Evaluation of Selected Processes

In this stage, the selected EOR processes are subjected to quick economic evaluation. The objective of this exercise is to check which processes are economically viable. Some of the data needed for the economic evaluation are:

1. Sources and costs of the gas, chemicals, and steam to be used for injection.
2. Estimates of expected hydrocarbon recoveries.
3. Market prices for the hydrocarbons recovered from the process.
4. Cost of processing produced fluids.
5. Cost of facilities for the EOR process.

Most of the required data can be estimated from analogue fields, if available. Alternatively, industry rule-of-thumb estimates can be used. The goal at this stage is to quickly eliminate processes that are clearly uneconomic and do not merit further evaluations.

17.9.3 Geologic and Reservoir Modeling of Selected Processes

Geologic and reservoir modeling of selected processes is a very important stage of the evaluation process. Geologic models of the reservoir are constructed from all available geologic data including structural, stratigraphic, seismic, rock properties, fluid data, etc. The geologic models should be reviewed in recognition of historical production data. The geological models should be characterized and populated with reservoir properties. Representative geologic models are then converted to reservoir models for simulation. The reservoir models should be subjected to a history-matching process, if the reservoir has historical production data. The history-matching process should serve as a test of the geologic models, and could be used to rank the geologic models in terms of how closely they match historical reservoir performance.

The most representative geologic model is then used to evaluate the selected EOR processes. This requires incorporating laboratory data that are specific to the EOR process in constructing the simulation model. Using the simulation models, each selected EOR processes is evaluated in terms of injection rates, production rates, displacement efficiency, cumulative recovery, etc. depending on the process design. The output from this stage is the used in an expanded economic evaluation in the next stage.

17.9.4 Expanded Economic Evaluation of Selected Processes

In the expanded economic evaluation, the outputs from the design of the simulated EOR processes are used. These outputs are primarily injection rates and production rates from process design. Development costs, such as the costs to install flood patterns, drill new wells (producers, injectors, and observation wells), install specialized facilities for fluid processing, etc. are included. At this stage, the cost of injectants and selling prices of produced hydrocarbons are upgraded to reflect market rates. All the relevant costs that could significantly impact the economic viability of the processes must be included in the analyses at this stage. At the conclusion of the stage, one process is chosen for pilot testing or field implementation, if pilot testing is not needed. Pilot testing may not be necessary, if the EOR process have been applied successfully in an analogue field or if it was decided that conducting a pilot test will not reduce the risks and uncertainties associated with the project. In that case, a phased field implementation of the project should be used to mitigate risks.

17.9.5 Pilot Testing

Pilot testing is very important in the evaluation of an EOR process. Pilot tests are typically designed to reduce the risks and uncertainties associated with an EOR process. Some of these risks and uncertainties may have been identified during simulation of the EOR process. Risks and uncertainties associated with EOR processes are typically elevated because of limited data on the impact of well spacing, pattern configuration, injection and production rates, injection strategy, vertical and areal sweep efficiencies, gravity override, viscous fingering, thief zones, mobility control, etc. Pilot testing can be used to assess the impact of these factors on the EOR process before embarking on a field-wide implementation. For a pilot test to be successful, the objectives of the test must be clearly defined.[101] Some key objectives of a pilot test that should be considered are:

1. Acquisition of geologic and performance data to improve the process simulation model developed in the earlier stage.
2. Evaluation of the impact of geology on volumetric sweep efficiency in the target area.
3. Assessment of the efficiency of the selected process in increasing hydrocarbon recovery.
4. Validation of predicted injection and production rates.
5. Testing processing facilities and identifying potential problems with handling injection fluids or processing produced fluids.

The pilot test should be designed with the simulation model in terms of selection of well type, well spacing, pattern configuration, well completion, injection strategy, injection and production rates, injection volumes, etc. The data acquired from the pilot should be compared with predictions from the simulation model. This comparison should serve as guide on the reliability of the simulation model, and the level of confidence that could be placed on its predictions.

Pilot tests can be classified according to well configurations.[101] The four common configurations are:

1. Non-producing pilot.
2. Small-scale unconfined pilot.
3. Small-scale confined pilot.
4. Multi-pattern producing pilot.

A non-producing pilot consists of one injector, and is used primarily for injectivity tests. If observation wells are added to the configuration, the system can be used for vertical conformance tests. Small-scale unconfined/confined and multi-pattern pilots have producing wells (and in many cases observation wells) in their configurations. These types of pilot tests yield valuable data on process efficiency, reservoir conformance, performance data, and overall direct assessment of the EOR process.[101]

Acquisition of accurate and reliable data during pilot testing is important for assessment of the success of the EOR process and future utilization of the data in field implementation. Frequent and regular monitoring of all wells involved in the pilot testing should include measurements of injection and production rates, bottomhole pressure surveys, running of injection and production profiles, pressure transient tests, temperature surveys, time-lapse logging, coring, use of chemical or radioactive tracers, etc. The nature and frequency of the well tests depend on the type of EOR process. For instance, installation of permanent downhole gauges for continuous monitoring of reservoir pressure may be desirable for miscible injection processes. The surveillance and monitoring program for the pilot test should be as extensive as necessary to acquire the data needed to meet the stated objectives for the test.

17.9.6 Upgrade Geologic and Reservoir Models with Pilot Test Data/Results

The geologic and reservoir models used in designing the pilot test should be upgraded with the data and results from the pilot test. In practice, this activity should be conducted almost simultaneously with the pilot test. As soon as data are collected from the pilot test, comparisons should be conducted between predictions from the simulation model and the results from the test. Severe divergence between model predictions and pilot test results should be investigated thoroughly. This may be an indication that the reservoir model is not a good representation of the actual reservoir, and should be reviewed. Actually, there are many other issues that could cause the difference between predictions and actual results other than the reservoir model. These issues must be resolved before using the model for future predictive work. Future predictive work should include extensive uncertainty analysis on critical parameters identified by the pilot test. This should be accompanied with sensitivity analysis on injectant costs and product prices.

17.9.7 Final Detailed Economic Evaluation

The final detailed economic evaluation incorporates the data from the pilot test and predicted rates from the upgraded simulation model. Further detailed costs of equipment and facilities, projected costs of injectants, future product prices, production costs, etc. should be included in the final economic analysis. This is the last economic evaluation of the entire project before proceeding to field implementation.

17.9.8 Field-Wide Project Implementation

Commercial field implementation of an EOR process is capital intensive. It demands careful planning and budgeting so that the project is installed in a timely manner. It is advantageous to use the team that installed and managed the pilot test for the field implementation. Also, field personnel trained in the operation of the pilot should be retained for the field implementation phase.

17.9.9 EOR Process Project Management

Proper management of EOR projects requires an intensive program of monitoring and surveillance of reservoir performance. The monitoring and surveillance practices that were proven during the pilot testing phase should be adopted for the field implementation of the EOR process. Well surveillance and monitoring should be conducted regularly by a team dedicated to this activity. The data gathered by the team should be used in frequent well and project reviews.

The simulation model history-matched with the results from the pilot tests should be used in the management of the project. Data gathered from field surveillance of the project should be used to update the model as necessary. The simulation model should be used in a predictive mode to adjust injection and production rates, pattern configuration, well locations, injection strategy, and other activities which will optimize the management of the EOR process.

Nomenclature

$M_{C_{5+}}$	molecular weight of C_{5+} fraction, lbm/lb-mole
$M_{C_{7+}}$	molecular weight of C_{7+} fraction, lbm/lb-mole
T	reservoir temperature, °F
T_{ci}	critical temperature of component i in the gas mixture
γ_o	specific gravity of stock tank oil

Abbreviations

AIP	air injection process
ASP	alkali-surfactant-polymer
CGI	continuous gas injection

CSS	cyclic steam stimulation
CV	combined condensing and vaporizing
DTS	distributed temperature sensing
EOS	equation of state
EOR	enhanced oil recovery
ES-SAGD	expanding solvent steam assisted gravity drainage
FCM	first-contact miscibility
HC	hydrocarbon
HCPV	hydrocarbon pore volume
HPAI	high pressure air injection
IFT	interfacial tension
ISC	in-situ combustion
MCM	multiple-contact miscibility
MEOR	microbial enhance oil recovery
MME	minimum miscibility enrichment
MMP	minimum miscibility pressure
OOIP	original oil in place
OSR	oil-steam ratio
RBA	rising-bubble apparatus
SAGD	steam assisted gravity drainage
SD	steam drive
SWAG	simultaneous-water-and-gas
WAG	water-alternating-gas
WASP	water-alternating-steam process
XSAGD	cross steam assisted gravity drainage

References

1. Moritis, G.: "California Steam EOR Produces Less; Other EOR Continues," *Oil and Gas Journal* (April 15, 2002) 100, 15.
2. Moritis, G.: "SPECIAL REPORT: More US EOR Projects Start but EOR Production Continues Decline," *Oil and Gas Journal* (April 21, 2008) 106, 15.
3. Energy Information Administration—Annual Energy Review: "Crude Oil Domestic First Purchase Prices, 1949–2007," http://www.eia.doe.gov/emeu/aer/petro.html.
4. Energy Information Administration—Annual Energy Review: "Crude Oil Production and Crude Oil Well Productivity, 1954–2007," http://www.eia.doe.gov/emeu/aer/petro.html.
5. Koottungal, L: "2008 Worldwide EOR Survey," *Oil and Gas Journal* (April 21, 2008) 47–59.
6. Awan, A.R., Teigland, R., and Kleppe, J.: "A Survey of North Sea Enhanced–Oil–Recovery Projects Initiated During the Years 1975 to 2005," *SPEREE* (June 2008) 497–512.

7. Lazar, I: "MEOR Field Trials Carried Out Over the World During the Last 35 Years," International Conference on MEOR, Norman, Oklahoma, May 27–June 1, 1990.

8. Bryant, R.S., and Lindsey, R.P.: "World-Wide Applications of Microbial Technology for Improving Oil Recovery," paper SPE/DOE 35356 presented at the SPE/DOE Improved Oil Recovery Symposium, Tulsa, Oklahoma, April 21–24, 1998.

9. Hutchinson, C.A., and Braun, P.H.: "Phase Relations of Miscible Displacement in Oil Recovery," *AICHEJ* (March 1961) 7, 64–72.

10. Benham, A.L., Dowden, W.E., and Kunzman, W.J.: "Miscible Fluid Displacement—Prediction of Miscibility," *Trans.,* AIME (1960) 219, 229–237.

11. Stalkup, F.I.: "Status of Miscible Displacement," *JPT* (April 1983) 815–826.

12. Zick, A.A.: "A Combined Condensing/Vaporizing Mechanism in the Displacement of Oil by Enriched Gases," paper SPE 15493 presented at the SPE 61st Annual Technical Conference and Exhibition, New Orleans, Louisiana, October 5–8, 1986.

13. Stalkup, F.I.: "Displacement Behavior of the Condensing/Vaporizing Gas Drive Process," paper SPE 16715 presented at the SPE 62nd Annual Technical Conference and Exhibition, Dallas, Texas, September 27–30, 1987.

14. Glasø, Ø: "Generalized Minimum Miscibility Pressure Correlation," *SPEJ* (December 1985) 927–934.

15. Firoozabadi, A., and Aziz, K.: "Analysis and Correlation of Nitrogen and Lean-Gas Miscibility Pressure," *SPERE* (November 1986) 575–582.

16. Emera, M.K., and Sarma, H.K: "Use of Genetic Algorithm to Estimate CO_2-Oil Minimum Miscibility Pressure—A Key Parameter in Design of CO_2 Miscible Flood," *Journal of Petroleum Science & Engineering* (2005) 46, 37–52.

17. Yuan, H., Johns, R.T., Egwuenu, A.M., and Dindoruk, B.: "Improved MMP Correlations for CO_2 Floods Using Analytical Gasflooding Theory," *SPEREE* (October 2005) 418–425.

18. Johnson, J.P., and Pollin, J.S.: "Measurement and Correlation of CO_2 Miscibility Pressures," paper SPE 9790 presented at the SPE/DOE Enhanced Oil Recovery Symposium, Tulsa, Oklahoma, April 5–8, 1981.

19. Alston, R.B., Kokolis, G.P., and James, C.F.: "CO_2 Minimum Miscibility Pressure: A Correlation for Impure CO_2 Streams and Live Oil Systems," *SPEJ* (April 1985) 268–274.

20. Sebastian, H.M., Wenger, R.S., and Renner, T.A.: "Correlation of Minimum Miscibility Pressure for Impure CO_2 Streams," *JPT* (November 1985) 2076–2082.

21. Kovarik, F.S.: "A Minimum Miscibility Pressure Study Using Impure CO_2 and West Texas Oil Systems: Data Base, Correlations, and Compositional Simulation," paper SPE 14689 presented at the SPE Production Technology Symposium, Lubbock, Texas, November 11–12, 1981.

22. Orr, F.M., Jr., and Silva, M.K.: "Effect of Oil Composition on Minimum Miscibility Pressure—Part 2: Correlation," *SPERE* (November 1987) 479–491.

23. Enick, R.M., Holder, G.D., and Morsi, B.I.: "A Thermodynamic Correlation for the Minimum Miscibility Pressure in CO_2 Flooding of Petroleum Reservoir," *SPERE* (February 1988) 81–92.

24. Eakin, B.E, and Mitch, F.J.: "Measurement and Correlation of Miscibility Pressures of Reservoir Oils," paper SPE 18065 presented at the SPE Annual Technical Conference and Exhibition, Houston, Texas, October 2–5, 1988.

25. Dong, M.: Potential of Greenhouse Gas Storage and Utilization Through Enhanced Oil Recovery—Task 3: Minimum Miscibility Pressure Studies. Final Report (SRC Publication No. P-110-468-C-99), September 1999.

26. Emera, M. K., and Sarma, H.K.: "A Reliable Correlation to Predict the Change in Minimum Miscibility Pressure When CO_2 Is Diluted with Other Gases," *SPEREE* (August 2006) 366–373.

27. Egwuenu, A.M, and Johns, R.T.: "Improved Fluid Characterization for Miscible Gasfloods," *SPEERE* (August 2008) 655–665.

28. Yuan, H., Chopra, A., Marwah, V., and Stalkup, F.: "Fluid Characterization for Miscible Floods," paper SPE 114913 presented at the SPE Annual Technical Conference and Exhibition, Denver, Colorado, September 21–24, 2008.

29. Monroe, W.W., Silva, M.K., Larsen, L.L., and Orr, F.M., Jr.: "Composition Paths in Four-Component Systems: Effect of Dissolved Methane on 1D CO_2 Flood Performance," *SPERE* (August 1990) 423–432.

30. Johns, R.T., Dindoruk, B., and Orr, F.M., Jr.: "Analytical Theory of Combined Condensing/Vaporizing Gas Drive," *SPE Advanced Technical Series* (1993) 1, No. 2.

31. Johns, R.T., and Orr, F.M., Jr.: "Miscible Gas Displacement of Multicomponent Oils," *SPEJ* (March 1996) 39–50.

32. Wang, Yun, and Orr, F.M., Jr.: "Calculation of Minimum Miscibility Pressure," paper SPE 39683 presented at the SPE/DOE Improved Oil Recovery Symposium, Tulsa, Oklahoma, April 19–22, 1998.

33. Jessen, K., Michelsen, M.L., and Stenby, E.H.: "Effective Algorithm for Calculation of Minimum Miscibility Pressure," paper SPE 50632 presented at the SPE European Petroleum Conference, Hague, The Netherlands, October 20–22, 1998.

34. Yuan, H., and Johns, R.T.: "Simplified Method for Calculation of Minimum Miscibility Pressure or Enrichment," *SPEJ* (December 2005) 416–425.

35. Ahmadi, K., and Johns, R.T.: "Multiple Mixing-Cell Method for MMP Calculations," paper SPE 116823 presented at the SPE Annual Technical Conference and Exhibition, Denver, Colorado, September 21–24, 2008.

36. Orr, F.M., Jr., Silva, M.K., Lien, C.L., and Pelletier, M.T.: "Laboratory Experiments to Evaluate Field Prospects for CO_2 Flooding," *JPT* (April 1982) 888–898.

37. Randall, T.E., and Bennion, D.B.: "Laboratory Factors Influencing Slim Tube Results," *The Journal of Canadian Petroleum Technology* (July–August, 1989) 28, 4, 60–70.

38. Flock, D.L., and Nouar, A.: "Parametric Analysis on the Determination of the Minimum Miscibility Pressure in Slim Tube Displacements," *The Journal of Canadian Petroleum Technology* (September–October, 1984) 80–88.

39. Novosad, Z., Sibbald, L.R., and Costain, T.G.: "Design of Miscible Solvents for a Rich Gas Drive—Comparison of Slim Tube and Rising Bubble Tests," *The Journal of Canadian Petroleum Technology* (January–February, 1990) 29, 1, 37–42.

40. Christiansen, R.L., and Haines, H.K.: "Rapid Measurement of Minimum Miscibility Pressure with the Rising-Bubble Apparatus," *SPERE* (November 1987) 523–527.

41. Elsharkawy, A.M., Poettmann, F.H., and Christiansen, R.L.: "Measuring Minimum Miscibility Pressure: Slim-Tube or Rising-Bubble Method?," paper SPE/DOE 24114 presented at the SPE/DOE Eight Symposium on Enhanced Oil Recovery, Tulsa, Oklahoma, April 22–24, 1992.

42. Thomas, F.B., Zhou, X.L., Bennion, D.B., and Bennion, D.W.: "A Comparative Study of RBA, P-x, Multicontact, and Slim Tube Results," *The Journal of Canadian Petroleum Technology* (February, 1994) 33, 2, 17–26.

43. Zhou, D., and Orr, F.M., Jr.: "Analysis of Rising–Bubble Experiments to Determine Minimum Miscibility Pressures," *SPEJ* (March 1998) 19–25.

44. Bon, J., Sarma, H.K., and Theophilos, A.M.: "An Investigation of Minimum Miscibility Pressure for CO_2-Rich Injection Gases with Pentanes-Plus Fraction," paper SPE 97536 presented at the SPE International Improved Oil Recovery Conference in Asia Pacific, Kuala Lumpur, Malaysia, December 5–6, 2005.

45. Orr, F.M., Jr.: "Storage of Carbon Dioxide in Geologic Formations," SPE 88842, Distinguished Author Series (September 2004) 90–97.

46. Qi, R., Laforce, T.C., and Blunt, M.J.: "Design of Carbon Dioxide Storage in Oilfields," paper SPE 115663 presented at the Annual Technical Conference and Exhibition, Denver, Colorado, September 21–24, 2008.

47. Wood, D.J., Lake, L.W., Johns, R.T., and Nunez, V.: "A Screening Model for CO_2 Flooding and Storage in Gulf Coast Reservoirs Based on Dimensionless Groups," *SPEERE* (June 2008) 513–520.

48. Flaaten, A.K., Nguyen, Q.P., Pope, G.A., and Zhang, J.: "A Systematic Laboratory Approach to Low-Cost, High Performance Chemical Flooding," paper SPE 113469 presented at the 2008 SPE/DOE Improved Oil Recovery Symposium, Tulsa, Oklahoma, April 19–23, 2008.

49. Manrique, E.J., Muci, V.E., and Gurfinkel, M.E.: "EOR Field Experiences in Carbonate Reservoirs in the United States," *SPEREE* (December 2007) 667–686.

50. Arihara, N., Yoneyama, T., Akita, Y., and XiangGuo, L.: "Oil Recovery Mechanisms of Alkali-Surfactant-Polymer Flooding," paper SPE 54330 presented at the SPE Asia Pacific Oil and Gas Conference and Exhibition, Jakarta, Indonesia, April 20–22, 1999.

51. Vargo, J., Turner, J., Vergnani, B., Pitts, M.J., Wyatt, K., Surkalo, H., and Patterson, D.: "Alkaline-Surfactant-Polymer Flooding of the Cambridge Minnelusa Field," *SPEREE* (December 2000) 552–558.

52. Hernández, C., Chacón, L.J., Anselmi, L., Baldonedo, A., Qi, J., Dowling, P.C., and Pitts, M.J.: "ASP System Design for an Offshore Application in La Salina Field, Lake Maracaibo," *SPEREE* (June 2003) 147–156.

53. Jiechang, C., Xu, D., and Bai, W.: "Commercial ASP Flood Test in Daqing Oil Field," paper SPE 117824 presented at the Abu Dhabi International Petroleum Exhibition and Conference, Abu Dhabi, United Arab Emirates, November 3–6, 2008.

54. Zhang, D.L., Liu, S., Yan, W., Puerto, M., Hiraski, G.J., and Miller, C.A.: "Favorable Attributes of Alkali-Surfactant-Polymer Flooding," paper SPE 99744 presented at the SPE/DOE Symposium on Improved Oil Recovery, Tulsa, Oklahoma, April 22–26, 2006.

55. Zubari, H.K., and Babu, V.C.: "Single Well Tests to Determine the Efficiency of Alkaline–Surfactant Injection in a Highly Oil–Wet Limestone Reservoir," paper SPE 81464 presented at the SPE Middle East Oil Show and Exhibition, Bahrain, April 5–8, 2003.

56. Taber, J.J., Martin, F.D., and Seright, R.S.: "EOR Screening Criteria Revisited—Part 1: Introduction to Screening Criteria and Enhanced Recovery Field Projects," *SPERE* (August 1997) 189–198.

57. Wang, D., Jiechang, C., Junzheng, W., and Gang, W.: "Experiences Learned After Production More Than 300 Million Barrels of Oil by Polymer Flooding in Daqing Oil Field," paper SPE 77693 presented at the SPE Annual Technical Conference and Exhibition, San Antonio, Texas, September 29–October 2, 2002.

58. Wang, D., Seright, R.S., Shao, Z., and Wang, J.: "Key Aspects of Project Design for Polymer Flooding at the Daqing Oil Field," *SPEREE* (December 2008) 1117–1124.

59. Shouliang, Z., Yitang, Z., Shuhong, W., Shangqi, L., Xiuluan, L., and Songlin, L.: "Status of Heavy-Oil Development in China," Paper SPE 97844 presented at the 2005 SPE International Thermal Operations and Heavy Oil Symposium, Calgary, Alberta, November 1–3, 2005.

60. Matthews, C.S.: "Steamflooding," *JPT* (March 1983) 465–471.

61. Blevins, T.R.: "Steamflooding in the U.S.: A Status Report," *JPT* (May 1990) 548–554.

62. Nasr, T.N., and Ayodele, O.R.: "Thermal Techniques for the Recovery of Heavy Oil and Bitumen," paper SPE 97488 presented at the SPE International Improved Oil Recovery Conference in Asia Pacific, Kuala Lumpur, Malaysia, December 5–6, 2005.

63. Jimenez, J.: "The Field Performance of SAGD Projects in Canada," paper SPE 12860 presented at the International Petroleum Technology Conference, Kuala Lumpur, Malaysia, December 3–5, 2008.

64. Das, S.: "Improving the Performance of SAGD," paper SPE 97921 presented at the SPE International Thermal Operations and Heavy Oil Symposium, Calgary, Alberta, Canada, November 1–3, 2005.

65. Stalder, J.L.: "Thermal Efficiency and Acceleration Benefits of Cross SAGD (XSAGD)," paper SPE 117244 presented at the SPE International Thermal Operations and Heavy Oil Symposium, Calgary, Alberta, Canada, October 20–23, 2008.

66. Gupta, S.: "Unlocking the Planet's Heavy Oil and Bitumen Resources—A Look at SAGD," SPE 108799, Distinguished Lecture Series, 2005.

67. Marx, J.W., and Langenheim, R.H.: "Reservoir Heating by Hot Fluid Injection," *Trans.* AIME (1959) 216, 312–315.

68. Boberg, T.C., and Lantz, R.B.: "Calculation of the Production Rate of a Thermally Stimulated Well," *JPT* (December 1966) 1613–1623.

69. Mandl, G., and Volek, C.W.: "Heat and Mass Transport in Steam-Drive Processes," paper SPE 1896 presented at the 42nd SPE Annual Fall Meeting, Houston, Texas, October 1–4, 1967.

70. Farouq Ali, S.M.: "Graphical Determination of Oil Recovery in a Five-Spot Steamflood," paper SPE 2900 presented at the SPE Rocky Mountain Regional Meeting, Casper, Wyoming, June 8–9, 1970.

71. Myhill, N.A., and Stegemeier, G.L.: "Steam-Drive Correlation and Prediction," *JPT* (February 1978) 173–182.

72. Gomaa, E.E.: "Correlations for Predicting Oil Recovery by Steamflood," *JPT* (February 1980) 325–332.

73. Rhee, S.W., and Doscher, T.M.: "A Method for Predicting Oil Recovery by Steamflooding Including the Effects of Distillation and Gravity Override," *SPEJ* (August 1980) 249–266.

74. Jones, J: "Steam Drive Model for Hand-Held Programmable Calculators," *JPT* (September 1981) 1583–1598.

75. Aydelotte, S.R., and Pope, G.A.: "A Simplified Predictive Model for Steamdrive Performance," *JPT* (May 1983) 991–1002.

76. van Lookeren, J.: "Calculation Methods for Linear and Radial Steam Flow in Oil Reservoirs," *SPEJ* (June 1983) 427–439.

77. Vogel, J.V.: "Simplified Heat Calculations for Steamfloods," *JPT* (July 1984) 1127–1136.

78. Neuman, C.H.: "A Gravity Override Model of Steamdrive," *JPT* (January 1985) 163–169.

79. Miller, M.A., and Leung, W.K.: "A Simple Gravity Override Model of Steamdrive," paper SPE 14241 presented at the 60th SPE Annual Technical Conference and Exhibition, Las Vegas, Nevada, September 22–25, 1985.

80. Chandra, S., and Mamora, D.D.: "Improved Steamflood Analytical Model," *SPEREE* (December 2007) 638–643.

81. Shutler, N.D.: "Numerical Three-Phase Simulation of the Linear Steamflood Process," *SPEJ* (June 1969) 232–246.

82. Shutler, N.D.: "Numerical Three-Phase Model of the Two-Dimensional Steamflood Process," *SPEJ* (December 1970) 405–417.

83. Abdalla, A., and Coats, K.H.: "A Three-Phase, Experimental and Numerical Simulation Study of the Steamflood Process," paper SPE 3600 presented at the SPE-AIME 46th Annual Fall Meeting, New Orleans, Louisiana, October 3–6, 1971.

84. Weinstein, H.G., Wheeler, J.A., and Woods, E.G.: "Numerical Model for Steam Stimulation," paper SPE 4759 presented at the SPE–AIME Third Improved Oil Recovery Symposium, Tulsa, Oklahoma, April 22–24, 1974.

85. Coats, K.H., George, W.D., Chu, Chieh, and Marcum, B.E.: "Three-Dimensional Simulation of Steamflooding," *SPEJ* (December 1974) 573–592.

86. Coats, K.H.: "Simulation of Steamflooding with Distillation and Solution Gas," *SPEJ* (October 1976) 235–247.

87. Coats, K.H.: "A Highly Implicit Steamflood Model," *SPEJ* (October 1978) 369–383.

88. Grabowski, J.W., Vinsome, P.K., Lin, R.C., Behie, A., and Rubin, B.: "A Fully Implicit General Purpose Finite–Difference Thermal Model for In Situ Combustion and Steam," paper SPE 8396 presented at the 54th Annual Fall Technical Conference and Exhibition, Las Vegas, Nevada, September 23–26, 1979.

89. Aziz, K., Ramesh, A.B., and Woo, P.T.: "Fourth SPE Comparative Solution Project: Comparison of Steam Injection Performance," *JPT* (December 1987) 1576–1584.

90. Kumar, M., Satik, C., and Hoang, V.: "New Developments in Steamflood Modeling," paper SPE 97719 presented at the 2005 SPE International Thermal Operations and Heavy Oil Symposium, Calgary, Alberta, Canada, November 1–3, 2005.

91. Shi, X., Chang, Y., Muller, M., Obi, E., and Lim, K.: "A General Unstructured Grid, Parallel, Thermal Simulator and its Application for Large Scale Thermal Models," paper SPE 119172 presented at the 2009 SPE Reservoir Simulation Symposium, The Woodlands, Texas, February 2–4, 2009.

92. Johnson, D.O., Sugianto, R., Mock, P.H., and Jones, C.H.: "Identification of Steam-Breakthrough Intervals with DTS Technology," *SPEPF* (February 2004) 41–48.

93. Nath, D.K., Sugianto, R., and Finley, D.: "Fiber-Optic Distributed-Temperature-Sensing Technology Used for Reservoir Monitoring in an Indonesian Steamflood," *SPEDC* (June 2007) 149–156.

94. Hanzlik, E.J., and Mims, D.S.: "Forty Years of Steam Injection in California—The Evolution of Heat Management," paper SPE 84848 presented at the SPE International Improved Oil Recovery Conference in Asia Pacific, Kuala Lumpur, Malaysia, October 20–21, 2003.

95. Ziegler, V.M., Crookston, R.B., Sanford, S.J., and Merrell, J.M.: "Recommended Practices for Heat Management of Steamflood Projects," paper SPE 25808 presented at the International Thermal Operations Symposium, Bakersfield, California, February 8–10, 1993.

96. Turta, AT., and Singhai, A.K.: "Reservoir Engineering Aspects of Light-Oil Recovery by Air Injection," *SPEREE* (August 2001) 336–344.

97. Fassihi, M.R., Yannimaras, D.V., and Kumar, V.K.: "Estimation of Recovery Factor in Light-Oil Air-Injection Projects," *SPERE* (August 1997) 173–178.

98. Kumar, V.K., Gutierrez, D., Moore, R.G., and Mehta, S.A.: "Case History and Appraisal of the West Buffalo Red River Unit High-Pressure Air Injection Project," paper SPE 107715 presented at the 2007 SPE Hydrocarbon Economics and Evaluation Symposium, Dallas, Texas, April 1–3, 2007.

99. Gutierrez, D., Taylor, A.R., Kumar, V.K., Ursenbach, M.G., Moore, R.G., and Mehta, S.A.: "Recovery Factors in High-Pressure Air Injection Projects Revisited," *SPEREE* (December 2008) 1097–1106.

100. Hite, J.R., Avasthi, S.M., and Bondor, P.L.: "Planning EOR Projects," paper SPE 92006 presented at the 2004 SPE International Petroleum Conference, Puebla, Mexico, November 8–9, 2004.

101. Teletzke, G.F., Wattenbarger, R.C., and Wilkinson, J.R.: "Enhanced Oil Recovery Pilot Testing Best Practices," paper SPE 118055 presented at the 2008 Abu Dhabi International Petroleum Exhibition and Conference, Abu Dhabi, UAE, November 3–6, 2008.

General Reading

- Jensen, T.B., Harpole, K.J., and Østhus, A.: "EOR Screening for Ekofisk," paper SPE 65124 presented at the SPE European Petroleum Conference, Paris, France, October 24–25, 2000.
- Hammershaimb, E.C., and Kuuskraa, V.A.: "Recovery Efficiency of Enhanced Oil Recovery Methods: A Review of Significant Field Tests," paper SPE 12114 presented at the 58th Annual Technical Conference and Exhibition, San Francisco, California, October 5–8, 1983.
- Rathmell, J.J., Stalkup, F.I., and Hassinger, R.C.: "A Laboratory Investigation of Multiple Displacement by Carbon Dioxide," paper SPE 3483 presented at the SPE 46th Annual Fall Meeting, New Orleans, Louisiana, October 3–6, 1971.
- Holm, L.W., and Josendal, V.A.: "Mechanism of Oil Displacement by Carbon Dioxide," *JPT* (December 1974) 1427–1436.
- Yellig, W.F., and Metcalfe, R.S.: "Determination and Prediction of CO_2 Minimum Miscibility Pressure, *JPT* (January 1980) 160–168.
- Holm, L.W., and Josendal, V.A.: "Effect of Oil Composition on Miscible-Type Displacement by Carbon Dioxide," paper SPE 8814 presented at the SPE-DOE Enhanced Oil Recovery Symposium, Tulsa, Oklahoma, April 20–23, 1980.
- Stalkup, F.I.: "Carbon Dioxide Miscible Flooding: Past, Present, and Outlook for the Future," *JPT* (August 1978) 1102–1112.

- Stalkup, F.I.: "Status of Miscible Displacement," *JPT* (April 1983) 815–826.
- Stalkup, F.: "Predicting the Effect of Continued Gas Enrichment Above the MME on Oil Recovery in Enriched Hydrocarbon Gasfloods," paper SPE 48949 presented at the SPE Annual Technical Conference and Exhibition, New Orleans, Louisiana, September 27–30, 1998.
- Dindoruk, B., Orr, F.M., Jr., and Johns, R.T.: "Theory of Multicontact Miscible Displacement with Nitrogen," *SPEJ* (September 1997) 268–279.
- Johns, R.T., Yuan, H., and Dindoruk, B.: "Quantification of Displacement Mechanisms in Multicomponent Gasfloods," *SPEJ* (September 2004) 314–321.
- Wu, R.S, and Batycky, J.P.: "Evaluation of Miscibility from Slim Tube Tests," *The Journal of Canadian Petroleum Technology* (November–December 1990) 29, 6, 63–70.
- Taber, J.J., Martin, F.D., and Seright, R.S.: "EOR Screening Criteria Revisited—Part 2: Applications and Impact of Oil Prices," *SPERE* (August 1997) 199–205.
- Akin, S.: "Mathematical Modeling of Steam–Assisted Gravity Drainage," *SPEREE* (October 2005) 372–376.
- Gates, I.D., Kenny, J., Hernandez-Hdez, I.L., and Bunio, G.L.: "Steam–Injection Strategy and Energetics of Steam-Assisted Gravity Drainage," *SPEREE* (February 2007) 19–34.
- Collins, P.M.: "Geomechanical Effects on the SAGD Process," *SPEREE* (August 2007) 367–375.
- Chen, Q., Gerritsen, M.G., and Kovscek, A.R.: "Effects of Reservoir Heterogeneities on the Steam-Assisted Gravity-Drainage Process," *SPEREE* (October 2008) 921–932.
- Neuman, C.H.: "A Mathematical Model of the Steam Dive Process—Derivation," paper SPE 5495.
- Jones, J.: "Why Cyclic Steam Predictive Models Get No Respect," *SPERE* (February 1992) 67–74.
- Abou-Kassem, J.H., and Aziz, K.: "Sensitivity of Steamflood Model Results to Grid and Timestep Sizes," *SPEJ* (February 1984) 65–74.
- Satik, C., Kumar, M., Defrancisco, S., Hoang, V., and Basham, M.: "Important Modeling Parameters for Predicting Steamflood Performance," *SPEREE* (October 2007) 508–513.
- Hong, K.C.: "Recent Advances in Steamflood Technology," paper SPE 54078 presented at the 1999 International Thermal Operations and Heavy Oil Symposium, Bakersfield, California, March 17–19, 1999.

Chapter 18

Geologic Modeling and Reservoir Characterization

18.1 Introduction

Geologic modeling is the art of constructing a structural and stratigraphic model of a reservoir from analyses and interpretations of seismic data, log data, core data, fluid production data, etc., plus inclusion of knowledge on geologic structures acquired by geoscientists from basin sedimentology and outcrop studies of similar formations. Reservoir characterization can be described as the process by which various rock definitions and petrophysical properties (such as facies, flow units, porosity, permeability, net sand, saturations, etc.) are distributed in the geologic model in a consistent and logical manner to obtain a probable representation of the reservoir. Geologic modeling and reservoir characterization are processes that require extensive knowledge and experience before they can be applied properly on a reservoir. The tools and techniques used in geologic modeling and reservoir characterization are numerous, varied, and intricate. Due to these reasons, this chapter serves mainly as an introduction to these subjects.

The basic concepts in geologic modeling and reservoir characterization presented in this chapter include the types and sources of data, scale and integration of data, and data quality and control. This is followed with a general procedure for conducting geologic modeling and characterization of a reservoir. The chapter concludes with examples on the application of geologic modeling and characterization on two reservoirs.

18.2 Sources of Data for Geologic Modeling and Reservoir Characterization

There are many sources of data for geologic modeling and reservoir characterization. These depend on the stages of exploration, appraisal, and development of the reservoir. At the exploration stage, data sources are limited to seismic, analogue reservoirs, and outcrops. In the appraisal stage, data sources expand to include cores, logs, fluids, and other subsurface data acquired from wells drilled into the reservoir. During the development stage, additional data may be available from fluid production tests, and other tests designed to assess the extent and productive potential of the reservoir. These sources of data should be integrated at each stage to construct and characterize reservoir models that are based on **all** available data. A brief description of the various sources of data for geologic modeling and reservoir characterization is presented in this section. This list is not considered to be complete and exhaustive. It is provided as a guide to some of the data that are generally used for geologic modeling and reservoir characterization. Advancements in the technology for data acquisition from reservoirs and transformation of these data into quantities that are useful for geologic modeling and reservoir characterization are making significant progress in the industry. It is expected that the list of data sources presented in this book will continue to grow as more data sources become available in the future.

18.2.1 Seismic Data

One of the important sources of data for geologic modeling and reservoir characterization is seismic data. Seismic data can be used in geologic modeling to map the structure and stratigraphy of the reservoir, as well as designate the locations of faults, if any are present. The applications of seismic data in geologic modeling are especially important in deepwater reservoirs, where due to few well penetrations, seismic data are used almost exclusively in the construction of the geologic models.[1] Another important use of seismic data is in reservoir characterization. Seismic impedance data obtained from seismic inversion processes are widely used in reservoir characterization.[2,3] The seismic inversion data are typically integrated with well log and other petrophysical data to improve reservoir characterization.

18.2.2 Outcrop and Basin Studies

Knowledge and experience gained from outcrop and basin studies by the geoscientist generally aid in the construction of the geologic model and characterization of the reservoir. During field trips, geoscientists gain valuable visual description of the structure and stratigraphy of outcrops which could be analogous to subsurface structures that are the candidates for geologic modeling. Another practice that aids geologic modeling is basin studies. Generally, the depositional environment described in basin studies can aid in defining the structure and stratigraphy of geologic structures located in similar basins.[4,5]

18.2.3 Well Log Data

Well log data are obtained from running tools in wells drilled in the reservoir or analogue reservoirs. These include well log data obtained from logging-while-drilling (LWD), wireline logs, and any other means of running tools in wells. These log data may include gamma ray, resistivity, density, neutron, sonic, nuclear magnetic resonance (NMR), image logs, vertical seismic profile, etc.

The log data are correlated, processed, and interpreted. They are then used in the determination of the structure and stratigraphy of the reservoir in concert with other data, such as seismic data. Further petrophysical analyses of the log data provide data for facies description, and other rock properties data (such as porosity, permeability, water saturation, formation thickness, etc.) for reservoir characterization. Well log data are very important in the process of geologic modeling and reservoir characterization.

18.2.4 Core Data

Samples of the reservoir rock are obtained from whole and sidewall cores. From these core samples, direct data on many rock properties can be measured in the laboratory. Core plugs taken from whole core samples can be used in routine core analysis (RCA) programs to measure basic rock properties, such as porosity, permeability, and water saturation. Some of the core plugs are used in special core analysis (SCAL) programs to measure capillary pressures and relative permeability data. In addition, high density profile permeability data can be obtained from whole core samples at close intervals using the mini-permeameter tool. Other data obtained from core samples include detailed core descriptions, geomechanical data, x-ray diffraction data, sand particle size distribution, geochemical data, etc. All these data are integrated and used in geologic modeling and characterization of the reservoir.

18.2.5 Formation Pressures and Fluid Properties Data

Formation pressures and fluid samples are obtained from the reservoir using wireline formation tester (WFT) tools. More details on the use of WFT to measure formation pressures and obtain fluid samples are presented in Chapter 5, Section 5.5. Formation pressures can be used in pressure-depth plots to determine fluid contacts, identify existence of reservoir heterogeneities or permeability barriers, and evaluate the presence of reservoir compartments and/or multiple reservoirs. These data can supplement other data in geologic modeling of the reservoir. Fluid properties, such as bubble point pressure, formation volume factor, solution gas-oil ratio, oil viscosity, oil gravity, etc. can be measured from fluid samples. Vertical and spatial variation of fluid properties can be incorporated in the characterization of the reservoir.

18.2.6 Pressure Transient Test Data

Pressure transient test data include data obtained from drillstem tests (DST), and conventional drawdown and buildup tests. Procedures for the analysis of pressure transient test data are presented in Chapters 11 to 14. Analyses of pressure transient tests generally provide data on formation permeability, location of faults, and presence of reservoir boundaries. The permeability data obtained from pressure transient tests are at different scale from permeability data obtained from well logs and core measurements. These permeability data should be reconciled before they can be used in the characterization of the reservoir. Further discussions on scales of different data sources are presented in Section 18.4.

18.2.7 Reservoir Performance Data

Reservoir performance data include all fluid production and pressure data measured after the reservoir has been placed on production. These include data from well production rates, flowing and static bottomhole pressures, production logs, injection logs, etc. These data are complied and compared with assumptions made during construction of the geologic model and characterization of the reservoir. For instance, performance data can be used to verify whether a fault assumed to be a sealing fault in the geologic model is acting as a sealing fault based on actual performance data. If it appears that the fault is not sealed on the basis of performance data, the geologic model should be modified.

18.3 Data Quality Control and Quality Assurance

Data quality control and quality assurance (QC/QA) are essential for execution of successful geologic modeling and reservoir characterization programs. The data acquired or collected at all stages of exploration, appraisal, development, and fluid production should be subjected to rigorous QC/QA procedures. These data should be organized and stored in a data management system that is accessible to a multidisciplinary team. The QA/QC program applied to each data set varies with the type of data. It is important to emphasize that a comprehensive QA/QC program should be devised for each data type discussed earlier, and the program should be under the management of a professional with experience in the acquisition, collection, and processing of that particular type of data.

18.4 Scale and Integration of Data

Petrophysical data used in reservoir characterization are measured at different scales. For instance, petrophysical data measured on core plugs are at a different scale from similar data obtained from well logs. Typically, the dimensions of a core plug are two inches diameter by two inches length to give a core volume of 0.0036 $ft.^3$ Most modern well logs measure petrophysical properties as an average over an interval in the range from 2 to 10 feet along the wellbore. It is rather obvious then that the volume of rock sampled by well logs is several orders of magnitude larger than the volume of rock sampled by core plugs. The difference in scale is even higher when data from well tests and seismic interpretations are included. The dependence of petrophysical data on scale is described as the support effect. The support for core, well log, well test, and seismic data ranges from a few inches to thousands of feet. These differences in scale should be taken into account before petrophysical data from these sources are integrated and used for reservoir characterization.[6–8]

Many techniques have been used to reconcile petrophysical data with different scales or supports. Frykman and Deutsch[7] used volume-variance scaling techniques to integrate data of different scales. Al-Ali and Worthington[8] used averaging techniques to reconcile core plug and well log data. Shafer and Ezekwe[9] applied regression analysis in the integration of core and well log permeability data. The methods that have been reported to be successful appear to apply to specific data types and data sets. Consequently, no particular method for reconciling petrophysical data with different supports is recommended. Different methods should be tried and reviewed to find a method that is applicable to the data set. It is important to recognize that petrophysical data with different scales or supports should be integrated before using them in characterizing the reservoir.

18.5 General Procedure for Geologic Modeling and Reservoir Characterization

A general procedure for geologic modeling and reservoir characterization is presented in this section. This general procedure is designed to function as a guide in devising workflows for geologic modeling and reservoir characterization. The approach recognizes that the quantity, quality, and type of data available for the process vary from one reservoir to the other. For this reason, all the steps in the general procedure are described in generic terms, and may not apply in all cases. The workflow captures all the essential steps that are applicable in a geologic modeling and reservoir characterization process. The workflow could be modified and changed as necessary to meet specific requirements of a project. The general procedure recommended for geologic modeling and reservoir characterization is presented under the following processes:

1. Generation of geologic surfaces or horizons
2. Structural modeling
3. Stratigraphic modeling
4. Correlation and assignment of well log data
5. Property modeling
6. Uncertainty analysis
7. Upscaling of geologic model to reservoir flow model

The application of the general procedure for geologic modeling and reservoir characterization is illustrated with two examples at the end of the chapter.

18.5.1 Generation of Geologic Surfaces or Horizons

Geologic surfaces or horizons can be mapped from seismic data or correlation of formation tops from well logs. In some cases, seismic and well log data are used together to map geologic surfaces that define the geologic structure. The number of geologic surfaces generated depends on the complexity of the structure. In simple structures, two geologic surfaces representing the top and base of the structure may be sufficient. In complex geologic structures, additional surfaces may be added to capture geologic features (such as faults, permeability barriers, unconformities, truncations, etc.) that are necessary in the construction of the geologic model. Additional surfaces may also be added to serve as guides in defining the stratigraphy of the geologic model. The number of geologic surfaces used in developing the geologic model is typically determined by the geoscientists responsible for the construction of the model.

18.5.2 Structural Modeling

Structural modeling involves the design and definition of geologic features within the geologic model. Structural modeling generally includes modeling of faults and other geologic structures that were observed or mapped from seismic data, well log data, and performance data, if available. Fault modeling is a very important process under structural modeling. The faults represented in the geologic model may include vertical, inclined, curved, or reverse faults. The fault system may consist of a single fault, crossing faults, branching faults, truncated faults, eroded faults, etc. The complexity of the fault system is a major consideration in the determination of the appropriate fault models to use in the representation of the faults. The potential impact of the faults on reservoir performance is also a major consideration in the selection of the fault models. The capability of the software available for the fault modeling exercise is part of the consideration. Selection of fault models and modeling of faults are processes that are highly dependent on the particular geologic model, features available in the fault modeling software, and experience of the modeler. For these reasons, no specific fault modeling process is recommended. However the geoscientists and engineers should work closely together to devise fault models that represent the fault system structurally as envisioned by the geoscientists, while including the impact of the faults on reservoir performance as recognized by the engineers.

18.5.3 Stratigraphic Modeling

The stratigraphy of the geologic model is determined from seismic and well log data. Seismic data can be used to define the stratigraphy of the geologic structure, depending on the quality of the data. The most reliable means of defining the stratigraphy of the structure is correlation of formation tops determined from well logs. By correlating formation tops, the stratigraphy of the structure can be defined and categorized into zones. The geologic zones should represent geologic units separated by recognizable geologic features, such as correlated shale intervals. These geologic zones can be subdivided further into geologic layers either to represent flow units or facies. The use of stratigraphic modeling to create geologic zones which can be subdivided further into geologic layers should precede the definition of facies or flow (hydraulic) units during the reservoir characterization step.

18.5.4 Correlation and Assignment of Well Log Data

Well log data are processed with petrophysical models to generate rock properties, such as porosity, permeability, water saturation, formation thickness, etc. before they are loaded into the geologic model. Also, the well logs may have been used to define and assign facies or flow units to specific formation intervals. Classification of the formation into facies or flow units are based on well log data, studies of mud log data, and analogous studies of outcrops and depositional basins.

The processed well log data, comprising of rock property data and facies designations, are loaded into the geologic model. The processed well log data are tied to well tops previously used in defining the stratigraphy of the geologic model. Consequently, the processed well log data are correlated in depth to the geologic zones created in the geologic model. The processed well log data can then be used in characterizing the geologic model as discussed in property data modeling.

18.5.5 Property Data Modeling

In property data modeling, some of the methods that can be used to populate the geologic model with rock, fluid, or any property data are presented. There are many different property data modeling techniques. These include techniques as simple as direct assignment of properties to the gridblocks in the geologic model to sophisticated methods based on geostatistics. Geostatistics can be described as the application of mathematical methods based on statistics in analyzing, integrating, interpreting, and distributing geologic, petrophysical, or any other property-based data in a geologic model. Geostatistics is presented in this book as a methodology widely used in the petroleum industry for reservoir characterization. An introduction to basic concepts in geostatistics, such as variogram analysis, stochastic, and object modeling are presented. The reader is advised to consult text books[10–12] on geostatistics for more detailed and comprehensive treatments of these topics. Property modeling is generally based on the grid system applied on the geologic model. In this section, model gridding is presented followed by the application of geostatistical methods.

18.5.5.1 Model Gridding

One of the critical processes in property modeling is the design of the grid system for the geologic model. The design of the grid system for the model should be the joint effort of the geoscientists and reservoir engineers involved in the project. The key factors that should be considered in the design of the grid system are:

1. Resolution of the geologic and petrophysical data
2. Size of the static geologic model
3. Limitations on the size of the reservoir flow model
4. Potential uses of the reservoir model
5. Upscaling (also termed upgridding or scaleup) techniques to convert the static geologic model into a dynamic reservoir model

1. Resolution of the geologic and petrophysical data. One of the key factors in the design of the model grid system is the resolution or scale of the available geologic and petrophysical data. The grid system should be designed at a scale that reflects the scale of the data to be used in the construction of the geologic model. This approach enables data used in the modeling process to be captured at the scale of the original data, while avoiding potential "smoothing" effects caused by applying averaging techniques to the original data in order to fit the grid system. For instance, in designing the areal grid system, the grid dimensions should be selected to ensure that important geologic features, such as faults, permeability barriers, unconformities, etc. are represented adequately in the geologic model. If the areal grid dimensions are too large, these geologic features could be poorly represented or actually absent in the geologic model.

Similarly, the vertical grid dimensions should be as close as possible to the scale of the available vertical data. For instance, if log data are available at half-foot intervals, the vertical grid dimensions should be at this resolution or as close as possible to this scale to avoid excessive smoothing of the vertical data caused by averaging to fit a much larger vertical grid scale.

2. Size of the static geologic model. The selection of the grid system is sometimes determined by the desired size of the static geologic model. Some conditions that may cause the size of the geologic model to be limited include potential applications of the model, quality and quantity of available data, resources assigned to the project, and time constraints. For instance, if the geologic model is a conceptual model and very little data are available, there is little or no need to construct a geologic model with millions of gridblocks in this case, since a geologic model with a few thousand gridblocks is sufficient for the application. Similarly, if limited human resources and time are available for work on the project, a smaller geologic model with few thousand gridblocks could be adequate due to limited resources. The key point is that the grid system should be designed to fit the intended applications of the geologic model and at the same time accommodate utilization of available resources.

3. Limitations on the size of the flow reservoir model. In most cases, geologic models are converted into flow models for simulation of reservoir processes. This should be taken into consideration in designing a grid system for the geologic model. A large geologic model will convert into a large reservoir flow model, if excessive upscaling is not applied. Excessive upscaling will lead to loss of the detailed geologic features present in the geologic model. The key message is that if the objective is to obtain a reservoir flow model with a few thousand gridblocks, it is adequate to construct a geologic model that is at most five times the expected size of the reservoir flow model.

4. Potential uses of the reservoir model. Reservoir simulation models are typically designed to achieve specific objectives. These objectives generally define the potential uses of the reservoir model which may include investigations of reservoir drive mechanisms, effect of rock and fluid properties on hydrocarbon recovery, impact of faults on reservoir performance, incremental recoveries from secondary and/or enhanced oil recovery processes, etc. These potential uses of the reservoir model can directly influence the design of its grid system. For instance, a reservoir model designed to evaluate the impact of fluid properties on hydrocarbon recovery may require a finer grid system than a reservoir model designed to investigate the effects of faults on reservoir performance. Since reservoir models are based on geologic models, the potential uses of the reservoir model should be considered in the design of the grid system for the geologic models.

5. Upscaling (also termed upgridding or scaleup) techniques to convert the static geologic model into a reservoir flow model. A general rule of thumb is to minimize the extent of upscaling required to convert the static geologic model to a reservoir flow model. Upscaling generally results in some loss of heterogeneities built into the larger geologic model. This should be borne in mind in designing a grid system for the geologic model. Irrespective of the upscaling techniques used to convert the geologic model into a reservoir model, the objective should be to select a grid system for the geologic model that should be retained as much as feasible in the reservoir model. Additional details on upscaling are presented in this chapter as part of the workflow in geologic modeling and reservoir characterization.

18.5.5.2 Geostatistical Methods

Geostatistical methods are presented in this book as part of the workflow in geologic modeling and reservoir characterization. Some prior knowledge and exposure to geostatistical methods is assumed. It is recommended that the reader consult text books[10–12] for detailed presentations of geostatistical methods. In this section, an overview on spatial analysis using semivariograms (called variograms hereafter) is presented, followed by a brief presentation of geostatistical estimation and conditional simulation methods.

1. Spatial analysis with variograms. Many geostatistical methods are based on variograms. The variogram is a geostatistical tool widely used for spatial analysis of the variability of facies, porosity, permeability, and any other petrophysical and geologic data. The semivariogram function can be defined as one-half of the squared difference between random variables separated by a distance. It is represented mathematically as:

$$\gamma(\mathrm{h}) = \frac{1}{2N(\mathrm{h})}\sum_{i=1}^{N(\mathrm{h})}[z(x_i) - z(x_i + \mathrm{h})]^2 \tag{18.1}$$

In Eq. (18.1), $\gamma(\mathrm{h})$ = the semivariogram at the lag distance, h; $N(\mathrm{h})$ is the number of sample pairs at lag distance, h; and $z(x_i)$ and $z(x_i + \mathrm{h})$ are the sample values at locations x_i and $(x_i + \mathrm{h})$, respectively. Note that by definition, the semivariogram, $\gamma(\mathrm{h})$, is one-half of the variogram, $2\gamma(\mathrm{h})$. But the term variogram is used in spatial analysis generally instead of the more correct term semivariogram. In many practical applications of Eq. (18.1), the number of pairs at a given lag distance may not be statistically significant. To increase the number of sample pairs at any lag distance, a tolerance, $(\pm\Delta\mathrm{h})$, is often added to the lag distance. The modified semivariogram with the added tolerance is represented as:

$$\gamma(\mathrm{h} \pm \Delta\mathrm{h}) = \frac{1}{2N(\mathrm{h})}\sum_{i=1}^{N(\mathrm{h})}[z(x_i) - z(x_i + \mathrm{h})]^2 \tag{18.2}$$

As Eq. (18.2) indicates, sample pairs are collected within a band represented by the lag distance, $\mathrm{h} \pm \Delta\mathrm{h}$. This is illustrated in Figure 18.1 in which four sample pairs were formed by using the tolerance, $\Delta\mathrm{h}$.

A diagram showing all the features of a simple variogram is shown in Figure 18.2. The features shown in Figure 18.2 are the sill, the range, and the nugget. The point at which the variability between the sample pairs becomes flat is called the sill. The sill is equal to the variance of the sample data. The distance at which the sill is reached is called the range. Beyond the range, the sample data are no longer correlated. The nugget is the point at which the variogram intersects the y-axis above the origin. The nugget effect is caused by the combination of geologic microstructures and

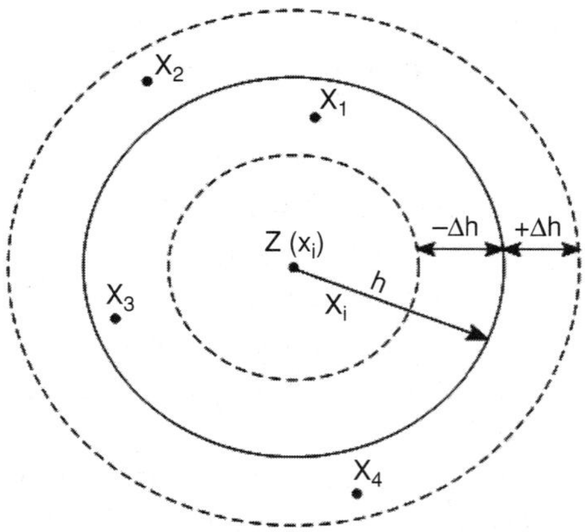

Figure 18.1 Sample pairs formed by lag distance h and lag tolerance Δh.

measurement errors. Sparse data may also cause nugget effects. The presence of nuggets in the variogram has significant effect on the simulation of distributed properties.[13]

By plotting the variance for each sample pair against the lag distance as calculated from Eq. (18.1) or Eq. (18.2), an experimental variogram is generated. If the experimental variogram represents spatial relationship in all directions, it is called an omnidirectional or isotropic variogram. In some cases, different experimental variograms define spatial relationships in different directions. These experimental variograms are called directional or anisotropic variograms.

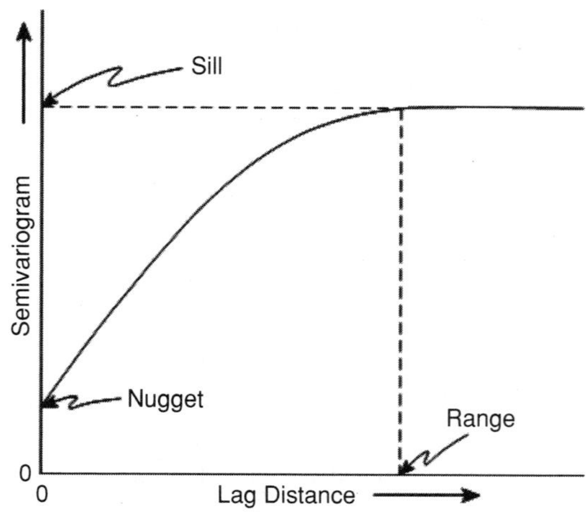

Figure 18.2 A variogram showing sill, range, and nugget.

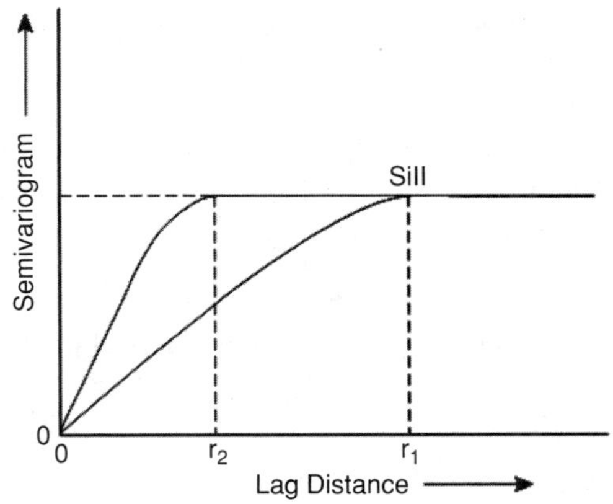

Figure 18.3 A variogram showing geometric anisotropy.

Anisotropic variograms are functions of lag distances as well as directions. There are two types of anisotropic variograms. These are geometric anisotropy and zonal anisotropy. Geometric anisotropy is observed where the variograms in different directions have different ranges but the same sill as shown in Figure 18.3. This type of anisotropy is generally observed when analyzing areal data in different directions. Zonal anisotropy is observed where the variograms in different directions have different ranges and sills as shown in Figure 18.4. This type of anisotropy is present where variograms in vertical direction are compared to horizontal variograms in the areal

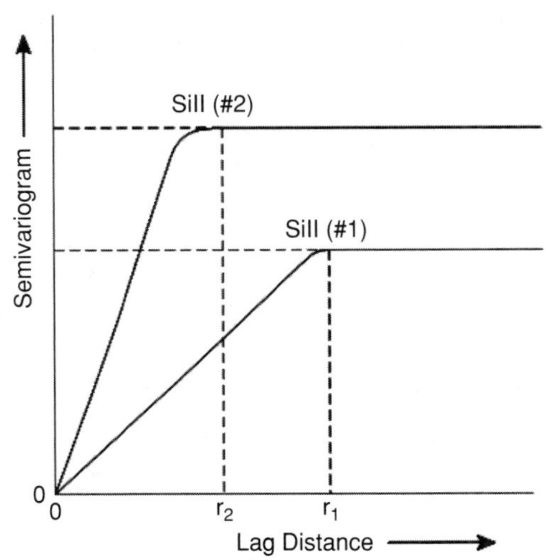

Figure 18.4 A variogram showing zonal anisotropy.

direction. Calculation of anisotropic variograms can be a very tedious process. However, most geostatistical software packages offer visual graphical support for the calculation of anisotropic variograms.

2. Variogram Models. Variogram models, expressed as mathematical functions, are divided into two groups. The first group is variogram models with sills, also called transition models. The second group is variogram models without sills, also called non-transition models.

A. Transition models. The spherical model is an example of a transition model. If the range is represented as r and the sill is normalized to 1, the spherical model can be expressed as:

$$\gamma(\mathrm{h}) = \begin{cases} \frac{3}{2}\frac{\mathrm{h}}{r} - \frac{1}{2}\left(\frac{\mathrm{h}}{r}\right)^3 & \text{if} \quad \mathrm{h} \leq r \\ 1 = \text{sill} & \quad\quad \mathrm{h} \geq r \end{cases} \tag{18.3}$$

Another transitional model is the exponential model. The exponential model is represented as:

$$\gamma(\mathrm{h}) = 1 - \exp\left(-\frac{3\mathrm{h}}{r}\right) \tag{18.4}$$

The Gaussian model is a transitional model represented as:

$$\gamma(\mathrm{h}) = 1 - \exp\left(-\frac{3\mathrm{h}^2}{r^2}\right) \tag{18.5}$$

In Eqs. (18.3), (18.4), and (18.5), h = lag distance; and the sill of the sample data has been normalized to 1. The shapes of the spherical, exponential, and Gaussian models are shown in Figure 18.5. Note that the spherical model is linear near the origin, while the exponential and Gaussian models are parabolic near the origin. For practical purposes, the ranges of these models are defined as the lag distances at which the variograms reach 95% of their respective sill values.

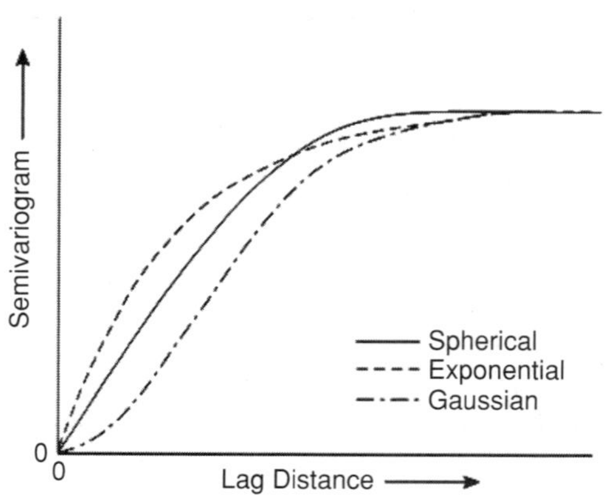

Figure 18.5 Shapes of spherical, exponential, and Gaussian variogram models.

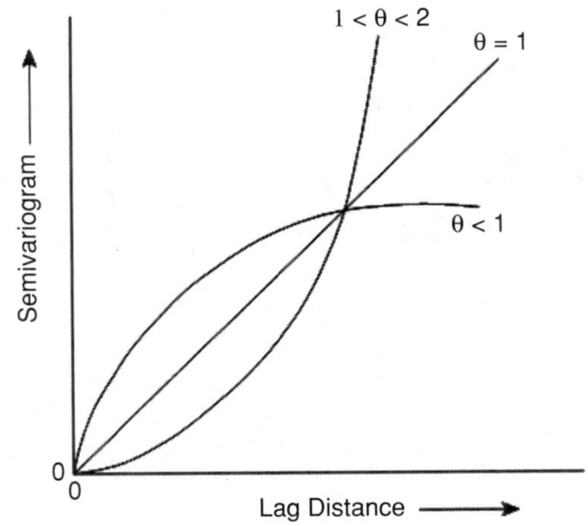

Figure 18.6 Shape of Power Law variogram model.

B. Non-Transition models. The most common example of a non-transition model is the Power Law model. The Power Law model is expressed as:

$$\gamma(h) = h^{\theta}, \tag{18.6}$$

where θ is between 0 and 2 excluding these limits.

The shapes of the Power Law model are shown in Figure 18.6. Other non-transition models include the cosine model and the log normal model. These models are rarely used in modeling experimental variograms.

C. Modeling of experimental varicograms. After the experimental variogram has been calculated with either Eq. (18.1) or (18.2), the next step in spatial analysis is modeling of the experimental variograms with one or combinations of the variogram models. The requirement for the use of any function as a variogram model is that the function must be conditionally positive definite. This condition ensures the existence of a unique solution for the estimation of a variable at an unsampled location. The transition and non-transition models described earlier satisfy this condition. In modeling experimental variograms, it is important to capture major features such as the sill, range, and nugget.[14] In some cases, any linear combinations of variogram models to create a nested model are permissible. However, it is better to use simple models and avoid nested structures as much as possible. In a sense, modeling of experimental variograms is a curve fitting exercise involving systematic fitting of several parameters, namely the sill, the range, and the nugget. Most geostatistical software packages provide visual and interactive graphical applications to assist in the modeling of experimental variograms.

The choice of variogram models used in modeling (or fitting) the experimental variogram have considerable impact on the resulting simulated realizations using the same sample data as input.[7] Figures 18.7a, 18.7b, and 18.7c show the 2D simulated realizations using exponential, spherical, and Gaussian variogram models with the same sill and range values.

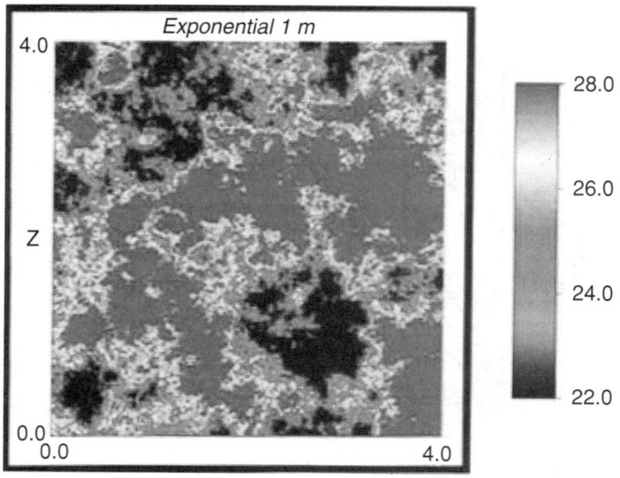

Figure 18.7a 2D simulated property using exponential variogram model (from Frykman and Deutsch[7]).

The exponential variogram model gives a realization that appears "noisy," while the Gaussian model produces a smoother realization.[7] The realization with the spherical variogram model lies in-between these two realizations. These figures are used to illustrate the impact of variogram models on simulated realizations, and emphasize that the choice of variogram models during modeling of experimental variograms should be guided by knowledge of spatial variability of the sample data.

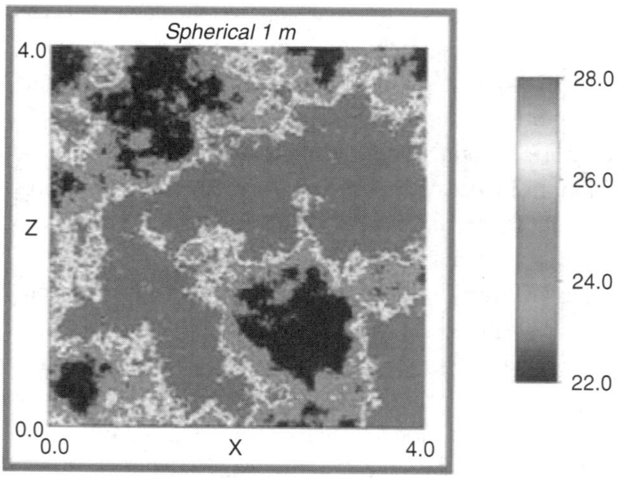

Figure 18.7b 2D simulated property using spherical variogram model (from Frykman and Deutsch[7]).

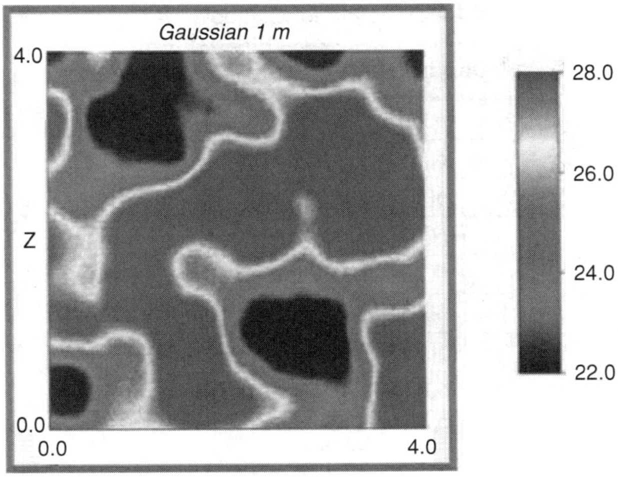

Figure 18.7c 2D simulated property using Gaussian variogram model (from Frykman and Deutsch[7]).

Example 18.1 Calculation of an Experimental Variogram as a Function of Lag Distance

Problem

Calculate an experimental variogram at lag distances of 1 to 12 feet for the data shown in Table 18.1.

Solution

The experimental variogram at all lag distances were calculated using Eq. (18.1).
At lag distance, h = 1 ft.:
Substituting Eq. (18.1) gives:

$$\gamma(1) = \frac{(19.0-20.2)^2+(20.2-21.4)^2+(21.4-19.7)^2+\cdots+(21.2-20.9)^2+(20.9-21.5)^2+(21.5-21.8)^2}{2\times 33}$$
$$= 0.5073$$

At lag distance, h = 2 ft.:
Substituting Eq. (18.1) gives:

$$\gamma(2) = \frac{(19.0-21.4)^2+(21.4-22.0)^2+(22.0-21.7)^2+\cdots+(20.8-21.0)^2+(21.0-21.2)^2+(21.2-21.5)^2}{2\times 32}$$
$$= 0.7458$$

The variograms calculated at higher lag distances are shown in Table 18.2. The results in Table 18.2 were plotted as shown in Figure 18.8. A simple computer program can be developed for the calculation of experimental variograms.

Table 18.1 Data for Example 18.1

Depth (feet)	Porosity (%)
6050	19.00
6051	20.20
6052	21.40
6053	19.70
6054	22.00
6055	21.00
6056	21.70
6057	21.40
6058	19.80
6059	19.30
6060	18.90
6061	19.70
6062	18.30
6063	19.50
6064	19.90
6065	20.20
6066	20.70
6067	20.90
6068	21.40
6069	18.20
6070	18.10
6071	18.40
6072	18.70
6073	19.10
6074	19.60
6075	20.40
6076	20.80
6077	21.40
6078	21.00
6079	21.50
6080	21.20
6081	20.90
6082	21.50
6083	21.80

Table 18.2 Variogram Data Calculated from Data in Table 18.1

Lag distance, h (ft)	Variogram, γ (h)
1	0.5073
2	0.7458
3	1.0752
4	1.4928
5	1.6659
6	1.9164
7	1.7024
8	1.7160
9	1.3986
10	1.3554
11	1.1913
12	1.2559

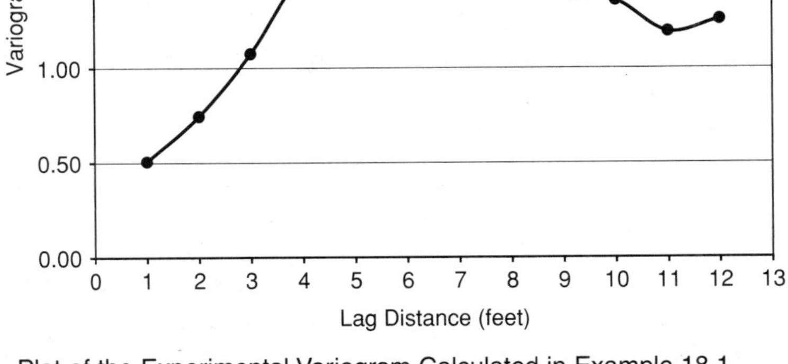

Figure 18.8 Plot of the Experimental Variogram Calculated in Example 18.1.

3. Geostatistical Estimation and Conditional Simulation Methods. They are many estimation and conditional simulation methods that can be used to assign values at unsampled locations. For reservoir characterization, kriging is the most applied estimation method. Kriging is different from conditional simulation in three key areas:

a. Kriging yields the best local estimate of the unsampled value without considering the resulting spatial variability. In conditional simulation, global features and spatial statistics take precedence over local accuracy.

b. Kriging tends to produce a "smoothing" effect unlike conditional simulation.

c. Kriging produces a single model (realization) of the attribute. Conditional simulation produces many alternate models (probable realizations) of the attribute.

The development of estimation and conditional simulation methods involve complex statistical and mathematical equations. Furthermore, implementations of these methods for analysis of real data require in-depth knowledge of geostatistics. Due to these reasons, the statistical theories and mathematics that are the bases of estimation and conditional simulation methods are not presented. These topics are available in several geostatistical books.[10–12] For the purposes of this book, estimation methods are limited to a listing of commonly used kriging methods. Similarly, conditional simulation methods are limited to a listing of methods commonly used in reservoir characterization.

A. Kriging methods. Kriging is a minimum variance estimator based on variograms. It is an unbiased estimator that accounts for the configuration of the sampled data. There are many variation of the kriging method. The kriging methods most used in reservoir characterization are:

- Ordinary kriging
- Kriging with external drift
- Cokriging
- Collocated cokriging
- Indicator kriging

B. Conditional simulation methods. Conditional simulation methods are stochastic processes of building alternative, equiprobable models (realizations) using the spatial distribution of the attribute. The simulation process is described as conditional if it produces realizations that honor sample data at their locations. The attributes simulated can be either continuous or categorical. Conditional simulation methods commonly used in reservoir characterization are:

- Sequential Gaussian simulation (SGS)
- Sequential Indicator simulation (SIS)
- Object-based modeling

18.5.6 Uncertainty Analysis

There is some degree of uncertainty inherent in any geologic model because of the differences (or errors) between the true and sampled values of data used in building the model. If the amount of sample data available for constructing the model is sparse, the level of uncertainty is expected to be high. Conversely, if the amount of sample data available is large and well distributed across the

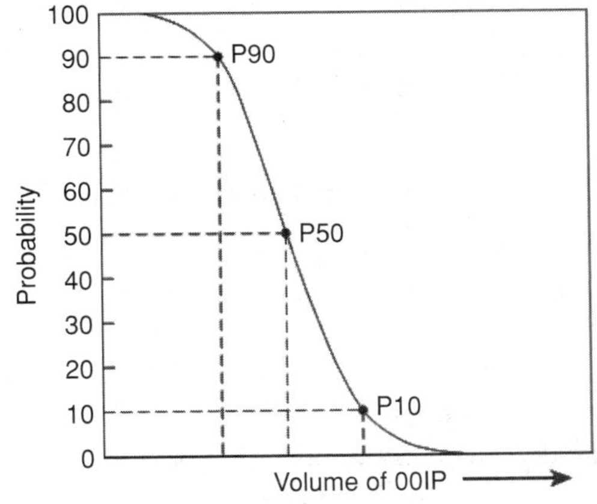

Figure 18.9 Cumulative probability curve for OOIP based on multiple realizations showing P10, P50, and P90 quantiles.

volume of interest, the level of uncertainty is comparatively less. In geologic model building, the two main sources of uncertainties are structural data and petrophysical data. Structural uncertainties may result from processing and interpretation of seismic data, mapping of faults, structural and stratigraphic correlation of log data, "picking" of structure tops from seismic and/or log data, description of the depositional environment, etc. Petrophysical uncertainties may derive from measurements of porosity, permeability, and water saturation data, determination of gross and net thickness data, definition and assignment of facies, location of fluid contacts, etc. Practically most of the data used in the construction of geologic models have some degree of uncertainty associated with them. Consequently, it is important to develop methodologies for assessing and quantifying the uncertainty associated with geologic models built from these data.

One of the methods that can be used to assess quantitatively the degree of uncertainty associated with a geologic model is by generation of multiple, equiprobable realizations of the model. Even though the realizations are equally probable (equiprobable), there are differences from one realization to the next realization. These differences are measures of the uncertainties present in the models. As a result, the realizations can serve as the input for uncertainty analysis. By application of statistical analysis, several measures of uncertainty can be generated from the set of multiple realizations. These statistical metrics may include the mean, standard deviation, and the cumulative probability distribution of variables such as total pore volumes and oil-in-place volumes which can be calculated from multiple realizations. A typical cumulative probability curve representing the distribution of oil-in-place volumes based on multiple realizations is shown in Figure 18.9. The nominal classification based on P10, P50, and P90 quantiles are shown for illustration purposes in Figure 18.9. For instance, the P10 quantile represents realizations with volumes of oil in place equal to or larger than the designated volume. Note that in some organizations, the designations of magnitudes of P10, P50, and P90 quantiles are reversed from the designations used in this book. From the cumulative probability curve calculated for a variable, the probability of achieving any level of outcome for that variable can be estimated. The cumulative probability curve can also be used to select specific realizations (geologic models) at the P10, P50, and P90 quantiles for further analyses in reservoir flow simulations.

18.5.7 Upscaling of Geologic Model to Reservoir Flow Model

The term upscaling (also called upgridding or scaleup) is used to describe the systematic process of building a coarse-gridded reservoir flow (simulation) model from a fine-gridded geologic model.[15] The objective in the upscaling process is to reduce the number of gridblocks in the geologic model by a factor of two or more in converting it into a reservoir flow model, and still retain important characteristics of the geologic model. Upscaling is sometimes necessary because geologic model with multimillion gridblocks are routinely constructed in many projects with the advent of advanced techniques for geologic modeling and reservoir characterization. The sizes of these geologic models are in some cases too large to simulate as reservoir models because of the need to reduce model run times, evaluate alternative reservoir management strategies, or conduct iterative simulation runs in a history matching process. Other reasons for scaling up geologic models include limitations on computer memory storage required for large simulation models, and the cost of processing the output from large models. However, it is relevant to observe that recent advances in computer hardware and simulation software technologies indicate that the practice of upscaling geologic models to simulation models is expected to decline with time. In other words, it is expected that geologic models will be converted into reservoir simulation models without the need for upscaling as computer hardware and simulation software technologies advance over time. This view is supported by the construction and simulation of billion-cell reservoir models as reported in the literature.[16–18]

18.5.7.1 Recommended Steps for Upscaling a Geologic Model to a Reservoir Model

The following steps are recommended for upscaling a geologic model to a reservoir model:

1. Grid construction for the reservoir model
2. Generation of properties for reservoir model gridblocks
3. Verification of the upscaling process

Step 1: Grid construction for the reservoir model

The goal in designing a grid system for the reservoir simulation model is to retain the major geologic features and heterogeneities built into the geologic model. It is also important to consider the potential uses of the reservoir model in designing the grid system. Major geologic features that should be preserved in the reservoir model include faults, fault geometry, permeability barriers, location of fluid contacts, stratigraphy of the geologic model, and variability of rock and fluid properties. Potential applications of the reservoir model that can depend on the grid system include monitoring rapid changes in fluid saturations and pressures, fluid flow around wells, and microscopic displacement processes as observed in enhanced oil recovery processes.

The areal grid system determines the location of geologic features in the *X*,*Y* coordinates. Consequently, it is important to select an areal grid system for the reservoir model that closely mirrors the areal grid system of the geologic model. This approach ensures that the location of major geologic features in the geologic model is maintained in the simulation model. A common practice that is recommended is to use the same areal grid system of the geologic model for the reservoir model.[19] If it is necessary to use a different areal grid system for the reservoir model, it is recommended that the new areal grid system should be devised to encompass consistent multiples of the geologic model gridblocks, and have the same orientation as the areal grid of the geologic model.

By using the same or similar areal grid system for the reservoir model as present in the geologic model, most of the upscaling exercise is focused on selecting the vertical grid system or layering for the reservoir model. In upscaling the vertical grids from the geologic model to the reservoir model, it is important to maintain the stratigraphy developed for the geologic model. The recommended practice is to reduce the number of layers in the reservoir model by combining similar layers in the geologic model **within** the same zone, while retaining geologic layers that mark the location of permeability barriers such as identifiable shale layers. Proper upscaling of the vertical grid system can be achieved by close inspection of the geologic model coupled with a thorough understanding of the stratigraphy used in its construction.

The reservoir engineers and the geoscientists who constructed the geologic model should work closely together to obtain a representative reservoir model from the geologic model. This cooperative effort should begin at the initial construction of the grid system for the geologic model. With proper input from the reservoir engineer, the areal grid system for the geologic model should be devised with the intention that the same grid system is retained for the reservoir model. Similarly, the vertical grid system for the geologic model should be devised to capture the high resolution of data at the geologic level, while retaining the ease of combining vertical layers that exhibit similar characteristics during the upscaling process.

Step 2: Generation of properties for reservoir model gridblocks

The reservoir model gridblocks are populated with rock properties, such as porosity, net sand thickness, water saturation, permeability, rock types or facies based on the properties of the gridblocks in the geologic model. The method of upscaling rock properties from the geologic model to the reservoir model depends on the type of rock property to be upscaled. Static rock properties such as porosity, net sand thickness, and water saturation are scaled-up using averaging techniques for the corresponding properties in the geologic model. For instance, gridblock porosities in the reservoir model can be upscaled as the bulk volume weighted averages of the porosities of corresponding geologic model gridblocks. Net sand thickness of gridblocks in the reservoir model can be upscaled as the sum of the net sand thickness of the corresponding gridblocks in the geologic model. Water saturations can be upscaled as pore volume weighted averages of the corresponding gridblock water saturations in the geologic model.

They are no simple techniques for upscaling rock types or facies, since a gridblock in the geologic model may represent more than one rock type or facies. An approach that is recommended is to assign rock type or facies based on the proportion of non-shale rock type or facies in the corresponding geologic model gridblocks.[15] The rock type or facies with the highest proportion is assigned to the corresponding reservoir model gridblock.

The best practice for upscaling permeability is by applying flow-based scale averaging (FBSA).[15] This method uses the Darcy flow equation to calculate the effective permeability due to the flow of a single phase fluid in the fine grid geologic model under set pressures, flow rates, and boundary conditions. The effective permeabilities assigned to the reservoir model gridblocks have been shown to achieve the same flow characteristic as the corresponding geologic model gridblocks.[20] Many geologic modeling software provide permeability upscaling techniques based on FBSA. It is strongly recommended that this technique should be used in upscaling permeabilities. Geometric and harmonic averaging techniques should not be used since these averaging techniques generally yield lower gridblock permeabilities in the reservoir model.

Step 3: Verification of the upscaling process

Verification of the upscaling process is a quality control step to ensure that the reservoir model upscaled from the geologic model is a **reasonable** representation of the geologic model. This process involves an intensive interrogation of the upscaled reservoir model. Every aspect of the reservoir model should be investigated and compared with the geologic model. Visual and graphic displays should be used as much as possible in comparing the two models.

The first activity in the verification process is to compare the two models and ascertain that major geologic features such as faults, permeability barriers, etc. present in the geologic model are correctly reproduced in the reservoir model. The location and geometry of all faults in the two models should be compared and shown to be correctly represented in the reservoir model. Other geologic heterogeneities and features should be compared in the same manner.

This should then be followed by comparisons of the gridblock properties, such as porosities, net sand thickness, saturations, permeabilities, etc. Cross-sections of the geologic model displaying these properties should be compared with the same cross-section of the reservoir model. The comparisons should be conducted in a systematic and thorough manner. Another way of comparing the petrophysical properties of the two models is by plotting the histograms of each property in the two models together. The histograms should show similar trends. Most geologic modeling software has applications that facilitate these comparisons. These applications can be used to calculate and compare statistical metrics such as mean, standard deviation, maximum and minimum values of any petrophysical parameter in the geologic and reservoir models.

It is also useful to calculate and compare gross rock volumes, pore volumes, and hydrocarbon-in-place volumes, of the geologic and reservoir models. These volumetric comparisons should also be conducted for major geologic zones as defined in the stratigraphy. The gross rock volume of the geologic and reservoir models should agree within a margin of 1 to 2%.[15] The hydrocarbon-in-place volumes between the two models should be in agreement within a margin of 1 to 5%.

At the end of the verification process, if no major deficiencies or errors are found in the upscaled reservoir model, then the upscaling effort can be considered to be successful. The upscaled reservoir model can then be used for flow simulation work.

Example 18.2 Geologic Modeling and Characterization of Reservoir "A"

Problem

Reservoir "A" consists of two main sand formations, designated here as 1A and 1B sands. The 1A sands are separated from the 1B sand by a shale interval as shown in Figure 18.10. The deepest sand formation designated as the A2 sand is not included in the model. The data available for building and characterizing the geologic model are:

1. Top and base structural contour maps for 1A and 1B sands
2. Net sand maps for 1A and 1B sands
3. Fault maps for the major faults
4. Processed log data from five wells
5. Core data from one well

Using the above data, build and characterize a geologic model for the reservoir.

Solution

Step 1: Generate geologic surfaces or horizons

From the top and base structural contour maps, surface maps were generated for the top of 1A sand and base of 1B sand using facilities available in a commercial geologic

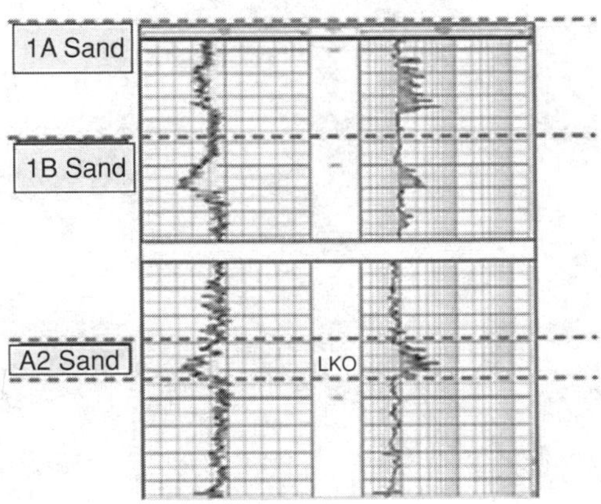

Figure 18.10 Type log of Reservoir "A" in Example 18.2.

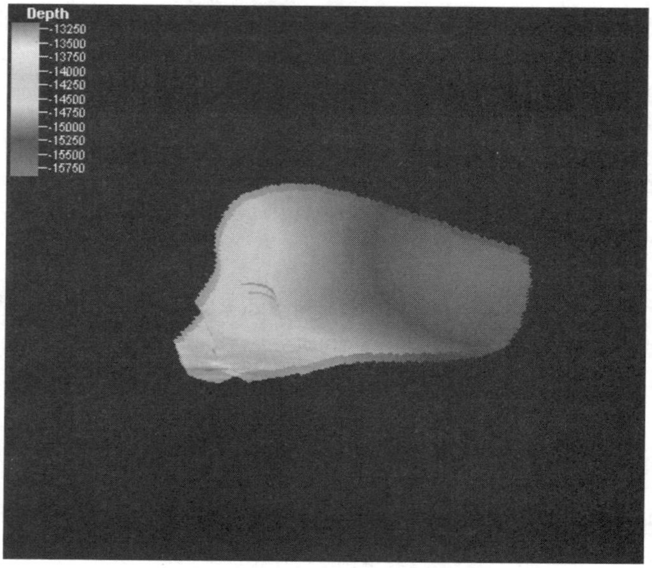

Figure 18.11 Top and base surface maps for Reservoir "A" in Example 18.2.

modeling software. The surfaces generated by this process are shown in Figure 18.11. Also net sand map surfaces were generated from net sand contour maps for 1A and 1B sands. These are shown as Figures 18.12 and 18.13, respectively.

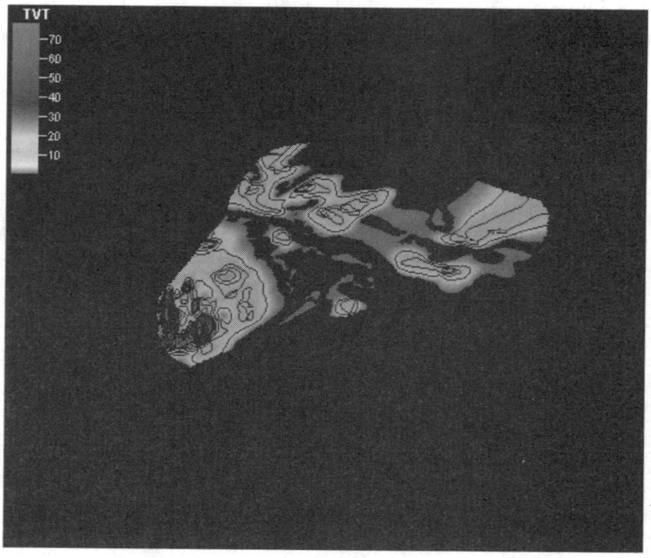

Figure 18.12 Net sand surface map of 1A sand in Example 18.2.

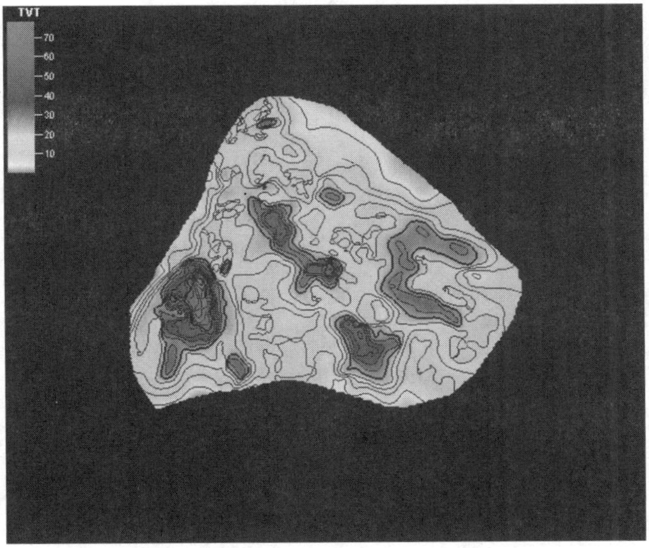

Figure 18.13 Net sand surface map of 1B sand in Example 18.2.

Step 2: Build the structural model

For this reservoir, the main geologic feature that must be represented in the geologic model is the main fault the divides the structure into a southwestern section and a northeastern section. The southwestern section has gas in the 1A sand and the 1B sand is wet. The northeastern section has oil in both the 1A and 1B sands. The juxtapositions of the sands across the main fault are shown in Figure 18.14. The main fault was considered to be a sealing fault at discovery.

The faults were digitized from structure maps into polygons. From the polygons, fault surfaces were generated and are shown in Figure 18.15. The fault surfaces were treated as vertical faults and used in the construction of the geologic model.

Step 3: Develop the model stratigraphy

The stratigraphy of the geologic model is relatively simple. It consists of three zones. Zone 1 is assigned to the 1A sand. Zone 2 designated the shale interval between 1A and 1B sands. And Zone 3 is assigned to the 1B sand.

Step 4: Load and assign well log data

Processed log data were available from five wells. The log data include porosity, water saturation, and permeability. The log data were calibrated with core data from one of the wells. The log data were uploaded and assigned to corresponding gridblocks in the geologic model.

Step 5: Conduct property modeling

To start property modeling, a grid system has to be created for the geologic model. The grid system designed for the model is 91 $\times$ 104 $\times$ 34 for a total of 321,776 gridblocks.

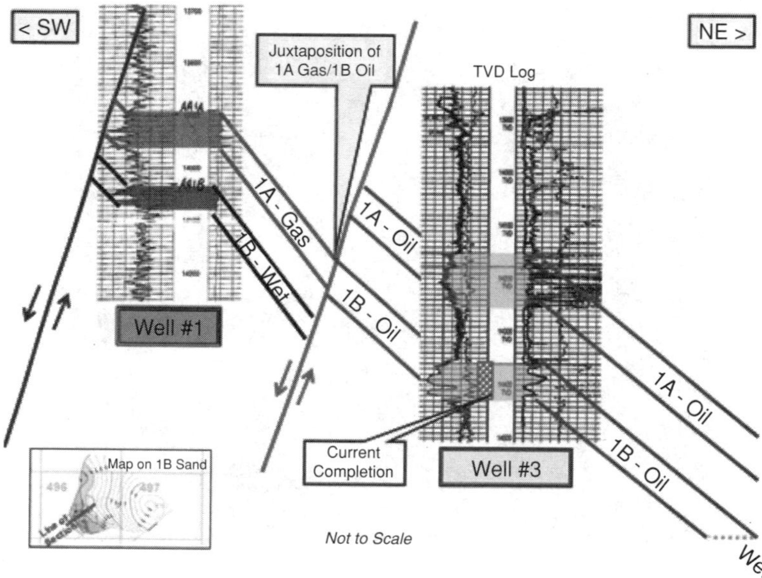

Figure 18.14 Cross-section through the main fault in Reservoir "A" of Example 18.2.

Figure 18.15 Fault surfaces in Reservoir "A" of Example 18.2.

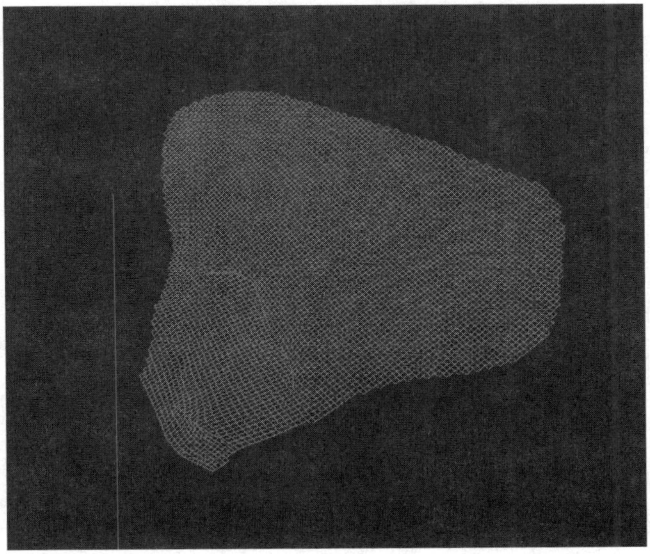

Figure 18.16 Areal grid system for the geologic model in Example 18.2.

The dimensions of the gridblocks on the average are 197.6 ft × 200.5 ft × 5.8 ft. The areal grid is shown as Figure 18.16. For the vertical layering, 10 layers were assigned to 1A sand and 20 layers were assigned to the 1B sand. The shale interval between the two sands was assigned 4 layers. This grid system was considered to be sufficiently fine to capture the heterogeneity of the rock properties and could still be used for flow simulation without the need to upscale the geologic model.

The next phase in property modeling is data analysis using variogram modeling. Data analysis was conducted on porosity, permeability, and water saturation data upscaled from well logs. A variogram model was fitted to the particular data for the corresponding sand. For example, a spherical variogram model was fitted to the porosity data for 1A sand as shown in Figure 18.17. Similar variogram models were obtained for other petrophysical data. These variogram models were then used in conditional simulation of the petrophysical data.

In conditional simulation of petrophysical data for the geologic model, porosity was distributed in the model using the method of Sequential Gaussian Simulation (SGS) with collocated cokriging on net sand data. Figure 18.18 shows the geologic model with the distribution of porosity generated by this process. Similar distributions of permeability and water saturation were generated using SGS with collocated cokriging on distributed porosity data.

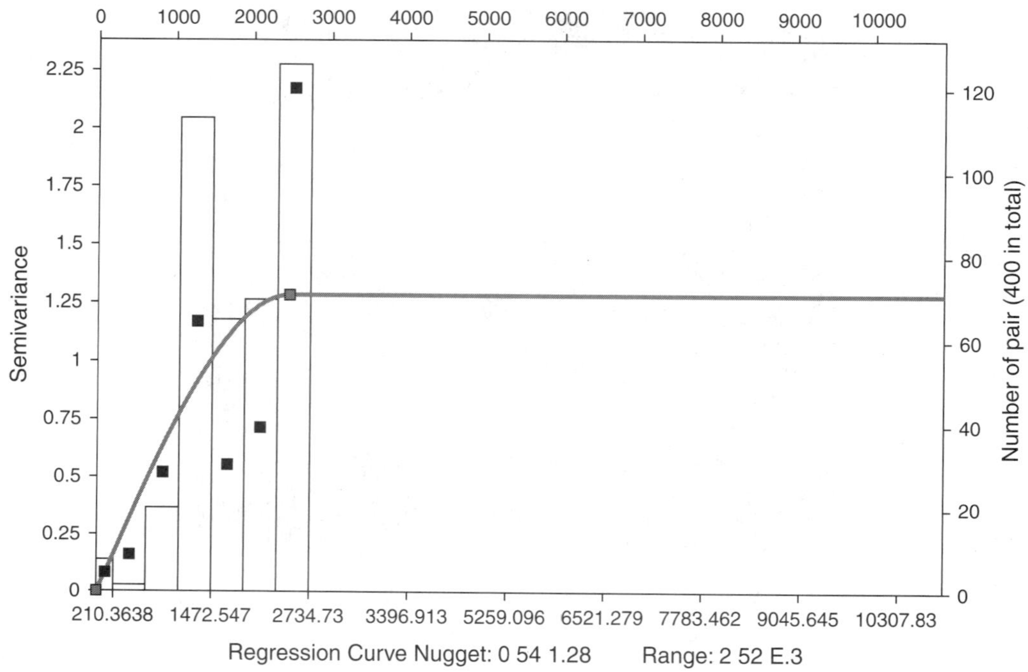

Figure 18.17 Spatial variogram (spherical model) of porosity in the 1A sand (nugget = 0, sill = 1.28, range = 2520 ft) of Example 18.2.

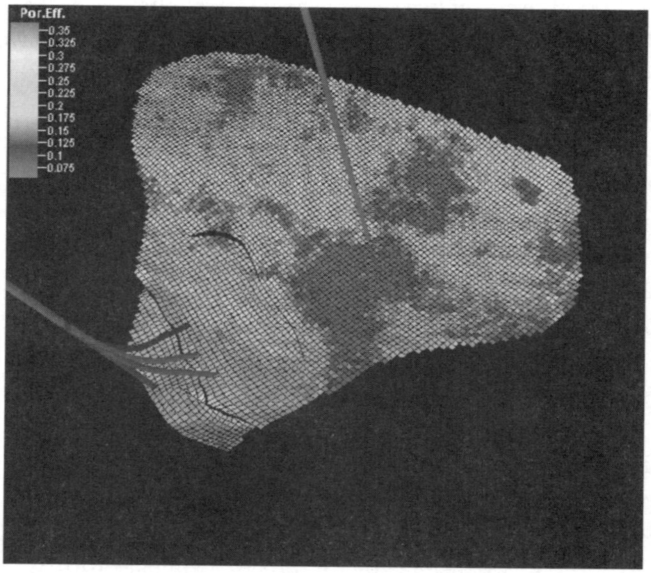

Figure 18.18 Porosity distribution for Reservoir "A" of Example 18.2.

Example 18.3 Geologic Modeling and Characterization of Reservoir "B"

Problem

The reservoir in this example is much larger and has more complex lithology than the reservoir in Example 18.2. The data available for building and characterizing the geologic model for Reservoir "B" are:

1. Seismic data
2. Horizons from seismic data
3. Fault maps from seismic data
4. Processed log data from three wells
5. Core data from two wells
6. Analogs from outcrop studies

Using the above data, build and characterize a geologic model for Reservoir "B."

Solution

Step 1: Generate geologic surfaces or horizons

Since this is a very large reservoir with over 2500 feet of gross thickness, several horizons were "picked" from seismic data. Only the surface maps generated for the top and base of the structure are shown in Figures 18.19 and 18.20.

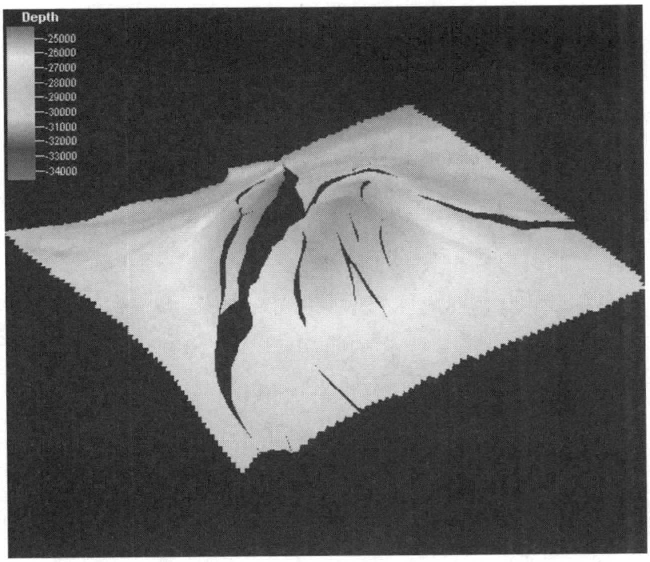

Figure 18.19 Horizon map on top of structure of Reservoir "B" in Example 18.3.

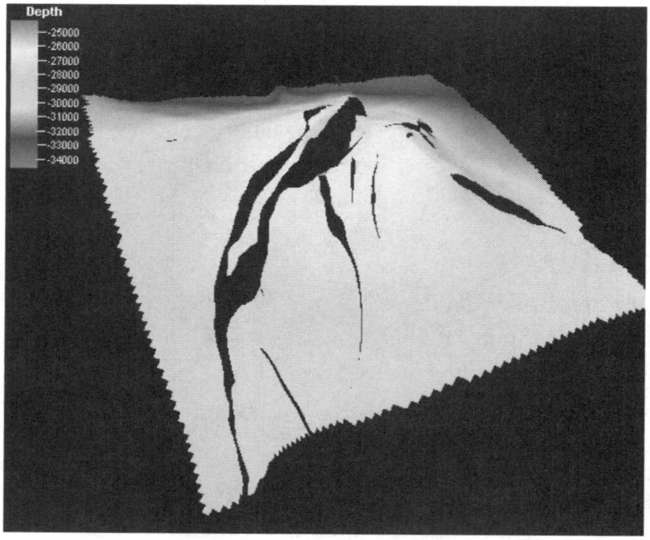

Figure 18.20 Horizon map on base of structure of Reservoir "B" in Example 18.3.

Step 2: Build the structural model

It is evident in the surface maps shown in Figures 18.19 and 18.20 that Reservoir "B" has a complex fault system. The challenge in constructing a structural model for this reservoir is proper representation of the faults in the geologic model. The faults were mapped from seismic data. The fault model used in the structural model is shown in Figure 18.21. All faults were treated as vertical faults in the fault model.

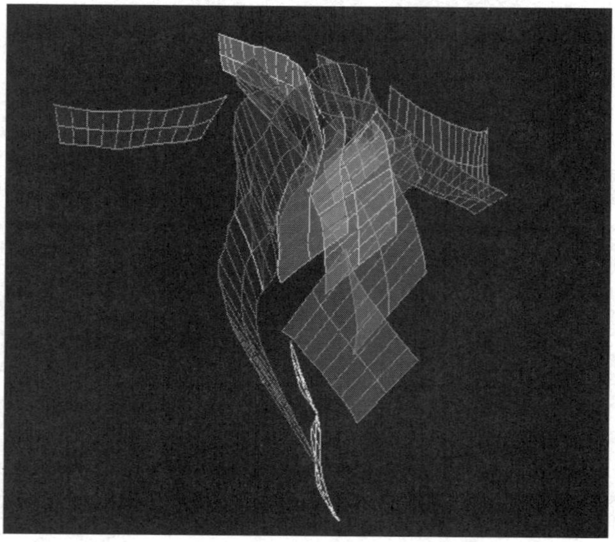

Figure 18.21 Fault surfaces of Reservoir "B" in Example 18.3.

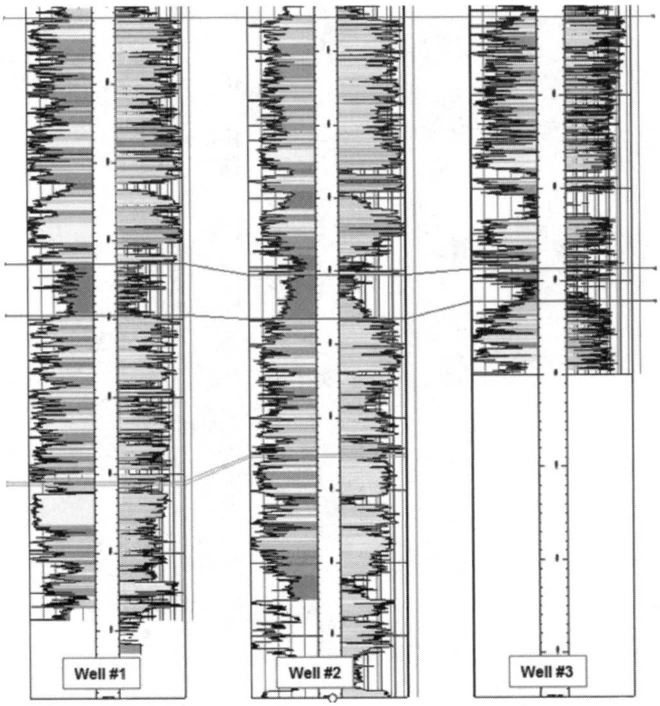

Figure 18.22 Correlation of well log data in Example 18.3.

Step 3: Develop the model stratigraphy
The stratigraphy of the reservoir was defined in the geologic model with 45 zones. This required the generation of 91 conformable horizons. It was necessary to refine the stratigraphy as much as possible so that vertical variations of lithology are captured in the geologic model.

Step 4: Load and assign well log data
Processed log data were available from three wells. An example of correlation of well log data is shown in Figure 18.22. The well log data were also used to define facies in the model. Five facies were designated for the various flow units present in the model. Figure 18.23 shows the classification of the facies in the model.

Step 5: Conduct property modeling
The grid system designed for the model is 113 × 112 × 557 for a total of 7.0 million gridblocks. The dimensions of the gridblocks on the average are 400 ft × 404 ft × 3.6 ft. The areal grid is shown as Figure 18.24. The areal grid as designed is expected to be retained in the flow model without further need for upscaling. The vertical resolution of the grid system was chosen so that well data are captured at the scale of the log data.

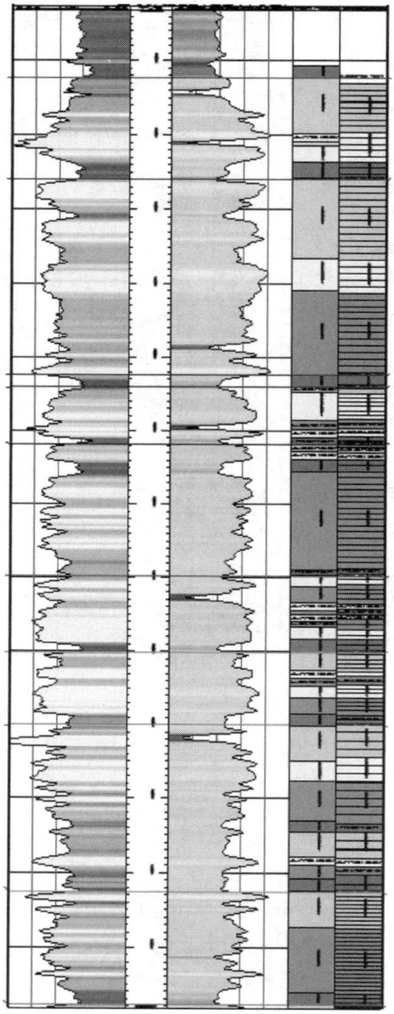

Figure 18.23 Facies definition and upscaling of a well in Example 18.3.

The variogram models were based on various data sources as follows:

- Directional variograms were based on seismic data. An example of variogram map based on seismic amplitude is shown as Figure 18.25.
- Vertical variogram models were based on well log data. Vertical trends and relative proportions of facies in the geologic model were derived from logs.
- Input variograms were based on published outcrop studies.

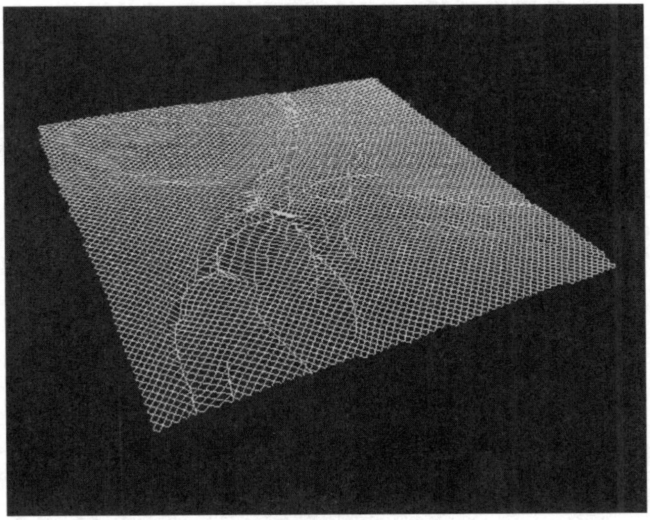

Figure 18.24 Areal grid system of the geologic model in Example 18.3.

The geologic model was populated stochastically using the variogram models derived as described earlier. Figure 18.26 shows a realization of facies distribution in one layer of the geologic model. Figure 18.27 is a cross-section showing vertical distribution of facies in the geologic model. Porosity distribution in the geologic model was by facies type. Figure 18.28 shows porosity distribution for a realization of the geologic model.

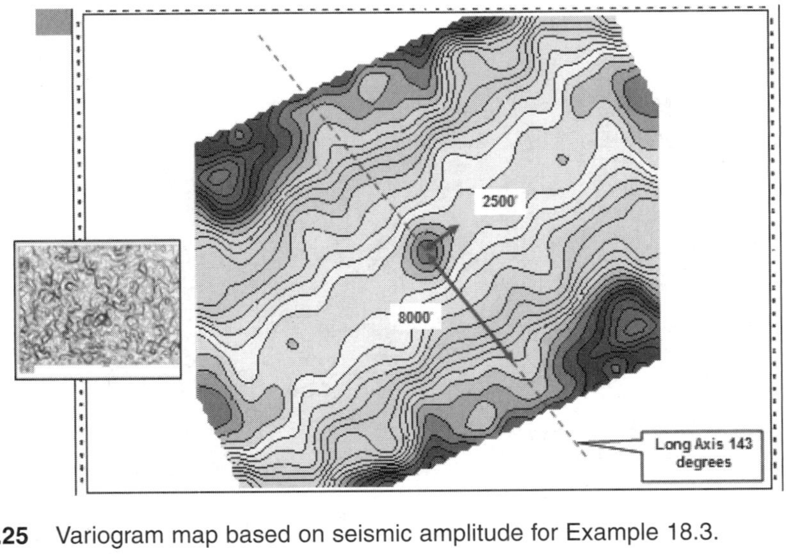

Figure 18.25 Variogram map based on seismic amplitude for Example 18.3.

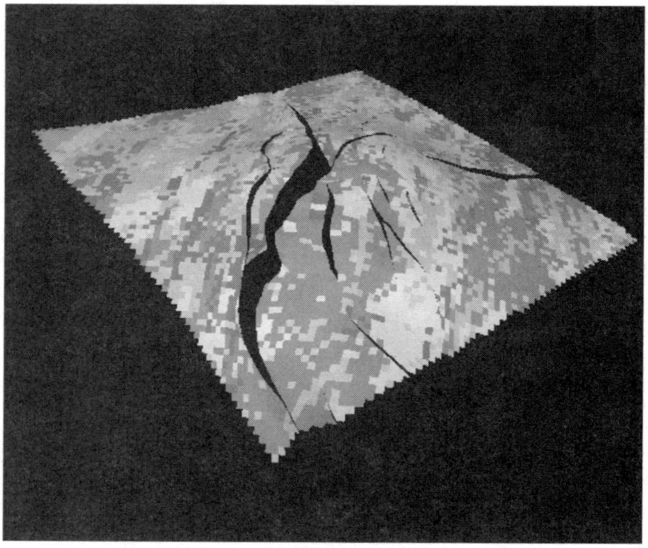

Figure 18.26 A realization of facies distribution in one layer of Reservoir "B" in Example 18.3.

Net-to-gross ratio was also distributed by facies. Permeability distribution by facies was based on cloud transform from porosity/permeability crossplots. The porosity/permeability crossplot for all the wells is shown in Figure 18.29. The permeability distribution for a realization of the geologic model is shown in Figure 18.30.

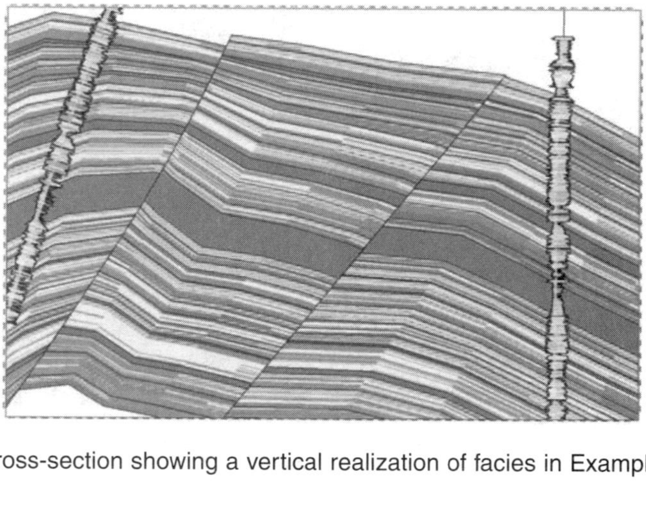

Figure 18.27 Cross-section showing a vertical realization of facies in Example 18.3.

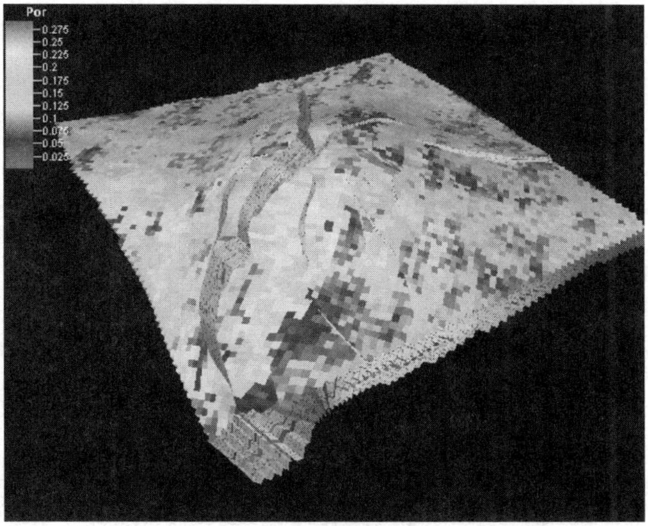

Figure 18.28 A porosity realization of Reservoir "B" in Example 18.3.

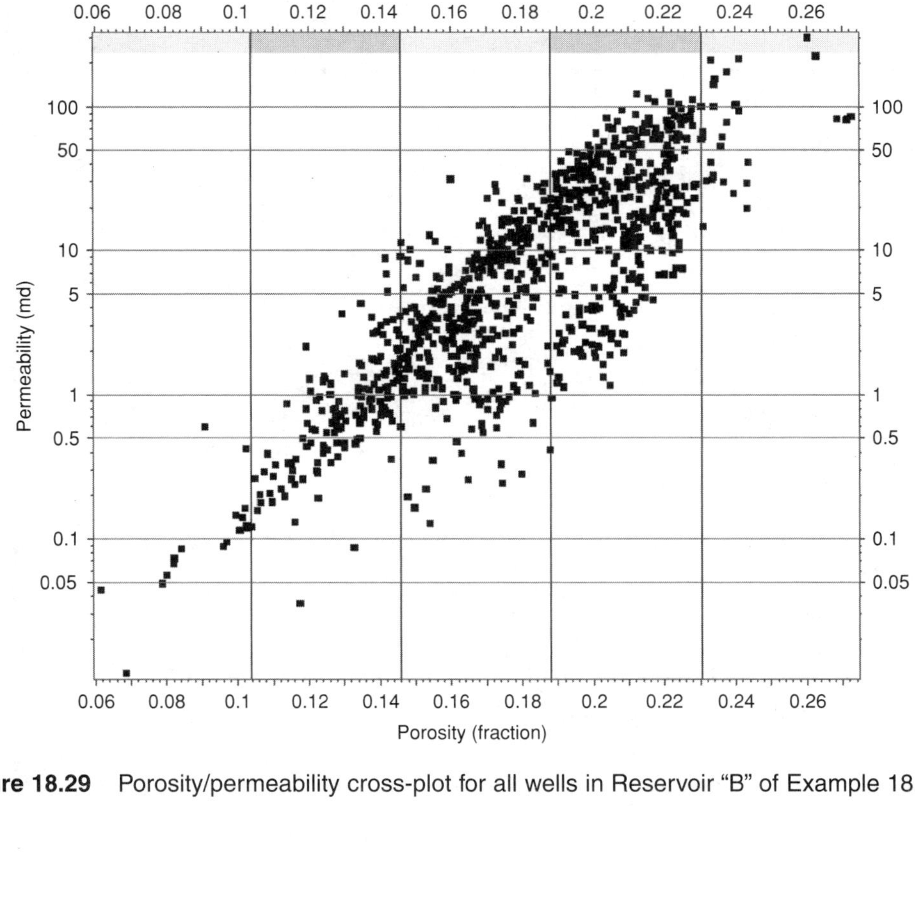

Figure 18.29 Porosity/permeability cross-plot for all wells in Reservoir "B" of Example 18.3.

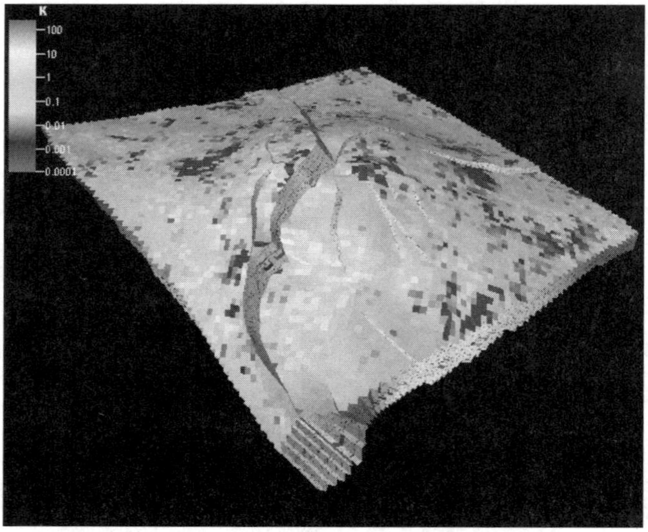

Figure 18.30 A permeability realization of Reservoir "B" in Example 18.3.

Nomenclature

h	lag distance
x_i	location of sample points
$z(x_i)$	sample values at location x_i
N(h)	number of sample pairs at lag distance, h
r	range of a semivariogram
γ(h)	semivariograms at lag distance, h
Δh	lag tolerance

Abbreviations

FBSA	flow-based scale averaging
LWD	logging-while-drilling
NMR	nuclear magnetic resonance
RCA	routine core analysis
SCAL	special core analysis
SGS	sequential Gaussian simulation
SIS	sequential indicator simulation
WFT	wireline formation tester

DST drill stem test
QA quality assurance
QC quality control

References

1. Ezekwe, J.N., and Filler, S.L.: "Modeling Deepwater Reservoirs," paper SPE 95066 presented at the Annual Technical Conference and Exhibition, Dallas, Texas, October 9–12, 2005.
2. McCarthy, P., Brand, J., Paradiso, B., Ezekwe, J., Wiltgen, N., Bridge, A., Willingham, R., and Bogaards, M.: "Using Geostatistical Inversion of Seismic and Borehole Data to Generate Reservoir Models for Flow Simulations of Magnolia Field, Deepwater Gulf of Mexico," *SEG Technical Program Abstracts,* (2005) 1351–1354.
3. Wu, Heyong, Wu, Haibo, Wang, D., Nian, J., Sun, Z., and Dusseault, M.: "Combined Seismic Inversion Methods in Reservoir Model Development," paper SPE 109626 presented at the 2007 Annual Technical Conference and Exhibition, Anaheim, California, November 11–14, 2007.
4. Wickens, H.D., and Bouma, A.H.: "The Tanqua Fan Complex, Karoo Basin, South Africa—Outcrop Analog for Fine-Grained, Deepwater Deposits," *in* A.H. Bouma and C.G. Stone, eds., *Fine-grained turbidite systems*, AAPG Memoir 72/SEPM Special Publication 68, Chapter 14, (2000).
5. Carr, M., and Gardner, M.H.: "Portrait of a Basin-Floor Fan for Sandy Deepwater Systems, Permian Lower Brushy Canyon Formation, West Texas," *in* A.H. Bouma, C.E. Stelting, and C.G. Stone, eds., *Fine-grained turbidite systems*, AAPG Memoir 72/SEPM Special Publication 68, Chapter 20, (2000).
6. Panda, M.N., Mosher, C., and Chopra, A.K.: "Reservoir Modeling Using Scale-Dependent Data," *SPEJ* (June 2001) 157–170.
7. Frykman, P., and Deutsch, C.V.: "Practical Application of Geostatistical Scaling Laws for Data Integration," *Petrophysics* (May–June 2002) 153–171.
8. Al-Ali, H.A., and Worthington, P.F.: "Applications of Petrophysical Scale Reconciliation to Saudi Arabian Reservoirs," paper IPTC 10726 presented at the International Petroleum Technology Conference, Doha, Qatar, November 21–23, 2005.
9. Ezekwe, N.J., and Shafer, J.L.: "Methods for Upscaling Diverse Rock Permeability Data for Reservoir Characterization and Modeling," paper SPE 109128 presented at the Annual Technical Conference and Exhibition, Anaheim, CA, November 11–14, 2007.
10. Isaaks, E.H., and Srivastava, R.M.: *An Introduction to Geostatistics,* Oxford University Press, Oxford, United Kingdom (1989).
11. Deutsch, C.: *Geostatistical Reservoir Modeling*, 1st Edition, Oxford University Press, New York (2002).

12. Kelkar, M., and Perez, G.: *Applied Geostatistics for Reservoir Characterization*, SPE, Richardson, Texas (2002).

13. Yarus, J.M., and Chambers, R.L.: "Practical Geostatistics—An Armchair Overview for Petroleum Reservoir Engineers," paper SPE 103357, Distinguished Author Series, *JPT* (November 2006) 78–87.

14. Gringarten, E., and Deutsch, C.V.: "Methodology for Variogram Interpretation and Modeling for Improved Reservoir Characterization," paper SPE 56654 presented at the 1999 SPE Annual Technical Conference and Exhibition, Houston, Texas, October 3–6, 1999.

15. Stern, D.: "Practical Aspects of Scaleup of Simulation Models," paper SPE 89032, Distinguished Author Series (September 2005) 74–81.

16. Dogru, A.H., Fung, L.S.K., Middya, U., Al-Shaalan, T.M., Pita, J.A., Kumar, K.H., Su, H.J., Tan, J.C.T., Hoy, H., Dreiman, W.T., Hahn, W.A., Al-Harbi, R., Al-Youbi, A., Al-Zamel, N.M., Mezghani, M., and Al-Mani, T.: "A Next-Generation Parallel Reservoir Simulator for Giant Reservoirs," paper SPE 119272 presented at the SPE Reservoir Simulation Symposium, The Woodlands, Texas, February 2–4, 2009.

17. Dogru, A.H, Fung, L.S.K., Al-Shaalan, T.M., Middya, U., and Pita, J.A.: "From Mega-Cell to Giga-Cell Reservoir Simulation," paper SPE 116675 presented at the SPE Annual Technical Conference and Exhibition, Denver, Colorado, September 21–24, 2008.

18. Saggaf, M.M.: "A Vision for Future Upstream Technologies," paper SPE 109323, Distinguished Author Series (March 2008) 94–98.

19. Brown, J.M.G., and Ezekwe, N.: "Improved Reservoir Management Using Geostatistical Reservoir Description, 24Z Stevens Reservoir, Elk Hills Oilfield, California," SPE 46232 presented at the Western Regional Meeting in Bakersfield, California, May 10–13, 1998.

20. Peaceman, D.W.: "Effective Transmissibilities of a Gridblock by Upscaling—Comparison of Direct Methods with Renormalization," paper SPE 36722, *SPEJ* (September 1997) 338–349.

General Reading

- Worthington, P.F.: "The Effect of Scale on the Petrophysical Estimation of Intergranular Permeability," SPWLA 44th Annual Logging Symposium, June 22–25, 2003.
- Slatt, R.M., Browne, G.H., Davis, R.J., Clemenceau, G.R., Colbert, J.R., Young, R.A., Anxionnaz, H., and Spang, R.J.: "Outcrop–Behind Outcrop Characterization of Thin-bedded Turbidites for Improved Understanding of Analog Reservoirs: New Zealand and Gulf of Mexico," paper SPE 49563 prepared for presentation at the 1998 SPE Annual Technical Conference and Exhibition, New Orleans, Louisiana, 27–30 September, 1998.
- Stewart, J., Dunn, P., Lyttle, C., Campion, K., Oyerinde, A., and Fischer, B.: "Improving Performance Prediction in Deep-Water Reservoirs: Learning From Outcrop Analogues,

Conceptual Models and Flow Simulation," paper IPTC 12892 presented at the International Petroleum Technology Conference, Kuala Lumpur, Malaysia, December 3–5, 2009.

- Bogan, C., Johnson, D., Litvak, M., and Stauber, D.: "Building Reservoir Models Based on 4D Seismic & Well Data in Gulf of Mexico Fields," paper SPE 84370 presented at the 2003 SPE Annual Technical Conference and Exhibition, Denver, Colorado, October 5–8, 2003.
- Aniekwena, A.U., McVay, D.A., Ahr, W.M., and Watkins, J.S.: "Integrated Characterization of the Thin-Bedded 8 Reservoir, Green Canyon 18, Gulf of Mexico": paper SPE 84051 presented at the 2003 SPE Annual Technical Conference and Exhibition, Denver, Colorado, October 5–8, 2003.
- Strebelle, S., Payrazyan, K., and Caers, J.: "Modeling of a Deepwater Turbidite Reservoir Conditional to Seismic Data Using Multiple-Point Geostatistics," paper SPE 77425 presented at the 2002 SPE Annual Technical Conference and Exhibition, San Antonio, Texas, September 29–October 2, 2002.
- Caers, J., Asveth, P., and Mukerji, T.: "Geostatistical integration of rock physics, seismic amplitudes and geological models in North-Sea turbidite systems," paper SPE 71321 presented at the 2001 SPE Annual Technical Conference and Exhibition, New Orleans, Louisiana, September 30–October 3, 2001.
- Shmaryan, L.E. and Deutsch, C.V.: "Object–Based Modeling of Fluvial/Deepwater Reservoirs with Fast Data Conditioning: Methodology and Case Studies," paper SPE 56821 presented at the 1999 SPE Annual Technical Conference and Exhibition, Houston, Texas, October 3–6, 1999.
- Shew, R.D., Tiller, G.M., Hackbarth, C.J., Rollins, D.R., and White, C.D.: "Characterization and Modeling of Channel and Thin-Bedded Turbidite Prospects in the Gulf of Mexico: Integration of Outcrops, Modern Analogs, and Subsurface Data," paper SPE 30535 presented at the 1995 SPE Annual Technical Conference and Exhibition, Dallas, Texas, October 22–25, 1995.
- Al-Aziz, S.A., Al-Rowyeh, A., AL-Enezi, T., Chetri, H.B., and AL-Anzi, E.: "Sabiriyah Upper Burgan Geological Model and Reservoir Characterization," paper SPE 120380 presented at the 2009 SPE Middle East Oil & Gas Show and Conference, Bahrain, March 15–18, 2009.
- Camey, S., Chai, S.N., Abu Bakar, M.F., Barr, D., and Leal, L.: "Reservoir Characterization of the Miocene Reservoirs of the North Malay Basin, Malaysia-Thailand Joint Development Area (MTJDA)", paper SPE 116311 presented at the 2008 SPE Asia Pacific Oil & Gas Conference and Exhibition, Perth, Australia, October 20–22, 2008.
- Al Maskeen, A.M.A.: "Advanced Geological Modeling and Uncertainty Analysis in a Complex Clastic Gas Reservoir from Saudi Arabia," paper SPE 109275 presented at the 2007 Asia Pacific Oil & Gas Conference and Exhibition, Jakarta, Indonesia, October 30–November 1, 2007.

- Lerat, O., Nivlet, P., Doligez, B., Lucet, N., Roggero, F., Berthet, P., Lefeuvre, F., and Vittori, J.: "Construction of a Stochastic Geological Model Constrained by High–Resolution 3D Seismic Data—Application to the Girassol Field, Offshore Angola," paper SPE 110422 presented at the 2007 SPE Annual Technical Conference and Exhibition, Anaheim, California, November 11–14, 2007.
- Christie, M.A.: "Upscaling for Reservoir Simulation," *JPT* (November 1996) 1004–1010.
- Christie, M.A.: "Tenth SPE Comparative Solution Project: A Comparison of Upscaling Techniques," *SPEREE* (August 2001) 308–317.
- Salazar, M.O., and Villa, J.R.: "Permeability Upscaling Techniques for Reservoir Simulation," paper SPE 106679 presented at the 2007 SPE Latin America and Carribean Petroleum Engineering Conference, Buenos Aires, Argentina, April 15–18, 2007.
- Delfiner, P.: "Three Statistical Pitfalls of Phi-K Transforms," paper SPE 102093 presented at the 2006 SPE Annual Technical Conference and Exhibition, San Antonio, Texas, September 24–27, 2006.

CHAPTER 19

Reservoir Simulation

19.1 Introduction

The term "reservoir simulation" is generally used to describe the activities involved in the building and execution of a model that represents the reservoir, such that the behavior of the model mirrors or "simulates" as much as possible the observed behavior of the reservoir. The model built in this case consists of sets of mathematical equations that represent material balance, fluid flow, and other physical processes occurring in the reservoir, subject to some defined constraints and conditions. Many forms of the mathematical model that represent the reservoir can be developed, depending on the nature of the reservoir, physical data available to describe the reservoir, processes occurring within the reservoir, and the objectives of simulating the behavior of the reservoir. In this book, three-dimensional, three-phase mathematical reservoir models (called black oil models) are used almost exclusively to describe the general principles and practice of reservoir simulation. This approach was taken to limit the scope of the material covered on reservoir simulation in this book, and also in recognition of the extensive body of work and books already published on reservoir simulation. These books[1–3] and literature[4–6] are recommended as additional sources of material for acquiring more knowledge on the principles and applications of petroleum reservoir simulation.

Why is it necessary to conduct numerical simulations of petroleum reservoirs? The answers to this cogent question can be approached in a variety of ways. But the simplest, most straightforward answer to the question was provided by Coats[7] with this statement: "Whereas the field can be produced only once—and at considerable expense—a model can be produced or 'run' many times at low expense over a short period of time." In other words, reservoir simulation provides a powerful tool for evaluating alternative reservoir management strategies, thereby making it possible to select an optimum management strategy, based on existing reservoir and operating conditions. Since the above statement was made by Coats[7] in 1969, the application of reservoir simulation as a tool for reservoir management, and for assessments of uncertainty and risks associated with reservoir development, has become widespread in the petroleum industry. There is scarcely any recent major field development or project that is not evaluated or assessed with some form of reservoir simulation. The reach of project evaluation with reservoir simulation is quite extensive from the development of deepwater reservoirs to the improvement of hydrocarbon recovery with secondary or enhanced oil recovery processes. The relevant question in the industry is no longer why simulate, but rather how to effectively and efficiently apply the technology of reservoir simulation to develop reservoir management strategies, optimize reservoir development, and improve hydrocarbon recovery.

Reservoir simulation is applied on the evaluation or assessment of key reservoir management issues, problems or operating questions such as:

1. What are the expected hydrocarbon recoveries under current and alternative reservoir management strategies? Can the expected hydrocarbon recovery be improved?
2. Is the reservoir a good candidate for pressure maintenance? Should gas or water injection be used for pressure maintenance?
3. For gas condensate reservoirs, will gas cycling improve total hydrocarbon recovery from the reservoir?
4. What is the optimum well count and well locations for development of new reservoirs?
5. Is the development of a new discovery economic under prevailing drilling and completion costs, operating costs, and product prices? What management strategies can be used to change or improve the economic outcomes of new projects?
6. What is the optimum producing rate for each well in the reservoir? What is the effect of producing rate on hydrocarbon recovery? Can the reservoir sustain production at levels necessary to maintain profitability?
7. Is the reservoir a good candidate for secondary recovery processes? If so, when is the proper time to initiate the secondary recovery process? What type of secondary recovery process to implement; for instance, pattern waterflooding versus peripheral (flank) waterflooding? What is the expected incremental hydrocarbon recovery from secondary recovery processes?
8. What types of enhanced oil recovery processes will be suitable for the reservoir? What is the ranking of various enhanced oil recovery processes when applied to the reservoir? What is the expected incremental recovery and profitability of the selected enhanced oil recovery process?
9. How is the reservoir performing under current operating strategy? Are they indications from current performance data of future problems? How can those problems be avoided or alleviated by modifying current practices and/or management strategies?
10. What are the uncertainties and risks associated with planned or current reservoir management strategies? Are those uncertainties and risks acceptable? How can those uncertainties and risks be improved by adopting alternative reservoir management strategies?

These are some of the problems of reservoir management that can be evaluated with reservoir simulation. Clearly, there are many more problems of reservoir management that can be addressed with reservoir simulation. Consequently, reservoir simulation has developed over many years to finally become an indispensible tool in modern reservoir engineering practice.

The material balance equations for three-dimensional, three-phase immiscible flow of oil, water, and gas were derived by Muskat.[8] These material balance equations are the fundamental equations used in black oil simulators. There are no analytical solutions of these non-linear, partial differential equations. Solutions of these equations are generally based on numerical methods. Advancement of numerical reservoir simulation has depended on the development of numerical methods to solve these partial differential equations, and availability of high speed digital computers for implementation of the solution techniques. Douglas et al.[9] proposed an Implicit (simultaneous) method for solving two-dimensional, two-phase flow problems. This method was used later by Coats et al.[10] on three-dimensional, two-phase flow problems, and by Coats[11,12] on three-dimensional, three-phase flow of incompressible and compressible fluids, respectively. An alternative solution technique described as Implicit-Pressure, Explicit Saturation (IMPES) method was proposed by Sheldon et al.[13] and Stone and Garder.[14] The method was later used on a two-dimensional, multiphase flow simulator by Fagin and Stewart[15] in 1966. The differences between these two solution methods (Implicit and IMPES) are discussed later. The above account is meant to provide a historical perspective and is not considered an accurate historical account of the development of numerical reservoir simulators. Many other pioneers contributed in various ways to the development of this technology. It is not possible to provide a complete list of these pioneers, but their contributions to the development of reservoir simulation is hereby acknowledged.

The use of numerical methods to solve the equations formulated to represent material balance and fluid flow in reservoirs leads to non-exact solutions. Moreover, application of different numerical solution techniques on the same set of equations will generally yield solutions that are slightly different, including associated errors. In addition, solutions can be achieved by making different sets of assumptions in the process of formulating the equations for solution. The point is that numerical simulators generally yield results that may vary when applied to the same reservoir problem, depending on the formulation of the equations and the numerical technique used in solving these equations. For many years, the non-uniqueness of solutions provided by numerical simulators undermined its acceptability as a valuable tool for reservoir analyses. These skepticisms on the applicability of numerical reservoir simulators have been erased gradually over the years in the petroleum industry through publication of sponsored comparative studies using reservoir simulators built by various groups,[16–25] and availability of a large body of work on reservoir simulation in the literature.[1–7, 26–29]

Numerical reservoir simulators are classified on the basis of reservoir type, reservoir process, and formulation of the simulator. The following types of reservoir simulators are commercially available in the petroleum industry:

1. Black oil simulators: oil, water, and gas are treated as separate immiscible phases. This simulator can be used on dry gas, and black oil reservoirs (as defined in Chapter 4), and immiscible recovery processes (Chapters 15 and 16). The black oil simulator is the "work-horse" of the petroleum industry with respect to reservoir simulation.

2. Compositional simulators: oil and gas are represented as hydrocarbon components. Water is present as a phase. This type of simulator is used on volatile oil, and gas condensate reservoirs (as defined in Chapter 4), and miscible gas enhanced oil recovery processes (Chapter 17).
3. Dual-porosity simulators: used on naturally fractured reservoirs that exhibit dual porosity behavior.
4. Thermal simulators: used for simulating thermal processes, such cyclic steam injection, steam flooding, and in-situ combustion (Chapter 17).
5. Chemical flood simulators: applicable to enhanced oil recovery processes using alkalis, surfactants, and polymers (Chapter 17).
6. Streamline simulators: This class of simulators is widely used in the industry for upscaling of large geologic models, modeling of waterflooding, and other uncomplicated reservoir processes. Streamline simulators can be used in conjunction with the other grid-based simulators, and are not considered a substitute for them.[30] Streamline simulators are not covered in this book.

Reservoir simulation is presented in this book with more emphasis on its use as a tool for reservoir analysis and evaluation. Theoretical development and formulation of equations are used where necessary to illustrate basic concepts, such as solution techniques, stability issues, material balance errors, etc. Detailed presentations of various formulations of the basic material balance equations and solution techniques are not covered in this book. These are available in other books written exclusively on reservoir simulation.[1–3] In this chapter, the basic material balance and flow equations are presented for a black oil simulator. This is then followed with presentation of basic terms, concepts, and solution methods encountered in the practice of reservoir simulation. The chapter concludes with the presentation of the basic structure of data input generally encountered in most commercial simulators.

19.2 Derivation of the Continuity Equation in Rectangular Form

Consider a rectangular block with fluid flow in three directions, x, y, and z as shown in Figure 19.1. During time interval, Δt, the principle of mass conservation can be applied on the block to obtain the following mass balance equation:

$$\begin{aligned}&\text{Mass flowing into the block in time, } \Delta t - \text{Mass flowing out of the block in time, } \Delta t\\&= \text{Mass accumulation in the block in time, } \Delta t\end{aligned} \tag{19.1}$$

Let mass flow rate per unit cross-sectional area normal to the direction of flow be denoted as F. Using this designation, fluid flow into the block in the x-direction at x is $(F_x)_x$, and flow out of the block at $x + \Delta x$ is $(F_x)_{x+\Delta x}$. Similarly, flow into the block in the y and z directions are designated as $(F_y)_y$, and $(F_z)_z$, respectively. And flow out of the block in the y and z directions

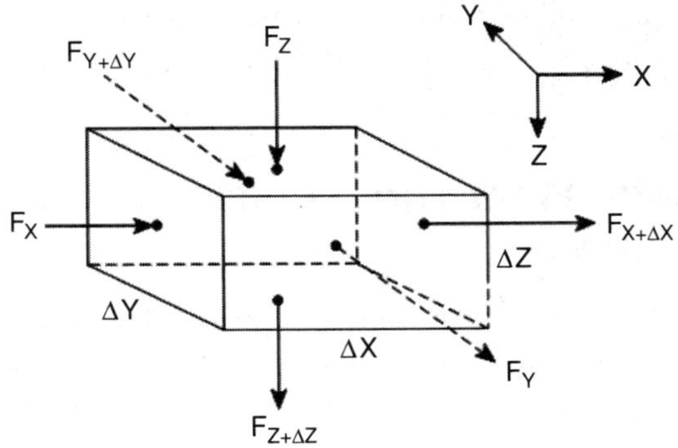

Figure 19.1 A gridblock showing fluid flow in three directions.

are designated as $(F_y)_{y+\Delta y}$, and $(F_z)_{z+\Delta z}$, respectively. The length, width, and height of the block are designated as Δx, Δy, and Δz, respectively, as shown in Figure 19.1. Using these definitions, it can be written from Eq. (19.1) that:

$$\text{Mass flowing into block in time, } \Delta t = [(F_x)_x \Delta y \Delta z + (F_y)_y \Delta x \Delta z + (F_z)_z \Delta x \Delta y]\Delta t \quad \textbf{(19.2)}$$

$$\begin{aligned}\text{Mass flowing out of block in time, } \Delta t &= [(F_x)_{x+\Delta x} \Delta y \Delta z + (F_y)_{y+\Delta y} \Delta x \Delta z + (F_z)_{z+\Delta z} \Delta x \Delta y]\Delta t \\ &\quad + q \Delta x \Delta y \Delta z \Delta t\end{aligned} \quad \textbf{(19.3)}$$

$$\text{Mass accumulation in block in time, } \Delta t = [(C_p)_{t+\Delta t} - (C_p)_t]\Delta x \Delta y \Delta z \quad \textbf{(19.4)}$$

In Eq. (19.3), q is a source/sink term that represents mass flow into or out of the block primarily through a well. For a well that is a producer, $q > 0$ (or positive), and for a well that is an injector, $q < 0$ (or negative) by convention. In Eq. (19.4), C_p is the concentration of any phase (oil, water, and gas) in the block defined in mass per unit volume.

Substituting Eqs. (19.2), (19.3), and (19.4) into Eq. (19.1) gives:

$$\begin{aligned}&[(F_x)_x \Delta y \Delta z + (F_y)_y \Delta x \Delta z + (F_z)_z \Delta x \Delta y]\Delta t \\ &- [(F_x)_{x+\Delta x} \Delta y \Delta z + (F_y)_{y+\Delta y} \Delta x \Delta z + (F_z)_{z+\Delta z} \Delta x \Delta y]\Delta t \\ &- q \Delta x \Delta y \Delta z \Delta t = [(C_p)_{t+\Delta t} - (C_p)_t]\Delta x \Delta y \Delta z\end{aligned} \quad \textbf{(19.5)}$$

Dividing Eq. (19.5) with $\Delta x \Delta y \Delta z \Delta t$ and collecting like terms yields:

$$-\frac{(F_x)_{x+\Delta x} - (F_x)_x}{\Delta x} - \frac{(F_y)_{y+\Delta y} - (F_y)_y}{\Delta y} - \frac{(F_z)_{z+\Delta z} - (F_z)_z}{\Delta z} - q = \frac{(C_p)_{t+\Delta t} - (C_p)_t}{\Delta t} \quad \textbf{(19.6)}$$

Taking limits as Δx, Δy, Δz, and Δt approach zero gives:

$$-\frac{\partial F_x}{\partial x} - \frac{\partial F_y}{\partial y} - \frac{\partial F_z}{\partial z} - q = \frac{\partial C_p}{\partial t} \quad \textbf{(19.7)}$$

Equation (19.7) is the continuity equation for fluid flow in rectangular coordinates (three directions). Eq. (19.7) is similar in form to Eq. (10.52) developed in Chapter 10 as the continuity equation in radial form.

19.3 Flow Equations for Three-Phase Flow of Oil, Water, and Gas

The mass flow rate per unit area in a given direction for any fluid is the product of the density of the fluid and its velocity in that direction. If oil, water, and gas are denoted with the subscripts o, w, and g, respectively, then $(\vec{F})_o$,$(\vec{F})_w$, and $(\vec{F})_g$, respectively, can be represented as:

$$(\vec{F})_o = \frac{\rho_{osc}}{B_o}\vec{v}_o \tag{19.8}$$

$$(\vec{F})_w = \frac{\rho_{wsc}}{B_w}\vec{v}_w \tag{19.9}$$

$$(\vec{F})_g = \frac{\rho_{gsc}}{B_g}\vec{v}_g + \frac{R_{so}\rho_{gsc}}{B_o}\vec{v}_o + \frac{R_{sw}\rho_{gsc}}{B_w}\vec{v}_w \tag{19.10}$$

In Eqs. (19.8), (19.9), and (19.10), $\vec{F}$ is the mass flow rate in x, y, and z directions; $\vec{v}$ is the fluid velocity in x, y, and z directions; ρ_{osc}, ρ_{wsc}, and ρ_{gsc} are oil, water, and gas densities at standard conditions, respectively; R_{so} and R_{sw} are gas solubilities in oil and water, respectively, scf/STB; and B_o, B_w, and B_g are formation volume factors of oil, water, and gas, respectively, RB/STB. For oil, water, and gas, the concentration of each phase, C_p, is defined, respectively, as:

$$C_o = \frac{\phi\rho_{osc}S_o}{B_o} \tag{19.11}$$

$$C_w = \frac{\phi\rho_{wsc}S_w}{B_w} \tag{19.12}$$

$$C_g = \phi\rho_{gsc}\left[\frac{S_g}{B_g} + R_{so}\frac{S_o}{B_o} + R_{sw}\frac{S_w}{B_w}\right] \tag{19.13}$$

In Eqs. (19.11), (19.12), and (19.13), ϕ is the porosity of the block; and S_o, S_w, and S_g are oil, water, and gas saturations, respectively, in the block. A constraint applied to fluid saturations in the block is the requirement that the sum of the fluid saturations is equal to unity:

$$S_o + S_w + S_g = 1 \tag{19.14}$$

Substituting Eqs. (19.8) to (19.13) into Eq. (19.7) with vector expansions gives a mass balance equation for each phase as:

For oil:

$$-\left[\frac{\partial}{\partial x}\left(\frac{\rho_{osc}}{B_o}v_{xo}\right) + \frac{\partial}{\partial y}\left(\frac{\rho_{osc}}{B_o}v_{yo}\right) + \frac{\partial}{\partial z}\left(\frac{\rho_{osc}}{B_o}v_{zo}\right)\right] - q_o = \frac{\partial}{\partial t}\left(\frac{\phi\rho_{osc}S_o}{B_o}\right) \tag{19.15}$$

For water:

$$-\left[\frac{\partial}{\partial x}\left(\frac{\rho_{wsc}}{B_w}v_{xw}\right)+\frac{\partial}{\partial y}\left(\frac{\rho_{wsc}}{B_w}v_{yw}\right)+\frac{\partial}{\partial z}\left(\frac{\rho_{wsc}}{B_w}v_{zw}\right)\right]-q_w=\frac{\partial}{\partial t}\left(\frac{\phi\rho_{wsc}S_w}{B_w}\right) \quad (19.16)$$

For gas:

$$\begin{aligned}&-\frac{\partial}{\partial x}\left(\frac{\rho_{gsc}}{B_g}v_{xg}+\frac{R_{so}\rho_{gsc}}{B_o}v_{xo}+\frac{R_{sw}\rho_{gsc}}{B_w}v_{xw}\right)-\frac{\partial}{\partial y}\left(\frac{\rho_{gsc}}{B_g}v_{yg}+\frac{R_{so}\rho_{gsc}}{B_o}v_{yo}+\frac{R_{sw}\rho_{gsc}}{B_w}v_{yw}\right)\\&-\frac{\partial}{\partial z}\left(\frac{\rho_{gsc}}{B_g}v_{zg}+\frac{R_{so}\rho_{gsc}}{B_o}v_{zo}+\frac{R_{sw}\rho_{gsc}}{B_w}v_{zw}\right)-q_g=\frac{\partial}{\partial t}\left[\phi\rho_{gsc}\left(\frac{S_g}{B_g}+\frac{R_{so}S_o}{B_o}+\frac{R_{sw}S_w}{B_w}\right)\right]\end{aligned} \quad (19.17)$$

Eqs. (19.15), (19.16), and (19.17) can be simplified by canceling out ρ_{osc}, ρ_{wsc}, and ρ_{gsc}, respectively, to get:

For oil:

$$-\left[\frac{\partial}{\partial x}\left(\frac{v_{xo}}{B_o}\right)+\frac{\partial}{\partial y}\left(\frac{v_{yo}}{B_o}\right)+\frac{\partial}{\partial z}\left(\frac{v_{zo}}{B_o}\right)\right]-\frac{q_o}{\rho_{osc}}=\frac{\partial}{\partial t}\left(\frac{\phi S_o}{B_o}\right) \quad (19.18)$$

For water:

$$-\left[\frac{\partial}{\partial x}\left(\frac{v_{xw}}{B_w}\right)+\frac{\partial}{\partial y}\left(\frac{v_{yw}}{B_w}\right)+\frac{\partial}{\partial z}\left(\frac{v_{zw}}{B_w}\right)\right]-\frac{q_w}{\rho_{wsc}}=\frac{\partial}{\partial t}\left(\frac{\phi S_w}{B_w}\right) \quad (19.19)$$

For gas:

$$\begin{aligned}&-\frac{\partial}{\partial x}\left(\frac{v_{xg}}{B_g}+\frac{R_{so}v_{xo}}{B_o}+\frac{R_{sw}v_{xw}}{B_w}\right)-\frac{\partial}{\partial y}\left(\frac{v_{yg}}{B_g}+\frac{R_{so}v_{yo}}{B_o}+\frac{R_{sw}v_{yw}}{B_w}\right)\\&-\frac{\partial}{\partial z}\left(\frac{v_{zg}}{B_g}+\frac{R_{so}v_{zo}}{B_o}+\frac{R_{sw}v_{zw}}{B_w}\right)-\frac{q_g}{\rho_{gsc}}=\frac{\partial}{\partial t}\left[\phi\left(\frac{S_g}{B_g}+\frac{R_{so}S_o}{B_o}+\frac{R_{sw}S_w}{B_w}\right)\right]\end{aligned} \quad (19.20)$$

In vector notation, Eqs. (19.18), (19.19), and (19.20) can be written as:

For oil:

$$-\nabla\cdot\frac{\vec{v}_o}{B_o}-\frac{q_o}{\rho_{osc}}=\frac{\partial}{\partial t}\left(\frac{\phi S_o}{B_o}\right) \quad (19.21)$$

For water:

$$-\nabla\cdot\frac{\vec{v}_w}{B_w}-\frac{q_w}{\rho_{wsc}}=\frac{\partial}{\partial t}\left(\frac{\phi S_w}{B_w}\right) \quad (19.22)$$

For gas:

$$-\nabla\cdot\left(\frac{\vec{v}_g}{B_g}+\frac{R_{so}\vec{v}_o}{B_o}+\frac{R_{sw}\vec{v}_w}{B_w}\right)-\frac{q_g}{\rho_{gsc}}=\frac{\partial}{\partial t}\left[\phi\left(\frac{S_g}{B_g}+\frac{R_{so}S_o}{B_o}+\frac{R_{sw}S_w}{B_w}\right)\right] \quad (19.23)$$

In vector notation,

$$\nabla \cdot \vec{v} = \frac{\partial}{\partial x} v_x + \frac{\partial}{\partial y} v_y + \frac{\partial}{\partial z} v_z \tag{19.24}$$

If the fluid velocities in Eqs. (19.21), (19.22), and (19.24) are based on Darcy's law, then the fluid velocities for each phase can be represented in terms of potential as:

$$\vec{v}_p = -\overleftrightarrow{K}\lambda_p \nabla \Phi_p \tag{19.25}$$

In Eq. (19.25), $\vec{v}$ is the fluid velocity of phase, p, in vector form; $\overleftrightarrow{K}$ represents permeability as a tensor; λ_p is the mobility of the phase, where $\lambda_p = k_{rp}/\mu_p$; and Φ_p is the potential of the phase where $\Phi_p = p_p - (\rho_p z)/144$, assuming $g = g_c$.

Substituting Eq. (19.25) into Eqs. (19.21), (19.22), and (19.23) gives:

For oil:

$$\nabla \cdot \frac{\overleftrightarrow{K}\lambda_o}{B_o} \cdot \nabla \Phi_o - \frac{q_o}{\rho_{osc}} = \frac{\partial}{\partial t}\left(\frac{\phi S_o}{B_o}\right) \tag{19.26}$$

For water:

$$\nabla \cdot \frac{\overleftrightarrow{K}\lambda_w}{B_w} \cdot \nabla \Phi_w - \frac{q_w}{\rho_{wsc}} = \frac{\partial}{\partial t}\left(\frac{\phi S_w}{B_w}\right) \tag{19.27}$$

For gas:

$$\nabla \cdot \overleftrightarrow{K}\left(\frac{\lambda_g}{B_g}\nabla \Phi_g + \frac{R_{so}\lambda_o}{B_o}\nabla \Phi_o + \frac{R_{sw}\lambda_w}{B_w}\nabla \Phi_w\right) - \frac{q_g}{\rho_{gsc}} = \frac{\partial}{\partial t}\left[\phi\left(\frac{S_g}{B_g} + \frac{R_{so}S_o}{B_o} + \frac{R_{sw}S_w}{B_w}\right)\right] \tag{19.28}$$

Equations (19.26), (19.27), and (19.28) are the basic equations for immiscible flow of oil, water, and gas, respectively. When written in difference form, these are the basic equations solved numerically in black oil simulators. The equations are presented in this book essentially as derived by Fanchi.[31]

The derivations of the continuity equation (Eq. (19.7)) and the flow equations (Eqs. (19.26), (19.27), and (19.28)) for a black oil simulator were presented to emphasize the point that the development of reservoir simulators are based on the fundamental principles of mass balances and the Darcy flow equation, which are the bedrock of most calculations in petroleum reservoir engineering. In other types of simulators, mass balances are applied on the components within the phases, and other fundamental flow equations may be included in the derivations of the relevant equations. For instance, in a compositional simulator, mass balances are applied to the defined components of the hydrocarbon phases (Chapter 6), and the equilibrium between components in the vapor and liquid phases are determined with an equation of state.[27,32] For thermal simulators, an energy balance on the system in addition to mass balances on water, steam, and oil are included in the relevant equations.[33,34] The point here is to reassure potential users of reservoir simulation that reservoir simulators are based on fundamental principles of mass balance, energy balance, Darcy flow equations, etc. that form the basis of most analytical reservoir engineering calculations. With proper geologic description and reservoir characterization, application of reservoir simulation should yield results that are comparable, if not superior, to analytical methods.

19.4 Basic Concepts, Terms, and Methods in Reservoir Simulation

In this section, basic concepts, terms, and methods that are commonly used in the reservoir simulation environment are presented. This section is intended for readers who have little or no prior knowledge of reservoir simulation. The subjects covered in this section are relatively basic, and readers with prior knowledge of reservoir simulation may elect to skip this section entirely.

19.4.1 Grid Systems

The discretization in space of the difference forms of the flow equations (such as Eqs. (19.26) to (19.28)) in reservoir simulators requires the use of grids. The key factors that influence the design of grid systems for reservoir simulation were discussed extensively in Chapter 18, Section 18.5.5.1 under Model Gridding. These factors are summarized here to re-emphasize their importance in the design of grid systems for reservoir simulation as follows:

1. Resolution of the geologic and petrophysical data
2. Size of the static geologic model
3. Limitations on the size of the reservoir flow model
4. Potential uses of the reservoir model
5. Upscaling (also termed upgridding or scaleup) techniques to convert the static geologic model into a reservoir flow model

In this section, the different types of grid systems and their potential effects on reservoir simulation results are presented.

19.4.1.1 Cartesian Grids

The most common type of grid systems used in reservoir simulation are based on Cartesian coordinates of x, y, and z. Cartesian grids can be based on a block-centered grid or a point-distributed grid systems (Figure 19.2). The block-centered grid is better for calculating mass accumulation terms in the flow equations, while the point-distributed grid is more accurate for calculating flows between the gridblocks.[35] The differences between the two grid systems (block-centered vs point-distributed) are insignificant, if the Cartesian grid systems are uniform. Most commercial simulators use block-centered grids. Cartesian grid systems can be classified as one-dimensional (1D), two-dimensional (2D), and three-dimensional (3D). Figure 19.3 shows a 1D grid system in the x-direction. Figure 19.4 shows a 2D grid system for an areal model (x, y), and a cross-sectional model (x, z). Figure 19.5 shows a 3D grid system for a reservoir model (x, y, z).

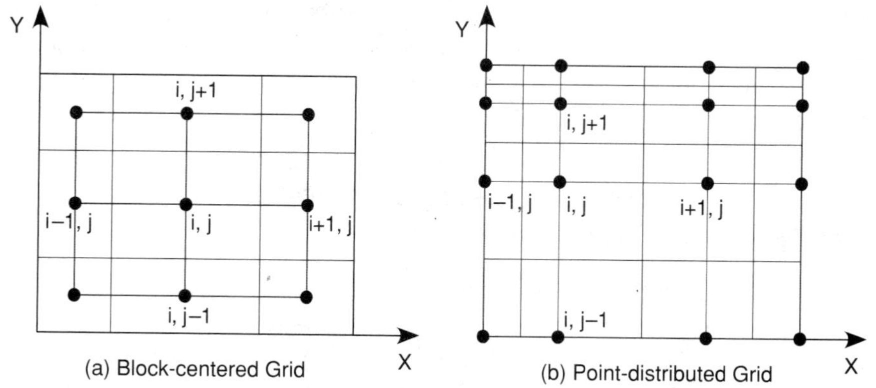

Figure 19.2 Block-centered and point-distributed grids (from Aziz[35] © 1993 SPE, Reproduced with permission).

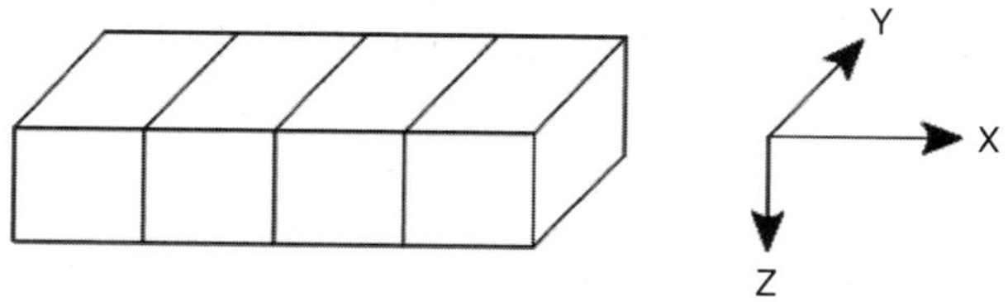

Figure 19.3 1D grid model.

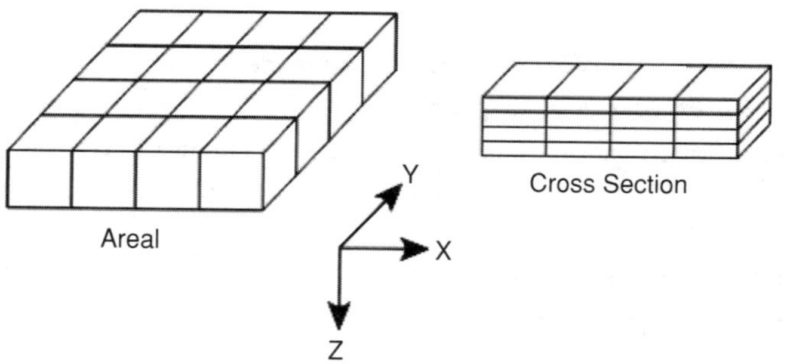

Figure 19.4 2D grid model.

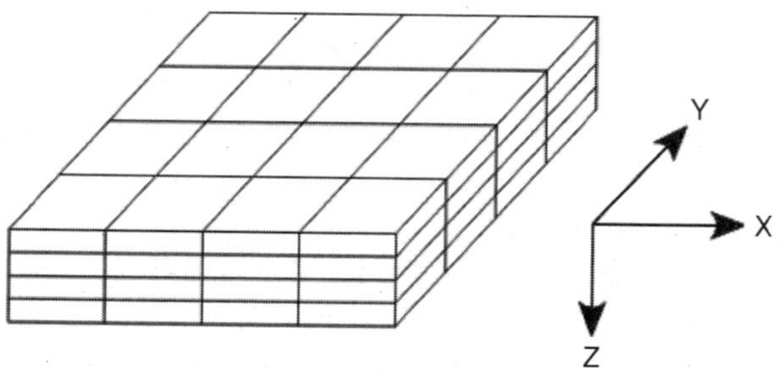

Figure 19.5 3D grid model.

The discretization in space of the simulator flow equations assumes that the gridblocks are orthogonal. This means that each column of gridblocks is at right angles to each row of gridblocks. In designing a Cartesian grid system for reservoir simulation, it is important to maintain the gridblocks orthogonal to each other. Otherwise, the results from the simulation may not be accurate, if the gridblocks are severely non-orthogonal. Most commercial simulators incorporate features to ensure orthogonality of the gridblocks for irregular grid systems, especially if the static geologic model is intended to be used in reservoir simulation.

19.4.1.2 Radial grids

Grid systems are also based on radial coordinates of r, θ, and z. An example of a grid system based on radial coordinates is shown in Figure 19.6. Radial grid systems are used mainly in the representation of wells in models that show rapid changes in fluid saturations near the wellbore as in gas cusping or water coning studies. In some applications, the wells are represented with radial grids while the remaining parts of the reservoir have Cartesian grids, thereby creating a **hybrid grid**[36] system as shown in Figure 19.7.

19.4.1.3 Corner-Point geometry

Corner-point geometry is based on specifying the locations of all the eight corners of each gridblock as shown in Figure 19.8. This gridding system is useful for representing complex geologic geometries such as faults, permeability barriers, irregular boundaries, sand channels, etc. However, they could generate highly irregular grids that may not be orthogonal. As pointed out earlier, non-orthogonal grids, especially from corner-point geometry, could yield erroneous flow simulation results.[35]

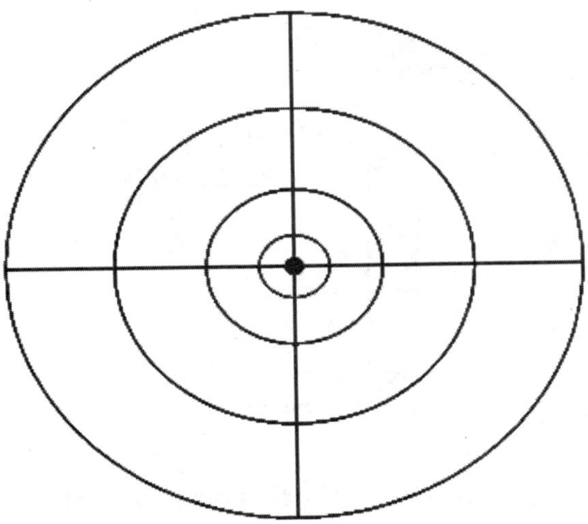

Figure 19.6 Radial grid model.

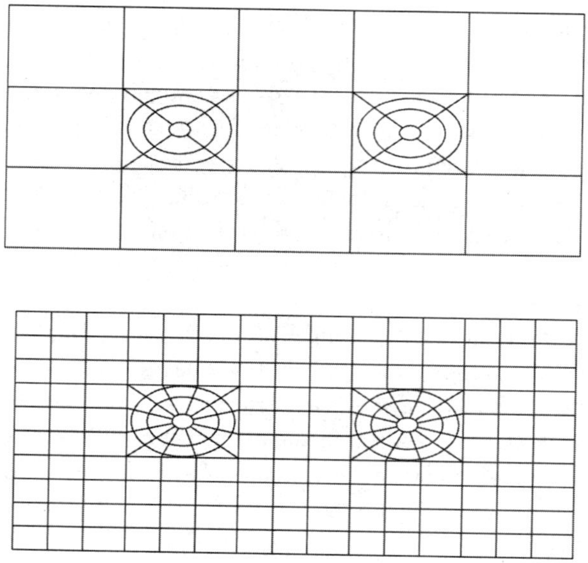

Figure 19.7 Hybrid grid model (from Aziz[35] © 1993 SPE, Reproduced with permission).

19.4.1.4 Voronoi or Perpendicular Bisection (PEBI) Grids

Voronoi grid, named after Voronoi,[37] is very flexible and locally orthogonal. It is also called perpendicular bisection (PEBI) grid in the petroleum industry. A Voronoi gridblock is defined as the region of space that is closer to its gridpoint than to any other gridpoint.[38] Voronoi grids are based on a generalization of the point-distributed grid system (Figure 19.2). Voronoi gridblocks are locally orthogonal because a line that joins gridpoints of any two connected gridblocks is perpendicular to the gridblock boundary between these two gridpoints and is divided into two equal

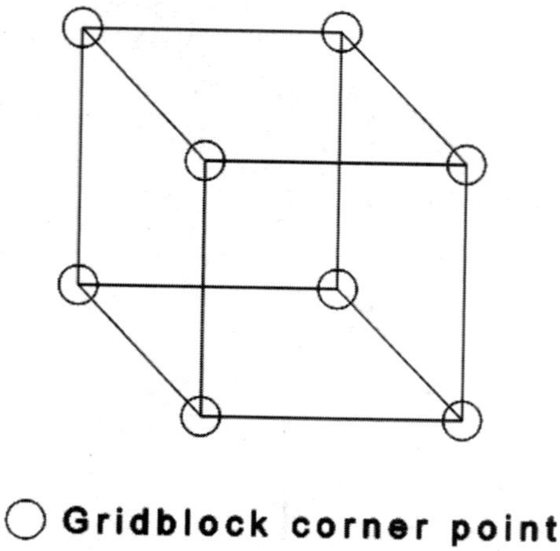

Figure 19.8 Gridblock showing corner-point geometry.

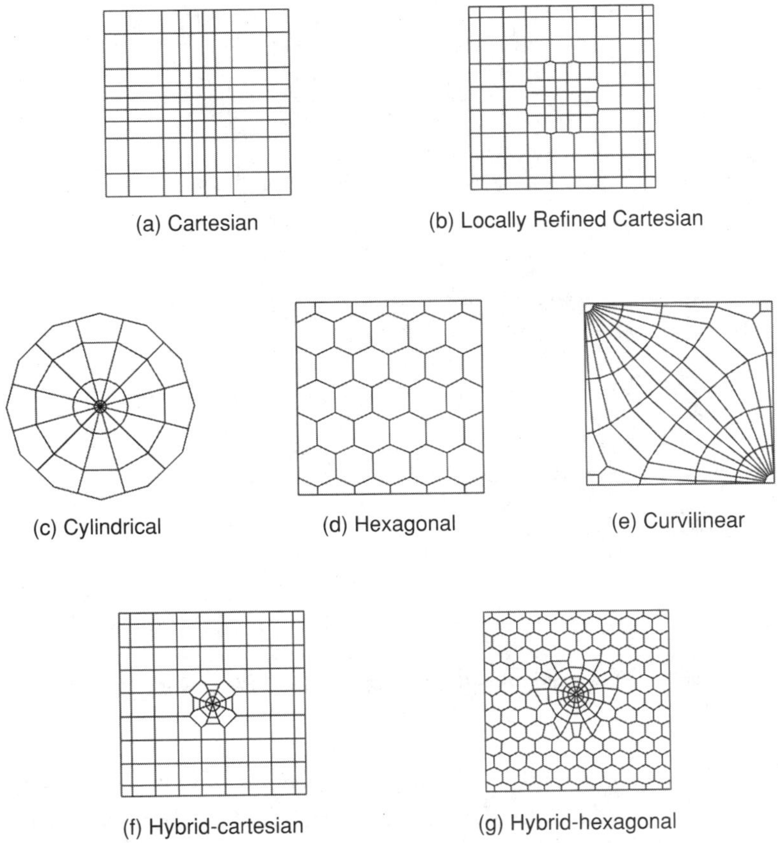

Figure 19.9 Examples of Voronoi grids (from Aziz[35] © 1993 SPE, Reproduced with permission).

parts by the boundary.[35] Voronoi grids are very flexible for representing complex well geometries, such as horizontal wells, and major geological features, such as faults, permeability barriers, etc. Examples of Voronoi grids are shown in Figure 19.9. The flexibility of Voronoi grid is illustrated in Figure 19.10. Wide applications of Voronoi grid in the petroleum industry are still limited primarily because many engineers are more familiar with using Cartesian-based grid systems, especially for analyses, visualization, and reporting of simulation results.

19.4.1.5 Local Grid Refinement (LGR)

Local grid refinement (LGR) is a technique used to increase the grid density in regions of the reservoir where rapid changes in saturations and pressures are occurring so as to represent those changes more accurately. LGR are used mainly around well locations or to monitor saturation changes at fluid boundaries. An example of LGR around well locations is shown in Figure 19.11 for Cartesian grids. LGR should be used with caution due to two main potentially adverse consequences from indiscriminate application of the technique. The first adverse result of LGR is

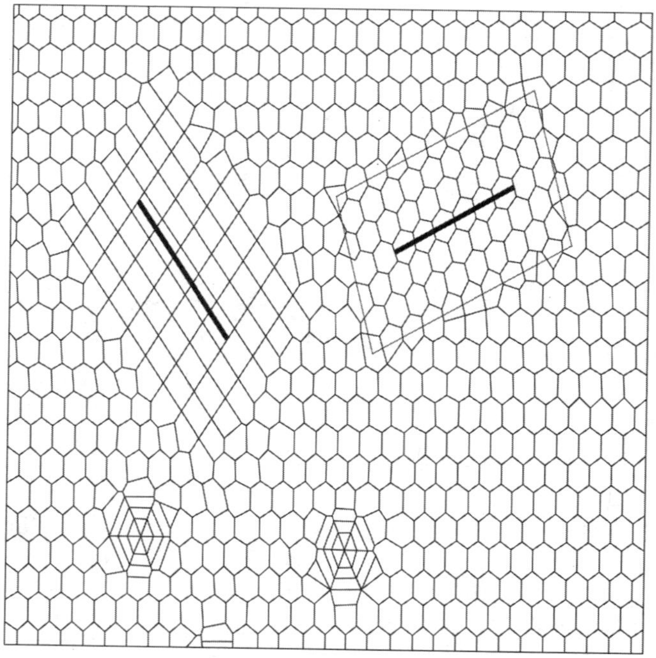

Figure 19.10 Flexibility of Voronoi grids (from Aziz[35] © 1993 SPE, Reproduced with permission).

fluid flow between a large gridblock and a small grid block in an LGR area, which could cause large oscillations in computed results. The second adverse effect is a significant increase in model run times, if many LGRs are present in a flow simulation model. A reverse form of LGR is **Grid Coarsening**. Grid coarsening involves reduction of the number of grid blocks in portions of the reservoir that do not have rapid changes in saturation or pressures, such as in aquifers surrounding the hydrocarbon portions of the reservoir (Figure 19.12). Grid coarsening is beneficial in terms of reducing computation and storage memory required for a given flow simulation model.

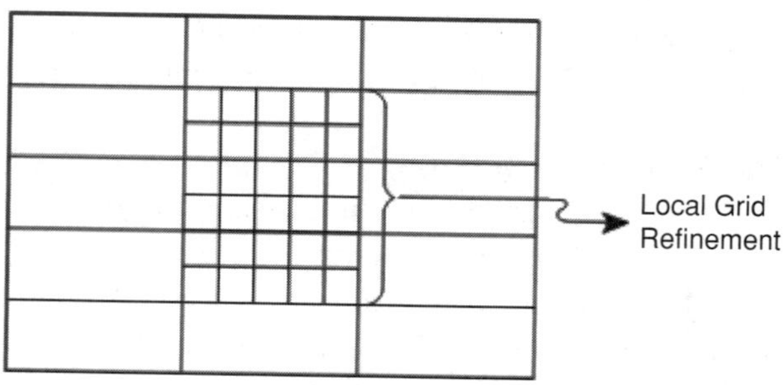

Figure 19.11 Local grid refinement.

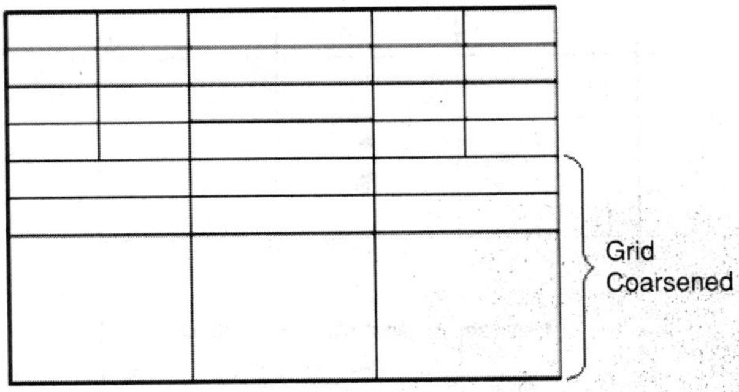

Figure 19.12 Grid coarsening.

19.4.1.6 Grid orientation effects

The orientation of Cartesian grids can have strong effects on the shapes of the displacement front and fluid breakthrough times. These effects are known as grid orientation effects. Strong grid orientation effects were reported by Todd et al.[39] for highly adverse mobility waterfloods and by Coats et al.[40] for pattern steamfloods. The impact of grid orientation effects can be reduced by using nine-point finite differencing[41] as shown in Figure 19.13 instead of five-point finite differencing shown in Figure 19.14. In nine-point finite difference formulation, flow is allowed between a gridblock and all its eight neighbor gridblocks including gridblocks diagonally adjacent to it (Figure 19.13). Five-point finite difference formulation allows flow between the gridblock and its four adjacent gridblocks (Figure 19.14). Note that nine-point formulation requires

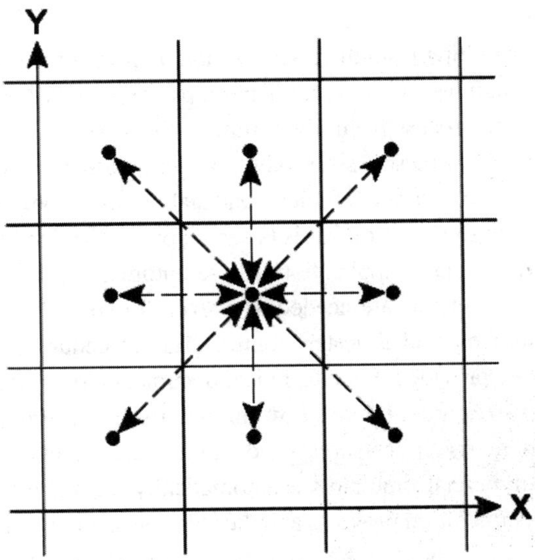

Figure 19.13 Nine-point differencing.

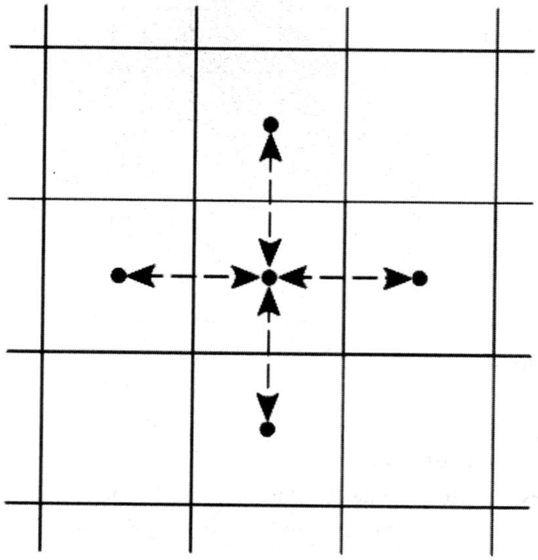

Figure 19.14 Five-point differencing.

more computation time than five-point formulation. Most commercial simulators provide the option of selecting either five-point or nine-point finite differencing schemes. The recommendation is to use nine-point finite difference formulation, if the process being simulated involves adverse mobility ratios, such as in waterfloods, gasfloods, or steamfloods.

19.4.2 Timesteps

The difference equations resulting from the discretization in time of the flow equations in reservoir simulators are solved in a number of timesteps. Timesteps are progressive increments of time used to advance simulation of the process from initial time to the stated end time. As the solution progresses in time, the selection of timestep sizes plays a very important role in ensuring that solutions of the difference equations are accurate, efficient, and stable. Most commercial simulators feature sophisticated automatic time step control that is based on pre-determined set of default parameters and tolerances. For many simulator applications, these automatic time step selection criteria are adequate and no further adjustments are needed. However, for specialized simulator applications, the user may prefer to use manual timestep control. The procedure for implementing manual timestep control is typically provided in the technical documentation of the commercial simulator.

Simulator timestep sizes are influenced by simulator startup, timing of input and output of data, and large changes in reservoir pressures, production rates, and injection rates. At startup, the timestep size for most commercial simulators is automatically set at about 1 day or less depending on the application. As the simulation progress and the solution stabilize, large time steps are taken, depending on other controls applied on the simulator. One of such controls applied on the simulator is data input and output. Most simulators adjust their timestep sizes as necessary to precisely comply with data input and/or output specified by time (or date). Frequent data input or output could slow down the simulator and increase run times, thereby increasing computation costs. Large changes in pressures, production rates, and injection rates may also cause reduction in timestep sizes. One technique that could be used to avoid sudden large changes in reservoir conditions (pressures, saturations, etc) is to incorporate additional times (or dates) for data input so as to "smooth-out" or reduce the impact of the large changes on the progress of the simulation.

Except for experienced users of simulators, it is recommended that automatic time step selection should be used as provided in commercial simulators. In general, automatic time step control in commercial simulator will yield reduced run times and computation costs.

19.4.3 Formulations of Simulator Equations

Three main methods are used in the formulation of simulator equations for solution. These are the **IMPES** (Implicit Pressure, Explicit Saturation) method, the **Sequential** method, and the **Implicit** method. The IMPES method was first applied to black oil systems by Sheldon et al.,[13] Stone and Garder,[14] and Fagin and Stewart.[15] For the IMPES formulation, the flow equations are combined to eliminate unknown saturations, thereby generating a single pressure equation. The pressure equation is solved implicitly for pressures at each gridblock at the current timestep designated as $n + 1$, using parameters at the previous (or old) timestep designated as n. This is then followed by the explicit substitution of the pressures into the corresponding flow equations to calculate saturations at the current timestep, $(n + 1)$, for each gridblock.[42] The advantages of the IMPES method are that it requires less computing time per timestep and less computer memory for storage. The main disadvantage of the IMPES method is that it is relatively unstable at large timesteps. There are techniques for improving the stability of the IMPES method as reported by Coats.[43,44] To improve the stability of IMPES formulations, MacDonald and Coats[45] introduced a modification of the IMPES formulation which was later named the **Sequential** method by Spillette et al.[46] The Sequential method is a two step solution process. In the first step, the pressure equation is solved implicitly for pressures at each gridblock as was done in the IMPES method. This is then followed by a second step in which the saturation equations derived with the aid of fractional flow equations are also solved implicitly. However, the Sequential method may have material balance problems in areas of the gridblocks with large saturation (or composition) changes. The formulation method that is widely used in many simulators (black oil, compositional, thermal, chemical, etc.) is the **Implicit** formulation. In the Implicit method, the flow equations are formulated to be solved simultaneously at the current timestep, $n + 1$. A fully implicit formulation was first published by Blair and Weinaug[47] in 1969. This was followed by authors that used Implicit formulations in black oil,[48] compositional,[32] and thermal[34,49] simulators. The Implicit formulation requires more computer memory for storage and higher computation costs than either the IMPES or the Sequential methods. However, the Implicit formulation is considerably more stable and can take larger timestep than either IMPES or Sequential methods, especially for flow problems with large saturation (or composition) changes in the gridblocks. It is noteworthy to discuss a hybrid solution method that cleverly combines the strengths of the IMPES and Implicit methods in one solution method. This hybrid method is called the **Adaptive**

Implicit Method (AIM). The AIM approach uses Implicit method in gridblocks or groups of gridblocks with rapid pressure and saturation changes, while the IMPES method is used in the rest of the gridblocks with slower (less rapid) changes in pressures and saturations.[50,51] The switch between IMPES and Implicit methods under AIM is automatic, and is applied on gridblocks or groups of gridblocks when needed and as long as required to achieve optimum balance on stability and cost of the computation process. However with the advancement of computer hardware, especially in terms of computation speed and cheap memory for storage, most commercial simulators use Implicit formulation as the preferred formulation for solution of most flow problems. Consequently, experienced users are required by most commercial simulators to specifically select the other solution methods (IMPES, SEQUENTIAL, AIM), if these methods are better suited for the solution of the flow problems.

19.4.4 Material Balance Errors and Other Convergence Criteria

One criterion used in all simulators to check the accuracy and stability of the solution at each timestep is the material balance error (MBE). A low value for material balance error by itself is not sufficient measure of the accuracy of the solution. Material balance error is used, in conjunction with other criteria, to determine convergence of the solution at each timestep. Commercial simulators use different methods to calculate and normalize material balance errors. Irrespective of the method used, the criterion applied in most simulators is that the material balance error for each gridblock at each timestep should be close to zero. One method for calculating material balance error is the summation of mass accumulation for each phase at the beginning and at the end of the timestep for all gridblocks. The net mass accumulation for each phase should be equal to net influx through the wells during the timestep. For a black oil simulator, the material balance error is calculated as follows:

For the oil phase:

$$\text{MBE}_o = \frac{\sum_{j=1}^{m}\left[PV\left(\frac{S_o}{B_o}\right)\right]_j^n - \sum_{j=1}^{m}\left[PV\left(\frac{S_o}{B_o}\right)\right]_j^{n+1}}{q_o \Delta t} - 1 \tag{19.29}$$

For the water phase:

$$\text{MBE}_w = \frac{\sum_{j=1}^{m}\left[PV\left(\frac{S_w}{B_w}\right)\right]_j^n - \sum_{j=1}^{m}\left[PV\left(\frac{S_w}{B_w}\right)\right]_j^{n+1}}{q_w \Delta t} - 1 \tag{19.30}$$

For the gas phase:

$$\text{MBE}_g = \frac{\sum_{j=1}^{m}\left[PV\left(\frac{S_g}{B_g}\right)\right]_j^n - \sum_{j=1}^{m}\left[PV\left(\frac{S_g}{B_g}\right)\right]_j^{n+1}}{q_g \Delta t} - 1 \tag{19.31}$$

In Eqs. (19.29) to (19.31), PV = pore volume of the gridblock, j; S_o, S_w, and S_g, are saturations of oil, water, and gas, respectively, in gridblock, j; B_o, B_w, and B_g are formation volume factors for oil, water, and gas, respectively, in gridblock, j; q_o, q_w, and q_g are net influx through wells for oil, water, and gas, respectively, in gridblock, j; Δt = length of timestep; m = total number of gridblocks in reservoir model; n = previous timestep; and $n + 1$ = current timestep. Many commercial simulators consider material balance errors to be sufficiently small for all the phases, if less than 1×10^{-7}.

Other measures that are used to check stability and accuracy of the solutions are changes in saturations of the phases in iteration at each timestep. The iteration is considered to have converged, if maximum changes in the saturation of each phase for each gridblock is less than a preset maximum saturation change for the timestep. Many commercial simulators set the largest convergence error for any phase in any grid block at less than 0.001. If the conditions for the material balance error for each phase and the convergence error for phase saturations are not met for the timestep size, the calculations and iterations are repeated with a reduction in the size of the timestep until convergence is achieved. The amount of reduction of timestep size applied automatically due to convergence failure varies among commercial simulators. Some commercial simulators automatically reduce the timestep size by half after a convergence failure. This can drastically slow down the speed of the simulation run. Note that the amount of cut applied to timestep size due to convergence failure can be specified by the user in most commercial simulators, in addition to changes on the convergence criteria. However, these options should be used as necessary with considerable caution and only by very experienced users of simulators because they can lead to simulation results that are unstable and erroneous.

19.4.5 Numerical Dispersion

Numerical dispersion is a term generally used to describe the false appearance of smeared spatial gradients of saturation or concentration arising from time and space discretizations of the flow equations in simulators based on finite differences.[52] In simpler terms, numerical dispersion describes the smearing of saturation or concentration of the displacing phase ahead of the actual front. Numerical dispersion generally increases with increases in the areal sizes of the model gridlocks, (Δx, Δy),and increases in timestep sizes, Δt. Consequently for the same areal grid sizes, numerical dispersion is likely to be larger with Implicit solution methods that use larger timestep sizes than IMPES solution methods that use smaller timestep sizes. For both IMPES and Implicit solution techniques, numerical dispersion can be reduced by increasing the number of areal gridblocks. This approach is expected to create larger simulation models. However, if accurate prediction of water, steam, or solvent breakthrough is an objective for simulation of the displacement process, then it is necessary to use as many areal gridblocks as necessary in order to minimize the effects of numerical dispersion. Other methods for reducing numerical dispersion include use of pseudofunctions as demonstrated by Kyte and Berry,[53] and mobility weighting techniques such as the two-point, upstream mobility weighting scheme proposed by Todd et al.[39]

19.4.6 Well Model

A well model is used in simulation to represent the flow between the well bore and the reservoir gridblock in which the well is located. Assuming the well is producing from a single gridblock, the flow rate of a phase, p (oil, water, or gas), into the well is given by:

$$q_{p,j} = \frac{WI_j}{\mu_p}(p_j - p_{wf}) \tag{19.32}$$

In Eq.(19.32), $q_{p,j}$ = the flow rate of phase, p, in gridblock, j; WI_j = well index; μ_p = viscosity of phase, p; p_j = gridblock pressure; and p_{wf} = flowing bottomhole pressure of the well. For the sign convention, the flow is considered positive from the gridblock into the well, and negative from the well into the gridblock. For Cartesian grids, the well index, WI_j (also called connection transmissibility factor), was derived by Peaceman[54] as:

$$WI_j = \frac{2\pi\sqrt{k_x k_y}h}{\ln\dfrac{r_o}{r_w} + s} \tag{19.33}$$

In Eq. (19.33), WI_j = well index for gridblock, j; k_x, k_y = permeabilities in the x and y directions, respectively; h = net thickness of the gridblock; r_o = equivalent radius of the gridblock; r_w = wellbore radius; and s = skin factor. The equivalent radius of the gridblock, r_o, is defined as the distance from the well at which the flowing wellbore pressure is equal the average pressure of the gridblock.[55,56] According to Peaceman,[56] equivalent wellbore radius, r_o, for an anisotropic reservoir in Cartesian grids is given by:

$$r_o = 0.28\frac{\sqrt{(k_y/k_x)^{1/2}\Delta x^2 + (k_x/k_y)^{1/2}\Delta y^2}}{(k_y/k_x)^{1/4} + (k_x/k_y)^{1/4}} \tag{19.34}$$

In Eq. (19.34), r_o = equivalent wellbore radius; k_x, k_y = permeabilities in the x and y directions, respectively; Δx, Δy = x and y dimensions of the gridblock, respectively. For an isotropic reservoir, the equivalent well bore radius, r_o, in Cartesian grids is given by:

$$r_o = 0.14\sqrt{(\Delta x^2 + \Delta y^2)} \tag{19.35}$$

It is important to note that the well index, WI_j, for a gridblock is not the same as the productivity (or injectivity) index of the well. The productivity index, J, of a well is defined as:

$$J = \frac{q_p}{\overline{p} - p_{wf}} \tag{19.36}$$

In Eq. (19.36), q_p = production rate of phase p (oil, water, or gas); $\overline{p}$ = average reservoir pressure or pressure at the drainage radius of the well; and p_{wf} = flowing bottomhole pressure of the well. Assuming steady-state, radial Darcy flow within the drainage radius of the well, a relationship between the well index and the productivity index of the well is expressed as:

$$J = \sum_j \left[\frac{WI_j}{\mu_p} \left\{ \frac{\ln(r_o/r_w) + s}{\ln(r_d/r_w) + s} \right\} \right] \quad (19.37)$$

In Eq. (19.37), WI_j = well index for gridblock, j; μ_p = viscosity of phase, p; r_o = equivalent wellbore radius; r_w = wellbore radius; s = skin factor; and r_d = drainage radius of the well. The summation is over all the gridblocks connected to the well. If the drainage radius of the well is assumed to be equal to equivalent wellbore radius, Eq. (19.37) reduces to:

$$J = \sum_j \left[\frac{WI_j}{\mu_p} \right] \quad (19.38)$$

Note that all the equations developed in this section for well models apply only to vertical wells located in rectangular gridblocks in Cartesian grids. They are not applicable to wells located in radial gridblocks or advanced wells. Advanced wells are described as wells with complex trajectories such as horizontal wells and multilateral wells, and wells with dowhhole control devices (sensors, flow control valves, etc.) called "smart" wells.[57] Peaceman[58] proposed a method for representing a horizontal well in reservoir models with Cartesian grids. The modeling of vertical and horizontal wells in unstructured grids was presented by Palagi and Aziz.[59] A technique for modeling an advanced well that is available in most commercial simulators consists of building a well model by subdividing the wellbore into segments as described by Stone et al.,[60] and Holmes et al.[61] Detailed presentations of these techniques for modeling advanced wells in reservoir simulators are not covered in this book. However, since many reservoirs are produced with horizontal wells, multilateral wells, or "smart" wells, it is important to note that these wells require advanced well models for accurate representation of their productivity.

Example 19.1 Calculation of Well Index

Problem

Calculate the well index for a vertical well located completely in a single rectangular gridblock in an areal Cartesian grid system. The dimensions and properties of the gridblock are given as follows:

Gridblock dimension, Δx	= 150 ft
Gridblock dimension, Δy	= 100 ft
Gridblock dimension, Δz	= 5 ft
Gridblock permeability, k_x	= 350 md
Gridblock permeability, k_y	= 225 md
Wellbore radius, r_w	= 0.5 ft
Skin factor, s	= 2.3

Assume the net thickness of gridblock, h, is equal to gridblock dimension, Δz.

Solution

Step 1. Calculate equivalent radius of the gridblock, r_o.
Using Eq. (19.34),

$$r_o = 0.28\frac{\sqrt{(k_y/k_x)^{1/2}\Delta x^2 + (k_x/k_y)^{1/2}\Delta y^2}}{(k_y/k_x)^{1/4} + (k_x/k_y)^{1/4}}$$

$$= 0.28\frac{\sqrt{(225/350)^{1/2}(150)^2 + (350/225)^{1/2}(100)^2}}{(225/350)^{1/4} + (350/225)^{1/4}}$$

$$= 24.3065 \text{ ft}$$

Step 2. Calculate well index for gridblock.
Using Eq. (19.33),

$$WI_j = \frac{2\pi\sqrt{k_x k_y}h}{\ln\frac{r_o}{r_w} + s}$$

$$= \frac{2\pi\sqrt{350 \times 225} \times 5}{\ln\frac{24.3065}{0.5} + 2.3}$$

$$= 1425.6513 \text{ md-ft}$$

19.4.7 Model Initialization

The initial conditions in the reservoir based on the input data are established during model initialization. These initial conditions are distributions of phase pressures and saturations for each gridblock in the reservoir model. Generally, two methods are used for model initialization. These are simulator generated initial conditions or user specified initial conditions. Under simulator generated initial conditions, the simulator uses the input data to generate equilibrated distributions of phase pressures and saturations at initial conditions. The input data consist of specification of pressure at a given datum depth, depth of fluid contacts (water-oil, gas-oil, or gas-water contacts), capillary pressure at the contacts, fluid densities at datum, and saturation endpoints and capillary pressure data in the relative permeability tables. For the user-specified initial conditions, the distributions of pressure and saturations at initial conditions for each gridblock are provided and set by the user. Note that user specified initial conditions may cause a non-equilibrated model at initial conditions. The user should verify that these conditions actually exist in the reservoir at initial conditions as a non-equilibrated model may lead to erroneous results.

After initializing the model, one of the key activities at this stage of model building is to check the quality of the model. A very important measure of model quality is to compare the in-place fluid volumes calculated by the simulator against volumes calculated volumetrically from the geologic models. The fluid-in-place volumes from the two methods should be in close agreement. A good rule-of-thumb is that the deviation between the two methods should be less than 2 percent.

The initialized model is typically the model which is used for simulation of the reservoir process. At this stage it should be reviewed completely before proceeding with the simulation. All the input data should be compared with model data for consistency and reliability. These include checking the model pressure and saturation data, gridblock properties (porosity, permeability, net thickness, etc.), well completion data, production, and injection data, if applicable. The objective of thoroughly checking the model at this stage is to ensure that the initialized model is an accurate representation of the reservoir model before proceeding with the time consuming process of history matching and predictions.

19.4.8 History Matching

History matching is the process of adjusting the properties and parameters of the reservoir model to match past actual performance data measured on the reservoir. The performance data may include pressure data (such as static bottomhole pressures, flowing bottomhole pressures, tubing head pressures, average reservoir pressures, etc.), production and injection data (such as oil, water, and gas production rates and water and gas injection rates), well completion data (such as perforated and isolated intervals), production and injection profile data (from production logs and injection logs), etc. A large variety of performance data can be used in the history matching process. It depends on reservoir type, and methods being used to deplete or improve production from the reservoir. The history matching process is a classical example of an inverse problem which leads to a non-unique solution or response. This means that different history matched models of a reservoir can be achieved by applying the same input data. The history matched models are non-unique solutions or responses to the applied performance data and may yield different predictions of the future performance of the reservoir.

The process of history matching a reservoir model varies from one practitioner to the other, and depends largely on experience and knowledge of reservoir engineering principles. The history matching process also depends considerably on the reservoir model type, the quality and quantity of available performance data, and the objectives of the study. For instance, if the reservoir model was constructed from sparse data, and limited performance data are available, it is reasonable to expect that the history matched model is likely to be limited in its application on prediction of the future performance of the reservoir.

Generally, the history matching process can be divided into two phases, namely the pressure match phase, and the saturation match phase. The pressure match phase normally precedes the saturation match phase. During the pressure match phase, the history matching process is focused on matching the net voidage (in reservoir barrels) from the reservoir with the aim of matching pressure distribution in the reservoir over the specified time period. In some cases, only a match of the trend in pressure distribution can be achieved during this phase. The goal during the pressure match phase is to obtain a reservoir model that has approximately the appropriate net volume of fluids in place over the duration of its known history. During the pressure match phase, minimal effort is applied in the validation of the distribution of phases or saturations in the model. At the end of the pressure match phase, the history matching process proceeds to the saturation match phase. During the saturation match phase, attention is then directed toward specifying the actual phases of fluids produced and/or injected into the reservoir. For instance, in history matching black oil reservoir models, oil production rates are typically specified and the model yields the corresponding water and gas production rates. The water and gas production rates generated from the model are then compared with the actual water and gas production rates. Also, pressures and other performance data generated from the model are compared with actual data. Different forms of the actual data may be used for this comparison during the history matching process. Some practitioners use production rates, cumulative production volumes, producing gas-oil ratios and water-oil ratios, flowing bottomhole pressures, tubing head pressures, average reservoir pressures, etc. The matching of data predicted by the model to actual data can be performed at the level of the field, well groups, or individual well. Again, the choice of actual data to match depends on the quality and quantity of the data, and the objectives of the simulation study. At the end, the saturation match phase generally produces a history matched model with appropriate distribution of pressures and saturations in the reservoir. Generally, the history matching process is terminated at the point where the team engaged in the effort decides that an acceptable history matched model has been obtained. History matched models are ruled to be acceptable, if they are adequate to meet the objectives of the simulation study, and further adjustments to the models do not result in significant changes on the history matched variables.

The following best practices are recommended to enhance the efficiency and effectiveness of the history matching process:

1. Make a log of all activities and changes made on the model during the history matching process. This log should be maintained preferably within the model. If the history match activities are conducted by a team, all members of the team should adhere to these guidelines to ensure that all changes made on the model are documented in this historical record. No changes made on the model should be left out of this record.
2. The changes made on the model during the history match process should be graduated and systematic. Avoid making changes in a haphazard manner. Clear and defined results expected from the changes must be understood before the adjustments to the model are made at each stage.
3. The results from previous changes or adjustments to the model should be analyzed and understood before proceeding with the next set of changes or adjustments. If a team is working cooperatively on the project, it is advantageous for members of the team to reach consensus on the next set of changes or adjustments to be made.
4. The least reliable input data should be modified first during the history match process. For instance, if permeability data are the least reliable data, then permeability data should be changed systematically at the inception of the history match process. If further changes to permeability data lead to unreasonable results, then the next set of input data considered to be unreliable should be changed in a systematic manner.
5. Recognition of the point in the history matching process when further changes on the input data cause insignificant improvement on the state of the history matched model is important. This is the point of diminishing return and should be recognized as the end of the history matching process to save money and resources.

19.4.9 Predictions

The future performance of the reservoir is predicted with the history matched models or with the initialized model in the case that the reservoir has no historical production. For a history matched model, it is usually necessary to adjust individual well productivity index so that a smooth transition is achieved from the history matching phase to the prediction phase. The process of tuning the well productivity indices to achieve the desired smooth transition may require an iterative process. For new reservoirs with no production history, initial well production rates may be adjusted with productivity index from drill stem tests, extended well tests, or production data from analogue reservoirs. The predicted performance of new reservoirs is subject to a much higher level of uncertainty due to lack of historical performance data from the reservoir to guide or condition predicted future performance as is the case with history matched models.[62]

The most effective application of reservoir models is to compare alternative management strategies or to rank competing management strategies. This is actually the understated power of reservoir simulation. The actual reservoir can be produced one time at considerable cost over an extended period of time in many cases, whereas a reservoir model can be "produced" (or run) as many times as needed at a tiny fraction of the cost.[7] With the aid of reservoir models, it is possible to investigate the potential outcome of many reservoir management strategies without conducting actual field or pilot tests. It is reasonable to assert that learned application of reservoir simulation technology should lead to better reservoir management practices when compared with cases where the technology was not applied.

For a typical comparison of alternative reservoir management strategies, it is useful to create a base case that is focused on current operating strategy. For instance, the base case may consist of current operations based on depletion drive mechanism. This can then be compared with other alternative strategies based on fluid injection. The injected fluid could be gas, and/or water. The important point is that predicted performances of the model under various alternative management strategies can be compared with the base case leading to the selection of an optimum management strategy.[63] Note that the use of reservoir models to rank competing strategies were presented in Chapter 16 for secondary recovery methods, and Chapter 17 for enhanced oil recovery methods.

19.4.10 Uncertainty Analysis

The presence of uncertainties in the geologic and petrophysical data used in constructing the geologic model was discussed in Chapter 18, Section 18.5.6. Uncertainty analysis of geologic models was performed by generation of multiple realizations and application of statistical analysis on the realizations to create a cumulative distribution curve. It was suggested then that the cumulative probability curve can be used to select models at the P10, P50, and P90 quantiles for use in reservoir flow simulations. The decision to select geologic models for flow simulation only at the P10, P50, and P90 quantiles is based largely on limitation of resources which can be devoted to the project. In other words, the team may choose to use additional geologic models including the P10, P50, and P90 models, if all the models can be processed at the same time with available resources.

The level of uncertainties present in the geologic models is increased by introducing additional variable parameters during the process of transforming the static geologic model into a reservoir simulation model. These additional variable parameters include fluid properties (oil viscosity, oil gravity, saturation pressures, solution gas-oil ratios, etc.), rock/fluid properties (relative permeability data, rock compressibility data, etc.), fluid contacts (oil-water contact, gas-oil contact, gas-water contact, etc.), well locations, and performance data (production rates, pressures, completion data, etc.). These uncertainties present in the reservoir model from geological data, petrophysical data, fluid properties data, rock/fluid properties data, etc. compound the problem of uncertainty analysis in reservoir simulation models.

One of the processes widely used in the petroleum industry to systematically evaluate the impact of uncertainty on flow simulation outcomes is called **Experimental Design** (ED). Experimental Design is also called **Design of Experiments** (DOE). ED is a statistically based procedure that systematically evaluates the effects of these uncertain variables on dependent outcomes such as hydrocarbon recovery. Commercial software based on Plackett–Burman or Latin Hypercube methods for designing the experiments are readily available in the industry. They are copious documentations of these methods available in the technical manual provided by the vendors of the commercial software. Interested readers are encouraged to consult these technical documents for further details on these methods. It is important to note that the commercial ED software have been configured to facilitate input for most commercial reservoir simulators.

19.5 General Structure of Reservoir Flow Models

The structures of input data in most reservoir flow simulators are remarkably similar. The general structure of reservoir simulation models is described here to acquaint the reader with sequence of data input in most simulators. Obviously, the structure of input data will vary slightly between different simulators. But in many cases, the differences are minor and can be quickly reconciled between simulators from different sources. The purpose of this section is to familiarize the reader with a readily available data structure that can be used to transfer data from one simulator to the other.

19.5.1 Definition of Model and Simulator

The entry of data into every simulator begins with definition of the size of the reservoir model and the type of simulator to be used for modeling the reservoir. The key entry that defines the size of the reservoir model is the number of gridblocks in the model. For instance, in the Cartesian system, the number of gridblocks in the x, y, and z directions for a 3D model is specified. Also defined in this data entry section is the number of wells in the model, the number of tabular data (PVT, relative permeability, equilibration regions, etc.), and the number of initialization regions, etc. The type of simulator to be used including the formulation (solution) type is specified in this section. For instance, a black oil model based on the Implicit formulation may be selected. Most important, the date for start of simulation is specified. This section may be considered as the section in which the scope of the simulation problem is specified for the simulator. In many ways, this section defines the amount of computer memory that will be required to run the reservoir model.

19.5.2 Geologic Model Data

All the structural and petrophysical data in the geological model are typically assembled as data input for the gridblocks in a section of the simulator. The structure of the geologic model is represented by geometrical data on the gridblocks in terms of location and dimensions. This is usually accompanied with separate specifications of the petrophysical data for each gridblock. The petrophysical data usually specified for each gridblock include porosity, permeability, and net sand thickness or net-to-gross ratio data. Initial fluid saturations for each gridblock may also be specified in some models. These data are then followed with modifications to the grid system (such as local grid refinement), and modifications to the petrophysical data specified for the gridblocks.

19.5.3 Fluid Properties Data

This section of the model data set contains data that represent the PVT properties of the fluids present in the reservoir. The PVT data are usually presented in a tabular form for black oil models. For compositional simulators, the PVT data are represented in a compatible form as output generated with an equation of state.

19.5.4 Rock/Fluid Properties Data

Rock/fluid properties data in the form of relative permeability data and capillary pressure data are represented in the model as functions of fluid saturations. These data are usually presented in the simulator in a tabular form. Note that the data in these tables are sometimes used by the simulator to establish initial conditions in the reservoir model, if the option for simulator generated initial conditions is selected.

19.5.5 Model Equilibration Data

Model equilibration data include fluid contact depths (oil-water contact, gas-oil contact, or gas-water contact), capillary pressures at the fluid contacts, and reservoir pressure at a selected datum depth. The model equilibration data are in some cases used by the simulator to establishing initial reservoir conditions.

19.5.6 Well Data

In the section of the model data for wells, the locations of the wells in the grid system are specified. Also, the gridblocks in which the wells are completed are specified. The production or injection rates of the wells including the type of fluid produced or injected are specified. The progression of the simulation in terms of time is defined in this section in the form of time steps, cumulative time, or dates. These time-based data are very important because the speed and duration of the simulation are controlled by these data. For reservoirs with production history, the production data are provided at specific time intervals which could be daily, monthly, quarterly, semi-annually, or annually. The frequency of production data entry is totally at the discretion of the user. However, note that higher frequency of production data specifications reduces the speed of the simulator during history match. Additional data that may be specified at progressive time periods include introduction of new wells, recompletion of existing wells, and changes to well fluid production or injection rates.

19.5.7 Simulator Data Output

All simulators are capable of generating large amounts of output data, especially for large models. This should be controlled by the user to avoid being overwhelmed by the simulator data input/output.[64] In most cases, the amount of simulator data output can be controlled by the user. The minimum output that can be specified by the user may include well performance data and distributions of pressures and saturations over time. Most simulators are equipped with powerful graphical software that can assists in the evaluation and presentation of the output from the simulator.

Nomenclature

B_g	gas formation volume factor
B_o	oil formation volume factor
B_w	water formation volume factor
C_g	concentration of gas, mass per unit volume
C_o	concentration of oil, mass per unit volume
C_w	concentration of water, mass per unit volume
F	mass flow rate per unit cross sectional area normal to the direction of flow
$\vec{F}$	mass flow rate per unit cross sectional area in the x, y, and z directions
h	net thickness of gridblock, j
J	productivity index
$\vec{K}$	permeability tensor
k_{rp}	relative permeability of phase, p
k_x	permeability in x-direction
k_y	permeability in y-direction
k_z	permeability in z-direction
n	timestep
p_j	pressure in gridblock, j
p_{wf}	flowing bottomhole pressure
$\overline{p}$	average reservoir pressure
q	mass flow rate per unit volume
r_d	well drainage radius
r_o	equivalent radius of gridblock
r_w	wellbore radius
R_{so}	gas solubility in oil

R_{sw} gas solubility in water
s skin factor
S_g gas saturation
S_o oil saturation
S_w water saturation
$\vec{v}_g$ gas velocity in x, y, and z directions
$\vec{v}_o$ oil velocity in x, y, and z directions
$\vec{v}_w$ water velocity in x, y, and z directions
Δx grid length in the x-direction
Δy grid width in the y-direction
Δz grid height in the z-direction
Δt time interval
λ_p mobility of phase, p
μ_p fluid viscosity of phase, p
ρ_{gsc} gas density at standard conditions
ρ_{osc} oil density at standard conditions
ρ_{wsc} water density at standard conditions
ϕ porosity
Φ_p potential of phase, p

Subscripts

g gas
o oil
w water
x x-direction
y y-direction
z z-direction
t time
sc standard conditions

Abbreviations

AIM Adaptive Implicit Method
DOE Design of Experiments
ED Experimental Design
IMPES Implicit Pressure Explicit Saturation
MBE Material Balance Error
PV Pore Volume
PVT Pressure Volume Temperature
WI Well Index

References

1. Aziz, K., and Settari, A: *Petroleum Reservoir Simulation*, Applied Science Publishers Ltd., London (1979).
2. Mattax, C.C., and Dalton, R.L.: Reservoir Simulation, SPE Monograph Volume 13, Society of Petroleum Engineers, Richardson, Texas (1990).
3. Erketkin, T., Abou-Kassem, J.H., and King, G.R.: "Basic Applied Reservoir Simulation," SPE Textbook Series, Vol. 7, Society of Petroleum Engineers, Richardson, TX (2001).
4. Coats, K.H.: "Reservoir Simulation: State of the Art," paper SPE 10020, Distinguished Author Series (August 1982) 1633–1642.
5. *Numerical Simulation II,* SPE Reprint Series No. 20, Society of Petroleum Engineers, Richardson, Texas (1986).
6. Dogru, A.H., Fung, L.S.K., Middya, U., Al-Shaalan, T.M., Pita, J.A., Kumar, K.H., Su, H.J., Tan, J.C.T., Hoy, H., Dreiman, W.T., Hahn, W.A., Al-Harbi, R., Al-Youbi, A., Al-Zamel, N.M., Mezghani, M., and Al-Mani, T.: "A Next-Generation Parallel Reservoir Simulator for Giant Reservoirs," paper SPE 119272 presented at the SPE Reservoir Simulation Symposium, The Woodlands, Texas, February 2–4, 2009
7. Coats, K.H.: "Use and Misuse of Reservoir Simulation Models," *JPT* (November 1969) 1391–1398.
8. Muskat, M.: *Physical Principles of Oil Production,* McGraw-Hill Book Co., New York (1949).
9. Douglas, J., Jr., Peaceman, D.W., and Rachford, H.H., Jr.: "A Method for Calculating Multi-Dimensional Immiscible Displacement," *Trans.* AIME (1959) 216, 297–308.
10. Coats, K.H., Nielsen, R.L., Terhune, M.H., and Weber, A.G.: "Simulation of Three-Dimensional, Two-Phase Flow in Oil and Gas Reservoirs," *SPEJ* (Dec. 1967) 377–388.
11. Coats, K.H.: "An Analysis for Simulating Reservoir Performance Under Pressure Maintenance by Gas and/or Water Injection," *SPEJ* (Dec. 1968) 331–340.
12. Coats, K.H.: "A Mathematical Model for Simulating Three-Dimensional Three-Phase Fluid Flow in Reservoirs," The University of Texas, Austin, Texas, November 1968.
13. Sheldon, J.W., Harris, C.D., and Bavly, D.: "A Method for General Reservoir Behavior Simulation on Digital Computers," paper 1521-G presented at SPE 35th Annual Fall Meeting, Denver, Colorado, October 2–5, 1960.
14. Stone, H.L., and Garder, A.O., Jr.: "Analysis of Gas Cap or Dissolved-Gas Drive Reservoirs," *SPEJ* (June 1961) 92–104.
15. Fagin, R.G., and Stewart, C.H., Jr.: A New Approach to the Two-Dimensional Multiphase Reservoir Simulator," *SPEJ* (June 1966) 175–182.
16. Odeh, A.S.: "Comparison of Solutions to a Three-Dimensional Black Oil Reservoir Simulation Problem," *JPT* (Jan. 1981) 13–25.

17. Chappelear, J.E., and Nolen, J.S.: "Second Comparative Solution Project: A Three-Phase Coning Study," *JPT* (March 1986) 345–353.

18. Kenyon, D.E., and Behie, A.: "Third SPE Comparative Solution Project: Gas Cycling of Retrograde Condensate Reservoirs," *JPT* (August 1987) 981–997.

19. Aziz, K., Ramesh, A.B., and Woo, P.T.: "Fourth SPE Comparative Solution Project: Comparison of Steam Injection Simulators," *JPT* (December 1987) 1576–1584.

20. Killough, J., and Kossack, C.: "Fifth SPE Comparative Solution Project: Evaluation of Miscible Flood Simulators," paper SPE 16000 presented at the Ninth SPE Symposium on Reservoir Simulation, San Antonio, Texas, February 1–4, 1987.

21. Firoozabadi, A., and Thomas, L.K.: "Sixth SPE Comparative Solution Project: Dual-Porosity Simulators," *JPT* (June 1990) 710–715, and 762–763.

22. Nghiem, L., Collins, D.A., and Sharma, R.: "Seventh SPE Comparative Solution Project: Modeling of Horizontal Wells in Reservoir Simulation," paper SPE 21221 presented at the 11th SPE Symposium on Reservoir Simulation, Anaheim, California, February 17–20, 1991.

23. Quandalle, P.: "Eighth SPE Comparative Solution Project: Gridding Techniques in Reservoir Simulation," paper SPE 25263 presented at the 12th SPE Symposium on Reservoir Simulation, New Orleans, Louisiana, February 28–March 3, 1993.

24. Killough, J.E.: "Ninth SPE Comparative Solution Project: A Reexamination of Black Oil Simulation," paper SPE 29110 presented at the 13th SPE Symposium on Reservoir Simulation, San Antonio, Texas, February 12–15, 1995.

25. Christie, M.A., and Blunt, M.J.: "Tenth SPE Comparative Solution Project: A Comparison of Upscaling Techniques," *SPEREE* (August 2001) 308–317.

26. Odeh, A.S.: "RESERVOIR SIMULATION...What is it?" *JPT* (November 1969) 1383–1388.

27. Coats, K.H., Thomas, L.K., and Pierson, R.G.: "Compositional and Black Oil Reservoir Simulation," *SPEREE* (August 1998) 372–379.

28. Varavei, A., and Sepehrnoori, K.: "An EOS-Based Compositional Thermal Reservoir Simulator," paper SPE 119154 presented at the 2009 SPE Reservoir Simulation Symposium, The Woodlands, Texas, February 2–4, 2009.

29. Mishev, I., Beckner, B., Fedorova, N., and Terekhov, S.: "Linear Solver Performance Optimization in Reservoir Simulation Studies, paper SPE 119266 presented at the 2009 SPE Reservoir Simulation Symposium, The Woodlands, Texas, February 2–4, 2009.

30. Datta-Gupta, A.: "Streamline Simulation: A Technology Update," paper SPE 65604, Distinguished Author Series (December 2000) 68–74.

31. Fanchi, J.R.: *Principles of Applied Reservoir Simulation,* Gulf Publishing Co., Houston, Texas (1997).

32. Coats, K.H.: "An Equation of State Compositional Model," *SPEJ* (October 1980) 363–376.

33. Coats, K.H., George, W.D., Chu, Chieh, and Marcum, B.E.: "Three-Dimensional Simulation of Steamflooding," *SPEJ* (December 1974) 573–592.

34. Coats, K.H.: "A Highly Implicit Steamflood Model," *SPEJ* (October 1978) 369–383.

35. Aziz, K.: "Reservoir Simulation Grids: Opportunities and Problems," *JPT* (July 1993) 658–663.

36. Pedrosa, O.A., Jr., and Aziz, K.: "Use of Hybrid Grid in Reservoir Simulation," *SPERE* (November 1986) 611–621.

37. Voronoi, G.: "Nouvelles applications des parameters continues a la teorie des formes quadratiques," *J. Reine Angew. Math.* (1908) 134, 198–287.

38. Palagi, C.L., and Aziz, K.: "Use of Voronoi Grid in Reservoir Simulation," paper SPE 22889 presented at the 1991 SPE Annual Technical Conference and Exhibition, Dallas, Texas, October 6–9, 1991.

39. Todd, M.R., O'Dell, P.M., and Hirasaki, G.J.: "Methods for Increased Accuracy in Numerical Reservoir Simulators," *SPEJ* (December 1972) 515–530.

40. Coats, K.H., George, W.D., Chu, C., and Marcum, B.E.: "Three-Dimensional Simulation of Steamflooding," *SPEJ* (December 1974) 573–592.

41. Yanosik, J.L., and McCracken, T.A.: "A Nine-Point, Finite-Difference Reservoir Simulator for Realistic Prediction of Adverse Mobility Ratio Displacements," *SPEJ* (August 1979) 253–262.

42. Coats, K.H.: "A Note on IMPES and Some IMPES-Based Simulation Models," *SPEJ* (September 2000), 245–251.

43. Coats, K.H.: "IMPES Stability: Selection of Stable Timesteps," *SPEJ* (June 2003) 181–187.

44. Coats, K.H.: "IMPES Stability: The CFL Limit," *SPEJ* (September 2003) 291–297.

45. MacDonald, R.C., and Coats, K.H.: "Methods for Numerical Simulation of Water and Gas Coning," *SPEJ* (December 1970) 425–436.

46. Spillette, A.G., Hillestad, J.G., and Stone, H.L.: "A High-Stability Sequential Solution Approach to Reservoir Simulation," paper SPE 4542 presented at the SPE 48th Annual Fall Meeting, Las Vegas, Nevada, September 30–October 3, 1973.

47. Blair, P.M., and Weinaug, C.F.: "Solution of Two-Phase Flow Problems Using Implicit Difference Equations," *SPEJ* (December 1969) 417–424.

48. Bansal, P.P., Harper, J.L., McDonald, A.E., Moreland, E.E., Odeh, A.S., and Trimble, R.H.: "A Strongly Coupled, Fully Implicit, Three Dimensional, Three Phase Reservoir Simulator," paper SPE 8329 presented at the 54th Annual Fall Technical Conference and Exhibition, Las Vegas, Nevada, September 23–26, 1979.

49. Coats, K.H.: "In-Situ Combustion Model," *SPEJ* (December 1980) 533–553.

50. Thomas, G.W., and Thurnau, D.H.: "Reservoir Simulation Using an Adaptive Implicit Method," *SPEJ* (October 1983) 759–768.

51. Thomas, G.W., and Thurnau, D.H.: "The Mathematical Basis of the Adaptive Implicit Method," paper SPE 10495 presented at the Sixth SPE Symposium on Reservoir Simulation, New Orleans, Louisiana, January 31–February 3, 1982.

52. Lantz, R.B.: "Quantitative Evaluation of Numerical Diffusion (Truncation Error)," *SPEJ* (September 1971) 351–320.

53. Kyte, J.R., and Berry, D.W.: "New Pseudo Functions to Control Numerical Dispersion," *SPEJ* (August 1975) 269–276.

54. Peaceman, D. W.: "A New Method for Calculating Well Indexes for Multiple Wellblocks with Arbitrary Rates in Numerical Reservoir Simulation," paper SPE 79687 presented at the SPE Reservoir Simulation Symposium, Houston, Texas, February 3–5, 2003.

55. Peaceman, D.W.: "Interpretation of Well-Block Pressures in Numerical Reservoir Simulation," *SPEJ* (June 1978) 183–194.

56. Peaceman, D.W.: "Interpretation of Well-Block Pressures in Numerical Reservoir Simulation with Nonsquare Grid Blocks and Anisotropic Permeability," *SPEJ* (June 1983) 531–543.

57. Holmes, J.A.: "Modeling Advanced Wells in Reservoir Simulation," paper SPE 72493, Distinguished Author Series, November 2001.

58. Peaceman, D.W.: "Representation of a Horizontal Well in Numerical Reservoir Simulation," paper SPE 21217, SPE *Advanced Technology Series*, Vol. 1, No. 1 (1993) 7–16.

59. Palagi, C.L., and Aziz, K.: "Modeling Vertical and Horizontal Wells with Voronoi Grid," *SPERE* (February 1994) 15–21.

60. Stone, T.W., Edmunds, N.R., and Kristoff, B.J.: "A Comprehensive Wellbore/Reservoir Simulator," paper SPE 18419 presented at the 1989 SPE Reservoir Simulation Symposium, Houston, Texas, February 6–8, 1989.

61. Holmes, J.A., Barkve, T., and Lund, O.: "Application of a Multisegment Well Model to Simulate Flow in Advanced Wells," paper SPE 50646 presented at the 1998 European Petroleum Conference, The Hague, The Netherlands, October 20–22, 1998.

62. Ezekwe, J.N.: "Comparison of Predictions from Reservoir Models to Actual Production Data: Field Examples," SPE 77972 presented at the Annual Technical Conference and Exhibition in San Antonio, Texas, September 29–October 2, 2002.

63. Echols, D.P., and Ezekwe, N.: "A Reservoir Management Model for 26R Reservoir, Elk Hills Oil Field, California," SPE 46231 presented at the Western Regional Meeting in Bakersfield, California, May 10–13, 1998.

64. Gencer, S.C., Ketcherside, B.P., Morrell, G.O., Mulkay, E.L., and Wiegand, K.D.: "Data Management in Reservoir Simulation," paper SPE 106075 presented at the 2007 SPE Reservoir Simulation Symposium, Houston, Texas, February 25–28, 2007.

General Reading

- Wong, T.W., Firoozabadi, A., Nutakki, R., and Aziz, K.: "A Comparison of Two Approaches to Compositional and Black Oil Simulation," paper SPE 15999 presented at the Ninth SPE Symposium on Reservoir Simulation, San Antonio, Texas, February 1–4, 1997.
- Coats, K.H.: "Simulation of Steanflooding With Distillation and Solution Gas," *SPEJ* (October 1976) 235–247.
- Coats, K.H.: "A Highly Implicit Steamflood Model," *SPEJ* (October 1978) 369–383.
- Vestergaard, H., Olsen, H., Sikandar, A.S., and Noman, R.: "The Application of Unstructured-Gridding Techniques for Full-Field Simulation of a Giant Carbonate Reservoir Developed with Long Horizontal Wells," *SPEREE* (December 2008) 958–967.
- Hammersley, R.P., and Ponting, D.K.: "Solving Linear Equations in Reservoir Simulation Using Multigrid Methods," paper SPE 115017 presented at the 2008 SPE Russian Oil & Gas Technical Conference and Exhibition, Moscow, Russia, October 28–30, 2008.
- Nolen, J.S., and Berry, D.W.: "Tests of the Stability and Time-Step Sensitivity of Semi–Implicit Reservoir Simulation Techniques," *SPEJ* (June 1972) 253–266.
- Mehra, R., Hadjitofi, M., and Donnelly, J.K.: "An Automatic Time-Step Selector for Reservoir Models," paper SPE 10496 presented at the Sixth SPE Symposium on Reservoir Simulation, New Orleans, Louisiana, January 31–February 3, 1982.
- Dogru, A.H.: "Megacell Reservoir Simulation," paper SPE 57907, Distinguished Author Series, May 2000.
- Mattax, C.C., and Dalton, R.L.: "Reservoir Simulation," *JPT* (June 1990) 692–695.
- Williamson, A.S., and Chappelear, J.E.: "Representing Wells in Numerical Reservoir Simulation: Part 1—Theory," *SPEJ* (June 1981) 323–338.
- Chappelear, J.E., and Williamson, A.S.: "Representing Wells in Numerical Reservoir Simulation: Part 2—Implementation," *SPEJ* (June 1981) 339–344.
- Abou-Kassem, J.H., and Aziz, K: "Analytical Well Models for Reservoir Simulation," *SPEJ* (August 1985) 573–580.
- Peaceman, D.W.: "A New Method for Calculating Well Indexes for Multiple Wellblocks with Arbitrary Rates in Reservoir Simulation," paper SPE 79687 presented at the SPE Reservoir Simulation Symposium, Houston, Texas, February 3–5, 2003.
- Najafabadi, N.F., Han, C., Delshad, M., and Sepehrnoori, K,: "Development of a Three Phase, Fully Implicit, Parallel Chemical Flooding Simulator," paper SPE 119002 presented at the 2009 SPE Reservoir Simulation Symposium, The Woodlands, Texas, February 2–4, 2009.
- Varavei, A., and Sepehrnoori, K.: "An EOS-Based Compositional Thermal Reservoir Simulator," paper SPE 119154 presented at the 2009 SPE Reservoir Simulation Symposium, The Woodlands, Texas, February 2–4, 2009.

- Iranshahr, D.V., Voskov, D.V., and Tchelepi, H.A.: "Phase Equilibrium Computations Are No Longer the Bottleneck in Thermal Compositional EoS Based Simulation," paper SPE 119166 presented at the 2009 SPE Reservoir Simulation Symposium, The Woodlands, Texas, February 2–4, 2009.
- Mishev, I., Beckner, B., Fedorova, N., and Terekhov, S.: "Linear Solver Performance Optimization in Reservoir Simulation Studies," paper SPE 119266 presented at the 2009 SPE Reservoir Simulation Symposium, The Woodlands, Texas, February 2–4, 2009.
- Fung, L.S.K., and Dogru, A.H.: "Parallel Unstructured-Solver Methods for Simulation of Complex Giant Reservoirs," *SPEJ* (December 2008) 440–446.
- Frost, R.J., Parker, M.J., and Ponting, D.K.: "Scalability and Usability Aspects of Reservoir Simulation," paper SPE 115015 presented at the 2008 SPE Russian Oil & Gas Technical Conference and Exhibition, Moscow, Russia, October 28–30, 2008.
- Rietz, D., and Usmani, A.: "Case Studies Illustrating the Use of Reservoir Simulation Results in the Reserves Estimation Process," *SPEREE* (February 2009) 149–158.
- Ezekwe, N., Echols, D., and Hudgens, P.: "Diagnostic Application of Reservoir Simulation to 31S Structure Stevens Reservoirs—Elk Hills Oil Field, California," SPE 46230 presented at the Western Regional Meeting in Bakersfield, California, May 10–13, 1998.

CHAPTER 20

Reservoir Management

20.1 Introduction

Reservoir management is a term which can be described as consisting of operational plans or strategies, based on analyses of current geologic, reservoir, and production data, designed to optimize the development and exploitation of a reservoir with the goal of achieving the maximum efficient **and** economic recovery of its hydrocarbons. The most important ingredient necessary for the creation of sound reservoir management strategies is the collection and analyses of key geologic, reservoir, and performance data.[1] For this reason, this book has emphasized in previous chapters, the importance of collection and analyses of basic geologic, rock, fluid properties and reservoir production data, the construction and characterization of geologic models with these data, and use of reservoir flow models in simulation to devise reservoir management strategies. But depending on the quantity and quality of available data, it is **not** always necessary to use reservoir models in devising effective reservoir management strategies. Other methods such as the general material balance equation presented in Chapter 7, and other analytical methods presented in Chapters 8 and 9 can be used in gaining knowledge and understanding of the important recovery mechanisms occurring in the reservoir. These methods can also be used to develop short term reservoir management strategies based on operational activities like curtailing production from high gas-oil ratio wells or high water-cut wells to reduce costs. As additional data are collected from the reservoir, models can then be constructed which could be used to generate long term management strategies for the reservoir which may include initiation of pressure maintenance by gas and/or water injection and other strategies.

The collection, analyses, and assimilation of data should be at the center of any sound reservoir management process. The traditional process of creating reservoir management strategies had relied on somewhat regular periods of collecting reservoir and well performance data. These data are then analyzed, and eventually used to generate or modify strategies for managing the reservoir. The petroleum field of current era is rapidly embracing the digital age. This has led to the installation of permanent downhole gauges and control devices in wells, thereby creating the so-called "smart" or "intelligent" wells. Also, petroleum fields are equipped with sensors, real-time data communication systems, massive data storage capacities, reservoir simulators, and data visualization centers to create the "intelligent" field.[2] The combination of "smart" wells and "intelligent" fields is leading the development of a new concept in reservoir management which has been termed "closed-loop reservoir management."[3,4] Closed-loop reservoir management can be described as a process that seeks to maximize reservoir performance in terms of hydrocarbon recovery or net present value for the remaining projected life of the reservoir by changing production operations based on near real-time processes. In applying closed-loop reservoir management, a combination of reservoir model-based optimization and data assimilation are used. As the use of flow control, flow monitoring, and flow optimization devices continue to spread throughout the petroleum industry, more reservoirs will become candidates for application of closed-loop reservoir management.

To assist reservoir management teams in crafting and implementing sound reservoir management strategies, five reservoir management principles are enunciated. The five reservoir management principles are:[1]

1. Conservation of reservoir energy.
2. Early implementation of simple, proven strategies.
3. Systematic and sustained practice of data collection.
4. Application of emerging technologies for improved hydrocarbon recovery.
5. Long term retention of staff in multi-disciplinary teams.

In this chapter, the concepts behind the five reservoir management principles are explored further. The chapter is concluded by presenting case histories of three reservoirs where by using these principles it is demonstrated that the applied reservoir management strategies can be pronounced as successful.

20.2 Reservoir Management Principles

The reservoir management principles proposed in this chapter are designed to guide reservoir management teams in the development, implementation, and monitoring of sound reservoir management strategies. The principles are simple, easy to understand, and can be applied to practically all reservoirs. The five principles can be used as a checklist by reservoir management teams to ensure that they are doing all the "right things" in managing their reservoirs. The five reservoir management principles are:

1. Conservation of reservoir energy.
2. Early implementation of simple, proven strategies.
3. Systematic and sustained practice of data collection.
4. Application of emerging technologies for improved hydrocarbon recovery.
5. Long term retention of staff in multi-disciplinary teams.

20.2.1 Conservation of Reservoir Energy

The central principle of the five reservoir management principles is conservation of reservoir energy. This principle applies to controlled and optimized use of the energy stored in the reservoir at discovery or at any stage of depletion to maximize economic and efficient recovery of its hydrocarbons. **The principle should not be interpreted to mean that the reservoir should not be produced until the "correct" strategies for managing it have been devised.** It does require, however, that management strategies applied on the reservoir should avoid depleting reservoir energy inefficiently, especially during the early stages of reservoir development and production. Common production practices that should be avoided because they could deplete reservoir energy include excessive production of gas from the gas cap of a saturated reservoir, high production rates due to excessive pressure drawdown, comingling of production from separate reservoirs, as examples. In applying this principle, the reservoir management teams should strive to achieve a balance between conserving reservoir energy and maximizing economic recovery of hydrocarbons from the reservoir.

20.2.2 Early Implementation of Simple, Proven Strategies

The principle of early implementation of simple, proven strategies is directed at supporting the concept of conserving reservoir energy. Simple, proven strategies are reservoir management practices that are known from industry experience to conserve reservoir energy. Some of these simple, proven strategies include some form of pressure maintenance by fluid injection, limited pressure drawdown at production wells, isolation of separate reservoirs at producers, optimal well spacing, and selective perforation of productive zones. **The costs of implementing these strategies should be weighed against their expected benefits by the reservoir management teams so that economic recovery of hydrocarbons is always maintained.**

20.2.3 Systematic and Sustained Practice of Data Collection

The collection, analysis, and assimilation of data can be considered as the foundation on which sound reservoir management strategies can be devised and implemented. The culture of data collection should start at discovery of the reservoir and maintained throughout its life, as long as it is economic. **This means that the value of information gained from the data should outweigh the cost of collecting and analyzing the data.** The data collection process should include geologic, geophysical, petrophysical, pressure, production/injection data, and any other data to support reservoir development and management. The objectives of the data collection program should target continuously improving knowledge on reservoir processes, and applying that knowledge on devising improved reservoir management strategies. The petroleum industry is rapidly adopting the practice of sustained data collection by installing monitored downhole gauges and flow control devices on key wells. This has led to substantial increase in the amount of data available on many reservoirs. It is expected that this practice will continue to increase in the industry with the ultimate result of key reservoir data available in real time for improved reservoir management decisions.

20.2.4 Application of Emerging Technologies for Improved Hydrocarbon Recovery

Reservoirs that have been properly managed on the basis of the first three principles of reservoir management discussed earlier are good candidates for application of emerging technologies for improved hydrocarbon recovery. In general, reservoirs whose energies have been conserved with simple, proven strategies and have ample data acquired are good candidates for application of emerging technologies for improved hydrocarbon recovery. These new technologies include advances in well architecture, drilling and completion of wells, applications of new chemicals for improved recovery, equipment for production operations, and any other technology that improves the hydrocarbon recovery process. For instance, an emerging technology in well architecture that is expected to substantially improve productivity is application of multilateral wells.[5–7] Multilateral wells have been used in many well-managed reservoirs with long production history to improve hydrocarbon recovery. **The reservoir management team should monitor the industry regularly for new emerging technologies, and assess the potential of these technologies for improving recovery from the reservoir.** The benefits of applying sound reservoir management strategies are realized by positioning the reservoir ready to take advantage of new technologies as they emerge in the industry.

20.2.5 Long Term Retention of Staff in Multi-Disciplinary Teams

In many organizations, reservoir management teams are composed of reservoir engineers, geologists, geophysicists, petrophysicists, production engineers, drilling engineers, facilities engineers, and other staff. These multi-disciplined teams have been recognized in the industry as being very effective in fashioning sound reservoir management strategies. **This final principle of reservoir management is necessary because it puts emphasis on the importance of keeping members of the team together long enough to gain critical understanding of the reservoir.** After this critical level of knowledge has been achieved, management teams function more efficiently in crafting new strategies or altering existing strategies for improved reservoir management. When staff changes are necessary or mandated, it is recommended that at least one or two key members of the reservoir management team should be retained to train and transfer knowledge and experience on the reservoir to the new team members.

20.3 Case Histories Demonstrating Applications of Reservoir Management Principles

The applications of the reservoir management principles can be demonstrated by reviewing the case histories of three reservoirs, and discussing the strategies used to manage the reservoirs on the basis of these principles. The case histories were selected as examples of reservoirs which have long histories of sound reservoir management practices. Many other case histories reported in the literature or constructed from the histories of reservoirs in many organizations can be used as examples.

20.3.1 The Case History of 26R Reservoir (1976–1996)

The reservoir management strategy for the 26R reservoir (1976–1996) can be summarized as follows:

1. Control excessive gas production
2. Maintain reservoir pressure by gas injection
3. Assist gravity drainage
4. Data collection
5. Improve oil recovery with horizontal wells

Geology. The 26R reservoir is located within the Elk Hills Field in the southern San Joaquin valley of central California.[8] The reservoir is contained within the steeply, dipping southwestern limb of the 31S Anticline (Figure 20.1). The structure map of the 26R reservoir is shown in Figure 20.2. The 26R structure is roughly three miles long and one mile wide. As shown in Figure 20.2, a non-sealing fault with several hundred feet of throw intersects the structure. The reservoir had an initial oil column of 1800 ft, and a net productive sand thickness of approximately 1150 ft. The 26R reservoir is an Upper Miocene, Stevens turbidite sand channel classified under the Monterey Formation. Theses sand channels consist of many thin turbidite beds ranging in thickness from a few inches to several feet. Within the beds, sand grain sizes range from very fine to very coarse with infrequent conglomerates. In clean portions of the reservoir, net-to-gross ratios could be as high as 90%. Significant correlatable shale layers exist between

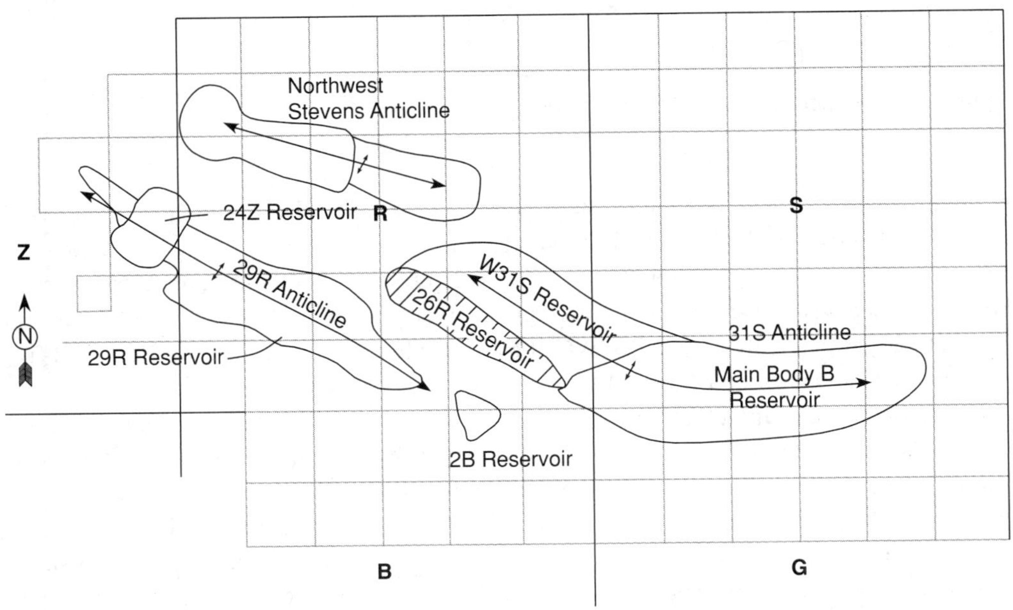

Figure 20.1 Map showing anticlines in Elk Hills Field.

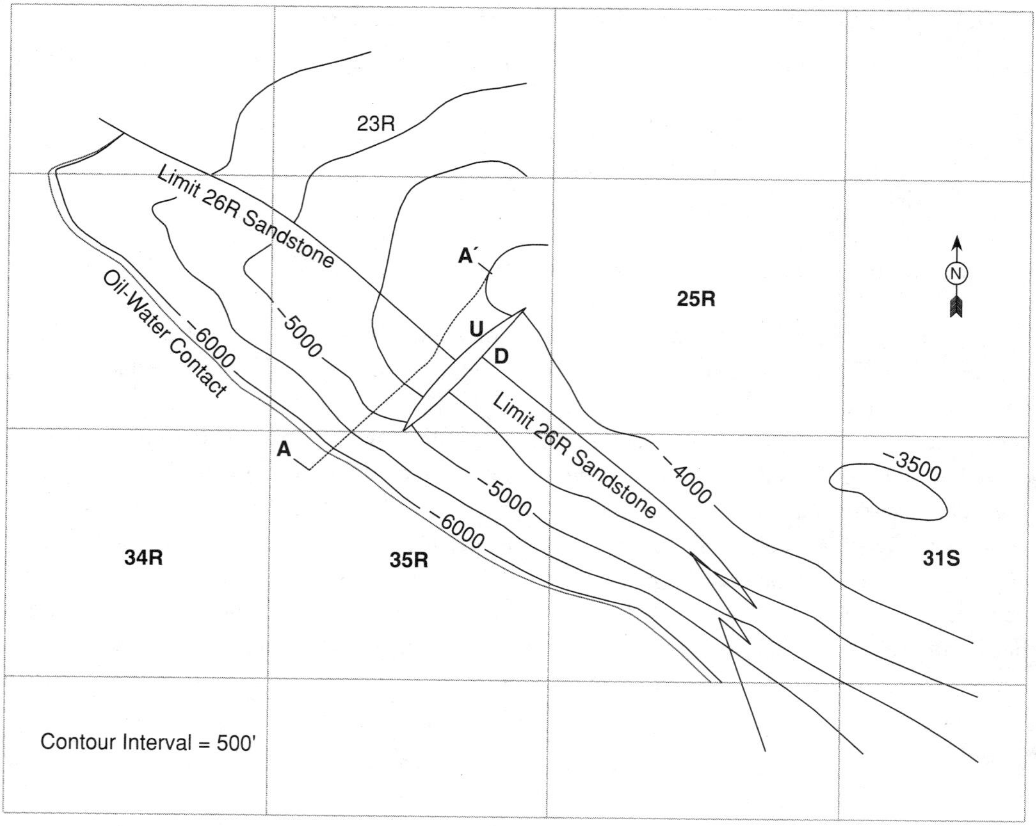

Figure 20.2 Structure map of 26R reservoir.

turbidite sequences. These continuous shale layers were used to classify five large intervals as mega-units, namely: A-C, C-F, F-K, K-N, and N-P, as shown in Figure 20.3. These mega-units are not considered to be separate reservoirs. Formation beds within the 26R reservoir generally dip from 23 degrees near the top of the structure to 70 degrees at the OWC. The OOIP was estimated at 424 million barrels with an OWC at 6010 ft subsea. The 26R reservoir rock and fluid properties are summarized in Table 20.1.

Production and Injection History. The 26R reservoir was discovered in 1950. The reservoir was barely produced from 1950 to 1976 because it was mothballed as part of the United States of America Naval Petroleum Reserve to supply oil for naval operations during national emergency. The production history of the 26R reservoir is shown in Figure 20.4. Oil production reached its peak between 1980 and 1983 at average rates of 50,000 BOPD. Gas and water production during this period averaged 72,809 Mcf/D and 1833 BWPD, respectively. Excessive gas production from the reservoir was controlled by performing gas isolation remedial work with multiple packers and/or shutting-in wells producing at high gas-oil ratios.

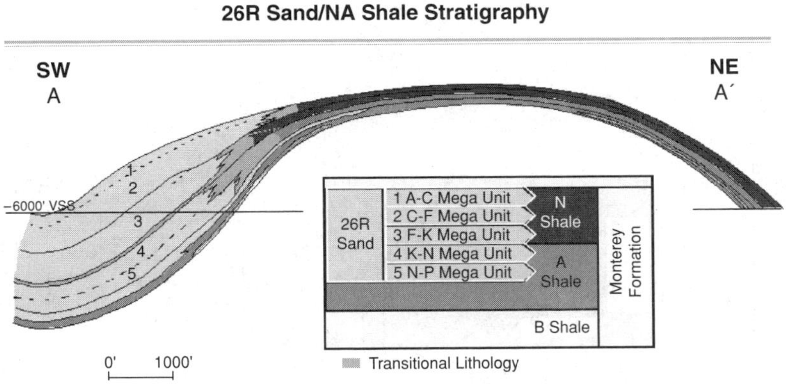

Figure 20.3 Cross-section of 26R reservoir (along A-A′ in Figure 20.2) showing mega-units.

Gas injection at the crest of the structure began in October 1976 barely three months after initial production started in July 1976. The management strategy of injecting gas into this reservoir was to balance reservoir voidage, thereby maintaining reservoir pressure. The average pressure history of the 26R reservoir is shown in Figure 20.5. As shown in Figure 20.5, reservoir pressure declined gradually from 1976, in spite of the pressure maintenance program. There is evidence that the inability to maintain pressure in the 26R reservoir was caused by the migration of some of the injected gas into the contiguous and overlaying shale reservoirs (Figure 20.3). However, the gas injection strategy minimized pressure decline and helped to stabilize reservoir pressure.

Displacement Mechanisms. The major fluid displacement mechanisms occurring within the 26R reservoir in order of prominence are: gravity drainage, liquids vaporization, oil displacement

Table 20.1 Summary of 26R Reservoir Rock and Fluid Properties

Average porosity	23.6%
Average permeability	88 md
Average water saturation	16%
Initial saturation pressure	3155 psia
Reservoir temperature	210°F
Reservoir oil viscosity	0.42 cp
Oil gravity	36°API
Water-oil contact	6010 feet subsea
Residual oil saturation to gas	49%
Residual oil saturation to water	26%
Estimated original oil in place	424 MMBO

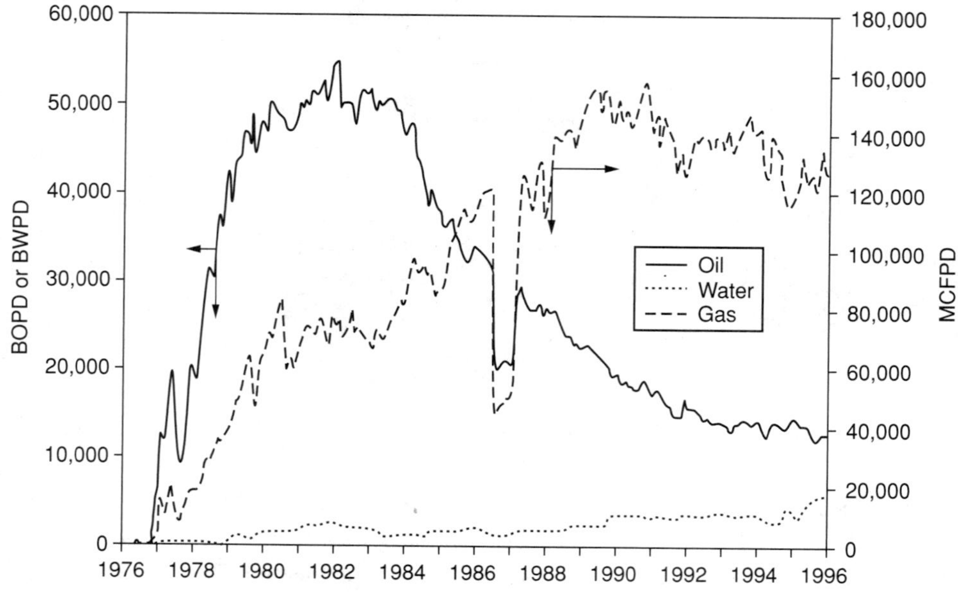

Figure 20.4 Historical production of 26R reservoir.

by gas, and solution gas drive. Gravity drainage in the 26R reservoir is promoted by high dip angles, moderate oil relative permeabilities, and low oil viscosity.[9,10] Liquids vaporization in the secondary gas cap are due to injection and cycling of dry gas as part of the gas injection program.[11] Oil recovery is further assisted by the downward displacement of oil by injected gas. And additional oil recovery is due to expansion of solution gas as reservoir pressure declined gradually.

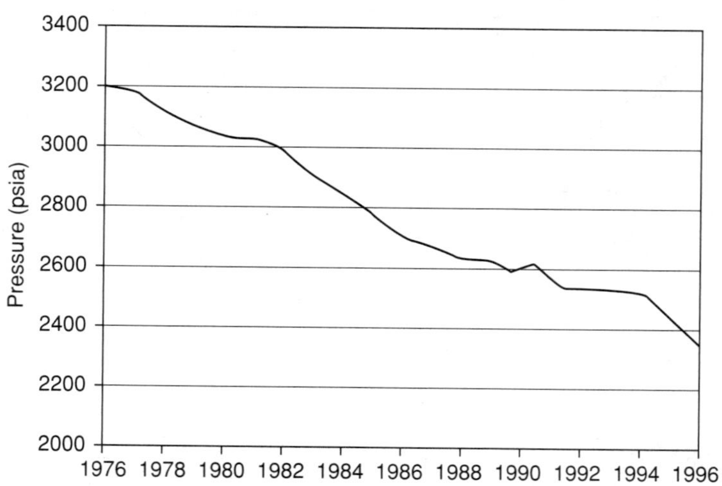

Figure 20.5 Historical average reservoir pressure of 26R reservoir.

These four mechanisms have worked together to achieve a very efficient system of oil displacement in the 26R reservoir.

20.3.2 Application of Reservoir Management Principles to 26R Reservoir

Conservation of Reservoir Energy. The reservoir management principle of conservation of reservoir energy was practiced in the 26R reservoir by controlling excessive gas production. This was achieved by installing multiple packers in wells located high on the structure. As the gas-oil contact moved downdip, high gas producing intervals straddled with packers were shut-off. Eventually, these upstructure wells were shut-in as the gas-oil contact reached the lowest packers in the well. The historical gas-oil ratio plot of the 26R reservoir is shown in Figure 20.6. A review of Figure 20.6 indicates that field GOR was controlled in this reservoir for over a 20-year period.

Early Implementation of Simple, Proven Strategies. Gas injection into the 26R reservoir for pressure maintenance was started just three months from initial production. Gas was injected into the reservoir at wells located along the crest of the structure. The volume of gas injected was calculated to balance reservoir voidage. Pressure maintenance by gas injection is a simple, proven strategy which can be implemented early in the life of a reservoir to reduce precipitous decline in reservoir pressure. In many reservoirs similar to 26R reservoir, reinjection of produced gas is efficient in terms of pressure maintenance and utilization of excess gas produced

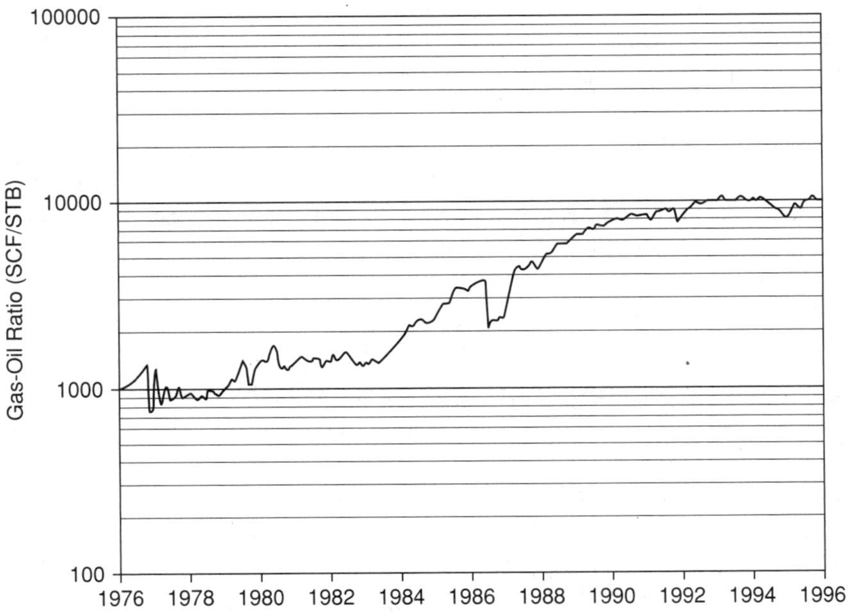

Figure 20.6 Historical GOR of the 26R reservoir.

from the reservoir. In the case of 26R reservoir, pressure maintenance by gas injection was not entirely successful because reservoir pressure continued to decline gradually over the entire period of gas injection (Figure 20.5). This was due primarily to gas migration out of the 26R reservoir into the overlying shale reservoirs (Figure 20.3). However, it is evident that the pressure decline in this reservoir would have been more severe, if the pressure maintenance program was not installed.

Systematic and Sustained Practice of Data Collection. The collection of reservoir pressure data was very important in the management of the 26R reservoir. Several key wells were designated in the gas cap and the oil leg as pressure monitoring wells. These wells in the gas cap were equipped with permanently installed capillary tubing for regular pressure monitoring. Buildup surveys were conducted on the oil band wells as part of a field wide pressure monitoring program. The field wide pressure monitoring program was conducted every six months. Consequently, substantial pressure data were available on the 26R reservoir. These pressure data were used to adjust gas injection volumes as part of the management strategy of maintaining reservoir pressure. In addition to the pressure data, log data, wireline formation tool pressure data, fluid samples, and whole core samples were collected from all new wells drilled in the reservoir. These data were used to monitor the progression of the gas injection strategy, measure residual oil saturation in the gas cap, and update existing geological and reservoir models. These models[12,13] were used to develop new reservoir management strategies or adjust current strategies.

Application of Emerging Technologies for Improved Hydrocarbon Recovery. The successful implementation of the first three principles of reservoir management positioned the 26R reservoir as a good candidate for application of horizontal well technology. Horizontal well technology was virtually non-existent in the petroleum industry in the 1950s when the 26R reservoir was developed with vertical wells. In the 1980s, as horizontal well technology was emerging in the industry, it was recognized that the 26R reservoir was ideally suited for this technology because of its steeply dipping formation that promoted gravity drainage. It was recognized that horizontal wells could be located within the oil column close to the oil-water contact to produce oil draining to the base of the structure by gravity.[14,15] The first horizontal well was drilled in the 26R reservoir in 1988. By 1996, twenty-two horizontal wells were drilled in the 26R reservoir (Figure 20.7). The horizontal wells were more efficient than the vertical wells because they produce at lower GOR, thereby requiring less gas injection for pressure maintenance. In 1998, the horizontal wells produced 70% of the total oil production with one-third of the GOR of the remaining vertical wells (Figures 20.8 and 20.9). In the 1980s, the application of horizontal well technology in the 26R reservoir was the most successful program of its type in the United States of America.

Long Term Retention of Staff in Multi-Disciplinary Teams. The 26R reservoir management team was composed of staff from geology, petrophysics, reservoir, production, facilities, and other disciplines. Over several years, the staff coalesced into a highly efficient team focused on optimizing hydrocarbon recovery from the reservoir. New reservoir management

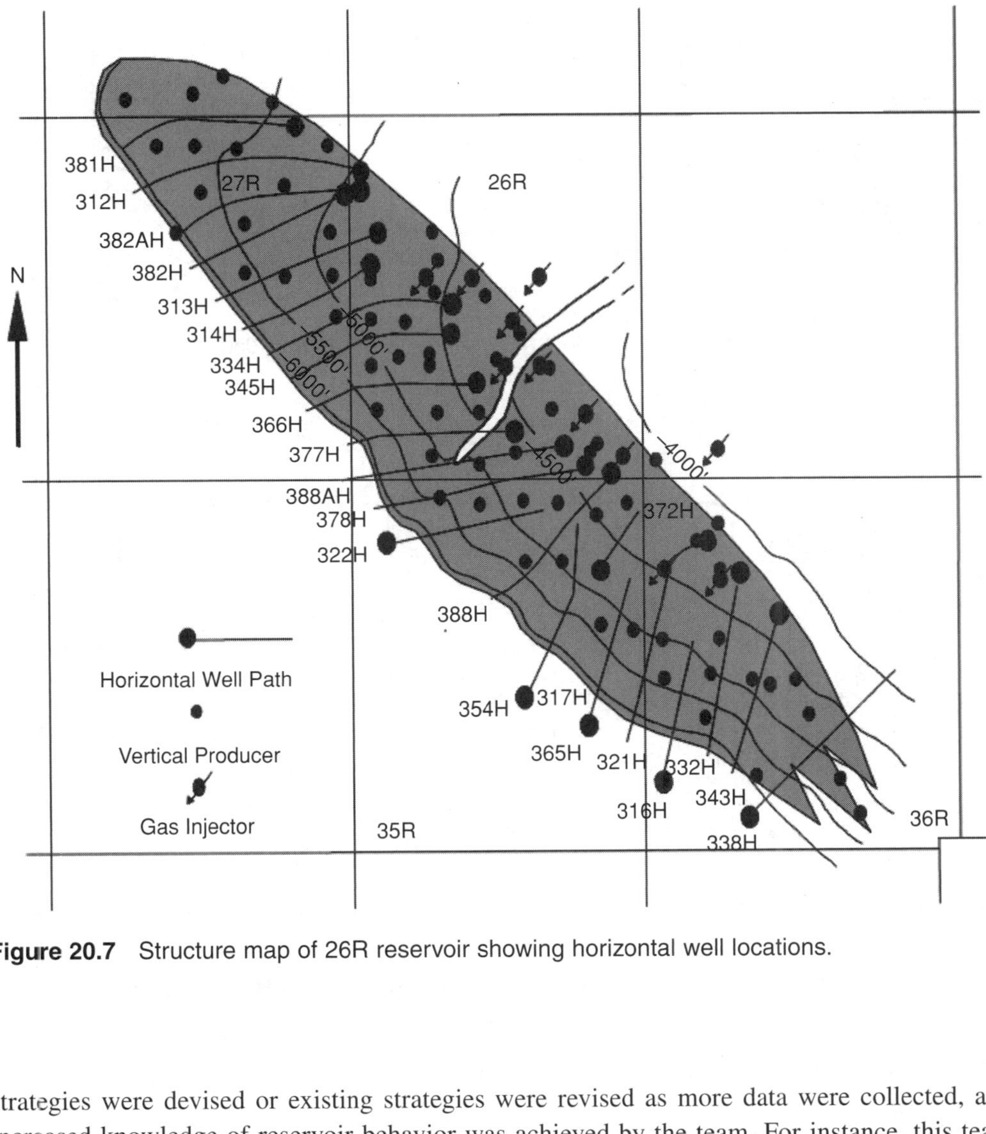

Figure 20.7 Structure map of 26R reservoir showing horizontal well locations.

strategies were devised or existing strategies were revised as more data were collected, and increased knowledge of reservoir behavior was achieved by the team. For instance, this team was able to adjust gas injection volumes weekly on the basis of pressure data from pressure monitoring wells located in the gas cap, and reported total field production volumes. Furthermore, the team was kept generally together for longer than five years. When new members were brought in, there were always legacy members who could convey their experience and knowledge of the reservoir to the new members. Keeping the management team of the 26R reservoir stable over a long period contributed to the success achieved in the management of the reservoir.

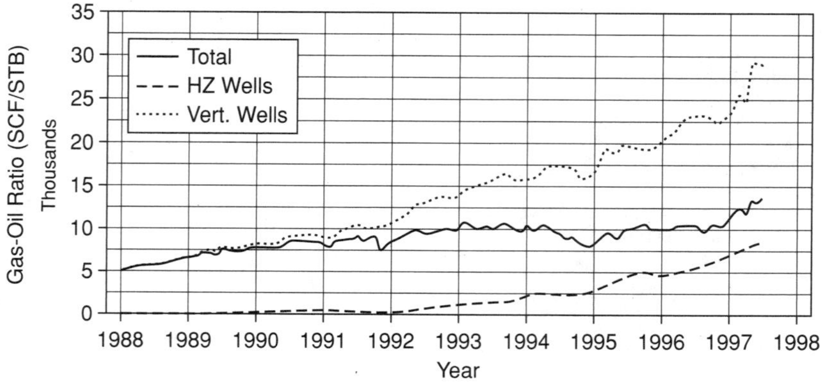

Figure 20.8 GOR comparison—horizontal versus vertical wells in the 26R reservoir.

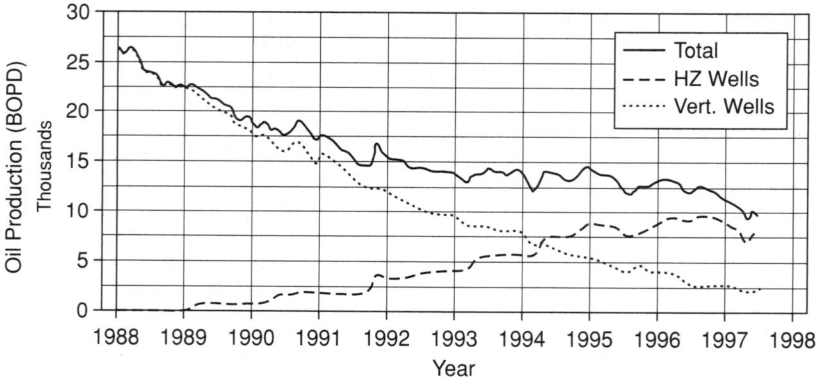

Figure 20.9 Oil production comparison—horizontal versus vertical wells in the 26R reservoir.

20.3.3 The Case History of MBB/W31S Reservoirs (1976–1999)

The reservoir management strategy that proved successful in the Main Body "B" (MBB) and Western 31S (W31S) reservoirs can be enumerated as follows:

1. Reduce production of free gas from the gas cap.
2. Maintain reservoir pressure by gas and peripheral water injection.
3. Install extensive program of data collection and monitoring of waterflood performance.
4. Exploit the oil bank created by the waterflood by drilling infill wells.

Geology. The largest of the three anticlines in the Elk Hills Oil Field is the 31S structure (Figure 20.1). The MBB and W31S reservoirs occupy the entire 31S structure which is about 9 miles long and 1.5 miles wide. The 31S structure is an elongated, doubly plunging anticline, which was slightly deflected at its mid region to form two distinct halves, called the MBB and the W31S

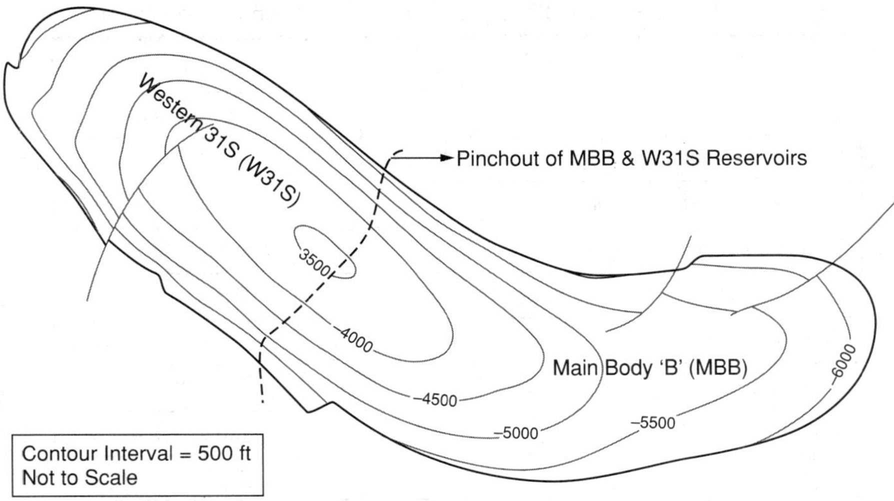

Figure 20.10 Structure map of MBB/W31S reservoirs.

reservoirs (Figure 20.10). The eastern half of the 31S anticline with formations dipping up to 30 degrees is the MBB reservoir, and the western half of the 31S anticline with formations dipping up to 60 degrees is the W31S reservoir. The two reservoirs are combined for management purposes and called the MBB/W31S reservoirs. The MBB/W31S reservoirs are turbidite, sandstone reservoirs consisting of feldspathic, clay-rich deposits. Core studies show that the turbidite beds are composed of medium to coarse grained dirty sands that grade upward first into fine grained sands, followed by silts and mudstones.[16,17] The complete "Bouma Sequence,"[18] consisting of fining upward cycle of coarse sand, fine sand, silt, and mud were created in a low-density sandy turbidity currents. The Bouma sequences are not complete in all depositional cycles of the MBB/W31S formations. However, where these complete and incomplete Bouma sequences are stacked vertically on top of each other, the resulting formations are very stratified and highly heterogeneous. The Dykstra-Parsons permeability variability coefficients of the MBB/W31S formations range from 0.7 to 0.85.

Several relatively thick shale layers can be correlated over long distances in the MBB and W31S formations. These shale layers have been used to subdivide the MBB and W31S formations into flow units for waterflood management. The flow units in the MBB reservoir are sheet-like sands, whereas the flow units in the W31S reservoir are channelized sands. Some of the flow units in the MBB reservoir can be correlated over distances as long as five miles, whereas the flow units in the W31S reservoir are more sporadic and extend about one-half mile. The MBB and W31S reservoirs overlap in the central portion of the 31S anticline (Figure 20.10).

The OOIP was calculated at 530 MMSTBO for the MBB reservoir, and 80 MMSTBO for the W31S reservoir for a total of 610 MMSTBO for the combined MBB/W31S reservoirs. The range of porosity is 11%–26%. The geometric mean air permeability from core samples is 32.3 md,

Table 20.2 Summary of MBB/W31S Reservoir Rock and Fluid Properties

Porosity range	11–26%
Air permeability range	10–250 md
Initial water saturation range	30–45%
Initial average reservoir pressure	3150 psia
Initial bubble point pressure	2965 psia
Reservoir temperature	210°F
Reservoir oil viscosity	0.40 cp
Oil gravity	36°API
Mobility ratio	0.6
Residual oil saturation to water	25%
Estimated original oil-in-place	610 MMBO

with a range of 10 to 250 md. The range of initial water saturation is 30%–45%. The mobility ratio for water displacement is favorable at 0.6, with residual oil saturation to water of 25%. The MBB/W31S reservoir rock and fluid properties are summarized in Table 20.2.

Production and Injection History. Before the energy crisis in mid-1970s, the MBB/W31S reservoirs were partially developed and preserved as part of the United States of America Naval Petroleum Reserve to supply oil for naval operations in case of national emergency. Cumulative oil production from initial production in January 1942 through December 1975 was just 1.15 MMBO. Consequently, oil production could be considered to have started in the MBB/W31S reservoirs in June 1976 under primary depletion. Total oil production rate rose rapidly to 25,900 BOPD in May 1977 but was later reduced to 10,765 BOPD towards the end of that year due to rapid decline of reservoir pressure. A pressure maintenance program by gas injection at the crest of the structure was initiated in October 1976. In June 1978, a pilot peripheral waterflood project was started on the southeastern nose of the MBB reservoir. The pilot waterflood project was designed to evaluate the effectiveness of peripheral water injection in displacing oil towards the first line of in-board producers. The pilot waterflood project was very successful because the in-board producers responded with higher oil production rates. This led to the expansion of the peripheral water injection around the entire MBB/W31S reservoirs from 1982 to 1985 (Figure 20.11). The gas injection program was ended in 1988 with cumulative gas injection volume of 117 BCF. Gas injection was terminated because increased water injection was considered sufficient for pressure maintenance.

Peripheral water injection resulted in the development of a significant oil bank within the MBB/W31S reservoirs (Figure 20.11). The oil bank was created updip in areas of the structure where a large secondary gas cap had developed because of the decline in reservoir pressure and

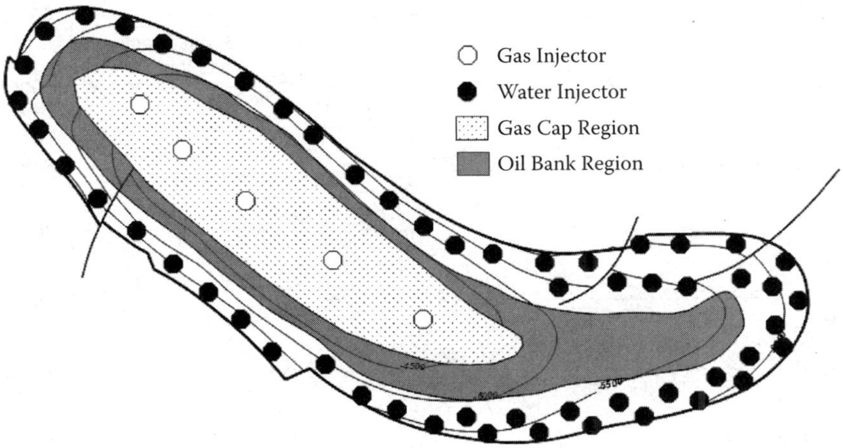

Figure 20.11 Map of MBB/W31S reservoirs showing the peripheral waterflood project.

the gas injection program. As the oil bank developed, infill wells were drilled on the leading edge of the oil bank. Most wells drilled in the oil bank produced at GORs that were close to solution GOR. Development drilling of infill wells reduced well spacing in the MBB/W31S reservoirs from 20 acres to 10 acres. The production history of MBB/W31S reservoirs is shown in Figure 20.12. The historical gas and water injection rates are shown in Figure 20.13. The historical gas-oil ratio

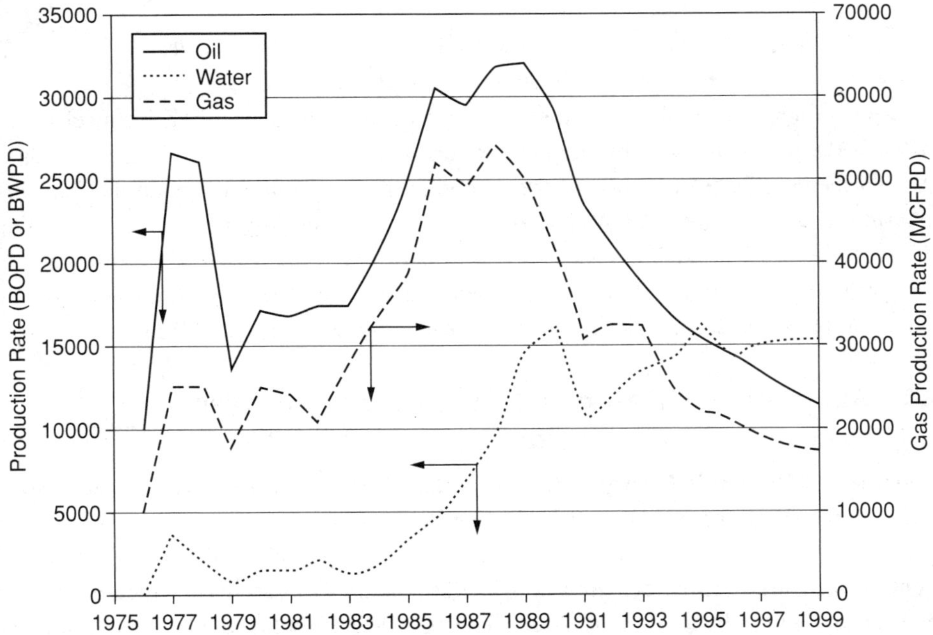

Figure 20.12 Historical production of MBB/W31S reservoirs.

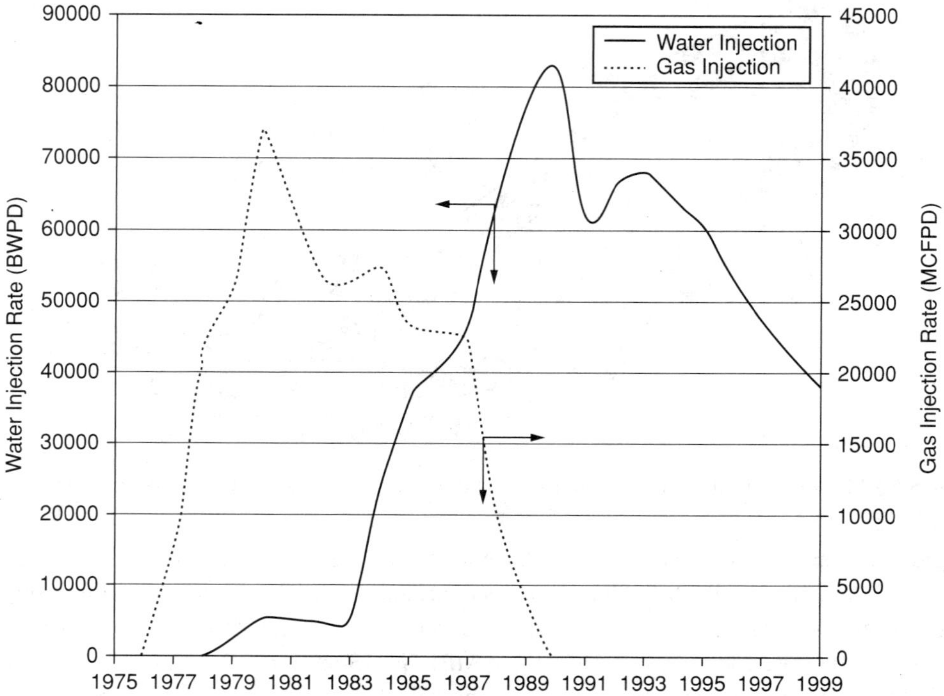

Figure 20.13 Historical gas and water injection in the MBB/W31S reservoirs.

and water-oil ratio for the reservoirs are shown in Figure 20.14. The historical average reservoir pressure of MBB/W31S reservoirs is shown in Figure 20.15.

Displacement Mechanisms. At the start of production, the dominant drive mechanisms in the MBB/W31S reservoirs were by fluid expansion and solution gas drive. These were assisted by gas displacement from expansion of the secondary gas cap. Later, the main mechanism that assisted oil recovery in the MBB/W31S reservoirs is oil displacement by water injected at the periphery of the reservoirs. Oil displacement by waterflood in the MBB/W31S reservoirs has a favorable mobility ratio of 0.6. Peripheral waterflooding in the MBB/W31S reservoirs over time created an oil bank and shrunk its secondary gas cap as shown in Figure 20.11.

20.3.4 Application of Reservoir Management Principles to MBB/W31S Reservoirs

Conservation of Reservoir Energy. After the MBB/W31S reservoirs were opened up for production in June 1976, there was a rapid decline in reservoir pressure. To conserve reservoir energy, oil production was curtailed from 25,900 BOPD to 10,765 BOPD in 1977. This early intervention measure reduced the growth of the secondary gas cap that formed after average reservoir pressure initially at 3150 psia declined about 450 psi below the bubble point pressure at 2950 psia. By shutting in high gas-oil ratio wells and increasing water injection, the producing gas-oil ratio of the reservoirs was maintained relatively stable as shown in Figure 20.14.

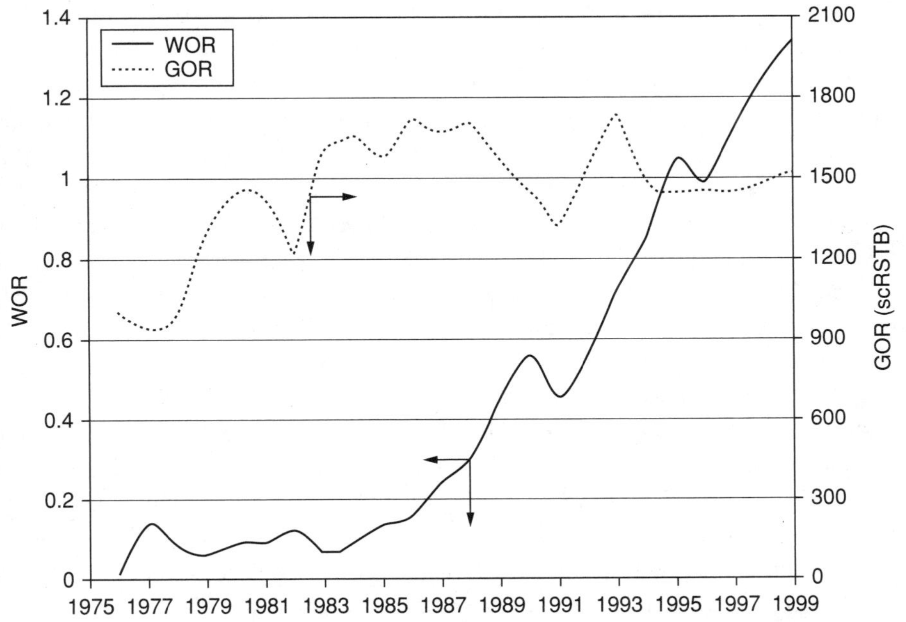

Figure 20.14 Historical GOR and WOR for MBB/W31S reservoirs.

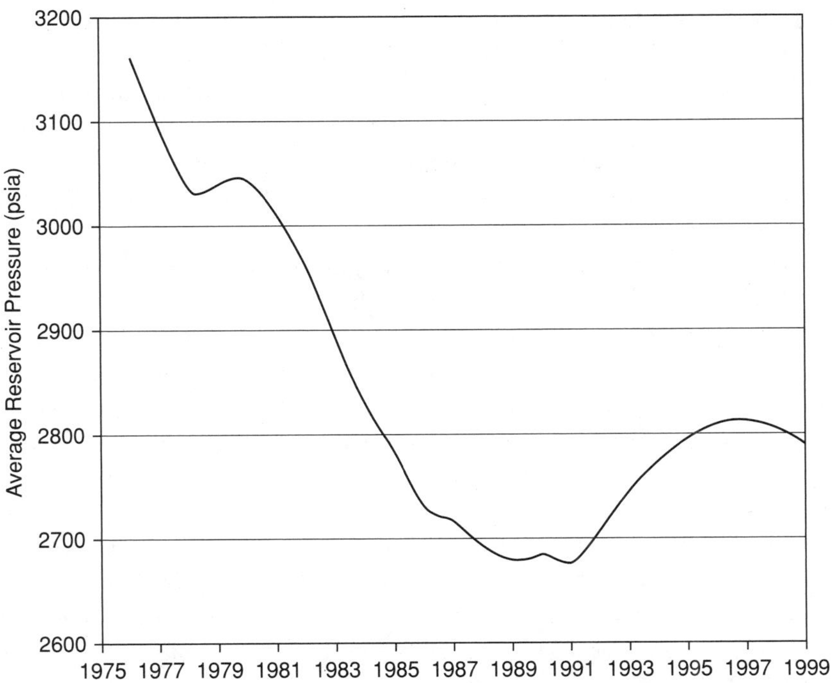

Figure 20.15 Historical average reservoir pressure of MBB/W31S reservoirs.

Early Implementation of Simple, Proven Strategies. Gas injection at the crest of the structure was started in October 1976. This was followed with the start of a pilot water injection project in June 1978. The gas injection project was primarily to maintain reservoir pressure. The pilot peripheral water injection was designed to evaluate oil displacement with water, and also to maintain reservoir pressure. Gas and water injection are simple, proven strategies that can be used to maintain reservoir pressure and improve oil recovery. In the case of MBB/W31S reservoirs, peripheral water injection strategy was selected because there were concerns about premature water breakthrough at the producers for the highly heterogeneous MBB/W31S sands. Also, peripheral water injection was cheaper to install compared with pattern water flood and could be discontinued quickly, if adverse results were observed. Early implementation of these two simple strategies improved oil recovery in the MBB/W31S reservoirs substantially, when compared with the performance of almost identical reservoirs in the area.[19,20]

Systematic and Sustained Practice of Data Collection. An extensive data collection program was undertaken as an integral part of the reservoir management strategy for the peripheral waterflood project in the MBB/W31S reservoirs. This data collection system included pressure monitoring in wells located in key areas of the reservoir, wireline formation test pressures, core samples, and complete log suites from new infill wells, production logs from producing wells, injection profile logs from water injectors, fluid entry surveys with downhole videos, and gamma ray surveys to identify layers showing significant water production. For instance as of 1996, the MBB/W31S reservoirs had a database consisting of conventional core analysis on 9000 samples, and special core analysis on 37 samples.[21] Included in this database are logs for 725 wells, including 413 wells with modern log suites.[22] In the area of pressure data, the MBB reservoir had about 7000 static pressure data, 80 buildups/falloffs, and 43 wells with wireline formation test points.[22] In addition, a well was drilled behind the floodfront for the sole purpose of evaluating the potential for bypassed reserves in the less permeable layers of the MBB reservoir in 1988. As a result of this massive data base, the management of the MBB/W31S waterflood project was constantly updated and improved to take advantage of the advancement of the floodfront into unflooded areas of the reservoir.

Application of Emerging Technologies for Improved Hydrocarbon Recovery. The emerging technologies that were used to improve hydrocarbon recovery in the MBB/W31S reservoirs were mainly in the area of monitoring and surveillance of the waterflood project. New tools for fluid entry surveys were routinely used to determine the layers that were being flooded. This assisted the reservoir management team to adjust the water injection profile by installation of multiple packers in water injection wells. For instance, downhole video cameras were especially useful in identifying water producing layers for remedial work to isolate or shut-off water production.

Long Term Retention of Staff in Multi-Disciplinary Teams. The successful peripheral waterflooding of the MBB/W31S reservoirs can be linked to the formation of a management team composed of staff from diverse disciplines. This team worked together on the management of this project for many years. The team was able to devise new management strategies or modify existing strategies based on analysis of extensive data collected on the waterflood project.

20.3.5 The Case History of the Shaybah Field

The Shaybah field in Saudi Arabia is a low permeability carbonate reservoir overlain by a large gas cap and underlain by a relatively weak aquifer. The Shaybah field was developed initially with massive application of single lateral horizontal wells. Later to improve well productivity, multilateral wells with maximized reservoir contact (MRC) were used to develop the reservoir. The use of horizontal wells and later multilateral wells with MRC as reservoir management strategies were successful in this reservoir due to the following reasons:

1. The formations in the reservoir have low permeability and porosity.
2. Ability to control and reduce producing gas-oil ratio.
3. Ability to control and reduce producing water cut.
4. The need to improve well productivity index.
5. The requirement of achieving production targets.
6. Reduction of unit development cost.
7. Use of smart completion technology.

Geology. The Shaybah field in the Rub' al-Khali desert of Saudi Arabia is about 40 miles long and 8 miles wide.[23] The structure of the field is described as a gently folded northeast/southwest-trending anticline. The formation consists primarily of Cretaceous age sandstones, shales, and carbonates with rudist buildups that change laterally into barrier and shelf slope facies.[23] The average matrix porosity is 25% with no lateral variations. The matrix permeability is facies-dependent and has spatial variations. In south Shaybah, matrix permeability has a range of 5 to 10 md, whereas in the north Shaybah the range of matrix permeability is from 50 to 200 md. Three-dimensional seismic data show that the Shu'aiba carbonate reservoir in the Shaybah field contains a number of faults. These faults have been identified from openhole logs and are more prevalent in the northern part of the reservoir. The Shu'aiba formation contains Arabian Extra Light oil with average API of 42°, and solution gas-oil ratio of 750 scf/STB. The Shaybah field was discovered in 1968 and placed on production in July 1998.

Development History. The surface terrain of the Rub' al-Khali desert, where the Shaybah field is located, comprises of salt flats area called *sabkhahs* and mountainous sand dunes which are as high as 650 feet. Due to the rugged character of the terrain, the Shaybah field was developed from the flat sabkhahs. This required the use of highly directional wells to reach formation targets. The Shaybah field was developed in 1996 with 1-km long, single-lateral horizontal wells designed to drain the oil effectively, while reducing cusping of gas from the gas cap. From 1998 to 2001, reservoir contact was increased by drilling single-lateral wells with lengths of 2 to 3 km based on well and field performance of the 1-km horizontal wells. The longer, single-lateral horizontal wells resulted in significant improvements in well performance, in terms of higher productivity indices, lower drawdowns, and further delays of gas cusping into wells.[23–25] To capitalize on the success of drilling these extra long, single-lateral horizontal wells, the concept of drilling multilateral (ML) and maximum reservoir contact (MRC) wells was developed in Saudi Aramco.[23,26,27]

A MRC well is defined as a well with a minimum aggregate reservoir contact of 5 km either as a single-lateral or a multilateral configuration.[25] The first MRC well was drilled in the Shaybah field in 2002. This was a trilateral well with total reservoir contact of 8.5 km. This was followed by two other MRC wells with reservoir contacts of 5.8 km and 12.3 km, respectively. These three MRC wells had openhole completions. By the end of 2007, a total of 50 ML/MRC wells have been completed in the Shaybah field.[25] MRC wells in the Shaybah field showed reductions in water cut and gas-oil ratio, increases in oil production and well productivity index, and reduction in unit development cost.[23,25]

Displacement Mechanisms. The Shaybah field was described as a low permeability carbonate reservoir overlain by a large gas cap and underlain by a relatively weak aquifer. From this description, it is evident that the primary drive mechanisms for hydrocarbon recovery are gas cap expansion, solution gas drive, and weak water drive from the aquifer. When these drive mechanisms are coupled with the low permeability conditions present in the reservoir, it becomes quite comprehensible that reservoir development with wells that can reduce drawdown, minimize free gas and water production, and achieve high target production rates are desirable. These observations undoubtedly led to the decision to use long single-lateral horizontal wells for the initial development of the reservoir in 1996.

20.3.6 Application of Reservoir Management Principles to the Shaybah Field

It is instructive to apply the five principles of reservoir management to the Shaybah Field and illustrate how the reservoir management strategies devised for this field may have contributed to the success reported for the field.

Conservation of Reservoir Energy. The reservoir energy in the Shaybah field was conserved by the management strategy of using horizontal wells instead of vertical wells for the initial development of the reservoir. The use of horizontal wells minimized production of free gas from the overlying primary gas cap, thereby conserving reservoir energy. It also reduced coning of water from the aquifer which could have reduced hydrocarbon recovery potentials of the wells. The use of horizontal wells minimized the potential for early gas breakthrough, while maintaining production at economic rates.[23]

Early Implementation of Simple, Proven Strategies. Horizontal well technology emerged in the petroleum industry in the early 1980s. In 1996 alone, about 3700 horizontal wells were drilled around the world and the total number of horizontal wells in the world was about 38,000.[28] The same year 1996, the reservoir management team chose to develop the Shaybah field with horizontal wells. This is a clear example of early implementation of simple, proven strategy to maximize hydrocarbon recovery. By 1996 when horizontal well technology was implemented for the development of the Shaybah field, the technology has been proven around the world to be efficient, effective, and economic for developing low permeability reservoirs with large gas caps.

Systematic and Sustained Practice of Data Collection. The database of the Shaybah field was initially created from core and log data of 29 delineation wells drilled before 1976. The database was used to build a facies-based geological model in 1995.[29] As horizontal wells were drilled during the development phase from 1996 to 2002, additional data were obtained from cores and logs leading to the update of the geological model in 1998, and in 2002. Consequently, it appears that a culture of systematic and sustained practice of data collection was a part of the management strategy of the Shaybah field. During drilling of MRC wells, the motherbore was logged with a logging-while-drilling (LWD) tool, and a measurement-while-drilling (MWD) log was run on the laterals. These tools enabled the use of real-time geosteering to improve the efficiency of drilling the ML/MRC wells.[26] Furthermore, the data collected from the closely spaced laterals of the ML/MRC wells improved the distribution of facies in the Shaybah geologic model.[29]

Application of Emerging Technologies for Improved Hydrocarbon Recovery. The Shaybah field is a very good example of the application of emerging technologies for improvement of hydrocarbon recovery. In this case, the management team of the Shaybah field developed the technology of ML/MRC wells to improve hydrocarbon recovery from the field.[23–28] ML/MRC wells were designed for the Shaybah field to take advantage of its low permeability and relatively low porosity, minimize cusping gas from the large overlying primary gas cap, and avoid coning water from the underlying weak aquifer, while maintaining desired target oil production rates. In comparison to single-lateral horizontal wells, ML/MRC wells produced at lower GORs (Figure 20.16), and higher oil rates resulting in much higher cumulative oil productions (Figures 20.17 and 20.18). The productive indices (PIs) of ML/MRC wells were higher than those of single-lateral horizontal wells as shown in Figure 20.19. The costs of ML/MRC wells defined in terms of unit-development costs (drilling + completion + hookup

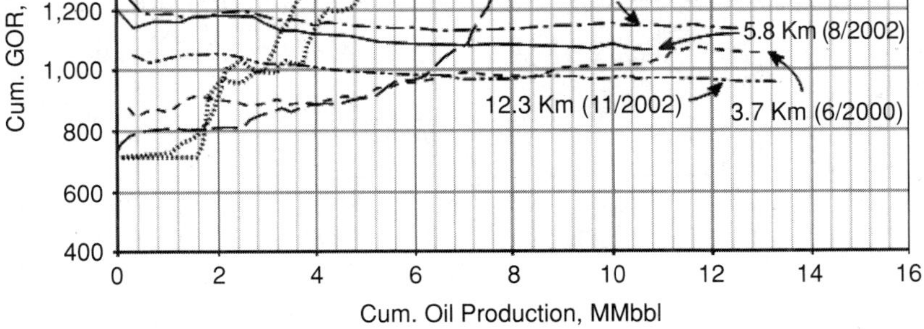

Figure 20.16 Shaybah Field GOR plot based on reservoir contact-South area wells (from Salamy et al.[25] © 2008 SPE, Reproduced with permission).

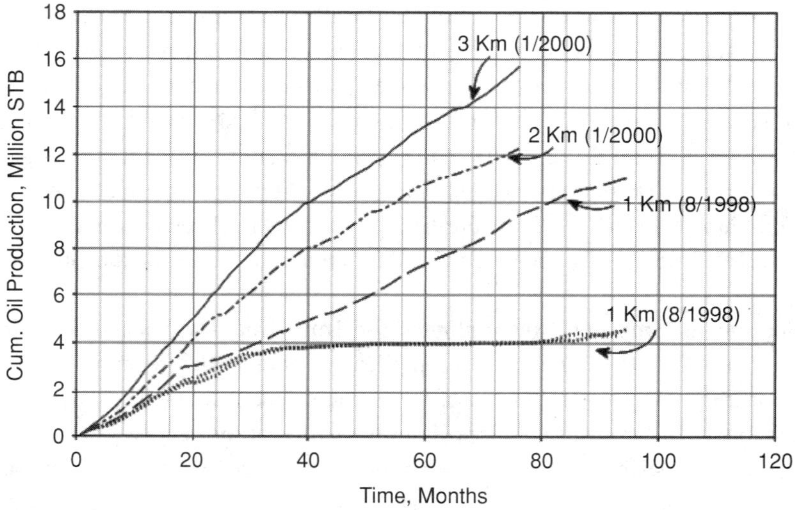

Figure 20.17 Shaybah Field cumulative oil production vs. time based on reservoir contact-North Area wells (from Salamy et al.[25] © 2008 SPE, Reproduced with permission).

costs per B/D of initial production) are lower when compared with single-lateral horizontal wells. Figure 20.20 shows normalized (relative to 1 km horizontal well) unit-development costs in terms of reservoir contact for single-lateral horizontal wells and ML/MRC wells in South Shaybah field. Consequently in terms of performance and costs, ML/MRC wells are considerably superior to single-lateral wells justifying the application of this advanced well technology for the development of the Shaybah field.

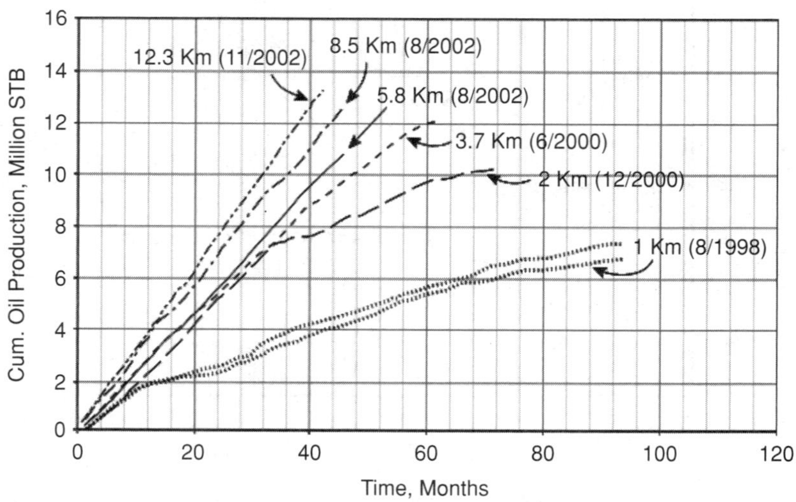

Figure 20.18 Shaybah Field cumulative oil production vs. time based on reservoir contact-South Area wells (from Salamy et al.[25] © 2008 SPE, Reproduced with permission).

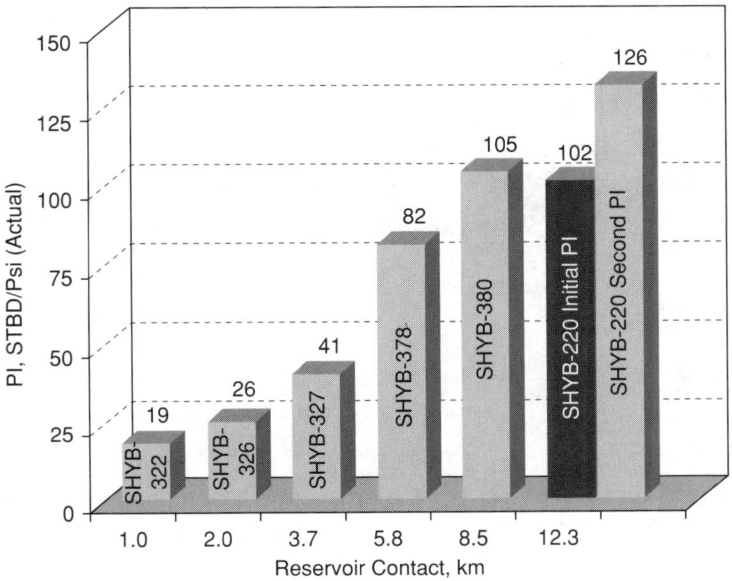

Figure 20.19 Productivity index versus reservoir contact in South Shaybah (from Saleri et al.[23] © 2004 SPE, Reproduced with permission).

The management team of the Shaybah field has continued to apply emerging technologies to improved hydrocarbon recovery.[30] This is reflected in the use of smart technologies to improve control of new ML/MRC wells and conversion of existing single-lateral horizontal wells into ML/MRC wells. The smart technologies deployed in the Shaybah field include downhole flow

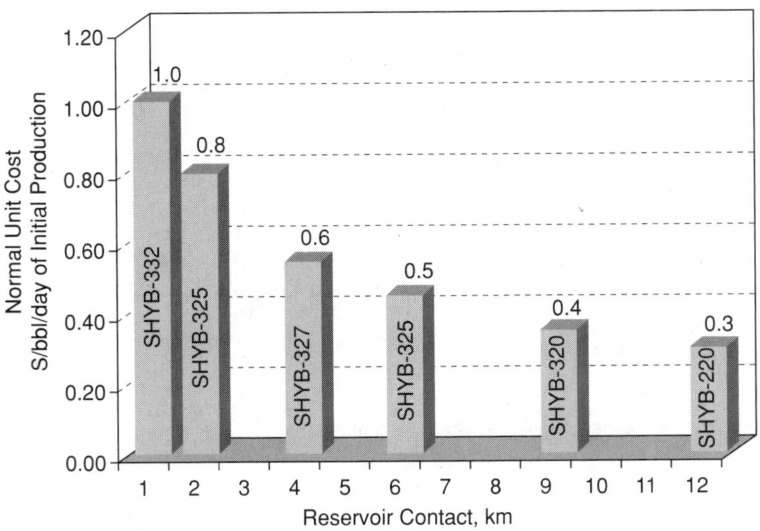

Figure 20.20 Normalized unit cost versus reservoir contact in South Shaybah (from Saleri et al.[23] © 2004 SPE, Reproduced with permission).

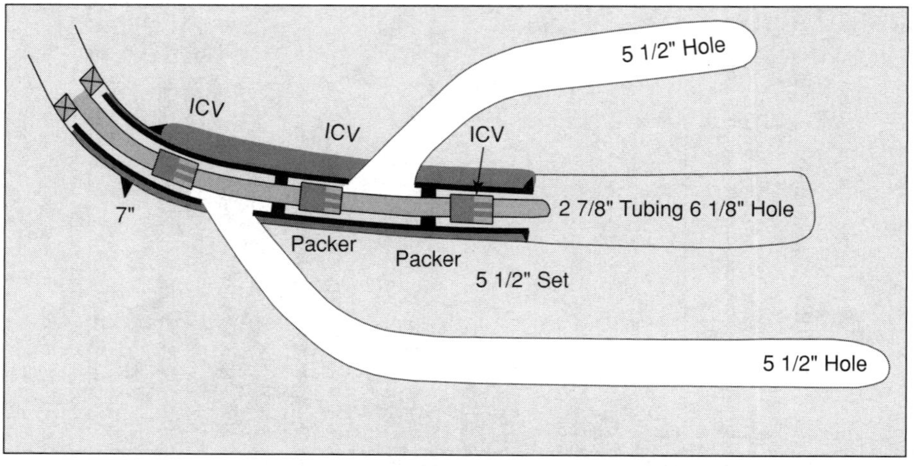

Figure 20.21 Smart completion inside 5 1/2" expandable liner (from Salamy et al.[25] © 2008 SPE, Reproduced with permission).

control systems (DHFCS) and permanent downhole monitoring systems (PDHMS). The DHFCS are surface-controlled hydraulic inflow control valves (ICV) and the PDHMS consists of surface-based pressure and temperature monitoring systems. Smart completions have been applied successfully with expandable liners as shown in Figure 20.21 and in open-holes with swellable packers as shown in Figure 20.22. In addition, production equalizer is being used in the Shaybah field as an advanced completion technology to control and reduce GORs of wells.[30] The application of these new technologies clearly demonstrate that the management team of the Shaybah field are proactive in using emerging new technologies to improve hydrocarbon recovery from the field.

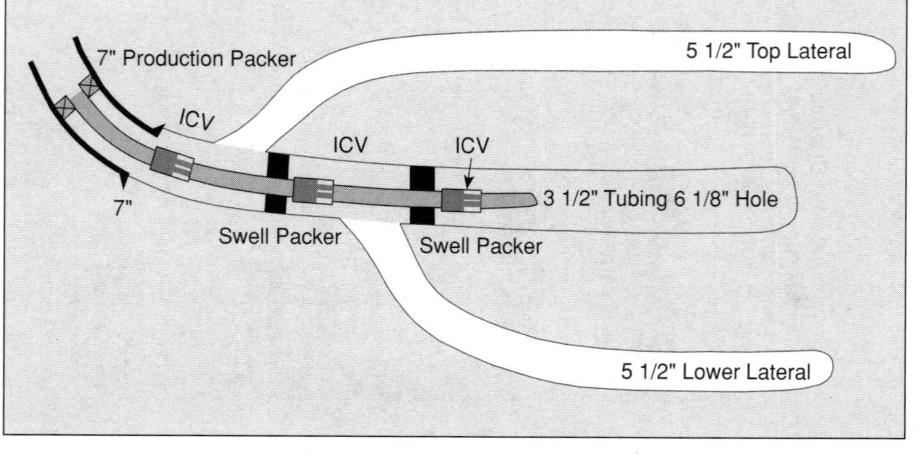

Figure 20.22 Swell-packer in openhole smart completion (from Salamy et al.[25] © 2008 SPE, Reproduced with permission).

Long Term Retention of Staff in Multi-Disciplinary Teams. Evidence of the application of this principle in the management of the Shaybah field is revealed by reviewing the technical literature published on the Shaybah field.[23–31] It is worthy to note that the names of the same authors have reoccurred in most of these publications. It can be readily deduced from these publications that these authors have worked consistently on the Shaybah field from the drilling of the single-lateral horizontal wells to drilling of ML/MRC wells and application of smart well technologies. There is little doubt that this long term retention of staff on the Shaybah field has contributed to its resounding success.

References

1. Ezekwe, J.N.: "Applied Reservoir Management Principles with Case Histories," paper SPE 84148 presented at the SPE Annual Technical Conference and Exhibition, Denver, Colorado, October 5–8, 2003.
2. Konopczynski, M.R., and Ajayi, A.: "Reservoir Surveillance, Production Optimisation and Smart Workflows for Smart Fields—A Guide for Developing and Implementing Reservoir Management Philosophies and Operating Guidelines in Next Generation Fields," paper SPE 112244 presented at the 2008 SPE Intelligent Energy Conference and Exhibition, Amsterdam, The Netherlands, February 25–27, 2008.
3. Mubarak, S.M.: "Real-Time Reservoir Management from Data Acquisition Through Implementation: Closed-Loop Approach," paper SPE 111717 presented at the 2008 SPE Intelligent Energy Conference and Exhibition, Amsterdam, The Netherlands, February 25–27, 2008.
4. Jansen, J.D., Douma, S.D., Brouwer, D.R., Van den Hof, P.M.J., Bosgra, O.H., and Heemink, A.W.: "Closed-Loop Reservoir Management," paper SPE 119098 presented at the 2009 SPE Reservoir Simulation Symposium, The Woodlands, Texas, February 2–4, 2009.
5. Meader, T., Allen, F., and Riley, G.: "To the Limit and Beyond—The Secret of World-Class Extended-Reach Drilling Performance at Wytch Farm," paper IADC/SPE 59204 presented at the 2000 IADC/SPE Drilling Conference, New Orleans, Louisiana, February 23–25, 2000.
6. Yaliz, A., Chapman, T., and Downie, J.: "Case Study of a Quad-Lateral Horizontal Well in the Lennox Field: A Triassic Oil-Rim Reservoir," paper SPE 75249 presented at the SPE/DOE Improved Oil Recovery Symposium, Tulsa, Oklahoma, April 13–17, 2002.
7. Mercado, O., Velez, J., and Fipke, S.: "Multilateral Wells in the Castilla Field of Eastern Colombia: A Case History of the Guadalupe Reservoir," paper SPE/IADC 119458 presented at the SPE/IADC Drilling Conference and Exhibition, Amsterdam, The Netherlands, March 17–19, 2009.

8. Wei, M.H., Yu, J.P., Moore, D.M., Ezekwe, N., Querin, M., and Williams, L.L.: "Case History of Pressure Maintenance by Crestal Gas Injection in the 26R Gravity Drainage Reservoir," paper SPE 24035 presented at the Western Regional Meeting, Bakersfield, California, March 20–April 1, 1992.

9. Essley, P.L., Jr., Hancock, G.L., Jr., and Jones, K.E.: "Gravity Drainage Concepts in a Steeply Dipping Reservoir," paper SPE 1029-G, SPE of AIME, Petroleum Conference on Production and Reservoir Engineering, Tulsa, Oklahoma, March 20–21, 1958.

10. Lewis, J.O.: "Gravity Drainage in Oil Fields," *Trans. AIME* (1944) 155, 133–154.

11. Fleckenstein, W.W., Bouck, L.S., Hudgens, P.D., Querin, M.E., and Williams, L.L.: "An Application of Oil Vaporization Evaluation Methods," paper SPE 24041 presented at the 1992 Western Regional Meeting, Bakersfield, California, March 30–April 1, 1992.

12. Echols, D.P., and Ezekwe, N.: "A Reservoir Management Model for 26R Reservoir, Elk Hills Oil Field, California," paper SPE 46231 presented at the 1998 SPE Western Regional Meeting, Bakersfield, California, May 10–13, 1998.

13. Ezekwe, N., Echols, D., and Hudgens, P.: "Diagnostic Application of Reservoir Simulation to 31S Structure Stevens Reservoirs—Elk Hills Oil Field, California paper SPE 46230 presented at the 1998 SPE Western Regional Meeting, Bakersfield, California, May 10–13, 1998.

14. Gangle, F.J., Schultz, K.L., McJannet, G.S., and Ezekwe, N.: "Improved Oil Recovery Using Horizontal Wells at Elk Hills, California," *SPE Drilling and Completion* (March 1995) 27–33.

15. Mut, D.L., Moore, D.M., Thompson, T.W., and Querin, E.M.: "Horizontal Well Development of a Gravity Drainage Reservoir, 26R Pool, Elk Hills Field, California," paper SPE 35710 presented at the 1996 SPE Western Regional Meeting, Anchorage, Alaska, May 22–24, 1996.

16. Ezekwe, N., Smith, S., Wilson, M., Humphrey, M., and Murphy, J.: "Performance of a Peripheral Waterflood Project in the Main Body 'B' Reservoirs (Stevens) at the Elk Hills Oil Field, California" paper SPE 21759 presented at the Western Regional Meeting, Long Beach, California, March 20–22, 1991.

17. Walker, T., Kerns, S., Scott, D., White, P., Harkrider, J., Miller, C., and Singh, T.: "Fracture Stimulation Optimization in the Redevelopment of a Mature Waterflood, Elk Hills Field, California," paper SPE 76723 presented at the SPE Western Regional/AAPG Pacific Section Joint Meeting, Anchorage, Alaska, May 20–22, 2002.

18. Bouma, A.H.: *Sedimentology of Some Flysch Deposits: A Graphical Approach to Facies Interpretation,* Elsevier, Amsterdam (1962).

19. Underdown, D.R., Schultz, D.J., Marino, A., and Miranda, R.: "Optimizing Production by Fluid/Rock Studies for the Stevens Sands, North Coles Levee Field, Kern County, California," paper SPE 21785 presented at the Western Regional Meeting, Long Beach, California, March 20–22, 1991.

20. MacAllister, D.J.: "Evaluation of CO_2 Flood Performance: North Coles Levee CO_2 Pilot, Kern County, California," paper SPE 15499 presented at the 61st Annual Technical Conference and Exhibition, New Orleans, Louisiana, October 5–8, 1986.

21. Starcher, M.G., Engineer, R., and Wilson, M.: "Case History of the 31S Peripheral Waterflood Project, Stevens Zone, Elk Hills Field, California," paper SPE 35673 presented at the 1996 SPE Western Regional Meeting, Anchorage, Alaska, May 22–24, 1996.

22. Hampton, T.J., Danielli, C., McChesney, D., and Singh, S.P.: "Fullfield Numerical Simulation Using Geostatistical Reservoir Model of Upper Main Body "B" Reservoir, Monterey Formation, Elk Hills, California," paper SPE 68879 presented at the SPE Western Regional Meeting, Bakersfield, California, March 26–30, 2001.

23. Saleri, N.G., Salamy, S.P., Mubarak, H.K., Sadler, R.K., Dossary, A.S., and Muraikhi, A.J.: "Shaybah-220: A Maximum-Reservoir-Contact (MRC) Well and its Implications for Developing Tight-Facies Reservoirs," *SPEREE* (August 2004) 316–321.

24. Saleri, N.G.: "'Learning Reservoirs': Adapting to Disruptive Technologies," *JPT* (March 2002) 57–60.

25. Salamy, S.P., Al-Mubarak, H.K., Ghamdi, M.S., and Hembling, D.: "Maximum-Reservoir-Contact-Wells Performance Update: Shaybah Field, Saudi Arabia," *SPE Production & Operation* (November 2008) 439–443.

26. Dossary, A.S., and Mahgoub, A.A.: "Challenges and Achievements of Drilling Maximum Reservoir Contact (MRC) Wells in Shaybah Field," paper SPE 85307 presented at the SPE/IADC Middle East Drilling Technology Conference and Exhibition, Abu Dhabi, UAE, October 20–22, 2003.

27. Saleri, N.G., Salamy, S.P., and Al-Otaibi, S.S.: "The Expanding Role of the Drill Bit in Shaping the Subsurface," *JPT* (December 2003) 53–58.

28. Salamy, S.P.: "A New Generation of Wells for Developing Tight Reservoir Facies," Paper SPE 108806, *SPE Distinguished Lecture Series Program,* 2004–2005.

29. Al-Awami, F., Hemanthkumar, K., Salamy, S., and Sadler, R.: "A Mega-Cell Simulation Model of the Shaybah Field," paper IPTC 10240 presented at the International Petroleum Technology Conference, Doha, Qatar, November 21–23, 2005.

30. Salamy, S.P., Al-Mubarak, H.K., Hembling, D.E., and Al-Ghamdi, M.S.: "Deployed Smart Technologies Enablers for Improving Well Performance in Tight Reservoirs—Case: Shaybah Field, Saudi Arabia," paper SPE 99281 presented at the SPE Intelligent Energy Conference and Exhibition, Amsterdam, The Netherlands, April 11–13, 2006.

31. Salamy, S.P., Al-Mubarak, S., Al-Mubarak, H., Al-Dawood, N., and Al-Alawi, A.: "Maximum Reservoir Contact Wells: Six Years of Performance-Lessons Learned and Best Practices," paper SPE 118030 presented at the 2008 Abu Dhabi International Petroleum Exhibition and Conference, Abu Dhabi, UAE, November 3–6, 2008.

General Reading

- Saputelli, L., Nikolaou, M., and Economides, M.J.: "Self-Learning Reservoir Management," *SPEREE* (December 2005) 534–547.
- Wang, C., Li, G., and Reynolds, A.C.: "Production Optimization in Closed-Loop Reservoir Management," paper SPE 109805 presented at the 2007 SPE Annual Technical Conference and Exhibition, Anaheim, California, November 11–14, 2007.
- Al-Hussainy, R., and Humphreys, N.: "Reservoir Management: Principles and Practices," *JPT* (December 1996) 1129–1135.
- Hill, A.D., Zhu, D., and Economides, M.J.: *Multilateral Wells,* SPE, Richardson, Texas (2008).
- Hembling, D., Salamy, S., Qatani, A., Carter, N., and Jacob, S.: "Swell Packers: Enabling Openhole Intelligent and Multilateral Well Completions for Enhanced Oil Recovery," paper IADC/SPE 100824 presented at the IADC/SPE Asia Pacific Drilling Technology Conference and Exhibition, Bangkok, Thailand, November 13–15, 2006.
- Mubarak, S., Dawood, N., and Salamy, S.: "Lessons Learned from 100 Intelligent Wells Equipped with Multiple Downhole Valves," paper SPE 126089 presented at the 2009 SPE Saudi Arabia Section Technical Symposium, Al Khobar, Saudi Arabia, May 9–11, 2009.

Index

A

B

C

D

E

F

G

H

K

L

M

N

O

P

Q

R

S

T

U

X

Y

Z

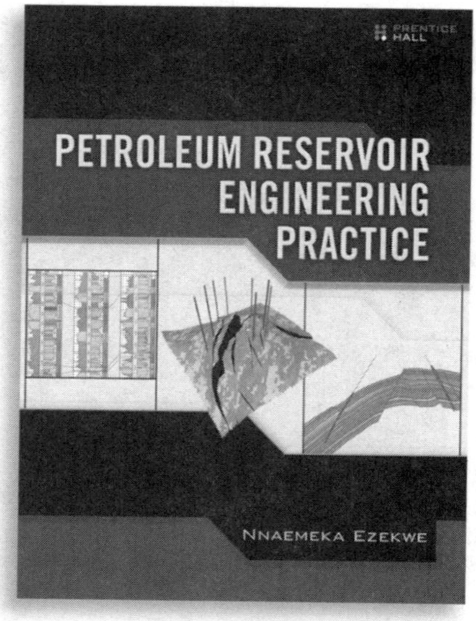
PRENTICE HALL
PETROLEUM RESERVOIR ENGINEERING PRACTICE
NNAEMEKA EZEKWE